THE SCIENTIFIC LETTERS AND PAPERS OF
JAMES CLERK MAXWELL

James Clerk Maxwell at Cambridge in autumn 1855, holding the colour top.

THE SCIENTIFIC
LETTERS AND PAPERS OF
JAMES CLERK MAXWELL

VOLUME I
1846–1862

EDITED BY

P. M. HARMAN

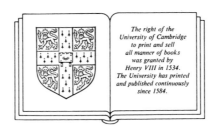

The right of the
University of Cambridge
to print and sell
all manner of books
was granted by
Henry VIII in 1534.
The University has printed
and published continuously
since 1584.

CAMBRIDGE UNIVERSITY PRESS

CAMBRIDGE

NEW YORK   PORT CHESTER

MELBOURNE   SYDNEY

Published by the Press Syndicate of the University of Cambridge
The Pitt Building, Trumpington Street, Cambridge CB2 1RP
40 West 20th Street, New York, NY 10011, USA
10 Stamford Road, Oakleigh, Melbourne 3166, Australia

First published 1990

Printed in Great Britain at the University Press, Cambridge

British Library cataloguing in publication data
Maxwell, James Clerk, 1831–1879
The scientific letters and papers of James Clerk Maxwell
Vol. 1, 1846–1862
1. Physics
I. Title II. Harman, P. M. (Peter Michael), 1943–
530

Library of Congress cataloging in publication data
Maxwell, James Clerk, 1831–1879.
The scientific letters and papers of James Clerk Maxwell/edited by P. M. Harman
p. cm.
Bibliography: p. Includes index. Contents: v. 1. 1846–1862
ISBN 0 521 25625 9
1. Electromagnetic theory. 2. Light, Wave theory of. 3. Molecular theory – Statistical methods.
I. Harman, P. M. (Peter Michael), 1943– . II. Title.
QC670.M385 1990
530.1′41–dc19 89–452 CIP

ISBN 0 521 25625 9

For Juliet

CONTENTS

PREFACE

James Clerk Maxwell's contributions to science, especially his theory of the electromagnetic field and electromagnetic theory of light, and his development of statistical molecular theory, have established his special place (with Isaac Newton and Albert Einstein) in the history of physics. When, from the 1870s, his major theoretical ideas were espoused, in varying measure, by Helmholtz, Boltzmann, Lorentz and Hertz, his physics achieved the pre-eminence in the 'classical' physics of the nineteenth century, which it has retained ever since. The revolution in the structure of physical theory which has occurred in the twentieth century has reinforced rather than qualified Maxwell's unique status. His conception of the physical field and his representation of molecular processes by statistical descriptions stand behind the relativity and quantum theories.

The pervasiveness of Maxwell's influence in physics, his creative power and fecundity of imagination, and the depth of his critical examination of the foundations of physics, give his writings an especial interest and importance. From the inclusion by his biographers Lewis Campbell and William Garnett of extracts from his correspondence and occasional essays in their *Life of Maxwell*, and Joseph Larmor's partial publication of his letters to George Gabriel Stokes and William Thomson (Lord Kelvin), up to the present day, the publication of his letters and manuscript papers has seemed important to commentators.

The present edition of Maxwell's scientific letters and manuscript papers, which will be completed in three volumes, is however the first comprehensive gathering of Maxwell's manuscripts. The primary intention of this edition is the reproduction of an accurate text of Maxwell's autograph manuscripts. As explained in the General introduction to this first volume of the edition, Maxwell's legacy of manuscript papers has suffered various vicissitudes and depletions during the past century; and this edition of his scientific letters and manuscript papers (which is augmented by supplementary documents drawn from printed sources) encompasses the autograph letters and papers now known to exist.

In undertaking this edition I am indebted in many ways to D. T. Whiteside. His advice on matters of editorial practice and on points of scholarship has been invaluable, but I am especially grateful for his friendly encouragement. I owe a special debt to Alan Shapiro for much friendly advice, especially on matters of editorial practice.

Much of the work on this volume has been carried out in the Cambridge University Library. A. E. B. Owen first introduced me to the collection of Maxwell manuscripts held by the Library, and I am grateful to him for this early help and for his continued assistance. I have benefited from the kindness and ready helpfulness of archivists and librarians in many libraries, and I am deeply grateful to them. I thank especially Godfrey Waller of the Manuscripts Room of the Cambridge University Library. He and his colleagues have, over the years, borne the brunt of my unceasing requests for books and manuscripts.

My work in Cambridge has been aided by my continued association with Clare Hall. I am grateful to the President and Fellows for first electing me to a Research Fellowship, and for welcoming me back as a regular visitor to this unique scholarly community. I thank Sir Brian Pippard especially for many stimulating discussions over the years, and for his continued interest in this work. Professor Alan Cook has taken a kind interest in the progress of the edition on behalf of the Syndics of the Cambridge University Press; and my editor at the Press, Richard Ziemacki, has been generous with advice and assistance. It is a pleasure to thank Susan Bowring for her careful work as copy editor.

I am especially grateful to David Wilson for generously allowing me to consult his transcript of the Stokes–Kelvin correspondence. Other friends and colleagues have been helpful in providing information about the location of manuscripts and in offering advice. I thank especially Stephen Brush, Jed Buchwald, David Dewhirst, Francis Everitt, Robert Fox, A. T. Fuller, Ivor Grattan-Guinness, Rupert Hall, Robert Kargon, Martin Klein, Ole Knudsen, David Kohn, David MacAdam, Nathan Reingold, Dan Siegel, Thomas Simpson, Carlene Stephens and Paul Theerman.

In April 1985 I was invited by Nathan Reingold to a conference on scientific editing held at the Rockefeller Archive Center, New York. The discussions I had on this occasion, especially with Reese Jenkins and John Stachel, were helpful in aiding me to clarify my approach in the preparation of this edition.

I am indebted to Ethel Dunkerley for typing the entire manuscript from my handwritten transcriptions, a task she has undertaken with enthusiasm and competence. I am very grateful for the excellent assistance I have had from Isabel Matthews in preparing the figures. Keith Papworth, of the Cavendish Laboratory, has been very helpful in preparing photographs of Maxwell's experimental apparatus.

I undertook the work on this edition without the advantage of regular access to a research library. That the task of preparing this volume for publication has not proved impossible to accomplish, I am grateful to the Council of the Royal Society for generously awarding me a succession of research grants; to the Leverhulme Trust for the generous award of a Research Fellowship in

1986–87; and to the Research Committee of the University of Lancaster, for a research grant and for the allowance of leave of absence, which enabled me to devote two years of uninterrupted research to this volume at a crucial stage in its preparation.

For kind permission to reproduce manuscripts in their keeping I am grateful to the Syndics of the Cambridge University Library; the Cavendish Laboratory, Cambridge; the Cambridge Observatory; the Master and Fellows of Trinity College, Cambridge; the Master and Fellows of Peterhouse, Cambridge; the Librarian of Glasgow University Library; Aberdeen University Library; St Andrews University Library; the Royal Society; the Royal Society of Edinburgh; the Institution of Electrical Engineers, London; the College Secretary, King's College, London; the Greater London Record Office; the Royal Greenwich Observatory; the Staatsbibliothek Preussische Kulturbesitz, Berlin; and the Smithsonian Institution Libraries, Washington.

I am grateful to the Cavendish Laboratory, Cambridge, for kindly providing photographs of Maxwell's apparatus; to the Master and Fellows of Trinity College, Cambridge for the photograph of Maxwell reproduced as the frontispiece to this volume; and to the Syndics of Cambridge University Library for photographs of documents in their keeping.

I offer my deepest thanks however to Juliet, Tim and Rosie for tolerating my continuing absorption in this work.

LIST OF TEXTS

GENERAL INTRODUCTION

On 10 November 1879, five days after Maxwell's death, his cousin and executor Colin Mackenzie drew up a 'Memorandum by the Executors of James Clerk Maxwell'. The 'Memorandum' was signed by Mackenzie himself, and by Maxwell's other two executors, George Gabriel Stokes, Lucasian Professor of Mathematics at Cambridge, and his physician George Edward Paget, Regius Professor of Physic.

The Executors, having in view that Professor Maxwell had directed in a Codicil to his Will dated 28 October 1879 that his Executors should have power to examine his whole papers & decide which should be destroyed & which prepared for publication unanimously requested Professor Stokes to undertake this duty, having the fullest discretionary powers in his hands. They further, at Professor Stokes request, deputed Mr William Garnett St John's College to arrange classify & inventory the whole papers.[1]

Stokes, burdened with duties as Secretary of the Royal Society, and already involved (along with John Couch Adams, H. R. Luard and G. D. Liveing) with the catalogue of the Portsmouth Collection of Newton's scientific papers, passed the papers to William Garnett, who had been Maxwell's Demonstrator at the Cavendish Laboratory. Garnett had effective custody of Maxwell's papers for some years.

On 14 February 1881 Mackenzie drew up a 'Memorandum as to Biography of James Clerk Maxwell' which was signed by Maxwell's widow, which set out the division of work between the biographers, Lewis Campbell (Maxwell's lifelong friend) and Garnett himself, who was 'to send Mr Campbell all diaries letters etc or copies of them in his hands bearing on the matter'.[2] Campbell and Garnett's *Life of Maxwell* was published in 1882, and followed by a revised edition in 1884 (which included the texts of a few additional Maxwell letters).[3]

The Maxwell papers were apparently returned to Mrs Maxwell. In 1903 Garnett wrote to Joseph Larmor in response to a query about the fate of the Maxwell papers, no doubt occasioned by Larmor's unavailing search for letters

---

(1) ULC Add. MSS 8385/1.

(2) ULC Add. MSS 8385/3.

(3) These letters are included in Robert Kargon's 'Appendix' to his 1969 New York reprint of the first edition of the *Life of Maxwell*.

written to Maxwell by George Gabriel Stokes, whose papers he was collecting for a memoir.[4] Garnett wrote that

All the Maxwell letters which Campbell and I had were either returned to their owners from whom they were received on loan or were placed with the other letters in a big deal box which was covered with wall paper, and which was in my possession while the work was in progress, but was afterwards returned to Mrs. Maxwell's house [in Cambridge]. I had the impression that the box was sent back to Glenlair [Maxwell's Galloway home], and was in charge of Mr. Wedderburn Maxwell [the heir to Maxwell's estate], but I know absolutely nothing of what happened at Mrs. Maxwell's death [in 1886].... Of course this box contained mainly letters received by Maxwell, and note-books and papers in his own handwriting, which remained his property. The great bulk of the letters written by him were returned, as I stated to their owners. Lord Kelvin probably possesses a good many. I do not know what became of the letters which were in the possession of P. G. Tait, who was perhaps Maxwell's most frequent correspondent....[5]

The collection of Maxwell's autograph papers which came into the possession of the Cavendish Laboratory in 1887[6] clearly derived from the manuscript material in the box described by Garnett. This collection, augmented by letters in the possession of Peter Guthrie Tait,[7] was transferred to the Cambridge University Library in 1964. In a preamble to his catalogue of this collection of the 'Papers of James Clerk Maxwell' (ULC Add. MSS 7655), A. E. B. Owen recalls that the papers were received 'mainly in a series of envelopes, in some confusion', a description which accords with Derek J. Price's reference to 'about 85 large file envelopes of manuscript material together with some 350 letters' in his review of the Cavendish Laboratory archives in 1953.[8]

There remains however a puzzle about the fate of the corpus of Maxwell papers which passed into Garnett's hands. In his 1903 letter to Larmor Garnett recollected that letters written by Maxwell were generally returned to their owners, the correspondents to whom they were addressed; while Maxwell's own papers and notebooks were returned to his widow. Comparison of the 'Papers of James Clerk Maxwell' held by Cambridge University Library with documents included in the *Life of Maxwell* reveals that numerous letters and papers, including Maxwell's letters to his father and wife, his essays for the Cambridge Apostles, and his father's papers, which were published in partial

---

(4) Larmor, *Correspondence*, **1**: v–vii.

(5) William Garnett to Joseph Larmor, 17 November 1903 (ULC Add. MSS 7655, III, b/1).

(6) J. J. Thomson, 'Survey of the last twenty-five years', in *A History of the Cavendish Laboratory 1871–1910* (London, 1910): 85.

(7) ULC Add. MSS 7655, I, b.

(8) Derek J. Price, 'The Cavendish Laboratory archives', *Notes and Records of the Royal Society*, **10** (1953): 139–47, on 139.

extract in the *Life of Maxwell* and which would presumably have been returned to his widow, are no longer extant as autograph documents.

According to the 'Memorandum as to Biography of James Clerk Maxwell', the 'writing of the Earlier life of James Clerk Maxwell, with a Sketch of the Writers recollection of him during his latter years is entrusted to Professor Lewis Campbell and shall be kept in a distinct & separate volume or division of a volume'; and 'Mr Garnett to compile & write an account of the later life, and of James Clerk Maxwell's public career...'.[9] In his preface to the first edition of the *Life of Maxwell*, Campbell stated that he himself had written the account of Maxwell's life, but that the 'substance' of the three chapters which covered the period 1860–79 was 'largely drawn from information obtained through [Garnett].'[10] This period of Maxwell's life is treated very cursorily in the biography. Campbell declared that 'from this point [1860] onward the interest of Maxwell's life (save things "wherewith the stranger intermeddles not") is chiefly concentrated in his scientific career'.[11] Garnett's treatment of Maxwell's scientific work, which formed the second part of the biography, was envisaged as providing the substantive biographical narrative of Maxwell's later life. While Campbell's account of the young Maxwell is rich and affectionate, and is embellished by numerous letters written to friends and family which provide insight into his character and personal development, the period after 1862 is very thinly illustrated by original documents. Garnett included few such materials in his account of Maxwell's scientific work. The documents reproduced in the *Life of Maxwell* were chosen to reveal Maxwell's character and philosophy, not his scientific thought; though in many cases, notably the letters to his Cambridge friends R. B. Litchfield and C. J. Monro and letters addressed to Campbell himself, and some of the Cambridge essays, they illuminate Maxwell's intellectual development and the pattern of his scientific work.

With the major exceptions of Maxwell's letters to Litchfield and Monro, which Campbell extracted in his biography, and the letters to Faraday which were included in the second edition of 1884, the documents which Campbell included in the *Life of Maxwell* are no longer extant as autograph originals. The numerous letters which Maxwell wrote to his friends and family, including the substantial set of letters addressed to Campbell himself which form the backbone of his biography, and the Cambridge essays on literary and philosophical subjects, can now with few exceptions only be consulted in the form in which they are printed in the *Life of Maxwell*. Campbell reproduced most of these texts in partial extract, frequently merely as fragments, and he certainly published

(9) ULC Add. MSS 8385/3.   (10) *Life of Maxwell*: vii.

(11) *Life of Maxwell*: 314.

documents on a selective basis; and for the period after 1862 the number of such documents made available in the *Life of Maxwell* is severely limited.

The corpus of Maxwell autograph documents now extant falls into three groups. First, the collections held by Cambridge University Library, numerically the most substantial. The 'Papers of James Clerk Maxwell' (Add. MSS 7655) consists mainly of drafts of scientific papers, notebooks, and scientific letters, including his correspondence with P. G. Tait. The Stokes and Kelvin papers (Add. MSS 7656 and 7342)[12] contain a substantial number of Maxwell letters; and there are smaller collections of letters catalogued separately. Second, the archive of the Royal Society holds Maxwell's referee's reports on papers communicated to the Society for publication, the autograph manuscripts of two of his papers published in the *Philosophical Transactions*, and some scientific letters. Third, there are scattered collections of letters held in university, public and private archives. These are almost exclusively scientific letters.[13]

The history of Maxwell's essays and family papers extracted by Campbell in the *Life of Maxwell*, and of other related materials which he did not choose to print, can only be conjectured. It is possible that these documents were separated from Maxwell's scientific papers on his wife's death in 1886. His books were sorted into two separate collections at that time, his scientific books being bequeathed to the Cavendish Laboratory while his other books were donated to the Cambridge Free Library, the city's public library.[14] Garnett recollected that he had the impression that Maxwell's papers may have been transferred to Glenlair; and it is possible that the nonscientific and family papers were handed over to the custody of Maxwell's heir. If so, the papers were no doubt destroyed when Glenlair was later ravaged by a fire which reduced the building to a shell. Lewis Campbell presumably retained in his own possession the letters which Maxwell had written to him; but with the exception of two letters which have found their way into other collections, the autographs of Maxwell's letters to Campbell are no longer extant. As a result, the *Life of Maxwell* remains the substantive and almost unique source for Maxwell documents encompassing matters other than his intellectual development, scientific work and public career.

---

(12) See David B. Wilson, *A Catalogue of the Manuscript Collections of Sir George Gabriel Stokes and Sir William Thomson, Baron Kelvin of Largs in Cambridge University Library* (Cambridge, 1976).

(13) The few extant 'personal' letters not reproduced here are listed in the Appendix §1.

(14) T. C. Fitzpatrick and W. C. D. Whetham in *A History of the Cavendish*: 8; *Cambridge Public Free Library. 32nd Annual Report. 1886–1887* (Cambridge, 1887); the inventory of Mrs K. M. Clerk Maxwell's effects, ULC Add. MSS 8385/4. Maxwell's scientific books are held by the Cavendish Laboratory, Cambridge.

The Cambridge collection of the 'Papers of James Clerk Maxwell' contains around 250 letters written to Maxwell by his scientific correspondents. Most of the letters held in the collection date from after 1871, when Maxwell returned to Cambridge as the University's first Professor of Experimental Physics. These letters form a very disparate collection. With the major exceptions of the substantial correspondence with Peter Guthrie Tait, and a series of letters exchanged with George Chrystal towards the end of Maxwell's life, this collection of letters written to Maxwell does not correlate with the extant Maxwell letters so as to form a coherent correspondence. There are however some letters in existence from Michael Faraday, J. D. Forbes and C. J. Monro. The most significant gaps are the missing letters from Stokes and William Thomson (Lord Kelvin). Judging by the letters Maxwell addressed to them, these men were (with P. G. Tait) his major correspondents; yet only a few short letters of the many they must have written to Maxwell are now extant. This contrasts with the substantial Maxwell–Tait correspondence; thus the history of the Stokes and Thomson letters remains obscure. Campbell's hope, expressed in the preface to the revised edition of the *Life of Maxwell*, that 'it is greatly to be desired that the *scientific correspondence* which Maxwell held from time to time with other eminent men might, before it is too late, be collected, sifted, and arranged',[15] cannot now be fully realised.

(15) *Life of Maxwell* (2nd edn) : vii.

EDITORIAL NOTE

The terms of reference of this edition of Maxwell's scientific letters and papers are shaped by the scope of the documents now known to exist (as described in the General introduction), and encompass his extant autograph letters and papers which form the main corpus of the texts here reproduced. These autograph manuscripts are supplemented by selections from documents which were published (almost invariably in partial extract and frequently in fragmentary form) in the *Life of Maxwell*, where these shed light on his scientific work and intellectual development; and by his shorter publications – letters, reviews, and abstracts of contributed and published papers – which were omitted from the memorial edition of his *Scientific Papers* published by Cambridge University Press in 1890.

The Maxwell texts are reproduced in chronological sequence. While Maxwell generally dated his letters, other manuscripts have to be dated on the basis of independent evidence. Because of the fragmentary and disordered nature of much of the material, I developed an independent chronology of the documents on the basis of available contemporary evidence, including the inspection of writing style and paper. In the case of texts drawn from the *Life of Maxwell* Campbell's dating has been generally followed, though always after critical scrutiny. In the case of abstracts of published papers the convention is adopted of citing the date when the paper was read; or, in the case of papers read to the Royal Society (of London), the date the paper was received as recorded by the Society's Secretary.

The primary intention of this edition is the reproduction of an accurate text of all Maxwell's scientific letters and substantive manuscript papers. The edition includes documents that shed light on his science in describing his general intellectual development and milieu, but does not contain all juvenilia, all sheets of data and calculations, or all manuscript fragments and jottings. Maxwell's undergraduate notes[1] are entirely derivative; these notes, student exercises, and questions he set for university examinations, are not generally reproduced. The few autograph letters which have not been reproduced are listed in the appendix, together with a list of the documents which have not been reprinted from the *Life of Maxwell*.

Letters written to Maxwell are listed in the appendix. Because of the limited

---

(1) See the Introduction notes (19), (22), (43), (44) and (48).

scope of these documents (as described in the General introduction), these letters are reproduced on a selective basis. In the case of his correspondence with P. G. Tait, the substantial series of letters written by Tait (dating from 1865–79) will be published. In most cases the letters are reproduced in selective extract, as annotations to the Maxwell texts. In some cases these letters have been published in the *Life of Maxwell* or in other editions. Third-party letters and other documents that contain information bearing on Maxwell's activities are likewise reproduced in selective extract.

In accordance with the principles of modern scholarship the reproduction of the texts faithfully follows the manuscripts in spelling, punctuation, capitalization, and in preserving contractions. Where the texts are reproduced from printed sources the style of the original is followed. Trivial cancellations have been omitted without comment, but corrections deemed significant have been recorded. Minor deletions are placed within angle brackets ⟨ ... ⟩ preceding the revised text; longer cancelled passages are reproduced by setting a double vertical bar against them in the left-hand margin. Appended passages are reproduced with a single vertical bar in the left-hand margin; appended phrases by corners ⌞ ... ⌟ which enclose the added words. Trivial mathematical corrections which Maxwell introduced for the sake of consistency (for example, the correction of coefficients of algebraic variables), which led to some modifications in the text, have been recorded. Annotations which were subsequently appended by Maxwell or by his correspondents are recorded. The name enclosed by brackets { ... } denotes the annotator. The very few editorial insertions to the text, which have been introduced for the sake of clarity, are enclosed within square brackets. In general I have attempted to preserve the layout of the manuscripts in their transformation to the printed page, but some necessary adjustments have been made for reasons of clarity.

This applies especially to the reproduction of figures, which, like the transcription of handwriting, requires editorial interpretation. Some of Maxwell's figures are clearly drawn, but most are rough sketches. The aim has been to elucidate Maxwell's intentions as determined by study of both the figure and the corresponding text. The aim has been clarity rather than the precise reproduction of Maxwell's figures. I have sought a compromise between a facsimile of Maxwell's rough figures and structurally ideal geometrical diagrams; and between his hasty sketches of apparatus and instruments and the polished draughtsman's figures in his published papers to which they sometimes have relation. Figure numbers have been added; the captions are Maxwell's.

The editorial commentary – the historical and textual notes and the Introduction – is intended to aid the reader in following Maxwell's arguments and his allusions to concepts, events and personalities. The Introduction provides a broad account of his intellectual development and career in the period covered by this volume, and an outline review of the texts here reproduced. In addition to clarifying obscurities in the texts, the historical notes seek to establish the context within which the documents were written, employing contemporary published as well as manuscript sources, including letters written to Maxwell, third-party correspondence, and fragmentary manuscript jottings not reproduced as texts.

LIST OF PLATES

ABBREVIATED REFERENCES

Ann. Chim. Phys.   *Annales de Chimie et de Physique* (Paris).

Ann. Phys.   *Annalen der Physik und Chemie* (Leipzig).

Camb. & Dubl. Math. J.   *Cambridge and Dublin Mathematical Journal* (Cambridge).

Camb. Math. J.   *Cambridge Mathematical Journal* (Cambridge).

Comptes Rendus   *Comptes Rendus Hebdomadaires des Séances de l'Académie des Sciences* (Paris).

DNB   *Dictionary of National Biography*. Ed. L. Stephen and S. Lee, 63 vols. and 2 supplements (6 vols.) (London, 1885–1912).

Electricity   Michael Faraday, *Experimental Researches in Electricity*, 3 vols. (London, 1839–55).

Electrostatics and Magnetism   William Thomson, *Reprint of Papers on Electrostatics and Magnetism* (London, 1872).

Larmor, *Correspondence*   *Memoir and Scientific Corrrespondence of the Late Sir George Gabriel Stokes, Bart.* Ed. J. Larmor, 2 vols. (Cambridge, 1907).

Larmor, 'Origins'   'The origins of Clerk Maxwell's electric ideas, as described in familiar letters to W. Thomson'. Communicated by Sir Joseph Larmor, in *Proc. Camb. Phil. Soc.*, **32** (1936): 695–750. Reprinted separately (Cambridge, 1937).

Life of Maxwell   Lewis Campbell and William Garnett, *The Life of James Clerk Maxwell. With a Selection from his Correspondence and Occasional Writings and a Sketch of his Contributions to Science* (London, 1882).

Life of Maxwell (2nd edn)   Lewis Campbell and William Garnett, *The Life of James Clerk Maxwell with Selections from his Correspondence and Occasional Writings*, new edition, abridged and revised (London, 1884).

Math. & Phys. Papers   William Thomson, *Mathematical and Physical Papers*, 6 vols. (Cambridge, 1882–1911).

OED   *The Oxford English Dictionary*. 12 vols. (Oxford, 1970).

Papers   George Gabriel Stokes, *Mathematical and Physical Papers*, 5 vols. (Cambridge, 1880–1905).

Phil. Mag.   *Philosophical Magazine* (London)

Phil. Trans.   *Philosophical Transactions of the Royal Society of London* (London).

Proc. Camb. Phil. Soc.   *Proceedings of the Cambridge Philosophical Society* (Cambridge).

Proc. Roy. Soc.   *Proceedings of the Royal Society of London* (London).

Proc. Roy. Soc. Edinb.   *Proceedings of the Royal Society of Edinburgh* (Edinburgh).

Scientific Memoirs	*Scientific Memoirs, Selected from the Transactions of Foreign Academies of Science and Learned Societies, and from Foreign Journals.* Ed. Richard Taylor, 5 vols. (London, 1837–52).
Scientific Papers	*The Scientific Papers of James Clerk Maxwell.* Ed. W. D. Niven, 2 vols. (Cambridge, 1890).
Trans. Camb. Phil. Soc.	*Transactions of the Cambridge Philosophical Society* (Cambridge).
Trans. Roy. Soc. Edinb.	*Transactions of the Royal Society of Edinburgh* (Edinburgh).
Treatise	James Clerk Maxwell, *A Treatise on Electricity and Magnetism,* 2 vols. (Oxford, 1873).
ULC	Manuscripts in the University Library, Cambridge.
Venn	*Alumni Cantabrigienses. A Biographical List of all Known Students, Graduates and Holders of Office at the University of Cambridge, from the Earliest Times to 1900.* Compiled by J. A. Venn. Part II. From 1752 to 1900, 6 vols. (Cambridge, 1940–54).

INTRODUCTION

This first volume of James Clerk Maxwell's scientific letters and manuscript papers opens with his first contribution to science in 1846 when he was a schoolboy of 14, a paper on the mechanical description of Cartesian ovals. The volume concludes early in 1862 following his triumphant announcement, in letters to Michael Faraday and William Thomson (later Lord Kelvin), of the first version of his most famous work, the electromagnetic theory of light. The reproduction of manuscript materials may not explain Maxwell's genius, but the documentation of his education and early scientific work, and of the period of his greatest scientific innovation, will contribute to an understanding of his intellectual development. These documents provide ample evidence of the cultural conditions which fostered Maxwell's scientific ideas and career, and of the way in which his work was moulded by the attitudes of his contemporaries. Maxwell was a man of wide reading and deep learning, a scholar as well as a scientist; and his friend and biographer Lewis Campbell, quoting a phrase of Maxwell's own, insisted that there were 'no water-tight compartments' in his mind.[1] His letters and papers provide evidence of the raw materials from which emerged the classic works on which his enduring fame as a physicist rests.

Edinburgh: education and early scientific interests

In autobiographical remarks written in 1873 Maxwell recollected his early delight in the forms of regular figures and curves and his view of mathematics as the search for harmonious and beautiful shapes.[2] An early interest in

(1) *Life of Maxwell*: 205, 433. There are two biographical studies by C. W. F. Everitt; his *James Clerk Maxwell. Physicist and Natural Philosopher* (New York, 1975), based on his article on 'Maxwell' in *Dictionary of Scientific Biography*, ed. C. C. Gillispie, 16 vols. (New York, 1970–80), **9**: 198–230; and his essay on 'Maxwell's scientific creativity', in *Springs of Scientific Creativity. Essays on Founders of Modern Science*, ed. R. Aris, H. T. Davis, R. H. Stuewer (Minneapolis, 1983): 71–141.

(2) V. L. Hilts has identified the replies sent by scientists, including Maxwell, in response to a questionnaire issued by Francis Galton in 1873 while gathering materials for his book *English Men of Science* (London, 1874); see V. L. Hilts. 'A guide to Francis Galton's *English Men of Science*', *Transactions of the American Philosophical Society*, **65** Part 5 (1975): 59. Hilts attributes the following statement to Maxwell: 'I always regarded mathematics as the method of obtaining the best shapes and dimensions of things; and this meant not only the most useful and economical, but chiefly the most harmonious and the most beautiful'. Galton edited his respondents' replies, so the wording may have been modified from Maxwell's autograph original.

geometry can be seen in a letter to his father written in June 1844, when Maxwell was just 13, describing his construction of 'a tetra hedron, a dodeca hedron, and 2 more hedrons that I don't know the wright names for'.[3] Writing with the benefit of personal acquaintance, his biographer William Garnett noted his fascination with the harmony of colours.[4] Maxwell's starting point in mathematics was geometry, leading to his first mathematical papers on Cartesian ovals and on rolling curves; and his first studies in physics were of the chromatic effects exhibited by strained glass in polarised light. If his interest in mathematics and physical science was initiated by his sense of harmony and beauty, his work was shaped by his education and by the environment in which he grew to maturity.

In 1831, the year of Maxwell's birth, his father, John Clerk Maxwell, published a paper on an automatic printing press;[5] and in the 1840s he was a prominent member of the Royal Scottish Society of Arts. During the winter of 1845–6, while Maxwell was still at school at the Edinburgh Academy, John Clerk Maxwell encouraged his son to accompany him to meetings of the Society of Arts and of the Royal Society of Edinburgh.[6] Maxwell became aware of the work of David Ramsey Hay, a prominent Edinburgh decorative artist. Hay was engaged in studies which aimed to reduce beauty in form and colour to mathematical principles, and his current attempt to construct 'the perfect egg-oval'[7] led Maxwell to his first mathematical paper, a method of drawing oval curvēs (Number 1). John Clerk Maxwell assiduously promoted his son's effort; his contact with James David Forbes, Professor of Natural Philosophy at Edinburgh University,[8] led Forbes to present the paper to the Royal Society of Edinburgh.[9] This early paper displays ingenuity, but Maxwell's development of his work on ovals during the following year demonstrates the burgeoning of the mathematician and physicist. Forbes' summary and his 'remarks' on the paper on ovals established the relation of Maxwell's description of ovals to the Cartesian ovals. Forbes' allusion to Descartes

(3) *Life of Maxwell*: 60.   (4) *Life of Maxwell*: 487–8n.

(5) John Clerk Maxwell, 'Outline of a plan for combining machinery with the manual printing-press', *The Edinburgh New Philosophical Journal*, **10** (1831): 352–7. There are reprints of this paper among the Maxwell manuscripts (ULC Add. MSS 7655, V, i/12). Compare Maxwell's comment, as quoted by Galton: '[Father.] Very great mechanical talent'; see Hilts, 'A guide to Francis Galton's *English Men of Science*': 59.

(6) *Life of Maxwell*: 73.

(7) D. R. Hay, 'Description of a machine for drawing the perfect egg-oval', *Transactions of the Royal Scottish Society of Arts*, **3** (1851): 123–7. See Number 1 note (1).

(8) See *DNB*; J. C. Shairp, P. G. Tait and A. Adams-Reilly, *Life and Letters of James David Forbes* (London, 1873); and *An Index to the Correspondence and Papers of James David Forbes (1809–1868). And also to some papers of his son, George Forbes* (St Andrews, 1968).

(9) *Scientific Papers*, **1**: 1–3. See Number 1 notes (2), (4) and (7).

prompted Maxwell to study the *Geometrie* with critical attention (Number 4); and drawing freely on Euclid's *Elements* he discussed the geometrical and optical properties of ovals, questions which had been expounded by Descartes and noted by Forbes (Number 3).

Maxwell was still at school at this time, but he had there formed a friendship with Peter Guthrie Tait[10] who later recalled that they discussed 'numerous curious problems' and exchanged manuscripts (Numbers 2, 3 and 6).[11] He continued his attendance at meetings of the Edinburgh Royal Society and the Society of Arts, which may have stimulated his outline of the principles of the conical pendulum and a pendulum clock (Number 6).[12] This interest in scientific instruments, fostered by John Clerk Maxwell's own enthusiasm for mechanical devices,[13] later led him to design a 'platometer', an instrument for measuring the areas of figures drawn on a plane, which he presented to the Society of Arts in January 1855 (Number 55). The design of the platometer involved problems of gearing, frictionless motion and rolling contact, and he had already explored some of the geometrical problems of rolling contact in his paper 'On the theory of rolling curves' (Numbers 11 and 18), on the geometry of curves generated like the cycloid by one curve rolling on another. He worked on this paper during the summer of 1848, at the end of his first session as a student at Edinburgh University. But while still at school his interest had been aroused in another pattern of inquiry. Once again family circumstances were to foster his interest. In April 1847 his uncle John Cay, who was a prominent member of the Royal Scottish Society of Arts (becoming President of the Society in 1848–9),[14] took Maxwell and his close school friend Lewis Campbell[15] to the laboratory of the Edinburgh experimental optician William Nicol, who introduced him to the study of polarised light.[16] The visit to Nicol's

---

(10) Born 1831, educ. Edinburgh Academy 1841, Edinburgh University 1847, Peterhouse, Cambridge 1848, Senior Wrangler 1852, Fellow of Peterhouse 1852, Professor of Mathematics, Queen's College Belfast 1854, Professor of Natural Philosophy, Edinburgh University 1860; see C. G. Knott, *Life and Scientific Work of Peter Guthrie Tait. Supplementing the Two Volumes of Scientific Papers published in 1898 and 1900* (Cambridge, 1911).

(11) P. G. Tait, 'James Clerk Maxwell', *Proc. Roy. Soc. Edinb.*, **10** (1879): 331–9; extract quoted in *Life of Maxwell*: 86–7.

(12) See Number 6 note (8).

(13) See Maxwell's comment, as quoted by Galton: 'Fond of mathematical instruments and delighted with the forms of regular figures and curves of all sorts. Strong mechanical power...'; in Hilts, 'A guide to Francis Galton's *English Men of Science*': 59.

(14) See the *Proceedings* of the Royal Scottish Society of Arts for 1848–9: 201, in *Transactions of the Royal Scottish Society of Arts*, **3** (1851).

(15) Born 1830, educ. Edinburgh Academy, Glasgow University 1847, Trinity College then Balliol College, Oxford 1849, Fellow of Queen's College, Oxford 1855, Professor of Greek, St Andrews 1863 (*DNB*).   (16) *Life of Maxwell*: 84, 487–8.

laboratory had a great effect on Maxwell,[17] leading him to study the chromatic effects of polarised light in crystals and strained glass (Numbers 9, 12, 13, 14, 19 and 20), work which led him to his first published paper on a physical subject (Number 26).

On entering the University of Edinburgh in the autumn of 1847 he attended Sir William Hamilton's class in Logic, Forbes' in Natural Philosophy, and Philip Kelland's class in Mathematics. Forbes divided his class into three divisions, and while Tait (who entered Edinburgh University at the same time as Maxwell but remained for only one session before going up to Cambridge) joined the first division where knowledge of the calculus was required, Maxwell more cautiously was content with the middle division of the class where a modest mathematical accomplishment was expected.[18] In 1847–8 Forbes' lectures paid special attention to mechanics, with some discussion of pneumatics, heat and the steam engine, and a substantial number of lectures on astronomy.[19] Maxwell was awarded the class prize by Forbes at the end of the session, the generous gift of a copy of the second edition of *Principia*.[20] His mathematical studies also prospered. In March 1848 he wrote a systematic exercise on analytical geometry,[21] and during the following summer he prepared a comprehensive memoir on the theory of rolling curves (Number 11). His draft differs most notably from the paper he published the following year (Number 18) in its organisation. In the draft he classifies his examples of rolling curves into orders, genera and species, using categories of classification which had been discussed by Hamilton in his logic lectures the previous session.

---

(17) See Maxwell's comment, as quoted by Galton: 'I was taken to see [William Nicol], and so, with the help of 'Brewster's Optics' and a glazier's diamond, I worked at polarization of light, cutting crystals, tempering glass, etc.'; in Hilts, 'A guide to Francis Galton's *English Men of Science*': 59. See David Brewster, *A Treatise on Optics* [in Dr Lardner's *Cabinet Cyclopaedia*] (London, 1831), and his 'Optics' in *The Edinburgh Encyclopaedia*, 18 vols. (Edinburgh, 1830), **15**: 460–662; see Number 14 notes (4) and (5).

(18) Knott, *Life of Tait*: 6.

(19) David B. Wilson, 'The educational matrix: physics education at early-Victorian Cambridge, Edinburgh and Glasgow Universities', in *Wranglers and Physicists. Studies on Cambridge Physics in the Nineteenth Century*, ed. P. M. Harman (Manchester, 1985): 12–48, esp. 21–2. Maxwell's notes on the lectures on astronomy are preserved in ULC Add. MSS 7655, V, m/4.

(20) This copy of *Principia*, preserved in the Cambridge City Library, bears the inscription: 'Juveni Meritissimo/Jacobo Clerk Maxwell/Inter provectiores/Qui in alma Academia Jacobi VI/Edinburgena/Physices Studio incumbebant/Hoc Praemium adjudicavit/Jacobus D. Forbes/MDCCCXLVIII.'

(21) 'ESSAY. The propositions in Wallace's Treatise on the Ellipse, Established Analytically', ULC Add. MSS 7655, V, d/2. The date 'March 1848' (in Maxwell's hand) was added later. See William Wallace, *A Geometrical Treatise on the Conic Sections; with an Appendix Containing Formulae for the Quadrature, &c* (Edinburgh, 1837) [especially 'Part II Of the Ellipse']: 32–66.

Maxwell had taken detailed notes on these lectures[22] which he considered the most substantial of the lectures he had attended (Numbers 7 and 8). After spending the winter months largely on formal study in Edinburgh, he was able to pursue his own original line of investigations more fully and at leisure during the summer at Glenlair, the family estate in Galloway. His letters to Lewis Campbell at this time (Numbers 9, 10 and 12) attest to his enthusiasm and to his growing confidence in his awakening scientific and mathematical powers.

During the following session at Edinburgh University he continued his study of mathematics, winning a prize in Kelland's class,[23] and now entered the first division of Forbes' class in Natural Philosophy. In the session 1848–9 Forbes again concentrated on mechanics, with some attention to the physical properties of bodies and the physics of heat, topics in which he himself had a research interest. In that session he also gave substantial attention to optics. First-division students studied texts such as Poisson's *Mechanics* and Airy's tract on the undulatory theory of light in his *Mathematical Tracts*.[24] A grasp of more advanced mathematical methods was required (Number 15) in the examinations on these texts. Maxwell also attended Hamilton's class on Metaphysics (Number 16). Hamilton formed a high estimate of his abilities, and these lectures left a lasting impression, encouraging his abiding concern to establish the conceptual rationale of his physics by appeal to philosophical argument.[25] This strand of Maxwell's thought is a recurrent theme in his writings, but appears most explicitly in his general essays and lectures and in some of his letters (Numbers 88, 105, 147 and 183). Maxwell's philosophical outlook was shaped by his diverse reading, but his philosophical reflections bear the stamp of his own originality of mind.

At the end of the university session he returned to his interest in the polarisation of light (Numbers 19 and 20). He directed his attention to the phenomenon of induced double refraction in strained glass which had been investigated by Sir David Brewster, a towering figure in optical science and in Scottish

---

(22) ULC Add. MSS 7655, V, m/1–3; and see Number 11 note (4).

(23) See Number 26 note (86).

(24) Wilson, 'The educational matrix': 22; see Number 26 note (87) and Number 14 esp. note (3); and *Life of Maxwell*: 134.

(25) Hamilton wrote a testimonial, dated 26 February 1856, for Maxwell's candidature for the Professorship of Natural Philosophy at Marischal College, Aberdeen (Glasgow University Library, Y1–h.18): 'In the class of Logic and Metaphysics (and here I speak from personal knowledge) – in this class, though differing in subject from his favourite pursuits, he laboured zealously, and received by general suffrage one of the honours.' On the philosophical themes in Maxwell's physics see P. M. Harman, 'Edinburgh philosophy and Cambridge physics: the natural philosophy of James Clerk Maxwell', in *Wranglers and Physicists*: 202–24; and on Maxwell's relation to Scottish philosophy see George Elder Davie, *The Democratic Intellect. Scotland and her Universities in the Nineteenth Century* (Edinburgh, ₂1964): 138–45, 192–7. See also Number 88 note (2).

scientific life at this time. Maxwell's studies broadened to include the measurement of the coefficients of elasticity of rods and wires (Number 24) and to consider the determination of the compressibility of water, a problem which had engaged the attention of leading experimental physicists such as Oersted and Regnault (Number 26 §3). Forbes had himself written papers on these problems, one of them (on the measurement of the extensibility of solids) having been presented to the Royal Society of Edinburgh the previous year.[26] Maxwell began to draft a systematic paper embracing the elasticity of solids and the chromatic effects of strained glass in polarised light (Numbers 21 to 23). The extraordinary thrust of the ensuing 'On the equilibrium of elastic solids' (Number 26) derives from his thorough grasp of complex material and his presentation of a mathematical theory of elasticity, which provides the theoretical framework of the explanation of the experimental data of elasticity and photo-elasticity. Maxwell's control of this material demonstrates the extent of his mathematical reading [27] and his confidence in tackling a subject at the forefront of research. The paper is remarkable in breadth of coverage and in its depth and specificity of analysis of a large number of special cases, an astonishing achievement for an 18 year-old. The mathematical theory of elasticity had been the subject of considerable recent work by the most notable French mathematicians, Navier, Poisson and Cauchy, and Maxwell shows some grasp of their work;[28] but his chief sources were a memoir by the Cambridge mathematician George Gabriel Stokes,[29] from which he derived his basic theoretical model, and a paper by Gabriel Lamé and Émile

(26) J. D. Forbes, 'On an instrument for measuring the extensibility of elastic solids', *Proc. Roy. Soc. Edinb.*, **2** (1848): 173–5. On Forbes' interest in the compressibility of water at this time see a letter from William Thomson to G. G. Stokes of 5 February 1848 (ULC Add. MSS 7656, K24), quoted in Number 26 note (63).

(27) Tait, 'James Clerk Maxwell', extract quoted in *Life of Maxwell*: 133–4; and Tait, 'Clerk-Maxwell's scientific work', *Nature*, **21** (1880): 317–21; 'His reading was very extensive. The records of the [Edinburgh] University Library show that he carried home for study, during these years, such books as Fourier's *Théorie de la Chaleur*, Monge's *Géometrie Descriptive*, Newton's *Optics*, Willis' *Principles of Mechanism*, Cauchy's *Calcul Différentiel*, Taylor's *Scientific Memoirs*, and others of a very high order. These were *read through*, not merely consulted. Unfortunately no list is kept of the books consulted in the Library'.

(28) See Number 26 notes (7), (43) and (93); and Isaac Todhunter, ed. and completed by Karl Pearson, *The History of the Theory of Elasticity and of the Strength of Materials, from Galileo to Lord Kelvin*, **1** (Cambridge, 1886).

(29) Born 1819, educ. Bristol College 1835, Pembroke College, Cambridge 1837, Senior Wrangler 1841, Fellow of Pembroke 1841, Lucasian Professor of Mathematics 1849, Secretary of the Royal Society 1854 (*DNB*); see *Memoir and Scientific Correspondence of the Late Sir George Gabriel Stokes, Bart.*, ed. J. Larmor, 2 vols. (Cambridge, 1907); and David B. Wilson, *Kelvin and Stokes. A Comparative Study in Victorian Physics* (Bristol, 1987).

Clapeyr…   …of the special cases which
make uμ   …ese include the compression
of spher   …a cylinder, and the theory
of Oersι   …f water; and he draws on
experim   …others. The discussion of
Brewste   …refraction in strained glass
forms aι   …ell added an account of his
experim   …ript (Number 26 §14), but
this exμ   …the published paper. The
manusc   …ιring prior to publication,
being re   …y criticised by Forbes.[31]

Duriι   …(in 1849–50) he continued
to folloν   …tudy, now attending classes
in Cheι   …5 and 28), but soon began
prepara   …Peterhouse after much discussion.   …ιdies jostled for his attention
with exμ   …ιg of 'On the equilibrium of
elastic s   …ιoral philosophy (Numbers
29, 30 a   …ed the meeting of the British
Associa   …Edinburgh. Papers on the
phenon   …'Haidinger's brushes' presented by Brewster and Stokes led him to immediately investigate the matter
experimentally (Number 32). He was already known to William Thomson,
Professor of Natural Philosophy at Glasgow,[33] with whom he established
an important relationship which shaped his development as a physicist
(Number 33).

Cambridge: the Mathematical Tripos

Maxwell arrived in Cambridge as an accomplished mathematician and physicist, whose intellectual development had been shaped by Edinburgh traditions
of experimental science, the mechanical arts and philosophy, and by his own

---

(30) Number 26 notes (7), (11) and (94).

(31) See Forbes' letter to Maxwell of 4 May 1850 (*Life of Maxwell*: 137–8), quoted in Number 28 note (2).

(32) See Number 30 note (2); and compare Forbes' letter to Whewell of 30 September 1850 (Trinity College, Add. MS. a.204[96]): 'I recommended him to Trinity, but some friendships amongst Scotch undergraduates at S[t] Peters prevailed'. Tait was at Peterhouse (see note (10)).

(33) Born 1824, educ. Glasgow University 1834, Peterhouse, Cambridge 1841, Second Wrangler 1845, Fellow of Peterhouse 1845, Professor of Natural Philosophy, Glasgow University 1846; see S. P. Thompson, *The Life of William Thomson, Baron Kelvin of Largs*, 2 vols. (London, 1910); and Wilson, *Kelvin and Stokes*.

intensive studies in physics and mathematics. According to Tait 'he brought to Cambridge, in the autumn of 1850, a mass of knowledge which was really immense for so young a man, but in a state of disorder appalling to his methodical private tutor ... William Hopkins'.[34] The opinion of his seniors was not free from some criticism. His Edinburgh mentor Forbes wrote to William Whewell, Master of Trinity College, recommending him to Whewell's attention but remarking that 'he is not a little uncouth in manners, but withal one of the most original young men I have ever met with, & with an extraordinary aptitude for physical enquiries'.[35] After Maxwell's migration to Trinity in December 1850[36] Forbes wrote again to introduce him to Whewell; and observed that 'he is a singular lad, & shy', but 'very clever and persevering'.[37] The course of study at Cambridge was exacting for those undergraduates with ambitions to achieve distinction in the competitive examinations. The first academic hurdle to be surmounted was the Previous Examination, which Maxwell sat in March 1852. In that year candidates were required to display familiarity with the Gospel of St Matthew (in the original Greek), Paley's *Evidences of Christianity*, the Old Testament (Genesis to Esther), the Iliad Book XXIII, the first book of Livy, Euclid's *Elements* Books I and II, and examples in the rules of arithmetic.[38] Not surprisingly, after the freedom and vitality of his Edinburgh years, Maxwell was not totally enamoured of Cambridge life

(34) Tait, 'James Clerk Maxwell'; passage quoted in *Life of Maxwell*: 133.

(35) Forbes to Whewell, 30 September 1850, see note (32); 'As you have been looking into our Transactions I may mention that the author of not the most inconsiderable or at least original paper in the Collection Mr James Clerk Maxwell (the paper is on the Mathematical Theory of Elastic Solids) is about to enter himself as a freshman at St. Peters'. The following extract from a letter from P. G. Tait to Lewis Campbell of 14 February 1849 may shed light on Maxwell's supposed uncouthness: 'Forbes' forms and the cruel men of his class will be the death of Dafty [Maxwell's nickname at school] – I saw Forbes catch him one day running up over the backs of the forms and presenting to his Professor the *broad disc* of a pair of striped trousers – I thought Forbes would have muttered, he wished the stripes on his skin for the Insolence –...' (Peterhouse, Tait MSS).

(36) *Admissions to Trinity College, Cambridge. Vol. IV 1801 to 1850*, ed. W. W. Rouse Ball and J. A. Venn (London, 1911): 666. Maxwell was admitted as a Pensioner on 14 December 1850.

(37) Forbes to Whewell, 18 February 1851 (Trinity College, Add. MS. a.204^{98}); in reply to Whewell's letter of 16 February (Forbes Papers, St Andrews University Library, 1851/21), reporting that 'I think I have seen young Clerk Maxwell already here', and expressing the hope that he would 'not allow himself to be misled by his own fancies (if he has any, of which I know nothing)'. Forbes had written a letter of introduction to Whewell on 23 January 1851 (Trinity College, Add. MS. a.204^{97}): 'Allow me to introduce to you my young friend Mr James Clerk Maxwell (a nephew of Sir George Clerk) and whom I mentioned to you some time ago as the author of some papers of great promise in the Edinburgh Transactions .... I think there can be very little doubt that he will do credit to his College'.

(38) *The Cambridge University Calendar for the Year 1851* (Cambridge, 1851): iv.

(Number 35). Nevertheless he progressed well enough to gain a scholarship at Trinity in April 1852, prompting the Olympian Whewell to report to Forbes on the progress of his protégé: 'he has undoubtedly an able mathematical head; but he is marvellously uncouth in bringing out the produce of it. I hope he will continue to get a little more culture of various kinds before he comes under our consideration again'.[39]

In October 1851 Maxwell became one of the pupils of William Hopkins, the pre-eminent mathematical coach of the period.[40] Hopkins' teaching rather than the lectures of the university professors,[41] provided the essential training for the Mathematical Tripos. Cambridge mathematical education had gone through various changes in the first half of the nineteenth century.[42] The Analytical Society, formed in 1812 by Charles Babbage, John Herschel and George Peacock, had aimed to replace Newtonian fluxions, justified by the method of first and last ratios of nascent and vanishing finite quantities approaching their limits, by appealing to Lagrange's theory of analytical functions in which the derivatives of a function are defined as the coefficients of the terms in its expansion in a Taylor series. Hopkins' notes on the 'Differential

(39) Whewell to Forbes, 20 April 1852 (Forbes Papers, St Andrews University Library, 1852/44a). Forbes replied on 2 May 1852 (Trinity College, Add. MS. a.204[103]): 'I was glad to see in the newspapers & also to hear from you that Clerk Maxwell got his scholarship. Pray do not suppose, though I take an interest in him, that I am not aware of his exceeding uncouthness, as well Mathematical as in other respects; indeed, as he has passed through my examinations, I have been a sufferer from it, & cannot flatter myself that I exercised almost any perceptible influence on him. I thought the Society & Drill of Cambridge the only chance of taming him, & much advised his going; but I have no idea that he will be Senior Wrangler. But he is most tenacious of physical reasonings of a mathematical class, & perceives there far more clearly than he can express. This (in my experience) is a rather rare characteristic, & I should think he might be a discoverer in those branches of knowledge, as indeed he already partly is'.

(40) *Life of Maxwell*: 154. See Wilson, 'The educational matrix': 14–19.

(41) George Gabriel Stokes, Lucasian Professor of Mathematics, lectured on 'Hydrostatics, pneumatics and optics, with particular reference to the physical theory of light'; James Challis, Plumian Professor Astronomy and Experimental Philosophy, lectured on 'Practical Astronomy'; George Peacock, Lowndean Professor of Astronomy and Geometry, lectured alternately on 'Astronomy' and 'Geometry and the general principles of mathematical reasoning'; and Robert Willis, Jacksonian Professor of Natural and Experimental Philosophy, lectured on 'Statics, dynamics, and mechanism, with their practical applications'. See *The Cambridge University Calendar for the Year 1850* (Cambridge, 1850): 132–8. Maxwell attended Stokes' lectures in 1853 (see Number 39) and Willis' in 1855 (see Number 79).

(42) Harvey Becher, 'William Whewell and Cambridge mathematics'. *Historical Studies in the Physical Sciences*, **11** (1980): 1–48; and I. Grattan-Guinness, 'Mathematics and mathematical physics from Cambridge, 1815–40: a survey of the achievements and of the French influences', in Harman, ed., *Wranglers and Physicists*: 84–111.

& Integral Calculus', copied by Maxwell and before him by Stokes,[43] reflect this approach.[44] Whewell had however successfully insisted on the educational value of traditional geometrical methods, and urged a more intuitive approach to the foundations of the calculus. The first part of the Tripos examination, devoted to the more elementary parts of mathematics, included (in the 1850s) 'the 1st, 2nd and 3rd Sections of *Newton's Principia*; the Propositions to be proved in Newton's manner'.[45] The second and more advanced part of the Tripos emphasised, along with the calculus and geometry, the study of 'mixed mathematics', including mechanics, hydrodynamics, astronomy and the theory of gravitation, and geometrical and physical optics. Whewell's *Treatise on Dynamics*, Pratt's *Mathematical Principles of Mechanical Philosophy*, and Airy's *Mathematical Tracts* were important Tripos texts.[46] In 1849 the Board of Mathematical Studies had recommended that 'the Mathematical Theories of Electricity, Magnetism, and Heat, be not admitted as subjected of examination'[47] in line with current practice, though questions on these topics were set in the papers for the Smith's Prizes for which the high 'wranglers' in the

(43) Maxwell's Cambridge notes for the Mathematical Tripos were apparently copied from texts supplied by Hopkins. There is a close similarity, often an identity, between Maxwell's notes and those transcribed by Stokes and Thomson a decade earlier. For comparison between Maxwell's notes on the calculus (ULC Add. MSS 7655, V, m/7) with Stokes' notes (Add. 7656, PA 2) see note (44); for comparison between Maxwell's notes on hydrodynamics (Add, 7655, V, m/8) with notes by Stokes (Add. 7656, PA 19) and Thomson (Add. 7342, PA 16) see Number 61 note (4); and for comparison between Maxwell's notes on dynamics (Add. 7655, V, m/10) and notes by Stokes (Add. 7656, PA 6) and Thomson (Add. 7342, PA 11) see Number 105 note (8). Thomson's notes on 'Hydrostatics' are endorsed: '(Manuscript given in Sep. & Oct. 1843, when I was at home. Commenced copying Oct. 20th 1843).' (Add. 7342, PA 16, f. 1), suggesting transcription from Hopkins' text.

(44) Maxwell's notebook 'Differential & Integral Calculus' (ULC Add. MSS 7655, V, m/7, ff. 1–10) and Stokes' notes 'Differential Calculus N°.1'(Add. 7656, PA 2). A discussion of Taylor's theorem leads to the definition of a derivative as the coefficients of the terms in the Taylor series; and there is subsequently a definition of a derivative by a limiting ratio, yet without a definition of limits. Compare the similar approach adopted by W. H. Miller, *An Elementary Treatise on the Differential Calculus* (Cambridge, 1833): 3. Cauchy's 'arithmetical' approach to the calculus, based on a rigorous definition of limits by finite differences made as small as desired, did not find favour in Cambridge. See George Peacock's discussion in his 'Report on the recent progress and present state of certain branches of analysis', *Report of the Third Meeting of the British Association for the Advancement of Science* (London, 1834): 185–352, esp. 247–8n. See also P. M. Harman, 'Newton to Maxwell: the *Principia* and British physics', *Notes and Records of the Royal Society*, **42** (1988): 75–96, esp. 80–82.

(45) *Cambridge Calendar for 1850*: 12.

(46) See Numbers 38 note (3) and 109 note (14); and Wilson, 'The educational matrix': 15–17.

(47) Report of the Board of Mathematical Studies, 19 May 1849 (Cambridge University Archives), quoted in Wilson, 'The educational matrix': 16.

Mathematical Tripos competed. The topics covered in Maxwell's Cambridge notes reflect Hopkins' concern with 'mixed mathematics'.[48] While Tait later recollected that '[Maxwell] to a great extent took his own way', formal study with Hopkins was demanding.[49] He did however write two short papers during these undergraduate years (Numbers 38 and 44), and made good progress on a third substantial paper on the geometry of surfaces (Numbers 45, 46 and 47).

Trinity however offered a wide range of friendships. In the winter of 1852–3 he was elected a member of the Apostles Club, and the essays he read to the Society between 1853 and 1856 (Numbers 42, 48, 62 and 88) shed light on his intellectual development. Two friendships in particular, with Richard Buckley Litchfield and Cecil James Monro,[50] led to a significant correspondence; and his letters to them (Monro having scientific interests as well) abound in allusions to his scientific work and to the activities of this generation of Cambridge Apostles. Along with other members of his circle he became active in F. D. Maurice's Christian Socialist movement,[51] and continued to teach evening classes for artisans in Aberdeen and London. His work in mathematics excited Hopkins' admiration,[52] but the strain of Cambridge work was not without its effect. In the summer of 1853 he became ill with 'a sort of brain fever'[53] which occasioned an emotional and religious crisis (Number 40), reinforcing a dimension to his thought[54] which appears most explicitly in some of his letters to Campbell and Litchfield (Numbers 35 and 147) and in his inaugural lectures (Numbers 105 and 183), where he presents a view of science inspired by religious values.

In the Tripos examination in January 1854 Maxwell graduated second wrangler to E. J. Routh, but was bracketed equal Smith's Prizeman with Routh in the examination for Smith's Prizes which followed. The questions in

(48) Notebooks on 'Astronomy', 'Differential Equations', 'Differential & Integral Calculus', 'Hydrostatics, Hydrodynamics, & Optics', 'Lunar Theory & Rigid Dynamics', 'Statics Dynamics' (ULC Add. MSS 7655, V, m/5–10). Hopkins follows Whewell's treatment in *A Treatise on Dynamics* (Cambridge, 1823): 4, in placing emphasis on Newton's laws of motion (rather than, with Lagrange, D'Alembert's principle and the principle of virtual velocities) as foundational to the science of mechanics. See Maxwell's notebook 'Statics Dynamics' (Add. 7655, V, m/10, ff. 58–60), and Harman, 'Newton to Maxwell': 85.

(49) Tait, 'James Clerk Maxwell', passage quoted in *Life of Maxwell*: 133; and see *ibid.*: 170–76.

(50) See Numbers 42 notes (1) and (9) and 56 note (1).

(51) See Numbers 42, 82 and 98; *Life of Maxwell*: 192, 194, 217, 218; and Number 82 note (3).

(52) See Number 92 note (2).

(53) *Life of Maxwell*: 170, 173.

(54) See Maxwell's comment, as quoted by Galton: 'given to theological ideas and not reticent about them', in Hilts, 'A guide to Galton's *English Men of Science*': 59. See also Paul Theerman, 'James Clerk Maxwell and religion', *American Journal of Physics*, **54** (1986): 312–17.

the Tripos papers were strongly geometrical and physical in emphasis, the questions for the Smith's Prizes (set by James Challis,[55] Whewell and Stokes) overwhelmingly so: Euclid, *Principia*, astronomy and optics figure prominently in the Smith's Prize examination.[56] The next step was to be elected to a Fellowship at Trinity, but he was unsuccessful at his first attempt in October 1854. Forbes wrote to Whewell of Maxwell's 'disappointment of a fellowship', and in reply Whewell observed that 'I am sorry that Maxwell is in anxiety. I consider his chance of a Fellowship here next to certain, though it would be well that he should attend to his classics more than he has done, and give some neatness and finish to his mathematics: but I should like to know that he would be willing to labour in College as a mathematical lecturer for some years when he is elected ... if you should happen to know Maxwell's intentions and plan of life, I should be obliged if you would tell me'[57] In the event he was elected a Fellow of Trinity in October 1855, and immediately found himself engaged in college lecturing (Numbers 73, 76, 77, 78 and 79).

Electricity and magnetism: Faraday's lines of force

During his years as an undergraduate Maxwell had maintained contact with William Thomson (Number 37); having been introduced to the study of magnetism by Thomson in 1850 (Number 33) he had followed Thomson's work on the subject (Number 36). Writing to Thomson in February 1854 (Number 45) he refers to a discussion they had had (presumably the previous summer) on the bending of surfaces; and also declares his intention 'to attack Electricity'. His paper on the representation of the geometry of surfaces by 'lines of bending' develops work by Gauss on the curvature of surfaces (Numbers 46 and 47); by the process that Maxwell later termed the 'cross-fertilization of the sciences'[58] this suggests links to his geometrical theory of Faraday's lines of force. The geometrical relations between lines and surfaces became a crucial element in Maxwell's field theory of electricity and magnetism, where the use of vectors, integral theorems and topology is central to his mathematical method.[59] In choosing electricity as a field of inquiry he

(55) See note (41).

(56) *The Cambridge University Calendar for the Year 1854* (Cambridge, 1854): 372–416.

(57) Forbes to Whewell, 29 December 1854 (Trinity College, Add. MS. a.204[112]); Whewell to Forbes, 10 March 1855 (Forbes Papers, St Andrews University Library, 1855/40a). Maxwell's letter to Campbell of 21 November 1865, recollecting his 'bad classics', will be published in Volume II.

(58) *Scientific Papers*, **2**: 744. See Everitt, 'Maxwell's scientific creativity': 122–3.

(59) P. M. Harman, 'Mathematics and reality in Maxwell's dynamical physics', in *Kelvin's Baltimore Lectures and Modern Theoretical Physics*, ed. R. Kargon and P. Achinstein (Cambridge, Mass./London, 1987): 267–97.

was selecting an area of physics at the forefront of current research; and he was naturally drawn to the work of Michael Faraday and Thomson.

Faraday's extraordinary series of experimental discoveries, of electromagnetic induction, the laws of electrochemistry, and magneto-optical rotation, had formed (in Maxwell's words) 'the nucleus of everything electric since 1830' (Number 144). In studying dielectric and electrolytic processes Faraday had emphasised the transmission of forces mediated by the action of contiguous particles of matter in the space between charged bodies. He explained paramagnetism and diamagnetism in terms of the propensity of lines of force to pass through different substances. The explanation of these magnetic phenomena became central to Maxwell's work (Number 84). Faraday's theory of the 'magnetic field' sought to explain magnetic and electromagnetic phenomena in terms of the action of lines of force in space. Faraday's view of electrostatics had been generally thought to be incompatible with the mathematical theory of electrostatics based upon direct action at a distance; but in 1845, in the first of a series of papers on the theory of electricity, Thomson established that Faraday's theory and mathematical electrostatics could be reconciled. Thomson's argument drew upon the analogy between electrostatics and heat conduction that he had himself proposed, an analogy which opened up important applications of the theorems of potential theory, suggesting links between heat conduction, electrostatics, and fluid flow. Thomson published a number of mathematical papers in the 1840s in which he derived theorems relevant to this framework of physical theory; and he deployed some of these ideas in a systematic paper on the theory of magnetism.[60]

Responding to Thomson's advice about a course of reading on electricity Maxwell made rapid progress (Number 50); and he wrote to Thomson in November 1854 to outline 'the confessions of an electrical freshman' (Number 51), speculations about electricity and magnetism which laid the foundations, albeit in preliminary form, for some of the main arguments of the paper 'On Faraday's lines of force' that he wrote in the winter of 1855–6. He announces his commitment to the Faraday–Thomson theory of 'magnetic lines of force' and the 'magnetic field'; his comprehensive conceptual and mathematical transformation of their ideas reinforced his basic adherence to this point of view. He perceived that Faraday's concept of lines of force could be employed as a purely geometrical representation of the structure of the field, while Thomson's 'analogy of the conduction of heat' provided a geometrical analogy between the flux of heat and the flow of electric force across isothermal

(60) On the term 'magnetic field' see Number 51 note (6). Faraday's papers are collected in his *Electricity*, Thomson's in his *Electrostatics and Magnetism*. See also Number 51 notes (3), (7) and (19), Number 66 note (21), Number 71 notes (12) and (18), and Number 84 notes (7), (14) and (37).

and equipotential surfaces. Drawing on Thomson's exposition of the theorems of potential theory and his statement of propositions in the theory of magnetism, Maxwell began to formulate a theory to represent the number of lines of magnetic force that pass through the surface enclosed by an electric circuit.[61] These ideas formed the basis of his subsequent work. This letter also documents his breadth of reading on the subject, including memoirs by Ohm, Neumann and Kirchhoff.

Thomson had noted the similarity between an expression for the distribution of magnetism and the equation of continuity for the flow of an incompressible fluid,[62] and Maxwell began to pursue the implications of this analogy (Numbers 61 and 63), drawing on the series of 'Notes on hydrodynamics' that Stokes and Thomson had published in the late 1840s.[63] On writing to Thomson in May 1855 (Number 66) he states theorems expressing the stability of fluid motion; these theorems form the basis of his subsequent analogy between lines of force and lines and tubes of motion in an incompressible fluid. He had also continued his study of the literature, and by this time had achieved a grasp of the electrodynamic theories of Ampère and Wilhelm Weber. In contrast to the field theory which Maxwell was striving to develop, these physicists explained electrodynamics in terms of the direct action between current elements.[64] In a further letter to Thomson written in September 1855 (Number 71) he gives a more complete account of the scope of his work, with detailed allusions to papers by Faraday (on lines of force and diamagnetism) and Thomson (including his 'allegorical representation' of electrostatics by heat conduction, and his theorem on Poisson's equation for the potential), which had helped to shape the structure of his evolving theory. Maxwell aimed to explain many of the magnetic and electromagnetic phenomena discovered by Faraday, but he insisted that his theoretical model should be understood as 'a collection of purely geometrical truths embodied in geometrical conceptions of lines, surfaces &c'.[65]

This emphasis on the physical geometry of fluid flow as an analogy for lines of force shapes his exposition in the first part of his paper 'On Faraday's lines of force' (Numbers 83 and 84), which he presented to the Cambridge Philosophical Society in December 1855 (Number 85). He applies the analogy to the theory of electrostatics, dielectrics, paramagnetism and diamagnetism, to Faraday's ideas on the magnetic properties of crystalline materials, and to

---

(61) Number 51 notes (3), (7), (10) to (13), and (19).
(62) Number 51 note (16).
(63) *Camb. & Dubl. Math. J.*, **2** (1847): 282–6; **3** (1848): 89–93, 121–7, 209–19; **4** (1849): 90–94, 219–40.
(64) Number 66 notes (3) and (5).   (65) Number 71 notes (7) to (10), (12) and (13).

electric currents (Number 84). The second part of the paper, concerned with the theory of electromagnetism, was read to the Society in February 1856 (Number 87). Drawing upon many of the ideas which he had outlined in his letters to Thomson, which were now reformulated and developed systematically, he proposes a theory based on his distinction between electric and magnetic 'quantities' (acting through surfaces) and 'intensities' (acting along lines).[66] The second part of the paper was not complete however. In a further letter to Thomson, written in February 1856 (Number 91), he is still considering the supplementary examples which were appended to the paper, and seeking to investigate by experiment the flux of lines of force passing through a circuit.[67] The final letter of the series, written in April 1856 (Number 100), shows him struggling to express the analytical theorems of the second part of 'On Faraday's lines of force' in their finished form, once again drawing on propositions derived by Thomson which he refashioned into his newly-wrought theory of lines of force. An essay on 'Analogies in Nature' (Number 88) which he read to the Apostles in February 1856 enabled him to expound his view of the principle of analogy which had helped to shape his approach to the theory of lines of force. The style of this essay is informal and discursive, and his argument cryptic, but the essay is important in indicating the scope of his reading (with echoes of Hamilton and Whewell) and his depth of thought on the epistemological problems bearing on his scientific work.[68] This early work on field theory contains the rudiments of some of Maxwell's most important ideas on the subject.[69] From Faraday the paper drew the reaction that 'I was at first almost frightened when I saw such mathematical force made to bear upon the subject and then wondered to see that the subject stood it so well',[70] leading Maxwell to respond with a wide-ranging exposition of the principles of field theory (Number 133).

Colour vision and optics

In reporting his conversation with Maxwell in December 1854 to Whewell, Forbes remarked that '[Maxwell] has made some most ingenious experiments & deductions about Combinations of Colours, a subject which once occupied

(66) See Number 51 notes (7) and (12).

(67) See also Number 87 esp. note (19).

(68) Harman, 'Edinburgh philosophy and Cambridge physics': 212–14.

(69) Everitt, *James Clerk Maxwell*: 87–93; M. N. Wise, 'The mutual embrace of electricity and magnetism', *Science*, **203** (1979): 1310–18; and Harman, 'Mathematics and reality in Maxwell's dynamical physics': 269–73.

(70) Faraday to Maxwell, 25 March 1857; see Number 133 note (5).

me a good deal'.[71] Forbes had in fact introduced Maxwell to experiments on colour mixing in the summer of 1849,[72] at a time when he had himself been engaged in a systematic review of the problem of the classification of colours.[73] Forbes' experiments consisted in observing the hues generated by adjustable coloured sectors fitted to a rapidly spinning disc, using tinted papers supplied by D. R. Hay, whose *Nomenclature of Colours* exhibited an elaborate system of colour plates which distinguished variations in colour.[74] Forbes adopted a triangular classification of colours, derived from schemes which had been used by Tobias Mayer and J. H. Lambert (and later by Thomas Young); assuming red, blue and yellow primary colours which are placed at the vertices of a triangle, Forbes showed that points within the triangle represent different colour combinations.[75] Maxwell's approach to colour vision was shaped by the attempts by Hay and Forbes to provide a nomenclature for the classification of colours, by Forbes' use of a colour triangle to represent colour combinations, and by Forbes' method of experimentation. He performed experiments on the mixture of coloured beams of light in August 1852,[76] and on graduating at Cambridge he continued this interest (Number 47), which had broadened to include the problem of colour blindness which had been investigated by another Edinburgh acquaintance, George Wilson.[77]

The study of colour vision had been significantly advanced in the early 1850s in major papers by Helmholtz and Grassmann,[78] work which Maxwell absorbed into his own theory. But the ideas of the two leading British colour theorists, Newton and Young, were especially important in shaping the development of his work. In the *Opticks* Newton explains colour mixing by his colour circle, each arc of the circle representing one of the seven principal colours. The circle enables the compound colour resulting from any mixture of colours to be determined.[79] While Newton distinguished his principal colours from the painters' triad of primary colorants (red, yellow and blue), he supposed the identity of the mixing rule for lights and pigments. As Maxwell recalled to Forbes in May 1855 (Number 64), Forbes himself had found that blending yellow and blue by the rapid rotation of coloured sectors on a disc did

---

(71) Forbes to Whewell, 29 December 1854; see note (57). See P. D. Sherman, *Colour Vision in the Nineteenth Century. The Young–Helmholtz–Maxwell Theory* (Bristol, 1981).

(72) See Number 64, esp. notes (2) and (5).

(73) J. D. Forbes, 'Hints towards a classification of colours', *Phil. Mag.*, ser. 3, **34** (1849): 161–78.

(74) See Numbers 54 notes (6) and (11) and 64 note (8).

(75) See Number 54 note (8).   (76) *Scientific Papers*, **1**: 144.

(77) See Number 54 note (1).   (78) Number 54 notes (3) and (5).

(79) Number 54 note (9); and see Alan E. Shapiro, 'The evolving structure of Newton's theory of light and colour', *Isis*, **71** (1980): 211–35, esp. 234–5.

not produce green.[80] The conventional mixing rule had been challenged by Helmholtz, who found that in experiments on the mixture of spectral colours 'yellow and blue do not furnish green'. Helmholtz explained this by maintaining that the mixture of coloured lights is an additive process, while pigment mixing is subtractive.[81]

In adopting this interpretation Maxwell also incorporated two further developments in colour theory. The first was Young's three-receptor theory of colour vision; his claim that red, green and violet should be considered as the primary constituents of white light; and the suggestion by Young and John Herschel that John Dalton's insensibility to red light was caused by the absence of one of the three receptors.[82] The second was Grassmann's theory that there are three variables of colour vision (spectral colour, intensity of illumination, and the degree of saturation); and his demonstration that this method of representing colour mixtures could be expressed by loaded points on Newton's colour circle.[83] In his studies of colour vision (Numbers 54 and 59) Maxwell argues that these colour variables could be represented on a colour diagram based on three primary colours. He follows Young in adopting red, green and violet as the primary colours; and his colour diagram (Number 54) incorporates Newton's colour circle and the triangular scheme for the classification of colours based on three primary colours. The relation between these two methods of representing colours then 'becomes a matter of geometry'.[84]

In making his experiments Maxwell modified the colour disc, adding a second set of sectors of smaller diameter so that colour comparisons could be made. He obtained observations for several groups of observers, generating colour equations which could be manipulated algebraically. He concluded his work by investigating the colour vision of colour-blind observers, suggesting that colour blindness could be explained (as Young and Herschel had suggested) using only two colour variables. To achieve greater accuracy in the observations he had an improved colour top made by an Edinburgh instrument-maker (Number 58), which he used to obtain systematic observations for his published paper (Number 59).[85]

(80) See Number 64 notes (2) and (5). But as Maxwell pointed out to Forbes (Number 64), in his 'Hints towards a classification of colours' Forbes had followed Newton in stating that yellow and blue rays mixed will produce green; see Number 64 note (7).

(81) H. Helmholtz, 'On the theory of compound colours', *Phil. Mag.*, ser. 4, **4** (1852): 519–34, esp. 528. See Number 64 note (2).   (82) Numbers 47 note (12) and 54 note (7).

(83) On the terminology employed in classifying these variables see Number 54 note (6).

(84) *Scientific Papers*, **1**: 135.

(85) Maxwell is holding this colour top in the photograph (taken in 1855 when he was elected a Fellow of Trinity) reproduced as the frontispiece. See Number 58 for a description of this colour top made by J. M. Bryson; and plate III.

Maxwell's early work on colour vision took the science of colorimetry to a new level of sophistication. He extended and synthesised the methods of his predecessors and established quantitative techniques. His interest in optics, manifest since his youthful endeavours, was not limited to work on colour vision. Since his Edinburgh days he had been interested in geometrical optics and optical instruments (Numbers 20 and 44), and around October 1855, encouraged by the suggestion that he should write a text on optics (Number 75), he began to study geometrical optics in earnest (Number 80). In a manner that is typical of Maxwell's scholarly research methods, he was led to classic work on geometrical optics, in particular to the account in Robert Smith's *Opticks* of a theorem due to Roger Cotes. Developing this work Maxwell advanced a new approach to the subject which improved on the contributions of Euler, Gauss and Möbius. He proposed a new theory of optical instruments in which theorems expressing geometrical relations between an object and image are separated from discussion of the dioptrical properties of lenses (Numbers 90, 91 and 93). This approach was subsequently developed independently by Ernst Abbe, and these methods have become standard in geometrical optics.[86] Following from his interest in optical instruments, an offshoot of this work was his design of a mirror (reflecting) stereoscope (Numbers 91, 94, 96 and 97).

Marischal College, Aberdeen

On taking up his fellowship in October 1855 Maxwell had probably anticipated staying at Trinity for some years as a lecturer in mathematics. But in February 1856 Forbes wrote to him to report that the Professorship of Natural Philosophy at Marischal College, Aberdeen was vacant.[87] Maxwell decided to apply, soon securing an impressive clutch of testimonials.[88] According to Campbell Maxwell decided to apply for the post to please his father: the arrangement of session and vacation time would enable him to spend the whole

(86) Numbers 91 notes (8), (9) and (10) and 93 note (5).

(87) Forbes to Maxwell, 13 February 1856 (*Life of Maxwell*: 250); see Number 92 note (2).

(88) Maxwell's testimonials for Marischal College were reprinted in a new set of testimonials when he applied for the Chair of Natural Philosophy at Edinburgh University in 1859; these are preserved in Glasgow University Library (Y1–h.18). He obtained testimonials from J. D. Forbes [see Number 92 note (2)], Sir William Hamilton [see note (25)], Philip Kelland and George Wilson (Edinburgh); William Thomson [see Number 94 note (5)], Sir John Herschel [see Number 97 note (2)], Sir David Brewster (eminent physicists); William Hopkins [see Number 92 note (2)] (Cambridge coach); the Master and Fellows of Trinity College, Arthur Thacker [see Number 92 note (5)], W. C. Mathison (Tutors of Trinity); William Whewell, James Challis, G. G. Stokes [see Number 97 note (4)] (Smith's Prize examiners 1854); Joseph Wolstenholme, William Walton, Percival Frost (Mathematical Tripos examiners and moderators 1854); Frederick Fuller [see Number 96 note (2)] (former Tutor of Peterhouse).

summer (April to October) at Glenlair. Even though John Clerk Maxwell died before the appointment was made, Maxwell decided to accept the post when it was offered.[89] He left Cambridge in early June 1856,[90] and commenced his duties in Aberdeen with an inaugural lecture delivered on 3 November 1856 (Number 105). In this lecture he presents a broad review of the scope of physics in relation to other forms of knowledge. He places special emphasis on the importance of a unified science of physics based on the programme of mechanical explanation, which sought to explain physical phenomena in terms of the structure and laws of motion of a mechanical system. This approach was associated in particular with the objectives of Cambridge 'mixed mathematics', but this view of science, and the scope of physics as Maxwell presents it, was also that of Forbes in his Edinburgh lectures.[91] Maxwell emphasises the relations of the 'physical sciences' (of elasticity, heat, optics, electricity and magnetism) to the 'mechanical sciences'. His philosophical remarks echo expressions in Whewell's *Philosophy of the Inductive Sciences* rather than the Baconian methodology presented by Forbes; and his view of the aims of science is strongly religious in tone.

Maxwell devoted considerable effort to his teaching duties at Marischal College. He probably began the course with a lecture on the 'Properties of Bodies' (Number 106); Forbes had commenced with this topic in 1847 when Maxwell had attended his Edinburgh lectures.[92] In early June 1857, perhaps taking advantage of a lull in his activities following the submission of his Adams Prize essay on Saturn's rings (Number 107) and the completion of his paper on the 'dynamical top' (Number 116),[93] he wrote a general summary of elementary mechanical principles (Numbers 120 to 124), presumably for his lectures in the next session. He gave information about his teaching during his first year at Marischal College in several of his letters (Numbers 108, 109, 112, 113, 114, 115, 118, 119). The course consisted largely of mechanics with some attention to the theory of heat and geometrical optics at the end of the session. This course of study was maintained during his following three sessions at Marischal College.[94] He prepared the text for an introductory lecture for the 1857–8

(89) *Life of Maxwell*: 247–57. See Number 99.

(90) Maxwell to R. B. Litchfield, 4 June 1856 (Trinity Add. MS. Letters. c.1[82]) (= *Life of Maxwell*: 256–7).

(91) Wilson, 'The educational matrix': 19–26.

(92) See Number 8.

(93) The essay on 'Saturn's rings' was completed by December 1856 (see Number 107 note (2)). The paper 'On a dynamical top' (Number 116) was presented to the Royal Society of Edinburgh on 20 April 1857; the paper as published contains revisions dated 7 May 1857 (*Scientific Papers*, **1**: 250).

(94) As recorded in his notebooks on 'Examinations in Natural Philosophy' and 'Exercises in Natural Philosophy Course' (ULC Add. MSS 7655, V, k/3, 4).

session (Number 132); Forbes and Thomson commenced with an introductory review.[95] The lecture gives a general introduction to the scope of the course, and is clearly intended only for the students attending during the session, being less philosophical and elevated in tone than the formal inaugural lecture of the previous year. In this session he began an advanced class, adapting the practice of Forbes, who divided his class into three divisions, and Thomson (at Glasgow), who divided his course into mathematical and experimental parts and allowed the abler students to continue in the class for a second year to do more advanced work. Maxwell's advanced students worked on physical astronomy, optics and electricity and magnetism, as well as studying the first three sections of Book I of *Principia* (Numbers 135, 136, 142, 144 and 146). This last topic formed part of the course of study for the Mathematical Tripos at Cambridge, and Maxwell clearly wished to introduce some mathematics into his lectures, again following Forbes' practice for his first-division students.

Saturn's rings

In a letter to Litchfield in July 1856 (Number 101) Maxwell remarked that he was engaged on the 'stiff... but curious subject' of the stability of Saturn's rings. This is his first reference to the subject of the Adams Prize for 1857 which had been formally announced in March 1855.[96] But a letter to Monro of October 1856 (Number 104) shows that he had made considerable progress with his prize essay (Number 107) by that date. Between April 1855 and April 1856 Maxwell was heavily occupied with his work on the theory of lines of force, but even had he been able to work on the Adams Prize problem at that time, correspondence on the subject would have been unlikely. Thomson, who was his chief scientific correspondent of the period, was one of the examiners for the prize (the others being James Challis, as Plumian Professor of Astronomy at Cambridge, and Stephen Parkinson, of St John's College, John Couch Adams' own college). Correspondence with Thomson on the subject before the award of the prize would have been impossible, though once the prize had been awarded to Maxwell no such restraint applied (Numbers 126, 128, 134, 137 and 143). The prize itself had been established in 1848 by members of St John's College in honour of Adams' prediction of the existence of the planet Neptune, and the subjects for the prize had been on celestial mechanics. But the prize had attracted few candidates on the three previous occasions it had been set, and had only been awarded once.[97] On this occasion Challis was determined

---

(95) Number 8; and Thompson, *Life of Thomson*, **1**: 190–2, 239–51.

(96) See Number 107 note (1).

(97) R. Peirson, 'The theory of the long inequality of Uranus and Neptune, depending on the near commensurability of their mean motions: an essay which obtained the Adams Prize for the

to excite greater interest among Cambridge graduates who alone were eligible to compete. Writing to Thomson in February 1855 he remarked that 'Cambridge mathematicians have no taste for investigations that require long mathematical calculations', and his suggestions for possible topics for the prize included an 'investigation of the perturbations of the forms of Saturn's Rings, supposing them to be fluid'.[98] After some discussion they reached agreement, and Challis and Parkinson drew up a draft notice on 'The mechanical stability of Saturn's Rings'. Thomson amended the title to 'The Motions of Saturn's Rings' on the grounds that 'it may perhaps be found that the Rings do not possess mechanical stability'.[99] This became the subject of the Adams Prize for 1857 as officially advertised.

The problem of the stability of Saturn's rings was, as Challis intimated to Thomson, consonant with 'the general tenor of Cambridge mathematics'.[100] Indeed, Whewell had set a Smith's Prize question in February 1854 (the year Maxwell was awarded the prize) requiring that candidates 'Shew that a fluid may revolve in a permanent annulus, like Saturn's Ring. How does it appear that Saturn's Ring is not a rigid body?'.[101] The second part of the question involved reference to Chapter 6 of Book III of Laplace's *Traité de Mécanique Céleste* where Laplace had established that the motions of a uniform solid ring were dynamically unstable; he concluded that the rings could be irregular solid bodies whose centres of gravity did not coincide with their geometrical centres.[102] The reference to the rotation of a fluid ring was probably prompted by Joseph Plateau's suggestion that the appearance of the rings of Saturn was analogous to the effect of rotation on a sphere of oil immersed in a mixture of alcohol and water, where the sphere is transformed into a ring.[103] The problem of Saturn's rings, as Challis was aware, had also excited some recent interest among astronomers. A dark 'obscure ring' interior to the two bright

year 1850, in the University of Cambridge', *Trans. Camb. Phil. Soc.*, **9** (1853): i–lxvii. The history of the prize subjects is described by Challis in a letter to William Thomson of 28 February 1855 (ULC Add. MSS 7342, C 76A), published in *Maxwell on Saturn's Rings*, ed. S. G. Brush, C. W. F. Everitt and E. Garber (Cambridge, Mass./London 1983): 6–7. Challis' 'Book of Minutes relating to the Adams Prize, kept by the Plumian Professor' (Cambridge Observatory Archives) is a record of the competition.

(98) Challis to Thomson, 28 February 1855, and a list of 'Subjects suggested for the Adams Prize' (ULC Add. MSS 7342, C 76A, C 76Aa), in *Maxwell on Saturn's Rings*: 6–7.

(99) Challis to Thomson, 14 and 23 March 1855 (ULC Add. MSS 7342, C 76B and C 76C), in *Maxwell on Saturn's Rings*: 8–10; and drafts relating to the Adams Prize competition on Saturn's rings (Cambridge Observatory Archives).

(100) Challis to Thomson, 14 March 1855 (note (99)).

(101) *Cambridge Calendar for 1854*: 413.

(102) Number 107 note (7).

(103) Number 107 notes (16) and (17).

rings had been first observed by George Phillips Bond in 1850;[104] and Otto Struve had recently claimed that the inner bright ring was approaching the planet.[105] Challis had remarked to Thomson that the problem of Saturn's rings acquired an interest because of Struve's 'singular conclusions', and that he himself considered Struve's 'evidence to be satisfactory'.[106] For these reasons the problem of the appearance and motion of Saturn's rings was a suitable subject for the prize.[107]

Maxwell's reasons for attempting the prize were no doubt mixed. The problem involved rigid body dynamics and questions of dynamical stability, and there was therefore a 'cross-fertilization of the sciences' with respect to his work on the 'dynamical top' at around the same time (Number 116).[108] The cash value of the prize, of 'about £130', was a not inconsiderable sum.[109] He may have wished to establish himself with a memoir on mathematical physics in the Cambridge style and which bore directly on Laplace's classical work on celestial mechanics. But the intrinsic interest of the subject will surely have weighed most heavily with him. The terms of the competition required the essays to be submitted by 16 December 1856. Maxwell met that deadline, and his essay (apparently the only one sent in) was first read by Challis, who received it back from Thomson in late April 1857. On 30 May 1857 Maxwell was awarded the Adams Prize, and his essay, bearing annotations by Challis and Thomson, was returned to him (Number 107).[110]

The essay submitted for the prize differs markedly from the memoir *On the Stability of the Motion of Saturn's Rings* that Maxwell published in 1859. These changes can be followed from the text of the essay, the annotations to this text by Challis and Thomson (Number 107), and from Maxwell's letters in the summer and autumn of 1857, especially his letters to Thomson and Challis (Numbers 126, 128, 134, 137 and 138). The essay is divided into two parts, the

(104) 'Inner ring of Saturn', *Monthly Notices of the Royal Astronomical Society*, **11** (1851): 20–27.

(105) *Ibid.*, **13** (1852): 22–4; and see Number 107 notes (1) and (80).

(106) Challis to Thomson, 28 February 1855 (note (98)).

(107) The Adams Prize notice on 'The Motions of Saturn's Rings' (Cambridge Observatory Archives) is reproduced in part in Number 107 note (1).

(108) For Maxwell's work on the 'dynamical top' in 1856 see Numbers 89 esp. note (4) and 101. For correspondences between the essay on Saturn's rings and 'On a dynamical top' see Numbers 107 note (76), 116 esp. note (5), and 128 esp. note (14).

(109) As stated in the Adams Prize notice (see note (107)). Compare Stokes' current salary of £155 p.a. as Lucasian Professor, as given in *The Cambridge University Calendar for the Year 1855* (Cambridge, 1855): 147.

(110) Challis to Thomson, 31 December 1856 (ULC Add. MSS 7342, C 76 D), in *Maxwell on Saturn's Rings*: 12; Challis to Thomson, 28 April 1857 (Add. 7342, C 76E); and Challis' 'Book of Minutes relating to the Adams Prize', entry for 30 May 1857. See Number 107 note (4).

first part being concerned with the motion of a rigid ring, the second with the motion of a fluid ring or a ring formed of disconnected particles. This division follows the terms suggested by the examiners. The mathematical argument rests on potential theory, Taylor's theorem and Fourier analysis, methods familiar to a Cambridge wrangler. He begins with the work of Laplace, seeking to determine the conditions under which the rotation of a solid ring would be stable, and concluding that such a ring would have to be so irregular 'as to be quite inconsistent with the observed appearance of the rings'. Nevertheless, his discussion of the stability conditions was not free from error, as Challis noticed and tried (without success) to remedy. Maxwell had erred in establishing the equations for the gravitational potential of the ring; and this entailed that a uniform solid ring would be stable, contrary to Laplace's demonstration.[111] Maxwell reworked the argument the following August, perceiving the source of his error (Number 128) and concluding, with Laplace, that a uniform solid ring would be unstable (Number 138).

On turning to the case of a fluid ring he established that if the ring were assumed to be at rest 'the whole ring would collapse into satellites'. If the ring were assumed to be in motion, however, he found that 'we are able to understand the possibility of the stable motion of a fluid ring'. The effect of any disturbance of the ring would be such as to produce waves in the fluid in the plane of the ring. He obtains a biquadratic equation for the angular velocity (relative to the rotation of the ring) with which a system of waves travels round the ring. The solution of this equation forms the core of the argument of the second part of the essay. On discussing the question (raised by Struve) of the possible change in form of the rings over time, he considers the effect of disturbing causes on the stability of the rings: these are the friction of the rings, and an external disturbing force due to the irregularities of the planet, the attraction of satellites, or the effect of the irregularities in neighbouring rings. He concludes that the result of a long-continued series of disturbances would 'make the outer rings extend farther from the planet and the inner rings come nearer to it'.

Maxwell's starting-point in commencing the revision of the essay was the annotations of Challis and Thomson to its text. He visited Cambridge to take his M.A. degree early in July 1857 (Numbers 118 and 119), shortly after the award of the Adams Prize, and took advantage of his visit to discuss his work with Challis and follow up some reading on the topic in the University Library.[112] He was also aided by letters from Thomson in commencing a sys-

---

(111) Maxwell failed to draw this conclusion as a result of a slip; see Part I Problem VI of Number 107, esp. note (31).

(112) See Number 126 esp. note (3).

tematic revision of the essay (Number 126). He soon corrected his treatment of the conditions of stability of a solid ring (Number 128), and then turned to the more difficult task of reconstructing the second part of the essay, on the conditions of stability of a fluid ring. He reported the results of his work to Thomson and Challis in November 1857 (Numbers 134 and 138), discussing the conditions of stability of possible systems of rings and concluding that the ring system consists of concentric rings of satellites: 'we are driven to a plurality of rings with independent angular velocities' (Number 138). Writing to Campbell in December 1857 he mentions a model which would exhibit the waves in a ring of satellites 'for the edification of sensible image worshippers' (Number 142); and describes the model in some detail in a letter to Thomson written early the following year (Number 143). After a respite of some months he recommenced work at Glenlair in late July 1858 (Number 151), completing the memoir for publication by early September (Numbers 152, 153 and 155).

The publication of the memoir *On the Stability of the Motion of Saturn's Rings* undoubtedly consolidated Maxwell's growing reputation as a mathematical physicist.[113] The essay was reviewed for the Royal Astronomical Society by the Astronomer Royal, George Biddell Airy, who concluded his report with the observation that 'the essay which we have abstracted is one of the most remarkable contributions to mechanical astronomy that has appeared for many years.'[114]

The kinetic theory of gases

While engaged in revising the essay on Saturn's rings Maxwell remarked to Thomson that the problem of determining the motions of 'a fortuitous concourse of atoms...is a subject above my powers at present' (Number 134), a point he repeated in the published memoir.[115] Some time between February and May 1859 he happened to notice the translation of a paper by Rudolf Clausius (published in the *Philosophical Magazine* of February 1859) on the kinetic theory of gases (Number 157). Clausius' paper led Maxwell to a study of the collisions of particles as a means of establishing the properties of gases.[116]

---

(113) For discussion of the memoir and its reception see Brush, Everitt and Garber, *Maxwell on Saturn's Rings*: 1–38.

(114) G. B. Airy, 'On the stability of the motion of Saturn's rings', *Monthly Notices of the Royal Astronomical Society*, **19** (10 June 1859): 297–304; reprinted in *Maxwell on Saturn's Rings*: 159–66.

(115) *Scientific Papers*, **1**: 354; see Number 134 note (17).

(116) Number 157 note (2). For discussion of the development of Maxwell's work on gas theory see *Maxwell on Molecules and Gases*, ed. E. Garber, S. G. Brush and C. W. F. Everitt (Cambridge, Mass./London, 1986): 1–63, esp. 4–20; S. G. Brush, *The Kind of Motion We Call Heat: a History of the Kinetic Theory of Gases in the Nineteenth Century*, 2 vols. (Amsterdam/New York, 1976); and Everitt, *James Clerk Maxwell*: esp. 131–7.

While his investigation of Saturn's rings had already alerted him to the problem of determining the motions of large numbers of colliding particles, two features of Clausius' work may have attracted his attention. In concluding his work on Saturn's rings Maxwell had drawn on data on gas viscosity, relevant to establishing the effect of friction in disturbing the stability of the rings, in a paper by Stokes on the damping of pendulums (Number 152). The study of the physical properties of gases was therefore not unfamiliar to him. The second significant feature of Clausius' paper arose in Clausius' explanation of the slow diffusion of gas molecules. Arguing that the molecules would repeatedly collide, he introduced a statistical argument to calculate the probability of a molecule travelling a given distance (termed the 'mean free path') without collision. Maxwell had been interested in probability theory as early as 1850 (Number 31). His interest may have been aroused by an essay by Sir John Herschel in the *Edinburgh Review* of July 1850 on Adolphe Quetelet's *Theory of Probabilities*; and he would have encountered the review when it was reprinted in Herschel's collected *Essays* which he read in the winter of 1857–8 (Numbers 142 and 144). In this review Herschel had responded to some criticisms of his own discussion of probabilistic arguments (made in connection with the distribution of the stars) by J. D. Forbes; and Forbes in turn then replied to Herschel.[117] Herschel's review was published when Maxwell was Forbes' student in Edinburgh, so it is likely that his comments on the logic of probability in a letter to Campbell in the summer of 1850 (Number 31) were occasioned by the review. There are in addition similarities between Maxwell's derivation of the velocity distribution law in his paper 'Illustrations of the dynamical theory of gases' and Herschel's proof of the law of least squares,[118] which suggests that Maxwell had studied Herschel's argument when reading his collected *Essays*. Even before he encountered Clausius' paper Maxwell had interests in the logic of probability, particle collisions, and gas viscosity.

In a letter to Stokes of May 1859 (Number 157) he outlines the main features of his theory of gases as subsequently published in 'Illustrations of the dynamical theory of gases'. Maxwell advanced on Clausius' procedure by introducing a statistical formula for the distribution of velocities among gas molecules, a distribution function which is identical in form to the distri-

---

(117) See Number 31 note (11). The impact of Herschel's review is suggested by C. C. Gillispie, 'Intellectual factors in the background of analysis by probabilities', in *Scientific Change*, ed. A. C. Crombie (London, 1963): 431–53; and for further discussion see the papers by Heimann, Garber and Porter cited in Number 157 note (16).

(118) J. F. W. Herschel, *Essays from the Edinburgh and Quarterly Reviews, with Addresses and other Pieces* (London, 1857): 398–400; *Scientific Papers*, **1**: 380–82; see Garber, Brush and Everitt, *Maxwell on Molecules and Gases*: 10.

bution formula in the theory of errors. This letter gives a full account of the genesis of his kinetic theory of gases. He emphasises that he had undertaken the study of the motions of particles as 'an exercise in mechanics'; but following Clausius' lead he immediately looked for confirmation of his argument in work on molecular science, specifically in experiments on gaseous diffusion. He hoped to be 'snubbed a little by experiments'. He presented an account of his preliminary results to the British Association meeting in Aberdeen in September 1859 (Number 160), and with his enthusiasm undampened (Number 163) he decided to publish his theory, of which he gave a more complete account at the Oxford meeting of the British Association the following year (Number 181).

Maxwell's description of physical processes by a statistical function was a major innovation in the science of physics, and one whose implications he was to pursue for the rest of his life. More immediately his treatment of heat conduction was criticised by Clausius in 1862,[119] and he perceived the need to modify his theory (Number 195), initiating revisions (Number 196) which ultimately were to lead him to a new formulation of the kinetic theory of gases, to discussions of the essentially statistical nature of the second law of thermo- dynamics, and to speculations about the limitations of statistical knowledge. In 1859 he was however concerned with formulating mechanical theorems expressing the regularity of the motions of gas molecules,[120] which suggests a 'cross-fertilization of the sciences' to his discussion of stability problems both in his treatment of stream lines in an incompressible fluid (Numbers 63 and 66) and in his theory of the motions of Saturn's rings (Number 107).

From Marischal College, to King's College, London; the Royal Society paper on colour vision

Two events, one private the other public, shaped the circumstances of Max- well's life at Marischal College. The first was his marriage to Katherine Mary Dewar, daughter of the Principal of the college, in June 1858 (Numbers 146, 147 and 151); the second was the appointment in 1857 of Commissioners to administer the Act of Parliament[121] which enacted the union of King's and Marischal Colleges to form the University of Aberdeen. The Commissioners had to decide whether to recommend either the union of the colleges (with the retention of existing classes) or their 'fusion', which would leave a single class

(119) See Number 195 note (8).

(120) See Theodore M. Porter, 'A statistical survey of gases: Maxwell's social physics', *Historical Studies in the Physical Sciences*, **12** (1981): 77–116, esp. 79, 97.

(121) 21 & 22 Vict. c. 83.

and professor in each subject and would have the consequence that redundant professorships would be abolished. Maxwell was at first strongly in favour of fusion (Numbers 140 and 142); but when the Commissioners reported early in 1860 recommending fusion they proposed that the Professorship of Natural Philosophy in the University of Aberdeen should be held by David Thomson, the King's College professor.[122] Writing to Monro in January 1860 Maxwell announced that 'I am to be turned out by the Commissioners' (Number 174). He declared his intention to fight his enforced redundancy, but all such efforts were unavailing and the colleges were fused. By this time he was a candidate for the Professorship of Natural Philosophy at Edinburgh University, which Forbes was resigning on his appointment to succeed Sir David Brewster as Principal of the United Colleges at St Andrews (Numbers 164, 166, 167, 168 and 174).[123] Maxwell was unsuccessful, his old school-fellow P. G. Tait being appointed.[124]

While all this was going on during the winter of 1859–60, Maxwell was heavily engaged in further work on colour vision. He had presented two papers, one on the mixture of the colours of the spectrum (Number 102) and the other developing his interest in the variations in colour sensitivity across the retina (Number 70), to the Cheltenham meeting of the British Association in 1856. These papers established the basis for his subsequent research on the problems of colour vision. Maxwell devised a 'colour box' to mix spectral colours, first using an apparatus of this type in 1852 and improving the design in 1855. With this latter version he was able to compare two different colour combinations, but the apparatus was unwieldy (Numbers 71 and 139). In July 1856 he constructed a smaller portable instrument which he displayed at the British Association meeting in August. This colour box used the method of folded optics, light being passed twice through a prism (Numbers 109 and 117); but

---

(122) See Number 174 note (6).

(123) Maxwell supplemented his testimonials for Marischal College (note (88)) with two additional sets of testimonials. He added testimonials from G. B. Airy, Astronomer Royal (see Number 167 note (2)), C. Piazzi Smyth, Astronomer Royal for Scotland (see Number 168 note (2)); and from several colleagues at Marischal College: John Cruickshank (Mathematics), the Rev. Robert J. Brown (Greek), the Rev. John Crombie Brown (Botany) John Macrobin (Medicine), Thomas Clark (Chemistry) and William Martin (Moral Philosophy); and from the Rev. James Smith, father of a pupil of Maxwell's (Glasgow University Library, Y1–h.18). On Maxwell's colleagues see [P. J. Anderson,] *Officers of the Marischal College and University of Aberdeen* (Aberdeen, 1897). A second series gave testimonials from John Webster (late Lord Provost of Aberdeen), Sir Thomas Blaikie (late Lord Provost and Dean of the Faculty of Marischal College), Robert Maclure (Professor of Humanity), F. Ogston (Medical Logic and Jurisprudence), and from the Very Rev. Daniel Dewar, (Principal of Marischal College and Maxwell's father-in-law) (see Number 174 note (8)) (ULC Add. MSS 7655, V, i/12).

(124) *Life of Maxwell*: 258–60, 277; and see Knott, *Life of Tait*: 16.

this instrument was only intended to exhibit spectral colour mixtures. As he informed Forbes in November 1857 (Number 139), he wished to compare spectral colour mixtures in juxtaposition, and this would require a colour box capable of yielding more precise observations. He had the improved colour box made the following year (Numbers 149 and 154), and had planned to do some experiments during the summer at Glenlair after his marriage in June 1858 (Numbers 148 and 149). But in the event the newly-weds were 'no great students' that summer, preferring the 'elemental influences of sun, wind & streams', though Maxwell did succeed in completing *Saturn's Rings* (Numbers 151 and 152). During the spring of 1859 he made some observations on the vision of a colour-blind student, using the colour top (Numbers 156 and 157); and he began to use the colour box for experiments on the mixture of the colours of the spectrum during the summer. Describing his work to the Aberdeen meeting of the British Association in September 1859 (Number 161), he explained that his new apparatus enabled mixtures of spectral colours to be directly compared with white light. The problem of differences in colour sensitivity across the retina also continued to interest him, and discussion of this question, especially the insensitivity of his own eyes to greenish-blue light, features prominently in his subsequent work (Number 170), especially in his letters to Stokes (Numbers 163 and 178).

In October 1859 he described his work to Stokes (Number 163), and explored the possibility of submitting a paper on colour vision to the Royal Society, of which Stokes was Secretary.[125] Meeting with Stokes' encouragement he completed the paper early in the new year (Numbers 169 to 172). He rounded off his work by re-modelling the portable colour box to compare colour combinations with white light (Numbers 168 and 169). He used this small colour box to obtain observations on colour-blind subjects (Numbers 175 and 180), writing up his conclusions as a 'postscript' to his Royal Society paper (Number 179). The substance of this was to reaffirm and develop his theory of colour vision and colour blindness. While the Royal Society paper was his last substantial contribution to the study of colour vision, he continued to engage in experiments on the subject. In 1862 he designed another colour box (Number 194), a more complex instrument which was based on an experimental arrangement described by Newton,[126] and which produced spectral hues of great purity. Maxwell used this instrument to study the problem of determining colour vision at different points of the retina.[127] His projection of the first trichromatic colour photograph in a lecture at the Royal

---

(125) Knowing that he had recently been nominated for a Royal Society Royal Medal for his work on colour vision: see Number 159 note (6).

(126) See Number 194 note (14).   (127) *Scientific Papers*, **2**: 230–32.

Institution in May 1861 (Number 184) gave a vivid demonstration of his theory of colour vision, even though the experiments succeeded through a combination of fortuitous circumstances.[128]

Like his memoir on *Saturn's Rings* the work on colour vision brought Maxwell rapid recognition. Writing to Stokes in January 1856 Thomson expressed doubts about Maxwell's early paper (Number 59), and in reply Stokes indicated disinterest in the subject; but in writing to Maxwell in November 1857 Stokes voiced enthusiastic approval of Maxwell's work.[129] At a meeting of the Council of the Royal Society in June 1859 Stokes and Whewell nominated Maxwell for a Royal Medal 'for his Mathematical Theory of the Composition of Colours, verified by quantitative experiments, and for his Memoirs on Mathematical and Physical subjects', and another unsuccessful nomination was made the following year. But a nomination in May 1860 by Stokes and the Cambridge Professor of Mineralogy W. H. Miller for the Rumford Medal (which was awarded especially for studies of heat and light), for his 'Researches on the Composition of Colours, and other Optical Papers', met with success.[130] In February 1860 he was appointed Bakerian Lecturer by the Council of the Royal Society, but did not read his paper 'On the theory of compound colours' as the Bakerian Lecture as he was found to be ineligible to do so, not being a Fellow of the Society.[131] Stokes had inquired in the past whether Maxwell wished to be a candidate for the Royal Society, and he had declined in January 1857 (Number 111); but in May 1860, perhaps because of the problem he had encountered over his appointment as Bakerian Lecturer, he asked Stokes to place his name on the list of candidates for the Society (Number 178), and he was elected the following year.[132]

Maxwell's failure to retain his post in Aberdeen and to be appointed at Edinburgh University had not discouraged his aim to continue a career as a university teacher. In July 1860 he was appointed Professor of Natural Philosophy at King's College, London, following the resignation of Thomas Minchin Goodeve.[133] Before taking up his professorship he fell ill with smallpox while at Glenlair, but was able to move to London and deliver his inaugural lecture at King's (Number 183). This lecture is similar in scope to his 1856 inaugural lecture at Marischal College (Number 105). The variations are due to differences between the two institutions; and to Maxwell's emphasis, in the

---

(128) See Number 184 note (12).

(129) Thomson to Stokes, 28 January 1856 (ULC Add. MSS 7656, K 89) and Stokes to Thomson, 4 February 1856 (ULC Add. MSS 7342, S 387), quoted in Number 72 note (7). See Stokes to Maxwell, 7 November 1857 (ULC Add. MSS 7655, II/12), reproduced in Number 139 note (2).

(130) See Number 176 note (6).   (131) See Numbers 175 note (2) and 176 note (6).
(132) See Number 178 note (9).   (133) See Number 183 note (3).

1860 lecture, on the status of the concept of energy and the principle of the conservation of energy as fundamental to the science of mechanics, reflecting developments in the scope of the science of physics during the 1850s, a conceptual transformation which Maxwell had outlined to Faraday in a letter of November 1857 (Number 133).

His duties at King's College required his attendance for three mornings a week, when he gave two lectures, one to first-year students and the other to second- and third-year students. He also taught an evening class on experimental physics, a course of 20 weekly lectures. In his first session at King's his syllabus for his elementary lectures followed Goodeve's practice. He announced lectures on mechanics, the properties of matter and heat, followed by a more detailed mathematical study of these problems; and the course concluded with experimental lectures on light. Maxwell's more advanced lectures were more mathematical in scope than Goodeve's, proposing the study of rigid body dynamics, the motion of an incompressible fluid and its application to electricity and magnetism, astronomy, and of waves and their application to sound and light, clearly bringing in some of his own interests in mathematical physics. In his second session (1861–2) he abandoned the more mathematical lectures for the first-year students.[134] The university year was longer than at Marischal College, stretching from early October to the end of June (Number 186), permitting a shorter summer in the relaxed environment of Glenlair.

Electricity and magnetism: molecular vortices, electromagnetism and light

Maxwell's major work in physics during his first two years at King's College was the preparation of his paper 'On physical lines of force', which was published in instalments in the *Philosophical Magazine* in March, April and May 1861 (Parts I and II) and in January and February 1862 (Parts III and IV). The basis of this physical theory of the electromagnetic field was a model of 'molecular vortices' oriented along magnetic field lines. In writing to Faraday in November 1857 (Number 133), Maxwell had looked to the development of his theory of lines of force in relation to Thomson's explanation of the Faraday effect, the rotation of the plane of polarisation of linearly polarised light by a magnetic field. Thomson supposed that this phenomenon was caused by the

---

(134) The duties of the professor are set out in the advertisement (King's College Archives, Special Committees Nᵒ. 2, f. 231). The syllabus for Goodeve's course is given in *The Calendar of King's College, London, for 1859–60*: 99–100; Maxwell's courses for his first two sessions are given in the *Calendar for 1860–61*: 115–17, and the *Calendar for 1861–62*: 113–15. See also C. Domb, 'James Clerk Maxwell in London 1860–1865', *Notes and Records of the Royal Society*, **35** (1980): 67–103.

rotation of molecular vortices in an ether, the axes of revolution of the vortices being aligned along the direction of the lines of force.[135] Thomson's paper was first published in 1856, and reprinted in the *Philosophical Magazine* the following March, and soon excited Maxwell's interest. Writing to Monro in May 1857 he remarked that he was working at 'a Vortical theory of magnetism & electricity which is very crude but has some merits' (Number 118). The problem of the rotation of molecular vortices in a fluid, of special interest to Thomson at the time,[136] is discussed in a letter to Thomson of November 1857 (Number 137); and in early 1858 (Number 143) he outlines an experiment on a freely rotating magnet which could establish the effect of revolving vortices within the magnet. He refers to this experiment again in letters to Faraday and Thomson of October and December 1861 (Numbers 187, 189 and 190), having had the apparatus constructed and having tried the experiment, though without success.[137] The theory of molecular vortices was therefore of long gestation.[138]

It is likely that Maxwell originally envisaged his paper 'On physical lines of force' as being in two parts, on the theory of molecular vortices as applied to magnetism and electric currents. But during the summer of 1861 'in the country' at Glenlair (Numbers 187 and 189), he developed his mechanical ether theory along new lines. He calculated the velocity of transverse elastic waves in a cellular ether, supposing the elastic properties of the ether to have electromagnetic correlates. He established 'the nearness between the two values of the velocity of propagation of magnetic effects and that of light' (Number 189). This discovery was apparently unexpected, leading him to conclude (in the third part of the published paper) that he could 'scarcely avoid the inference that *light consists in the transverse undulations of the same medium which is the cause of electric and magnetic phenomena*'.[139] To complete his theory of physical lines of force he wished to give a quantitative treatment of the Faraday effect in terms of the rotation of molecular vortices, and in October 1861 (Number 187) asked Faraday for information about experiments on the rotation of polarised light by magnets. It is possible that Faraday drew his attention to the work of Émile Verdet,[140] and Maxwell gave a preliminary account of his theory of the Faraday effect, which formed the substance of the

---

(135) See Number 133 note (16).

(136) Wilson, *Kelvin and Stokes*: 156–63. See Number 137 notes (2) and (5).

(137) See Number 187 note (20). The apparatus is preserved in the Cavendish Laboratory, Cambridge: see plate X.

(138) On Maxwell's theory of molecular vortices see the papers by Daniel Siegel cited in Number 189 notes (4) and (5).

(139) *Scientific Papers*, **1**: 500. See Numbers 187 note (15) and 189 note (18).

(140) See Faraday's annotation 'Verdet' to Number 187.

fourth part of 'On physical lines of force', in writing to Thomson in December 1861 (Number 189). Maxwell had unexpectedly established the basis for his 'electromagnetic theory of light';[141] and it was the development of this insight, first announced in some excitement to his two mentors in field theory Faraday and Thomson, that was to form the main thrust of his work during the next decade, and to lead him to write the *Treatise on Electricity and Magnetism*.

---

(141) Maxwell first uses the expression in his paper 'A dynamical theory of the electromagnetic field', *Phil. Trans.*, **155** (1865): 459–512, on 497 (= *Scientific Papers*, **1**: 577).

TEXTS

PAPER ON THE DESCRIPTION OF OVAL CURVES[1]

FEBRUARY 1846[2]

From the holograph copy by John Clerk Maxwell at the Royal Society of Edinburgh[3]

OBSERVATIONS ON CIRCUMSCRIBED FIGURES HAVING A PLURALITY OF FOCI, AND RADII OF VARIOUS PROPORTIONS[4]

by James Clerk Maxwell

Some time ago while considering the analogy of the Circle and Ellipsis – and the common method of drawing the latter figure by means of a cord of any given length – fixed by the ends of the foci – which rests on the principle,

---

(1) According to Lewis Campbell (*Life of Maxwell*: 73–4) Maxwell's interest was aroused by the work of the Edinburgh decorative painter D. R. Hay. See Hay, 'Description of a machine for drawing the perfect egg-oval; and of a method of producing curvilinear figures, on a principle whereby beauty of form may be imparted to ornamental vases and mouldings in architecture – to the works of the silversmith, brazier, and potter – equal to such works of the ancients', *Transactions of the Royal Scottish Society of Arts*, **3** (1851): 123–7 (read 9 March 1846). See also D. R. Hay, *First Principles of Symmetrical Beauty* (Edinburgh/London, 1846); Maxwell's copy of this work (Cavendish Laboratory, Cambridge) is endorsed: 'To Mr Clerk Maxwell Junr with the Author's Comp^{ts}. 23^d Feby 1847'.

(2) According to John Clerk Maxwell's 'Diary' for Wednesday 25 February 1846: 'Called on...Mr D. R. Hay at his house, Jordan Lane, and saw his diagrams and showed James's Ovals...'. Thursday 26 February: 'Call on Prof. Forbes at the College, and see about Jas. Ovals and 3-foci figures and plurality of foci. New to Prof. Forbes, and settle to give him the theory in writing to consider'. Monday 2 March: 'Wrote account of James's ovals for Prof. Forbes. Evening. – Royal Society with James, and gave the above to Mr. Forbes' (*Life of Maxwell*: 74). The ovals shown to Hay and Forbes are no doubt those reproduced here; the text of the manuscript is presumably the document prepared for Forbes by John Clerk Maxwell, no doubt a fair copy of a draft of Maxwell's.

(3) The text is in the hand of John Clerk Maxwell; the figures – and appended series of ovals – were undoubtedly drawn by Maxwell; see note (2).

(4) J. D. Forbes published an account of the paper, 'On the description of oval curves and those having a plurality of foci. By Mr Clerk Maxwell junior; with remarks by Professor Forbes', *Proc. Roy. Soc. Edinb.*, **2** (1846): 89–91 (= *Scientific Papers*, **1**: 1–3). The paper was presented on 6 April 1846, as John Clerk Maxwell recorded in his 'Diary': 'Royal Society with Jas. Professor Forbes gave acct. of James's ovals. Met with very great attention and approbation generally'. Writing to John Clerk Maxwell on 6 March 1846 Forbes declared the paper to be 'very ingenious – certainly very remarkable for his years; and, I believe, substantially new'; and having shown the paper to Philip Kelland he reported on 11 March that 'it is...we believe, a new way of considering higher curves with reference to foci'. Maxwell and his father visited Forbes on 17 March 'to discourse on the ovals' (*Life of Maxwell*: 75–6).

that the sum of the two lines drawn from the foci to any point in the circumference is a constant quantity,[5] it occurred to me that the *Sum* of the Radii being constant was the essential condition in all circumscribed figures, and that the foci may be of any number and the radii of various proportions.

This rule applies to the circle – here there is one focus & one length of radius. – In the Ellipsis there are two foci & to any point of the circumference two radii – the sum of which are constant – and the circle may be considered as drawn on the same principle supposing the two foci & radii to be conjoined.

Impressed with these views, I proceeded to put them to the practical test of tracing figures on the principle of the constant quantity of the radii – of various proportions. –

The most simple proportion of one to two was first tried with two foci – which I did merely by doubling the thread which was attached to one of the foci – *A* & *B* are the two foci.

A thread is fixed on the focus *B* at the other end is loop at *D* for the tracing point and the thread is passed round a pin in the focus *A* – the tracing point is then carried round the figure guided by the bight of the thread as shown by the red line[6] – by this arrangement the elliptical figure is modified by the preponderating influence of the double proportion of the Radius *AD* which enlarges that end of the figure and a fine oval form is produced. – The proportions of a figure produced in this way are affected by the distance between the foci.[7] The method I have adopted for giving a measure for fixing

---

(5)   The drawing of an ellipse by string 'employer par des Jardiniers', as described by Descartes in his *Dioptrique*; R. Descartes, *La Discours de La Methode...plus La Dioptrique, Les Meteores et La Geometrie qui sont des essais de cete Methode* (Leiden, 1637): 89–90.

(6)   There is no red line in Maxwell's figure.

(7)   In his published 'remarks' on Maxwell's paper Forbes notes 'the identity of the oval with the Cartesian oval', observing that 'the simplest analogy of all is that derived from the method of description'. In Maxwell's method a tracing pin describes an oval by moving so that m times its distance from one focus (ρ) together with n times its distance from another focus (ρ') is equal to a constant quantity; compare George Salmon, *A Treatise on the Higher Plane Curves* (Dublin, 1852): 119, where 'a Cartesian oval is defined as the locus of a point whose distances from two given foci $[\rho, \rho']$ are connected by the relation $m\rho + n\rho' = c$.' In Maxwell's description of ovals m and n are determined by the number of times the thread is wrapped round the pins placed at the two given foci; and Forbes observed that 'it probably has not been suspected that so easy and elegant a method exists of describing these curves by the use of a thread and pins whenever the powers of the foci $[m, n]$ are commensurable'. This restriction, that m/n is rational, does not limit Descartes' description of his first oval (using thread, pins and a ruler); see Descartes, *La Discours de La Methode...La Geometrie*: 356. Maxwell himself noted this feature of Descartes' description of ovals 20 years later, in a letter to J. J. Sylvester of 21 December 1866 (which will be reproduced in Volume II): '[Descartes] has overcome the difficulty of describing ovals with strings when the value of μ $[m/n]$ is incommensurable'.

that distance is to state the Angle at one of the foci *B* formed by the lines to the other focus & to a point in the circumference *C* the line *AC* being always drawn at a right angle to the Axis *AB* in this figure it is about 25° –

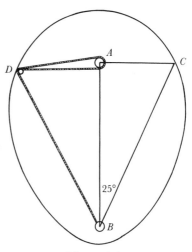

Figure 1,1

This figure may be defined either to be a figure with two foci in which the sum of one of the radii and twice the sum of the other is a constant quantity or it may be regarded as a figure of three foci – two of them being co-incident, and the sum of the three radii constant – as the circle may be said to be of any number of foci all coincident – for if in the figure, *A* & *B* were made to coincide then a circle would be traced with a triple radius.[8]

Having thus obtained the oval form I proceeded to draw it with the distances of the foci varied – on a separate paper are a series of examples.[9] It is obvious that if the distance of the foci is very small, the figure will differ but little from the circle. The first of the series is marked as drawn on an angle of 45° as explained above – then 30° – 27° – 24° – at this angle the focus of the single radius is in the circumference, & the figure is of a pointed form – at lower angles the focus is out of the figure & the small end again becomes more rounded as seen by the example of 22° 30′ – & 18° – in the last example the focus is supposed at an infinite distance – expressed by the parallel lines, in which case the figure is still of oval form – but the difference in the two ends is infinitely small.

Having investigated the varieties of form producible for the two foci with the radii – as one & two – I have drawn figures with the proportion of 1 to 3 – 1 to 4 – 2 to 3 – 3 to 4 – all done by lapping the tracing thread so many times round the pins placed in the foci. The greater the difference in the proportion such as 1 to 3 the figure of the Oval next to the preponderating focus approaches more nearly to the circle – and the less the difference between the power of the foci as in the case the radii being as 3 to 4 – the form is kept more nearly to the ellipsis – and by adjusting the distances between the foci – innumerable varieties of the oval form can be produced – but I have not had leisure to

---

(8) Compare Forbes' comment: 'we have a further generalization of the same kind as that so highly recommended by Montucla, by which Descartes elucidated the conic sections as particular cases of his oval curves'; see J. E. Montucla, *Histoire des Mathématiques*, 4 vols. (Paris, 1799–1802), **2**: 129; and Descartes, *Geometrie*: 357.

(9) See Maxwell's appended figures *infra*.

draw many examples – but a few are drawn on a separate paper.

The arrangement I have used for drawing figures with three separate foci is the following –

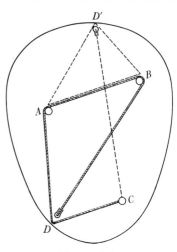

Let *A*, *B*, *C* be the three foci – let a thread be fixed at *C* passed around *A* & *B* with the tracing point in a loop at the end of the thread – with the tracer in the loop catch up a bight of the thread between *A* & *C* and stretch it to *D* – then trace the arc as far as the thread from *B* to *D* keeps clear of *A* on the one side and *C* on the other – the thread is then shifted to over the pin at *C*, and the other side traced – the dotted line shows the position of the threads in tracing the upper portion of the figure. The portion of the thread between *A* & *B* is

Figure 1,2

a constant quantity always subtracted from the whole length of the thread – the remainder is in all positions divided into the 3 radii from the point of the tracer to the three foci.

By stating the position of the foci as being placed in the angles of any kind of triangle – such as the equilateral – or isosceles of any degree of angle a definite symmetrical figure will be produced. If the triangle is scalene the figure is not symmetrical. The foci may be either within or without the figure. An innumerable variety of forms can be produced by varying the positions of the foci and the sum of the radii – and again by changing the proportions of the radii. Many of the forms are pleasing, and may be applicable for designs for vases cups & vessels of various kinds.[10] Some examples are given on a separate paper.

Symmetrical figures are also obtained when the three foci are in a straight line. The outer figure is drawn from the 3 foci *A*, *B* & *C* – with simple radii – the inner figure is drawn from the same foci with radii from *A* & *B* single and the Radius from *C* in double proportion.

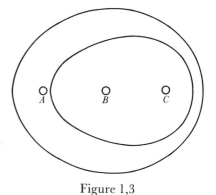

The following figure shows the arrangement of the thread for drawing a figure from four foci – the Radii being in simple proportion. The portion of the thread from *A* to *B* is a constant quantity to be subtracted from the whole thread. The management of the thread

Figure 1,3

(10) See note (1).

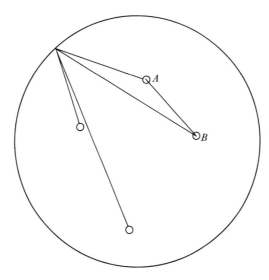

Figure 1,4

is exactly the same as in the three foci figure. The Threads from the foci have to be lifted over the pins of the other foci whenever their motion is interrupted.

The arrangement for five foci is similar only there are two portions of the thread subtracted viz. *A* to *B* and *C* to *D*. It is difficult to get so many threads to slip smoothly – and I have not gone beyond this number. The figures which may be produced are probably of little beauty of form: they appear to approach the nearer to the circle the more the foci are multiplied and have the effect of badly drawn circles.

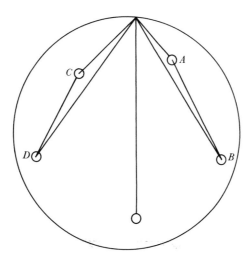

Figure 1,5

Series of Ovals – on two Foci & Radii in the proportion of one & two – from 45° to 18° – & on parallel lines.

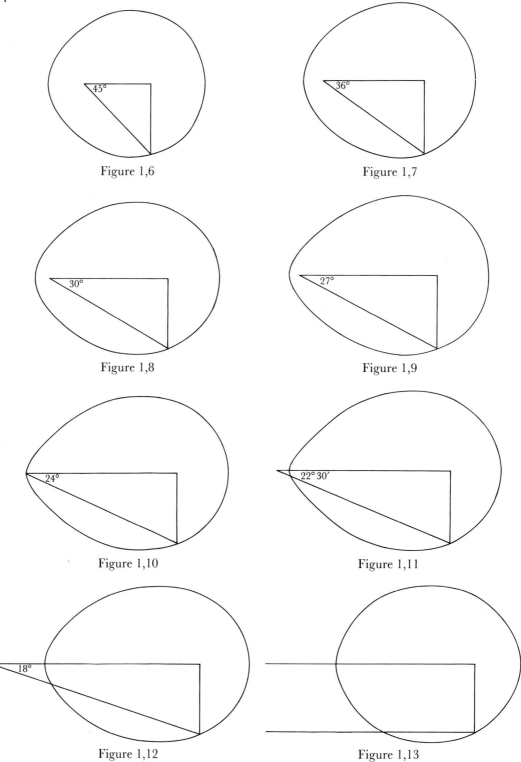

Figure 1,6

Figure 1,7

Figure 1,8

Figure 1,9

Figure 1,10

Figure 1,11

Figure 1,12

Figure 1,13

Figures from two foci – Radii as 3 to 4.

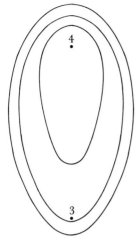

Figure 1,14

Figures on two foci – Radii as 2 to 3.

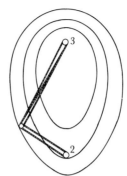

Figure 1,15

Figures on two foci – Radii as 1 to 4.

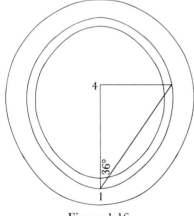

Figure 1,16

– Radii as 2 to 5.

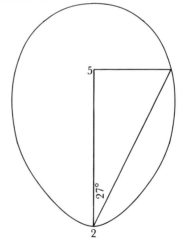

Figure 1,17

Figures from two foci – Radii in proportion of 1 to 3.

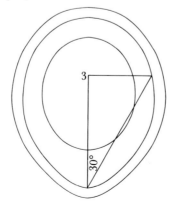

Figure 1,18

Figure from 3-foci – Radii equal proportions.

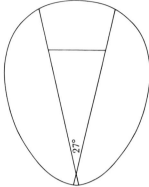

Figure 1,19

Figures drawn from three foci – Radii in single proportion.

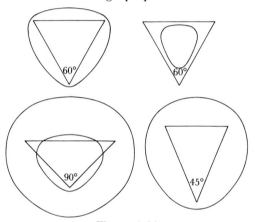

Figure 1,20

Figures from 3 foci – Radii in Single proportion.

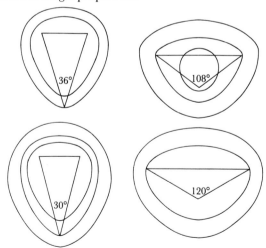

Figure 1,21

2

MANUSCRIPT ON TRIFOCAL CURVES[1]

MARCH 1847[2]

From the original in the University Library, Cambridge[3]

ON TRIFOCAL CURVES

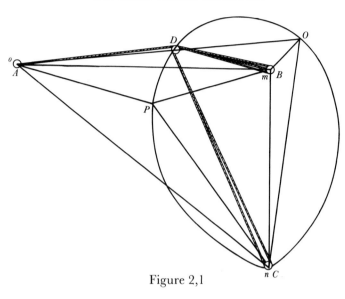

Figure 2,1

Def 1   Trifocal curves are those in which any point being taken m times its distance from one point $+ n$ times the distance from a second $+ o$ times the distance from a third $=$ constant quantity.[4]

---

(1) In his Obituary Notice 'James Clerk Maxwell', *Proc. Roy. Soc. Edinb.*, **10** (1879): 331–9 (on 332), Peter Guthrie Tait recalled his association with Maxwell during their last years at the Edinburgh Academy: 'I still possess some of the MSS we exchanged in 1846 and early in 1847. Those by Maxwell are on "The Conical Pendulum", "Descartes' Ovals", "Meloid and Apioid", and "Trifocal Curves". All are drawn up in strict geometrical form, and divided into consecutive propositions.' The manuscript 'On Trifocal curves', which is annotated in pencil by Tait, and dated 'March 1847', is clearly the paper referred to by Tait. The manuscript on the conical pendulum dated 25 May 1847 (Number 6), and the propositions on 'Oval' and 'Meloid and Apioid' (Number 3) which are reproduced from the *Life of Maxwell*, are possibly the other papers referred to, or are drafts of these papers. The papers 'On Trifocal curves', 'Oval' and 'Meloid and Apioid' are closely related in content; the order in which they are reproduced here may not be the chronological order of their composition.

(2) At the end of the MS there is a pencilled date 'March 1847'.

(3) ULC Add. MSS 7655, V, d/3.

(4) Compare J. D. Forbes' 'remarks' on Maxwell's paper 'On the description of oval curves', *Proc. Roy. Soc. Edinb.*, **2** (1846): 91 ($=$ *Scientific Papers*, 1: 3), where he discussed the equation $mr + nr' + or'' + \&c =$ constant for curves with more than two foci.

E ⊤

F ⊥

Figure 2,2

2 The 3 points are called the foci.

3 *m, n* and *o* are called the powers of the foci.[5]

4 When 2 of the foci are equidistant from the third and have the same power the curve is called isosceles.

Proposition 1 Problem

To describe a trifocal curve the foci their powers and the constant quantity being given.

Let A, B and C be the foci, 1, 2 and 3 the powers and EF the constant quantity to describe the curve.

Erect focal cylinders as in the oval[6] take a thread $= EF$ wrap it round the foci and a moveable cylinder D so that there may be 1 thread between A and D 2 between C and D and 3 between B and D. Move D keeping the threads tight. D will describe the curve.

For at any point D, $AD + 2CD + 3BD =$ constant quantity.

Proposition 2 Theorem

A focus is within on or without the curve according as the distances from the other foci multiplied respectively by their powers is less than equal to or greater than the constant quantity.

Let $oAB + nBO <$ constant quantity. B is within the curve. Bisect ABC by BO then

$$oAO + nCO + m \llcorner O \lrcorner B^{(a)} = EF$$

and

$$BO = \frac{oAO + nCO - oAB - nCB}{m}.$$

It is evident that C is in the curve when

$$oAO + mBC = EF.$$

Let $mBA + nCA > EF$ draw AF within $\llcorner BA \lrcorner C^{(b)}$

then

$$AP = \frac{mBA + nCA - mBP - nCP}{o}.$$

Proposition 3 Problem

The powers of the 3 foci any 2 must be $> 3^{\text{rd}}$ and the constant quantity being given to make a triangle such that if its angles be taken as foci and a trifocal

---

(a) {Tait} $(+mBO)$ *P. G. Tait.*

(b) {Tait} $+ (BAC)$ P.G.T.

(5) The term used by Forbes.

(6) Compare Number 3.

H

curve described with the said constant quantity all the foci will be in the curve and if a greater constant quantity be taken they will be within and if a less, without the curve.

Let HL be the constant quantity $= c$, and the powers of the foci m, n and o make a line

$$AB = \frac{(m+n+o)\,c}{2mn}$$ let A the one end be the focus whose power is m and B the focus whose power is n.

Complete the triangle so that $AC = \dfrac{(m+o-n)\,c}{2mo}$ and $BC = \dfrac{(n+o-m)\,c}{2no}$. $(1.23)^{(7)}$ ABC is the triangle.

For
$$nAB = \frac{(m+n-o)\,c}{2m} \quad \text{and} \quad oAC = \frac{(m+o-n)\,c}{2m}$$

and
$$nAB + oAC = \frac{(m+n-o+m+o-n)\,c}{2m} = \frac{2mc}{2m} = c$$

and
$$nAB + oAC = c \text{ therefore.}$$

A is in the curve in the same way prove C and $\llcorner B \lrcorner^{(c)}$ in the curve. It is evident from Prop 2 that the foci will be within or without the curve according as the constant quantity is greater or less than c.

Figure 2,3

E

Proposition 4 Problem

The powers of the foci and the constant quantity being given to cut an indefinite line in 3 points so that if the 3 points be taken as foci and a trifocal curve described with the said constant quantity the two extreme points may be in the curve.

Let $c = EF$ the constant quantity and m, n and o the powers of the foci in their order

take a line $\quad AB = \dfrac{(m+n-o)\,c}{mn+n^2+no}$

produce it so that

$$BC = \frac{(n+o-m)\,c}{mn+n^2+no} \quad \text{then} \quad AC = \frac{2c}{a+b+c}.$$

A, B, C are the foci.

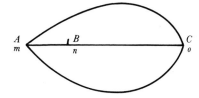

Figure 2,6

(c) {Tait} (B) P. G. Tait.

(7) Euclid, *Elements of Geometry*, Book I, Prop. xxiii (to construct a triangle on a straight line as base).

L

Figure 2,4

F

Figure 2,5

For
$$nAB = \frac{(m+n-o)\,c}{m+n+o} \quad \text{and} \quad oAC = \frac{(2o)\,c}{a+b+c}$$
$$\therefore nAB + oAC = \frac{(m+n+o)\,c}{m+n+o} = c$$

and A is in the curve in the same way C is in the curve.

Proposition 5 Problem

To determine the angle in a trifocal curve when the focus is in the curve.

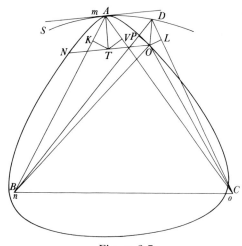

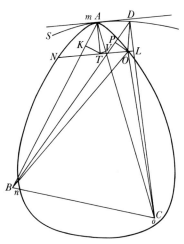

Figure 2,7   Figure 2,8

Let A be the focus in the curve with a power of m and let the powers of B and C be respectively n and o. With the same constant quantity and the same powers of B and C describe an Oval or Ellipse SAD it will pass through A. Take a point D indefinitely near A. Join DB, DC, AB, AC then because D is near A, $BAC = BDC$. Draw a tangent AD and AT perpendicular make $PDO = KAT \therefore TAV = ODL$.

[d] Join OB, OC then $nBO + oCO + mAO = nBD + oCD$. Cut off $BP = BO$ and $CL = CO$ then because the angles PBO, OCL are very small, PO, OL are perpendicular to DB, DC and also $nPD + oLD = mAO$.

Draw $TO \parallel AD$ and KT, TV perpendicular to AB, AC then $AT = DO$, $AK = PD$ and $AV = DL$.

$\therefore nAK + oAV = mAO$ and the same can be proved at the other side of A. $\therefore AN = AO$ and AON is a very small isosceles triangle coinciding with the curve.

Therefore having drawn the Oval or ellipse and its tangent draw AT perpendicular and NTO parallel to it. Draw TK, TV perpendicular to AB, AC and make $mAO = nAK + oAV$ make $NA = AO$. NAO is the angle.[e]

(d) {Tait} (*BP*). P. G. Tait.   (e) {Maxwell} March 1847.

3

PROPOSITIONS ON OVAL CURVES

circa 1847[1]

From Campbell and Garnett, *Life of Maxwell*[2]

[1]   OVAL[3]

Definition 1. – If a point move in such a manner that *m* times its distance from one point, together with *n* times its distance from another point, may be equal to a constant quantity, it will describe a curve called an Oval.[4]

Definition 2. – The two points are called the foci, and the numbers signified by *m* and *n* are called the powers of the foci.[5]

Definition 3. – The line joining the foci is called the axis.

Proposition 1 – Problem

To describe an oval with given foci, given multiples, and given constant quantity.

Let *A* and *B* be the given foci, 3 and 2 the multiples, and *EF* the constant quantity, it is required to describe an oval. At *A* and *B* erect two infinitely small cylinders. Take a perfectly flexible and inextensible thread, without breadth or thickness,[6] equal to *EF*; wind it round the focal cylinders and another movable cylinder *C*, so that the number of plies between *A* and *C* may be equal to *m*, that is 3, and the number between *B* and *C* equal to *n* or 2. Now move *C* in such a manner that the thread may be quite tight, and an oval will be described by the point.

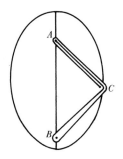

Figure 3,1

---

(1) See Number 2 note (1).   (2) *Life of Maxwell*: 91–104.   (3) See Number 2 note (1).

(4) On the identity of Maxwell's ovals with Cartesian ovals see Number 1 note (7) and Number 4. In his discussion of ovals Maxwell seems unaware that a Cartesian oval has three foci; Descartes' description of his first oval (*Geometrie*: 356) depends on this condition. In his letter to J. J. Sylvester of 21 December 1866 (see Number 1 note (7)) he states that 'I tried to establish a third focus when I began mathematics about the [year] 1846 and several times afterwards'. Referring to Descartes' description of his first oval, Maxwell notes that Descartes had 'found out a case of the 3rd focus'; he remarks that 'I daresay this proposition has been well known though I never understood it till now'. The issue had however been discussed by Michel Chasles, *Aperçu Historique sur l'Origine et le Développement des Méthodes en Géométrie*, in *Mémoires couronnés par l'Académie Royale des Sciences et Belle-Lettres de Bruxelles*, **11** (1837): 350–3 (Note XXI).

(5) See Number 2 note (5).

(6) Maxwell appreciates the practical difficulty of his method of the description of ovals.

For take any point C. There are m plies of thread between A and C, and n plies between B and C, which taken together make up the thread or the constant quantity, therefore $mAC + nBC = EF$.

The focus A, which has the greatest number of plies is called the greater focus, and B is called the less focus.

Proposition 2 – Theorem

The greater focus is always within the oval, but the less is within, on, or without the curve, according as the distance between the foci, multiplied by the power of the greater focus, is less equal or greater than, the constant quantity.

The greater focus is always within the Oval, for suppose it to be at A, Fig. [3,2], then $mAC + nBC =$ constant quantity $= mAD + mDC + nBC$, and $mAD + nDB =$ constant quantity $= mAD + nDC + nBC = mAD + mDC + nBC$, and $nDC = mDC$, and $m = n$, but $m > n$.

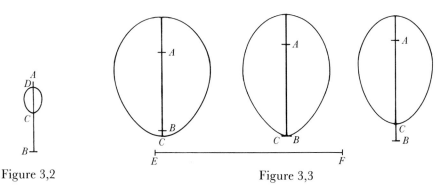

Figure 3,2   Figure 3,3

The less focus B is within the oval when $mAB < EF$ the constant quantity.

$$\text{For it is evident that } BC \text{ (Fig. [3,3])} = \frac{EF - mAB}{m + n}.$$

It is in the curve when $mAB = EF$, this is evident.

$$\text{It is without the curve when } mAB > EF, \text{ for } AC = \frac{EF - nAB}{m - n}.$$

Proposition 3 – Theorem

If a circle be described with a focus for a center, and the constant quantity divided by the power of that focus for a radius, the distance of any point of the oval from the other focus is to the distance from the circle as the power of the central focus is to the power of the other.

The circle *EHP* is described with the centre *B* and radius = constant quantity, divided by the power of *B*. At any point *C*, *AC*:*CH*::power of *B*:power of *A*.

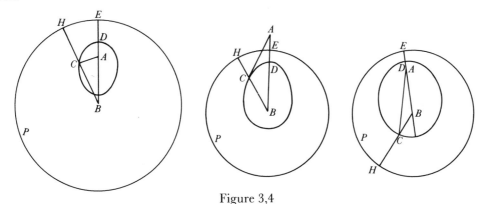

Figure 3,4

Let *m* = power of *A*, and *n* = power of *B*, *nBH* = constant quantity = *nBC*+*nCH*, and *nBC*+*mAC* = constant quantity = *nBC*+*nCH*, take away *nBC*, and *nHC* = *mAC*, therefore *HC*:*CA*::*m*:*n*.   QED.

Cor. 1. – When the powers of the foci are equal the curve is an ellipse.

Cor. 2. – When the less focus is at an infinite distance the curve is an ellipse, for the circle becomes a straight line; and

Cor. 3. – When the greater focus is at an infinite distance, the curve is an hyperbola for the same reason.

Proposition 4 – Theorem

When the less focus is in the curve, an angle will be formed equal to the vertical angle of an isosceles triangle, of which the side is to the perpendicular on the base, as the power of the greater focus to that of the less.

For let a circle be described, as in Prop. 3, it is evident that it will pass through *B*. Take indefinitely small arcs *CB* = *BD*, join *CA* and *DA*, join *EH*. *EC*:*EB* = power of *B*:power of *A*, and *EC* = *BO*, therefore *EB*:*BO*, power of *A*:power of *B*.

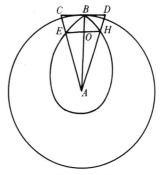

Figure 3,5

Proposition 5 – Problem

A point *A*, and a point *B* in the line *BC* being given, to find a point in the line as *D*, so that *mAD*+*nBD* may be a minimum.

Take a line HP, raise HX perpendicular, and from X as a centre describe a circle with a radius $= \dfrac{mXH}{n}$, so that $m:n::XP:XH$.

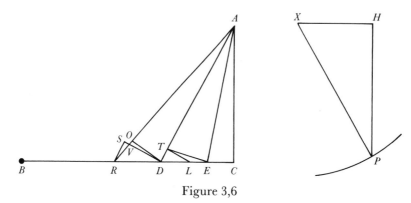

Figure 3,6

Draw AC perpendicular to BC, and make an angle $CAD = HPX$. D is the required point. If not, take any point E on the opposite side of D from B, make $AT = AE$. Join TE, and make $DTL = $ right angle, then $DTE > $ right angle DTL, because ATE is less than a right angle (1.16),[7] therefore L is within E. And in the triangle DTL, $DTL = PHX$ and $TDL = HXP$, therefore the triangles are similar, and $TD:DL = (XH:XP) = n:m$; and $nDL = mDT$, then $nDE > mDT$, add nBD and mTA or mEA, then $nBD + nDE + mEA > nBD + mDT + mTA$ and $nBD + mDA < nBE + mAE$.

Now take a point R on the other side. Join AR, and cut off $AO = AD$, make $SRD = HXP = ADC$. S will be beyond AR. Draw DS perpendicular to AD, then it will be perpendicular to RS, and below the line OD, then RSD similar to XHP and $RS:RD = (XH:XP =) n:m$, and $mRS = nRD$, but $SR < RV < RO$ and $nRD (= mRS) < mRO$, add nBR and mOA or mDA, and $nBR + nRD + mDA < nBR + mRO + mOA$ and $nBD + mDA < nBR + mRA$. QED.

Proposition 6 – Problem

To draw a tangent to an oval from a focus without:

Take m for the power of the greater focus, and n for that of the less, and find

---

(7) Book I, Prop. XVI of Euclid's *Elements*. From his citations (see note (9)) Maxwell is apparently using the edition by John Playfair, *Elements of Geometry; containing the First Six Books of Euclid, with a Supplement on the Quadrature of the Circle and the Geometry of Solids* (Edinburgh, 1795). A copy, with John Clerk Maxwell's bookplate and annotations on the flyleaf by Maxwell, is in Maxwell's library (Cavendish Laboratory, Cambridge). (Theorem) 'If one side of a triangle be produced, the exterior angle is greater than either of the interior, and opposite angles' (*Elements*: 20).

the angle *ADB* (Prop. 5). Upon *AB* describe a segment of a circle containing an equal angle. Join *B* and *C*, the point where it cuts the oval, *BC* is a tangent. For take any point *O*, join *AO*, $mAO + nOB > mAC + nCB$, therefore *O* is without the oval.

Proposition 7 – Problem

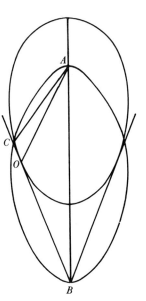

Figure 3,7

To draw a tangent to a given Oval, the foci and the ratio being given:

It is required to draw a tangent at *C* to the oval *CNT*, ratio $m:n$. From *B* the less focus describe a circle as in (Prop. 3). Join *CA*, join *CB*, and produce to *D*, then $DC:CA::m:n$. Join *AD* and produce it. Bisect the angle *DCA* by the line *CO*. $DO:OA::(DC:CA::)\ m:n$. Draw *CK* perpendicular to *CO*, and describe the circle *OCLK*. Because *OCK* is a right angle, $OCK = DCO + KCB = OCA + ACK$. But $OCA = DCO \therefore KCB = ACK$ and $(6.3)^{(8)}$ $KA:KD::CA:CD::AO:OD \therefore KA:KD::AO:OD$, and at any point *L* in the circle, $AL:DL::AO:OD$ (6 F),[9] and $AL:DL::n:m$; and the circle is wholly without the oval, and can only touch it in the two points *C* and *P*, where *DB* cuts the oval; for suppose the circle coincided with the oval at *L*,

$$\left.\begin{array}{ll}\text{Join } BL \text{ and produce to } X, \text{ then } AL:DL::n:m \\ \text{And by Prop. 3} \qquad . \qquad . \qquad AL:XL::n:m\end{array}\right\} \therefore DL = XL;$$

but $DL > XL\ (3.7)$,[10] therefore the oval is within the circle. Therefore draw

---

(8) Book VI, Prop. III of Euclid's *Elements*: (Theorem) 'If the angle of a triangle be bisected by a straight line which also cuts the base; the segments of the base shall have the same ratio which the other sides of the triangle have to one another: And if the segments of the base have the same ratio which the other sides of the triangle have to one another; the straight line drawn from the vertex to the point of section, bisects the vertical angle' (*Elements*: 159).

(9) A proposition appended by Playfair 'from the *Loci Plani* of Apollonius' (*Elements*: 384): (Theorem.) 'If two points be taken in the diameter of a circle, such that the rectangle contained by the segments intercepted between them and the centre of the circle be equal to the square of the semidiameter; and if from these points two straight lines be drawn to any point whatsoever in the circumference of the circle, the ratio of these lines will be the same with the ratio of the segments intercepted between the two first mentioned points and the circumference of the circle' (*Elements*: 198).

(10) Book III, Prop. VII of Euclid's *Elements*: (Theorem.) 'If any point be taken without a circle, and straight lines be drawn from it to the circumference, whereof one passes through the centre; of those which fall upon the concave circumference, the greatest is that which passes

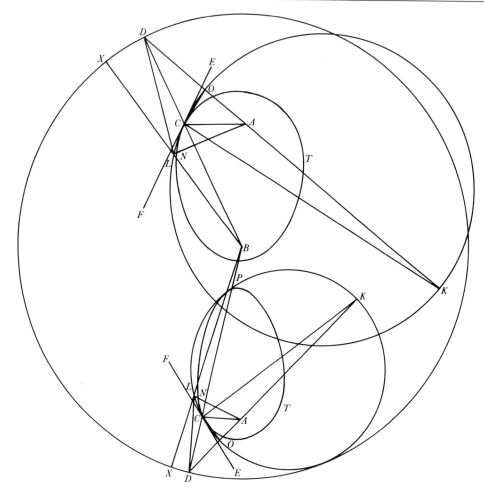

Figure 3,8

ECF a tangent to the circle, and as it is without the circle it is without the oval.

Scholium. – Join *AC*, join *BC*, and produce till *AC*:*CD*::*n*:*m*. Join *AD*, bisect *ACD* by *CO*. Draw *OE* perpendicular to *AO*. Make *OCE* = *EOC*. It is evident from the proposition that *ECF* is the tangent.

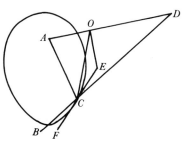

Figure 3,9

---

through the centre; and of the rest, that which is nearer to that through the centre is always greater than the more remote: But of those which fall upon the convex circumference, the least is that between the point without the circle, and the diameter; and of the rest, that which is nearer to the least is always less than the more remote: And only two equal straight lines can be drawn from the point unto the circumference, one upon each side of the least' (*Elements*: 77).

Proposition 8 – Theorem

If a line AE be cut in C and B, so that $AB:BC::AE:CE$, and two semi-circles BOE, BXE be described on BE, and AO, CO be drawn to O in the circumference, and perpendiculars DH, DL, be drawn from the centre, $DH:DL::AB:BC$.

For $AB:BC::AE:CE$ $\therefore AB:AE::BC:CE$ $\therefore (AE+AB) = (2AB+2BD)$ $= 2AD:AB::(CE+BC =)\, 2BD:BC$ and $AD:AB::BD:BC$ $\therefore (AD-AB =)$ $BD:AB::(BD-BC =)\, CD:BC$ $\therefore BD:CD::AB:BC$.

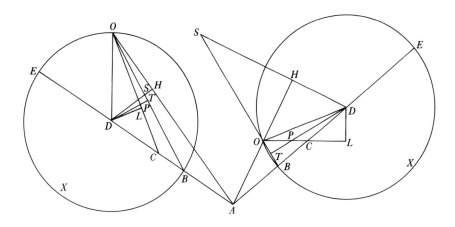

Figure 3,10

Draw DT perpendicular to OB, then as it bisects the base it bisects the angle ODB and $ODT = BDT$, then in the triangles OTP, DPL, $OTP = DLP$ and $OPT = DPL$ $\therefore PDL = POT$. But $POT = BOA$ (6 F Cor.)[11] and $BOA = SOH = PDL$, and in the triangles SOH, STD, $SHO = STD$, and $OSH = TSD$ $\therefore SOH = SDT$ and $SDT = PDL$. But $ODT = BDT$ $\therefore CDL = ODH$ and

(11) The corollary to Prop. F appended by Playfair to Book VI of the *Elements* (see note (9)). 'Cor. If AB be drawn, because $FB:BE::FA:AE$ [by Prop. F], the angle FBE is bisected by AB. Also, since $FD:DC::DC:DE$, by comparison, $FC:DC::CE:ED$, and since it has been shown that $FA:AD(DC)::AE:ED$ [Book V, Prop. XVIII], therefore, ex æquo, $FA:AE::FC:CE$. But $FB:BE::FA:AE$, therefore $FB:BE::FC:CE$; so that if FB be produced to G, and if BC be drawn, the angle EBG is bisected by the line BC.' (*Elements*: 199).

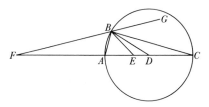

Figure 3,11

$DHO = DLC \therefore HOD$, CDL are equiangular, and $HD:DL::(DO =) BD:CD$, but $BD:CD::AB:BC \therefore HD:DL::AB:BC$.

Cor. Sine HOD : Sine $DOL::AB:BC$.

Proposition 9 – Theorem

If lines be drawn from the foci to any point in the oval, the sines of the angles which they make with the perpendicular to the tangent are to one another as the powers of the foci.[12]

Sine DCE : sine ACE :: power of A : power of B.

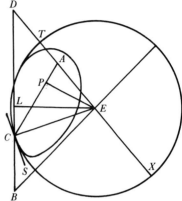

For describe the circle CXT as in Prop. 7, so that $DT:TA::DX:AX$, then SC the tangent to the circle is a tangent to the oval, and CE the radius is perpendicular to it; then by Prop 8 Corollary, sine DCE : sine $ACE::(DT:TA::)$ power of A : power of B.

As the ratio of the sines of incidence and refraction is invariable for the same medium, and as this ratio is as 2 to 3 in glass; and as in the oval at Fig. [3,12] the powers of the foci are as 2 to 3; if the oval were made of glass, rays of light from B would be refracted to A. If now a circle be described from A, the case will not be altered, for the rays are perpendicular to the circle as in Fig. [3,13].

Figure 3,12

The ellipse and hyperbola, Figs. [3,14] and [3,15] cause parallel rays to converge, for (Prop. 3) they are ovals with one of the foci at an infinite distance.

Fig. [3,16] is a combination of 2 hyperbolas, and rays from A are refracted to B.

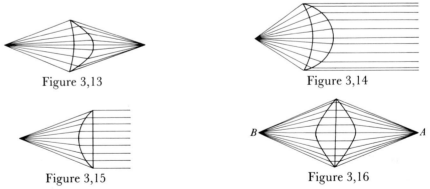

Figure 3,13   Figure 3,14

Figure 3,15   Figure 3,16

(12) In his 'remarks' on Maxwell's paper 'On the description of oval curves' Forbes had noted that the 'demonstration of the optical properties [of ovals] was given by Newton in the *Principia*,

[2]   MELOID AND APIOID[13]

If a point C move so that $mCA \sim nCB =$ a constant quantity, it will describe a curve called a meloid when $mCA > nCB$, and an apioid when $nCB > mCA$. The points A and B are called the foci.

Proposition 1 – Problem

To describe a Meloid, Fig. [3,17], or Apioid, Fig. [3,18].[14]

Take up a rigid straight rod AD, centred at A. At D erect a small cylinder upon the rod, and also another at B on the paper. Then take a flexible and inextensible thread of such a length that $mAD \sim n$ times the thread $=$ constant quantity; then wrap the thread round B and D and a movable cylinder C, so that there may be n plies between B and C, and m plies between C and D; if now AD be turned round and the thread kept tight, the point C will describe a meloid or apioid.

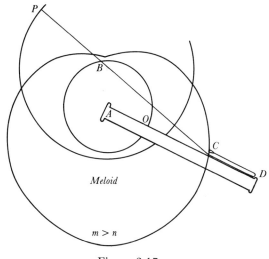

Meloid

$m > n$

Figure 3,17

Book I, prop. 97, by the law of sines' (*Scientific Papers*, **1**: 3). See Isaac Newton, *Philosophiae Naturalis Principia Mathematica* (Cambridge, ₂1713): 208, where he discusses the refraction of corpuscles at a surface converging them from a point A to a point B: 'Faciendo autem ut puntum A vel B nunc abeat in infinitum…habebuntur Figuræ illæ omnes, quas *Cartesius* in Optica & Geometria ad Refractiones exposuit. Quarum inventionem cum *Cartesius* maximi fecerit & studiose celaverit, visum fuit hac propositione exponent.' See Descartes, *Discours de La Methode…plus La Dioptrique, Les Meteores et La Geometrie* (Leiden, 1637): 89–121, 357–69.

(13) Possibly the MS 'Meloid and Apioid' mentioned by Tait, 'James Clerk Maxwell', *Proc. Roy. Soc. Edinb.*, **10** (1879): 332; see Number 2 note (1). The terms 'meloid' and 'apioid' for the Cartesian conjugate ovals seem unknown in the literature; compare F. Gomes Teixeira, *Traité des Courbes Spéciales Remarquables*, 2 vols. (Coimbra, 1908–9), **1**: 218–33; and G. Loria, *Spezielle Algebraische und Transzendente Ebene Kurven*, 2 vols. (Leipzig/Berlin, ₂1910–11), **1**: 174–83. The terms are very likely Maxwell's own, as implied by P. G. Tait, 'Clerk-Maxwell's scientific work', *Nature*, **21** (1880): 317–21, esp. 317; 'apioid' and 'meloid' are perhaps from ἄπιον (pear) and μῆλον (apple).

(14) Compare Descartes' description of his first oval, using thread, pins, and a ruler, in his *Geometrie*: 356.

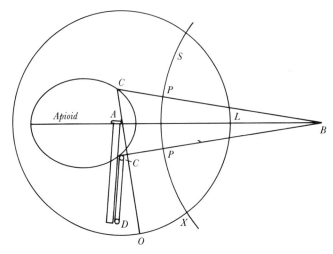

Figure 3,18

For take any point C, $mDC + nCB =$ thread, and $mDC + mAC = mAD$; therefore take away mDC, and $nCB \sim mAC =$ thread $\sim mAD =$ constant quantity. As $m > n$, A is called the greater and B the less focus.

Proposition 2 is the same as proposition 2 of the Oval.

Proposition 3 – Theorem

If a circle be described with a focus for a center, and the constant difference divided by the power of that focus for a radius, the distance of any point in the curve from the other focus is to the distance from the circle as the power of the central focus is to the power of the other.

Let TLO, PXS be the circles: at any point C, $BC : CO :: m : n$ and $PC : CA :: m : n$.

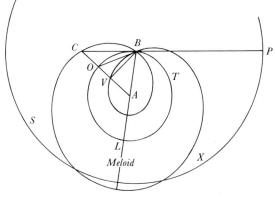

Figure 3,19

For $mCA \sim nCB =$ constant difference, but $mOA =$ constant difference $\therefore mCO = nCB$ and $BC : CO :: m : n$. QED.

And $nBP =$ constant difference $\therefore mCA = nCP$ and $PC : CA :: m : n$.

Cor. 1. – If the constant difference $= 0$ the curve is a circle.

Cor. 2. – If an oval be described with a thread $=$ constant difference, with the same foci and the same powers as the meloid, and any line be drawn from A, and CB, OB, VB be joined, the angle $CBO = VBO$.

For $BC:CO::m:n$ and $BV:VO::m:n \therefore BC:CO::BV:VO \therefore BC:BV::CO:VO$ and $CBO = VBO$ (6.3).[15]

Proposition 4 – Theorem

When the less focus is in the curve, an angle will be formed = that in the Oval (Prop. 4).

For take an indefinitely small arc DB in the circle, $CBD = DBE$ (3. Cor. 2), and $LBT = TBP \therefore CBL = EBP$.

Or it may be proved as in the oval. If the greater focus A is at an infinite distance the figure will

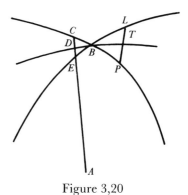

Figure 3,20

appear thus:

Figure 3,21

Proposition 5 – Theorem

If the distance between the greater focus and the point where the axis cuts the meloid, be to the distance between that point and the less focus, in a greater proportion than the power of the greater focus to that of the less, the curve is convex toward the greater focus at that point, but if the proportion is less, concave.

Let $AD:DB::p:q$.

If $AD:DB > m:n$, CDE is convex towards A; but if $AD:DB < m: n$, it is concave towards A.... Take C and E near D, and $CD = DE$.

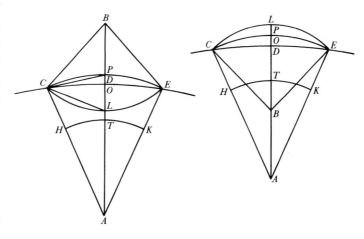

Figure 3,22

---

(15) Book VI, Prop. III of Euclid's *Elements* (see note (8)).

Join CE, CE cuts the axis in O. Draw the circle CLE from B, and CPE from A. Also draw HTK as in Prop. 3 –

Then $BD:DT::m:n::BC:CH$, but $BC = BL$, and $CH = PT$ $\therefore BD:DT::BL:PT$ $\therefore BD:BL::DT:PT$ $\therefore BD:BL-BD::DT:PT-DT$ $\therefore BD:DL::DT:DP$ $\therefore BD:DT::DL:DP$ $\therefore DL:DP::m:n$.

As E and C are very near D, $AD:BD::AC:BC$, but $PE = LE$ and $PCE:2$ right angles $::PE:$ circumference of CPE, and $LCE:2 \llcorner ::LE:$ circumference of CLE, but circ. $CPE:$ circ. $CLE::AC:BC::AD:BD::p:q$ and $PE = LE$ $\therefore PCE:2 \llcorner ::qPE:q$ circ. CPE, and $LCE:2 \llcorner ::pPE:(p$ circ. CLE or$)$ q circ. CPE $\therefore PCE:LCE::qPE:pPE::q:p$ $\therefore PCO:LCO::q:p$ and $PO:OL::q:p$; and if $p:q > m:n$, $OL:OP > DL:DP$, and D is nearer to A than the line COE, and CDE is convex toward A; but if $p:q < m:n$, D is on the opposite side, and it is concave.

Proposition 6 – Problem

To draw a tangent to a meloid or apioid from a focus without:

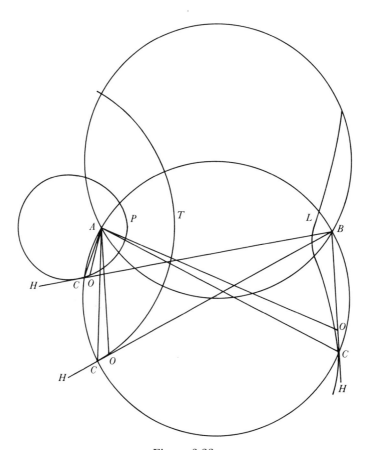

Figure 3,23

Take m for the power of the greater focus, and n for that of the less, and find the angle ADC (Prop. 5 of the Oval) upon AB describe a segment $ACCCB$ containing an equal angle. Join BC and produce to H, BH is a tangent, for suppose H to be the end of the rod, and take any point O, $nCH + mCA < nOH + nOA$, therefore O is without the curve.

CP is an apioid, CT is a circle, and CL is a meloid, with A and B as foci.

Proposition 7 – Problem

To draw a tangent to an apioid from any point in the same, the foci and the ratio being given.

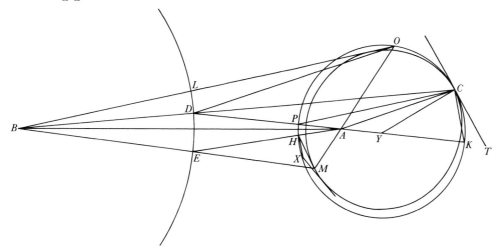

Figure 3,24. Apioid.

It is required to draw a tangent to the apioid at the point C. Join BC, and draw a circle as in Prop. 3. Join AD, and produce it. Make $AP:PD::n:m$. Join CP, and draw CK at right angles to PC. Describe a circle through P, C, K, and it was proved in Prop. 7 of the Oval that if any point O be taken and DO, AO joined, $DO:AO::m:n$. Suppose O to be both in the circle and in the apioid, join BO, then $LO:AO::m:n$, but $DO:AO::m:n$ ∴ $LO = DO$, but $LO < DO$ ∴ the circle is without the apioid; therefore a tangent CT to the circle at C is a tangent to the apioid.

Axiom 1

It is possible for a circle to be described touching any given curve internally.

Proposition 8

To draw a tangent to a meloid at any point C.

Case 1. – Let the curve be concave towards B. Describe the circle $POCK$ as in Prop. 7: it will be wholly within the meloid. At C draw a tangent to the circle: it is also a tangent to the meloid. For let RN be the tangent to the meloid, it must cut the circle (3.16),[16] and therefore cuts the meloid.

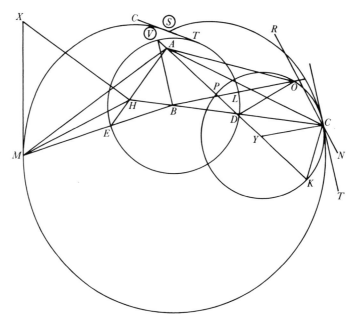

Figure 3,25. Meloid.

Case 2. – When the curve is convex towards B, draw a circle V as before. Draw CT a tangent to the circle; it is also a tangent to the meloid. For suppose a circle S drawn touching the curve internally, it must touch V and also CT, and any other line would cut S.   QED.

Scholium (see Figs. [3,24] and [3,25]). – Let M be the point. Join AM, BM, cut off ME, so that $AM:ME::$ power of B: power of A. Join AE, bisect AME by MH, make XH perpendicular, make $XMH = XHM$, XM is a tangent.

Proposition 9 – Theorem

If lines be drawn from the foci to any point in a meloid or apioid, the sines of the angles which they make with the perpendicular to the tangent are to one another as the powers of the foci. See Figs. [3,24, 3,25].

---

(16) Book III, Prop. XVI of Euclid's *Elements*: (Theorem) 'The straight line drawn at right angles to the diameter of a circle, from the extremity of it, falls without the circle; and no straight line can be drawn between that straight line and the circumference from the extremity so as not to cut the circle' (*Elements*: 85).

For *CY* is the perpendicular to the tangent and (Prop. 8 Oval Cor) Sine ACY : Sine BCY :: $(AP : PD)$ :: $m : n$.

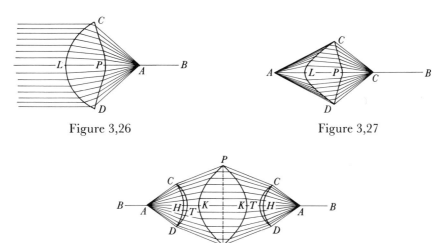

Figure 3,26   Figure 3,27

Figure 3,28

Fig. [3,26] *CLD* is an ellipse converging parallel rays to *B*, *CPD* is a meloid foci *A*, *B* converging the rays to *A*.

Fig. [3,27] *CLD* is an oval foci *A*, *B* converging rays from *A* to *B*. *CPD* is a meloid foci *C*, *B* refracting the rays to *C*.

Fig. [3,28] *CHD* is a circle not altering rays from *A*; *CTD* is a circle as in Prop 8 of the oval refracting the rays as if these had come from *B*. *PKLK* is a lens of hyperbolas refracting from *B* to *B* the whole 3 lenses refract from *A* to *A*.

FROM A LETTER TO JOHN CLERK MAXWELL

circa APRIL 1847

From Campbell and Garnett, *Life of Maxwell*[1]

[April] 1847

I have identified Descartes' ovals with mine.[2] His first oval is an oval with one of its foci outside; the second is a meloid with a focus outside. It also comprehends the circle and apioid.[3] The third is a meloid with both foci inside; and the fourth is an oval with both foci inside. He says with regard to the last that rays from *A* are *reflected* to *B*, which I can disprove.[4]

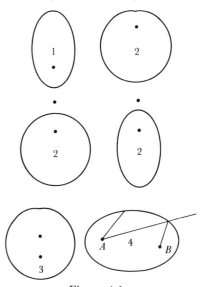

Figure 4,1

---

(1) *Life of Maxwell*: 88. The date is suggested by Campbell.

(2) See Number 1 esp. note (7) and Number 3 esp. note (4). Descartes considers ovals as falling in four classes; see Descartes, *Discours de La Methode...La Geometrie* (Leiden 1637): 352–6. In the 'References to memoirs on scientific subjects' appended to his testimonials for the Professorship of Natural Philosophy at Edinburgh in 1860 (ULC Add. MSS 7655, V, i/12) Maxwell gave the title of his 1846 paper as 'On the mechanical description of Descartes' ovals'.

(3) On the terms 'meloid' and 'apioid' see Number 3 note (13).

(4) The oval being supposed as the concave surface of a mirror; see Descartes, *Geometrie*: 358–9. Descartes' claim that the oval 'sert toute aux reflexions' rests on his supposition that 'les angles de la reflexion seroient inesgaus, aussy bien que sont ceux de la refraction, & pourroient estre mesurés en mesme sorte'. He supposes an incident ray to be reflected at the angle with which it would be refracted were the oval supposed as the surface of a lens; hence Maxwell's objection. Descartes requires the surface of the mirror to be 'de telle matiere qu'il diminuast la force de ces rayons' in a given proportion. In his letter to Sylvester of 21 December 1866 (to be reproduced in Volume II) Maxwell again refers to Descartes' 'erroneous statements about reflexion'.

FROM A LETTER TO JOHN CLERK MAXWELL

1847

From Campbell and Garnett, *Life of Maxwell*[1]

[1847]

I have made a map of the world, conical projection –

ACD is the one cone, *BCD* the other; the side *AC* = diameter of the base *CD*; therefore, when unrolled, they are each semicircles.

To make the map take $OD = OD$ and $OA = \frac{3}{4}$ of OD, and describe a circle from O, with radius OA; this is the circle of contact. Let the longitude of a point $P = 60°$; take AOC in the map $= 30°$; let its latitude be $65°$; subtract $30°$, and take the tangent of 35, make $AB = $ tangent. B is the point.

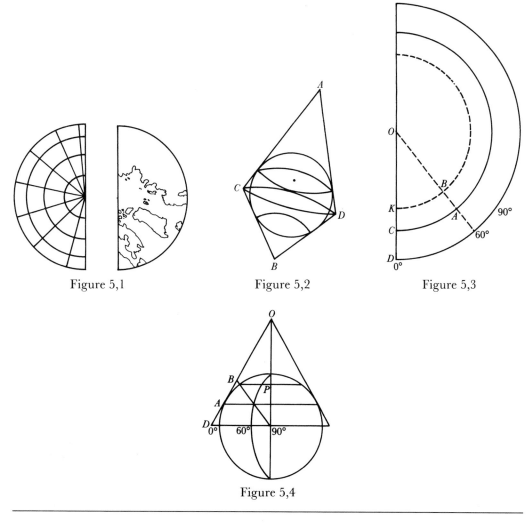

Figure 5,1   Figure 5,2   Figure 5,3

Figure 5,4

(1) *Life of Maxwell*: 89. The date is suggested by Campbell.

MANUSCRIPT ON THE CONICAL PENDULUM
AND A REVOLVING PENDULUM CLOCK[1]

25 MAY 1847

From the original in the University Library, Cambridge[2]

[CONICAL PENDULUM]

Proposition 1 Theorem[3]

Centrifugal force increases directly as the radius and inversely as the square of the time of revolution.

For let a body move from A to D and let AD be a nascent arc and the angle ASD very small and let AD be the space passed over in a time p and AC the direction in which it would go if free, then the disturbing force is represented by CD and the centrifugal force of B revolving in the same time with a radius BS is represented by EH and $SE:SC::SB:SA$.

$$\therefore SE - SB : SC - SA :: SB : SA$$

but

$$SE - SB = EH \text{ and } SC - SA = CD,$$

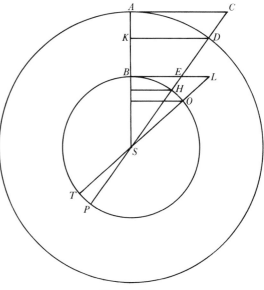

Figure 6,1

the centrifugal forces of weights moving in equal times round S with radii SB, SA.

And let the times of revolution of two bodies revolving round S with the radius SB be as BL to BE then the one will get to L when the other is at H, and

---

(1) See Number 2 note (1).

(2) ULC Add. MSS 7655, V, d/1. Previously published by A. T. Fuller, 'James Clerk Maxwell's Cambridge manuscripts: extracts relating to control and stability – I', *International Journal of Control*, **35**, (1982): 786–91.

(3) Compare Thomas Young's similar theorems on centrifugal force (in his 'Mathematical elements of natural philosophy') in Young, *A Course of Lectures on Natural Philosophy and the Mechanical Arts*, 2 vols. (London, 1807), **2**: 30.

BE and *BL* represent the velocities then $BE^2 = EP \cdot EH$ and $BL^2 = TL \cdot OL$[4] but as *HE* and *OL* are very small and there is very little difference between *EP* and *TL*

$BE^2 : BL^2 :: EH : OL$ and the centrifugal force is inversely[5] as the square of the velocity. QED.

Proposition 2 Theorem[6]

Bodies suspended by a thread from a fixed point and revolving in the same horizontal plane in equilibrio revolve in the same time; and as to those revolving in different planes, the squares of their times are as the perpendiculars on the planes from the fixed point.

1st let *S* and *E* revolve in the plane to which *AC* is perpendicular suspended from it they will revolve in the same time.

Figure 6,2

Let *ET* = force of gravity then *AE* is the thread and the force tending towards $C = ER$. In the same way the force of *S* toward $C = SX$.

Then by last proposition centrifugal force is as $\dfrac{\text{distance}}{\text{square of time}}$ and centripetal force is here equal to centrifugal.

$$\therefore \frac{XS}{CS} = \frac{1}{\text{square of time of } S} \text{ and } \frac{ER}{CE} = \frac{1}{\text{square of time of } E}.$$

But $AC : CS :: NX : XS$ and $AC : CE :: (OR) = NX : RE$.

$$\therefore CS : XS :: CE : RE \quad \therefore \frac{XS}{CS} = \frac{ER}{CE}$$

$$\therefore \text{time of } S = \text{time of } E.$$

---

(4) From Euclid, *Elements of Geometry*, Book III, Prop. xxxvi.
(5) Read: directly.   (6) Compare Young, *Lectures on Natural Philosophy*, **2**: 35.

2nd Let M and S be bodies revolving in different planes find the centrifugal force as before and as $LM = XS$ the centrifugal force of $M =$ that of S and

$$\frac{BM}{\text{square of time of } M} = \frac{CS}{\text{square of time of } S}.$$

$\therefore BM : CS :: (\text{time } M)^2 : (\text{time } S)^2$

but $BM : CS :: AB : AC$

$\therefore AB : AC ::$ square of time of M : square of time of N.[7]   QED.

Proposition 3 Theorem

If a revolving body be suspended from a point by a thread which lengthens proportionally to the force extending it it will always move in a plane the distance of which from the point of suspension is equal to the length of the thread when extended by the weight of the body.

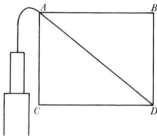

Figure 6,3

For let AC represent the force of gravity then CD will be the centripetal force and $AD =$ tension of thread.

$\therefore$ the thread will lengthen out to be equal to AD and will always move in the plane.

REVOLVING PENDULUM CLOCK[8]

In this clock Wheel 6 carries the weight and the hour hand wheel 4 the minuit hand and wheel 2 the second hand wheel 2 drives wheel 1 and this drives P the pendulum which is thus made.

C is a cylinder of iron suspended in a cylindrical vessel Q of quicksilver, P is a hollow ball attached to C by a string of such a length that, when at rest and in equilibrio P will be at O then if any force be applied to draw P down it will be drawn down proportionally to the force applied as the cylinder rises out of the quicksilver.

---

(7) Read: S.

(8) On 12 April 1847 James M'Ewan presented a paper 'Description of a new regulating index for the pendulum' to the Royal Scottish Society of Arts: see *Proceedings* of the Society (1846–7): 104, in *Transactions of the Royal Scottish Society of Arts*, **3** (1851), the paper being awarded a prize for the session (*Proceedings*: 126). M'Ewan's paper, in which he suggested an index hand to indicate the raising and lowering of the pendulum bob, may have occasioned Maxwell's manuscript, which is dated 25 May 1847. Compare Young, *Lectures on Natural Philosophy*, **1**: 47–8 and Plate II, Figs. 26–8, opp. page 759, on the conical pendulum as a governor for speed regulation.

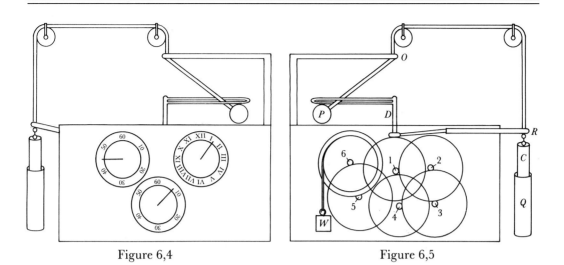

Figure 6,4   Figure 6,5

Now if P be loaded till it reaches D (the distance OD being 9.788 inches) it will be in the condition of the body in Prop 3 and will always swing round in the same time (1 second).

Therefore this Clock will keep true time.

When P swings out too far the drag R comes into play and retards the machinery.

To regulate the clock drop weights into P and the time will vary as square root of the weight.

JAS C. MAXWELL
25th May 1847

FROM A LETTER TO LEWIS CAMPBELL

16 NOVEMBER 1847[1]

From Campbell and Garnett, *Life of Maxwell*[2]

27 Heriot Row
[Edinburgh]
Tuesday [16 November 1847]

In Kelland[3] we find the value of expressions in numbers as fast as we can, the values of the letters being given; light work. In Forbes[4] we do Lever, which is all in Potter,[5] no notes required, only read Pottery ware (light reading). Logic needs long notes.[6] On Monday, Wednesday, Friday, I read Newton's *Fluxions*[7] in a sort of way, to know what I am about in doing a prop. There is no time of reading a book better than when you need it, and when you are on the point of finding it out yourself if you were able.[8]

---

(1) Given by Campbell.

(2) *Life of Maxwell*: 115; abridged.

(3) The Mathematics class at Edinburgh University (Philip Kelland).

(4) The Natural Philosophy class (James David Forbes).

(5) Richard Potter, *An Elementary Treatise on Mechanics* (London, 1846): 62–8.

(6) The Logic class (Sir William Hamilton).

(7) Isaac Newton, *A Treatise of the Method of Fluxions and Infinite Series, With its Application to the Geometry of Curve Lines* (London, 1737) (= *The Mathematical Works of Isaac Newton*, ed. D. T. Whiteside, 2 vols. (London/New York, 1964), **1**: 29–137).

(8) On Forbes, Hamilton, Kelland and the cultural milieu of Edinburgh University, see George Elder Davie, *The Democratic Intellect. Scotland and her Universities in the Nineteenth Century* (Edinburgh, ₂1964).

FROM A LETTER TO LEWIS CAMPBELL

NOVEMBER 1847

From Campbell and Garnett, *Life of Maxwell*[1]

31 Heriot Row
[Edinburgh]
November 1847

As you say, sir, I have no idle time. I look over notes and such like till 9.35, then I go to Coll., and I always go one way and cross streets at the same places; then at 10 comes Kelland. He is telling us about arithmetic and how the common rules are the best. At 11 there is Forbes, who has now finished introduction and properties of bodies, and is beginning Mechanics in earnest.[2] Then at 12, if it is fine, I perambulate the Meadows; if not, I go to the Library and do references. At 1 go to Logic. Sir W. reads the first $\frac{1}{2}$ of his lecture, and commits the rest to his man, but reserves to himself the right of making remarks. To-day was examination day, and there was no lecture. At 2 I go home and receive interim aliment, and do the needful in the way of business. Then I extend notes, and read text-books, which are Kelland's *Algebra*[3] and Potter's *Mechanics*.[4] The latter is very trigonometrical, but not deep; and the Trig. is not needed. I intend to read a few Greek and Latin beside. What books are you doing?... In Logic we sit in seats lettered according to name, and Sir W. takes and puts his hand into a jam pig[5] full of metal letters (very classical), and pulls one out and examines the bench of the letter. The Logic lectures are far the most solid and take most notes.[6]

Before I left home I found out a prop for Tait (P. G.); but he *will* not do it. It is 'to find the algebraical equation to a curve which is to be placed with its axis vertical, and a heavy body is to be put on any part of the curve, as on an inclined plane, and the horizontal component of the force, by which it is actuated, is to vary as the nth power of the perpendicular upon the axis'.

---

(1) *Life of Maxwell*: 115–16.

(2) See J. D. Forbes' 'Record of lectures delivered to class, 1833–57' (St Andrews University Library, Forbes Papers, Box IX, 5); in the session 1847–8 Forbes delivered 4 lectures 'Introduction and general' followed by 31 on 'Statics' and 21 on 'Dynamics'. See D. B. Wilson, 'The educational matrix: physics education at early-Victorian Cambridge, Edinburgh and Glasgow Universities', in *Wranglers and Physicists. Studies on Cambridge Physics in the Nineteenth Century*, ed. P. M. Harman (Manchester, 1985): 12–48, on 22.

(3) Philip Kelland, *The Elements of Algebra* (Edinburgh, 1839).

(4) See Number 7 note (5).   (5) A jar.

(6) Three notebooks, 'Logic Nº2', 'Logic Nº3' and 'Logic Nº4' (ULC Add. MSS 7655, V, m/1, 2, 3) record Maxwell's attention in 1847–8 to Hamilton's lectures. Compare William Hamilton, *Lectures on Metaphysics and Logic*, ed. H. L. Mansel and J. Veitch, 4 vols. (Edinburgh/London, 1859–60), **3**: 45–468, and **4**: 1–225. Maxwell's notes closely follow Hamilton's lectures as subsequently published.

FROM A LETTER TO LEWIS CAMPBELL

26 APRIL 1848

From Campbell and Garnett, *Life of Maxwell*[1]

Glenlair
26 April 1848

On Saturday, the natural philosophers ran up Arthur's Seat with the barometer. The Professor set it up at the top and let us pant at it till it ran down with drops. He did not set it straight, and made the hill grow fifty feet; but we got it down again.

We came here on Wednesday by Caledonian. I intend to open my classes next week after the business is over. I have been reading Xenophon's *Memorabilia* after breakfast; also a French collection book. This from 9 to 11. Then a game of the Devil,[2] of whom there is a duality and a quaternity of sticks, so that I can play either conjunctly or severally. I can jump over him and bring him round without leaving go the sticks. I can also keep him up behind me.

Then I go in again to science, of which I have only just got the books by the carrier. Hitherto I have done a prop on the slate on polarised light. Of props I have done several.

1. Found the equation to a square.
2. The curve which Sir David Brewster sees when he squints at a wall.[3]
3. A property of the parabola. ...
4. The same of the Ellipse and Hyperbola. ...[4]

I can polarise light now by reflection or refraction in 4 ways, and get beautiful but evanescent figures in plate glass by heating its edge. I have not yet unannealed any glass. ...[5]

---

(1) *Life of Maxwell*: 116–17; abridged.

(2) The devil on two sticks: a wooden toy in the form of a double cone, made to spin in the air by means of a string attached to two sticks held in the hands (*OED*).

(3) David Brewster, 'On the knowledge of distance given by binocular vision', *Trans. Roy. Soc. Edinb.*, **15** (1844): 663–75, esp. 663–5 (= *Phil. Mag.*, ser. 3, **30** (1847): 305–18). The visual illusion on viewing a regularly patterned surface, leading to a false estimate of its distance; 'the surface seems slightly convex towards the eye'. On binocular vision see Number 20.

(4) Campbell cut the letter.

(5) See Numbers 12, 13 and 14; and *Life of Maxwell*: 84, 487–8, on Maxwell's visit, with Campbell and his uncle John Cay, to the laboratory of William Nicol, inventor of Nicol's polarising prism, in April 1847. On Nicol's polarising prism see Number 19, esp. note (3).

FROM A LETTER TO LEWIS CAMPBELL

5 AND 6 JULY 1848

From Campbell and Garnett, *Life of Maxwell*[1]

Glenlair
5 July 1848

I was much glad of your letter, and will be thankful for a repetition. I understand better about your not coming. I have regularly set up shop now above the wash-house at the gate, in a garret. I have an old door set on two barrels, and two chairs, of which one is safe, and a skylight above, which will slide up and down.

On the door (or table), there is a lot of bowls, jugs, plates, jam pigs,[2] etc., containing water, salt, soda, sulphuric acid, blue vitriol, plumbago ore; also broken glass, iron, and copper wire, copper and zinc plate, bees' wax, sealing wax, clay, rosin, charcoal, a lens, a Smee's Galvanic apparatus, and a countless variety of little beetles, spiders, and wood lice, which fall into the different liquids and poison themselves. I intend to get up some more galvanism in jam pigs; but I must first copper the interiors of the pigs, so I am experimenting on the best methods of electrotyping. So I am making copper seals with the device of a beetle. First, I thought a beetle was a good conductor, so I embedded one in wax (not at all cruel, because I slew him in boiling water in which he never kicked), leaving his back out; but he would not do. Then I took a cast of him in sealing wax, and pressed wax into the hollow, and black-leaded it with a brush; but neither would that do. So at last I took my fingers and rubbed it, which I find the best way to use the black lead. Then it coppered famously. I melt out the wax with the lens, that being the cleanest way of getting a strong heat, so I do most things with it that need heat. To-day I astonished the natives as follows. I took a crystal of blue vitriol and put the lens to it, and so drove off the water, leaving a white powder. Then I did the same to some washing soda, and mixed the two white powders together; and made a small native spit on them, which turned them green by a mutual exchange, thus: 1. Sulphate of copper and carbonate of soda. 2. Sulphate of soda and carbonate of copper (blue or green).

With regard to electro-magnetism you may tell Bob[3] that I have not begun the machine he speaks of, being occupied with better plans, one of which is rather down cast, however, because the machine when tried went a bit and

---

(1) *Life of Maxwell*: 117–20.   (2) Jars.

(3) Lewis Campbell's brother Robert, born 1832, educ. Edinburgh Academy and University, Caius 1850 and Trinity Hall 1852 (Venn). See Number 28.

stuck; and I did not find out the impediment till I had dreamt over it properly, which I consider the best mode of resolving difficulties of a particular kind, which may be found out by thought, or especially by the laws of association. Thus, you are going along the road with a key in your pocket. You hear a clink behind you, but do not look round, thinking it is nothing particular; when you get home the key is gone; so you dream it all over, and though you have forgotten everything else, you remember the look of the place, but do not remember the locality (that is, as thus, 'Near a large thistle on the left side of the road' – nowhere in particular, but so that it can be found). Next day comes a woman from the peats who has found the key in a corresponding place. This is not 'believing in dreams', for the dream did not point out the place by the general locality, but by the lie of the ground.

Please to write and tell how Academy matters go, if they are coming to a head. I am reading Herodotus, *Euterpe*, having taken the turn; that is to say, that sometimes I can do props, read diff. and Int. Calc., Poisson,[4] Hamilton's dissertations,[5] etc. Off, then I take back to experiments, history of what you may call it, make up leeway in the newspapers, read Herodotus, and draw the figures of the curves above. O deary, 11 P.M.! Hoping to see you *before* October.... I defer till to-morrow.

July 6. To-day I have set on to the coppering of the jam pig which I polished yesterday.

I have stuck in the wires better than ever, and it is going on at a great rate, being a rainy day, and the skylight shut and a smell of Hydrogen gas. I have left it for an hour to read Poisson, as I am pleased with him

Figure 10,1

to-day. Sometimes I do not like him, because he pretends to give information as to calculations of sorts, whereas he only tells how it might be done if you were allowed an infinite time to do it in, as well as patience. Of course he never stoops to give a particular example or even class of them. He tells lies about the way people make barometers, etc.

I bathe regularly every day when dry, and try aquatic experiments.

I first made a survey of the pool, and took soundings and marked rocky places well, as the water is so brown that one cannot see one's knees (pure peat, not mud). People are cutting peats now. So I have found a way of swimming round the pool without knocking knees. The lads are afraid of melt-

(4) Siméon Denis Poisson, *Traité de Mécanique*, 2 vols. (Paris, ₂1833); (trans. H. H. Harte), *A Treatise of Mechanics*, 2 vols. (London, 1842).

(5) Sir William Hamilton's 'Supplementary Dissertations' to his edition of *The Works of Thomas Reid, D.D. Now Fully Collected, with Selections from his Unpublished Letters* (Edinburgh, 1846).

ing, except one. No one here would touch water if they could help it, because there are two or three eels in the pool, which are thought near as bad as adders.

I took down the clay gun and made a centrifugal pump of it; also tried experiments on sound under water, which is very distinct, and I can understand how fishes can be stunned by knocking a stone.

We sometimes get a rope, which I take hold of at one end, and Bob Fraser the other, standing on the rock; and after a flood, when the water is up, there is sufficient current to keep me up like a kite without striking at all.

The thermometer ranged yesterday from 35° to 69°.

I have made regular figures of 14, 26, 32, 38, 62, and 102 sides of cardboard.

Latest intelligence – Electric Telegraph. This is going so as to make a compass spin very much. I must go to see my pig, as it is an hour and half since I left it; so, sir, am your afft. friend,

JAMES CLERK MAXWELL.

11

DRAFT OF PAPER 'ON THE THEORY OF ROLLING CURVES'

SUMMER 1848[1]

From the original in the University Library, Cambridge[2]

ON ROLLING CURVES[3]

Let there be a curve CAS, of which the pole is C.

Let $DCA = \theta_1$ and $CA = r_1$ and let $\theta_1 = \phi_1 r_1$.   (1)

Let this curve remain fixed to the paper.   (2)

Let there be another curve BAT whose pole is B.

Let $MBA = \theta_2$ and $BA = r_2$ and let $\theta_2 = \phi_2 r_2$.   (3)

Let this curve roll along CAS (4) without slipping (5) so that its pole B describes a third curve whose pole is C.

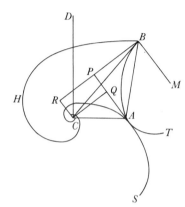

Figure 11,1

Let $DCB = \theta_3$ and $CB = r_3$ and let $\theta_3 = \phi_3 r_3$.   (6)

It is required from any two of the functions $(\phi_1 \phi_2 \phi_3)$ to find the third.

Here we have six unknown quantities $\theta_1 \theta_2 \theta_3 r_1 r_2 r_3$ we have [three] equations of the form $(\theta = \phi r)$ therefore we must obtain three more, from the enunciation.

Now the enunciation may be divided into six articles which contain the six equations, but of the three (1) (3) and (6) one is supposed not given.

Let us examine the other articles (2) (4) and (5) in an inverse order.

---

(1) See Number 12.   (2) ULC Add. MSS 7655, V, d/6.

(3) A draft of 'On the theory of rolling curves', *Trans. Roy. Soc. Edinb.*, **16** (1849): 519–40 (= *Scientific Papers*, **1**: 4–29). Likely sources for Maxwell's knowledge of the geometry of curves are the introductory text by John Leslie, *Geometrical Analysis and Geometry of Curve Lines, being Volume Second of a Course of Mathematics, and designed as an Introduction to the Study of Natural Philosophy* (Edinburgh, 1821): 323–438; and the more advanced text by D. F. Gregory, *Examples of the Processes of the Differential and Integral Calculus* (Cambridge, 1841): 127–95, esp. 133–41. Maxwell made reference to Leslie's discussion of rolling curves in 'On the theory of rolling curves': 529 (= *Scientific Papers*, **1**: 16); 'The operation of finding the fixed curve from the rolled curve is what Sir John Leslie calls "divesting a curve of its radiated structure".' Compare Leslie, *Geometrical Analysis*: 420, 428. For works on rolling curves cited by Maxwell see Number 18.

Article (5) 'Without slipping'. This shows that at any instant the curves CAS, BAT are equal in length at the point of contact or in the form of an equation

$$\int_{\theta_1} \sqrt{r_1^2 + \frac{dr_1^2}{d\theta_1^2}} = \int_{\theta_2} \sqrt{r_2^2 + \frac{dr_2^2}{d\theta_2^2}}.$$

Article (4) 'Let this curve roll along CAS'. This shows that the curves touch each other without cutting or they have a common tangent.

Let PA be the tangent, draw BP, CQ perpendiculars from B, C.

Let $BP = p_2 = \dfrac{r_2^2}{\sqrt{r_2^2 + \dfrac{dr_2^2}{d\theta_2^2}}}.$ Let $CQ = p_1 = \dfrac{r_1^2}{\sqrt{r_1^2 + \dfrac{dr_1^2}{d\theta_1^2}}}.$

Complete the rectangle $QPRC$ then

$$r_3^2 = BC^2 = BR^2 + RC^2 = (BP + PR)^2 + PQ^2 = (BP + CQ)^2 + (PA - AQ)^2$$
$$r_3^2 = BP^2 + 2BP \cdot CQ + CQ^2 + PA^2 - 2PA \cdot AQ + AQ^2$$
$$r_3^2 = (BP^2 + AP^2) + (AQ^2 + CQ^2) + 2BP \cdot CQ - 2PA \cdot AQ$$
$$r_3^2 = r_2^2 + r_1^2 + 2p_1 p_2 - 2\sqrt{(r_1^2 - p_1^2)(r_2^2 - p_2^2)}$$
$$r_3^2 = r_1^2 + r_2^2 + 2\frac{r_1^2 r_2^2 \, d\theta_1 \, d\theta_2 - r_1 \, r_2 \, dr_1 \, dr_2}{\sqrt{r_1^2 r_2^2 \, d\theta_1^2 \, d\theta_2^2 + r_1^2 \, dr_2^2 \, d\theta_1^2 + r_2^2 \, dr_1^2 \, d\theta_2^2 + dr_1^2 \, dr_2^2}}.$$

Article (2) 'Let this curve remain fixed to the paper' shews how to find the angle θ_3 in terms of the other quantities, for $\theta_3 = \theta_1 - BCQ - ACQ$ and

$$\tan BCQ = \frac{CR}{BR}$$

$$\tan BCQ = \frac{AP - AQ}{BP + CQ}$$

$$-\tan BCQ = \frac{r_1 \, dr_1 \sqrt{r_2^2 \, d\theta_2^2 + dr_2^2} - r_2 \, dr_2 \sqrt{r_1^2 \, d\theta_1^2 + dr_1^2}}{r_1^2 \, d\theta_1 \sqrt{r_2^2 \, d\theta_2^2 + dr_2^2} + r_2^2 \, d\theta_2 \sqrt{r_1^2 \, d\theta_1^2 + dr_1^2}}$$

$$\therefore \theta_3 = \theta_1 + \tan^{-1} \frac{r_1 \, dr_1 \sqrt{r_2^2 \, d\theta_2^2 + dr_2^2} - r_2 \, dr_2 \sqrt{r_1^2 \, d\theta_1^2 + dr_1^2}}{r_1^2 \, d\theta_1 \sqrt{r_2^2 \, d\theta_2^2 + dr_2^2} + r_2^2 \, d\theta_2 \sqrt{r_1^2 \, d\theta_1^2 + dr_1^2}} - \tan^{-1} \frac{dr_1^2}{r_1^2 \, d\theta_1^2}.$$

Thus we have found the other three equations and now we propose to give examples of their use but first let us classify the different kinds of problems.[4]

Now we may have two of the functions ($\phi_1 \, \phi_2 \, \phi_3$) given to find the third. (We may call this Order I.)

---

(4) Hamilton had discussed the classification of concepts into genus and species in his Logic lectures in 1847–8, recorded by Maxwell in notes dated 'Jan. 10 [1848]' in his Edinburgh notebook 'Logic N°. 2' (ULC Add. MSS 7655, V, m/1). Compare William Hamilton, *Lectures on Metaphysics and Logic*, ed. H. L. Mansel and J. Veitch, 4 vols. (Edinburgh/London, 1859–60), **3**: 191–2.

Or we may have one function given and the other two stated to be identical (Order II).

Or we may have all three identical (Order III).

In the first Order there will be three Genera according to the given functions. In each Genus there will be two species the given functions being either different or identical. In the first species the principal varieties will be those in which one of the curves is a straight line, a circle or some other notable curve, the individuals are those in which both curves are given.

In the second Order there will also be three Genera but only one species and the rest individuals.

In the third Order there is only one Genus Species and individual.

We will give rules for the Genera Species etc and examples of individuals.

Order I

Given two functions find the third.

Genus I

Given ϕ_1 and ϕ_2 find ϕ_3.

Rule. From the five equations of the rolling curves eliminate $\theta_1 \, \theta_2 \, r_1 \, r_2$ and we have the polar equation of the curve required.

Species 1

Let ϕ_1 be different from ϕ_2.

Variety 1

Let $\phi_1 = \sec^{-1}$ that is let $\theta = \sec^{-1} \dfrac{r}{a}$ or $r = a \sec \theta$. This is the equation to a straight line but it is better in this case to consider the straight line as the axis of y and find the equation of the curve produced in terms of rectangular co-ordinates. To do this let the equation to the rolling curve be $\theta_2 = \phi_2 r_2$, find $\dfrac{dr_2^2}{d\theta_2^2}$ in terms of r_2 substitute $x \sqrt{\dfrac{dx^2}{dy^2} + 1}$ for r in this expression and equate it to

$$x^2 \frac{dx^4}{dy^4} + x^2 \frac{dx^2}{dy^2}$$ or in other words let $\dfrac{dr_2^2}{d\theta_2^2} = fr_2$ then $x^2 \dfrac{dx^4}{dy^4} + x^2 \dfrac{dx^2}{dy^2} = fx \sqrt{\dfrac{dx^2}{dy^2} + 1}$.

Solving this equation we have the value of $\dfrac{dx}{dy}$ in terms of x from which we find y in terms of x. This is the most convenient for easy problems, but we may do it otherwise from the five equations thus. Make the fixed pole a point in the fixed line then we have

$$r_1 = \int_{\theta_1} \sqrt{r_1^2 + \frac{dr_1^2}{d\theta_1^2}} = \int_{\theta_2} \sqrt{r_2^2 + \frac{dr_2^2}{d\theta_2^2}}$$

$$p_1 = 0^{(5)}$$

thus we obtain

$$r_3^2 = r_2^2 + r_1^2 - \frac{2\frac{dr_2}{d\theta_2} r^2}{\sqrt{r_2^2 + \frac{dr_2^2}{d\theta_2^2}}} r_1$$

$$\theta_3 = \tan^{-1} \frac{r_2^2}{r_1 \sqrt{r_2^2 + \frac{dr_2^2}{d\theta_2^2}} - \frac{dr_2}{d\theta_2} r_2}$$

$$r_1 = \frac{r_3\, dr_3}{r_3\, d\theta_3 \sin\theta_3 + dr_3 \cos\theta_3}, \quad r_2 = r_3 \tan\theta_3 \frac{\sqrt{r_3^2\, d\theta_3^2 + dr_3^2}}{r_3\, d\theta_3 \tan\theta_3 + dr_3}.$$

If the rectangular equation of the rolling curve is given we have the following equations

equation to rolling curve $x_2 = \phi_2 y_2$

equation to traced curve $x_3 = \phi_3 y_3$

$$s_2 = \int_{x_2} \sqrt{1 + \frac{dy^2}{dx^2}}^{(6)}$$

$$x_3 = \frac{y_2\, dx_2 - x_2\, dy_2}{ds_2}$$

$$y_3 = \frac{s_2\, ds_2 - x_2\, dx_2 - y_2\, dy_2}{ds_2}$$

$$\frac{dx_3}{dy_3} = \frac{x_2\, dx_2 + y_2\, dy_2}{y_2\, dx_2 - x_2\, dy_2}.$$

We will now give examples of the use of these equations.

---

(5) Here and in sequel p denotes the perpendicular from the pole on the tangent.

(6) Here and in sequel s denotes the length of the curve from the pole.

Take the polar equation to the cardioid $r = 2a(1 - \cos\theta)$ or

$$\theta = \cos^{-1}\left(1 - \frac{r}{2a}\right), \quad \frac{d\theta}{dr} = \frac{1}{\sqrt{r^2 + 2ar}}$$

$$\frac{dr^2}{d\theta^2} = -r^2 + 2ar$$

$$\frac{dx^4}{dy^4} + \frac{dx^2}{dy^2} = 2\frac{a}{x}\sqrt{1 + \frac{dx^2}{dy^2}} - 1 - \frac{dx^2}{dy^2}$$

$$\left(\frac{dx^2}{dy^2} + 1\right)^{\frac{3}{2}} = 2\frac{a}{x}$$

$$\frac{dx^2}{dy^2} = \left(2\frac{a}{x}\right)^{\frac{2}{3}} - 1$$

$$\frac{dy}{dx} = \frac{1}{\sqrt{\left(2\dfrac{a}{x}\right)^{\frac{2}{3}} - 1}}$$

the integration of which gives the equation to the curve.

As an example of the second method, take the involute to the circle, here

$$\theta_2 = \frac{\sqrt{r^2 - a^2}}{a} - \sec^{-1}\frac{r}{a}$$

$$r_1 = \int_{\theta_2}\sqrt{r_2^2 + \frac{dr_2^2}{d\theta_2^2}} = \frac{r_2^2 - a^2}{2a}$$

$$\therefore r_3^2 = r_2^2 + \frac{r_2^4}{4a^2} + \frac{a^2}{4} - \frac{r_2^2}{2} - r_2^2 + a^2$$

$$\theta_3 = \tan^{-1}\frac{2a\sqrt{r^2 - a^2}}{r^2 - 3a^2}$$

$$\left\langle \therefore \theta_3 = \cos^{-1}\left(\frac{2a}{r} - 1\right)\right\rangle \quad \text{the} \quad \text{polar} \quad \text{equation} \quad \text{of} \quad \text{the} \quad \text{parabola}$$

$$r = \frac{2a}{\tan\theta}\sqrt{1 + \frac{1}{\tan^2\theta}} = \frac{2a}{\sin\theta\tan\theta}.$$

Example 3rd

The equation to the parabola, the focus being the origin is

$$y_2^2 = 4a^2 + 4ax_2$$

$$\frac{dx_2}{ds_2} = \frac{\sqrt{a^2 + ax_2}}{\sqrt{2a^2 + ax_2}}$$

$$\frac{dy_2}{ds_2} = \frac{a}{\sqrt{2a^2 + ax_2}}$$

$$s_2 = \sqrt{x_2^2 + 3ax_2 + 2a^2} + \frac{a}{2}\log\left(\frac{2x_2 + 3a + 2\sqrt{x_2^2 + 3ax_2 + 2a^2}}{a}\right)$$

then by the equations of the third method

$$x_3 = \frac{2\sqrt{4a^2 + 2ax_2}}{\sqrt{2a^2 + ax_2}} - \frac{ax_2}{\sqrt{2a^2 + ax_2}} = \sqrt{2a^2 + ax_2}$$

$$\therefore x_3^2 = 2a^2 + ax_2$$

$$\therefore x_2 = \frac{x_3^2 - 2a^2}{a}$$

$$y_3 = s - \frac{x\sqrt{a^2 + ax} + 2a\sqrt{a^2 + ax}}{\sqrt{2a^2 + ax}}$$

$$= s - \sqrt{x^2 + 3ax + 2a^2}$$

$$\therefore y_3 = \frac{a}{2} \log\left(2x_2^2 + 3a^2 + 2\sqrt{x_2^2 + 3ax_2 + 2a^2}\right)$$

$$y_3 = \frac{a}{2} \log\left(\frac{2x_3^2 - a^2 + 2x_3\sqrt{x_3^2 - a^2}}{a^2}\right) = \frac{a}{2} \log\left(\frac{(x + \sqrt{x^2 - a^2})^2}{a}\right)$$

$$y_3 = a \log\left(\frac{x + \sqrt{x^2 - a^2}}{a}\right)$$

$$e^{\frac{y}{a}} = \frac{x + \sqrt{x^2 - a^2}}{a}$$

$$a^2 e^{\frac{2y}{a}} + 2axe^{\frac{y}{a}} + x^2 = x^2 - a^2$$

$$x = \frac{a}{2}\left(e^{\frac{y}{a}} + e^{-\frac{y}{a}}\right). \text{[7]}$$

Variety 2

Let $\phi_2 = \sec^{-1}$ that is let the rolling curve be a straight line. Here we have

$$p_2 = a \qquad s_2 = a\tan\theta_2 = \sqrt{r_2^2 - a^2}$$

$$\therefore \int_{\theta_1} \sqrt{r_1^2 + \frac{dr_1^2}{d\theta_1^2}} = \sqrt{r_2^2 - a^2}$$

$$r_3^2 = \frac{2ar_1^2 - 2\sqrt{r_2^2 - a^2}\, r_1 \dfrac{dr_1}{d\theta_1}}{\sqrt{r_1^2 + \dfrac{dr_1^2}{d\theta_1^2}}} + r_2^2$$

$$\theta_3 = \theta_1 - \tan^{-1}\frac{dr_1}{r_1\, d\theta_1} - \tan^{-1}\frac{\sqrt{r^2 - a^2}\sqrt{r^2 + \dfrac{dr^2}{d\theta^2}} + r\dfrac{dr}{d\theta}}{r^2 + a\sqrt{r^2 + \dfrac{dr^2}{d\theta^2}}}.$$

---

(7) The equation of the catenary; compare 'On the theory of rolling curves': 525–6 (= *Scientific Papers*, **1**: 11–12).

When the fixed curve is expressed in rectangular coordinates we have

$$x_3 = \frac{x_2\, dx_2\, dy_2 - s_2\, dy_2 - a\, dx_2}{dx_2\, dy_2}$$

$$y_3 = \frac{y_2\, dx_2\, dy_2 - s_2\, dx_2 + a\, dy_2}{dx_2\, dy_2}$$

$$\frac{dy_3}{dx_3} = \frac{a\, dx_2 - s_2\, dy_2}{s_2\, dx_2 + a\, dy_2}.$$

Erratum in these three equations – for $x_2\, y_2$ read $x_1\, y_1$.

As an example let us find what curve will be traced by a point distant a from a line which rolls along a Catenary whose parameter is a.

Here
$$s_1 = a\frac{dx_1}{dy_1}$$

$$\therefore \frac{dy_3}{dx_3} = \frac{a-a}{a\dfrac{dx}{dy}+a\dfrac{dy}{dx}} = 0$$

therefore the locus required is a straight line parallel to y and distant from it by $-a$.

In this variety, when $a = 0$, the curve is the involute of the fixed curve, this is treated of in most works.[8]

If tangents of a certain length are drawn to a curve their extremities will be in a certain other curve, which in relation to the former may be called its tramitory[9] whilst the former is called the tractory of the latter.

Thus in the last example we see that the involute of the catenary is the tractory of the straight line.[10]

In the discussion of the next variety we shall give an example which belongs to both varieties, and exemplifys the first or polar method given here and shows a polar tractory and tramitory.

Variety 3

Let the fixed curve be a circle radius a.

Here
$$s_1 = s_2 = \int_{\theta_2}\sqrt{r_2^2 + \frac{dr_2^2}{d\theta_2^2}} = a\theta_1$$

$$p_1 = a$$

---

(8) See Leslie, *Geometrical Analysis*: 396.   (9) Maxwell's term.
(10) The tractory or tractrix; see Leslie, *Geometrical Analysis*: 393, 396–8.

$$r_3^2 = 2a\frac{r_2^2}{\sqrt{r_2^2 + \dfrac{dr_2^2}{d\theta_2^2}}} + a^2 + r_2^2$$

$$\theta_3 = \theta_1 - \tan^{-1}\frac{r_2\dfrac{dr_2}{d\theta_2}}{r_2^2 + a\sqrt{r_2^2 + \dfrac{dr_2^2}{d\theta_2^2}}}.$$

Ex. 1 Let the rolling line be $r = a\sec\theta$ then the equations become

$$a\tan\theta_2 = a\theta_1 \quad \therefore \tan\theta_2 = \theta_1$$
$$r_3^2 = -2a^2 + a^2 + r_2^2 = r_2^2 - a^2 = a^2\tan^2\theta_2$$
$$r_3 = a\tan\theta_2 = a\theta_1$$

$$\theta_3 = \theta_1 - \tan^{-1}\infty = \theta_1 - \frac{\pi}{2} \quad \therefore \theta_1 = \theta_3 + \frac{\pi}{2} \quad \therefore r_3 = a\theta_3 + a\frac{\pi}{2}$$

therefore the curve is the spiral of Archimedes and the involute of the circle is a tractory of the spiral of Archimedes and vice versâ.

Let the rolling curve be the logarithmic spiral

$$\theta = \log\frac{x}{b} \qquad\qquad \frac{d\theta}{dr} = \frac{1}{r}$$
$$a\theta_1 = \sqrt{2}r$$

$$r_3^2 = \sqrt{2}ar_2 + a^2 + r_2^2 \quad \therefore \sqrt{r_3^2 - \frac{a^2}{2}} = r_2 + \frac{a}{\sqrt{2}}$$

$$\therefore r_2 = \sqrt{r_3^2 - \frac{a^2}{2}} - \frac{a}{\sqrt{2}}$$

$$\theta_3 = \theta_1 - \tan^{-1}\frac{\sqrt{2}r_2}{\sqrt{2}r_2 + a} = \sqrt{\frac{2r_3^2}{a^2} - 1} - 1 - \tan^{-1}1 - \frac{a}{\sqrt{2r^2 - a^2}}$$

this is the equation of the involute of the circle.

Let $ALPMN$ be the log spiral.

Let $PTOXQ$ be the fixed circle.

Let $XTBCX$ be the curve traced which is the involute to the inner circle.

The log. spiral rolls along the outer circle till it comes to T where the involute cuts the circle, the spiral has made an

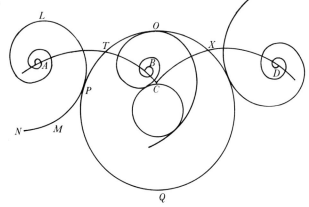

Figure 11,2

infinity of revolutions and has come to the end. It now enters on the other branch changing the direction of revolution and rolling on the inner side of the circle the radius of curvature being less than that of the circle.

When the pole arrives at C the cusp of the involute the radius of curvature becomes equal to that of the circle and the spiral begins to roll on the outside of the circle but still curved inwards and so describes the other branch of the involute.

Variety 4

Let the rolling curve be a circle and let its centre describe the curve required. Let the radius $= a$.

Here $r_3^2 = r^2 + 2ap + a^2$

$$\theta_3 = \theta_1 - \tan^{-1} \frac{dr_1}{r_1 \, d\theta_1} + \tan^{-1} \frac{\sqrt{r_1^2 - p_1^2}}{r_1 + a}$$

or

$$x_3 = x_2 - a \frac{dy_2}{dx_2}, \quad y_3 = y_2 + a \frac{dx_2}{dy_2}$$

$$\frac{dx_3}{dy_3} = \frac{dx_2}{dy_2}$$

$$\therefore x_2 = x_3 + a \frac{dy_3}{dx_3}, \quad y_2 = y_3 - a \frac{dx_3}{dy_3}$$

$$y_3 = \phi_2 \left(x_3 + a \frac{dy_3}{dx_3} \right) + a \frac{dx_3}{dy_3}.$$

Example. Take the log. spiral

$$r_3^2 = r_1^2 + a\sqrt{2}r_1 + a^2 \quad \therefore r_1 = \sqrt{r_3^2 - \frac{a^2}{2}} - \frac{\sqrt{2}}{2} a$$

$$\theta_3 = \log \frac{r_2}{a} - \tan^{-1} 1 + \tan^{-1} \frac{\sqrt{2} r_2}{\sqrt{2} r_2 + a}$$

$$\theta_3 = \log \left(\sqrt{\frac{r_3^2}{a^2} - \frac{1}{2}} - \frac{\sqrt{2}}{2} \right) - \frac{\pi}{4} + \tan^{-1} \frac{\sqrt{2} \sqrt{r_3^2 - \frac{a^2}{2}} - a}{\sqrt{2} \sqrt{r_3^2 - \frac{a^2}{2}}}$$

this equation is that of the curve which cuts all tangents to a circle whose centre is the pole and radius $\dfrac{a}{\sqrt{2}}$ at an angle of $\dfrac{\pi}{4}$ or $45°$.

$$\theta_3 = \log \left(\sqrt{\frac{r_3^2}{a^2} - \frac{1}{2}} - \frac{1}{\sqrt{2}} \right) - \frac{\pi}{4} + \tan^{-1} \left(1 + \frac{1}{\sqrt{\frac{2r^2}{a^2} - 1}} \right).$$

As an example of a geometrical investigation of a problem of this genus let us take the following. Let *SPOT* be the involute of the circle *SCK*. Let a logarithmic spiral roll upon it and trace out the curve *SRQ*. Let *R* and *Q* be consecutive positions of the pole of the spiral and *P, O* the points of contact then by the laws of rolling *RP, QO* are normals to *SRQ* therefore the centre of curvature at *Q* is at *D* the intersection of the normals.

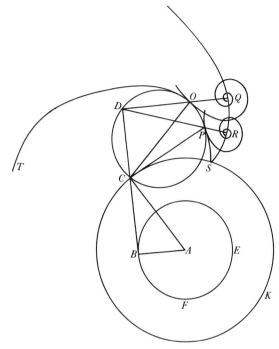

Now since the spiral is equiangular the angle *DOC = DPC* for *OC* and *PC* are normals to the involute.

∴ *ODP = OCP* and *OPCD* may be inscribed in a circle but because *O* is very near *P* the circle touches at *O* and *ODC* is a right angle but *DOC* = 45° ∷ *DCO* = 45°. Draw

Figure 11,3

$AB \perp DC$ then $AB = \dfrac{AC}{\sqrt{2}}$ and constant, therefore the normal of the evolute of *SRQ* always touches a circle, it is therefore the involute of the circle and *SRQ* is the second involute of the circle *BEF*.

In the same way we may prove that if a log. spiral be rolled on any involute to a circle the curved traced will be the next involute to a circle the square of whose radius is half the square of the radius of the first. Let any involute be denoted by θ^n and evolute by θ^{-n}. We have proved this proposition with regard to the θ^0 and θ^1 of a circle and it is evident for the θ^{-1} and we shall soon prove it for the θ^∞.

⌐Note. This is true of any log. spiral, besides that of which the constant angle is = 45°.⌐

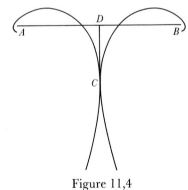

Figure 11,4

Species 2

Let one curve roll upon another equal and similar curve so that similar parts touch each other.

Join the poles AB and draw the tangent CD and it is evident that the locus of D may be found thus.

$$r_3 = 2p_1 \quad \theta_3 = \theta_1 - \tan^{-1}\frac{dr_1}{rd\theta_1}$$

and

$$p_3^2 = \frac{4p_1^4}{r_1^4} = \frac{r_3^4}{4r_1^2} \quad p_3 = \frac{r_3^2}{2r_1}.$$

If this curve again rolls on itself and produces a curve $r_4 = \phi_4\theta_4$ then

$$2p_3 = r_4 = \frac{r_3^2}{2r_1} = \frac{2p_1^{2\,(11)}}{r_2}$$

$$\theta_4 = \theta_3 - \tan^{-1}\frac{dr_3}{rd\theta_3} = \theta_1 - \tan^{-1}\frac{dr_1}{rd\theta_1} - \tan^{-1}\frac{dr_3}{rd\theta_3}.$$

Similarly it may be proved that

$$r_m = 2p_{m-1} = \frac{r_{m-1}^2}{r_{m-2}} = \frac{r_{m-2}^3}{r_{m-3}^2} \text{ etc.}$$

$$\therefore r_m = \frac{r_2^{m-1}}{r_1^{m-2}} = r_2\left(\frac{r_2}{r_1}\right)^{m-2} = r_1\left(\frac{r_2}{r_1}\right)^{m-1}$$

therefore

$$\frac{r_{m+1}}{2r_m} = \frac{r_2}{2r_1}$$

therefore

$$\theta_m = \theta_1 - \cos^{-1}\frac{r_2}{2r_1} - \cos^{-1}\frac{r_3}{2r_2} \text{ etc.}$$

$$\theta_m = \theta_1 - m\cos^{-1}\frac{r_2}{2r_1} = \theta_1 - m\cos^{-1}\left(\frac{r_m}{r_2}\right)^{\frac{1}{m-2}}.$$

Let $m = \infty$ then we have three cases, $\frac{r_2}{2r_1}$ is either 0, ∞ or finite. If it is $= 0$ all the curves are circles if it is ∞ they are straight lines but if it is finite then for every finite value of θ_m, $\cos^{-1}\frac{r_2}{2r_1}$ will be constant, for if it was to vary by any finite quantity $= h$ hm would be infinitely changed, but θ_m is finitely changed, therefore θ_1 is infinitely changed, therefore when $\cos^{-1}\frac{r_2}{2r_1}$ is finitely changed θ_1 is infinitely changed therefore when θ_1 is finitely changed, $\cos^{-1}\frac{r_2}{2r_1}$ does not vary at all that is it is constant therefore the angle at which the curve cuts the radius vector is constant or, it is the equiangular spiral, now we shall prove that the equiangular spiral, being rolled on itself, produces itself therefore since, by this erroneous supposition the primary curve $(\theta_1 = \phi_1 r_1)$ is the equilateral

---

(11) Maxwell's amendments introduce an inconsistency.

spiral, the curve in question $(\theta_m = \phi_m r_m)$ is so too therefore we have the following reasoning.

The curve is either the equiangular spiral or it is not. If it is, it is; but if it is not, we have proved that it must be so.

Therefore the curve is the equiangular spiral.

Example 1. Find the curve produced when an equiangular spiral rolls on itself.

Here $$p_1 = mr_1 \; \therefore r_3 = 2mr_1 \quad p_3 = \frac{r_3^2}{2r_1} = \frac{r_3^2}{\dfrac{r_3}{m}} = mr_3$$

therefore the curve is the equiangular spiral.

Example 2. Find the curve when a circle rolls on itself and traces a curve with a point in its circumference also the curve produced in the same way by this etc.

Here $$r_1 = 2a\cos\theta_1 \quad p_1 = \frac{r_1^2}{2a} \quad \therefore \frac{r^2}{r_1} = \frac{r_1}{a} = 2\cos\theta$$

$$\therefore r_m = r_1\left(\frac{r_2}{r_1}\right)^{m-1} = a(2\cos\theta_1)^m = (2)^m a(\cos\theta_1)^m$$

$$\theta_m = m\theta_1 \quad \therefore \theta_1 = \frac{\theta_m}{m}$$

$$\therefore r_m = 2^m a\left(\cos\frac{\theta_m}{m}\right)^m$$

when $m = 2$ the equation is that of a cardioid

$$r_2 = 4a\left(\cos\frac{\theta_2}{2}\right)^2.$$

When we have a rectangular equation we use the following equations

$$y_1 = \phi_1(x_1) \quad \frac{dy_1}{dx_1} = f_1(x_1)$$

$$y_3 - 2\phi_1(x_1) = f_1(x_1)(x_3 - 2x_1)$$

$$y_3 = -2f_1(x_1)x_3.$$

Genus 2 Species 1

Given the fixed curve and the curve traced, find the rolling curve.

Variety 1

Let the fixed line be a straight line.
Let the curve traced have a rectangular equation.

Let $\dfrac{dx_3^2}{dy_3^2} = \psi_3\, x_3$ then we have the equation

$$\frac{dr_2^2}{d\theta_2^2} = r_2^2\, \psi_3 \frac{r_2^2}{\sqrt{r_2^2 + \dfrac{dr_2^2}{d\theta_2^2}}}.$$

Example 1. Let the curve produced be the tractory to the straight line.

Here $\qquad \dfrac{dx_3}{dy_3} = \dfrac{x_3}{\sqrt{a^2 - x_3^2}} \qquad\qquad \dfrac{dx_3^2}{dy_3^2} = \dfrac{x_3^2}{a^2 - x_3^2}$

$$\frac{dr^2}{d\theta^2} = r^2 \frac{r^4}{a^2 r^2 + a^2 \dfrac{dr^2}{d\theta^2} - r^4}$$

$$\frac{dr^4}{d\theta^4} + \left(r^2 - \frac{r^4}{a^2}\right)\frac{dr^2}{d\theta^2} + \frac{r^4}{4a^4}(a^2 - r^2)^2 = \frac{r^4}{4a^4}(a^4 + 2a^2 r^2 + r^4)$$

$$\frac{dr^2}{d\theta^2} = \frac{r^2}{2a^2}\left((a^2 + r^2) - (a^2 - r^2)\right) = \frac{r^4}{a^2}$$

$$\frac{dr}{d\theta} = \frac{r^2}{a},\ \frac{d\theta}{dr} = \frac{a}{r^2},\ \theta = -2\frac{a}{r},\ r = -\frac{2a}{\theta}$$

therefore the rolling curve is the hyperbolic spiral.

In the same way we may do the following examples which are worked in another place.

curve produced	rolling curve
$\dfrac{y^2}{b^2} \pm \dfrac{x^2}{a^2} = 1$	$\dfrac{d\theta}{dr} = \dfrac{\sqrt{b^2 r^2 - a^4}}{ar\sqrt{a^2 \pm r^2}}$
$xy = a^2$	$\dfrac{dr^3}{d\theta^3} + r^2 \dfrac{dr}{d\theta} = \dfrac{r^5}{a^2}$
$ax = y^2$	$\dfrac{dr^6}{d\theta^6} + r^2 \dfrac{dr^4}{d\theta^4} = \dfrac{16 r^8}{a^2}$
$x^2 = ay$	involute of circle
$y^2 - x^2 = a^2$	$\theta = \log\!\left(\dfrac{\sqrt{r^4 - a^4} + r^2}{a^2}\right) - \log\!\left(\dfrac{\sqrt{r^4 - a^4} + a^2}{r^2}\right)$
$y = \dfrac{a}{2}\left(e^{\frac{x}{a}} + e^{-\frac{x}{a}}\right)$	$r = \dfrac{2a}{1 + \cos\theta}$
logarithmic curve	$\dfrac{d\theta}{dr} = \dfrac{\sqrt{a}\sqrt{a^2 + r^2} + a^2}{r^2}.$

Variety 2

Let the line produced be straight.[12] Here we have

$$y = \phi_1 x_1 \qquad \frac{dy}{dx} = \psi_1 x_1 \qquad \frac{d\theta_2}{dr_2} = \frac{1}{r_2} \psi_1 r_2$$

Ex. 1. Let

$$y_1 = \frac{x_1^2}{4a} \qquad \therefore \frac{dy_1}{dx_1} = \frac{x_1}{2a}$$

$$\therefore \frac{d\theta_2}{dr_2} = \frac{1}{r_2}\frac{r_2}{2a} = \frac{1}{2a} \qquad \therefore \theta = \frac{r}{2a}$$

therefore the rolling curve is the spiral of Archimedes.

2. If the fixed curve is a parabola $y_1 = 2\sqrt{ax_1 - a^2}$ the rolling curve is an equal parabola.

3. If the fixed line is straight, $y_1 = \dfrac{a}{b}x_1 + c$ the rolling curve is the equiangular spiral, if $a = 0$ a circle. If $b = 0$ a straight line.

4. If the fixed line is an ellipse $\left(\dfrac{y^2}{b^2} + \dfrac{x^2}{a^2} = 1\right)$ the rolling curve is $r = a\cos\dfrac{a\theta}{b}$.

5. For an hyperbola $\left(\dfrac{y^2}{b^2} - \dfrac{x^2}{a^2} = 1\right)$ it is $r = \dfrac{a}{2}\left(e^{\frac{a\theta}{b}} - e^{-\frac{a\theta}{b}}\right)$.

6. For an hyperbola $\left(y = \dfrac{a^2}{x}\right)$ it is $\theta = -\dfrac{a^2}{3r^2}$ or $3r^2\theta = -a^2$ (lituus).

7. For the tractory $\dfrac{dy}{dx} = \dfrac{\sqrt{a^2 - x^2}}{x}$ $\quad \theta = -\dfrac{\sqrt{a^2 - r^2}}{r} + \cos^{-1}\left(\dfrac{r}{a}\right)$.

8. For the log. curve $\dfrac{dy}{dx} = \dfrac{a}{x}$ $\quad r = -\dfrac{a}{\theta}$ (hyperbolic spiral).

9. For the catenary $\dfrac{dy}{dx} = \dfrac{a}{\sqrt{x^2 - a^2}}$ $\quad r = a\sec\theta$ (straight line).

10. For the cycloid to produce the base it is the cardioid.

11. To produce the line touching the vertex

$$\theta = -\operatorname{versin}^{-1}\left(\frac{r}{a}\right) - 2\sqrt{\frac{2a}{r} - 1}. \text{[13]}$$

The rule for this variety enables us to find a curve which, being set on a given curve will be in neutral equilibrio however much displaced.

---

(12) Compare 'On the theory of rolling curves': 536–8 (= *Scientific Papers*, **1**: 24–7).

(13) For a definition of a versed sine, where $\operatorname{versin}\theta = 1 - \cos\theta$, see G. B. Airy's article on 'Trigonometry', in *Encyclopaedia Metropolitana, or Useful Dictionary of Knowledge. First Division. Pure Sciences*, **1** (London, 1829): 672–3.

Variety 3

Let the fixed curve be a circle, then

$$s_2 = a\theta_1 \quad r_2 = \sqrt{r_3^2 - p_3^2} \mp \sqrt{a^2 - p_3^2},$$

$$\frac{r_2^2\, d\theta_2^2}{dr_2^2} = \frac{a^2\, dr^2}{r^4\, d\theta^2} + \frac{a^2}{r^2} - 1.$$

Example 1. Let the curve produced be the spiral of Archimedes $r = a\theta$, here we have two solutions

$$p = \frac{ar}{\sqrt{a^2 - r^2}} \quad \text{and} \quad p = a$$

the first is the hyperbolic spiral, the second the straight line.

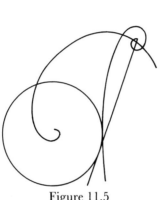
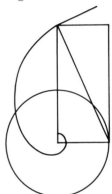

Figure 11,5   Figure 11,6

2. Let the line produced be the diameter of the circle the rolling line is a circle of half the size.

3. Let the curve produced be an ellipse, the rolling curve is a circle.

Variety 4

Let the curve traced be a circle.

$$\text{Let } \frac{d\theta_1}{dr_1} = \psi_1 r_1 \text{ then } \frac{d\theta_2}{dr_2} = \frac{r_2^2 + a}{r_2}\psi_1(r_2 + a).$$

Example 1. Let the fixed line be straight and touch the circle then

$$\frac{d\theta_1}{dr_1} = \frac{a}{r\sqrt{r^2 - a^2}}$$

$$\frac{d\theta_2}{dr_2} = \frac{a}{r^2\sqrt{r^2 + 2ar}}, \quad \theta = -\frac{\sqrt{2ar - r^2}}{r}$$

$$r = \frac{2a}{\theta^2 + 1} \qquad \theta = \sqrt{2\frac{a}{r} - 1}$$

Example 2. Let the fixed curve be $r_1 = 2a \cos \theta_1$

$$\theta = \text{versin}^{-1} \frac{r}{a} - \frac{\sqrt{2ar - r^2}}{r}.$$

3. Let the fixed curve be an ellipse or hyperbola whose major axis is a and whose focus is at the centre the rolling curve is an ellipse or hyperbola.

4. For the spiral of Archimedes the curve is

$$\theta = \frac{r}{a} + \log r.$$

5. For the logarithmic spiral we have a similar logarithmic spiral.

6. For the Cardioid it is

$$\theta = \sin^{-1} \frac{r}{a} + \log \frac{r}{\sqrt{a^2 - r^2} + a}.$$

This variety is the same as the following problem. Given the form of a wheel, find that of another which will work on the former without friction.

Genus 3

Given the rolling curve and the curve traced to find the fixed curve.

Species 1 Variety 1

Let the rolling curve be a straight line $r_2 = a \sec \theta_2$.

Here

$$y_1 = y_3 - r_2 \frac{dx_3}{\sqrt{dx_3^2 + dy_3^2}}$$

$$x_1 = x_3 + r_2 \frac{dy_3}{\sqrt{dx_3^2 + dy_3^2}}$$

$$s_1 = \sqrt{r_2^2 - a^2}$$

$$y_3 = \phi_3 x_3.$$

When $a = 0$ then the problem is that of finding the evolute to a curve.

Variety 2

Let the line produced be a straight line, then

$$\frac{d\phi_2}{dr_2} = \psi_2 r_2 \quad \frac{dy_1}{dx_1} = x_1 \psi_1(x_1).$$

Example 1. Let the rolling line be straight $r_2 = a \sec \theta_2$

$$\frac{d\theta_2}{dr_2} = \frac{a}{r \sqrt{r^2 - a^2}} \quad \frac{dy}{dx} = \frac{a}{\sqrt{x^2 - a^2}} \quad y = a \log \left(\frac{\sqrt{x^2 - a^2} + x}{a} \right)$$

the equation to the catenary.

2. Let the rolling curve be $x^2 + y^2 = a^2$, the fixed curve is $x^2 + y^2 = 4a^2$ a circle of twice the diameter of the former.

3. For the Cardioid it is the curve whose equation is $y = a \operatorname{versin}^{-1} \dfrac{x}{a} - \sqrt{2ax - x^2}$ or the Cycloid.

4. For the log. spiral it is a straight line.

5. For the involute of the circle it is the orthogonal trajectory of the catenary
$$\frac{dy}{dx} = \frac{\sqrt{x^2 - a^2}}{a}, \quad y = \frac{x\sqrt{x^2 - a^2}}{2a} + \frac{a}{2}\log\left(\frac{\sqrt{x^2 - a^2} + x}{a}\right).$$

6. For the spiral $\theta = \dfrac{n^m}{a^m}$ $\quad y = \dfrac{m}{m+1}\dfrac{x^{m+1}}{a^m}$.

Thus for the spiral of Archimedes it is a parabola
for the hyperbolic spiral it is the logarithmic curve
for the lituus it is a rectangular hyperbola.

Variety 3

Let the rolling curve be a circle. This variety is the same as Variety 4 of Genus 1.

Variety 4

Let the curve produced be a circle, then let
$$\frac{d\theta_3}{dr_3} = \psi_3 r_3 \qquad \frac{d\theta_1}{dr_1} = \frac{r_1 - a}{r_1}\psi_3(r_1 - a).$$

This variety is the same as Genus 2 Variety 4, if we change a into $-a$ or suppose the curve to trace the other side of the circle.

Order II

Genus 1

Let the fixed curve be the same as the rolling curve and let the curve traced be given.[14]

Let $\qquad\qquad \dfrac{p_3}{r_3} = \psi_3 r_3$ then $\dfrac{p_1}{r_1} = \psi_3(2p_1)$.

Example 1. Let the line produced be straight $(r = a \sec \theta)$
$$\frac{p_3}{r_3} = \frac{a}{r_3} \qquad \frac{p_1}{r_1} = \frac{a}{2p_1} \qquad 2p_1^2 = ar$$

---

(14) Compare 'On the theory of rolling curves': 539–40 (= *Scientific Papers*, **1**: 28–9).

22

Let $\theta_1 = \varphi_1 n_1$ be the equation of a curve

$\theta_0 = \varphi_0 n_0$ that of the curve which when rolled on itself produces $\theta_1 = \varphi_1 n_1$ etc

$\theta_{-1} = \varphi_{-1} n_{-1}$ the next or third

$\theta_{1-m} = \varphi_{1-m} n_{1-m}$, the mth

then

$$n_{1-m} = n_1 \left(\frac{n_1}{2\,\theta_1}\right)^m$$

$$\theta_{1-m} = \theta_1 + m \cos^{-1}\left(\frac{\theta_1}{n_1}\right)$$

therefore by performing this operation an infinite number of times we arrive at the equiangular spiral

Genus 2

Given the fixed curve similar to the ~~a thing~~ curve traced, and the equation to the rolling curve

We cannot give a rule for this genus but we may do a few examples

Example 1 Let the rolling curve be a straight line whose pole is in itself

This problem is the same as the following

Find the curve which is the same as its involute

Now the involution may always commence at the same point or it may begin where the last one stopped

thus ⌐ or ⌐ Also it is proved in the Journal of the Polytechnic School Number 18 page 431 that if any curve be developed an infinite number of times the curve produced will, in the first case, be an equiangular spiral

whence we obtain

$$\frac{d\theta}{dr} = \frac{a}{r(2r-a)}$$

$$\therefore \theta = \cos^{-1}\frac{a-r}{r}$$

$$\text{or } r = \frac{a}{1+\cos\theta}$$

the equation to a parabola whose parameter $= 2a$.

2. For a circle it is an ellipse or hyperbola.

3. For the spiral of Archimedes it is the involute of the circle.

Let $\quad\quad\theta_1 = \phi_1 r_1$ be the equation of a curve.

$\theta_0 = \phi_0 r_0$ that of the curve which when rolled on

itself produces $\theta_1 = \phi_1 r_1$ etc.

$\theta_{-1} = \phi_{-1} r_{-1}$ the next or third

$\theta_{1-m} = \phi_{1-m} r_{1-m}$, the mth

then $\quad\quad r_{1-m} = r_1\left(\frac{r_1}{2p_1}\right)^m$

$$\theta_{1-m} = \theta_1 + m\cos^{-1}\left(\frac{p_1}{r_1}\right)$$

therefore by performing this operation an infinite number of times we arrive at the equiangular spiral.

Genus 2

Given the fixed curve similar to the curve traced, and the equation to the rolling curve.

We cannot give a rule for this genus but we may do a few examples.

Example 1. Let the rolling curve be a straight line whose pole is in itself.

This problem is the same as the following.

Find the curve which is the same as its involute. Now the involution may always commence at the same point or it may begin where the last one stopped

thus 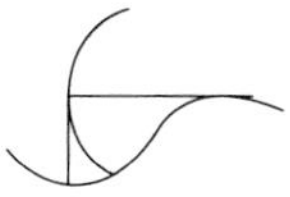 or .[15] Also it is proved in the Journal of the

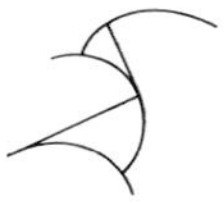

Figure 11,7   Figure 11,8

(15) See D. F. Gregory's discussion of the successive involution of a curve, so that 'the successive involutes approach continually nearer and nearer to the cycloid, and ultimately do not differ sensibly from that curve', in his *Differential and Integral Calculus*: 455; and compare Gregory's Fig. 21, which is identical to figures given by Johann Bernoulli, Euler and Legendre; see note (17).

Polytechnic School Number 18 page 431[16] that if any curve be developed an infinite number of times the curve produced will, in the first case, be an equiangular spiral and in the second, either a cycloid or an epicycloid.[17] Now if any curve except the above answer the conditions the resulting curve will be that curve but it is not therefore no other curve answers the conditions. Example 2. Let the rolling curve be a circle.

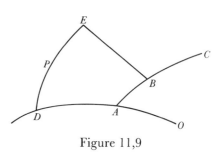

Figure 11,9

Let ABC be the curve required, let DPE be the equidistant curve which is to be the same as ABC, let DAO be the evolute of ABC then it is evident that DPE is an involute of DAO by a line longer by BE therefore $DA = BE$. But DPE is the same as ABC therefore its involute is the same therefore in the curve DAO the point D is similar to A etc or the curve recurs at intervals $= BE$.

If we suppose BE not given but taken anyhow the curve DAO must be the same at every point that is, a circle, straight line, or point therefore ABC is an involute of the circle a straight line or a circle, but in the last case the curves are similar not equal.

Genus 3

Given the rolling curve similar to the curve produced and the equation to the fixed curve.

As an example we may take the catenary upon which a straight line being

---

(16) S. D. Poisson, 'Mémoire sur la manière d'exprimer les fonctions par les séries de quantités périodiques, et sur l'usage de cette transformation dans la résolution de différens problèmes', *Journal de l'École Polytechnique*, **11**, cahier 18 (1820): 417–89, esp. 431–42 ('Développemens successifs des courbes planes').

(17) On this second case (the curve generated being an arc of a cycloid) see Gregory, *Differential and Integral Calculus*: 135–6 and Fig. 21. Gregory and Poisson ('Mémoire': 440) cite works by Johann Bernoulli, Euler and Legendre on this proposition. See Johann Bernoulli, 'De evolutione successiva et alternante curva cujuscunque in infinitum continuata, tandem cycloidem generante; schediasma cyclometricum', *Opera Omnia*, 4 vols. (Lausanne/Geneva, 1742), **4**: 98–108 and Tab. LXXIX, Nº. CLXV (page 124); L. Euler, 'Demonstratio theorematis Bernoulliani quod ex evolutione curvae cujuscunque rectangulae in infinitum continuata tandem cycloides nascantur', *Novi Commentarii Academiae Scientiarum Imperialis Petropolitanae*, **10** (1764): 179–98, esp. 180–1 and Tab. II, Fig. 2; A. M. Legendre, *Exercices de Calcul Intégral sur divers ordres de transcendantes et sur les quadratures*, 3 vols. (Paris, 1811–17), **2**: 541–4 and Fig. 34. See also V. Puiseux, 'Problèmes sur les développées et les développantes des courbes planes', *Journal de Mathématiques Pures et Appliquées*, **9** (1844): 377–99, esp. 397–9; and F. Gomes Teixeira, *Traité des Courbes Spéciales Remarquables*, 2 vols. (Coimbra, 1908–9), **2**: 147–9.

rolled produces a straight line or a circle on the internal circumference of which a circle of half the diameter being rolled produces itself.

Order III

Find the curve which, being rolled on itself produces itself.

Now if any curve has this property if it is rolled on itself and the curve thus produced rolled on itself and so on to infinity it will remain the same.

But we have shewn in Species 2 of Genus 1 Order I that any curve when this operation is performed an infinite number of times becomes the equiangular spiral. Therefore the equiangular spiral, and no other curve possesses this property.

Note. The following are some of the properties of this curve connected with rolling.

1   When rolled on itself back to back it produces itself. Fig. [11,10(1)].

Figure 11,10

2 When one is in advance of the other it is the curve of Order I Genus 1 Var. 4 Example 1 (Fig. [11,10(2)]).

3 When they roll as in Figs. [11,10(3)] and [11,10(4)] the curve is a circle.

4 When it rolls on a straight line it produces a straight line.[18]

5 When it rolls on a circle or any involute of the circle it produces an involute of a circle.[19]

6 When a straight line rolls on it it produces itself.[20]

7 When the straight line is in advance it is the curve of [11,10(2)].

8 When a circle rolls on it it is the curve of [11,10(8)].

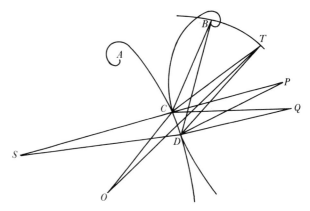

Figure 11,11

[Radii of curvature]

Relations between the radii of curvature of rolling curves.

Let ACD be the fixed curve let $SC = R_1$ = radius of curvature let $CD = h$ let BCD be the rolling curve, let $CP = R_2$ and $BC = r_2$ let BT be the curve traced and let $BO = R_3$

then $\quad CSD = \dfrac{h}{R_1} \quad CPD = \dfrac{h}{R_2} \quad CBD = \dfrac{dr_2^2}{r_2\sqrt{r_2^2 + \dfrac{dr_2^2}{d\theta_2^2}}}$

$$\frac{BT}{R_3} = BOT = BCT - CTD = PDQ - CTD = CSD + CPD - CBD$$

$$\frac{BT}{R_3} = \frac{h}{R_1} + \frac{h}{R_2} - \frac{h}{r_2\sqrt{r_2^2 + \dfrac{dr_2^2}{d\theta_2^2}}}$$

also $\quad BT = r_2 \left(\dfrac{h}{R_1} + \dfrac{h}{R_2} \right)$

$$\therefore \frac{r_3}{R_3} = 1 - \frac{R_1 R_2}{R_1 + R_2} \cdot \frac{1}{\sqrt{r_2^2 + \dfrac{dr_2^2}{d\theta_2^2}}} \cdot$$

---

(18) Fig. 11,10(5).   (19) Fig. 11,10(7).   (20) Fig. 11,10(6).

When the fixed line is straight $R_1 = \infty$

$$\frac{r_2}{R_3} = 1 - \frac{R_2}{\sqrt{r_2^2 + \dfrac{dr_2^2}{d\theta_2^2}}}.$$

When the rolling line is straight $r_2 = a \sec \theta_2$

$$R_3 = \frac{r_2^3}{r_2^2 - aR_1}.$$

When $a = 0$ $R_3 = r_2$.

When the line traced is straight

$$R_1 R_2 = (R_1 + R_2) \sqrt{r_2^2 + \frac{dr_2^2}{d\theta_2^2}}.$$

Let there be three curves, A, B, and C.

Let the curve A, when rolled on itself, produce the curve B and when rolled on a straight line, produce C then if B is rolled on C it will produce a straight line.

If the curve A when rolled on the curve B produces a circle then B rolled on A produces a circle.

12

FROM A LETTER TO LEWIS CAMPBELL

22 SEPTEMBER 1848

From Campbell and Garnett, *Life of Maxwell*[1]

Glenlair
22 September 1848

When I waken I do so either at 5.45 or 9.15, but I now prefer the early hour, as I take the most of my violent exercise at that time, and thus am *saddened down*, so that I can do as much still work afterwards as is requisite, whereas if I was to sit still in the morning I would be yawning all day. So I get up and see what kind of day it is, and what field works are to be done; then I catch the pony and bring up the water barrel. This barrel used to be pulled by the men, but Pa caused the road to be gravelled, and so it became horse work to the men, so I proposed the pony; but all the men except the pullers opposed the plan. So I and the children not working brought it up, and silenced vile insinuators. Then I take the dogs out, and then look round the garden for fruit and seeds, and paddle about till breakfast time; after that take up Cicero and see if I can understand him. If so, I read till I stick; if not, I set to Xen. or Herodt. Then I do props, chiefly on rolling curves,[2] on which subject I have got a great problem divided into Orders, Genera, Species, Varieties, etc.[3]

One curve rolls on another, and with a particular point traces out a third curve on the plane of the first, then the problem is: Order I. Given any two of these curves, to find the third.

Order II. Given the equation of one and the identity of the other two, find their equation.

Order III. Given all three curves the same, find them. In this last Order I have proved that the equi-angular spiral possesses the property, and that no other curve does.[4] This is the most reproductive curve of any. I think John Bernoulli had it on his tombstone, with the motto *Eadem mutata resurgo*.[5] There

---

(1) *Life of Maxwell*: 120–3.   (2) See Number 11.   (3) See Number 11, esp. note (4).

(4) See Number 11; and Maxwell, 'On the theory of rolling curves', *Trans. Roy. Soc. Edinb.*, **16** (1849): 519–40, esp. 532 (= *Scientific Papers*, **1**: 19).

(5) The epitaph, symbolising perpetual resurrection, on the tombstone of Jakob Bernoulli; misquoted from Bernoulli's epitaph in his paper on the logarithmic spiral by D. F. Gregory, *Examples of the Processes of the Differential and Integral Calculus* (Cambridge, 1841): 141. According to the preliminary *Vita* to Bernoulli's *Opera*, 2 vols. (Lausanne/Geneva, 1744), **1**: 30, his tombstone bears the epigraph 'EADEM MUTATA RESURGEM', in imitation of Archimedes who, according to Plutarch (*Marcellus*, Chap. 17), 'sepulchro suo cylindrum sphaera comprehensum ab amicis imponi voluit'. In his paper on the logarithmic spiral, 'Lineae cycloidales, evolutae, anti-evolutae, causticae, anti-causticae, peri-causticae. Earum usus & simplex relatio ad se invicem.

are a great many curious properties of curves connected with rolling. Thus, for example –

If the curve *A* when rolled on a straight line produces a curve *C*, and if the curve *A* when rolled up on itself produces the curve *B*, then the curve *B* when rolled upon the curve *C* will produce a straight line.[6]

Thus, let the involute of the circle be represented by *A*,

     the spiral of Archimedes by *B*,

     and the parabola by *C*,

then the proposition is true.[7] Thus the parabola rolled on a straight line traces a Catenary with its focus, an easy way to describe the Catenary.[8] Professor Wallace just missed it in a paper in the Royal Society.[9]

After props come optics, and principally polarised light.

Do you remember our visit to Mr. Nicol?[10] I have got plenty of unannealed glass of different shapes, for I find window glass will do very well made up in bundles. I cut out triangles, squares, etc., with a diamond, about 8 or 9 of a kind, and take them to the kitchen, and put them on a piece of iron in the fire one by one. When the bit is red hot, I drop it into a plate of iron sparks to cool, and so on till all are done. I have got all figures up to nonagons, triangles of all kinds, and irregular chips. I have made a pattern for a tesselated window of unannealed glass in the proper colours, also a delineation of triangles at every principal inclination. We were at Castle-Douglas yesterday, and got crystals of salt Peter, which I have been cutting up into plates to-day, in hopes to see rings.[11] There are very few crystals which are not hollow-hearted or filled up with irregular crystals. I have got a few cross cuts like ⬡ free of

---

Spira mirabilis. Alique', *Acta Eruditorum* (May 1692): 207–13 (= *Opera*, **2**: 491–502, esp. 502) Bernoulli wrote in a peroration: 'Aut, si mavis, quia Curva nostra mirabilis in ipsa mutatione semper sibi constantissime manet similis & numero eadem, poterit esse vel fortitudinis & constantiae in adversitatibus; vel etiam Carnis nostrae post varias alterationes & tandem ipsam quoque mortem, ejusdem numero resurrecturae symbolum; adeo quidem, ut si *Archimedem* imitandi hodienum consuetudo obtineret, libenter Spiram hanc tumulo meo juberem incidi cum Epigraphe: *Eadem numero mutata resurget*'.

(6) See Number 11, and 'On the theory of rolling curves': 535 (= *Scientific Papers*, **1**: 22).

(7) See Number 11 (Order I, Genus 2, Species 1, Varieties 1 and 2; Order II, Genus 1); and 'On the theory of rolling curves': 535–40 (= *Scientific Papers*, **1**: 22–9).

(8) See Number 11 (Order I, Genus I, Species 1, Variety 1, 'Example 3rd'); and compare 'On the theory of rolling curves': 525–6 (= *Scientific Papers*, **1**, 11–12).

(9) William Wallace, 'Solution of a functional equation, with its application to the parallelogram of forces, and to curves of equilibration', *Trans. Roy. Soc. Edinb.*, **14** (1840): 625–76, esp. 654–5.

(10) See Number 9 note (5).

(11) See Number 13.

irregularities and long [wedge-shaped][12] cuts for polarising plates. One has to be very cautious in sawing and polishing them, for they are very brittle.

Figure 12,1

I have got a lucifer match box fitted up for polarising, thus. The rays suffer two reflections at the polarising angle from glasses *A* and *B*. Without the lid it does for an analysing plate. In the lid there is set a plate of mica, and so one observes the blue sky, and turns the box round till a particular colour appears, and then a line on the lid of the box points to the sun wherever he is. Thus one can find out the time of day without the sun. These are a few of the figures one sees in unannealed glass.[13]

Pray write soon and tell when, how, and where by, you intend to come, that you may neither on the other hand fall upon us at unawares, nor on the one hand break and not come at all. I suppose when you come I will have to give up all my things of my own devizing, and take Poisson,[14] for the time is short, and I am very nearly unprepared in actual reading, though a great deal more able to read it.

I hope not to write any more letters till you come. I seal with an electrotype of the young of the ephemera. So, sir, I was, etc.

---

(12) Campbell's insertion.   (13) Campbell omitted these figures: but see Number 14.
(14) See Number 10 note (4).

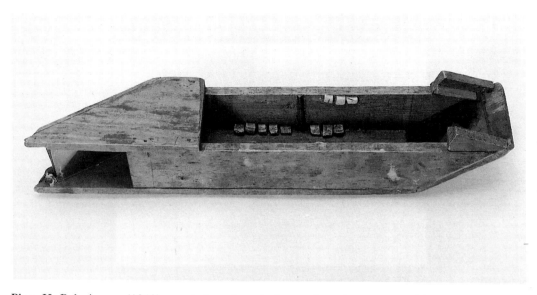

Plate II. Polariscope (1848), consisting of a polariser and analyser mounted in a wooden frame which has slots into which cardboard lens holders could be fitted (Number 13).

13

MANUSCRIPT ON THE OBSERVATION OF CRYSTALS USING POLARISED LIGHT

circa OCTOBER 1848[1]

From the original in the University Library, Cambridge[2]

AN INSTRUMENT FOR OBSERVING THE RINGS IN CRYSTALS SEEN BY POLARISED LIGHT[3]

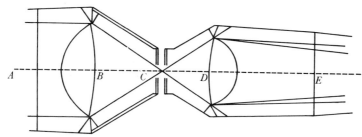

Figure 13,1

Let polarised light enter at *A*. It will be brought to a focus at *C* by the lens *B*, let the crystal be introduced at *C* with its axis in the line *AE*, then the light which passes through *C* making various angles with the axis will be variously coloured and be refracted at various parts of the lens *D* towards the eye in the direction of *E* where it is analysed.[4]

Thus the eye will see the rings traced on the lens *D* as it were.

This instrument possesses the following advantages.

1   All the light that reaches *C* is polarised completely.
2   This light comes in various directions so that the whole of the rings may be seen.
3   A very small piece of crystal will suffice.

---

(1) See Numbers 12, 14 and 19.   (2) ULC Add. MSS 7655, V, b/l.

(3) For discussion of the 'brilliant colours' of coloured rings produced by transmitting polarised light through doubly-refracting substances, see David Brewster, *A Treatise on Optics* [in Lardner's *Cabinet Cyclopædia*] (London, 1831) : 183–210. See also Number 19; and on the apparatus see *Life of Maxwell*: 487. See plate II.

(4) Compare Airy's discussion of the coloured rings produced by interposing a doubly-refracting substance between polarising and analysing plates, §§ 144–8 and Plate 5 Fig. 24 of his 'On the Undulatory Theory of Optics', in George Biddell Airy, *Mathematical Tracts on the Lunar and Planetary Theories, the Figure of the Earth, Precession and Nutation, the Calculus of Variations, and the Undulatory Theory of Optics* (Cambridge, ₃1842) : 355–8.

4   The rings may be seen at the distance of most distinct vision.

5   We may employ sun or candle light or light from any luminary however small.

6   We may polarise the light by reflexion and analyse it in the same way and there is no necessity for holding the crystal and analysing plate very near the eye.

14

MANUSCRIPT ON THE CHROMATIC EFFECTS OF POLARISED LIGHT ON UNANNEALED GLASS

10 OCTOBER 1848

From the original in the University Library, Cambridge[1]

TO FIND THE FORM OF THE CENTRAL BARS SEEN BY POLARISED LIGHT IN PIECES OF UNANNEALED GLASS[2]

Since unannealed glass is a doubly refracting substance these bars are the locus of the point where the principal plane is parallel or perpendicular to the plane of polarisation. See Airys Undulatory theory of Optics (148).[3]

Also the principal plane at any point is perpendicular to the line of equal density of the extraordinary medium passing through that point. See Edinburgh Encyclopaedia Article Optics 'On the cause of Double refraction'.[4]

But these lines of equal density seem to be those of equal heat while the piece of glass is cooling.

Therefore find the expression for the heat at a given point.

$$z = \phi_1(x, y)$$
$$\therefore y = \phi_2(x, z)$$
$$\therefore \frac{dy}{dx} = \phi_3(x, z) = \phi_3(x, \phi_1(x, y)) = p.$$

Make $p =$ tangent of the angle between the axis of x and the plane of polarisation and make $y = \phi_4(x)$ then we have the equation of one of the bars and we may get the other by making $p = \dfrac{1}{p'}$.

If we suppose $z = \phi$ (product of distances of point (x, y)) from the sides of the figure we get results which do not differ much from the truth.

Thus for the triangle whose equation is

$$3a^2y - 3x^2y - 2\sqrt{3}ay^2 + y^3 = 0$$

if the heat at any point $= z$ and

$$z = \phi_1(3a^2y - 3x^2y - 2\sqrt{3}ay^2 + y^3)$$

(1) ULC Add. MSS 7655, V, b/2.

(2) See Numbers 12 and 13.

(3) §148 of Airy's 'On the Undulatory Theory of Optics' in his *Mathematical Tracts* (Cambridge, 31842): 357–8. See Number 13 note (4).

(4) David Brewster, 'Optics', *The Edinburgh Encyclopaedia*, 18 vols. (Edinburgh, 1830), **15**: 460–662, esp. 612–13. Compare also Airy, *Mathematical Tracts*: 329–39.

then the equation to the isothermal lines is

$$x^2 = a^2 - \frac{2}{\sqrt{3}}ay + \frac{y^2}{3} - \frac{c^3}{3y} \qquad c \text{ being constant}$$

$$\frac{dx}{dy} = \frac{\frac{2y}{3} + \frac{c^3}{3y^2} - \frac{2a}{\sqrt{3}}}{2x} = p$$

but

$$c^3 = 3a^2y - 3x^2y - 2\sqrt{3}ay^2 + y^3$$

$$\therefore 2px + y^2 + x^2 + 2\sqrt{3}ay = a^2$$

this is the equation to the central bars and varies as p varies.

Let $p = \tan\alpha$ then $\alpha = $ angle of inclination of the axis of x with the plane of polarisation and let $p_1 = \tan\alpha_1$ $\therefore \alpha_1 = 90 - \alpha$.

1. Let $\alpha = 0$, $60°$ or $120°$ then the bar is the third part of a circle whose radius $= \dfrac{2a}{\sqrt{3}}$ and whose centre is the point $\left(x = 0, y = -\dfrac{a}{\sqrt{3}}\right)$

$$\left(x = -a, \quad y = \frac{2a}{\sqrt{3}}\right) \quad \left(x = a, \quad y = \frac{2a}{\sqrt{3}}\right).$$

Then $\alpha_1 = 90$, 150 or 30 and the other bar will be a straight line passing through the centre of the triangle and making an angle of 90, 150 or 30 with this axis of x, together with that side of the triangle which it bisects.

2. Let $\alpha = 15°, 45°, 75°, 105°, 135°$ or $165°$ the bar is a parabola which passes through the centre of the triangle and the angle A, C, C B, B or A whose axis is parallel to the plane of polarisation and whose tangent at the centre of the triangle is inclined to the diameter at an angle of $45° + \tan^{-1}\frac{2}{3}$.

3. In all other positions of the plane of polarisation the bars are ellipses or hyperbolas.

The centre of the curve is situate at the point

$$x = \frac{-pa}{\sqrt{3}(p^2-1)} \qquad y = \frac{a}{\sqrt{3}(p^2-1)}$$

Figure 14,1

its axes are inclined to those of x and y at an angle of 45 and are equal to

$$\frac{\sqrt{3p^3 + 3p^2 - 6p - 4}}{\sqrt{3}(p^2-1)}a \quad \text{and} \quad \frac{\sqrt{3p^3 + 3p^2 - 6p - 4}}{(p-1)\sqrt{p^2-1}}a.$$

In the unannealed glass the dark lines cannot well be seen not being sharply terminated in thin pieces and in thick pieces there is that uniform tint which is produced by the axes in the plane of the plate.

See Article 'Optics' 'On rectangular plates of glass with two axes of Polarisation' Edinburgh Encyclopaedia.[5]

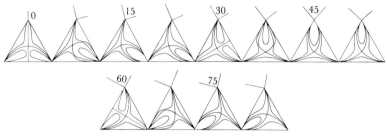

Figure 14,2

This however ought to be remedied by the interposition of a plate of mica. The above figures are those which the triangle produces according to the equation, and for thin pieces they appear perfectly correct.[6]

For the rectangle whose sides are $2a$ and $2b$ we have

$$y = \frac{\sqrt{x^4 + x^2(b^2 - 2p^2a^2) + a^4 + p^2(a^2 - x^2)}}{x}.$$

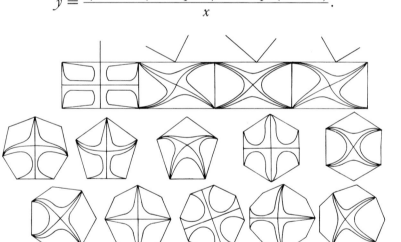

Figure 14,3

(a)

JAMES CLERK MAXWELL
10 October 1848

(a) {Maxwell} Big Lie J. C. M.

(5) David Brewster, 'Optics', *Edinburgh Encyclopaedia*, **15**: 603–4. On Brewster's substantive work on induced double refraction in heated and strained glass see Number 26 notes (19), (72) and (78).

(6) Compare Plate III, *Life of Maxwell*: 487, copied from Maxwell's water-colours depicting the chromatic effects of unannealed glass in polarised light; and see Number 26, Figs. 2 and 3.

15

EXERCISE ON THE CATENARY FOR J. D. FORBES' NATURAL PHILOSOPHY CLASS, EDINBURGH UNIVERSITY

circa 1848–49[1]

From the original in the University Library, St Andrews[2]

[ON THE CATENARY][3]

To find the equation to the catenary of uniform strength, that is, a curve of equilibrium of which the weight at any point is proportional to the secant, and the whole weight from the vertex to the tangent of elevation.

Let O be the origin $Ox\ Oy$ the axes of x and y. Let $AB = dx$, $BC = dy$ then $AC = ds$ the element of the curve, $\dfrac{dy}{dx}$ the tangent of elevation proportional to the weight from O, let $\dfrac{dy}{dx} = \dfrac{w}{a^3}$. $\qquad$ (1)

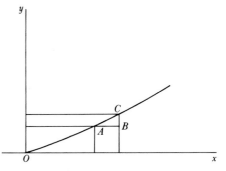

Figure 15,1

$\dfrac{ds}{dx}$ is the secant of elevation proportional to the weight at the point, therefore let

$$\frac{ds}{dx} = \frac{dw}{a^2 ds}$$

$$\therefore \frac{ds^2}{dx^2} = \frac{dw}{a^2 dx}$$

but (1) $\qquad \dfrac{d^2y}{dx^2} = \dfrac{dw}{a^3 dx} = \dfrac{ds^2}{a\,dx^2} = a\left(1 + \dfrac{dy^2}{dx^2}\right).$

Let $\qquad \dfrac{dy}{dx} = p \quad$ then $\quad \dfrac{dp}{dx} = a(1 + p^2)$

$$\therefore \frac{dx}{dp} = \frac{1}{a(1 + p^2)}$$

$$\therefore x = \frac{1}{a}\tan^{-1}p$$

$$\therefore p = \frac{dy}{dx} = \tan\frac{x}{a}$$

---

(1) The date is uncertain; but see note (4).

(2) St Andrews University Library, Forbes MSS, Box IX, 17.

(3) On the catenary compare William Whewell, *Analytical Statics. A Supplement to the Fourth Edition of An Elementary Treatise on Mechanics* (Cambridge, 1833): 66–80.

the quantity a in this expression is the cube root of a volume of the material of the chain whose mass represents unity or the square root of the sectional area at the vertex. By taking this for unity we have

$$\frac{dy}{dx} = \tan x \quad \therefore y = -a \log \cos \frac{x}{a}$$

which is the equation required.

From this equation it appears that when $x = 0, y = 0$ when $\frac{x}{a} = \frac{\pi}{2}$ $y = \infty$ when $\frac{x}{a} > \frac{\pi}{2} < \pi, y$ is impossible etc. We may transform the equation thus

$$y = -a \log \cos \frac{x}{a} \qquad \therefore e^{\frac{y}{a}} = -\cos \frac{x}{a}, \qquad \frac{x}{a} = -\cos^{-1} e^{\frac{y}{a}}$$

$$x = -a \cos^{-1} e^{\frac{y}{a}}.$$

$\frac{dy}{dx} = \tan \frac{x}{a}$. When $x = 0$ $\frac{dy}{dx} = 0$. When $x = a \frac{\pi}{2}$ $\frac{dy}{dx} = \infty$ therefore the curve has an asymptote whose equation is $x = \frac{\pi a}{2}$.

$$\frac{d^2y}{dx^2} = \sec^2 \frac{x}{a} \quad \therefore R = \frac{a\left(1 + \tan^2 \frac{x}{a}\right)^{\frac{3}{2}}}{\sec^2 \frac{x}{a}}$$

$$= \frac{a \sec^3 \frac{x}{a}}{\sec^2 \frac{x}{a}} = a \sec \frac{x}{a}.$$

$$\therefore \text{radius of curvature} = -a \sec \frac{x}{a}.$$

To find the evolute of the curve.
Let x', y' be the coordinates of B x, y those of C.

$$BD = a \quad BC = a \sec \frac{x}{a} = R', \quad CD = a \tan \frac{x}{a}.$$

Now $\qquad y = y' + a \quad \therefore y' = y - a.$

Also $\qquad x = x' - CD = x' - \tan \frac{x}{a}$

$$= -a \cos^{-1} e^{\frac{y}{a}} + a\sqrt{e^{\frac{-2y}{a}} - 1}.$$

$$\therefore x = -a \cos^{-1} e^{\frac{y}{a}-1} + a\sqrt{e^{1-\frac{2y}{a}} - 1}$$

the equation to the evolute.

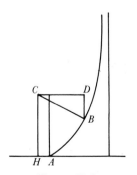

Figure 15,2

To find the length of any portion of the curve.

$$\frac{dy}{dx} = \tan\frac{x}{a} \quad \therefore \frac{ds}{dx} = \sec\frac{x}{a} \quad \therefore s = \log\tan\left(\frac{\pi}{4} + \frac{x}{2a}\right).$$

To make a series of spheres such that if they were strung on a thread they would approximately hang in this curve.[4]

It is evident that if equal spheres are strung on a string their weight is equal to that of a cylinder of equal diameter and a third of the length, so that the weight of any small part is proportional to its length and the square of the diameter of the balls.

Therefore in this chain, if we stop at the centre of any sphere the weight of half the first and last and the whole of the intermediate spheres will represent the total weight and will be proportional to the tangent of elevation while the square of the diameter of the last ball will be proportional to the secant of elevation.

Therefore let b^3 represent the weight of the chain up to a certain point, let x^3 be that of the next ball then $\dfrac{x^2}{a^2}$ = secant of elevation and $b^3 + \dfrac{x^3}{2}$ = whole weight.

$$\therefore \frac{b^3}{a^3} + \frac{x^3}{2a^3} = \text{tangent of elevation}$$

but $\qquad \sqrt{1+\tan^2} = \sec, \quad \therefore \sqrt{1 + \dfrac{b^6}{a^6} + \dfrac{b^3 x^3}{a^6} + \dfrac{x^6}{4a^6}} = \dfrac{x^2}{a^2}.$

$$\therefore a^6 + b^6 + b^3 x^3 + \frac{x^6}{4} = a^2 x^4,$$

$$\therefore x^6 - 4a^2 x^4 + 4b^3 x^3 + 4a^6 + 4b^6 = 0$$

which equation being solved, the value of x will be the cube root of the weight of the ball.

---

(4) Compare William Wallace, 'Solution of a functional equation, with its application to the parallelogram of forces, and to curves of equilibrium', *Trans. Roy. Soc. Edinb.*, **14** (1840): 625–76, esp. 661–76, on the theory of equilibrated curves and the construction of arches, computing a table of coordinates for the construction of a catenary arch; and see Numbers 12 and 26, esp. note (87). Compare also J. H. Pratt, *The Mathematical Principles of Mechanical Philosophy, and their application to the Theory of Gravitation* (Cambridge, 1836): 109–19. In his letter to Tait of 3 December 1856 (Number 108) Maxwell refers to Forbes' models of catenaries; the present manuscript may relate to the construction of one such model. Tait later recalled, in a conversation with C. G. Knott, that he had himself (while attending Forbes' class in 1847–8) made calculations of the sizes of discs to be strung together to represent a catenary; see C. G. Knott, *Life and Scientific Work of Peter Guthrie Tait* (Cambridge, 1911): 7. Tait attended the more mathematical first division of Forbes' class in 1847–8, while Maxwell was in the first division in the following session; see Knott, *Tait*: 6.

Thus we may begin at the beginning and find the weight of each ball by putting the weight of the preceding balls for b in the equation.

Since $R = a \sec \dfrac{x}{a}$ and the diameter of a ball $=$ $d \sec \frac{1}{2}\dfrac{x}{a}$ and since the angle $OCQ = OQP$

$$\therefore OP = \frac{d^2}{2a}.$$

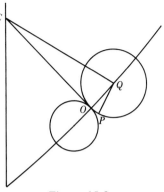

Therefore in piercing the spheres make the one orifice not directly opposite the other but distinct from it by the constant arc $OP = \dfrac{d^2}{2a}$.

Figure 15,3

The following is a practical method for describing the curve.

Suppose the curve described from A to P and that S is the centre of curvature

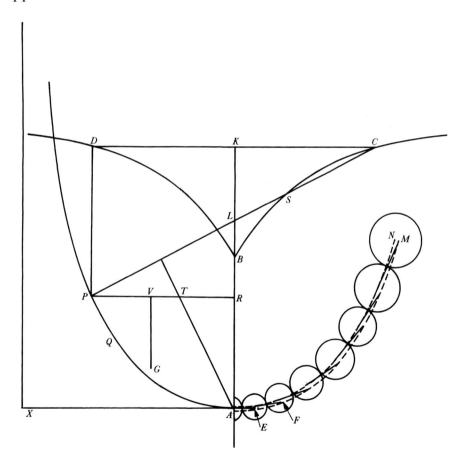

Figure 15,4

at a point Q near P. Join PS and produce. Make $PD = AB = a$ and $DC \parallel x$ then C is the next centre of curvature, and so on.

To find the centres and sizes of the spheres expeditiously go over the curve again enlarging the radii of curvature by a small quantity $MN = \dfrac{d^2}{2a}$.

Draw a tangent at A, then the point of section E is next the centre of the second sphere. Draw EF and F is the next centre and so on.

To find the centre of gravity of any arc of this curve (considering the curve not as a uniform line but as it really is) and first, to find its coordinate in x or x_1.

$$x_1 = \frac{\displaystyle\int x\,\frac{dw}{dx}}{w} = \frac{a^2 \displaystyle\int x\sec^2\frac{x}{a}}{a^3 \tan\frac{x}{a}}$$

$$= \frac{a^3 x \tan\dfrac{x}{a} + a^4 \log\cos\dfrac{x}{a}}{a^3 \tan\dfrac{x}{a}}$$

$$\therefore x_1 = x + \frac{a\log\cos\dfrac{x}{a}}{\tan\dfrac{x}{a}}$$

$$= x - \frac{a^3 y}{w} = x - y\cot\frac{x}{a}$$

and to find y_1

$$y_1 = \int_x y\,\frac{dw}{dy} = \frac{-a^3 \displaystyle\int \log\cos\frac{x}{a}\sec^2\frac{x}{a}}{a^3 \tan\dfrac{x}{a}}$$

$$= \frac{-a^4 \log\cos\dfrac{x}{a}\tan\dfrac{x}{a} - x^4\tan\dfrac{x}{a} + a^3 x}{a^3 \tan\dfrac{x}{a}}$$

$$\therefore y_1 = y - a + \frac{a^3 x}{w} = y - a + x\cot\frac{x}{a}.$$

These expressions afford a geometrical method of finding the centre of gravity of any arc from the vertex as PA.

Draw the ordinate PR and AT perpendicular to PC make $PV = TR$ and $VG = KL$, G is the centre of gravity.

To hang a chain from one point to another on the same level with the least possible material.

First it is evident that the chain must be of uniform strength for if not we might take away material from the thick parts.

Let $2x$ be the distance of the points and a the parameter of the catenary to be found then $y = -a \log \cos \dfrac{x}{a}$ the whole material $= ka^3 \tan \dfrac{x}{a}$. As all catenaries of uniform strength are similar we may begin with finding how to obtain the greatest span in proportion to the material.

or Let $\dfrac{2x}{a^3 \tan \dfrac{x}{a}}$ be a maximum

take the differential and equate it with zero

$$\frac{\tan \dfrac{x}{a} - \dfrac{x}{a} \sec^2 \dfrac{x}{a}}{\tan^2 \dfrac{x}{a}} = 0.$$

Now we may have either $\tan^2 \dfrac{x}{a} = \infty$ or $\dfrac{x}{a} = \dfrac{\pi}{2}$ or $\tan \dfrac{x}{a} - \dfrac{x}{a} \sec^2 \dfrac{x}{a} = 0$.

In the first case we have evidently a minimum. In the second

$$\tan \frac{x}{a} = \frac{x}{a} \tan^2 \frac{x}{a} + \frac{x}{a}$$

$$\tan^2 \frac{x}{a} - \frac{a}{x} \tan \frac{x}{a} + \frac{a^2}{4x^2} = \frac{a^2}{4x^2} - \frac{x}{a}$$

$$\tan \frac{x}{a} = \sqrt{\frac{a^2}{4x^2} - \frac{x}{a}} + \frac{a}{2x}.$$

Let $\qquad \dfrac{x}{a} = p$ then, $\tan p = \dfrac{\sqrt{1 - 4p^3} + 1}{2p}$

which equation being solved gives $p = .3069$ nearly.

$$\therefore a = \frac{x}{.3069} \quad y = \frac{x}{.3069} \log \cos .3069$$

$$= \frac{.048x}{.3069} = .15x$$

the whole weight $= a^3 \tan \dfrac{x}{a} = a^3 .316$ the angle of elevation at top $= 17°.35'$ nearly, the secant of this is 1.049 or nearly $1 + \frac{1}{20}$ therefore for a coarse approximation to the figure of this curve draw an arc of a circle subtending 35°.10′ at the centre this will be sufficient for ordinary purposes.

16

EXERCISE ON THE PROPERTIES OF MATTER, FOR SIR WILLIAM HAMILTON'S METAPHYSICS CLASS, EDINBURGH UNIVERSITY

1848–49[1]

From Campbell and Garnett, *Life of Maxwell*[2]

ON THE PROPERTIES OF MATTER[3]

These properties are all relative to the three abstract entities connected with matter, namely, space, time and force.

1. Since matter must be in some part of space, and in one part only at a time, it possesses the property of locality or position.

2. But matter has not only position but magnitude; this property is called extension.

3. And since it is not infinite it must have bounds, and therefore it must possess figure.

These three properties belong both to matter and to imaginary geometrical figures, and may be called the geometric properties of matter.[4] The following properties do not necessarily belong to geometric figures.

4. No part of space can contain at the same time more than one body, or no two bodies can coexist in the same space; this property is called impenetrability. It was thought by some that the converse of this was true, and that there was no part of space not filled with matter. If there be a vacuum, said they, that is empty space, it must be either a substance or an accident.

If a substance it must be created or uncreated.

If created it may be destroyed, while matter remains as it was, and thus

---

(1) See *Life of Maxwell*: 108–9; the MS was found among Hamilton's papers.

(2) *Life of Maxwell*: 109–13.

(3) Compare *Lectures on Metaphysics and Logic by Sir William Hamilton, Bart.*, ed. H. L. Mansel and J. Veitch, 4 vols. (Edinburgh/London, 1859–60), **2**. In the session 1848–9 (Maxwell's second year of attendance) Forbes gave 16 lectures 'Introduction and properties of bodies'; see David B. Wilson, 'The educational matrix: physics education at early-Victorian Cambridge, Edinburgh and Glasgow Universities', in *Wranglers and Physicists. Studies on Cambridge Physics in the Nineteenth Century*, ed. P. M. Harman (Manchester, 1985): 12–48, esp. 21–6.

(4) Compare Dugald Stewart's denotation of 'Extension and Figure by the title of the *mathematical affections of matter*', in his *Philosophical Essays* (Edinburgh, $_3$1818): 153; cited by Sir William Hamilton in a 'Supplementary Dissertation' in his edition of *The Works of Thomas Reid* (Edinburgh, 1846): 843 (and compare his *Lectures on Metaphysics*, **2**: 112). Maxwell read Hamilton's 'Dissertations' in July 1848 (see Number 10).

length, breadth and thickness would be destroyed while the bodies remain at the same distance.

If uncreated, we are led into impiety.

If we say it is an accident, those who deny a vacuum challenge us to define it, and say that length, breadth and thickness belong exclusively to matter.

This is not true, for they belong also to geometric figures, which are forms of thought and not of matter;[5] therefore the atomists maintain that empty space is an accident, and has not only a possible but a real existence, and that there is more space empty than full. This has been well stated by Lucretius.

5. Since there is a vacuum, motion is possible; therefore we have a fifth property of matter called mobility.

And the impossibility of a body changing its state of motion or rest without some external force is called *inertia*.

Of forces acting between two particles of matter there are several kinds.

The first kind is independent of the quality of the particles, and depends solely on their masses and their mutual distance. Of this kind is the attraction of gravitation and that repulsion which exists between the particles of matter which prevents any two from coming into contact.

The second kind depends on the quality of the particles; of this kind are the attractions of magnetism, electricity, and chemical affinity, which are all convertible into one another and affect all bodies.[6]

The third kind acts between the particles of the same body, and tends to keep them at a certain distance from one another and in a certain configuration.

When this force is repulsive and inversely as the distance, the body is called gaseous.

When it does not follow this law there are two cases.

There may be a force tending to preserve the figure of the body or not.

When this force vanishes the body is a liquid.

When it exists the body is solid.

If it is small the body is soft; if great it is hard.

If it recovers its figure it is elastic; if not it is inelastic.

The forces in this third division depend almost entirely on heat.

---

(5) Compare Hamilton's statement that 'extension... is one of our necessary notions, – in fact, a fundamental condition of thought itself'; he went on to assert that 'the analysis of Kant...has placed this truth beyond the possibility of doubt'; *Lectures on Metaphysics*, **2**: 113. Compare Stewart's discussion of Kant's concept of space as 'the *general form* of our external senses'; *Philosophical Essays*: 155.

(6) On Forbes' views on analogies in physics see his 'Dissertation' for the eighth edition of the *Encyclopaedia Britannica*; J. D. Forbes, *A Review of the Progress of Mathematical and Physical Science in more recent times, and particularly between the years 1775–1850* (Edinburgh, 1858).

The properties of bodies relative to heat and light are –

Transmission, Reflection, and Destruction,

and in the case of light these may be different for the three kinds of light,[7] so that the properties of colour are –

Quality, Purity, and Integrity; or
Hue, Tint, and Shade.[8]

We come next to consider what properties of bodies may be perceived by the senses.

Now the only thing which can be directly perceived by the senses is Force, to which may be reduced heat, light, electricity, sound, and all the things which can be perceived by any sense.

In the sense of sight we perceive at the same time two spheres covered with different colours and shades. The pictures on these two spheres have a general resemblance, but are not exactly the same; and from a comparison of the two spheres we learn, by a kind of intuitive geometry, the position of external objects in three dimensions.[9]

Thus, the object of the sense of sight is the impression made on the different parts of the retina by three kinds of light. By this sense we obtain the greater part of our practical knowledge of locality, extension, and figure as properties of bodies, and we actually perceive colour and angular dimension.

And if we take time into account (as we must always do, for no sense is instantaneous), we perceive relative angular motion.

By the sense of hearing we perceive the intensity, rapidity, and quality of the vibrations of the surrounding medium.

By taste and smell we perceive the effects which liquids and aeriform bodies have on the nerves.

By touch we become acquainted with many conditions and qualities of bodies.

1 The actual dimensions of solid bodies in three dimensions, as compared with the dimensions of our own bodies.
2 The nature of the surface; its roughness or smoothness.
3 The state of the body with reference to heat.

---

(7) Compare Thomas Young's trichromatic theory of vision, in his *A Course of Lectures on Natural Philosophy and the Mechanical Arts*, ed. P. Kelland, 2 vols. (London, 1845), **1**: 344–5. See Numbers 54 and 64.

(8) Compare D. R. Hay, *A Nomenclature of Colours, Hues, Tints and Shades* (Edinburgh/London, 1845). See Number 54.

(9) Compare Hamilton's discussion of the principle of abstraction: 'comparison is supposed in every, the simplest act of knowledge'; *Lectures on Metaphysics*, **2**: 277–90, on 279.

To this is to be referred the sensation of wetness and dryness, on account of the close contact which fluids have with the skin.

By means of touch, combined with pressure and motion, we perceive –

1 Hardness and softness, comprehending elasticity, friability, tenacity, flexibility, rigidity, fluidity, etc.
2 Friction, vibration, weight, motion, and the like.

The sensations of hunger and thirst, fatigue, and many others, have no relation to the properties of bodies.

Lucretius on Empty Space.[10]

Nec tamen undique corporeâ stipata tenentur
Omnia naturâ, namque est in rebus Inane.
Quod tibi cognôsse in multis erit utile rebus
Nec sinet errantem dubitare et quærere semper
De summâ rerum; et nostris diffidere dictis.
Quapropter locus est intactus Inane Vacansque
Quod si non esset, nullâ ratione moveri
Res possent; namque officium quod corporis extat
Officere atque obstare, id in omni tempore adesset
Omnibus. Haud igitur quicquam procedere posset,
Principium quoniam cedendi nulla daret res,
At nunc per maria ac terras sublimaque cæli
Multa modis multis variâ ratione moveri
Cernimus ante oculos, quæ, si non esset Inane,
Non tam sollicito motu privata carerent
Quam genita omninô nullâ ratione fuissent,
Undique materies quoniam stipata quiesset.

---

(10) *De Rerum Natura*, Book I, 329–45. The text cited suggests that Maxwell quoted from Gilbert Wakefield's edition, *T. Lucretii Cari De Rerum Natura Libri Sex*, 4 vols. (Glasgow, 1813), **1**: 77–80. Compare *Titi Lucretii Cari De Rerum Natura Libri Sex*, ed. and trans. H. A. J. Munro, 2 vols. (Cambridge, 1864), **1**: 15–16. Munro, following Bentley and Lachmann, considers l.334 spurious (*ibid.*, **2**: 31), and omits it in translating the passage. 'And yet all things are not on all sides jammed together and kept in by body: there is also void in things. To have learned this will be good for you on many accounts; it will not suffer you to wander in doubt and be to seek in the sum of things and distrustful of our words. If there were not void, things could not move at all; for that which is the property of body, to let and hinder, would be present to all things at all times; nothing, therefore, could go on since no other thing would be the first to give way. But in fact throughout seas and lands and the heights of heaven, we see before our eyes many things move in many ways for various reasons, which things, if there were no void, I need not say would lack and want restless motion: they never would have been begotten at all, since matter jammed on all sides would be at rest' (*ibid.*, **1**: 15–16).

17

MEMORANDUM ON A
PHYSICO-MATHEMATICAL SOCIETY

circa 1849[1]

From the original in the University Library, Cambridge[2]

[PHYSICO MATHEMATICAL SOCIETY]

At the preliminary meeting of the Physico Mathematical Society it was agreed that the members of the provisional Committee should each suggest some subjects deserving of the attention of the Society. It seems to me that the work of the Society might be arranged thus in order of importance.

1　The formation of committees of investigation into various subjects either by experiment or by consulting books in concert.

2　Observations experiments calculations and suggestions and essays by individuals either original or supposed by the Society to be so.

3　Communications of new discoveries and interesting facts connected with Math or Physics not generally known by the Members.

4　Observations on new books on Nat. Phil and Mathematics.

5　SHORT Abstracts of the principal contents of books to be kept for the use of members.

6　Speculative papers on theories of a Physical kind.

7　Problematic Challenges? NONE OF THEM PRAY.

1a　A member should be appointed to receive the key and list of the instruments of the Society from Mr T. Cleghorn and the first attention of the Society should be to the investigations experiments and observations which can be made with these instruments.

b　The whole society should be appointed to observe and note down the natural phenomena which they may have an opportunity of witnessing during the year. A list of these should be made as follows.

　　Haloes round the sun or moon (with the angular magnitude if possible). Rainbows (with internal bands or other peculiarities). The Magnetic Light and Aurora Borealis Crepuscular rays and polar bands etc. etc. etc. (Newspaper reports are not to be noted unless confirmed by witness.)

c　Subjects for reading connected with the business should be proposed and reports made on the results.

---

(1) See Number 29.　　　　(2) ULC Add. MSS 7655, V, i/3.

d   Committees to inquire into subjects started in discussions and not settled.

e   etc. + etc.

2   Examples of Calculations etc. which I am willing to undertake.[3]

a   Practical construction of Curves on paper.

b   On Certain interesting curves and surfaces.

c   Equations to squares triangles cubes etc.

d   Properties of curves in space.

e   Teeth of Wheels Moments of Inertia centrifugal clock.

f   To find the radii of curvature and the refractive power of lenses from 3 simple observations.

g   On the Iriscope, Whirling discs Gutta percha Isinglass wax and rosin etc. optical and mechanical properties.

h   On achromatic polarization by reflection Aireys apparatus Fresnels rhomb.[4]

i   Coloured glasses liquids etc.

j   Unannealed and compressed solids by polarized light with observations.

k   On the Eye,[5] and Haidingers aigrette.[6]
    ⌐Absorption of light and reflection by coloured bodies.⌐

NB   No member shall be made to study any subject for which he has an unconquerable aversion.

I have reported and am Sirs
Your humble Commissioner
Jas Clerk Maxwell

Solom Sec. P-MSE

---

(3)   Compare Numbers 1, 2, 3, 6, 11, 13, 14, 22, 23, 24 and 32.

(4)   For a description of 'Fresnel's rhomb', a parallelepiped of glass whose acute angles are 54° 37′ which produces circularly-polarised light by internal reflection, see G. B. Airy, *Mathematical Tracts on the Lunar and Planetary Theories...and the Undulatory Theory of Optics* (Cambridge, ₃1842): 346–52, esp. 349, and Plate 5 Figs. 24 and 33. In a manuscript fragment dated 'Aug 21, 1849' (ULC Add. MSS 7655, V, g/1) Maxwell drew a diagram of 'Fresnel's rhomb'.

(5)   A manuscript note in which Maxwell calculates the foci of the eye (ULC Add. MSS 7655, V, b/21) is endorsed by J. D. Forbes. In a letter of 19 April 1849 (ULC Add. MSS 7655, II/1) Forbes requested Maxwell 'will you be so kind as to re-calculate the Foci of the Eye...'. In a subsequent letter of 1 May 1849 (Add. 7655, II/2) he states: 'I have repeated the whole of your calculations about the Eye. Those for 10 inches are wrong (seriously) from the first Step. Might I ask you to repeat these as a verification of mine. I enclose the paper with the formulæ – *which please to return*. I hope you got your proof sheets.' He was here alluding to the proofs of Maxwell's paper 'On the theory of rolling curves'; see Number 18 note (3). Forbes' letter of 19 April 1849 is endorsed by Maxwell with some preliminary calculations of focal lengths.

(6)   See Number 32, and Campbell's recollection of Maxwell's early interest in 'Haidinger's brushes' (*Life of Maxwell*: 84).

ABSTRACT OF 'ON THE THEORY OF ROLLING CURVES'

[19 FEBRUARY 1849][1]

From the *Proceedings of the Royal Society of Edinburgh*[2]

This paper[3] commenced with an outline of the nature and history of the problem of rolling curves, and it was shewn that the subject had been discussed previously, by several geometers, amongst whom were De la Hire and Nicolè in the *Mémoires de l'Academie*,[4] Euler,[5] Professor Willis, in his *Principles of Mechanism*,[6] and the Rev. H. Holditch in the *Cambridge Philosophical Transactions*.[7]

None of these authors, however, except the two last, had made any application of their methods; and the principal object of the present communication was to find how far the general equations could be simplified in particular cases, and to apply the results to practice.

Several problems were then worked out, of which some were applicable to the generation of curves, and some to wheelwork; while others were interesting as shewing the relations which exist between different curves; and, finally, a collection of examples was added, as an illustration of the fertility of the methods employed.

---

(1) The date the paper was presented (by Philip Kelland) to the Royal Society of Edinburgh.

(2) *Proc. Roy. Soc. Edinb.*, **2** (1849): 222 (= *Scientific Papers*, **1**: 29).

(3) James Clerk Maxwell, 'On the theory of rolling curves', *Trans. Roy. Soc. Edinb.*, **16** (1849): 519–40 (= *Scientific Papers*, **1**: 4–29).

(4) Philippe de La Hire, 'Traité des roulettes', *Mémoires de l'Académie Royale des Sciences* (année 1706): 340–88; François Nicole, 'Methode generale pour determiner la nature des courbes formées par la roulement de toutes sortes de courbes sur une autre courbe quelconque', *ibid*, (année 1707): 81–97; and also Pierre Varignon, 'Nouvelle formation de spirales', *ibid*. (année 1704): 69–131, cited by Maxwell in 'On the theory of rolling curves': 519–20 (= *Scientific Papers*, **1**: 5).

(5) Leonhard Euler, 'De aptissima figura rotarum dentibus tribuenda', *Novi Commentarii Academiae Scientiarum Imperialis Petropolitanae*, **5** (1755): 299–316. Euler's paper is concerned with gearing, to obtain uniform frictionless motion from toothed wheels.

(6) Robert Willis, *Principles of Mechanism* (Cambridge, 1841): 31–61, 239–64, 287–95.

(7) Hamnet Holditch, 'On rolling curves', *Trans. Camb. Phil. Soc.*, **7**, (1842): 61–86.

FROM A LETTER TO LEWIS CAMPBELL

6 JULY [1849][1]

From Campbell and Garnett, *Life of Maxwell*[2]

Bannavie
6 July [1849]

But to the point. Perhaps you remember going with my Uncle John Cay (7th Class), to visit Mr. Nicol in Inverleith Terrace. There we saw polarised light in abundance. I purposed going this session but was prevented. Well, sir, I received from the aforesaid Mr. Cay a 'Nicol's prism', which Nicol had made and sent him. It is made of calc-spar, so arranged as to separate the ordinary from the extraordinary ray.[3] So I adapted it to a camera lucida, and made charts of the strains in unannealed glass.[4]

I have set up the machine for showing the rings in crystals, which I planned during your visit last year.[5] It answers very well. I also made some experiments on compressed jellies in illustration of my props on that subject. The principal one was this: The jelly is poured while hot into the annular space contained between a paper cylinder and a cork: then, when cold, the cork is twisted round, and the jelly exposed to polarised light, when a transverse cross, ×, not +, appears with rings as the inverse square of the radius, all which is fully verified. Hip! etc. QED.[6]

Figure 19,1

But to make an *abrupt transcision*, as Forbes says,[7] we set off to Glasgow on Monday 2d; to Inverary on 3d; to Oban by Loch Awe on 4th; round Mull, by Staffa and Iona, on (5th), and here on 6th. To-morrow we intend to get to Inverness and rest there. On Monday perhaps? to the land of Beulah, and afterwards back by Caledon. Canal to Crinan Canal, and so to Arran, thence

---

(1) Following Campbell. (2) *Life of Maxwell*: 123–4; abridged.

(3) A Nicol's prism consists of a prism of Iceland spar (which exhibits double refraction) cut into two along a diagonal plane; the segments are cemented together by Canada balsam; see William Nicol, 'On a method of so far increasing the divergence of the two rays in calcareous spar, that only one image may be seen at a time', *Edinburgh New Philosophical Journal*, **7** (1829): 83–4; and Nicol, 'Notice concerning an improvement in the construction of the simple vision prism of calcareous spar', *ibid.*, **27** (1839): 332–3.

(4) See Numbers 12, 14 and 26 §§5, 6. (5) See Number 13.

(6) Compare Case I of 'On the Equilibrium of Elastic Solids' (Number 26, §8). See Plate IV in *Life of Maxwell*: 491, copied from Maxwell's water-colour (preserved in the Cavendish Laboratory, Cambridge) depicting the chromatic effects exhibited by a hollow cylinder of gelatine subjected to a torsional shear and viewed by polarised light.

(7) Compare Number 28 note (2).

to Ardrossan, and then home. It is possible that you may get a more full account of all these things (if agreeable), when I fall in with a pen that will spell; my present instrument partakes of the nature of skates, and I can hardly steer it.

There is a beautiful base here for measuring the height, etc., of Ben Nevis. It is a straight and level road through a moss for about a mile that leads from the inn right to the summit.

It is proposed to carry up stones and erect a cairn 3 feet high, and thus render it the highest mountain in Scotland.

During the session Prof. Forbes gave as an exercise to describe a cycloid from top of Ben Nevis to Fort William and slide trees down it. We took an observation of the slide, but found nothing to slide but snow.

I think a body *deprived* of friction would go to Fort William in a cycloid in 49.6 seconds, and in 81 on an inclined plane.

FROM A LETTER TO LEWIS CAMPBELL

OCTOBER 1849

From Campbell and Garnett, *Life of Maxwell*[1]

[Edinburgh]
October 1849

Since last letter, I have made some pairs of diagrams representing solid figures and curves drawn in space; of these pictures one is seen with each eye by means of mirrors, thus...[2]

This is Wheatstone's Stereoscope,[3] which Sir David Brewster has taken up of late with much violence at the British Association. (The violence consists in making two lenses out of one by breaking it.) (See Report.)[4] Last winter he exhibited at the Scottish Society of Arts Calotype pictures of the statue of Ariadne and the beast seen from two stations, which, when viewed properly, appeared very solid.[5]

Since then I have been doing practical props on compression, and writing out the same that there may be no mistake.[6] The nicest cases are those of spheres and cylinders. I have got an expression for the hardness of a cricket ball made of case and stuffing. I have also the equations for a spherical cavity in an infinite solid,[7] and this prop: Given that the polarised colour of any part of a cylinder of unannealed glass is equal to the square of the distance from the centre (as determined by the observation),[8] to find – 1st, the state of strain at each point; 2d, the temperature of each.[9]

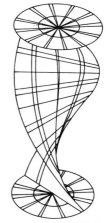

Figure 20,1. This is the likeness of a skew screw surface.

I have got an observation of the latitude just now with a saucer of treacle, but it is very windy.

---

(1) *Life of Maxwell*: 125–6; abridged.

(2) Campbell's deletion. On stereoscopic figures see Numbers 91 and 94.

(3) Charles Wheatstone, 'On some remarkable and hitherto unobserved phenomena of binocular vision', *Phil. Trans.*, **128** (1838): 371–94.

(4) See the report on Brewster's 'An account of a new stereoscope' using two semilenses in *The Athenæum* (22 September 1849): 960. On the lenticular stereoscope see David Brewster, 'Description of several new and simple stereoscopes for exhibiting, as solids, one or more representations of them on a plane', *Transactions of the Royal Scottish Society of Arts*, **3** (1851): 247–59.

(5) David Brewster, 'Account of a binocular camera, and of a method of obtaining drawings of full length and colossal statues, and of living bodies, which can be exhibited as solids by the stereoscope', *Transactions of the Royal Scottish Society of Arts*, **3** (1851): 259–64.

(6) See Numbers 21, 22, 23 and 24.

(7) See J. C. Maxwell, 'On the equilibrium of elastic solids', *Trans. Roy. Soc. Edinb.*, **20** (1850): 87–120, esp. 106 (= *Scientific Papers*, **1**: 55).

(8) See David Brewster, 'On the laws which regulate the distribution of the polarising force in plates, tubes and cylinders of glass that have received the polarising structure', *Trans. Roy. Soc. Edinb.*, **8** (1818): 353–72, esp. 353–4.   (9) See Number 23 and Number 26 §§5, 11.

DRAFT ON THE EQUATIONS OF COMPRESSION OF AN ELASTIC SOLID

circa OCTOBER 1849[1]

From the original in the University Library, Cambridge[2]

[EQUATIONS OF COMPRESSION]

The Laws of ⟨Compression⟩ ⌐elasticity⌐ express the relation between the changes of the dimensions of a body and the forces which produce them.

These forces may be called Pressures and their effects Compressions.

Pressures may be estimated in pounds on the square inch, and Compressions in fractions of the dimension compressed.

Having thus distinguished between Pressure and Compression we must consider the nature of Compression which is purely geometrical.

Let the position of points in space be expressed by their coordinates x, y, and z then any change in the dimensions of a body is expressed by giving to these coordinates the variations $\delta x, \delta y, \delta z, \; \delta x, \delta y, \delta z$ being functions of x, y, z. Here $\delta x, \delta y, \delta z$ represent the absolute motion of the point in the direction of its three coordinates, but as compression depends not on absolute but on relative displacement, we must transfer the origin of coordinates to the new position of the point.

Thus the quantities which we have to deal with are

$$\frac{d\delta x}{dx} \quad \frac{d\delta x}{dy} \quad \frac{d\delta x}{dz}$$
$$\frac{d\delta y}{dx} \quad \frac{d\delta y}{dy} \quad \frac{d\delta y}{dz}$$
$$\frac{d\delta z}{dx} \quad \frac{d\delta z}{dy} \quad \frac{d\delta z}{dz}.$$

Since the number of these quantities is nine, if we can find any other nine independent quantities of the same kind, the one set may be found in terms of the other. The quantities we shall assume for this purpose are

1   Three compressions, $\dfrac{\delta\alpha}{\alpha}, \dfrac{\delta\beta}{\beta}, \dfrac{\delta\gamma}{\gamma}$ in the directions of three axes α, β, and γ.

2   The nine direction-cosines of these axes, with the six connecting equations.

3   Three small angles of rotation of this system of axes about the axes of x, y, z.

---

(1) Compare Number 26 §7.   (2) ULC Add. MSS 7655, V, g/1.

The cosine of the angle between the axis of x and α will be denoted by a_1

Between $\quad x$ and β by b_1

$\qquad x$ and γ by c_1

$\qquad y$ and α, β, γ by a_2, b_2, c_2

$\qquad z$ and α, β, γ by a_3, b_3, c_3.

These quantities are connected by the six equations

$$\left.\begin{aligned} a_1^2+b_1^2+c_1^2 &= 1 & a_1 a_2+b_1 b_2+c_1 c_2 &= 0 \\ a_2^2+b_2^2+c_2^2 &= 1 & a_1 a_3+b_1 b_3+c_1 c_3 &= 0 \\ a_3^2+b_3^2+c_3^2 &= 1 & a_2 a_3+b_2 b_3+c_2 c_3 &= 0 \end{aligned}\right\} \quad (1)$$

The rotation round the axis of x from y to z is a very small angle denoted by $\delta\theta_1$.

The angle round $\quad y$ from x to $z = -\delta\theta_2$

$\qquad\qquad\qquad z$ from x to $y = \delta\theta_3$.

The equations connecting the two systems of axes are

$$\left.\begin{aligned} x &= a_1 \alpha + b_1 \beta + c_1 \gamma \\ y &= a_2 \alpha + b_2 \beta + c_2 \gamma \\ z &= a_3 \alpha + b_3 \beta + c_3 \gamma \end{aligned}\right\} \quad (2)$$

and from these we obtain the following expressions for the compressions

$$\left.\begin{aligned} \frac{d\delta x}{dx} &= \frac{\delta\alpha}{\alpha}a_1^2 + \frac{\delta\beta}{\beta}b_1^2 + \frac{\delta\gamma}{\gamma}c_1^2 \\ \frac{d\delta y}{dy} &= \frac{\delta\alpha}{\alpha}a_2^2 + \frac{\delta\beta}{\beta}b_2^2 + \frac{\delta\gamma}{\gamma}c_2^2 \\ \frac{d\delta z}{dz} &= \frac{\delta\alpha}{\alpha}a_3^2 + \frac{\delta\beta}{\beta}a_3^2 + \frac{\delta\gamma}{\gamma}c_3^2 \end{aligned}\right\} \quad (3)$$

$$\left.\begin{aligned} \frac{d\delta x}{dy} &= \frac{\delta\alpha}{\alpha}a_1 a_2 + \frac{\delta\beta}{\beta}b_1 b_2 + \frac{\delta\gamma}{\gamma}c_1 c_2 + \delta\theta_3 \\ \frac{d\delta x}{dz} &= \frac{\delta\alpha}{\alpha}a_1 a_3 + \frac{\delta\beta}{\beta}b_1 b_3 + \frac{\delta\gamma}{\gamma}c_1 c_3 + \delta\theta_2 \\ \frac{d\delta y}{dx} &= \frac{\delta\alpha}{\alpha}a_2 a_1 + \frac{\delta\beta}{\beta}b_2 b_1 + \frac{\delta\gamma}{\gamma}c_2 c_1 - \delta\theta_3 \\ \frac{d\delta y}{dz} &= \frac{\delta\alpha}{\alpha}a_2 a_3 + \frac{\delta\beta}{\beta}b_2 b_3 + \frac{\delta\gamma}{\gamma}c_2 c_3 + \delta\theta_1 \\ \frac{d\delta z}{dx} &= \frac{\delta\alpha}{\alpha}a_3 a_1 + \frac{\delta\beta}{\beta}b_3 b_1 + \frac{\delta\gamma}{\gamma}c_3 c_1 - \delta\theta_2 \\ \frac{d\delta z}{dy} &= \frac{\delta\alpha}{\alpha}a_3 a_2 + \frac{\delta\beta}{\beta}b_3 b_2 + \frac{\delta\gamma}{\gamma}c_3 c_2 - \delta\theta_1 \end{aligned}\right\} \quad (4)$$

If we wish to consider only the plane xy, $a_3\, b_3\, c_3\, c_1\, c_2$ are left out and we have

$$\left.\begin{aligned}
\frac{d\delta x}{dx} &= \frac{\delta\alpha}{\alpha}\, a_1^2 + \frac{\delta\beta}{\beta}\, b_1^2 \\[4pt]
\frac{d\delta y}{dy} &= \frac{\delta\alpha}{\alpha}\, a_2^2 + \frac{\delta\beta}{\beta}\, b_2^2 \\[4pt]
\frac{d\delta x}{dy} &= \frac{\delta\alpha}{\alpha}\, a_1\, a_2 + \frac{\delta\beta}{\beta}\, b_1\, b_2 + \delta\theta_3 \\[4pt]
\frac{d\delta y}{dx} &= \frac{\delta\alpha}{\alpha}\, a_1\, a_2 + \frac{\delta\beta}{\beta}\, b_1\, b_2 - \delta\theta_3
\end{aligned}\right\} \tag{5}$$

also

$$\left.\begin{aligned}
a_1^2 + b_1^2 = 1 \qquad a_1\, a_2 + b_1\, b_2 = 0 \\
a_2^2 + b_2^2 = 1 \qquad a_1\, b_1 + a_2\, b_2 = 0
\end{aligned}\right\}.$$

$$\therefore a_2^2 = b_1^2 = 1 - a_1^2, \qquad b_2 = -a_1$$

$$2\delta\theta_3 = \frac{d\delta x}{dy} - \frac{d\delta y}{dx}. \tag{6}$$

22

DRAFT ON THE TWISTING OF A PLATE TO FORM A SKEW SCREW SURFACE

circa OCTOBER 1849[1]

From the original in the University Library, Cambridge[2]

[A SKEW SCREW SURFACE][3]

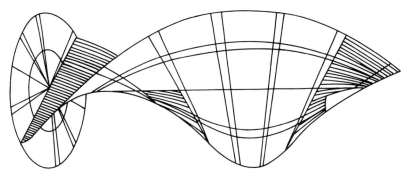

Figure 22,1

Let the original form of the solid be a very thin plate bounded by two parallel sides and of infinite length and let it be twisted into the form of a skew screw surface of which the equation is

$$a\theta = x$$

the distance from the axis of any point $= r$ and the thickness $= t$.

Let us consider the element bounded by ds and dr and t acted on by the forces p and q.

From the equation of the surface we obtain

$$ds = \sqrt{dx^2 + (rd\theta)^2} = \sqrt{1 + \frac{r^2}{a^2}}\, dx^{(4)}$$

and the curvature of the helix at any point

$$= \frac{1}{R} = \frac{r}{a^2 + r^2}$$

then the equations of the equilibrium of the element will be

$$q\frac{r}{a^2+r^2} - p\frac{r}{a}\frac{1}{\sqrt{a^2+r^2}} - \frac{dp}{dr}\frac{\sqrt{a^2+r^2}}{a} = 0$$

---

(1) See Number 20. (2) ULC Add. MSS 7655, V, g/1.
(3) Compare Number 26 § 10. (4) See Number 26 §§ 9 and 10.

and the equations of compression

$$\frac{d\delta s}{ds} = \left(\frac{1}{9\mu} + \frac{2}{3m}\right)q + \left(\frac{1}{9\mu} - \frac{1}{3m}\right)p. \text{(5)}$$

As the solid will be shortened by being twisted let this shortening be expressed by c so that $\dfrac{\delta x}{x} = c$

then
$$\frac{d\delta s}{ds} = \sqrt{1 + \frac{r^2}{a^2}(1-c)} - 1.$$

By combining these three equations we obtain this linear diff eq^n

$$p + \frac{\dfrac{1}{9\mu} + \dfrac{2}{3m}}{\dfrac{2}{9\mu} + \dfrac{1}{3m}}\left(\frac{a^2}{r} + r\right)\frac{dp}{dr} = \frac{1 - c - \left(1 + \dfrac{r^2}{a^2}\right)^{-\frac{1}{2}}}{\dfrac{2}{9\mu} + \dfrac{1}{3m}}.$$

---

(5) Maxwell terms μ and m 'the moduli of cubical and linear elasticity'; see Number 26 §1.

23

DRAFT ON THE OPTICAL PROPERTIES OF COMPRESSED SOLIDS

circa OCTOBER 1849[1]

From the original in the University Library, Cambridge[2]

ON THE OPTICAL PROPERTIES OF COMPRESSED SOLIDS[3]

It is observed that when a transparent elastic solid is compressed it possesses the properties of a doubly refracting crystal[4] whose axes are those of compression.[5] An axis of compression is negative like Iceland spar.

Let μ_1 = the refractive index of the substance, then since

$$\frac{\mu_1^2 - 1}{\text{spec. gravity}} = \text{constant for one substance}[6]$$

$$\frac{\delta\mu_1}{\left(\dfrac{\delta V}{V}\right)} = \frac{\mu_1^2 - 1}{2\mu_1}.$$

When a ray of light passes in the direction of one of the principal axes of compression it is divided into two portions of which that which is polarized in the plane of greatest compression is most retarded. Let T = the quantity by which the one is retarded more than the other and let z be the thickness of the solid then

$$T = zV\left(\frac{\delta\alpha}{\alpha} - \frac{\delta\beta}{\beta}\right)^{[7]}$$

$$= zV\sqrt{\left(\frac{d\delta x}{dx} - \frac{d\delta y}{dy}\right)^2 + \left(\frac{d\delta x}{dy} + \frac{d\delta y}{dx}\right)^2}$$

V being a coefficient to be determined by experiment for each substance but it is probably a function of μ and m.

---

(1) See Number 20.   (2) ULC Add. MSS 7655, V, g/1.   (3) Compare Number 26 §6.

(4) David Brewster, 'On the communication of the structure of doubly-refracting crystals to glass, muriate of soda, fluor spar and other substances, by mechanical compression and dilatation', *Phil. Trans.*, **106** (1816): 156–78.

(5) In his paper 'On the equilibrium of elastic solids', *Trans. Roy. Soc. Edinb.*, **20** (1850): 87–120, esp. 90, 119 (= *Scientific Papers*, **1**: 34, 70–71) Maxwell explains induced double refraction in strained glass by supposing that 'in homogeneous singly-refracting substances exposed to pressures, the principal axes of pressure coincide with the principal axes of double refraction'. See Number 26 notes (32) and (77).

(6) As stated by David Brewster, *A Treatise on Optics* (London, 1831): 26.

(7) $\dfrac{\delta\alpha}{\alpha}$ and $\dfrac{\delta\beta}{\beta}$ are the compressions in the directions of the principal axes of pressure α and β.

These equations may be applied to determine the state of strain in a cylinder of unannealed glass in which it is observed that the tint or interval of retardation varies directly as the square of the distance from the centre,[8] that is

$$q - p = c_1 r^2 = r \frac{dp}{dr}^{(9)}$$

$$c_1 r = \frac{dp}{dr}$$

$$p = \frac{c_1}{2} r^2 + c_2.$$

Let a be the radius of the cylinder then $p = 0$ when $r = a$

$$p = \frac{c_1}{2} (r^2 - a^2)$$

$$q = \frac{c_1}{2} (3r^2 - a^2)$$

$$\frac{d\delta r}{dr} = \frac{c_1}{9\mu} (2r^2 - a^2) - \frac{c_1}{6m} (r^2 + a^2)$$

$$\frac{\delta r}{r} = \frac{c_1}{9\mu} (2r^2 - a^2) + \frac{c_1}{6m} (5r^2 - a^2).$$

Integrate the value of $\dfrac{d\delta r}{dr} dr$ and divide by r

$$\frac{\int \frac{d\delta r}{dr} dr}{r} = \frac{c_1}{9\mu} (\tfrac{2}{3} r^2 - a^2) - \frac{c_1}{6m} (\tfrac{1}{3} r^2 + a^2)$$

$$\frac{\delta r}{r} - \frac{\int \frac{d\delta r}{dr} dr}{r} = \left(\frac{4}{27\mu} + \frac{8}{9m} \right) c_1 r^2 = c_2 r^2$$

$$\frac{d}{dr} \frac{\delta r}{r} r - \frac{d\delta r}{dr} = \left(\frac{4}{9\mu} + \frac{8}{3m} \right) c_1 r^2 = 3c_2 r^2.$$

The cooling of the cylinder goes on thus. It is taken from the fire in a viscous state and soon forms a crust round the circumference. This crust while forming has the temperature Θ but it cools to the temperature $\Theta - d\theta$ and contracts before a new layer forms by a quantity which as we see by the equations on the last page is proportional to the square of the distance from the centre.

---

(8) Compare Case X of the draft 'On the Equilibrium of Elastic Solids' (Number 26 §§5, 11). See David Brewster, 'On the laws which regulate the distribution of the polarising force in plates, tubes and cylinders of glass that have received the polarising structure', *Trans. Roy. Soc. Edinb.*, **8** (1818): 353–72, esp. 353–4.

(9) q and p are the tangential and radial pressures, r is the radius.

At the time when the centre becomes solid the temperature of any point $=$ $\Theta - \phi(r)^2$ therefore as the outer parts are coldest they will contract least when the whole is brought to a uniform temperature. If the cylinder remain entire after being cooled the outer rings will be forcibly contracted and the inner ones expanded the neutral distance being equal to $\dfrac{a}{\sqrt{3}}$. But there will also be a forcible separation between one ring and another which produces the force p.

Thus the general appearance will be that which was proposed to be explained, namely a series of rings whose tint is as the square of the distance and which are of the opposite kind to those seen in calcareous spar and negative crystals.

DRAFT ON THE MEASUREMENT OF COEFFICIENTS OF ELASTICITY

circa OCTOBER 1849[1]

From the original in the University Library, Cambridge[2]

ON THE DETERMINATION OF THE COEFFICIENTS m AND μ[3]

The coefficient m may be determined at once by means of experiments on the torsion of rods and wires. These experiments may be performed in different ways.

1st Let a uniform cylindrical rod of the substance be fixed at one end while it is twisted at the other by means of a wheel and weight or any other contrivance.[4] The rod passes through the centres of two graduated circles which are connected with each other but not with the rod or the rest of the apparatus. The position of the rod with respect to these circles may be observed either by means of marks on the rod or by pointers fastened to it. The circles are first read off and then a weight is applied to the wheel and the circles read off again and the difference of the revolution of the two circles is noted let it $= \delta\theta$.
Let the distance between the circles $= x$.

Let the radius of the wheel $= a$
————————————— rod $= b$
weight of the wheel $= W$

$$m = \frac{2aWx}{\pi b^4 \delta\theta}.$$

The quantity m may likewise be obtained from the time of oscillation of a body suspended by a wire of the substance,[5] but it appears that the first method is the best.

---

(1) See the draft 'On the Equilibrium of Elastic Solids' (Number 26 §§ 13, 14).

(2) ULC Add. MSS 7655, V, g/1.

(3) Maxwell terms μ and m 'the moduli of cubical and linear elasticity'; see Number 26 § 1.

(4) Compare Félix Savart, 'Mémoire sur la réaction de torsion des lames et des verges rigides', *Ann. Chim. Phys.*, ser. 2, **41** (1829): 373–97. See Number 26 note (133).

(5) Coulomb's torsion balance; see Charles Augustin Coulomb, 'Recherches théoriques et expérimentales sur la force de torsion, & sur l'élasticité des fils de métal', *Mémoires de l'Académie Royale des Sciences* (année 1784): 229–69.

The most direct way of finding the coefficient μ is by exposing the substance to hydrostatic pressure.

Let the content of the spherical or cylindrical vessel used in Œrsted's machine for the compression of liquids[6] be denoted by V and its cubical compressibility by $\dfrac{1}{\mu}$ and the sectional area of the capillary tube by c^2 and the change of level when the whole is exposed to a hydrostatic pressure h by l

$$\frac{lc^2}{Vh} = \frac{1}{\mu} - \frac{1}{\mu_2}$$

$\dfrac{1}{\mu_2}$ being the cubical compressibility of the contained liquid which is supposed unknown but may be found in the following way. Since the expression for the change of internal capacity of a sphere b being the internal radius, is[7]

$$\frac{\delta V}{V} = \tfrac{3}{2}a^3 \frac{h-k}{a^3-b^3}\frac{1}{m} + \frac{a^3h-b^3k}{a^3-b^3}\frac{1}{\mu}$$

the term involving $\dfrac{1}{\mu}$ may be made to vanish by making $a^3h = b^3k$ or $h = \dfrac{b^3}{a^3}k$ and thus

$$\frac{\delta V}{V} = -\tfrac{3}{2}k\frac{1}{m}.$$

Therefore if the hydrostatic pressure on the inner and outer surfaces of the hollow sphere be made inversely proportional to the cubes of the radii

$$\frac{lc^2}{Vk} = -\tfrac{3}{2}\frac{1}{m} - \frac{1}{\mu_2}$$

$$\frac{1}{\mu_2} = -\tfrac{3}{2}\frac{1}{m} - \frac{lc^2}{Vk}$$

and m has been found by experiments on torsion therefore μ_2 may be found.

Having thus determined the compressibility of water or any other liquid we may find that of any solid by using it as a containing vessel in Œrsted's apparatus or by introducing it into that vessel.

---

(6) H. C. Oersted, 'On the compressibility of water', *Report of the Third Meeting of the British Association for the Advancement of Science; held at Cambridge in 1833* (London, 1834): 353–60. In Oersted's piezometer the liquid is contained in a vessel with a narrow neck; after the application of pressure, the fall in the level of the liquid in the tube is observed: the difference between the change in volume of the liquid and the change in internal capacity of the containing vessel.

(7) Compare Maxwell's discussion in 'On the equilibrium of elastic solids', *Trans. Roy. Soc. Edinb.*, **20** (1850): 105 (= *Scientific Papers*, **1**: 52–3). See Number 26 §8; h and k are the external and internal pressures, a the external radius.

These results may be tested by straining a cylindric rod by weight and measuring the increase of length between two fixed points.[8] Then if

x = length of the rod between the points

δx = change of length

c^2 = area of section of the rod

W = the weight applied

$$\frac{1}{9\mu} + \frac{2}{3m} = \frac{\delta x}{x} \frac{c^2}{W}.$$

As the coefficient $\dfrac{1}{9\mu} + \dfrac{2}{3m}$ occurs frequently let it $= \dfrac{1}{n}$.[9]

This coefficient may be obtained from the bending of beams.

Let b = breadth of a beam

$\quad d$ = depth

$\quad l$ = length

$\quad R$ = radius of curvature at any point

$\quad M$ = moment of bending force at that point

$\dfrac{1}{n} = \dfrac{bd^3}{24MR}$ when the beam is rectangular

$\dfrac{1}{n} = \frac{3}{16}\pi \dfrac{a^4}{MR}$ when a is the radius of a cylinder

$\dfrac{1}{n} = \frac{1}{96} \dfrac{bd^3}{MR}$ when the beam is a rhomb. $\diamondsuit$

Figure 24,1

Let the rod be fixed by the middle and loaded at both ends with a weight $= w$. Let the weight of half the rod $= W$ and let the deflexion from the straight line at each end $= y$ then in the case of a rectangular beam

$$n = 3\frac{l^3}{bd^3}\frac{1}{y}\left\{\frac{W}{8} + \frac{w}{3}\right\}$$

---

(8) On 20 March 1848 J. D. Forbes read a paper to the Royal Society of Edinburgh 'On an instrument for measuring the extensibility of elastic solids', *Proc. Roy. Soc. Edinb.*, **2** (1848): 173–5, employing an instrument which was 'almost a faithful reproduction of S'Gravesande's apparatus' employed to verify Hooke's law; see Willelm Jacob van 'sGravesande, *Physices Elementa Mathematica experimentis confirmata, sive Introductio ad Philosophiam Newtonianam*, 2 vols. (Leiden, ₃1742), **1**: 373–7, and Fig. 4 Tab. 43. Maxwell refers to 'the apparatus of S'Gravesande' in 'On the equilibrium of elastic solids': 103 (= *Scientific Papers*, **1**: 50). In the apparatus employed by Forbes and 'sGravesande, a rod is secured at both ends, a weight applied at the middle of the rod, and the extension of the rod measured by the deflection of an indicator dial connected to the weight. See also Number 31.

(9) Maxwell subsequently writes E for n; see Number 26 §13; and 'On the equilibrium of elastic solids': 106 (= *Scientific Papers*, **1**: 54).

when the section of the rod is a rhombus

$$n = 12 \frac{l^3}{bd^3} \frac{1}{y} \left\{ \frac{W}{8} + \frac{w}{3} \right\}$$

and when the rod is a cylinder whose radius $= a$

$$n = \frac{2}{3\pi} l^3 \frac{1}{y} \left\{ \frac{W}{8} + \frac{w}{3} \right\} \frac{1}{a^4}.$$

25

FROM A LETTER TO LEWIS CAMPBELL

5 NOVEMBER 1849

From Campbell and Garnett, *Life of Maxwell*[1]

31 Heriot Row
Edinburgh, N.B.
Monday Nov. 5 1849

I go to Gregory[2] to Chemistry at 10, Morale Phil. at 12, and Pract. Chem. at 13, finishing at 14, unless perhaps I take an hour at Practical Mechanics at the School of Arts. I do not go to Sir W. H. logic, seeing I was there before.[3] Langhorne has got your Buchananic notes.[4] Why do you think that I can endure nothing but Mathematics and Logic, the only things I have plenty of? and why do you presuppose my acquaintance with your preceptors, professors, tutors, etc.?...

I don't wonder at your failing to take interest in the exponential theorem, seeing I dislike it, although I know the use and meaning of it. But I never *would have*, unless Kelland had explained it....

In your next letter you may give an abstract of Aristotle's *Rhet.*, for I do not attend Aytoun,[5] and so I know not what Rhet is. I know Logic only by Reid's account of it.[6] I will tell you about Wilson's Moral Philosophy,[7] provided always you want to know, and signify your desire.

---

(1) *Life of Maxwell*: 126–7.

(2) William Gregory, Professor of Chemistry at Edinburgh University (*DNB*).

(3) See Number 7 note (6).

(4) Robert Buchanan, Professor of Logic at Glasgow University (*DNB*). Campbell was a student at Glasgow.

(5) William Edmonstoune Aytoun, Professor of Rhetoric and Belles Lettres at Edinburgh University (*DNB*).

(6) See Number 10 note (5).

(7) John Wilson: see Number 28 note (8).

DRAFTS OF PAPER 'ON THE EQUILIBRIUM OF ELASTIC SOLIDS'[1]

AUTUMN 1849–SPRING 1850[2]

From the originals in the University Library, Cambridge[3]

[1][4]　　　ON THE EQUILIBRIUM OF ELASTIC SOLIDS[a][5]

There are few parts of mechanics in which theory has differed more from observed facts than in the theory of Elastic Solids. The mathematicians,[b] setting out from very plausible assumptions with respect to the constitution of bodies and the laws of molecular action, came to conclusions which were shewn to be erroneous by the observations of the[c] experimental philosophers.

The experiments of Œrsted[6] overturned[d] the mathematical theories of Navier, Poisson and Lamé and Clapeyron,[7] and apparently deprived this practically important branch of mechanics all assistance from mathematics.

(a) {Forbes} By Mr James Clerk Maxwell (Received 11 Dec 1849 JDF).

(b) {Forbes} Mathematicians.

(c) {Forbes} ⟨the⟩.

(d) {Forbes} ⟨overturned⟩ proved to be at variance with.[8]

(1) Published as: 'On the equilibrium of elastic solids', *Trans. Roy. Soc. Edinb.*, **20** (1850): 87–120 (= *Scientific Papers*, **1**: 30–73).

(2) The paper is dated 10 December 1849, and was submitted to J. D. Forbes for the Royal Society of Edinburgh, and marked 'Received 11 Dec 1849' by Forbes. The Appendix is dated 15 February 1850. The paper was read to the Royal Society of Edinburgh on18 February 1850 (see Number 27). In a letter to Campbell of 14 March 1850 (Number 28) Maxwell states that he had 'begun to write Elastic Equilibrium', presumably a revision, subsequently submitted to the Royal Society of Edinburgh; see a letter from Forbes to Maxwell of 4 May 1850 (Number 28 note (2)).

(3) ULC Add. MSS 7655, V, g/1, 2, 3. The original version of the paper (dated 10 December 1849) and its two addenda (V, g/2), and the Appendix (V, g/3), are annotated by Forbes; the manuscript is corrected by Maxwell. These annotations and corrections are recorded.

(4) In presenting a continuous text from an incomplete and corrected manuscript, the text is divided into sections; §7 is taken from the published paper, as the corresponding portion of the draft is missing from the manuscript (see note (88)).　　　(5) ULC Add. MSS 7655, V, g/2.

(6) Oersted summarised his experiments on the compressibility of water in a letter to William Whewell of 18 June 1833; see H. C. Oersted, 'On the compressibility of water', *Report of the Third Meeting of the British Association for the Advancement of Science; held at Cambridge in 1833* (London, 1834), part 2: 353–60.

(7) Claude Louis Marie Henri Navier, 'Mémoire sur les lois de l'équilibre et du mouvement des corps solides élastiques', *Mémoires de l'Académie Royale des Sciences de l'Institut de France*, **7** (1827): 375–93; Siméon Denis Poisson, 'Mémoire sur l'équilibre et le mouvement des corps élastiques', *ibid.*, **8** (1829): 357–570, 623–7; and Poisson, 'Mémoire sur les équations générales de l'équilibre et du mouvement des corps solides élastiques et des fluides', *Journal de l'École Polytechnique*, **13**, cahier 20 (1831): 1–174; Gabriel Lamé and Émile Clapeyron, 'Mémoire sur l'équilibre intérieur des corps solides homogènes', *Journal für die reine und angewandte Mathematik*, **7** (1831): 145–69, 237–52, 381–413.

The assumption on which these theories were founded may be stated thus –
'Solid bodies are composed of distinct molecules which are kept at a certain distance from each other by the opposing principles of attraction and heat. When the distance between two molecules is changed, they will attract or repel each other with a force equal to the change of distance multiplied into a function of the distance which must vanish when that distance becomes sensible, and this force acts in the line joining the two molecules.'[9]

I shall not here enquire how far this assumption is true. Perhaps it is not sufficient that the function vanishes when the distance becomes sensible, perhaps there may be some tangential force, or some polarity in the molecules. Whatever be the reason the results are not confirmed by experiment, and those who attempt to form a useful theory must leave the difficulty to be explained by the supporters of the doctrine of molecules.

In order to understand why these molecular theories will not explain the phenomena, let us consider the difference between a solid and a fluid. It requires no effort to retain a fluid in any form provided its volume is not altered, but where the form of a solid is changed, a force is excited tending to restore its former figure, and this constitutes the difference between elastic solids and fluids. Both tend to recover their volume but fluids do not tend to recover their shape.[e]

Now since there are in nature bodies which are in every intermediate state from perfect solidity to perfect fluidity, the two elastic powers cannot exist in every body in the same proportion, therefore the equations of elasticity must contain two coefficients,[10] one common to liquids and solids and the other peculiar to solids.

(e) {Forbes} *volume, shape.*

(8) In 'On the equilibrium of elastic solids': 87 (= *Scientific Papers*, **1**: 30) Maxwell adopts Forbes' modification. Compare also §5.

(9) In 'On the equilibrium of elastic solids': 87 (= *Scientific Papers*, **1**: 31) Maxwell states that this is 'the assumption of Navier'. Compare Navier, 'Mémoire sur les lois de l'équilibre et du mouvement des corps solides élastiques': 375–7; 'On regarde un corps solide élastique comme un assemblage de molécules matérielles placées às des distances extrêment petites. Ces molécules exercent les unes sur les autres deux actions opposées, savoir, une force propre d'attraction, et une force de répulsion due au principe de la chaleur....Nous regardons d'aileurs les actions moléculaires dont il s'agit comme ne subsistant qu'entre des molécules très-voisines, et comme ayant des valeurs qui décroissent très-rapidement, suivant une loi inconnue, pour des molécules de plus en plus éloignées l'une de l'autre.'

(10) Poisson and Navier assume an invariable ratio between the linear rigidity and the compressibility of elastic solids; see Navier, 'Mémoire sur les lois de l'équilibre et du mouvement des corps solides élastiques': 382–4; and Poisson, 'Mémoire sur l'équilibre et le mouvement des corps élastiques': 381. Augustin Louis Cauchy introduced two coefficients of elasticity in his memoir 'Sur les équations qui expriment les conditions d'équilibre ou les lois du mouvement

These coefficients will be called the moduli of cubical and linear elasticity,[11] and will be defined when we come to treat of them more particularly.

In the mean time before giving an account of the different theories it may be premised, that in the following pages, pressures will always be represented by the number of units of weight to the unit of surface, if in English measure in pounds to the square inch.

Compression will denote the change of any dimension which is the effect of pressure. It will be represented by the quotient of the change of dimension divided by the dimension compressed.[f]

Pressure will be understood to indicate tension, and compression dilatation, pressure and compression being reckoned positive.

Elasticity is the force which opposes pressure, and the equations of elasticity are those which express the relation of pressure to compression.[12]

(f) {Forbes} $\left(\text{as } \dfrac{\delta \alpha}{\alpha} \right)$.

intérieur d'un corps solide, élastique ou non élastique', *Exercices de Mathématiques*, **3** (Paris, 1828) : 160–87 (= *Oeuvres Complètes d'Augustin Cauchy*, ser. 2, **8** (Paris, 1890) : 190–226). Cauchy obtains general equations of elastic equilibrium (see note (93)) supposing two coefficients of elasticity, a measure of linear rigidity k and a coefficient of the compressibility of the volume of a body K. Referring to his equations of elastic equilibrium, he notes that: 'on suppose $k = 2K$, elles coincident avec elles que M. Navier a données pour déterminer l'équilibre ou le mouvement des corps élastique...et que M. Poisson a reproduite en les établissement d'une autre manière' (*Exercices de Mathématiques*, **3**: 182). See §3 and notes (43) and (44).

(11) Which Maxwell denotes by the symbols μ and m, respectively. Compare George Gabriel Stokes, 'On the theories of the internal friction of fluids in motion, and of the equilibrium and motion of elastic solids', *Trans. Camb. Phil. Soc.*, **8** (1845) : 287–319, esp. 288, 311, 313 (= *Papers*, **1**: 75–129). Stokes establishes equations for the equilibrium of elastic solids (see note (94)) which suppose two independent coefficients of elasticity, which he terms constants of 'cubical compressibility' B and 'extensibility' A. He notes that 'the equations at which Poisson arrives contain only *one* unknown constant k, whereas the equations...of this paper contain *two*'; and he observes that his own equations of elastic equilibrium 'reduce themselves to Poisson's equations for a particular relation between A and B...this relation is $A = 5B$'. Compare Maxwell's 'moduli of cubical and linear elasticity'; the terms 'compressibilité ou dilatabilité cubique' and 'compressibilité ou dilatabilité linéaire' had been used by Guillaume Wertheim, 'Mémoire sur l'équilibre des corps solides homogènes', *Ann. Chim. Phys.*, ser. 3, **23** (1848) : 52–95, on 75, to which Maxwell refers (§3). The relations between Maxwell's, Stokes' and Cauchy's coefficients are:

$$k = m, \qquad K = \mu - \tfrac{1}{3}m$$
$$A = 3\mu, \qquad B = \tfrac{1}{2}m.$$

See §3 and 'On the equilibrium of elastic solids': 89 (= *Scientific Papers*, **1**: 32–3). On the relation between Cauchy's and Stokes' assumptions see note (44).

(12) In rejecting any appeal to molecular forces, Maxwell follows the method of Stokes, 'On the theories of the internal friction of fluids in motion, and of the equilibrium and motion of elastic solids': 288–9, 310–11; see §§3 and 6.

(a)[2] [13] As the ⟨geometrical⟩ mathematical laws of compressions and pressures have been very thoroughly investigated, and as they are demonstrated with great elegance in the very complete and elaborate memoir of MM. Lamé and Clapeyron, I shall state as briefly as possible their results.

Let a solid be subjected to compression or pressure, then if through any given point a line be drawn whose length is proportional to the magnitude of the part of the compressions or pressures which may be resolved in the direction of the line, then the extremities of such lines will lie in the surface of an ellipsoid whose centre is the given point.

The properties of the system of compressions or pressures are analogous to those of the ellipsoid.

There are three diameters having perpendicular ordinates, these are called the principal axes of the ellipsoid. Similarly there are always three directions in compressed solids in which there is no tangential action or tendency of the parts to slide on each other. These directions are called the principal axes of compression or of pressure and the compression or pressure in any direction is equal to the sum of the compressions or pressures in each of the principal axes multiplied into the squares of the cosines of the angles which they respectively make with that direction.[14]

In order to connect pressure with compression I have assumed Hookes law of elasticity as established by experiment. Hookes law 'Ut tensio sic vis'[15] means that within the limits of perfect elasticity, compressions are proportional to pressure.

Now we have seen that the effect of pressure is twofold for it produces both change of form and change of volume let us consider its effect on a portion of an elastic solid the principal axes of compression and pressure being those of x, y and z and the pressures being $p_1 \, p_2$ and p_3 while the compressions are $\dfrac{\delta x}{x} \dfrac{\delta y}{y} \dfrac{\delta z}{z}$, and the change of volume $= \dfrac{\delta V}{V} = \dfrac{\delta x}{x} + \dfrac{\delta y}{y} + \dfrac{\delta z}{z}$.

(a) {Maxwell} To be inserted at Page 3. {Forbes}
[page] 5 or rather a note at the end of the paper.

(13) An addendum (ULC Add. MSS 7655, V, g/2). The first four paragraphs constitute (with minor emendations) 'Note A' appended to 'On the equilibrium of elastic solids': 119 (= *Scientific Papers*, **1**: 70–1).

(14) G. Lamé and É. Clapeyron, 'Mémoire sur l'équilibre intérieur des corps solides homogènes': 165–9, 237–48. See note (32).

(15) Robert Hooke, *Lectures de Potentia Restituva, or of Spring Explaining the Power of Springing Bodies* (London, 1678): 1; '*Ut tensio sic vis*; That is, The Power of any Spring is in the same proportion with the Tension thereof.' Compare Thomas Young, *A Course of Lectures on Natural Philosophy and the Mechanical Arts*, 2 vols. (London, 1807), **2**: 46; 'A substance perfectly elastic is initially extended and compressed in equal degrees by equal forces, and proportionally by proportional forces.'

Let us first find the effect of the pressure p_1. The first part of its effect is to produce a compression $\dfrac{\delta_1 x}{x}$ in the axis of x and by Hookes law

$$\frac{\delta_1 x}{x} = p_1 \frac{1}{a}.$$

The second part of its effect is to produce a diminution of volume $\dfrac{\delta_1 V}{V}^{(b)}$

and

$$\frac{\delta_1 V}{V} = p_1 \frac{1}{b} = \frac{\delta_1 x}{x} + \frac{\delta_1 y}{y} + \frac{\delta_1 z}{z}$$

from these equations we get

$$\frac{\delta_1 y}{y} + \frac{\delta_1 z}{z} = p_1\left(\frac{1}{b} - \frac{1}{a}\right)$$

but since

$$\frac{\delta_1 y}{y} = \frac{\delta_1 z}{z} = p_1 \tfrac{1}{2}\left(\frac{1}{b} - \frac{1}{a}\right).$$

Adding the parts of $\dfrac{\delta x}{x}$ produced by the three pressures

$$\frac{\delta x}{x} = \frac{\delta_1 x}{x} + \frac{\delta_2 x}{x} + \frac{\delta_3 x}{x} = p_1 \frac{1}{a} + p_2 \tfrac{1}{2}\left(\frac{1}{b} - \frac{1}{a}\right) + p_3 \tfrac{1}{2}\left(\frac{1}{b} - \frac{1}{a}\right)$$

$$^{(c)}\frac{\delta x}{x} = p_1 \frac{1}{ax} + \frac{p_2 + p_3}{2}\left(\frac{1}{b} - \frac{1}{a}\right)$$

$$\frac{\delta y}{y} = p_2 \frac{1}{ax} + \frac{p_3 + p_1}{2}\left(\frac{1}{b} - \frac{1}{a}\right)$$

$$\frac{\delta z}{z} = p_3 \frac{1}{ax} + \frac{p_1 + p_2}{2}\left(\frac{1}{b} - \frac{1}{a}\right).$$

From these equations we find

$$\frac{\delta V}{V} = \frac{\delta x}{x} + \frac{\delta y}{y} + \frac{\delta z}{z} = (p_1 + p_2 + p_3)\frac{1}{b}$$

$$\frac{\delta x}{x} - \frac{\delta y}{y} = (p_1 - p_2)\left(\tfrac{3}{2}\frac{1}{a} - \tfrac{1}{2}\frac{1}{b}\right).$$

Or as I have stated it elsewhere$^{(d)(17)}$
The sum of the Compressions is proportional to the sum of the Pressures, and
The difference of the Compressions in two directions is proportional to the difference of the Pressures in those directions.

(b) {Forbes} Hooke's law says nothing about Volume – This is a new assumption.$^{(16)}$

(c) {Forbes} The printer cannot read these.

(d) {Forbes} (refer).

(16) Maxwell generalises Hooke's law; compare his discussion of Hooke's law in §7.

(17) See §7.

It is convenient to substitute $\dfrac{1}{3\mu}$ for $\dfrac{1}{b}$ and $\dfrac{1}{m}$ for $\left(\dfrac{3}{2}\dfrac{1}{a} - \dfrac{1}{2}\dfrac{1}{b}\right)$ and thus to distinguish cubical from linear elasticity.

In the following account of the theories of Elastic Solids I have stated the different ratios which have been supposed to exist between these quantities.[18]

By confining ourselves to the consideration of normal pressures we may dispense altogether with tangential forces which as they can always be resolved into normal pressures need not be introduced at all.

A tangential force is equivalent to two pressures of contrary sign inclined 45° to the direction of the tangential force.

The effect of this force is a displacement of shifting the best example of this force is in a twisted wire hence it is sometimes called a force of torsion.

It is observed that unequal pressures communicate to transparent elastic solids the property of double refraction.[19]

Let I be the difference of retardation of the oppositely polarised rays for a thickness z when a pressure p_1 acts in the axis of x and p_2 in the axis of y then if $(p_1 - p_2)z\dfrac{1}{I} = \omega$ the quantity ω will indicate the relation of pressure to refractive power for the substance. This quantity is probably related to m perhaps $\omega^2 = m$.[20]

This quantity I is sometimes called the intensity of the action on light.

[3] The mathematicians who have treated of elasticity are MM. Navier, Poisson, Mdlle Sophie Germain Lamé and Clapeyron, Cauchy Stokes, and Wertheim.

The investigations of Navier are to be found in *Tom VII* of the *M*emoirs of the Institute, Page 373. It was read 14 May 1821.[21] He deduces the equations

---

(18) On Maxwell's coefficients of cubical elasticity μ and linear elasticity m see note (11).

(19) David Brewster, 'On the effects of simple pressure in producing that species of crystallization which forms two oppositely polarized images, and exhibits the complementary colours of polarized light', *Phil. Trans.*, **105** (1815): 60–4; Brewster, 'On the communication of the structure of doubly-retracting crystals to glass, muriate of soda, fluor spar and other substances, by mechanical compression and dilatation', *ibid.*, **106** (1816): 156–78, esp. 160–1. See note (78).

(20) See 'Note C' of 'On the equilibrium of elastic solids': 120 (= *Scientific Papers*, **1**: 72).

(21) Navier, 'Mémoire sur les lois de l'équilibre et du mouvement des corps solides élastiques', cited in note (7). See also Poisson, 'Mémoire sur l'équilibre et le mouvement des corps élastiques'; Sophie Germain, *Recherches sur la Théorie des Surfaces Élastiques* (Paris, 1821); Lamé and Clapeyron, 'Mémoire sur l'équilibre intérieur des corps solides homogènes'; Cauchy, 'Sur les équations qui expriment les conditions d'équilibre ou les lois du mouvement interieur d'un corps solide, élastique ou non élastique'; Stokes, 'On the theories of the internal friction of fluids in motion, and of the equilibrium and motion of elastic solids'; Wertheim, 'Mémoire sur l'équilibre des corps solides homogènes'; see notes (7), (10) and (11).

from the common assumption, and therefore has but one coefficient making $\mu = \frac{5}{6}m$.[22] He does not solve any particular cases.

There is also a memoir by Navier on elastic laminæ in the *Annales de Chimie et de Physique*, *2ᵉ série*, *XV.264*. This however and *Tom.I* of *L'Application de la Mécanique*, I have not been able to procure.[23]

Poisson, *Mém de l'Institut* VIII.429. 14 April 1828.

This memoir is denoted principally to the calculation of the nodal lines in vibrating plates. An abstract of it is given in *Ann de Ch. 2ᵉ série* XXXVII.337 and a communication to Œrsted in [*Ann de Ch. 2ᵉ série*] XXXVIII.330 and the results for rods in [*Ann de Ch. 2ᵉ série*] XXXVI.334[24] with an account of the experiments of Cagniard-Latour and Savart.[25]

Mr Stokes refers to a memoir by Poisson in the twentieth *Cahier* of the *Journal de l'Ecole Polytechnique*, entitled *Mémoire sur les equations générales de l'Equilibre et du Mouvement des Corps solides élastiques et des Fluides*.[26] Mr Stokes then gives an account of his theory. 'Poisson supposes the molecules to act on each other with forces of which the main part is a force in the direction of the line joining the centres of gravity, and varying as some function of the distance of these points, and the remainder a secondary force, or it may be two secondary forces

---

(22) In his paper 'On the theories of the internal friction of fluids in motion, and of the equilibrium and motion of elastic solids': 313, Stokes states that his own equations and Poisson's (and by implication Navier's) equations would be equivalent, when the relation between Stokes' coefficients of elasticity is $A = 5B$. As $A = 3\mu$ and $B = \frac{1}{2}m$, then $\mu = \frac{5}{6}\mu$. See note (11).

(23) 'Rapport fait à l'Académie des Sciences le lundi 4 Septembre 1820, sur un Mémoire de M. Navier qui traite de la flexion des lames élastiques', *Ann. Chim. Phys.*, ser. 2, **15** (1820): 264–79; C. L. M. H. Navier, *Résumé des Leçons données à l'École des Ponts et Chaussées sur l'Application de la Mécanique à l'établissement des Constructions et des Machines. Première Partie* (Paris, 1826).

(24) S. D. Poisson, 'Mémoire sur l'équilibre et le mouvement des corps élastiques', cited in note (7); S. D. Poisson, 'Mémoire sur l'équilibre et le mouvement des corps élastiques', *Ann. Chim. Phys.*, ser. 2, **37** (1828): 337–55; Poisson, 'Note sur la compression d'une sphère', *ibid.*, **38** (1828): 330–5; Poisson, 'Note sur l'extension des fils et des plaques élastiques', *ibid.*, **36** (1827): 384–7.

(25) Charles Cagniard-Latour, 'On the elasticity and change of volume of metallic strings while in a state of vibration', *Edinburgh Journal of Science*, **8** (1828): 201–3. Cagniard-Latour's experiments on the extension of wires were also described by G. Wertheim, 'Mémoire sur l'équilibre des corps solides homogènes': 52–4, to which Maxwell refers in §3. In his 'Note sur l'extension des fils et des plaques élastiques' Poisson made no reference to work by Félix Savart, but Maxwell may have had in mind Savart's 'Mémoire sur la réaction de torsion des lames et des verges rigides', *Ann. Chim. Phys.*, ser. 2, **41** (1829): 373–97, where he claimed experimental support for Poisson's theory of the torsion of cylindrical bars. Compare Maxwell's experiments on the torsion of rods; see §14, and Number 24.

(26) Poisson's memoir is cited in note (7).

depending on the molecules not being mathematical points. He supposes that it is on these secondary forces that the solidity of solid bodies depends.'[27]

The equations at which Poisson arrives contain only one unknown constant K and he finds $A = 5B$, or $\mu = \frac{5}{6}m$.[28]

In Crelle's Mathematical Journal vol. VII there is a memoir by Lamé and Clapeyron, entitled *Mémoire sur l'équilibre interieur des corps solides homogènes*.

It is accompanied with a report by MM. Poinsot and Navier 29 Sept. 1828.[29]

This report after noticing the want of generality and exactness in those parts of the works of Leibnitz Jac Bernoulli Euler and Lagrange,[30] explains the nature of the researches which the authors had undertaken, the first part consisting of the general equations and the second of particular solutions, and remarks that although the authors had pursued their investigations with success, they had not mentioned the previous researches of Navier, though their methods and results were nearly the same.[31]

They then describe the method of representing pressures and compressions by the radii of an ellipsoid, and refer to the works of Cauchy and Fresnel for the complete investigation of functions of this form.[a][32]

(a) {Forbes} (See note at the end of this paper).

(27) Stokes, 'On the theories of the internal friction of fluids in motion, and of the equilibrium and motion of elastic solids': 312.

(28) See note (22).

(29) Lamé and Clapeyron, 'Mémoire sur l'équilibre intérieur des corps solides homogènes' (cited in note (7)), prefaced by a 'Rapport fair par MM. Poinsot et Navier á l'académie des sciences de Paris dans la séance du 29 Septembre 1828': 145–9, apparently written by Navier.

(30) Lamé and Clapeyron make no reference to any earlier work on elastic solids. Navier presumably had in mind: Gottfried Wilhelm Leibniz, 'Demonstrationes novae de resistentia solidorum', *Acta Eruditorum*, (July 1684): 319–25; Jacob Bernoulli, 'Curvatura laminae elasticae', *Acta Eruditorum*, (June 1694): 262–76; Leonhard Euler, 'Genuina principia doctrinae de statu aequilibrii et motu corporum tam perfecte flexibilium quam elasticorum', *Novi Commentarii Academiae Scientiarum Imperialis Petropolitanae*, **15** (1770): 381–413; and Joseph Louis Lagrange, 'Sur la force des ressorts pliés', *Mémoires de l'Académie des Sciences de Berlin*, **25** (1769): 167–203. See C. Truesdell, *The Rational Mechanics of Flexible or Elastic Bodies 1638–1788*, in *Leonhardi Euleri Opera Omnia*, ser. 2, **11** part 2 (Zurich, 1960).

(31) Poinsot and Navier, 'Rapport': 146, referring to Navier's 'Mémoire sur les lois de l'équilibre et du mouvement des corps solides élastiques'.

(32) Navier refers to the second volume of Cauchy's *Exercices de Mathématiques*; see A. L. Cauchy, 'De la pression ou tension dans un corps solide', and 'Sur la condensation et la dilatation des corps solides', *Exercices de Mathématiques*, **2** (Paris, 1827): 42–59, 60–9 (= *Oeuvres Complètes d'Augustin Cauchy*, ser. 2, **7** (Paris, 1889): 60–93). Cauchy developed a theory of elasticity in which strain is expressed in terms of three axes of pressure at right angles to each other which correspond to the axes of an ellipsoid (*Exercices de Mathématiques*, **2**: 52). While Lamé and Clapeyron make no reference to Cauchy, they suppose that the strain of isotropic substances can be represented by 'un ellipsoide ayant pour demi-axes les trois pressions normales principales'; see their 'Mémoire sur

The memoir itself contains four sections, differential equations, Theorems relative to pressure, particular cases and general cases.

In the first section we find the equations of elasticity deduced from the assumption that when two molecules are separated from each other by a very small quantity Δz they begin to attract each other with a force equal to the product of Δz into a function of their original distance z this function vanishing when z becomes sensible.

As the equations derived from this principle contain only one coefficient they will be true only in one particular case where $\mu = \frac{5}{6}m$.[33]

In the remainder of this section which treats of the equations of compression and in the theorems relative to pressures, the authors show that pressures and compressions may be represented[b] in magnitude and direction by the radii of an ellipsoid.

In the third section which treats of simple cases the first is that of an infinitely long prism subjected to a strain in the direction of its length.

The second relates to a body exposed to fluid pressure.

The third is Case IV of this paper.[34] In this case the authors find that for a given substance and a given interior diameter there is a limit to the pressures which the interior surface can sustain, and that if the pressure be increased the cylinder will be ruptured, however thick it may be. The numerical result does not seem to be confirmed by experiment, for examples are given in the same memoir in which the pressure was increased beyond the calculated limit.[35] It is somewhat difficult to form a sufficiently exact idea of the nature of rupture and the circumstances under which it takes place, without a practical acquaintance with the subject which is not commonly possessed by those who undertake to determine its laws.

The fourth case is Case II of this paper.[36] From a comparison of the values

---

(b) {Forbes} in what instances?.

l'équilibre intérieur des corps solides homogènes: 239, and §2. Navier makes no direct reference to Fresnel, but see A. J. Fresnel, 'Mémoire sur la double réfraction', *Mémoires de l'Académie des Sciences*, **7** (1827): 45–176. To explain double refraction in biaxial crystals Fresnel represents the velocities of propagation of the ordinary and extraordinary rays by the intersection of the wave surface with the semi-axes of an ellipsoid.

(33) Lamé and Clapeyron, 'Mémoire sur l'équilibre intérieur des corps solides homogènes': 153–4; see note (22).

(34) Lamé and Clapeyron, 'Mémoire sur l'équilibre intérieur des corps solides homogènes': 382–9 (a hollow cylinder).

(35) Jacob Perkins, 'On the progressive compression of water by high degrees of force, with some trials of its effect on other fluids', *Phil. Trans.*, **116** (1826): 541–7.

(36) Lamé and Clapeyron, 'Mémoire sur l'équilibre intérieur des corps solides homogènes': 389–93 (a cylinder subjected to torsion).

of Lamé's coefficient A given by extension and torsion, Mr Stokes has found that $\frac{\mu}{m} = \frac{16}{3}$ for Iron and $= 8.8$ for Brass.[37]

The fifth case is that of a sphere all whose points attract each other with a force inversely proportional to the square of their distance. Although this case is a beautiful example of the methods of these ingenious mathematicians, it can never actually occur for the particles cannot be formed into a solid sphere before they begin to ⟨attract⟩ gravitate.

The sixth is Case V of this paper.[38] In this case also there is a limit to the interior pressure even in an infinite solid.

The fourth section concerns the general cases.

The first case relates to an indefinite plane exposed to indeterminate pressures. The authors arrive at this remarkable theorem[39] –

When a normal pressure is applied at a point of the plane surface of an indefinitely extended solid, the part of the solid which lies next to the surface at that point is compressed by a quantity equal to the quotient of the pressure applied referred to unity of surface divided by twice the coefficient A.

The second case is that of a solid contained between two parallel planes and acted on by any forces. The equations do not lead to any definite results.

The third is the general case of an indefinite cylinder acted on by any forces.

In these general cases the equations cannot be brought to be of practical utility without a knowledge of a particular species of definite integrals which had not as yet been treated of by mathematicians, but MM. Lamé and Clapeyron propose to treat of these problems in a forthcoming work.

In Annales de Chimie et de Physique 3ᵉ Série XXIII.74 M. Wertheim mentions the more general hypotheses of Cauchy[40] in these words.

'M. Cauchy a fait voir que l'on peut remplacer l'hypothèse fondamentale de Navier par une hypothèse plus générale : au lieu de supposer chaque traction ou pression principale proportionelle à la dilatation ou compression linéaire dans le sens de la force, on peut supposer qu'elle se compose de deux parties dont l'une est proportionelle au changement linéaire, tandis que l'autre est proportionelle à la dilatation ou condensation du volume.'[41]

---

(37) Stokes, 'On the theories of the internal friction of fluids in motion, and of the equilibrium and motion of elastic solids': 316.

(38) Lamé and Clapeyron, 'Mémoire sur l'équilibre intérieur des corps solides homogènes': 396–400 (a hollow sphere).

(39) Thus described by Lamé and Clapeyron, 'Mémoire sur l'équilibre intérieur des corps solides homogènes': 403.

(40) See note (10).

(41) G. Wertheim, 'Mémoire sur l'équilibre des corps solides homogènes': 74; see note (10).

From the examples given by M. Wertheim Cauchy's two coefficients appear to be
$$k = m \quad K = \mu - \tfrac{1}{3}m. \text{[42]}$$

This is the only statement of M. Cauchy's hypothesis which I have met with and the date of his investigations is not given.[43] Wertheim's Memoir was read 10th Feb 1848.

The first theory of elastic solids in which two coefficients are introduced seems to be that of Mr G. G. Stokes read to the Cambridge Philosophical Society April 14 1845, and published in vol. VIII Part 3 of their Transactions.[44]

He states his general principles thus –

'The capability which solids possess of being put into a state of isochronous vibration shows that the pressures called into action by small displacements depend on homogeneous functions of those displacements of one dimension. I shall suppose moreover, according to the general principle of the superposition of small quantities, that the pressures due to different displacements are superimposed, and consequently that the pressures are linear functions of the displacements. Since squares of $\alpha\,\beta\,\gamma$ (the displacements) are neglected, these pressures may be referred to a unit of surface in the natural state or after compression indifferently, and a pressure which is normal to any surface after displacement may be regarded as normal to the original position of that surface.

Let $-A\delta$ be the pressure corresponding to a uniform linear dilatation δ when the solid is in equilibrium, and suppose that it becomes $-mA\delta$ in consequence of the heat developed when the solid is in a state of rapid vibration. Suppose also that a displacement of shifting parallel to the plane xy for which $\alpha = kx$,

---

(42) Wertheim, 'Mémoire sur l'équilibre des corps solides homogènes': 74–5; see note (11).

(43) Maxwell had not read Cauchy's *Exercices de Mathématiques* when writing this draft. But see note (44).

(44) Stokes, 'On the theories of the internal friction of fluids in motion, and of the equilibrium and motion of elastic solids': 288, where he acknowledges that Cauchy had obtained similar equations for the equilibrium of elastic solids; as recognised by Maxwell in 'On the equilibrium of elastic solids': 88–9, 95 (= *Scientific Papers*, **1**: 32, 40); see §7, and notes (10), (11), (93) and (94). Unlike Cauchy, Stokes makes no reference to a molecular model; because no force law is assumed between the molecules, the two arbitrary constants cannot be reduced to one. In a 'Remark on the theory of homogeneous elastic solids', *Camb. & Dubl. Math. J.*, **3** (1848): 130–1, Stokes observes that in his paper 'The equations of equilibrium or motion of a homogeneous elastic solid are given with the two arbitrary independent constants which they must contain…'. He continues: 'there is no other work, of which I am aware, in which these equations are *insisted on* as being those which it is absolutely necessary to adopt; but the equations have been *obtained* by M. Cauchy…'; and notes that Cauchy allowed 'the reduction of the two constants to one'.

$\beta = ky$, $\gamma = 0$ calls into action a pressure $-Bk$ on a plane perpendicular to the axis of x and a pressure Bk on a plane perpendicular to the axis of y; the pressures on these planes being equal and of opposite signs, that on a plane perpendicular to the axis of z being zero and the tangential forces on those planes being zero. It may also be shewn that it is necessary to suppose B positive in order that the equilibrium of the solid medium may be stable, and it is easy to see that the same must be the case with A for the same reason.'[45]

Thus the coefficients are

$$A = 3\mu, \quad B = \frac{m}{2}.^{[46]}$$

Mr Stokes does not enter into the solution of his equations but gives the results in some simple cases. These cases are

1   A body exposed to a uniform pressure on all sides.
2   A rod extended in the direction of its length.
3   A cylinder twisted by a couple as in Case 2.

He then points out the method of finding the coefficients A and B from the two last cases.

While explaining why the equations of motion of the luminiferous ether are the same as those of elastic solids, Mr Stokes considers the property called *plasticity* or the tendency which a compressed body has to relieve itself from a state of constraint by its molecules assuming new positions of equilibrium. This property is opposed to linear elasticity and these two properties exist in all bodies but in variable ratio.[47]

M. Wertheim in the same memoir in which he mentions the hypothesis of Cauchy has given the results of some experiments on caoutchouc from which he finds that

$$K = k \text{ or } \mu = \tfrac{4}{3}m$$

and concludes that $K = k$ for all substances. In his equations μ is accordingly made equal to $\tfrac{4}{3}m$.[48]

---

(45)   Stokes, 'On the theories of the internal friction of fluids in motion, and of the equilibrium and motion of elastic solids': 311.

(46)   See note (11).

(47)   Stokes, 'On the theories of the internal friction of fluids in motion, and of the equilibrium and motion of elastic solids': 315–19.

(48)   Wertheim, 'Mémoire sur l'équilibre des corps solides homogènes': 75; and see also 'On the equilibrium of elastic solids': 89 (= *Scientific Papers*, **1**: 33). In a later paper 'Sur la double réfraction temporairement produite dans les corps isotropes, et sur la relation entre l'élasticité mécanique et entre l'élasticité optique', *Ann. Chim. Phys.*, ser. 3, **40** (1854): 154–221, on 206n, Wertheim protested 'contre la manière inexacte dont M. Maxwell a rapporté mes expériences'; Maxwell had not recorded that these results were 'designée comme préliminaires'.

For accounts of experimental researches on the value of the coefficients see

Œrsted   Mem Royal Society of Copenhagen IV
                           Annales de Chimie 2ᵉ série XXI, XXII, XXXVIII
Colladon and Sturm   —————————————— XXXVI
Aimé   3ᵉ série VIII.[49]

A communication from Œrsted is given in Transactions of the British association for 1833.[50]

M. Regnault, in his researches on the theory of the steam engine, has given an account of the experiments which he made in order to determine with accuracy the compressibility of mercury. He considers the mathematical formulæ very uncertain because the theories of the molecular forces from which they are derived are probably far from the truth. But even were these equations true, there would be much uncertainty in the coefficients obtained by the ordinary method of measuring the elongation of a wire for it is difficult to be certain that the wire is of the same material as the piezometer.

He therefore uses the piezometer only, taking three observations in one of which the interior and exterior pressures are equal while in the other two the pressure acts on the interior or exterior alone.[51]

It will be afterwards shown that by experiments with a sphere only two unknown quantities can be determined, so that without experiments on the substance in another form the two coefficients for the material of the piezometer and the coefficient for the liquid cannot be found; but as M. Regnault has adopted the equations of Lamé where μ is given equal to $\frac{5}{6}m$ he finds no difficulty in determining the coefficients thus reduced to two.[52]

The calculations now laid before the Society were undertaken at first with the desire of explaining the beautiful phenomena of heated and unannealed

---

(49) H. C. Oersted, 'Instrument pour mesurer la compression de l'eau', *Ann. Chim. Phys.*, ser. 2, **21** (1822): 99–100; Oersted, 'Sur la compressibilité de l'eau', *ibid.*, **22** (1823): 192–8; Oersted, 'Sur la compression de l'eau dans des vases de matières différentes', *ibid.*, **38** (1828): 326–9; D. Colladon and J. C. F. Sturm, 'Mémoire sur la compression des liquides', *ibid.*, **36** (1827): 113–59; G. Aimé, 'Mémoire sur la compression des liquides', *ibid.*, ser. 3, **8** (1843): 257–80. Oersted did not publish the paper cited as published in the *Memoirs of the Royal Society of Copenhagen*, but made reference to it in his papers 'Sur la compression de l'eau dans des vases de matières différentes': 329, and 'On the compressibility of water': 353.

(50) H. C. Oersted, 'On the compressibility of water'; see note (6).

(51) Henri Victor Regnault, 'De la compressibilité des liquides, et en particulier de celle du mercure', *Mémoires de l'Académie Royale des Sciences de l'Institut de France*, **21** (1847): 429–64. The piezometer, first constructed by Jacob Perkins, is an instrument for measuring the compressibility of liquids under varying pressures; see Perkins, 'On the compressibility of water', *Phil. Trans.*, **110** (1820): 324–30. For Regnault's discussion of his apparatus, see his paper 'De la compressibilité des liquides': 429–30; and see Number 24 note (6).   (52) See §§ 5, 13; and Number 24.

glass, but it soon became evident that as the equilibrium of unannealed glass depends on the laws of the cooling and solidification of a heated body combined with the laws of equilibrium of elastic solids, the calculation of the state of the body would be so difficult that the action and magnitude of the compressing forces must in most cases be determined by actual experiment. But though results have not yet been obtained in the case of angular solids acted on by heat there are many cases in which the equations are very simple and of easy solution. Some of these have been selected as affording the means of determining by experiment the coefficients which occur in the equations of equilibrium.

[4] In cases I, II and III a cylinder is subjected to forces causing displacements of shifting.[a][53]

In Case I this shifting is in a plane perpendicular to the axis.

In Case II the plane of shifting is parallel to the axis, and passes through the tangent.

In Case III it passes through the axis and the radius. It is easy to verify Case I by a simple experiment.

Let an elastic transparent jelly be poured between two consecutive cylinders. When it is cold let one of the cylinders be twisted round its axis while the other is fixed. The effects of the unequal pressures may be observed by transmitting polarized light through the jelly, where rings will be seen whose tint is proportional to the inverse square of the distance from the centre, traversed by a black cross the arms of which are inclined 45° to the plane of primitive polarization.[54] ⌞Case III may be verified in the same way.⌟

In Case IV the solid is a hollow cylinder bounded by two plane surfaces. These plane surfaces and the interior and exterior cylindric surfaces sustain different pressures which are given, and the internal pressures are deduced from them.

Case V relates to a hollow sphere, and Case VI to a beam bent in a uniform manner.

The expression found in Case VII may be of use in determining the effect of pressure applied on one surface of an elastic plate.

Case VIII relates to the twisting of a flat bar.

In Case IX the solid is acted on by centrifugal force, and in Case X by heat. Case XI is an inverse problem of the same kind.

(a) {Forbes} (What is shifting?).

(53) A term derived from Stokes, 'On the theories of the internal friction of fluids in motion, and of the equilibrium and motion of elastic solids': 311, 314, denoting elastic constraint in a plane. See the passages Maxwell quotes from Stokes.   (54) See Number 19; and Case I, §9.

Case XII exemplifies the principle of the superposition of compressions and in Case XIII the principle is applied to the determination of the flexure of beams in a few simple cases.[55] The theory of beams is generally more fully developed in works on Mechanics.[56]

[5] [a][57] I have first deduced from the principles of ⟨solid⟩ geometry of three dimensions the mathematical expressions for compressions. Compression is represented by the differential coefficient of the expression for the displacement of a point. The compressions in any axes x, y, z, are found in terms of the compressions in the principle axes α, β, γ.

I next find three equations which must be satisfied for every point in order that it may remain in equilibrium.

The equations between the pressures which act on a point and the compressions which they produce are found from Hookes law of elasticity first

---

(a) {Maxwell} At Page 12.

(55) In revising the draft by adding the addendum (§5), Maxwell changed the numbering of some of these special cases. The table indicates the differences in numbering in the draft, the revisions following the addendum (§5), and further changes in the published paper.

Draft	Revise	Published paper[d]
I	I	I
II	II	II
III	III ⎫	III
IV	IV ⎭	
V	V	IV
VI[a]	VI[a]	V
VII	VII	VI
VIII	deleted	—
IX	VIII	VIII
X	IX	IX
XI	X	X
XII	[XIII][b]	XIII[b]
XIII	XI	XII
[XIV][c]	[XIV]	XIV[c]

(a) Case VI is missing from the manuscript.
(b) There is discussion of this example in §5.
(c) This case is discussed in §6.
(d) Cases VII and XI of the published paper were added to its text.

(56) See for example: Henry Moseley, *Mechanical Principles of Engineering and Architecture* (London, 1843): 486–585.

(57) An addendum (ULC Add. MSS 7655, V, g/2) intended to replace §4.

with respect to the principal axes α, β, γ, and second with respect to any axes x, y, z.[58]

By finding the pressures in terms of the compressions and substituting their values in the equations of equilibrium we arrive at the general equations of elastic solids in the form in which they are given by Stokes, and by making $\mu = \frac{5}{6}m$ they become those of Navier Poisson and Lamé and Clapeyron.[59]

These general equations are not employed in the solution of particular cases for it is better to find the equations of equilibrium and of elasticity separately and to employ polar coordinates when the case requires them.

The first four cases relate to the cylinder and as the pressures are applied equally on all sides of the axis we need only consider the distance of a point from the axes and ⟨at a distance⟩ from the extremity of the cylinder.

The first case relates to an elastic solid bounded by two concentric cylindric surfaces and two parallel planes perpendicular to the axis. One of these cylindric surfaces is made to turn round its axis through a small angle while the other is fixed. This may be easily done by using as the elastic solid a transparent jelly poured while hot between two concentric cylinders. When any of these cylinders is twisted round its axis it causes a displacement of shifting which may be seen directly by the motion of the jelly. The effect of the torsion is best seen by transmitting polarized light through the jelly in the direction of the axis, rings are seen whose colours are those of thin plates whose thickness is inversely proportional to the square of the radius of the ring. These are intersected by two dark bands inclined at an angle of 45° to the plane of polarization.

In the second case a long cylinder is supposed to be held fast at one end while it is twisted at the other and a relation is found between the dimensions of the cylinder the force applied and the angle of rotation.

From this relation we may find the coefficient of linear elasticity by twisting a cylindric rod with a known force and measuring the difference of the angles of rotation at two points whose distance is known.[60]

The third case is when the interior cylinder in Case I instead of being twisted is pressed in the direction of the axis.

Cases IV and V refer to the compression of a liquid contained in a cylindric or spherical vessel.

The only known method of finding the compressibility of water and other liquids is by confining them in a vessel with a narrow neck then applying pressure and observing the change of level of the liquid in the neck. This gives the difference between the change of the volume of the liquid and the change of internal capacity of the vessel.

Now since the substance of which the vessel is made is compressible we cannot be certain that the interior capacity will not change.

If the pressure be applied only to the contained water it is evident that the vessel will be distended and the compressibility will appear too great.

The pressure therefore is commonly applied externally and internally at the same time, the result however is still affected by the compressibility of the vessel.[61]

As the value of the compressibility of water is an important quantity and as Oersted had invented an apparatus by which pressure might be conveniently applied all those who have developed theories of elastic solids have investigated the problem of a hollow sphere or cylinder exposed to normal interior and exterior pressures.

The results of Poisson were communicated to Oersted who applied them to the case in which the vessel is made of lead using the coefficient for the extension of a rod of lead as found by Tredgold and finding the cubical elasticity by the formula of Poisson which supposes $\mu = \frac{5}{6}m$. Thus Oersted found that by the theory of Poisson the apparent compressibility of water in a lead vessel would be negative but the observed compressibility was positive as it seems that the ratio of μ to m for lead is nearly four times greater than that assumed by Poisson.[62]

A similar experiment was made by Professor Forbes who used a vessel of caoutchouc, and as the cubical compressibility of caoutchouc is nearly equal to that of water the apparent compressibility vanishes.[63]

Some who reject the mathematical theories as unsatisfactory, have conjectured that if the vessel be sufficiently thin the pressure on both sides being equal its compressibility will not affect the result.[64]

It will be afterwards shown that the apparent compressibility depends on

---

(61) See §3 and the experiments cited in note (49).

(62) Oersted, 'On the compressibility of water': 358.

(63) Compare a letter from William Thomson to G. G. Stokes of 5 February 1848 (ULC Add. MSS 7656, K 24): 'Professor Forbes of Edinburgh, who is at present on a visit with us, has just...told me of an experiment he made a long time ago, & exhibited to his class 10 years ago (without publishing it in any other manner). In Œrsted's exper^t for the compressibility of water, he substituted india rubber for glass, as the bottle which, along with the water in the external vessel, is compressed; & he found that no water is forced into the bottle of india rubber, nor expelled from it, by the compression. This shows that the compressibility of india rubber differs from that of water by an amount inappreciable in an experiment capable of rendering very sensible the difference betw. the compressibility of water & glass....' For his published results on the compressibility of water, see J. D. Forbes, 'Note respecting the application of the compressibility of water to practical purposes', *Edinburgh New Philosophical Journal*, **19** (1835): 36–9.

(64) Regnault, 'De la compressibilité des liquides, et en particulier de celle du mercure': 456–7.

the compressibility of the vessel and is independent of the thickness[b] when the pressures are equal.[65]

In cases IV and V I have found the internal pressures and compressions at any point of a hollow cylinder or sphere the particular cases are

1   When all the pressures on the free surface are equal the result is then independent of m.[66]
2   When the result is independent of μ.[c][67]
3   When the cylinder or sphere is solid.
4   When its radius is rendered invariable by being confined in a strong vessel.
5   When it becomes a cylindric or spherical hollow in an infinite solid.[68]

In the sixth case I have found the internal pressures in a rectangular beam bent so as to have a uniform curvature. As the values of these pressures are not sufficiently simple I have neglected the pressures perpendicular to the length of the beam which are very small when the radius of curvature is large compared with the depth of the beam. The force with which the beam tends to recover its form is then obtained by integrating the effect of the longitudinal pressures and the ratio of this force to the curvature produced or the stiffness of the beam is found for beams having different transverse sections.[69]

In Case VII a thin plate is rendered hollow by the action of a uniform pressure acting on one of its surfaces and a relation is found between this pressure and the curvature produced. This case is exemplified in the concave mirrors made by[d]   who exhausted the air from behind a plate of glass so as to form a burning mirror.[70]

Case VIII is an example of an elastic solid in which variable forces act on the parts of the solid. A cylinder is made to revolve round its axis and its parts are thus acted on by the centrifugal force as in the case of a grindstone.

In Case IX heat is taken into account. The interior surface of a hollow cylinder is kept at a constant temperature different from that of the exterior

---

(b) {Forbes} *compressibility, independent of the thickness.*   (d) {Forbes} M[r] James Naysmith.
(c) {Forbes} (When is that?).

(65) See Case IV, §8; and 'On the equilibrium of elastic solids': 101, 105 (= *Scientific Papers*, **1**: 48, 52–3).   (66) See note (64).

(67) Missing from the MS (§8); but see 'On the equilibrium of elastic solids': 101, 105 (= *Scientific Papers*, **1**: 48, 53). See Number 24.

(68) See Case IV, §8; and 'On the equilibrium of elastic solids': 101–2, 106–7 (= *Scientific Papers*, **1**: 48–9, 55).

(69) Missing from the MS; but see 'On the equilibrium of elastic solids': 107–8 (= *Scientific Papers*, **1**: 55–7).

(70) James Nasmyth, 'On the bending of silvered plate glass with mirrors', *Report of the Ninth Meeting of the British Association for the Advancement of Science; held at Birmingham in 1839* (London, 1840), part 2: 7.

surface till the flow of heat becomes uniform as in the case of a pipe having hot water inside and cold air without. The internal pressures and compressions are then found and the expressions are not rendered much more complex than where the heat is uniform. The expression for the difference of the principal pressures perpendicular to the axis is the same as that in cases I and IV with the addition of a constant quantity.[71]

Case X is an example of an inverse problem. Sir David Brewster has observed (Edinburgh Transactions Vol VIII) that when a cylinder of glass is suddenly heated at the edges the difference of retardation of oppositely polarized rays of light is proportional to the square of the distance from the centre.[72] From this I find that the difference of the temperature at any point and the temperature at the centre is proportional to the square of the distance from the centre.

In Case XI I have given an example of the superposition of the effects of different pressures in a beam bent by weights.[73]

When a long beam is bent into the form of a circular ring as in Case VI all the pressures act either parallel or perpendicular to the direction of the length of the beam and as there are no forces tending to cause one part of the beam to slide on another part, the beam might be divided into planks without diminishing its stiffness.

But if the circular ring is not complete this is no longer the case the planks slide on each other and the beam loses its stiffness. In the case of a real beam these planks are united connected and the elasticity of the substance allows them little freedom yet the additional deflection from this cause may sometimes be considerable and I have here taken it into account for a beam bent by its own weight and forces acting at the middle and the extremities.

There are many other interesting cases of the flexure of beams to which this principle may be applied but they belong to a different branch of the subject the calculation of the stiffness and strength of framework. See Youngs Mathematical principles of Natural Philosophy and Duleau on the resistance of Iron Ann. de Ch. 2e série XII and Moseley's Mechanical principles of Engineering.[74]

---

(71) See §8; on the numbering of these Cases see note (55).

(72) David Brewster, 'On the laws which regulate the distribution of the polarising force in plates, tubes, and cylinders of glass that have received the polarising structure', *Trans. Roy. Soc. Edinb.*, **8** (1818): 353–72, esp. 353–4.

(73) The substance of Case XI is missing from the manuscript (§8); but see §13, and 'On the equilibrium of elastic solids': 114–15 (= *Scientific Papers*, **1**: 65–6).

(74) Thomas Young, *A Course of Lectures on Natural Philosophy and the Mechanical Arts*, 2 vols. (London, 1807), **2**: 46–51. A. Duleau, *Essai Théorique et Expérimental sur la Résistance du Fer Forgé* (Paris, 1820); reviewed in *Ann. Chim. Phys.*, ser. 2, **12** (1819): 133–48; Henry Moseley, *Mechanical Principles of Engineering and Architecture* (London, 1843): 486–585.

I have applied the principle of the superposition of compressions to the modification of the experiment illustrative of Case I in which two equal cylinders are twisted with the same force and in the same direction in an indefinitely extended elastic plate.[75]

In this case I find the following construction for the intensity of the pressures at any point.

Make the centres of the two cylinders the opposite angles of a square, and call them *A* and *B*, and the two other angles *C* and *D*; then the intensity of pressure at any point is proportional to the product of its distance from *C* and *D* divided by the square of the product of its distances from *A* and *B*.

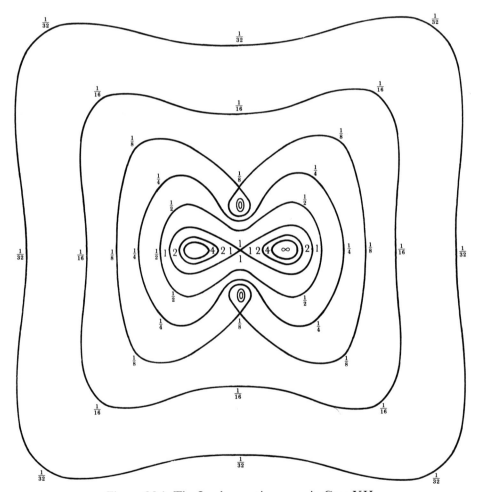

Figure 26,1. The Isochromatic curves in Case XII.

(75) See §12 *infra* [Case XII]; and compare 'On the equilibrium of elastic solids': 115–17 (= *Scientific Papers*, **1**: 66–7).

The curves of equal pressure thus found have a more complicated form than the common lemniscates. They are easily seen in transparent jelly when it is properly twisted by transmitting polarized light parallel to the axes of the twisted cylinders.[e][76]

[6] In the cases in which the solid is bounded by two parallel plane surfaces, the action on polarized light has been determined.

The magnitude and direction of the compressing forces may be determined by observation in the following manner.

The direction of the principal axes[a] of pressure at any point may be found by transmitting polarized light through the solid, then the parts which have no action on the light are those which have their principal axes parallel and perpendicular to the plane of polarization.[77]

The difference of pressure in these axes may be found by using circularly polarized light when the lines of equal tint will be seen without interruption. Experiments of this sort may be easily made with unannealed glass. The results thus obtained show the correctness of Sir John Herschells ingenious explanation of the phenomena of heated and unannealed glass.[78]

(e) {Maxwell} Page 13. (a) {Forbes} *principal axes*, Define.[79]

(76) ULC Add. MSS 7655, V, g/1. The figure (Fig. 26,1) is labelled 'The Isochromatic curves in Case XII'. Compare Fig. 1 of 'On the equilibrium of elastic solids': 116 (= *Scientific Papers*, **1**: 67). The term '*Isochromatic lines*, or lines of equal tint' had been used by David Brewster; see Brewster, 'On the distribution of the polarising force': 356. The isochromatic lines are represented in the figure; the values of *I* the optical effect of pressure (see Case I §8) are indicated.

(77) See Case XIV of 'On the equilibrium of elastic solids': 117–19 (= *Scientific Papers*, **1**: 68–70). Compare F. E. Neumann, 'Die Gesetze der Doppelbrechung des Lichts in comprimirten oder ungleichförmig erwärmten unkrystallinischen Körpern', *Abhandlungen der Königlichen Akademie der Wissenschaften zu Berlin*, Aus dem Jahre 1841, Zweiter Theil (Berlin, 1843): 1–254, esp. 35, 37. Neumann supposes 'optischen Elastizitätsaxen' to coincide in the direction with 'Hauptdruckaxen', the principal axes of the ellipsoid of elasticity: 'Die Richtungen der drei Axen dieser optischen Elastizitätsfläche fallen nothwendig zusammen mit den Richtungen der Hauptdruckaxen des gleichförmig comprimirten Körpers'. Maxwell was apparently unaware of Neumann's memoir.

(78) J. F. W. Herschel, 'On the effects of heat and mechanical violence in modifying the action of media on light, and on the application of the undulatory theory to their explanation', in his treatise on 'Light', *Encyclopaedia Metropolitana, or Useful Dictionary of Knowledge, Second Division: Mixed Sciences. Vol. II* (London, 1830): 341–586, esp. 562–8. Herschel reported Brewster's work on the effects of heat and mechanical compression in imparting doubly-refracting properties to substances such as glass; 'As the unusual heating or cooling of glass...is well known to produce in the parts heated or cooled a corresponding inequality of bulk, and thus to bring the parts adjacent into a state of strain in all respects analogous to that arising from mechanical violence...we have little hesitation in regarding the inequality of temperature as merely the remote, and the mechanical tension or condensation of the medium as the proximate cause of the phenomena in question,

Figure 26,2. The Isochromatic lines or curves in which the difference of the principal pressures is constant in a triangle of unannealed glass.

As an example of this method I have delineated the curves of equal intensity ⌐of action on polarised light⌐ and of equal inclination[b] ⌐of the principal axes of pressure⌐ in a triangle of unannealed glass.[80] In heated and unannealed

---

(b) {Forbes} *intensity, of equal inclination*, Define.[81]

and are very little disposed to call in the agency of a peculiar crystallizing fluid, endowed with properties analogous to those of magnetism, electricity, &c., to account for the phenomena, still less to regard media under the influence of heat or pressure as in any way thereby rendered more crystalline than in their natural state.... [The effect of] every external force applied to a solid is accompanied with a condensation of its particles in the direction of the force and a dilatation in a perpendicular direction...Hence [the effect of] dilatation will be the converse of that of compression'. Herschel remarked that this interpretation was 'in perfect accordance' with the experimental results reported by David Brewster, 'On the communication of the structure of doubly-refracting crystals to glass': 160–1; see note (19). On the work of Fresnel and Brewster see Number 110 esp. notes (7) and (8).   (79) See Maxwell's addendum (§2).

(80) ULC Add. MSS 7655, V, g/1. These figures are: 'The Isochromatic lines or curves in which the difference of the principal pressures is constant in a triangle of unannealed glass' (Fig.

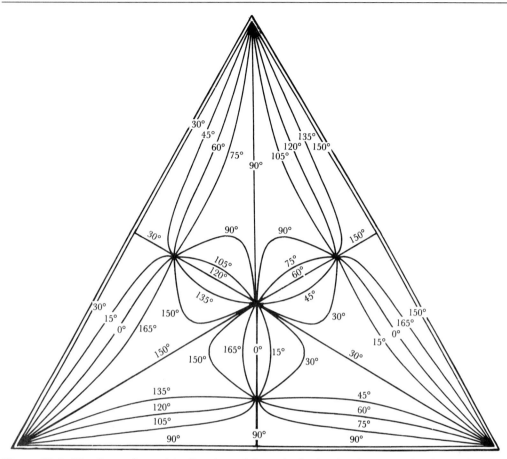

Figure 26,3. The lines of equal inclination of the principal axis of compression in a triangle of unannealed glass.

glass the pressures vary from one point to another and a uniform pressure in one direction cannot exist in every part of a homogeneous solid without an external pressure. But if two substances be mixed together one of which is much more transparent and elastic than the other, and if a uniform pressure in one direction be applied to the mixture, the whole will be compressed, and the

26,2), where the isochromatic lines and the intensity of the action of circularly polarised light at different points of a triangle of unannealed glass are indicated; and 'The lines of equal inclination of the principal axis of compression in a triangle of unannealed glass' (Fig. 26,3), where the direction of the principal lines of pressure is indicated for different directions of inclination of plane polarised light. Compare Figures 2 and 3 in Case XIV of 'On the equilibrium of elastic solids': 117–18 (= *Scientific Papers*, **1**: 68–9); see *Life of Maxwell*: 487, Plate III, copied from Maxwell's water colour of the isochromatic lines in a triangle of unannealed glass; and see Number 14.

(81) Compare 'On the equilibrium of elastic solids': 117 (= *Scientific Papers*, **1**: 68).

particles of the inelastic substance will have their shape permanently altered so that when the pressure is removed the particles of the elastic substance will be prevented from regaining their former shape, and will therefore be still in a state of compression, and capable of acting on light in a uniform manner like a crystal.

An experiment of this sort is described by Sir David Brewster in the Philosophical Transactions for 1815 and 1830.[82] A mixture of wax and resin is pressed into a thin plate. It has no action on polarized light at a perpendicular incidence but when inclined shews the segments of coloured rings.[83] This property belongs to every part of the plate, which acts like a uniaxial crystal, and I find that it belongs to the positive class. In this experiment the wax is less elastic and transparent than the resin, and retains it in a state of constraint. Gutta Percha possesses similar properties. If this substance is heated to the temperature of boiling water and cooled it has very little elasticity and may be drawn out in any direction without recovering its form. When it is drawn out to a little more than double its original length it becomes very elastic and cannot again be permanently extended. This property seems to have been long known to the manufacturers of the substance. ⟨It was noticed last year at the Royal Scottish Society of Arts.⟩[84]

If the substance thus drawn out be heated it suddenly returns to its original shape.

When a thin film is viewed by polarized light, the tints which are seen are like those of a piece of glass dilated in the direction in which the gutta percha was drawn out. These tints are permanent and belong to every part of the film and are visible in the smallest fragment of it.

The films are best prepared by pressing the substance while hot between two glass plates and drawing it out when cold. The large sheets which are sold are generally too thick to shew the tints of the second and third orders.

In the following pages I have endeavoured to apply a theory which is the same as that of Mr Stokes[85] to the solution of problems which have been selected on account of the possibility of fulfilling the conditions.

---

(82) David Brewster, 'On the effects of simple pressure in producing that species of crystallization which forms two oppositely polarized images, and exhibits the complementary colours by polarized light', *Phil. Trans.*, **105** (1815): 60–4; Brewster, 'On the laws of the polarization of light by refraction', *ibid.*, **120** (1830): 133–44.

(83) See Number 23.

(84) See James S. Torrop, 'Notice of a peculiar property of gutta percha' (read to the Royal Scottish Society of Arts on 26 March 1849), in the Society's *Proceedings* (1848–9): 214–15, in *Transactions of the Royal Scottish Society of Arts*, **3** (1851).

(85) Making no reference to a molecular model; see note (12).

I have not attempted to extend the theory to the case of imperfectly elastic bodies or to the laws of ⌊permanent⌋ bending and heating. The solids here considered are those which when compressed within the limits of perfect elasticity return to their original state when the pressure is removed.

The equations for the transformation of coordinates may be found in Gregory's Solid Geometry.[86]

The theorem for the composition of pressures may be derived from the proof of the rule for the composition of forces given by Dan Bernoulli, and improved by D'Alembert, Poisson, Whewell and Wallace.[87]

I have denoted the displacements by δx, δy, δz. They are generally denoted by α, β, γ, but as I had employed these letters to signify the three principal axes at any point, and as this had been done throughout the paper, I did not alter the notation which to me appears natural and intelligible.[c]

[7] [88]The laws of elasticity express the relation between the changes of the dimensions of a body and the forces which produce them.

These forces are called Pressures, and their effects Compressions. Pressures are estimated in pounds on the square inch, and compressions in fractions of the dimensions compressed.

Let the position of material points in space be expressed by their coordinates x, y, and z, then any change in a system of such points is expressed by giving to these coordinates the variations δx, δy, δz, these variations being functions of x, y, z.

The quantities δx, δy, δz, represent the absolute motion of each point in the directions of the three coordinates; but as compression depends not on

---

(c) {Forbes} (Space here).

(86) D. F. Gregory, concluded by W. Walton, *A Treatise on the Application of Analysis to Solid Geometry* (Cambridge, 1845): 48–57. Maxwell's copy, preserved in the Cavendish Laboratory, Cambridge, is inscribed: 'Presented to James Clerk Maxwell as a Prize in the third class of Mathematics University of Edinburgh April 12 1849 P. Kelland Prof.'

(87) Daniel Bernoulli, 'Examen principiorum mechanicae, et demonstrationes geometricae de compositione et resolutione virium', *Commentarii Academiae Scientiarum Imperialis Petropolitanae*, **1** (1726): 126–42; Jean Lerond D'Alembert, *Opuscules Mathématiques ou Mémoires sur différens Sujets de Géométrie, d'Optique, d'Astronomie &c*, 8 vols. (Paris, 1761–80), **1**: 169–79, **6**: 360–70; Siméon Denis Poisson, *Traité de Mécanique*, 2 vols. (Paris, ₂1833), **1**: 43–72; (trans. H. H. Harte) *A Treatise of Mechanics*, 2 vols. (London, 1842), **1**: 35–57; William Whewell, *Analytical Statics. A Supplement to the Fourth Edition of An Elementary Treatise on Mechanics* (Cambridge, 1833): 1–26; William Wallace, 'Solution of a functional equation, with its application to the parallelogram of forces, and to curves of equilibration', *Trans. Roy. Soc. Edinb.*, **14** (1840): 625–76, esp. 630–1 where Wallace gave these references.

(88) Missing from the MS, and reproduced from 'On the equilibrium of elastic solids': 90–5 (= *Scientific Papers*, **1**: 34–41). Compare Number 21.

absolute, but on relative displacement, we have to consider only the nine quantities –

$$\frac{d\delta x}{dx} \quad \frac{d\delta x}{dy} \quad \frac{d\delta x}{dz}$$

$$\frac{d\delta y}{dx} \quad \frac{d\delta y}{dy} \quad \frac{d\delta y}{dz}$$

$$\frac{d\delta z}{dx} \quad \frac{d\delta z}{dy} \quad \frac{d\delta z}{dz}.$$

Since the number of these quantities is nine, if nine other independent quantities of the same kind can be found, the one set may be found in terms of the other. The quantities which we shall assume for this purpose are –

1. Three compressions, $\dfrac{\delta\alpha}{\alpha}, \dfrac{\delta\beta}{\beta}, \dfrac{\delta\gamma}{\gamma}$, in the directions of three principal axes α, β, γ.

2. The nine *direction-cosines* of these axes, with the *six connecting equations*, leaving three independent quantities. (See Gregory's *Solid Geometry*).[89]

3. The small angles of rotation of this system of axes about the axes of x, y, z.

The cosines of the angles which the axes of x, y, z make with those of α, β, γ are –

$$\cos(\alpha 0 x) = a_1, \quad \cos(\beta 0 x) = b_1, \quad \cos(\gamma 0 x) = c_1,$$
$$\cos(\alpha 0 y) = a_2, \quad \cos(\beta 0 y) = b_2, \quad \cos(\gamma 0 y) = c_2,$$
$$\cos(\alpha 0 z) = a_3, \quad \cos(\beta 0 z) = b_3, \quad \cos(\gamma 0 z) = c_3.$$

These *direction-cosines* are connected by the six equations,

$$a_1^2 + b_1^2 + c_1^2 = 1 \quad a_1 a_2 + b_1 b_2 + c_1 c_2 = 0$$
$$a_2^2 + b_2^2 + c_2^2 = 1 \quad a_2 a_3 + b_2 b_3 + c_2 c_3 = 0$$
$$a_3^2 + b_3^2 + c_3^2 = 1 \quad a_3 a_1 + b_3 b_1 + c_3 c_1 = 0.$$

The rotation of the system of axes α, β, γ, round the axis of

$$x, \text{ from } y \text{ to } z, = \delta\theta_1,$$
$$y, \text{ from } z \text{ to } x, = \delta\theta_2,$$
$$z, \text{ from } x \text{ to } y, = \delta\theta_3.$$

By resolving the displacements $\delta\alpha, \delta\beta, \delta\gamma, \theta_1, \theta_2, \theta_3$, in the directions of the axes x, y, z, the displacements in these axes are found to be

$$\delta x = a_1 \delta\alpha + b_1 \delta\beta + c_1 \delta\gamma - \theta_2 z + \theta_3 y$$
$$\delta y = a_2 \delta\alpha + b_2 \delta\beta + c_2 \delta\gamma - \theta_3 x + \theta_1 z$$
$$\delta z = a_3 \delta\alpha + b_3 \delta\beta + c_2 \delta\gamma - \theta_1 y + \theta_2 x.$$

---

(89) D. F. Gregory, *A Treatise on the Application of Analysis to Solid Geometry*: 52–3.

But $$\delta\alpha = \alpha\frac{\delta\alpha}{\alpha}, \quad \delta\beta = \beta\frac{\delta\beta}{\beta}, \text{ and } \delta\gamma = \gamma\frac{\delta\gamma}{\gamma},$$

and $\alpha = a_1 x + a_2 y + a_3 z$, $\beta = b_1 x + b_2 y + b_3 z$, and $\gamma = c_1 x + c_2 y + c_3 z$.

Substituting these values of $\delta\alpha$, $\delta\beta$, and $\delta\gamma$ in the expressions for δx, δy, δz, and differentiating with respect to x, y, and z, in each equation, we obtain the equations –

$$\left.\begin{aligned}
\frac{d\delta x}{dx} &= \frac{\delta\alpha}{\alpha} a_1^2 + \frac{\delta\beta}{\beta} b_1^2 + \frac{\delta\gamma}{\gamma} c_1^2 \\
\frac{d\delta y}{dy} &= \frac{\delta\alpha}{\alpha} a_2^2 + \frac{\delta\beta}{\beta} b_2^2 + \frac{\delta\gamma}{\gamma} c_2^2 \\
\frac{d\delta z}{dz} &= \frac{\delta\alpha}{\alpha} a_3^2 + \frac{\delta\beta}{\beta} b_3^2 + \frac{\delta\gamma}{\gamma} c_3^2
\end{aligned}\right\} \qquad (1)$$

Equations of compression.

$$\left.\begin{aligned}
\frac{d\delta x}{dy} &= \frac{\delta\alpha}{\alpha} a_1 a_2 + \frac{\delta\beta}{\beta} b_1 b_2 + \frac{\delta\gamma}{\gamma} c_1 c_2 + \delta\theta_3 \\
\frac{d\delta x}{dz} &= \frac{\delta\alpha}{\alpha} a_1 a_3 + \frac{\delta\beta}{\beta} b_1 b_3 + \frac{\delta\gamma}{\gamma} c_1 c_3 - \delta\theta_2 \\
\frac{d\delta y}{dz} &= \frac{\delta\alpha}{\alpha} a_2 a_3 + \frac{\delta\beta}{\beta} b_2 b_3 + \frac{\delta\gamma}{\gamma} c_2 c_3 + \delta\theta_1 \\
\frac{d\delta y}{dx} &= \frac{\delta\alpha}{\alpha} a_2 a_1 + \frac{\delta\beta}{\beta} b_2 b_1 + \frac{\delta\gamma}{\gamma} c_2 c_1 - \delta\theta_3 \\
\frac{d\delta z}{dx} &= \frac{\delta\alpha}{\alpha} a_3 a_1 + \frac{\delta\beta}{\beta} b_3 b_1 + \frac{\delta\gamma}{\gamma} c_3 c_1 + \delta\theta_2 \\
\frac{d\delta z}{dy} &= \frac{\delta\alpha}{\alpha} a_3 a_2 + \frac{\delta\beta}{\beta} b_3 b_2 + \frac{\delta\gamma}{\gamma} c_3 c_2 - \delta\theta_1
\end{aligned}\right\} \qquad (2)$$

Equations of the equilibrium of an element of the solid

The forces which may act on a particle of the solid are:

1 Three attractions in the direction of the axes, represented by X, Y, Z.
2 Six pressures on the six faces.
3 Two tangential actions on each face.

Let the six faces of the small parallelopiped be denoted by x_1, y_1, z_1, x_2, y_2, and z_2, then the forces acting on x_1 are:

1 A normal pressure p_1 acting in the direction of x on the area $dy\,dz$.
2 A tangential force q_3 acting in the direction of y on the same area.
3 A tangential force q_2^1 acting in the direction of z on the same area, and so on for the other five faces, thus:

Forces which act in the direction of the axes of

	x	y	z
On the face x_1	$-p_1\,dy\,dz$	$-q_3\,dy\,dz$	$-q_2^1\,dy\,dz$
... ... x_2	$\left(p_1+\dfrac{dp_1}{dx}\,dx\right)dy\,dz$	$\left(q_3+\dfrac{dq_3}{dx}\,dx\right)dy\,dx$	$\left(q_2^1+\dfrac{dq_2^1}{dx}\,dx\right)dy\,dz$
... ... y_1	$-q_3^1\,dz\,dx$	$-p_2\,dz\,dx$	$-q_1\,dz\,dx$
... ... y_2	$\left(q_3+\dfrac{dq_3^1}{dy}\,dy\right)dz\,dx$	$\left(p_2+\dfrac{dp_2}{dy}\,dy\right)dz\,dx$	$\left(q_1+\dfrac{dq_1}{dy}\,dy\right)dz\,dx$
... ... z_1	$-q_2\,dx\,dy$	$-q^1\,dx\,dy$	$-p_3\,dx\,dy$
... ... z_2	$\left(q_2+\dfrac{dq_2}{dz}\,dz\right)dx\,dy$	$\left(q^1+\dfrac{dp^1}{dz}\,dz\right)dx\,dy$	$\left(p_3+\dfrac{dp_3}{dz}\,dz\right)dx\,dy$
Attractions,	$\rho X\,dx\,dy\,dz$	$\rho Y\,dx\,dy\,dz$	$\rho Z\,dx\,dy\,dz$

Taking the moments of these forces round the axes of the particle, we find

$$q_1^1 = q_1 \quad q_1^2 = q_2 \quad q_3^1 = q_3;$$

and then equating the forces in the directions of the three axes, and dividing by dx, dy, dz, we find the equations of pressures,

$$\left.\begin{aligned}
\frac{dp_1}{dx}+\frac{dq_3}{dy}+\frac{dq_2}{dz}+\rho X &= 0 \\
\frac{dp_2}{dy}+\frac{dq_1}{dz}+\frac{dq_3}{dx}+\rho Y &= 0 \\
\frac{dp_3}{dz}+\frac{dq_2}{dx}+\frac{dq_1}{dy}+\rho Z &= 0
\end{aligned}\right\} \quad \text{Equations of pressures.} \tag{3}$$

The resistance which the solid opposes to these pressures is called Elasticity, and is of two kinds, for it opposes either change of *volume* of change of *figure*. These two kinds of elasticity have no necessary connection, for they are possessed in very different ratios by different substances. Thus *jelly* has a cubical elasticity little different from that of water, and a linear elasticity as small as we please; while *cork*, whose cubical elasticity is very small, has a much greater linear elasticity than jelly.[90]

Hooke discovered that the elastic forces are proportional to the changes that excite them, or, as he expressed it, 'Ut tensio sic vis.'

To fix our ideas, let us suppose the compressed body to be a parallelopiped, and let pressures P_1, P_2, P_3 act on its faces in the direction of the axes α, β, γ, which will become the principal axes of compression, and the compressions will be $\dfrac{\delta\alpha}{\alpha}$, $\dfrac{\delta\beta}{\beta}$, $\dfrac{\delta\gamma}{\gamma}$.

---

(90) Compare §1.

The fundamental assumption from which the following equations are deduced is an extension of Hooke's law, and consists of two parts.[91]

I. The sum of the compressions is proportional to the sum of the pressures.

II. The difference of the compressions is proportional to the difference of the pressures.

These laws are expressed by the following equations:

$$\text{I.} \quad (P_1 + P_2 + P_3) = 3\mu \left(\frac{\delta\alpha}{\alpha} + \frac{\delta\beta}{\beta} + \frac{\delta\gamma}{\gamma} \right). \tag{4}$$

$$\left. \begin{aligned} (P_1 - P_2) &= m\left(\frac{\delta\alpha}{\alpha} - \frac{\delta\beta}{\beta} \right) \\ \text{II.} \quad (P_2 - P_3) &= m\left(\frac{\delta\beta}{\beta} - \frac{\delta\gamma}{\gamma} \right) \\ (P_3 - P_1) &= m\left(\frac{\delta\gamma}{\gamma} - \frac{\delta\alpha}{\alpha} \right). \end{aligned} \right\} \tag{5}$$

Equations of elasticity.

The quantity μ is the coefficient of cubical elasticity, and m that of linear elasticity.

By solving these equations, the values of the pressures P_1, P_2, P_3, and the compressions $\frac{\delta\alpha}{\alpha}, \frac{\delta\beta}{\beta}, \frac{\delta\gamma}{\gamma}$ may be found.

$$\left. \begin{aligned} P_1 &= (\mu - \tfrac{1}{3}m)\left(\frac{\delta\alpha}{\alpha} + \frac{\delta\beta}{\beta} + \frac{\delta\gamma}{\gamma} \right) + m\frac{\delta\alpha}{\alpha} \\ P_2 &= (\mu - \tfrac{1}{3}m)\left(\frac{\delta\alpha}{\alpha} + \frac{\delta\beta}{\beta} + \frac{\delta\gamma}{\gamma} \right) + m\frac{\delta\beta}{\beta} \\ P_3 &= (\mu - \tfrac{1}{3}m)\left(\frac{\delta\alpha}{\alpha} + \frac{\delta\beta}{\beta} + \frac{\delta\gamma}{\gamma} \right) + m\frac{\delta\gamma}{\gamma}. \end{aligned} \right\} \tag{6}$$

$$\left. \begin{aligned} \frac{\delta\alpha}{\alpha} &= \left(\frac{1}{9\mu} - \frac{1}{3m} \right)(P_1 + P_2 + P_3) + \frac{1}{m}P_1 \\ \frac{\delta\beta}{\beta} &= \left(\frac{1}{9\mu} - \frac{1}{3m} \right)(P_1 + P_2 + P_3) + \frac{1}{m}P_2 \\ \frac{\delta\gamma}{\gamma} &= \left(\frac{1}{9\mu} - \frac{1}{3m} \right)(P_1 + P_2 + P_3) + \frac{1}{m}P_3. \end{aligned} \right\} \tag{7}$$

From these values of the pressures in the axes α, β, γ, may be obtained the equations for the axes x, y, z, by resolution of pressures and compressions.*[92]

For
$$p = a^2 P_1 + b^2 P_2 + c^2 P_3$$
and
$$q = aaP_1 + bbP_2 + ccP_3$$

* See the Memoir of Lamé and Clapeyron.

(91) See notes (15) and (16).

(92) See Lamé and Clapeyron, 'Mémoire sur l'équilibre intérieur des corps solides homogènes': 165–9.

$$\left.\begin{aligned}
p_1 &= (\mu - \tfrac{1}{3}m)\left(\frac{d\delta x}{dx} + \frac{d\delta y}{dy} + \frac{d\delta z}{dz}\right) + m\frac{d\delta x}{dx} \\
p_2 &= (\mu - \tfrac{1}{3}m)\left(\frac{d\delta x}{dx} + \frac{d\delta y}{dy} + \frac{d\delta z}{dz}\right) + m\frac{d\delta y}{dy} \\
p_3 &= (\mu - \tfrac{1}{3}m)\left(\frac{d\delta x}{dx} + \frac{d\delta y}{dy} + \frac{d\delta z}{dz}\right) + m\frac{d\delta z}{dz}
\end{aligned}\right\} \quad (8)$$

$$\left.\begin{aligned}
q_1 &= \frac{m}{2}\left(\frac{d\delta y}{dz} + \frac{d\delta z}{dy}\right) \\
q_2 &= \frac{m}{2}\left(\frac{d\delta z}{dx} + \frac{d\delta x}{dz}\right) \\
q_3 &= \frac{m}{2}\left(\frac{d\delta x}{dy} + \frac{d\delta y}{dx}\right)
\end{aligned}\right\} \quad (9)$$

$$\left.\begin{aligned}
\frac{d\delta x}{dx} &= \left(\frac{1}{9\mu} - \frac{1}{3m}\right)(p_1 + p_2 + p_3) + \frac{1}{m}p_1 \\
\frac{d\delta y}{dy} &= \left(\frac{1}{9\mu} - \frac{1}{3m}\right)(p_1 + p_2 + p_3) + \frac{1}{m}p_2 \\
\frac{d\delta z}{dz} &= \left(\frac{1}{9\mu} - \frac{1}{3m}\right)(p_1 + p_2 + p_3) + \frac{1}{m}p_3
\end{aligned}\right\} \quad (10)$$

$$\left.\begin{aligned}
\frac{d\delta x}{dy} - \delta\theta_3 &= \frac{d\delta y}{dx} + \delta\theta_3 = \frac{1}{m}q_3 \\
\frac{d\delta y}{dz} - \delta\theta_1 &= \frac{d\delta z}{dy} + \delta\theta_1 = \frac{1}{m}q_1 \\
\frac{d\delta z}{dx} - \delta\theta_2 &= \frac{d\delta x}{dz} + \delta\theta_2 = \frac{1}{m}q_2
\end{aligned}\right\} \quad (11)$$

By substituting in equations (3) the values of the forces given in equations (8) and (9), they become

$$\left.\begin{aligned}
(\mu + \tfrac{1}{6}m)\left(\frac{d}{dx}\left(\frac{d\delta x}{dx} + \frac{d\delta y}{dy} + \frac{d\delta z}{dz}\right)\right) + \frac{m}{2}\left(\frac{d^2}{dx^2}\delta x + \frac{d^2}{dy^2}\delta y + \frac{d^2}{dz^2}\delta z\right) + \rho X = 0 \\
(\mu + \tfrac{1}{6}m)\left(\frac{d}{dy}\left(\frac{d\delta x}{dx} + \frac{d\delta y}{dy} + \frac{d\delta z}{dz}\right)\right) + \frac{m}{2}\left(\frac{d^2}{dx^2}\delta x + \frac{d^2}{dy^2}\delta y + \frac{d^2}{dz^2}\delta z\right) + \rho Y = 0 \\
(\mu + \tfrac{1}{6}m)\left(\frac{d}{dz}\left(\frac{d\delta x}{dx} + \frac{d\delta y}{dy} + \frac{d\delta z}{dz}\right)\right) + \frac{m}{2}\left(\frac{d^2}{dx^2}\delta x + \frac{d^2}{dy^2}\delta y + \frac{d^2}{dz^2}\delta z\right) + \rho Z = 0
\end{aligned}\right\} \quad (12)$$

These are the general equations of elasticity, and are identical with those of M. Cauchy, in his *Exercises d'Analyse*, vol. iii., p. 180, published in 1828, when

k stands for m, and K for $\mu - \dfrac{m}{2}$,[93] and those of Mr Stokes, given in the *Cambridge Philosophical Transactions*, vol. viii., part 3, and numbered (30); in his equations $A = 3\mu$, $B = \dfrac{m}{2}$.[94]

If the temperature is variable from one part to another of the elastic solid, the compressions $\dfrac{d\delta x}{dx}$, $\dfrac{d\delta y}{dy}$, $\dfrac{d\delta z}{dz}$, at any point will be diminished by a quantity proportional to the temperature at that point. This principle is applied in Cases X and XI.[95] Equations (10) then become

$$
\left.
\begin{aligned}
\frac{d\delta x}{dx} &= \left(\frac{1}{9\mu} - \frac{1}{3m}\right)(p_1 + p_2 + p_3) + c_3\,v + \frac{1}{m}p_1 \\
\frac{d\delta y}{dy} &= \left(\frac{1}{9\mu} - \frac{1}{3m}\right)(p_1 + p_2 + p_3) + c_3\,v + \frac{1}{m}p_2 \\
\frac{d\delta z}{dz} &= \left(\frac{1}{9\mu} - \frac{1}{3m}\right)(p_1 + p_2 + p_3) + c_3\,v + \frac{1}{m}p_3
\end{aligned}
\right\}
\tag{13}
$$

$c_3\,v$ being the linear expansion for the temperature v.[96]

---

(93) Read: *Exercices de Mathématiques*, vol. iii., p. 180; and: $K = \mu - \dfrac{m}{3}$ (see §3). Cauchy states equations of elastic equilibrium:

$$
\frac{k}{2\Delta}\left(\frac{\partial^2 \xi}{\partial x^2} + \frac{\partial^2 \xi}{\partial y^2} + \frac{\partial^2 \xi}{\partial z^2}\right) + \frac{k + 2K}{2\Delta}\frac{\partial v}{\partial x} + X = 0,
$$

and similar equations for the other components of the force X, Y, Z at a point x, y, z; where ξ, η, ζ, are functions of x, y, z which measure the displacement of a molecule at the point, k and K are the coefficients of linear rigidity and compressibility, Δ the density of the solid, and v measures the change in volume of the molecule at the point; Cauchy, *Exercices de Mathématiques*, **3**: 180 (see note (10)). Maxwell was now familiar with Cauchy's work: compare §3 and note (43).

(94) Stokes, 'On the theories of the internal friction of fluids in motion, and of the equilibrium and motion of elastic solids': 311, stating the equation of elastic equilibrium:

$$
\rho X + \tfrac{1}{3}(A + B)\frac{d}{dx}\left(\frac{d\alpha}{dx} + \frac{d\beta}{dy} + \frac{d\gamma}{dz}\right) + B\left(\frac{d^2\alpha}{dx^2} + \frac{d^2\alpha}{dy^2} + \frac{d^2\alpha}{dz^2}\right) = 0,
$$

and indicating similar equations for the other components of the force X, Y, Z at the point x, y, z in the solid; α, β, γ are the increments of these coordinates, ρ the density of the solid, and A and B the coefficients of 'extensibility' and 'cubical compressibility'; see note (11).

(95) See Cases IX and X, §11.

(96) Maxwell was apparently not familiar with the work of Duhamel on the thermoelasticity of isotropic substances; see J. M. C. Duhamel, 'Mémoire sur le calcul des actions moléculaires développées par les changements de température dans les corps solides', *Mémoires Présentes par Divers Savants à l'Académie des Sciences de l'Institut de France*, **5** (1838): 440–85; Duhamel, 'Second mémoire sur les phénomènes thermo-mécaniques', *Journal de l'École Polytechnique*, **15** cahier 25 (1837): 1–57.

[8] [97]Having found the general equations of ⟨compression⟩ ⌐elasticity⌐ I proceed to work some examples of their application which afford the means of determining the coefficients μ and m and of calculating the stiffness of solid figures. Instead of the coordinates $x\,y\,z$ other quantities may be assumed more suitable to particular cases. Thus in the case of a cylinder let

$dx = dx$ measured along the axis
$dy = dr$ ———————————— radius
$dz = rd\theta$ perpendicular to radius and axis
$p_1 = o =$ pressure in the direction of the axis x
$p_2 = p =$ ———————————— radius r
$p_3 = q =$ ———————————— tangent $rd\theta$.

Equations ⌐(9)⌐[98] become

$$q_1 = \frac{m}{2}r\frac{d\delta\theta}{dr}$$

$$q_2 = \frac{m}{2}r\frac{d\delta\theta}{dx}$$

$$q_3 = \frac{m}{2}\frac{\delta dx}{dr}.$$

Case I [99]The first equation is applicable to the case of the solid contained between two cylindric surfaces one of which is twisted through a small angle $\delta\theta$ by a couple whose moment is M.

The force q_1 acts on the surface of a cylinder whose radius is r and the moment of the whole of this force is equal to M that is $q_1(2\pi r)\,(x)\,r = M$

$$\frac{d\delta\theta}{dr}m\pi r^3 x = M$$

$$\frac{d\delta\theta}{dr} = \frac{M}{\pi m r^3 x}$$

$$\delta\theta = +\frac{M}{2\pi m r^2 x}$$

$$m = \frac{M}{2\pi r^2 x\delta\theta}.$$

---

(97) ULC Add. MSS 7655, V, g/2.

(98) See §7. Maxwell originally wrote the equations following as $q_1 = mr\,d\delta\theta/dr,\dots$. On correcting these equations by the factor of $\frac{1}{2}$, he modified the equations in Cases I, II and III in accordance with this correction. These changes are not further recorded, the corrected version being given.

(99) Case I of 'On the equilibrium of elastic solids': 95–8 (= *Scientific Papers*, **1**: 42–4).

Let $\langle T = \text{the tint} \rangle \lfloor I = \text{the optical effect of pressure} \rfloor$[100] at the distance r from the centre, as seen by polarized light in a transparent elastic solid thus constrained

$$\langle T \rangle \lfloor I \rfloor = Vr\frac{d\delta\theta}{dr} = V\frac{M}{\pi mr^2 x} = \omega\frac{M}{\pi r^3}$$

$$\omega = -\frac{m}{\pi r^2 I}.$$

The general appearance is therefore a system of rings arranged oppositely to Newtons rings and becoming closer as they approach the centre their diameters being inversely as the squares of the tints. To complete the contrast they are traversed by a black cross whose arms are inclined to the planes of polarization by an angle of 45°. This experiment has been tried and the phenomena are as described.[101]

Case II[102]

The second equation relates to the case of a cylinder held fast at one end and twisted at the other. In this case

$$M = \int_0^r \int_0^{2\pi} rq_2\, r\, dr\, d\theta$$

and

$$q_2 = \frac{m}{2}r\frac{d\delta\theta}{dx}.$$

$$\therefore M = \int_0^r \int_0^{2\pi} \frac{m}{2}\frac{d\delta\theta}{dx}r^3\, dr\, d\theta$$

$$= \int_0^r m\frac{\delta d\theta}{dx}\pi r^3\, dr.$$

$$M = m\frac{\pi}{4}r^4\frac{\delta d\theta}{dx}$$

$$m = \frac{4}{\pi}\frac{M}{r^4\,\delta\dfrac{d\theta}{dx}}. \quad \text{(a)}$$

(a) {Forbes}?

(100) That is: 'the difference of retardation of the oppositely polarized rays at any point in the solid'; see 'On the equilibrium of elastic solids': 97 (= *Scientific Papers*, **1**: 42) and §2. See also Number 23.

(101) See Number 19.

(102) Case II of 'On the equilibrium of elastic solids': 98–9 (= *Scientific Papers*, **1**: 44–5). Compare Lamé and Clapeyron, 'Mémoire sur l'équilibre intérieur des corps solides homogènes': 389–93.

⌞Let n be the number of degrees in θ and x the length of the rod

$$m = \frac{720}{\pi^2} \frac{x}{r^4} \frac{M}{n}.⌟^{(103)}$$

Every part of this solid has one of its optic axes or resultant axes of no polarization in the direction of the axis x and the other in the direction of the radius, therefore light passing parallel to ṭhe axis will not be affected ⟨except perhaps in a way hereafter to be explained⟩ ⌞any more than by passing along the optic axis of a biaxial crystal⌟.

Case III

From the third equation[104] may be deduced the state of a hollow cylinder one of whose surfaces is pressed in the direction of the axis by a force W the radius of the surface being $= a$.

$$q_3 \, x 2\pi r = W$$

$$m\frac{\delta dx}{dr}\pi rx = W \quad \text{or} \quad \frac{\delta dx}{dr} = \frac{W}{\pi mrx}$$

$$\delta x = c \log r.$$

Case IV[105]

[b]Equations ⌞(10)⌟ become when applied to the cylinder

$$\frac{d\delta x}{dx} = \left(\frac{1}{9\mu} - \frac{1}{3m}\right)(o+p+q) + \frac{1}{m}o$$

$$\frac{d\delta r}{dr} = \left(\frac{1}{9\mu} - \frac{1}{3m}\right)(o+p+q) + \frac{1}{m}p$$

$$\frac{\delta r}{r} = \left(\frac{1}{9\mu} - \frac{1}{3m}\right)(o+p+q) + \frac{1}{m}q.$$

Differentiating this last equation and equating the result with the former value of $\dfrac{d\delta r}{dr}$ we have

$$\left(\frac{1}{9\mu} - \frac{1}{3m}\right)\left(\frac{do}{dr} + \frac{dp}{dr} + \frac{dq}{dr}\right) + \frac{1}{m}\frac{dq}{dr} + \frac{q-p}{rm} = 0.$$

(b) {Forbes} Enunciate.[106]

(103) For a similar result see Lamé and Clapeyron, 'Mémoire sur l'équilibre intérieur des corps solides homogènes': 391; Stokes, 'On the theories of the internal friction of fluids in motion, and of the equilibrium and motion of elastic solids': 315; and Moseley, *Mechanical Principles of Engineering and Architecture*: 580–3.   (104) See *supra*.

(105) Case III of 'On the equilibrium of elastic solids': 99–103 (= *Scientific Papers*, **1**: 45–50). Compare Lamé and Clapeyron, 'Mémoire sur l'équilibre intérieur des corps solides homogènes': 382–9.

(106) Compare Maxwell's slight verbal amplificatioṇ in 'On the equilibrium of elastic solids': 99 (= *Scientific Papers*, **1**: 45).

The second equation of (11) becomes in the case of the cylinder

$$q-p = r\frac{dp}{dr}.$$

Substituting this value of $q-p$ in the preceding equation multiplying by dr and integrating

$$\left(\frac{1}{9\mu}-\frac{1}{3m}\right)o+\left(\frac{1}{9\mu}+\frac{2}{3m}\right)(p+q) = c_1.$$

Since o is constant we may assume

$$c_2 = \frac{c_1-\left(\dfrac{1}{9\mu}-\dfrac{1}{3m}\right)o}{\dfrac{1}{9\mu}+\dfrac{2}{3m}} = p+q.$$

$$\therefore r\frac{dp}{dr} = q-p = c_2-2p$$

$$\therefore p = c_3\frac{1}{r^2}+\frac{c_2}{2}.$$

To determine the coefficients c_2 and c_3 let p be given equal to h when r is equal to a and p equal to k when r is equal to b ⌊that is to say let the solid be a hollow cylinder whose external and internal radii are a and b and let it be acted on by normal pressures h and k on its cylindric surfaces⌋.

$$p_{r=a} = h$$

$$p_{r=b} = k.$$

From these equations we get

$$p = \frac{a^2h-b^2k}{a^2-b^2}-\frac{a^2b^2}{r^2}\frac{h-k}{a^2-b^2}$$

$$q = \frac{a^2h-b^2k}{a^2-b^2}+\frac{a^2b^2}{r^2}\frac{h-k}{a^2-b^2}$$

$$q-p = 2\frac{a^2b^2}{r^2}\frac{h-k}{a^2-b^2}$$

$$\langle T\rangle\llcorner I\lrcorner = 2\omega\frac{a^2b^2}{r^2}\frac{h-k}{a^2-b^2}x.\,^{(c)(107)}$$

This shews that the size and order of the rings seen by polarized light will be the same as described in the first example but the arms of the cross will in this

---

(c) {Forbes} Define T.

(107) Compare Case I; and 'On the equilibrium of elastic solids': 100 ($=$ *Scientific Papers*, **1**: 46–7).

case be in their normal position, that is, parallel and perpendicular to the plane of polarization.

$$\frac{\delta x}{x} = \left(\frac{1}{9\mu} - \frac{1}{3m}\right)\left(o + 2\frac{a^2h - b^2k}{a^2 - b^2}\right) + \frac{o}{m} \quad {}^{(108)}$$

$$\frac{\delta r}{r} = \left(\frac{1}{9\mu} - \frac{1}{3m}\right)\left(o + 2\frac{a^2h - b^2k}{a^2 - b^2}\right) + \frac{q}{m}$$

$$\frac{\delta V}{V_{r=a}} = \frac{1}{3\mu}\left(o + 2\frac{a^2h - b^2k}{a^2 - b^2}\right) + \frac{2}{m}b^2\frac{h - k}{a^2 - b^2}.$$

When $o = h = k$ then $o = p = q$ [109]

$$\frac{\delta x}{x} = \frac{\delta r}{r} = \frac{1}{3\mu}h \quad {}^{(110)}$$

$$\frac{\delta V}{V} = \frac{1}{\mu}h. \quad {}^{(d)}$$

When $b = 0$ the equations apply to a solid cylinder and they become

$$p = q = h$$

$$\frac{\delta x}{x} = \frac{1}{9\mu}(o + 2h) + \frac{2}{3m}(o - h)$$

$$\frac{\delta r}{r} = \frac{1}{9\mu}(o + 2h) - \frac{1}{3m}(o - h)$$

$$\frac{\delta V}{V} = \frac{1}{3\mu}(0 + 2h).$$

When $o = h$ or when the pressure is equal on all sides

$$\frac{\delta x}{x} = \frac{\delta r}{r} = \frac{o}{3\mu}.$$

When $h = 0$ or when the cylinder is pressed on the plane ends only

$$\frac{\delta x}{x} = \left(\frac{1}{9\mu} + \frac{2}{3m}\right)o$$

$$\frac{\delta r}{r} = \left(\frac{1}{9\mu} - \frac{1}{3m}\right)o$$

$$\frac{\delta V}{V} = \frac{1}{3\mu}o.$$

(d) {Forbes} V?.[111]

(108) See the equations for the compression of a solid cylinder.   (109) See §5.

(110) See 'On the equilibrium of elastic solids': 101 (= *Scientific Papers*, **1**: 48); 'The compression of a cylindrical vessel exposed on all sides to the same hydrostatic pressure is therefore independent of m, and it may be shewn that the same is true for a vessel of any shape'.

(111) Compare Maxwell's modifications in 'On the equilibrium of elastic solids': 101–2 (= *Scientific Papers*, **1**: 48–9).

When $o = 0$ or when it is pressed only on the cylindric surface

$$\frac{\delta x}{x} = 2h\left(\frac{1}{9\mu} - \frac{1}{3m}\right)$$

$$\frac{\delta r}{r} = h\left(\frac{2}{9\mu} + \frac{1}{3m}\right)$$

$$\frac{\delta V}{V} = \tfrac{2}{3}\frac{h}{\mu}.$$

When $\dfrac{\delta x}{x} = 0$ or when the ends of the cylinder are retained in their place

$$\frac{o}{h} = -2\frac{\dfrac{1}{9\mu} - \dfrac{1}{3m}}{\dfrac{1}{9\mu} + \dfrac{2}{3m}}$$

$$\frac{\delta r}{r} = \frac{3h}{m + 6\mu}$$

$$\frac{\delta V}{V} = \frac{6h}{m + 6\mu}.$$

When $\dfrac{\delta r}{r} = 0$ or when the radius cannot change

$$\frac{h}{o} = -\frac{\dfrac{1}{9\mu} - \dfrac{1}{3m}}{\dfrac{2}{9\mu} + \dfrac{1}{3m}}$$

$$\frac{\delta x}{x} = \frac{3o}{2m + 3\mu} = \frac{\delta V}{V}.$$

When p becomes infinite the solid becomes an infinitely extended plate and in this case

$$p = k + \frac{a^2}{r^2}(h - k)$$

$$q = k - \frac{a^2}{r^2}(h - k)$$

$$W = \frac{2\omega}{m}\frac{a^2 x}{r^2}(h - k)$$

$$\frac{\delta x}{x} = \frac{1}{9\mu}(o + 2k) + \frac{2}{3m}(o - k)$$

$$\frac{\delta r}{r} = \frac{1}{9\mu}(o + 2k) + \frac{1}{3m}(4k - 3h - o).$$

Case V[(e)(112)]

In the case of a sphere it is convenient to change the coordinates thus

$$dx = rd\phi$$
$$dy = dr$$
$$dz = rd\theta$$
$$p_1 = q = p_3$$
$$p_2 = p.$$

Equations ⌐(10)⌐ become ⌐for normal pressures⌐

$$\frac{d\delta r}{dr} = \left(\frac{1}{9\mu} - \frac{1}{3m}\right)(p+2q) + \frac{1}{m}p$$
$$\frac{\delta r}{r} = \left(\frac{1}{9\mu} - \frac{1}{3m}\right)(p+2q) + \frac{1}{m}q.$$

Differentiating this equation we obtain a value of $\dfrac{d\delta r}{dr}$ which when compared with the former gives

$$r\left(\frac{1}{9\mu} - \frac{1}{3m}\right)\left(\frac{dp}{dr} + 2\frac{dq}{dr}\right) + \frac{1}{m}\left(r\frac{dq}{dr} + q - p\right) = 0.$$

The second equation of (11) becomes

$$q - p = \frac{r}{2}\frac{dp}{dr}$$

which value of $q - p$ being substituted in the preceding equation

$$\frac{dp}{dr} + 2\frac{dq}{dr} = 0$$
$$p + 2q = c_1.$$

From this equation and the second equation of (11)

$$\frac{dp}{dr} + 3\frac{p}{r} + \frac{c_1}{r} = 0$$
$$p = c_2\frac{1}{r^3} + \frac{c_1}{3}. \text{ (f)(113)}$$

(e) {Forbes} Explain the physical condition of the sphere.[(114)]

(f) {Forbes} r^3.

(112) Compare Case IV of 'On the equilibrium of elastic solids': 103–7 (= *Scientific Papers*, **1**: 50–5).

(113) The remainder of the text of Case V, and the text of Case VI, are missing. But see §5.

(114) See Maxwell's addendum §5, incorporated into the text of 'On the equilibrium of elastic solids': 103, 105 (= *Scientific Papers*, **1**: 50–1, 53).

[9]   *Case VII*[115]

To find an expression for the curvature of a flat circular elastic plate produced by a hydrostatic pressure acting on one side of it.

Let t be the thickness of the plate, which must be small.

Let h be the pressure on one side, and k that on the other.

Let the form of the plate after displacement be expressed by an equation between r, the distance of any point from the axis or line passing perpendicularly through the centre of the plate, and x the distance of the point from the plane in which the middle of the plate originally was and let $ds = \sqrt{dx + dr}$.[116]

Let p and q be the pressures at any point of the plate, p acting in the line of intersection of the plane of the plate and the plane passing through the axis and q acting in a perpendicular direction.

Equations (3) become in this case

$$+ tp\frac{dr}{ds}\frac{dx}{ds} + tr\frac{dp}{ds}\frac{dx}{ds} + trp\frac{d^2x}{ds^2} + r(h-k)\frac{dr}{ds} = 0$$

$$+ tp\left(\frac{dr}{ds}\right)^2 + tr\frac{dp}{ds}\frac{dr}{ds} + trp\frac{d^2r}{ds^2} - (h-k)\,r\frac{dx}{ds} - qt = 0.$$

By making $p = 0$ when $r = 0$ in the first equation we have

$$p = -\frac{hr}{2t}\frac{ds}{dx}$$

and substituting this value in the second

$$q = -\frac{(h-k)}{t}r\left(\frac{dr}{ds}\frac{dr}{dx} + \frac{dx}{ds}\right) + \frac{(h-k)}{2t}r^2\left(\frac{dr}{dx}\frac{ds}{dx}\frac{d^2x}{ds^2} - \frac{ds}{dx}\frac{d^2r}{ds^2}\right).$$

Equations (10) become

$$\frac{d\delta s}{ds} = \left(\frac{1}{9\mu} - \frac{1}{3m}\right)\left(p + q + \frac{h+k}{2}\right) + \frac{p}{m}$$

$$\frac{\delta r}{r} = \left(\frac{1}{9\mu} - \frac{1}{3m}\right)\left(p + q + \frac{h+k}{2}\right) + \frac{q}{m}.$$

Differentiating

$$\frac{d\delta r}{dr} = \left(\frac{1}{9\mu} - \frac{1}{3m}\right)\left(p + q + \frac{h+k}{2}\right) + \frac{q}{m} + r\left(\frac{1}{9\mu} - \frac{1}{3m}\right)\left(\frac{dp}{dr} + \frac{dq}{dr}\right) + \frac{r}{m}\frac{dq}{dr}.$$

But

$$\frac{d\delta r}{dr} = 1 - \frac{dr}{ds} + \frac{dr}{ds}\frac{d\delta s}{ds}.$$

---

(115) Case VI of 'On the equilibrium of elastic solids': 108–10 (= *Scientific Papers*, **1**: 57–9).

(116) Read: $ds = \sqrt{(dx)^2 + (dr)^2}$.

Substituting for $\dfrac{d\delta s}{dr}$ and equating the values of $\dfrac{d\delta r}{dr}$ we obtain

$$\left(1-\frac{dr}{ds}\right)\left(\frac{1}{9\mu}-\frac{1}{3m}\right)\left(p+q+\frac{h+k}{2}\right)+\frac{q}{m}+\frac{dr}{ds}\frac{p}{m}$$
$$+r\left(\frac{1}{9\mu}-\frac{1}{3m}\right)\left(\frac{dp}{dr}+\frac{dq}{dr}\right)+\frac{r}{m}\frac{dq}{dr}-1+\frac{dr}{ds}=0.$$

To obtain an expression for the curvature of the plate at the vertex let a be the radius of curvature then as an approximation to the equation of the surface

$$x=\frac{r^2}{2a}.$$

By substituting this value of x in the values of p and q and in the last equation the approximate value of a is found to be

$$a=\frac{t}{h-k}\frac{\dfrac{h+k}{2}\left(\dfrac{1}{9\mu}-\dfrac{1}{3m}\right)-1}{5\left(\dfrac{1}{9\mu}-\dfrac{1}{3m}\right)-\dfrac{9}{2m}}$$

or

$$a=\frac{t}{h-k}\frac{-18m\mu}{10m+57\mu}+t\frac{h+k}{h-k}\frac{m-3\mu}{10m+51\mu}$$

the last term being small compared with the first.

[10]   *Case VIII*[117]

Let the original form of the elastic solid be a very thin plate bounded by two parallel infinite lines and let it be twisted into the form of a skew screw surface whose equation is

$$\theta=\frac{x}{a}.$$

Let the thickness of the plate be denoted by t and the distance of any point from the axis by r.

Let the force p act perpendicularly to the axis and in the plane of the plate and let q act perpendicularly to p and in the direction of a tangent to the helix.

From the equation to the surface we find

$$ds=\sqrt{1+\frac{r^2}{a^2}}\,dx$$

---

(117) See Numbers 20 and 22.

and the curvature of the helix at any point

$$= \frac{1}{R} = \frac{r}{a^2 + r^2}.$$

The equations of the equilibrium of the element become

$$q\frac{r}{a^2 + r^2} - p\frac{r}{a}\frac{1}{\sqrt{a^2 + r^2}} - \frac{dp}{dr}\frac{\sqrt{a^2 + r^2}}{a} = 0$$

and by the equations of compression

$$\frac{d\delta s}{ds} = \left(\frac{1}{9\mu} - \frac{1}{3m}\right)(p + q) + \frac{1}{m}q$$

as the plate will be shortened by being twisted let the compression in the axis be denoted by c then $\frac{\delta x}{x} = c$.

$$\frac{d\delta s}{ds} = \sqrt{1 + \frac{r^2}{a^2}(1 - c)} - 1.$$

By combining these three equations we obtain

$$\left(\frac{2}{9\mu} + \frac{1}{3m}\right)p + \left(\frac{1}{9\mu} + \frac{2}{3m}\right)\left(\frac{a^2}{r} + r\right)\frac{dp}{dr} + \left(1 + \frac{r^2}{a^2}\right)^{-\frac{1}{2}} + c - 1 = 0.$$

As the solution of this linear equation would be too much encumbered with exponentials to be of practical utility and as this case is not applicable to the determination of coefficients, and since the value of p depends on the square and higher powers of $\frac{r}{a}$ we may neglect p and the value of q becomes

$$q = \frac{9m\mu}{m + 6\mu}\frac{d\delta s}{ds}$$

$$q = \frac{9m\mu}{m + 6\mu}\left(\sqrt{1 + \frac{r^2}{a^2}(1 - c)} - 1\right)$$

$$\int qt\frac{dx}{ds}dr = \frac{9m\mu}{m + 6\mu}t\left\{r(1 - c) - a\log\left(\frac{r}{a} + \sqrt{1 + \frac{r^2}{a^2}}\right) + c\right\}.$$

This integral taken from $r = b$ to $r = -b$ gives the total tension in the direction of the axis.

$$\int qtr\frac{d\theta}{ds}r\,dr = \frac{9m\mu}{m + 6\mu}t\left\{\frac{r^3}{3a}(1 - c) - ar\sqrt{1 + \frac{r^2}{a^2}} + \frac{a^2}{2}\log\left(\frac{\sqrt{1 + \frac{r^2}{a^2}} - \frac{r}{a}}{\sqrt{1 + \frac{r^2}{a^2}} + \frac{r}{a}}\right)\right\}.$$

This integral taken from $r = 0$ to $r = b$ gives the moment of torsion or the magnitude of the statical couple requisite to retain the plate in its position.

[11]   *Case ⌞VIII⌟*

Let the elastic solid be a cylinder and let it revolve round its axis, to find the pressures and compressions resulting from the centrifugal force.
  Equations 3 become in this case

$$q - p = r\frac{dp}{dr} - \frac{c_1}{t^2}r^2$$

t being the time of revolution and c_1 a constant depending on the specific gravity of the substance.

  The values of q and $\frac{dq}{dr}$ being substituted in the equation

$$\left(\frac{1}{9\mu} - \frac{1}{3m}\right)\left(\frac{do}{dr} + \frac{dp}{dr} + \frac{dq}{dr}\right) + \frac{1}{m}\frac{dq}{dr} + \frac{1}{m}\frac{q-p}{r}\left[= 0\right]^{(118)}$$

we obtain a differential equation the solution of which is

$$p = c_2\frac{1}{r^2} + c_3 r^2 + c_4$$

$$c_3 = \tfrac{1}{4}\frac{c_1}{t^2}\left(1 - \frac{1}{\frac{2}{9}\frac{m}{\mu} + \frac{4}{3}}\right).$$

c_2 vanishes when the cylinder is solid. In this case

$$p = c_3(r^2 - a^2)$$

$$q - p = \left(2c_3 - \frac{c_1}{t^2}\right)r^2$$

$$q = \left(3c_3 - \frac{c_1}{t^2}\right)r^2 - c_3 a^2.$$

                            Case ⌞IX⌟[(119)]

Let the elastic solid be an infinitely long hollow cylinder and let the interior and exterior surfaces be exposed to two constant temperatures till the flow of heat is uniform or

$$r\frac{dv}{dr} = c_1$$

then $v = c_1 \log r + c_2$.
  This equation gives the law of the temperature at any point.

---

  (118) See Case IV, and compare Case VIII of 'On the equilibrium of elastic solids': 111–12
(= *Scientific Papers*, **1**: 60–2).
  (119) Case IX of 'On the equilibrium of elastic solids': 112–13 (= *Scientific Papers*, **1**: 62–3).

The equations of elasticity become for the cylinder[120]

$$\frac{d\delta x}{dx} = \left(\frac{1}{9\mu} - \frac{1}{3m}\right)(o+p+q) + \frac{o}{m} + \frac{v}{c_3} = c_4 \text{ or constant}$$

$$\frac{d\delta r}{dr} = \left(\frac{1}{9\mu} - \frac{1}{3m}\right)(o+p+q) + \frac{p}{m} + \frac{v}{c_3}$$

$$\frac{\delta r}{r} = \left(\frac{1}{9\mu} - \frac{1}{3m}\right)(o+p+q) + \frac{q}{m} + \frac{v}{c_3}$$

combining these with the following equations

$$\frac{d}{dr}\frac{\delta r}{r} r = \frac{d\delta r}{dr}$$

$$q = p + r\frac{dp}{dr}$$

$$0 = \frac{c_4 - \dfrac{v}{c_3} - \left(\dfrac{1}{9\mu} - \dfrac{1}{3m}\right)\left(2p + r\dfrac{dp}{dr}\right)}{\dfrac{1}{9\mu} + \dfrac{1}{3m}}$$

$$v = c_1 \log r + c_2$$

we find the following differential equation

$$\frac{c_1}{c_3} + \frac{3r}{m}\frac{dp}{dr} + \frac{r^2}{m}\frac{d^2p}{dr^2} = 0$$

the solution of which is

$$p = \frac{c_1}{2c_3 m} \log \frac{r}{c_5} + \frac{c_6}{r^2}.$$

By giving particular values to the pressures and temperatures of the inside and outside surfaces the constants may be determined.

Let

$$p = A \log \frac{r}{B} + \frac{c}{r^2}$$

then

$$q - p = A - 2\frac{c}{r^2}$$

$$q = A + A \log \frac{r}{B} - \frac{c}{r^2}$$

$$I = \omega\left(A - 2\frac{c}{r^2}\right).$$

---

(120) See equations (13) §7; and note (96).

Case ⌐X¬[121]

In a solid cylinder unequally heated it is observed that

$$q - p = c_1 r^2$$

$$\therefore r\frac{dp}{dr} = c_1 r^2$$

$$\therefore p = \frac{c_1}{2}r^2 + c_2.$$

If $p = 0$ when $r = a$, $c_2 = -\dfrac{c_1 a^2}{2}$

$$p = \frac{c_1}{2}\left(r^2 - a^2\right)$$

$$q = \frac{c_1}{2}\left(3r^2 - a^2\right).$$

$$\frac{d\delta r}{dr} = \left(\frac{1}{9\mu} - \frac{1}{3m}\right)\frac{c_1}{2}\left(4r^2 - 2a^2\right) + \frac{1}{m}\frac{c_1}{2}\left(r^2 - a^2\right) = \frac{v}{c_3}$$

$$= c_1\left(\frac{2}{9\mu} - \frac{1}{6m}\right)r^2 - c_1\left(\frac{1}{9\mu} + \frac{1}{6m}\right)a^2 + \frac{v}{c_3}.$$

$$\frac{\delta r}{r} = \left(\frac{1}{9\mu} - \frac{1}{3m}\right)c_1\left(2r^2 - a^2\right) + \frac{1}{m}\frac{c_1}{2}\left(3r^2 - a^2\right) + \frac{v}{c_3}$$

$$= c_1\left(\frac{2}{9\mu} + \frac{5}{6m}\right)r^2 - c_1\left(\frac{1}{9\mu} + \frac{1}{6m}\right)a^2 + \frac{v}{c_3}.$$

$$\frac{d}{dr}\frac{\delta r}{r}r = 3c_1\left(\frac{2}{9\mu} + \frac{5}{6m}\right)r^2 - c_1\left(\frac{1}{9\mu} + \frac{1}{6m}\right)a^2 + \frac{v}{c_3} + \frac{r}{c_3}\frac{dv}{dr}.$$

$$\therefore c_1\left(\frac{4}{9\mu} + \frac{8}{3m}\right)r^2 + \frac{r}{c_3}\frac{dv}{dr} = 0$$

$$v = c_1 c_3\left(\frac{2}{9\mu} + \frac{4}{3m}\right)r^2 + c$$

therefore the ⟨temperature⟩ ⌐difference of the temperature of any point from the central temperature¬ is as the square of the radius.[122]

---

(121) Case X of 'On the equilibrium of elastic solids': 113–14 (= *Scientific Papers*, **1**: 63–4).
(122) See §5 and note (72).

[12]　　　　　　　　　*Case XII*
　　　　　　　Combination of Compressions

When the values of the compressions have been found for pressures acting on a solid in two different ways separately, the compressions when the forces act at the same time may be found by adding them together, for since the displacements are small, the coordinates will not be sensibly altered and therefore the compressions which are functions of these coordinates will be independent of each other.

　It has been shown that in a cylinder

$$c_1 + \frac{c_2}{r^2} = \frac{d\delta r}{dr}$$

$$c_1 - \frac{c_2}{r^2} = \frac{\delta r}{r}.$$

By equations　　　　(123) these become

$$\frac{d\delta x}{dx} = c_1 + c_2 \frac{x^2 - y^2}{(x^2 + y^2)^2}$$

$$\frac{d\delta x}{dy} = c_1 - c_2 \frac{x^2 - y^2}{(x^2 + y^2)^2}$$

$$\frac{d\delta x}{dy} + \frac{d\delta y}{dx} = 4c_2 \frac{xy}{(x^2 + y^2)^2}.$$

　These equations may be applied to find the pressure at any point of an infinite plate in which the pressures emanate from two points in the axis of x whose coordinates are $+a$ and $-a$. Substitute $x+a$ and $x-a$ for x in the equations and add the results.[a]

[13]　　　　*On the determination of the coefficients m and μ*[124]

The coefficient m may be determined with great accuracy from experiments on the torsion of cylindrical rods or wires.

　The value of m has been found in Case II to be

$$m = \frac{4M}{\pi r^4 \dfrac{d\delta\theta}{dx}}.$$

(a) {Forbes} to be omitted.

(123)　Left blank in the manuscript.　　　　(124)　Compare Number 24.

The value of $\dfrac{9m\mu}{m+6\mu}$ may be deduced from the proportional elongation of the same rod corresponding to a given tension as in Case IV.[125]

$$\frac{9m\mu}{m+6\mu} = \frac{\,0\,}{\dfrac{\delta x}{x}} = E.$$

By substituting the previous value of m, we find

$$\mu = \tfrac{1}{3}\frac{Em}{3m-2E}.$$

Let a rod be placed between two supports, and let a weight $= w_2$ be suspended at the middle, then by ⌊Case XI⌋ $w_2 = -(2W+2w_1)$

and

$$\delta y = -W\left\{\tfrac{5}{24}\frac{l^3}{A}+\tfrac14\frac{l}{B}\right\}-w_2\left\{\frac{l^3}{6A}+\frac{l}{2B}\right\}.\text{[126]}$$

A and B may be found varying w_2 and l and from these the values of μ and m may be found.

The instrument employed by Oersted for determining the compressibility of liquids[127] may be employed for determining the moduli of elasticity. Let the modulus of cubical elasticity of the contained liquid be μ_2 and let the volume of the liquid which falls in the capillary tube[128] be denoted by ω then by the equation in Case V for a sphere

$$\frac{\omega}{V} = \frac{a^3h-b^3k}{a^3-b^3}\frac{1}{\mu}+\tfrac32 b^3\frac{h-k}{a^3-b^3}\frac{1}{m}-h\frac{1}{\mu_2}$$
$$= \left\{\frac{a^3}{a^3-b^3}\frac{1}{\mu}+\tfrac32\frac{b^3}{a^3-b^3}\frac{1}{m}-\frac{1}{\mu_2}\right\}h-\frac{b^3}{a^3-b^3}\left(\frac{1}{\mu}+\tfrac32\frac{1}{m}\right)k.\text{[129]}$$

If one of the three quantities μ m or μ_2 be known the other two may be found from these equations by taking different values of h and k, and the coefficient m seems to be that which can be determined with most accuracy.

Having found the modulus of cubical elasticity for the envelope μ_2 may be found by making $h = k$

then

$$\frac{\omega}{V}\frac{1}{h} = \frac{1}{\mu}-\frac{1}{\mu_2}$$

and this is independent of the form of the containing vessel so that the vessel

---

(125) See §8.

(126) See Case XII of 'On the equilibrium of elastic solids': 114–15 (= *Scientific Papers*, **1**: 65–6).

(127) See Oersted, 'On the compressibility of water'.   (128) See §5.

(129) Compare 'On the equilibrium of elastic solids': 105 (= *Scientific Papers*, **1**: 52–3).

may be made of such a form that solid bodies may be introduced. Let V_3 be the volume of the solid V_1 the content of the envelope μ_3 the modulus of cubical elasticity for the solid

$$\omega = h\left(\frac{V_1}{\mu} - \frac{V_1 - V_3}{\mu_2} - \frac{V_3}{\mu_3}\right).$$

When the vessel is a cylinder terminated by hemispheres the expression for the descent of the drop of mercury[130] is

$$\omega = \left(\frac{a^3 h - b^3 k}{a^3 - b^3}\frac{1}{\mu} + \tfrac{3}{2}b^3\frac{h-k}{a^3 - b^3}\frac{1}{m}\right)a^{\frac{34}{3}}\pi$$
$$+ \left(\frac{a^2 h - b^2 k}{a^2 - b^2}\frac{1}{\mu} + 2b^2\frac{h-k}{a^2 - b^2}\frac{1}{m}\right)a^2 c\pi.$$

Here a = internal radius of the sphere and cylinder
  b = external
  c = length of the cylinder.

The quantity ω, which gives the optical effect produced on a transparent elastic solid by a given compression, may be best determined for very soft solids by an experiment founded on Case I. For hard substances direct compression or bending would perhaps answer better.

As glass is a very elastic substance, and therefore well fitted for experiments, it is very convenient to have the means of determining the tension at any point by simple inspection, as we may thus ascertain when the glass is near its breaking point.

The quantity V is probably a function of m and μ, and the same function for different substances. If this function were known, some explanation might be given of the doubly refracting power communicated by pressure.

JAMES CLERK MAXWELL
Edinburgh 10th December 1849.

[14][131]   *Appendix*

In order to obtain an approximation to the numerical values of the moduli of linear and cubical elasticity which enter into all the equations of the equilibrium of elastic solids, I made some rough experiments on the twisting and bending of cylindric rods.[132]

---

(130) In Regnault's experiments; see note (51).

(131) ULC Add. MSS 7655, V, g/3. Compare Number 24.

(132) J. D. Forbes' experiments on elasticity, reported to the Royal Society of Edinburgh on 20 March 1848, may well have stimulated Maxwell's interest in determining the coefficients of elasticity; see Forbes, 'On an instrument for measuring the extensibility of elastic solids', *Proc. Roy. Soc. Edinb.*, **2** (1848): 173–5; and see Number 24 note (8).

In the first experiment the rods were fixed at one end while a twisting force was applied at the other by means of a lever and weight, two indices were fastened to the rod at a known distance apart, so that the angle of torsion might be observed at two distant points on the axis of the rod. Weights were then applied to the lever and the arc through which each index moved was measured.[133]

The modulus of linear elasticity was then determined by the formula

$$m = \frac{720}{\pi^2} \frac{x}{r^4} \frac{aw}{n} \quad [134]$$

where m is the modulus of linear elasticity

x is the distance of the indices in inches

r the radius of the cylindric rod ———

a the arm of the lever[a] ——————

w the weight acting on the lever in pounds

n the difference of the angle of rotation in degrees.

The modulus of linear elasticity[b] may be defined to be the reciprocal of the fraction of an inch which denotes the difference of the compression in the axis of pressure and in a perpendicular axis when a pressure of a pound is applied to two opposite faces of a cubic inch of the substance.

The rod was next placed between two supports and weights were applied at the middle of the rod and the deflection was observed for every additional weight.[135] The quantity E was then found from the expression for the deflection of a beam given in Case XI[136]

$$E = \frac{2}{3\pi} \frac{l^3}{r^4} \frac{w}{\delta y}$$

(a) {Forbes} in inches.

(b) {Forbes} $Aa - Cc = \dfrac{1}{m}$ in inches.

Figure 26,4

(133) Compare Félix Savart, 'Mémoire sur la réaction de torsion des lames et des verges rigides', *Ann. Chim. Phys.*, ser. 2, **41** (1829): 373–97; and experiments by Duleau reported in *ibid.*, **12** (1819): 133–48, to which Maxwell makes reference in §5 (see note (74)).

(134) See Case II, §8, and note (103).

(135) Compare the method employed by Forbes and 'sGravesande; see note (132), Number 24 note (8), and Number 31.

(136) Case XII of 'On the equilibrium of elastic solids': 114–15 (= *Scientific Papers*, **1**: 65–6).

where E is the reciprocal of the fraction of an inch by which the axis of pressure of a cubic inch of the substance is compressed when the pressure is the weight of a pound.

l is half the length of the rod in inches

r is the radius of the rod ————————

δy is the deflection produced by the addition of a weight

w is the weight in pounds.

Since
$$E = \frac{9m\mu}{m + 6\mu} \quad (137)$$

$$\mu = \frac{mE}{9m - 6E}.$$

So that μ may be found from m and E. The ratio of m to μ which varies from one substance to another is expressed by

$$\frac{\mu}{m} = \frac{1}{9\dfrac{m}{E} - 6} \qquad \frac{m}{\mu} = 9\frac{m}{E} - 6$$

so that any error in the determination of $\dfrac{m}{E}$ or $\langle E \rangle$ will produce an error nine times greater in the value of $\dfrac{m}{\mu}$, these ratios however will not be affected by any error in the measurement of the radius of the rod if the radius of the rod does not vary from one point on the axis to another.

μ, the modulus of cubical elasticity is the reciprocal of the fraction of a cubic inch which represents the change of volume when a cubic inch of the substance is acted on by pressures each equal to a pound acting on all its sides.

The quantity $\dfrac{15}{\mu}$ represents the proportional change of volume produced by the pressure of the atmosphere or 15 pounds to the square inch.

The compressibility of water is usually stated in this way.

Thus Oersted states the apparent compressibility of water in a glass vessel to be 46.1 millionths per atmosphere. If the compressibility of glass be 1.81 millionths per atmosphere that of water is 47.91.[138]

The results of my experiments are given in the following table.[139]

---

(137) See § 13.

(138) Oersted, 'On the compressibility of water': 359. The correction factor is Maxwell's: see his table.

(139) These experiments were subsequently mentioned by William Thomson, 'On the thermo-elastic and thermo-magnetic properties of matter', *Quarterly Journal of Pure and Applied Mathematics*, **1** (1855): 55–77, on 70 (= *Math. & Phys. Papers*, **1**: 291–313, on 305).

The first column contains the results of experiments on an iron wire, but as the texture of the wire is very fibrous, it can hardly be said to be truly homogeneous, and the results will be different on account of the hardening of the surface of the wire by the operation of drawing. The same may be said of the experiments on a brass wire. The modulus m deduced from the glass rod No. 3 differs from the result of No. 2 by a fraction of the probable error of observation, but No. 2 was broken in the determination of E. The columns headed 'Giulio' are deduced from his experiments on the torsion of metallic wires and the elasticity of helical springs in the Mémoires de l'Académie des Sciences de Turin, Série II Tome IV.[140] In these experiments m was determined by the oscillations of a suspended beam and E by the elongation of a helical spring.

	Iron 1[a]	Brass 1[a]	Glass 3[a]	Glass 2[a]	Iron Giulio
r radius of rod	0.069	$\dfrac{1.73}{8}$	$\dfrac{0.62}{8}$	$\dfrac{0.66}{8}$	
x distance of indices	36	41	34	36	
$aw = \frac{1}{4}$					
$n =$	1° 17′	1° 7′	2° 45′	2° 33′	
$m =$	21,800,000	9,748,000	6,251,000	6,280,000	20,470,000

(b)

	Iron	Brass	Glass 3	Brass (Giulio)	Iron (Giulio)
$2l =$	48	51	44		
$w = \frac{1}{16},\ \delta y =$	0.273	0.234	0.47		
$E =$	29,950,000	13,680,000	8,330,000	12,910,000	23,634,500

(c)

	Iron 1	Brass 1	Glass 3	Iron (Giulio)	Water (Oersted)
(d)					
$\mu =$	40,220,000	23,590,000	8,291,000	11,410,000	306,000
$\dfrac{\mu}{m} =$	1.848 = 1848	2.420	1.326	0.5573	

(a) {Forbes} m.
(b) {Forbes} Linear El.

(c) {Forbes} Mod. Elast[y].
(d) {Forbes} Cubic Elast[y].

(140)   Ignace Giulio, 'Sur la torsion des fils métalliques et sur l'élasticité des ressorts en hélices', *Memorie della Reale Accademia delle Scienze di Torino*, ser. 2, **4** (1842): 329–83.

	Iron 1	Brass 1	Glass 3	Iron (Giulio)	Water (Oersted)
$\dfrac{m}{E}$	0.7268[e]				
$\dfrac{m}{\mu}$	0.5412				
$1{,}000{,}000\,\dfrac{15}{m}$	$\langle 0.655 \rangle$	1.54	2.4		
$1{,}000{,}000\,\dfrac{15}{E}$	0.607	1.09	1.8		
$1{,}000{,}000\,\dfrac{15}{\mu}$	0.747	0.636	1.81		47.91

Jᴀᴍᴇs Cʟᴇʀᴋ Mᴀxᴡᴇʟʟ
15 Feb. 1850

(e) {Forbes} (On Poisson's theory $\frac{5}{6}$ const).[141]

(141) See note (22).

27

ABSTRACT OF PAPER 'ON THE EQUILIBRIUM OF ELASTIC SOLIDS'

[18 FEBRUARY 1850][1]

From the *Proceedings of the Royal Society of Edinburgh*[2]

This paper[3] commenced by pointing out the insufficiency of all theories of elastic solids, in which the equations do not contain two independent constants deduced from experiments. One of these constants is common to liquids and solids, and is called the modulus of *cubical* elasticity. The other is peculiar to solids, and is here called the modulus of *linear* elasticity. The equations of Navier, Poisson, and Lamé and Clapeyron,[4] contain only one coefficient; and Professor G. G. Stokes of Cambridge, seems to have formed the first theory of elastic solids which recognised the independence of cubical and linear elasticity,[5] although M. Cauchy seems to have suggested a modification of the old theories, which made the ratio of linear to cubical elasticity the same for all substances.[6] Professor Stokes has deduced the theory of elastic solids from that of the motion of fluids, and his equations are identical with those of this paper, which are deduced from the two following assumptions.

In an element of an elastic solid, acted on by three pressures at right angles to one another, as long as the compressions do not pass the limits of perfect elasticity –

1*st*, The sum of the pressures, in three rectangular axes, is proportional to the sum of the compressions in those axes.

2*d*, The difference of the pressures in two axes at right angles to one another, is proportional to the difference of the compressions in those axes.[7]

Or, in symbols:

$$(P_1 + P_2 + P_3) = 3\mu\left(\frac{\delta x}{x} + \frac{\delta y}{y} + \frac{\delta z}{z}\right) \tag{1}$$

---

(1) The date the paper was presented to the Royal Society of Edinburgh (see note (2)). Writing to William Thomson on 19 February 1850 J. D. Forbes reported that 'I did my best last night to render Clerk Maxwell's paper intelligible...' (ULC Add. 7342, F203A). Compare Number 28 note (2).

(2) *Proc. Roy. Soc. Edinb.*, **2** (1850): 294–6 (= *Scientific Papers*, **1**: 72–3).

(3) James Clerk Maxwell, 'On the equilibrium of elastic solids', *Trans. Roy. Soc. Edinb.*, **20** (1850): 87–120 (= *Scientific Papers*, **1**: 30–72). See Number 26.

(4) See Number 26 note (7).

(5) See Number 26 notes (11) and (44).

(6) See Number 26 notes (10), (44) and (93).

(7) See Number 26 §2.

$$
\left.
\begin{aligned}
(P_1 - P_2) &= m\left(\frac{\delta x}{x} - \frac{\delta y}{y}\right) \\
(P_2 - P_3) &= m\left(\frac{\delta y}{y} - \frac{\delta z}{z}\right) \\
(P_3 - P_1) &= m\left(\frac{\delta z}{z} - \frac{\delta x}{x}\right)
\end{aligned}
\right\}
\tag{2}
$$

μ being the modulus of *cubical*, and m that of *linear* elasticity.[8]

These equations are found to be very convenient for the solution of problems, some of which were given in the latter part of the paper.

These particular cases were –

That of an elastic hollow cylinder, the exterior surface of which was fixed, while the interior was turned through a small angle. The action of a transparent solid thus twisted on polarized light, was calculated, and the calculation confirmed by experiment.[9]

The second case related to the torsion of cylindric rods, and a method was given by which m may be found. The quantity $E = \dfrac{9mn}{m + 6n}$[10] was found by elongating, or by bending the rod used to determine m, and μ is found by the equation,

$$
\mu = \frac{Em}{9m - 6E}. \tag{11}
$$

The effect of pressure on the surfaces of a hollow sphere or cylinder was calculated, and the result applied to the determination of the cubical compressibility of liquids and solids.[12]

An expression was found for the curvature of an elastic plate exposed to pressure on one side; and the state of cylinders acted on by centrifugal force and by heat was determined.[13]

The principle of the superposition of compressions and pressures was applied to the case of a bent beam, and a formula was given to determine E from the deflection of a beam supported at both ends and loaded at the middle.[14]

The paper concluded with a conjecture, that as the quantity ω, (which expresses the relation of the inequality of pressure in a solid to the doubly-refracting force produced) is probably a function of m; the determination of these quantities for different substances might lead to a more complete theory of double refraction, and extend our knowledge of the laws of optics.[15]

---

(8) See Number 26 note (11).   (9) Case I of Number 26 §8.   (10) For n read μ.
(11) See Number 26 Case II (§5) and §§13 and 14.   (12) Cases IV and V of Number 26 §8.
(13) Cases VII, VIII and IX of Number 26 §§9 and 11.   (14) See Number 26 §§5 and 13.
(15) See Number 26 §2; and the supplementary 'Note C' of 'On the equilibrium of elastic solids': 120 (= *Scientific Papers*, **1**: 72).

28

FROM A LETTER TO LEWIS CAMPBELL

14 MARCH 1850

From Campbell and Garnett, *Life of Maxwell*[1]

[Edinburgh]
14 March 1850

As I am otherwise engaged, I take this opportunity of provoking you to write a letter or two. I have begun to write Elastic Equilibrium,[2] and I find that I must write you a letter in order *at least, indeed,* to serve *on the one hand* as an excuse to myself for sticking up, and *on the other hand* as a sluice for all the nonsense which I would have written. I therefore propose to divide this letter as follows: – My say naturally breaks up into – 1. Education; 2. Notions; 3. Hearsay.

1. Education – Public.

10–11. – Gregory[3] is on Alloys of Metals just now. Last Saturday I was examined, and asked how I would do if the contents of a stomach were submitted to me to detect arsenic, and I had to go through the whole of the preparatory processes of chopping up the tripes, boiling with potash, filtering, boiling with HCL and $KOCLO_5$,[4] all which Kemp the *Practical*[5] says are useless and detrimental processes, invented by chemists who want something to do. 11–12. – Prof. Forbes is on Sound and Light day about, as Bob[6] well knows, and can tell you if he chooses. He (R.C.) has written an essay on Probabilities, with very grand props in it; everything original, but no signs of

---

(1) *Life of Maxwell*: 127–9.

(2) The drafts (annotated by Forbes) are reproduced in Number 26. The paper was presented to the Royal Society of Edinburgh on 18 February 1850; see Number 27. On 4 May 1850 Forbes wrote to Maxwell: 'My dear Sir, Professor Kelland, to whom your paper was referred by the Council R.S., reports favourably upon it, but complains of the great obscurity of several parts, owing to the *abrupt transitions* and want of distinction between what is *assumed* and what is *proved* in various passages, which he has marked in pencil, and which I trust that you will use your utmost effort to make plain and intelligible. It must be perfectly evident that it must be useless to publish a paper for the use of scientific readers generally, the steps of which cannot, in many places, be followed by so expert an algebraist as Prof. Kelland; – if, indeed, they be *steps* at all and not assumptions of theorems from other writers who are not quoted. You will please to pay particular attention to clear up these passages, and return the MS by post to Professor Kelland, West Cottage, Wardie, Edinburgh, so that he may receive it by Saturday the 11th, as I shall then have left town. – Believe me, yours sincerely, James D. Forbes'. (*Life of Maxwell*: 137–8.)

(3) See Number 25 note (2).   (4) As in *Life of Maxwell*.

(5) Alexander Kemp, Teacher of Practical Chemistry at Edinburgh, and Assistant to William Gregory, Professor of Chemistry; see *Phil. Mag.*, ser. 3, **27** (1845): 350; and *Edinburgh New Philosophical Journal*, **48** (1850): 354.

(6) Robert Campbell: see Number 10 note (3).

reading, I guess.[7] It was all written in a week. He has despaired of Optics. –
12-1. Wilson,[8] after having fully explained his own opinions, has proceeded
to those of other great men: Plato, Aristotle, Stoics, Epicureans. He shows that
Plato's *proof* of immortality of the soul, from its immateriality, if it be a proof,
proves its pre-existence, the immortality of beasts and vegetables, and why not
transmigration? (Do you remember how Raphael tells Adam about meats and
drinks in *Paradise Lost?*)[9] (Greek Iambics, if you please.) He quarrels with
Aristotle's doctrine of the Golden Mean, – 'a virtue is the mean between two
vices', – not properly understanding the saying. He chooses to consider it as a
pocket rule to find virtue, which it is not meant to be, but an apophthegm or
maxim, or dark saying, signifying that as a hill falls away on both sides of the
top, so a virtue at its maximum declines by excess or defect (not of virtue but)
of some variable quantity at the disposal of the will. Thus, let it be a virtue to
give alms with your own money, then it is a greater virtue to pay one's debts to
the full. Now, a man has so much money: the more alms he gives up to a certain
point, the more virtue. As soon as it becomes impossible to pay debts, the virtue
of solvency decreases faster than that of almsgiving increases, so that the giving
of money to the poor becomes a vice, so that the variable is the sum given away,
by excess or defect of which virtue diminishes, say I; so that Wilson garbles
Aristotle, – but I bamboozle myself. I say that some things are virtues, others
are virtuous or generally lead to virtue. Substitute *goods* for virtues, and it will
be more general: thus, Wisdom, Happiness, Virtue, are *goods*, and cannot be in
excess; but Knowledge, Pleasure, and – what? (please tell me, Is it Propriety,
Obedience, or what is it?) lead to the other three, and are not so much goods
as tending to good; whereas particular knowledges, pleasures, and obediences
may be in excess and lead to evils. I postpone the rest of my observations to my
Collection of the Metaphysical principles of Moral Philosophy founded on the
three laws of Liberty, Equality, Fraternity, thus expressed:

1. That which can be done is that which has been done; that is, that the
possibility (with respect to the agent) of an action (as simple) depends on the
agent having had the sensation of having done it.

2. That which ought to be done is that which (under the given conditions)

---

(7) See also Robert Campbell, 'On the probability of uniformity in statistical tables', *Report of
the Twenty-ninth Meeting of the British Association* (London, 1860), part 2: 3–4; and Campbell, 'On
a test for ascertaining whether an observed degree of uniformity, or the reverse, in tables of statistics,
is to be looked upon as *remarkable*', *Phil. Mag.*, ser. 4, **18** (1859): 359–68.

(8) John Wilson (Christopher North), Professor of Moral Philosophy at Edinburgh (*DNB*).
See The *Works of Professor Wilson*, ed. J. Ferrier, 12 vols. (Edinburgh/London, 1855–8); and for
Maxwell's comments, see *Life of Maxwell*: 135–6, 145.

(9) John Milton, *Paradise Lost*, Book V, 405–33.

produces, implies, or tends to the greatest amount of good (an excess or defect in the variables will lessen the good and make evil).

3. Moral actions can be judged of only by the principle of exchange; that is (1), our own actions must be judged by the laws we have made for others; (2), others must be judged by putting ourselves in their place.

29

FROM A LETTER TO LEWIS CAMPBELL

22 MARCH [1850][1]

From Campbell and Garnett, *Life of Maxwell*[2]

[Edinburgh]
22 March [1850]

At Practical Mechanics I have been turning Devils of sorts. For private studies I have been reading Young's *Lectures*, Willis's *Principles of Mechanism*, Moseley's *Engineering and Mechanics*, Dixon on *Heat*, and Moigno's *Répertoire d'Optique*.[3] This last is a very complete analysis of all that has been done in the optical way from Fresnel to the end of 1849, and there is another volume a-coming which will complete the work. There is in it besides common optics all about the other things which accompany light, as heat, chemical action, photographic rays, action on vegetables, etc.

My notions are rather few, as I do not *entertain* them just now. I have a notion for the torsion of wires and rods, not to be made till the vacation; of experiments on the action of compression on glass, jelly, etc., numerically done up;[4] of papers for the Physico-Mathematical Society (which is to revive in earnest next session!);[5] on the relations of optical and mechanical constants, their desirableness, etc., and suspension-bridges, and catenaries, and elastic curves. Alex. Campbell, Agnew, and I are appointed to read up the subject of periodical shooting stars, and to prepare a list of the phenomena to be observed on the 9th August and 13th November. The Society's barometer is to be taken up Arthur's Seat at the end of the session, when Forbes goes up, and All students are invited to attend, so that the existence of the Society may be recognised.

I have notions of reading the whole of *Corpus Juris* and Pandects in no time at all; but these are getting somewhat dim, as the Cambridge scheme has been howked up from its repose in the regions of abortions, and is as far forward as an inspection of the Cambridge *Calendar* and a communication with Cantabs.

Mr. Bob is choosing his college.[6] I rejected for him all but Peter's, Caius, or Trinity Hall, the last being, though legal, not in favour, or lazy, or something. Caius is populous, and is society to itself. Peter's is select, and knows the

---

(1) Suggested by Campbell.   (2) *Life of Maxwell*: 129–30.

(3) Thomas Young, *A Course of Lectures on Natural Philosophy and the Mechanical Arts*, 2 vols. (London, 1807); rev. edn, ed. P. Kelland, 2 vols. (London, 1845); Robert Willis, *Principles of Mechanism* (Cambridge, 1841): Henry Moseley, *The Mechanical Principles of Engineering and Architecture* (London, 1843); R. V. Dixon, *A Treatise on Heat, Part I. The thermometer; dilatation; change of state; and laws of vapours* (Dublin, 1849); Francois Napoléon Marie Moigno, *Répertoire d'Optique Moderne*, 4 vols. (Paris, 1847–50).

(4) See Numbers 26 and 31.   (5) See Number 17.   (6) See Number 10 note (3).

University. Please give me some notions on these things, both for Bob and me. I postpone my answer to you about the Gorham business[7] till another time, when also I shall have read Waterland on *Regeneration*,[8] which is with Mrs. Morrieson,[9] and some Pusey books[10] I know. In the meantime I admire the *Judgement*[11] as a composition of great art and ingenuity.

---

(7) The appeal by the evangelical Rev. George Cornelius Gorham to the Judicial Committee of the Privy Council against the Bishop of Exeter's refusal to institute him into the living of Brampton-Speke, on the grounds that he denied that spiritual regeneration is always conferred by baptism. For evangelicals (and Calvinists) regeneration was associated with conversion. See Owen Chadwick, *The Victorian Church*, **1** (London, 1967): 250–62.

(8) Daniel Waterland, *Regeneration stated and explained according to Scripture and Antiquity* (London, 1740).   (9) Lewis Campbell's mother; see *Life of Maxwell*: 132.

(10) Edward Bouverie Pusey, Tractarian theologian (*DNB*); see his *Tracts for the Times. Scriptural Views of the Holy Baptism* (Oxford, 1835).

(11) The Judgement of the Judicial Committee of the Privy Council in favour of Gorham's appeal; see the report in *The Times* (9 March 1850), where the Judgement is printed.

FROM A LETTER TO LEWIS CAMPBELL

26 APRIL 1850

From Campbell and Garnett, *Life of Maxwell*[1]

Glenlair
26 April 1850

As I ought to tell you of our departure from Edinburgh and arrival here, so I ought to tell you of many other things besides. Of things pertaining to myself there are these: The tutor of Peterhouse has booked me, and I am booked for Peterhouse,[2] but will need a little more booking before I can write Algebra like a book.

I suppose I must go through Wrigley's problems[3] and Paley's Evidences[4] in the same sort of way, and be able to translate when required Eurip. *Iph. in Aulid.*[5] In the meantime I have my usual superfluity of plans.

1. Classics – Eurip. Iφ. εν Αυλ. for Cambridge. (I hope no Latin or Greek verses except for honours.) Greek Testament, Epistles, for my own behoof, and perhaps some of Cicero *De Officiis* or something else for Latin.

2. Mathematics – Wrigley's Problems, and Trig. for Cambridge; properties of the Ellipsoid and other solids for practice with Spher. Trig. Nothing higher if I can help it.

3. Nat. Phil. – Simple mechanical problems to produce that knack of solving problems which Prof. Forbes has taught me to despise. Common Optics at length; and for experimental philosophy, twisting and bending certain glass and metal rods, making jellies, unnealed glass, and crystals, and dissecting eyes – and playing Devils.[6]

4. Metaphysics – Kant's *Kritik of Pure Reason* in German, read with a determination to make it agree with Sir W. Hamilton.

5. Moral Philosophy – Metaphysical principles of moral philosophy. Hobbes' *Leviathan*, with his moral philosophy, to be read as the only man who

---

(1) *Life of Maxwell*: 134–5. The letter is cut from the version printed by Campbell, which includes discussion of poetry and aesthetics (*ibid.*: 135–7).

(2) See *Admissions to Peterhouse or S. Peter's College in the University of Cambridge. A Biographical Register*, compiled by Thomas Alfred Walker (Cambridge, 1912): 499, where for the year 1850 it is recorded: 'Aprilis 22. Jacobus Clerk Maxwell, Johannis C. Maxwell de Dumfries filius, ad mensam pensionariorum admittitur'.

(3) A. Wrigley, *A Collection of Examples and Problems in Pure and Mixed Mathematics, with answers and occasional hints* (London, ₂1847).

(4) William Paley, *A View of the Evidences of Christianity* (London, 1794).

(5) Euripides, 'Iphigenia in Aulis'.

(6) See Number 9 note (2).

has decided opinions and avows them in a distinct way.[7] To examine the first part of the seventh chapter of Matthew in reference to the moral principles which it supposes,[8] and compare with other passages.

But I question if I shall be able to overtake all these things, although those of different kinds may well be used as alternate studies.

---

(7) Maxwell was apparently writing an essay on Hobbes at this time. There is a fragment on 'Leviathan' (ULC Add. MSS 7655, V, 1/1). See also Number 31.

(8) See Number 31.

FROM A LETTER TO LEWIS CAMPBELL

circa JULY 1850[1]

From Campbell and Garnett, *Life of Maxwell*[2]

[Glenlair]

As there has been a long truce between us since I last got a letter from you, and as I do not intend to despatch this here till I receive Bob's answer with your address, I have no questions to answer, and any news would turn old by keeping, so I intend briefly to state my country occupations (otherwise preparation for Cambridge, if you please). I find that after breakfast is the best time for reading Greek and Latin, because if I read newspapers or any of those things, then it is dissipation and ruin; and if I begin with props, experiments, or calculations, then I would be continually returning on them. At first I had got pretty well accustomed to regular study with a Dictionary, and did about 120 lines of *Eurip.* a day, namely, 40 revised, 40 for to-day, and 40 for to-morrow, with the looking up of to-morrow's words. As I am blest with Dunbar's *Lexicon*, it is not very highly probable that I will find my word at all; if I do, it is used in a different sense from Dunbar's (so much the better), and it has to be made out from the context (either of the author or the Dictionary).[3] So much for regular study, which I have nearly forgot, for when I had got to the end of the first chorus I began to think of the rods and wires that I had in a box. They have entirely stopped *Eurip.*, for I found that if I spent the best part of the day on him, and took reasonable exercise, I could not much advance the making of the apparatus for tormenting these wires and rods. So the rods got the better of the Lexicon. The observations on the rods are good for little till they are finished; they are of three kinds, and are all distinguished for accuracy and agreement among themselves.

Thus – a rod bent by a weight at a middle takes the form of a curve, which is calculated to be one of the fourth order. Let ACB be the rod bent by a weight at C. Mirrors fastened to it at A and B make known the changes of the inclination of the tangent to the rod there, and a lens at C projects an image of a copper scale of inches and parts from A to B,

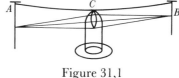

Figure 31,1

---

(1) The date June 1850 is suggested by Campbell; but July is likely (see notes (4) and (11)).

(2) *Life of Maxwell*: 138–44.

(3) George Dunbar, *A New Greek and English and English and Greek Lexicon, with an appendix explanatory of scientific terms, &c.* (Edinburgh, ₂1844, ₃1850); on Maxwell's reading of Euripides for Cambridge see Number 30.

where it is observed, and so the deflection of the rod at C becomes known. Now the calculated value of the elasticity deduced from the deflection differs from that deduced from the observations on the mirrors by about $\frac{1}{140}$ of either, and as the deflection at C was about $\frac{1}{4}$ inch, the difference of the observed and calculated deflections is about $\frac{1}{280}$ of an inch, which is near enough for home-made philosophical instruments to go.[4]

Thus you see I would run on about rods and wires, and weights, angles, and inches, and copper and iron, and silvered glass, and all sorts of practicalities. Where is now *Eurip.*? – Ay, where? On the top of the Lexicon, and behind bundles of observations and calculations. When will he come out? for he was a good soul after all, and wise (beg his pardon, *wiser*). For the rest I have been at Shakespeare and Cowper. I used to put Thomson and Cowper together (why?), and Thomson first; now they are reversed and far asunder.[5]

As I suppose my occupations are not very like yours, I pray you send me an account of what Oxford notions you have got, either from Oxonians, books, or observation; and as, if I was to question you, you could but answer my questions, I leave you to question yourself and send me some of the answers.

The only regular College science that I have thought of lately is Moral Philosophy. Whether it is an Oxford science I know not; but it must be, if not taught, at least interesting; so I purpose to fill up this letter with unuttered thoughts (or crude), which, as they are crammed into words, may appear like men new waked from sleep, who leap in confusion into one another's breeches, hardly fit to be seen of decent men. Then think not my words mad if their clothes fit them not, for they have not had an opportunity of trying them on before.

There are some Moral Philosophers whose opinions are remarkable for their general truth and good sense, but not for their utility, fixity, or novelty.

They tell you that in all your actions you ought to be virtuous, that benevolence is a virtue, that lawful rulers ought to be obeyed, that a man should give ear to his conscience.

Others tell you of unalterable laws of right and wrong, of Eternal truth and the Everlasting fitnesses of things. Others of the duty of following nature, of every virtue between two vices (Aristot.), and of the golden mean. That a man should do what is best on the whole (1) for himself; (2) for other men only, and *not* himself; (3) for the whole universe, including himself, and so on. Now I

---

(4) Compare Number 26 §14, esp. note (132). In his personal copy of 'On the equilibrium of elastic solids' reprinted from *Trans. Roy. Soc. Edinb.*, **20** (1850): 87–120 (Cavendish Laboratory, Cambridge), Maxwell recorded 'Results of Experiments July 1850' on the bending of iron wires.

(5) The poets William Cowper and James Thomson (author of 'The Seasons' (1730)).

think that the answers to the following questions should be separate parts of M. Ph.:

1. What is man? This is the introduction, and is called statical or proper Metaphysics.

2. What are the laws of human action? Action being all that man does – thought, word, deed.

3. What are the motives of human actions?

4. What actions do men perform in preference to what others, and why?

5. What is the principle by which men judge some actions right, others wrong?

6. What do particular men think of this principle? What are their doctrines?

7. What is the best criticism of right and wrong, or what (to us) is absolute right?

8. What are the best motives of human actions?

9. How are these motives to be implanted without violating the laws of human action?

10. What might, or rather what *will*, mankind become after this has been effected?

Moral Philosophy differs from Nat. Phil. in this, that the more new things we hear of in Nat. Phil. the better; but in Mor. Phil. the old things are best, so that a common objection to Mor. Phil. is that everybody knows it all before. If a man tells you that tyranny and anarchy are bad things, and that a just and lawful government is a good thing, it sounds very fine, but only means that when men think the government bad from excess or defect they give it the name of tyranny and anarchy. The ancient virtue of Tyrannicide was a man's determination to kill the king whenever he displeased him. Thus it is easy to call a dog a bad name to beat him for. But there are other parts of Mor. Phil. in which there are differences of opinion, such as the nature of selfishness, self-love, appetites, desires, and affections, disinterestedness (what a word for a rush at!), which belong to the first three questions, and so on. I have told you something[6] of three laws which I have been considering. In all parts of Mor. Phil. these three laws seem to meet one, and in each system of Morals they take a different form. Now, that I might not deceive myself in thinking that I was safe out of the hands of the philosophers who argue these matters, I have been looking into the books of Moralists the most opposed to one another, to see what it is that makes them differ, and wherein they agree. The three principles concerning the nature of man are continually changing their shape, so that it is not easy to catch them in their best shape. Nevertheless:

---

(6) See Number 28.

Lemma: Metaphysics. – A man thinks, feels, and wills, and therefore Meta-physicians give him the three faculties of cognition, feeling, and conation.

Cognition is what is called Understanding, and is most thought of generally.

Feelings are pleasures, pains, appetites, desires, aversions, approval and disapproval, love, hate, and all affections.

Conations are acts of will, whatever they be.

Now to move a man's will it is necessary to move his affections. (How? Wait!) For no convictions of the understanding will do, for a man does what he likes to do, not what he believes to be best for himself or others. The feelings can only be moved by notions coming through the understanding, for cognition is the only inlet of thoughts. Therefore, although it can be proved that self-love leads to all goodness, or, in other words, that goodness is happiness, and *self* loves happiness, yet it can also be proved that men are not able to act rightly from pure self-love; so that though self-love is a very fine theoretical principle, yet no man can keep it always in view, or act reasonably upon it. Now, most moralists take for granted that the end which men, good or bad, pursue is their own happiness, and that happiness, false or true, is the motive of every action, and that it is the only right motive. Others say that benevolence is the only virtue, and that any action not done expressly for the good of others is entitled to no praise.

Most of the ancients, and Hobbes among the moderns, are of the first opinion. Hutcheson and Brown (I think) are of the second, and call the first selfish Philosophers and the selfish school.[7] A few consider benevolence to the whole universe as the proper motive of every action, but they all (says Macintosh)[8] confound men's motives with the criterion of right and wrong, the reason why a thing is right, and that which actually causes a man to do it. In every book on Moral Philosophy some reference is made to that precept or maxim, which is declared to be the spirit of the law and the prophets (see Matt. vii. 12), and the application of it is a good mark of the uppermost thoughts or mode of thinking of the author.

Hobbes lays down as the first agreement of men to secure their safety, that a man should lay down so much of his natural liberty with respect to others, as he wishes that other men should to him. Hobbes having shown that men, in what

---

(7) Thomas Hobbes, *Leviathan, or The Matter, Forme, & Power of a Common-wealth Ecclesiasticall and Civill* (London, 1651); Francis Hutcheson, *An Inquiry into the Original of our Ideas of Beauty and Virtue* (Glasgow, ₄1772); Hutcheson, *A System of Moral Philosophy*, 2 vols. (Glasgow/London, 1755); for Thomas Brown's Edinburgh lectures on ethics see his *Lectures on the Philosophy of the Human Mind* (Edinburgh, ₇1833).

(8) James Mackintosh, *Dissertation on the Progress of Ethical Philosophy, chiefly during the Seventeenth and Eighteenth Centuries*, with a preface by William Whewell (Edinburgh, 1836).

the poets and moralists call a state of nature (that is, of equality and liberty, and without government), must be in a state of war, every man against every other, and therefore of danger to every man, deduces the obligation of obeying the powers that be from the necessity of Power to prevent universal war.[9] Adam Smith's theory of Moral Sentiments (which is the most systematic next to Hobbes) is that men desire others to sympathise with them, and therefore do those things which may be sympathised with; that is, as Smith's opponents say, men ought to be guided by the desire of esteem and sympathy. Not so. Smith does not leave us there, but I suppose you have read him, as he is almost the only Scotch Moral Philosopher.[10]

As it is Saturday night I will not write very much more. I was thinking to-day of the duties of [the] cognitive faculty. It is universally admitted that duties are voluntary, and that the will governs understanding by giving or with-holding Attention. They say that Understanding ought to work by the rules of right reason. These rules are, or ought to be, contained in Logic; but the actual science of Logic is conversant at present only with things either certain, impossible, or *entirely* doubtful, none of which (fortunately) we have to reason on. Therefore the true Logic for this world is the Calculus of Probabilities, which takes account of the magnitude of the probability (which is, or which ought to be in a reasonable man's mind).[11] This branch of Math., which is generally thought to favour gambling, dicing, and wagering, and therefore highly immoral, is the only 'Mathematics for Practical Men', as we ought to be. Now, as human knowledge comes by the senses in such a way that the existence of things external is only inferred from the harmonious (not similar) testimony of the different senses, Understanding, acting by the laws of right reason, will

---

(9) See Number 30 note (7).

(10) Adam Smith, *The Theory of Moral Sentiments* [1759], new edn, with a biographical and critical memoir of the author by Dugald Stewart (London, 1846). For Maxwell's further comments on Smith's moral philosophy see Number 62: Appendix.

(11) On Maxwell's reference to probabilistic arguments at this time compare [John Herschel,] 'Quetelet on probabilities', *Edinburgh Review*, **92** ([July] 1850): 1–57, esp. 8, a review of Adolphe Quetelet, *Letters addressed to H.R.H. the Grand Duke of Saxe-Cobourg and Gotha on the Theory of Probabilities as applied to the Moral and Political Sciences*, trans. O. G. Gregory (London, 1849). See also J. D. Forbes' critique of Herschel's discussion of John Michell's use of probabilistic arguments in connection with the distribution of the stars: J. D. Forbes, 'On the alleged evidence for a physical connexion between stars forming binary of multiple groups, arising from their proximity alone', *Phil. Mag.*, ser. 3, **35** (1849): 132–3; John Herschel, *Outlines of Astronomy* (London, 1849): 564–5; and John Michell, 'An inquiry into the probable parallax and magnitude of the fixed stars, from the quantity of light which they afford us, and the particular circumstances of their situation', *Phil. Trans.*, **57** (1767): 234–64. Forbes developed his argument in his paper 'On the alleged evidence for a physical connexion between stars forming binary or multiple groups, deduced from the doctrine of chances', *Phil. Mag.*, ser. 3, **37** (1850): 401–27.

assign to different truths (or facts, or testimonies, or what shall I call them) different degrees of probability. Now, as the senses give new testimonies continually, and as no man ever detected in them any real inconsistency, it follows that the probability and *credibility* of their testimony is increasing day by day, and the more a man uses them the more he believes them. He believes them. What is believing? When the probability (there is no better word found) in a man's mind of a certain proposition being true is greater than that of its being false, he believes it with a proportion of faith corresponding to the probability, and this probability may be increased or diminished by new facts. This is faith in general. When a man thinks he has enough of evidence for some notion of his he sometimes refuses to listen to any additional evidence *pro* or *con*, saying, 'It is a settled question *probatis probata*; it needs no evidence; it is certain.' This is knowledge as distinguished from faith. He says, 'I do not believe; I know.' 'If any man thinketh that he knoweth, he knoweth yet nothing as he ought to know.' This knowledge is a shutting of one's ears to all arguments, and is the same as 'Implicit faith' in one of its meanings. 'Childlike faith', confounded with it, is not credulity, for children are not credulous, but find out sooner than some think that many men are liars. I must now to bed, so good night; only please to write when you get this, if convenient, and state the probability of your coming here. We perhaps will be in Edinburgh when the Wise men are there.[12] Now you are invited in a corner of a letter by

JAMES CLERK MAXWELL.

---

(12) The meeting of the British Association for the Advancement of Science at Edinburgh in July and August 1850.

MANUSCRIPT ON EXPERIMENTS ON THE CAUSE
OF HAIDINGER'S BRUSHES

5 AUGUST 1850

From the original in the University Library, Cambridge[1]

EXPERIMENTS ON THE CAUSE OF HAIDINGERS BRUSHES[2]

AUGUST 5th 1850[3]

As the structure of the human eye has been minutely studied by many anatomists and opticians, the unexpected discovery of a polarizing apparatus in the living eye would lead us to think that some important part of the eye had escaped the observation of the oculists.

Since the phenomena of polarized light as seen by the lower animals are unknown to man, all our experiments must be made on the human eye, and it is well known that the dissection of the human eye requires all the skill of the most expert anatomists. I therefore made my own eye the subject of my experiments, and these experiments have induced me to think that a peculiar structure exists at a certain part of the retina ⟨   [4]⟩ a series of inductive experiments in order that those who are conversant with the anatomy of the eye may have an opportunity of testing the truth of my conjecture.

The appearance of Haidingers brushes has been described by Sir David Brewster[5] and Professor Stokes.[6] When white polarized light is used they

---

(1) ULC Add. MSS 7655, V, b/3.

(2) The phenomenon of vision reported in 1844 by Wilhelm Haidinger: The appearance, when looking at a source of polarized light, of pale yellow patches bounded on either side by curved arcs; see Wilhelm Haidinger, 'Ueber das directe Erkennen des polarisirten Lichts und der Lage der Polarisationsebene', *Ann. Phys.*, **63** (1844): 29–39; Haidinger, 'Beobachtungen der Lichtpolarisationsbüschel in geradlinig polarisirten Lichte', *ibid.*, **68** (1846): 73–87; and Haidinger, 'Beobachtungen der Licht-Polarisations-Büschel auf Flächen, welche das Licht in zwei senkrecht auf einander stehenden Richtungen polarisiren', *ibid.*, **68** (1846): 305–18.

(3) These experiments were apparently occasioned by papers by Brewster and Stokes at the meeting of the British Association for the Advancement of Science in July and August: see notes (5) and (6). See also William Swan's recollection of Maxwell's intervention at the meeting (*Life of Maxwell*: 489–90).   (4) An illegible deletion.

(5) David Brewster, 'On the polarizing structure of the eye', *Report of the Twentieth Meeting of the British Association for the Advancement of Science; held at Edinburgh in July and August 1850* (London, 1851), part 2: 5–6. Brewster attributed the 'brushes' to a 'polarizing structure existing in the cornea and crystalline lens, as well as in the tissues which lie in front of the sensitive layer of the retina'. Compare 'Experiment IV'.

(6) George Gabriel Stokes, 'On the Haidinger's brushes', *Report of the...British Association*

appear to my eye as blue and yellow spaces ranged round a central point so that a line drawn through the centre of the yellow spaces always coincides with the plane of polarization.

Experiment I

The first thing to be observed about the brushes is that though they are very faint their position is well defined. Now we know that [it] is necessary to distinct vision that an image should be distinctly pictured on the retina, and it may be shown that as we cannot distinctly see objects near the eye, so objects placed within the eye cannot be distinctly seen unless they are very near the retina. The apparatus which produces the brushes if within the eye at all must therefore be very near the retina.

In order to determine the distance of the structure from the retina, I made two parallel slits on a piece of metal, so near each other that both lay within the aperture of the pupil of the eye. When objects which are not within the limits of distinct vision are seen through these two slits at once they appear double and the apparent distance of the images of an object depends on its distance from the position of distinct vision.

When I examined Haidingers brushes in this way I found that they were not doubled but remained as distinct as before. I therefore concluded that the structure whatever it is must either be in the retina or close to it.

Experiment II

While observing the brushes seen in the polarized light reflected from a polished table I turned my eyes inwards so that all external objects appeared double, the brushes however still remained single. From this we learn that the centres of the polarizing structures in both eyes lie on the corresponding points of the retina. Now corresponding points on the retina in order to be symmetrical must lie in a vertical plane passing through the axis of the eye, for if two points were both to the right or left of the central plane of each eye they could not be symmetrical.

Definition

When we look steadily at any object our eyes are said to be turned towards it and fixed upon it. When this is the case the image of the object falls on a particular point on the retina let us call this point the *centre of vision*.

(London, 1851), part 2: 20–21. Stokes discussed the action of different colours in producing the brushes, finding the phenomenon could only be seen in blue light: 'the tint of the brushes ought to be made up of red, yellow, and perhaps a little green, the yellow predominating, on account of its greater brightness in the solar spectrum'. Compare 'Experiment VI'. See also Maxwell's letter to Stokes of 3 October 1853 (Number 43).

Experiment III

Standing before the same polished table I turned my attention to the upper part of Haidingers brushes. Immediately my eyes turned upwards the consequence of which was that the brushes moved upwards too and disappeared. I thus found that the only way to keep the brushes from going out of the field of view is to turn the attention on their centre. This experiment proves that the centre of the brushes corresponds with the centre of sight.

Experiment IV

It is well known that the foramen centrale of Sömmering[7] can be seen by the eye itself by alternately admitting and cutting off light from the eye.

Standing before the polished table I moved my fingers rapidly before my eye. I soon perceived the circular boundary of the foramen centrale. I then turned my head still looking in the same direction and immediately saw the brushes. The centre of the brushes corresponded with that of the foramen but the coloured spaces did not extend beyond its boundary.

I have thus I think proved

I　　That the polarizing structure lies in the retina.

II　　That the centres of this structure in the two eyes are corresponding points of the retina.

III　　That the centre of the brushes is the centre of sight.

IV　　That the centre of the brushes is that of the foramen centrale.[8]

Having determined the position of the polarizing structure I proceeded to investigate its optical properties.

Experiment V

On the position of the brushes

When we observe the brushes by means of a Nicoll's Prism[9] the yellow brush is always seen in the plane of polarization and the blue brush in the perpendicular plane but the shape or colour of the brushes is not changed by turning the prism round its axis.

---

(7) The *fovea centralis* (a depression of the retina of the eye) discovered by Samuel Thomas Soemmering in 1791.

(8) Compare Maxwell's paper 'On the unequal sensibility of the foramen centrale to light of different colours', *Report of the Twenty-sixth Meeting of the British Association for the Advancement of Science; held at Cheltenham in August 1856* (London, 1857), part 2: 12 (= *Scientific Papers*, **1**: 242).

(9) See Number 9 note (5) and Number 19 note (3).

The direction of the brushes has relevance only to the position of the prism and not to any fixed line in the eye, the optical structure therefore whatever it is is circular that is it is uniformly disposed round the axis of the eye.

Experiment VI

On the Colours of the brushes

I next examined the brushes seen in light which had first passed through a Nicoll's prism and then through the axis of a crystal of quartz.[10] The brushes were now red and green but the colours were impure.[11] By comparing the brushes thus seen with the colours seen when the light was analysed by another prism I found that the colour of any radius of the brush is that which is seen when the light is analysed in the plane of that radius.

These two experiments make it probable that the circular polarizing structure of the eye acts by analysing the light in the plane of the radius, we must next inquire how this can be done.

Light can be analysed in two ways.

1. When light falls obliquely on a transparent polished surface part of the reflected light is polarized in the plane of incidence and an equal part of the transmitted light is polarized in the opposite plane.

As there are no reflectors in the eye the light cannot be analysed by reflection.

If it is analysed by transmission through the membranes of the eye it must fall on them in an oblique direction but it does not appear that any of the membranes of the eye near the axis have a sufficient degree of obliquity and if they were disposed round the axis like a series of cones they would analyse the light in a plane *perpendicular* to the radius. We must therefore turn to the other method of polarizing light.

2. When light enters a doubly refracting substance it is divided into two pencils polarized in opposite planes. One of these pencils must be got rid of and this may be done in two ways the first by turning it aside as in Nicoll's prism and the second by absorbing it.

It is evident that there are no Nicoll's prisms in the eye. Several natural substances possess the property of absorbing one of the pencils. Of these I may mention tourmaline and agate. Now it is evident that by cutting tourmalines into sectors of a circle and building them up together a circle could be made which would have the required property of analysing the light in the plane of its radius.

---

(10) Compare Number 43, esp. note (6).		(11) Compare Stokes, 'On Haidinger's brushes'.

No tourmalines however can be found in the eye. But Sir David Brewster has described a method of making *artificial singly polarizing crystals.*[12]

When light enters a doubly refractive substance the two pencils into which it is divided have a different refrangibility. Now when a rough transparent substance is placed in a fluid of the same refractive power all its roughnesses disappear and even the form of the body can hardly be distinguished. If the rough transparent plate have two different refractive indices and if one of them be the same as that of the surrounding medium while the other is different light polarized in one plane will be freely transmitted while the rest of the light is stopped by the rough surface. Thus one of the refractive indices for nitre is nearly that of alcohol while the other corresponds to that of resinous substances if a rough plate of nitre cut parallel to the axis is placed in alcohol it will transmit only light polarized in a plane perpendicular to the axis but if it be placed in turpentine or Canada balsam the light transmitted will be polarized in a plane parallel to the axis.

Sir David Brewster has also proved that the property of double refraction may be communicated to transparent substances by pressure and tension.[13]

So high a degree of doubly refracting power may be communicated to gutta percha by tension[14] that a thin film of this substance may be substituted for the nitre in the preceding experiment and by forming a circle of small sectors so that the direction of the tension always be in the radius, an analyser is formed which possesses all the properties required for the reproduction of experiments V and VI.

Conclusion

The structure which I suppose to be the origin of Haidingers brushes may be thus described.

The polarizing structure lies within the boundary of the foramen centrale and consists of one or more thin membranes or tissues of fibre having their principal tensions at any point in the direction of the radius. I suppose these membranes to be immersed in a fluid of a refractive power such that the

---

(12) David Brewster, 'On the effects of simple pressure in producing that species of crystallization which forms two oppositely polarized images, and exhibits the complementary colours by polarized light', *Phil. Trans.*, **105** (1815): 60–64.

(13) David Brewster, 'On the communication of the structure of doubly-refracting crystals to glass, muriate of soda, fluor spar and other substances, by mechanical compression and dilatation', *Phil. Trans.*, **106** (1816): 156–78. See Number 26 §2.   (14) See Number 26 §6.

membranes are rendered more transparent when the incident light is polarized in the plane of the radius than when it is polarized in the opposite plane.

I leave it to the anatomists to determine whether such a structure actually exists in the human eye. If it does exist it may be the organ of this new sense whose object is polarized light. If no such structure can be found the other parts of the eye must be examined for it is probable that where a new sense has been discovered its organ will soon be found.[15]

JAMES CLERK MAXWELL

(15) See Hermann Helmholtz, *Handbuch der physiologischen Optik* (Leipzig, 1867): 421–3, for general agreement with Maxwell's conclusions.

FROM A LETTER TO LEWIS CAMPBELL

16 SEPTEMBER 1850

From Campbell and Garnett, *Life of Maxwell*[1]

Glenlair
16 September 1850

Professor W. Thomson[2] has asked me to make him some magne-crystallic preparations which I am now busy with.[3] Now, in some of these bismuth is required, which is not to be found either in Castle-Douglas or Dumfries. I have, therefore, thought fit to request you, and do now request you, during your transit through Edinburgh on your way here, to go either to Mr. Kemp's establishment in Infirmary Street, beside the College,[4] or to some other dealer in metals, and there purchase and obtain two ounces of metallic bismuth (called Regulus of Bismuth)[5] either powder or lumpish – all one. Thus you may perceive that the end of this letter is in two ounces of Regulus of Bismuth, that is, the metal bismuth, which if you do bring it with you, will please me well. Not that I am turned chemist. By no means; but common cook. My fingers are abominable with glue and chalk, gum and flour, wax and rosin, pitch and tallow, black oxide of iron, red ditto and vinegar. By combining these ingredients, I strive to please Prof. Thomson, who intends to submit them to Tyndall and Knoblauch, who, by means of them, are to discover the secrets of nature, and the origin of the magne-crystallic forces.[6]

---

(1) *Life of Maxwell*: 144–5; abridged.

(2) Maxwell was known to Thomson prior to the Edinburgh meeting of the British Association. See Thomson's letter to Stokes of [18 and 19 July 1850] (ULC Add. MSS 7656, K 42B); Forbes' letter to Thomson of 19 February 1850 (see Number 27 note (1)); and *Life of Maxwell*: 65.

(3) At the Edinburgh meeting of the British Association Thomson had presented a paper 'On the theory of magnetic induction in crystalline substances', *Report of the Twentieth Meeting of the British Association for the Advancement of Science; held at Edinburgh in July and August 1850* (London, 1851), part 2: 23. See Number 71 note (21).   (4) See Number 28 note (5).

(5) The term 'regulus' to denote a metal, especially 'regulus of antimony' for metallic antimony ('antimony' denoting one ore stibnite), was common in alchemy and early chemistry; see Thomas Thomson, *A System of Chemistry of Inorganic Bodies*, 2 vols. (London/Edinburgh, $_7$1831), **1**: 315.

(6) John Tyndall and Hermann Knoblauch, 'On the deportment of crystalline bodies between the poles of a magnet', *Phil. Mag.*, ser. 3, **36** (1850): 178–83.

PAPER ON A THEOREM BY RANKINE ON THE ELASTICITY OF SOLID BODIES[1]

AFTER MAY 1851[2]

From the original in the library of the Cavendish Laboratory[3]

RANKINE ON ELASTICITY

Theorem IV[4]

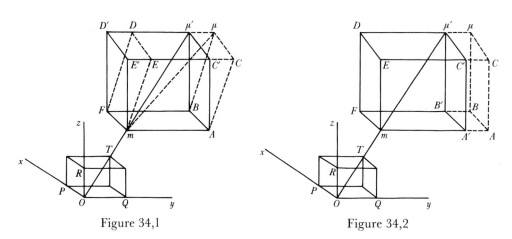

Figure 34,1   Figure 34,2

In each of the coordinate planes of elasticity of a perfect solid the two coefficients of lateral elasticity and the coefficient of rigidity are all equal to each other.[5]

---

(1) See W. J. Macquorn Rankine, 'Laws of the elasticity of solid bodies', *Camb. & Dubl. Math. J.*, **6** (1851): 47–80, esp. 63 (Theorem IV). The paper was presented to the Edinburgh meeting of the British Association in August 1850, a meeting which Maxwell attended (see Numbers 31, 32, and 33).

(2) The MS is bound in with a reprint of Rankine's note on 'Mr Clerk Maxwell's paper "On the equilibrium of elastic solids"', published in the May 1851 number of the *Camb. & Dubl. Math. J.*, **6** (1851): 185–6.

(3) Maxwell collection, Cavendish Laboratory, Cambridge.

(4) As editor of the *Cambridge and Dublin Mathematical Journal* William Thomson appended a note to Rankine's paper on the 'Laws of the elasticity of solid bodies': 80, remarking that in stating Theorem IV of the paper Rankine had ignored 'the effect of alterations of direction in the lines joining pairs of particles'. In response Rankine published a supplement (*ibid.*: 178–81) claiming that Theorem IV 'depends on the principle that the elastic force... *is sensibly the resultant of the variations of distance* [between particles] *only, the variations of relative direction producing no appreciable effect*'.

(5) Theorem IV of Rankine, 'Laws of the elasticity of solid bodies': 63. He explains that the

Let m and μ be two particles and let μ be displaced to μ' parallel to the axis of y.

On the diagonal $m\mu'$ construct a parallelopiped having its sides parallel to the axes of x, y, z.

By comparing the figures it will be seen that with respect to m and μ it is the same whether the plane μCED be transferred to $\mu'C'E'D'$ Fig [34,1] or μCAB to $\mu'C'A'B'$ Fig [34,2].

In the 1st case there is a displacement $\mu\mu'$ parallel to y between 2 planes parallel to xy whose distance from each other is $\mu'B$.

In the 2nd there is a displacement $\mu\mu'$ parallel to y between 2 planes parallel to xz whose distance is $\mu'D$ after displacement.

Since the Solid is Perfect the force produced acts in the direction $\mu'm$ and is equal in both cases. Let it be represented by the line $\mu'm$ then the resolved part parallel to y is represented by $\mu'D$ and that parallel to z by $\mu'B$.

Therefore the force in the direction of y caused by a displacement $\mu\mu'$ parallel to y between two planes parallel to xy whose distance is $\mu'B$ is to the force in the direction of z caused by a displacement $\mu\mu'$ parallel to y between two planes parallel to xz whose distance is $\mu'D$ is as $\mu'D$ to $\mu'B$.

Therefore, since the displacement between two parallel planes in a solid uniformly strained is proportional to their distance the state of strain remaining the same

The force parallel to y (due to the action of the particles m and μ) caused by a displacement $\mu\mu'\left(\dfrac{\mu'D}{\mu'B}\right)$ between two planes parallel to xy whose distance is $\mu'D$ is to that caused by a displacement $= \mu\mu'$ parallel to y between two planes parallel to xz whose distance is $\mu'D$ is as $\mu'D$ to $\mu'B$, or, as $\mu\mu'\left(\dfrac{\mu'D}{\mu'B}\right)$ is to $\mu\mu'$ that is as the displacements.

Therefore the coefficients of elasticity in the two cases are equal.

'term *perfect solid* is used to denote a body whose elasticity is due entirely to the mutual attractions and repulsions of atomic centres of force'; compare Maxwell's rejection of molecular theories of elasticity (Number 26 §1). In each plane there are two coefficients of '*lateral elasticity*' and a coefficient of '*transverse or tangential elasticity*, or of rigidity' ('Laws of the elasticity of solid bodies': 60).

35

FROM A LETTER TO LEWIS CAMPBELL

9 NOVEMBER 1851

From Campbell and Garnett. *Life of Maxwell*[1]

8 King's Parade
[Cambridge]
9 Nov. 1851

I began a letter last week, but stopped short for want of matter. I will not send you the abortion. Facts are very scarce here. There are little stories of great men for minute philosophers. Sound intelligence from Newmarket for those that put their trust in horses, and Calendristic lore for the votaries of the Senate-house. Man requires more. He finds x and y innutritious, Greek and Latin indigestible, and undergrads nauseous. He starves while being crammed. He wants man's meat, not college pudding. Is truth nowhere but in Mathematics? Is Beauty developed only in men's elegant words, or Right in Whewell's *Morality*?[2] Must Nature as well as Revelation be examined through canonical spectacles by the dark-lantern of Tradition, and measured out by the learned to the unlearned, all second-hand. I might go on thus. Now do not rashly say that I am disgusted with Cambridge and meditating a retreat. On the contrary, I am so engrossed with shoppy things that I have not time to write to you. I am also persuaded that the study of x and y is to men an essential preparation for the intelligent study of the material universe. That the idea of Beauty is propagated by communication, and that in order thereto human language must be studied, and that Whewell's *Morality* is worth reading, if only to see that there *may be* such a thing as a system of Ethics.

That few will grind up these subjects without the help of rules, the awe of authority, and a continued abstinence from unripe realities, etc.

I believe, with the Westminster Divines and their predecessors *ad Infinitum* that 'Man's chief end is to glorify God and to enjoy him for ever.'[3]

That for this end to every man has been given a progressively increasing power of communication with other creatures. That with his powers his susceptibilities increase.

That happiness is indissolubly connected with the full exercise of these powers in their intended direction. That Happiness and Misery must inevi-

---

(1) *Life of Maxwell*: 157–60.

(2) William Whewell, *The Elements of Morality, including Polity*, 2 vols. (London, 1845).

(3) The Westminster Confession: the profession of faith by the Westminster Assembly in 1846, based on the Calvinist doctrine of election, and approved by the Church of Scotland in 1647; the basis of Presbyterian faith.

tably increase with increasing Power and Knowledge. That the translation from the one course to the other is essentially *miraculous*, while the progress is natural. But the subject is too high. I will not, however, stop short, but proceed to Intellectual Pursuits.

It is natural to suppose that the soul, if not clothed with a body, and so put in relation with the creatures, would run on in an unprogressive circle of barren meditation. In order to advance, the soul must converse with things external to itself.

In every branch of knowledge the progress is proportional to the amount of facts on which to build, and therefore to the facility of obtaining data. In the Mathematics this is easy. Do you want a quantity? Take x; there it is! – got without trouble, and as good a quantity as one would wish to have. And so in other sciences, – the more abstract the subject, the better it is known. Space, time, and force come first in certainty. These are the subjects in Mechanics.

Then the active powers, Light, Heat, Electricity, etc. = Physics.

Then the differences and relations of Matter = Chemistry, and so on.

Here the order of advancement is just that of abstractedness and inapplicability to the actual. What poor blind things we Maths. think ourselves! But see the Chemists! Chemistry is a pack of cards, which the labour of hundreds is slowly arranging; and one or two tricks, – faint imitations of Nature, – have been played. Yet Chemistry is far before all the Natural History sciences; all these before Medicine; Medicine before Metaphysics, Law, and Ethics; and these I think before Pneumatology and Theology.[4]

Now each of these makes up in interest what it wants in advancement.

There is no doubt that of all earthly creatures Man is the most important to us, yet we know less of him than any other. His history is more interesting than natural history; but nat. history, though obscure, is much more intelligible than man's history, which is a tale half told, and which, even when this world's course is run, and when, as some think, man may compare notes with other rational beings, will still be a great mystery, of which the beginning and the end are all that can be known to us while the intermediate parts are perpetually filled up.

So now pray excuse me if I think that the more grovelling and materialistic sciences of matter are not to be despised in comparison with the lofty studies of Minds and Spirits. Our own and our neighbours' minds are known but very imperfectly, and no new facts will be found till we come in contact with some minds other than human to elicit them by counterposition. But of this more anon.

---

(4) Compare William Whewell, *Philosophy of the Inductive Sciences, founded upon their History*, 2 vols. (London, ₂1847), **1**: 78–81.

36

MANUSCRIPT ON THE POTENTIAL FUNCTION

circa 1851[1]

From the original in the University Library, Cambridge[2]

MATHEMATICAL THEORY OF POLAR FORCES[3]

When we examine the infinite variety of actions which take place between material bodies we find that the relative position of the bodies is an important element in the determination of the effect. If we confine our attention to the action on one of the bodies and conceive the rest to be fixed and unchangeable, then if a displacement of that body in a given direction produces a given effect then an equal small displacement in the opposite direction will produce the opposite effect, that is, an effect, which, if it could be combined with the first effect would neutralize it.

The way in which such actions have been generally investigated has been by an analysis of the action of the bodies as we find them by which that action is considered as the resultant of the action between every pair of elements in the bodies acting and acted on, and in this way the theory of various attractive forces has been established.

There is however another obvious method of treating the same subject in which the action of the fixed bodies on the body considered, being always the same for the same position of the body is regarded not as directly dependent on the fixed bodies but as the immediate consequence of its particular position in space, that space having been endued with certain properties by the action of the fixed bodies.

When properties of this kind are attributed to portions of space whether occupied by matter or not, it must be carefully remembered that such properties imply nothing more than what is expressed in their definitions and that they are to be considered for the present as mere mathematical abstractions introduced to facilitate the study of certain phenomena.

The most remarkable example of the use of a mathematical abstraction of this kind in mechanical science is the function used by Laplace in the theory of attractions and to which Green in his Essay on Electricity has given the name

---

(1) Handwriting suggests a date no later than the early 1850s; and see note (3).

(2) ULC Add. MSS 7655, V, c/1.

(3) Compare William Thomson, 'A mathematical theory of magnetism', *Phil. Trans.*, **141** (1851): 243–85, on 247 ($=$ *Electrostatics and Magnetism*: 347); '[there is] a very important branch of physical mathematics, which might be called "A Mathematical Theory of Polar Forces"'.

of the Potential Function.[4] We have no reason to believe that anything answering to this function has a physical existence in the various parts of space, but it contributes not a little to the clearness of our conceptions to direct our attention to the potential function as if it were a real property of the space in which it exists.

In the attraction of gravitation there does not yet appear to be anything equal & opposite to the attraction of a particle of matter in the direction of the increase of the potential but in the case of electrified & magnetic bodies we find that if a particle electrified positively tends to move in one direction another equally electrified negatively if placed in the same position would tend to move in the opposite direction with equal force & in the same way a north & a south magnetic pole of equal intensity will if placed at the same point in succession be acted on by forces equal & opposite.

This opposition of properties in opposite directions constitutes the polarity of the element of space. It is measured by the intensity of those equal & opposite forces & the direction of the polarity is determined by that of the force which is assumed to be *positive*.

---

(4)   Compare Thomson, 'A mathematical theory of magnetism': 259 (= *Electrostatics and Magnetism*: 363); 'The determination of the resultant force at any point is, as we shall see, much facilitated by means of a method first introduced by Laplace in the mathematical treatment of the theory of attraction, and developed to a very remarkable extent by Green in his "Essay on the Application of Mathematical Analysis to the Theories of Electricity and Magnetism" (1828) ... The term "potential", defined in connection with it, was first introduced by Green in his Essay (1828).' See P. S. Laplace, *Traité de Mécanique Céleste*, 5 vols. (Paris, An VII [1799]–1825), **1**: 136–7; and George Green, *An Essay on the Application of Mathematical Analysis to the Theories of Electricity and Magnetism* (Nottingham, 1828): 8. See Number 51 note (19).

LETTER TO LEWIS CAMPBELL

5 JUNE 1852

From the original in the Staatsbibliothek Preussischer Kulturbesitz, Berlin[1]

Trin. Coll. June 5th
1852

Dear Lewis

I have now settled my plans for my little vacation. Our Ex[ns] are over all but one paper on Monday. On Monday I discharge duties of sorts. On Tuesday I start by the Bedford Coach at 9.30 p.m. I suppose as you went by it you know when you arrive at Oxford. If I do not see you before I will come to Balliol and ask for you. Bob is going to London on Monday and will be at Oxford on Wednesday so he will be present. He is first in the Trin Hall *May*.

I have engaged with Tait to be at Cambridge on Tuesday the 15th to go to Lowestoft or some sea place and so start afresh on the 21st with Hopkins.[2] He poor man is now with Joule at Manchester experimenting on Spermaceti Lead and Rocks to determine their laws of melting.[3]

Our Ex[ns] are going on this week but now (Saturday evening) I have got through all the paying papers, done two big math papers well two ill no classics to swear by, some morality and there remains Roman History of which I intend to know nothing. Thomson as I think I told you is at Peterhouse. I bathe with him in the morning when there is not ex[n].

M[c]Lennan[4] is going to be first in his year at Trin. at Peterhouse Stewart Gardiner and Glynn are bracketed Gardiner slightly ahead Stewart having done very badly this time, best men got only 400 marks out of 1400.[5]

NB I am not at Lings.[6] My last two letters have been dated from Trinity.

Yrs truly
JAS C. MAXWELL

---

(1) Staatsbibliothek Preussischer Kulturbesitz, Berlin, Sammlung Darmstaedter Fle 1860 (3).

(2) William Hopkins, Peterhouse 1823, 7th Wrangler 1827, Maxwell's mathematics coach at Cambridge (Venn). See David Wilson, 'The educational matrix: physics education at early-Victorian Cambridge, Edinburgh and Glasgow Universities', in *Wranglers and Physicists*, ed. P. M. Harman (Manchester, 1985): 12–48, esp. 15–18, 33–4.

(3) William Hopkins, 'Experimental researches on the conductive powers of various substances, with the application of the results to the problem of terrestrial temperature', *Phil. Trans.*, **147** (1857): 805–49; this work was instigated by James Prescott Joule.

(4) John Ferguson McLennan, Trinity 1851 (Venn).

(5) Allan Duncan Stewart (an Edinburgh friend, see *Life of Maxwell*: 106), William Dundas Gardiner, Albert Glynn, all Peterhouse 1849 (Venn).

(6) A Cambridge lodging house; see *Life of Maxwell*: 149, 155.

DRAFT OF PAPER ON THE DESCENT OF A BODY IN A RESISTING MEDIUM[1]

13 AUGUST 1852

From the original in the University Library, Cambridge[2]

[FRICTION IN A RESISTING MEDIUM]

Take a slip of paper three inches long and half an inch broad and let it fall in still air, the direction of its length being horizontal.

In whatever way it may be projected the motion will soon become regular. It will descend with a tremulous but uniform motion, revolving round an axis in the direction of its length, and its path will not be vertical but its general direction will be in the plane of rotation and inclined to the vertical at an angle nearly constant.

This deviation will always be on that side of the vertical towards which the *under* side of the body turns while revolving round its axis.

The motion then consists of a succession of similar parts. The same velocity direction and rate of rotation must recur at every revolution.

Now the opposing forces here are gravity and the resistance of the air. These might maintain a uniform motion of translation but according to the ordinary hypothesis the resistance would tend continually to diminish the velocity of rotation while gravity can have no effect on it. It must therefore soon cease to rotate.

But the motion of rotation is in fact uniform and determinate and if it be accidentally less it will soon regain its former value. There must therefore be some force tending to increase the velocity of rotation.

Now the normal resistance of air may be shewn to have a continual tendency to stop rotation. The friction of the air against the surface acts in a direction passing through the axis (the paper being considered indefinitely thin) and therefore cannot influence the velocity of rotation.

But this friction acts on the air as well as the body. In this way it may alter the normal pressure and so act indirectly on the body.

When the surface moves obliquely through the air the friction presses the air near the body in the direction of the transverse motion of the surface and so increases the normal pressure on that side of the axis.

---

(1) J. C. Maxwell, 'On a particular case of the descent of a heavy body in a resisting medium', *Camb. & Dubl. Math. J.*, **9** (1854): 145–8 (= *Scientific Papers*, **1**: 115–18).

(2) ULC Add. MSS 7655, V, e/1. Endorsed: 'Friction in a Resisting Medium'.

To ascertain whether this indirect action would actually maintain the motion some calculation is necessary.

Let p_0 be the pressure of the air at rest referred to unit of surface. Let this become $p_0 + k_1 v^2$ where v is the velocity of the surface against the air and $p_0 - k_2 v^2$ where v is the velocity of the surface leaving the air behind.[3]

Let the friction referred to unit of surface be

$$f = c_1 pu$$

where p is the normal pressure u the transverse velocity and c_1 a constant to be determined by experiment. Since the first power of u only is involved this function is not necessarily discontinuous when u changes sign.

To explain the action of a resisting medium on a thin parallelogram moving through it.

Let AB be the projection of the parallelogram on the plane of motion. Let G be the origin of coordinates GB the axis of x GY that of y, let the velocity of G be v in the direction GY & u in the direction GB and let the angular velocity about $G = \omega$ from x to y.

Let P be a point distant r from G. Let the pressure on that point be p then the friction will be approxi-

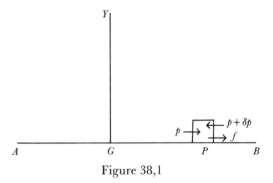

Figure 38,1

---

(3) Compare Maxwell's notes in his Cambridge notebook (dated March 1852) 'Statics Dynamics' (ULC Add. MSS, V, m/10): 'Rectilinear Motion of a Material point in a Resisting Medium. Whewell Dynamics Part II p. 118. The most important case is that in which a body is acted on by a constant force (as that of gravity) and in which the resistance varies as kv^2 (k being constant), this being found to be an approximation to the resistance when a body moves through a fluid medium' (f. 114). The reference is to William Whewell, *On the Motion of Points Constrained and Resisted, and on the motion of a Rigid Body. The Second Part of a new edition of A Treatise on Dynamics* (Cambridge, 1834); and compare Whewell, *A Treatise on Dynamics* (Cambridge, 1823): 171–2. See also S. D. Poisson, *Traité de Mécanique*, 2 vols. (Paris, $_2$1833), **1**: 400–401, commenting on the force of resistance R impeding the motion of a projectile in air: 'Je supposerai aussi, conformément aux hypothèses généralement admises, la force R proportionelle au carré de la vitesse du centre de gravité, à la surface du projectile, et à la densité de l'air'. In his *Discours de la Pesanteur* (appended to his *Traité de la Lumière* (Leyden, 1690): 125–80, esp. 169–76) Huygens referred to experiments that demonstrated that the resistance of air and water to the motion of a projectile was proportional to the square of the velocity. Subsequent discussion of the problem of determining the trajectory of a projectile in the case where the resistance is as the square of the instantaneous velocity focussed

mately $f = c_1 pu$ where c_1 is a small coeff^t and u is the transverse velocity of the body wh. may be taken for that of the particle as the coeff^t is small.

Again, the centrifugal force $C = \rho(r\omega^2 + v\omega)$. Therefore equating the forces acting on an element of the fluid at P we have

$$\frac{dp}{dr} - c_1 up - \rho(r\omega^2 + v\omega) = 0$$

whence $p = Ce^{c_1 ur} - \rho\dfrac{r\omega^2 + v\omega}{c_1 u} - \rho\dfrac{\omega^2}{c_1^2 u^2}.$

Let p_0 be the undisturbed pressure of the fluid then we have as the result of experiment when $r = 0$

$$p = p_0 + k_1 v^2$$

where k_1 is the ordinary coeff^t of normal resistance.

$$\therefore p_0 + k_1 v^2 = C - \rho\left(\frac{v\omega}{c_1 u} + \frac{\omega^2}{c_1^2 u^2}\right)$$

whence C. Putting for $e^{c_1 ur}$ its approximate value

$$p = p_0 + k_1 v^2 + \{c_1 u(p_0 + k_1 v^2) + \rho v\omega\}r + \&c.$$

Let the whole pressure $= P$ normal and F tangential and let the moment of the resistance about $G = P\bar{r}$ by integrating from $r = a$ to $r = -a$

$$P = 2a(p_0 + k_1 v^2)$$
$$F = 2ac_1 u(p_0 + k_1 v^2)$$
$$P\bar{r} = \tfrac{2}{3}a^3\{c_1 u(p_0 + kv^2) + \rho v\omega\}.$$

These eq^{ns} show that P is approximately the same as when found on the ordinary hypothesis

That $F = c_1 uP$

And that the couple is due partly to the resistance of the air to the actual rotation of the body and partly to the friction of the fluid which increases the pressure at B.

This part of the couple is independent of ω and always tends to turn the surface perpendicular to the direction of motion.

13th August 1852

---

on Proposition X in the *Principia*'s second book, and on Newton's recasting of the argument in the second edition; see Isaac Newton, *Philosophiae Naturalis Principia Mathematica* (Cambridge, $_2$1713): 232–40. For a full account of the controversy see *The Mathematical Papers of Isaac Newton*, ed. D. T. Whiteside, 8 vols. (Cambridge, 1967–81), **8**: 48–61, 312–424.

APPENDIX: DRAFT OF PAPER ON THE DESCENT OF A HEAVY BODY IN A RESISTING MEDIUM

EARLY 1853[4]

From the original in the University Library, Cambridge[5]

ON A PARTICULAR CASE OF THE DESCENT OF A HEAVY BODY IN A RESISTING MEDIUM

Everyone must have observed that when a slip of paper falls through the air, its motion being undecided at first sometimes becomes regular.[6] Its general path is not in the vertical direction but inclined to it at an angle nearly constant and its fluttering appearance will be found to be due to a rapid rotation round a horizontal axis. If this rotation be in the positive direction that is opposite to that of the hands of a watch, the path will incline to the right hand and if the direction of rotation be changed the horizontal part of the motion will be in the opposite direction.

These effects are commonly attributed to some peculiarity in the form of the paper, but a few experiments with a rectangular slip of paper (about two inches long and one broad) will prove that the direction of rotation is determined not by the accidental irregularities of the paper but by the initial circumstances of projection and that the symmetry of the form of the paper greatly increases the distinctness of the phenomena. We may therefore assume that if the form of the body were accurately that of a plane rectangle the same effects would be produced.

The following investigation is intended as a general explanation of the true cause of the phenomenon.

I suppose the resistance of the air to the motion of the plane to be in the direction of the normal and to vary as the square of the velocity estimated in that direction. Now though this be a sufficiently near approximation to the magnitude of the resisting force on the plane taken as a whole, the pressure on any element of the plane will vary with its position so that the resultant pressure will not pass through the centre of gravity.

It is found by experiment that the centre of pressure is on that side of the

---

(4) The published paper (to which this draft fairly closely corresponds) is dated 'April 5, 1853'; *Camb. & Dubl. Math. J.*, **9** (1854): 148 (= *Scientific Papers*, **1**: 118).

(5) ULC Add. MSS 7655, V, e/1.

(6) Compare Maxwell's comment (quoting Horace) in his letter to William Thomson of 24 November 1857 (Number 137).

centre of gravity towards which the tangential part of the motion is directed and that its distance from the centre of gravity increases with the tangential motion.

It is easy to point out the reason of this, but I am not aware that it has yet been subjected to mathematical calculation.

If we examine a plane surface placed obliquely in a current we may observe that

(1) On the anterior side of the plane the relative velocity of the fluid increases as it moves along the surface from the first to the second edge and since the pressure diminishes as the velocity increases the pressure must be greatest where the fluid first meets the plane.

(2) The motion of the fluid behind the plane is very unsteady but may be observed to consist of eddies diminishing in rapidity as they pass behind the plane and therefore relieving the pressure most at the edge of the plane which first cuts the fluid.

Both these causes tend to make the resistance greatest at the edge which first meets the fluid and therefore to bring the centre of pressure nearest to that edge.

Hence the moment of the resistance about the centre of gravity will always tend to turn the plane towards a position perpendicular to the direction of its motion. It will be shown that it is this moment which maintains the rotatory motion of the falling paper. When the plane has a motion of rotation the resistance will be modified by the unequal velocities of different parts of the surface. The magnitude of the resistance will not be much altered if the velocity of any point due to angular motion be small compared with that due to translation of the centre of gravity but the part of the moment of the resistance round the centre of gravity depending on the motion of rotation will vary directly as the normal velocity and the angular velocity together and will always act in the direction opposite to that of rotation.

The part of the moment of the resistance round the centre of gravity caused by the obliquity of the motion will remain nearly the same as before.

We are now prepared to give a general explanation of the motion of the slip of paper.

We may consider the disturbances of the motion small compared with the motion disturbed.

The mean motion would be one of uniform rotation and uniform descent along some line inclined at a given angle to the vertical.

Consider first the effect of the resistance of the air on the slip of paper if it were to descend in a vertical direction with a uniform motion of rotation.

The resolved part of the resistance in the vertical direction will always act upwards being greatest when the plane of the paper is horizontal and vanishing when it is vertical. When the motion has become regular the effect of this force during a whole rotation will be equal & opposite to that of gravity during the same time.

If the rotation be supposed to commence when the plane of the paper is horizontal since the resistance has its maxima when $\theta = n\pi$ and its minima when $\theta = (n+\frac{1}{2})\pi$ the maximum velocity will be attained in the second and fourth quadrant and the minimum velocity in the first and third.

The resolved part of the resistance in the horizontal direction will act in the negative direction in the first and third quadrants and in the positive direction in the second and fourth, but since the resistance increases with the velocity the whole effect during the second and fourth quadrants will be greater than the whole effect during the first & third. Therefore the horizontal part of the resistance will act on the whole in the positive direction and will therefore cause the general path of the body to incline towards that direction.

That part of the moment of the resistance about the centre of gravity wh. depends on the angular velocity will vary in magnitude but will always act in the negative direction. The other part wh. depends on the obliquity of the plane of the paper to the direction of motion will be negative in the first and third quadrants and positive in the second and fourth, but as its magnitude increases with the velocity the positive effect will be greater than the negative.

This excess in the positive direction will cause the angular velocity of the body to increase till it is balanced by the retarding force due to the angular velocity itself.

The motion will then consist of a succession of equal and similar parts performed in the same manner each part corresponding to half a revolution of the paper.

FROM A LETTER TO JANE CAY

29 MAY 1853[1]

From Campbell and Garnett, *Life of Maxwell*[2]

[Cambridge]

I have been attending Sir James Stephen's[3] lectures upon the causes of the first French Revolution.[4] They are now done, so I look in upon Stokes' dealing with light.[5]

---

(1) The letter is headed: 'Trin. Coll., Feast of St. Charles II'; Charles II born 29 May 1630, Restoration Day 1660. The letter is addressed to Maxwell's aunt.

(2) *Life of Maxwell*: 185; abridged.

(3) Regius Professor of Modern History at Cambridge 1849 (Venn, *DNB*).

(4) Stephen's published *Lectures on the History of France*, 2 vols. (London, ₂1852, ₃1857) terminate with the reign of Louis XIV.

(5) George Gabriel Stokes, Pembroke 1837, Lucasian Professor of Mathematics 1849 (Venn, *DNB*). Stokes regularly lectured on 'Hydrostatics, pneumatics, and optics' in the Easter term; see *The Cambridge University Calendar for the Year 1853* (Cambridge, 1853): 142. According to Stokes' lecture fee notebook (ULC Add. MSS 7656, NB 1) Maxwell attended his lectures in 1853, paying a fee of £2.2.0. According to his lecture outline for 1852 (ULC Add. MSS 7656 NB 2) Stokes lectured on optics during May. Beginning with an account of the properties of fluids and of the equations of hydrodynamics, he discussed waves in fluids and the theory of sound, followed by lectures on the undulatory theory of light. His lectures were regularly attended by prospective high wranglers in the Mathematical Tripos. See David B. Wilson, *Kelvin and Stokes. A Comparative Study in Victorian Physics* (Bristol, 1987): 42–53.

LETTER TO CHARLES BENJAMIN TAYLER[1]

8 JULY 1853

From Campbell and Garnett, *Life of Maxwell*[2]

Trin. Coll.
8 July 1853
Evening post

My dear Friend

Your letter was handed to me by the postman as I was taking a walk after morning chapel. As I was engaged then, I thought I might wait till the evening. I breakfasted with Macmillan the publisher,[3] who has a man called Alexander Smith with him, who published a volume of poems in the beginning of the year which have been much read here, and, indeed, everywhere, for 3000 copies have been sold already.[4] He is a designer of patterns for needlework, and he refuses to be made celebrated or to leave his trade. He speaks strong Glasgow, but without affectation, and is well-informed without the pretence of education, commonly so called. People would not expect from such a man a book in which the author seems to transfer all his own states of mind to the objects he sees. But he is young and may get wiser as he gets older. He sees and can tell of the beauty of things, but he connects them artificially. He may come to prefer the real and natural connection, and after that he may perhaps stir us all up by bringing before us real human objects of interest he has only dimly seen in the solitude of his youth.

I told you how I meant to go to Hopkins. He was not in. I had a talk with him on Sunday; he recommended light work for a while, and afterwards he would give me an opportunity of making up what I had lost by absence. Yesterday I did a paper of his on the Differential Calculus without fatigue, and as well as usual. Ask George how Mr. Hughes has arranged about Examinations. I will write to him soon, and send him a mass of papers in an open packet, to be taken twice a week, or not so often.

You dimly allude to the process of spoiling which has gone on during the last 2 years. I admit that people have been kind to me, and also that I have seen more variety than in other years; but I maintain that all the evil influences that I can trace have been internal and not external, you know what I mean – that

---

(1) Rector of Otley, Suffolk (*DNB*), uncle of George Wood Henry Tayler, Trinity 1850 (Venn).

(2) *Life of Maxwell*: 188–9. See Campbell's recollection that Maxwell subsequently referred to his experience during this visit as having 'given him a new perception of the love of God' (*ibid.*: 170); see also Number 42.   (3) Daniel Macmillan (*DNB*).

(4) Alexander Smith, *Poems* (London, 1853); see Numbers 68 and 119.

I have the capacity of being more wicked than any example that man could set me, and that if I escape, it is only by God's grace helping me to get rid of myself, partially in science, more completely in society, – but not perfectly except by committing myself to God as the instrument of His will, not doubtfully, but in the certain hope that that Will will be plain enough at the proper time. Nevertheless, you see things from the outside directly, and I only by reflexion, so I hope that you will not tell me you have *little* fault to find with me, without finding that little and communicating it.

In the *Athenæum* of the 2d there is Faraday's account of his experiments on Table-turning, proving mechanically that the table is moved by the unconscious pressure of the fingers of the people wishing it to move, and proving besides that Table-turners may be honest.[5] The consequence has been that letters are being written to Faraday boastfully demanding explanations of this, that, and the other thing, as if Faraday had made a proclamation of Omniscience. Such is the fate of men who make real experiments in the popular occult sciences, – a fate very easy to be borne in silence and confidence by those who do not depend on popular opinion, or learned opinion either, but on the observation of Facts in rational combination. Our anti-scientific men here triumph over Faraday.[6]

I hope the Rectory has flourished during the absence of you and Mrs. Tayler. I had got into habits with you of expecting things to happen, and if I wake at night I think the gruel is coming.

Macmillan was talking to me to-day about elementary books of natural science, and he had found the deficiency, but had a good report of 'Philosophy in Sport made Science in Earnest', which I spoke of with you. When I am settled I will put down some first principles and practicable experiments on Light for Charlie, who is to write to me and answer questions proposed; but this in good time. – Your affectionate friend,   J. C. MAXWELL.

---

(5) 'Professor Faraday on table-moving', *The Athenæum* (2 July 1853): 801–3 (= Michael Faraday, *Experimental Researches on Chemistry and Physics* (London, 1859): 385–91).

(6) Compare Faraday's letter (of 28 June 1853) on 'table-turning' in *The Times*, 30 June 1853 (= *Chemistry and Physics*: 382–5; and in *The Selected Correspondence of Michael Faraday*, ed. L. P. Williams, 2 vols. (Cambridge, 1971), **2**: 690–2).

41

FROM A LETTER TO LEWIS CAMPBELL

14 JULY 1853

From Campbell and Garnett, *Life of Maxwell*[1]

Trin. Coll.
14 July 1853

Faraday's experiments on Table-turning, and the answers of provoked believers and the state of opinion generally, show what the state of the public mind is with respect to the *principles* of natural science.[2] The law of gravitation and the wonderful effects of the electric fluid are things which you can ascertain by asking any man or woman not deprived by penury or exclusiveness of ordinary information. But they believe them just as they believe history, because it is in books and is not doubted. So that facts in natural science are believed on account of the number of witnesses, as they ought! I believe that tables are turned; yea! and by an unknown force called, if you please, the vital force, acting, as believers say, thro' the fingers. But how does it affect the table? By the *mechanical* action of the sideward pressure of the fingers in the direction the table ought to go, as Faraday has shown. At this last statement the Turners recoil.

---

(1) *Life of Maxwell*: 191; abridged. (2) See Number 40.

LETTER TO RICHARD BUCKLEY LITCHFIELD[1]

23 AUGUST 1853

From the original in the library of Trinity College, Cambridge[2]

Bankground
Coniston
Ambleside
Aug 23, 1853

Dear Litchfield

Pomeroy[3] tells me you intend to write to me, so I must tell you where I am and intend to be for a week or so. I came here with Campbell of Trin Hall to meet his brother & another Oxford man called Christie.[4] We are all in a house just above the lake, recreating ourselves and reading a little. Pomeroy is off to Ireland. I have seen a good deal of him & we have read 'at the same time successively' 'Vestiges of Creation'[5] and Maurices 'Theological Essays'.[6] Both have excited thought & talk. I have not yet seen Pye Smith's book[7] wh. you mention, but I intend to.

I suppose you may have heard that I was down after the May with Taylers uncle in Suffolk and that I was taken ill there.[8]

I was there made acquainted with the peculiar constitution of a well regulated family, consisting entirely of nephews & nieces, and educated entirely by the uncle & aunt. There was plenty of willing obedience but little diligence, much mutual trust & little self reliance. They did not strike out for them selves in different lines according to age sex & disposition but each so excessively sympathised (bona fide, of course) with the rest that one could not be surprised at hearing any one take part in criticising his own action.

In such a case some w^d recommend 'a little wholesome neglect'. I w^d suggest something like the scheme of self emancipation for slaves. Let each member of the family be allowed some little province of thought work or study wh. is not to be too much inquired into or sympathised with or encouraged by the rest,

---

(1) Trinity 1849, Inner Temple 1854 (Venn).

(2) Trinity College, Cambridge, Add. MS Letters c.1[78]. Published in extract in *Life of Maxwell*: 191–2.

(3) Robert Henry Pomeroy, Trinity 1850 (Venn).

(4) William Christie or Richard Copley Christie; see Joseph Foster, *Alumni Oxonienses*, 4 vols. (Oxford, 1888), 1: 251.

(5) [Robert Chambers,] *Vestiges of the Natural History of Creation* (London, 1844, ₁₀1853).

(6) Frederick Denison Maurice, *Theological Essays* (Cambridge, 1853).

(7) Perhaps: John Pye Smith, *The Relation between the Holy Scriptures and some parts of Geological Science* (London, ₄1848).   (8) See Number 40.

and let the limits of this be enlarged till he has a wide free field of independent action, which increases the resources of the family so much the more, as it is peculiarly his own.

I see daily more & more reason to believe that the study of the 'dark sciences' is one wh. will repay investigation. I think that what is called the proneness to superstition in the present day is much more significant than some make it.[9] The prevalance of a misdirected tendency proves the misdirection of a prevalent tendency. It is the nature & object of this tendency that calls for examination.

I had a letter from MacLennan some time ago. He is going to leave the 'Star'. Do you know why? I dont.[10] Harvey & Suter have been at Cambridge.[11] Suter is now in Scotland with 2 pupils. I have been very sleepy since I left Cambridge and must give up writing & reading for some time.

In about a week my address will change to

> Glenlair
>
> Kirkpatrick Durham
>
> Dumfries,

where I hope to stay till Sept. 24th when Hopkins begins again.

Yours truly

J. C. MAXWELL

APPENDIX I: FROM AN ESSAY FOR THE APOSTLES ON 'IDIOTIC IMPS'[12]

SUMMER 1853[13]

From Campbell and Garnett, *Life of Maxwell*[14]

The first question I would ask concerning a spiritual theory would be, Is it favourable or adverse to the present developments of Dark Science? The Dark

---

(9) See Maxwell's essay for the Apostles Essay Society on 'Idiotic Imps', reproduced as Appendix I. Maxwell was elected to the Society of Apostles in the winter of 1852–3 (*Life of Maxwell*: 165). Litchfield was also a member; see [Henrietta Litchfield,] *Richard Buckley Litchfield. A memoir written for his friends by his wife* (Cambridge, 1910): 6.

(10) John Ferguson McLennan; see Number 37 note (4). McLennan left Cambridge in 1853 (*DNB*).

(11) William Woodis Harvey, Trinity 1848; Andrew Burn Suter, Trinity 1849 (Venn).

(12) Fragments of Maxwell's Apostles essay on 'Idiotic Imps'. According to Campbell, the essay, which bore the alternative title 'Ought the Discovery of Plurality of Intelligent Creators to weaken our Belief in an Ultimate first Cause?', a third title 'Does the Existence of Causal Chains

Sciences...while they profess to treat of laws which have never been investigated, afford the most conspicuous examples of the operation of the well-known laws of association...in imitating the phraseology of science, and in combining its facts with those which must naturally suggest themselves to a mind unnaturally disposed. In the misbegotten science thus produced we have speciously sounding laws of which our first impression is that they are truisms, and the second that they are absurd, and a bewildering mass of experimental proof, of which all the tendencies lie on the surface and all the data turn out when examined to be heaped together as confusedly as the stores of button-makers...and those undigested narratives which are said to form the nutriment of the minute philosopher. ... The most orthodox system of metaphysics may be transformed into a dark science by its phraseology being popularised, while its principles are lost sight of.

[Three phases of dark science are described:]

(1.) At first they were or pretended to be physical sciences. Their language was imitated from popular physics, and their professed aim was to explain occult phenomena by means of new and still more occult material laws. Experiments in animal magnetism were always performed with the nose carefully turned towards the north. In electrobiology a scrupulous system of insulation was practised at first, and afterwards, when galvanism became more popular than statical electricity, circuits were formed of alternate elements, those of one sex being placed between those of the other. ... The fluid which in former times circulated through the nerves under the form of animal spirits, is in our day expanded so as to fill the universe, and is the invisible medium through which the communion of the sensitive takes place.

(2.) The next phase of the dark sciences is that in which...the phraseology of physics is exchanged for that of psychology. In this stage we hear much of the

---

prove an Astral Entity or Cosmothetic Idealism?' being added later, was occasioned by Maxwell's reading [Isaac Taylor,] *Physical Theory of Another Life* (London, 1836). Taylor's argument was based on a principle of 'analogy', that 'the vastness of the visible universe...may be taken as a sort of gauge of the vastness of that range of intellectual and moral existence of which the visible universe is the platform', and a 'rule of symmetry', that 'within the space encircled by the sidereal revolutions, there exists and moves a second universe...elaborate in structure and replete with life' (*ibid.*: 290, 301, 224). Maxwell commented on Taylor's book: 'the perusal of it has a tendency rather to excite speculation than to satisfy curiosity, and the author obtains the approbation of the reader while he fails to convince him of the soundness of his views' (*Life of Maxwell*: 227). On Maxwell's acquaintance with Taylor in February 1852 see *Life of Maxwell*: 176–7. For Taylor's views on the habitation of the universe see M. J. Crowe, *The Extraterrestrial Life Debate 1750–1900* (Cambridge, 1986): 229–31.

(13) As given by Campbell, *Life of Maxwell*: 227.   (14) *Life of Maxwell*: 228–30.

power of the will. The verb to will acquires a new and popular sense, so that every one now is able to will a thing without bequeathing it. People can will not to be able to do a thing, then try and not succeed; while those of stronger minds can will their victims out of their wits and back again.

(3.) The third or pneumatological phase begins by distrusting, as it well may, the explanations prevalent during the former stages of apparitions, distant intercourse, etc. It suggests that different minds may have some communion, though separated by space, through some spiritual medium. Such a suggestion if discreetly followed up might lead to important discoveries, and would certainly give rise to entertaining meditations. But the cultivators of the dark sciences have done as they have ever delighted to do. Their spirits are not content with making themselves present

'Where all the nerves of sense are numb,
Spirit to spirit, ghost to ghost.'

but they become the familiar spirits of money-making media, and rap out lies for hours together for the amusement of a promiscuous 'circle'. . . . While the believers sit round the table of the medium and form one loop of the figure 8, the spiritual circle enclosing the celestial mahogany forms the upper portion of the curve, the medium herself constituting the double point. But who shall say of the dark sciences that they have reached the maximum of darkness? Men have listened to the toes of a medium as to the voice of the departed. Let them now stand about her table as about the table of devils. If one spirit can wrap itself in petticoats, why may not another dance with three legs? A most searching question truly! And accordingly the powerful analysis of Godfrey[15] has led him to the conclusion that a table of which the plane surface is touched by believing fingers may be transformed into a diaboloid of revolution. . . . Will there be an interminable series of such expressions of belief, each more unnatural than its predecessor, and gradually converging towards absolute absurdity?

---

(15)   The Rev. Nathaniel Stedman Godfrey, author of *Table-Turning, the Devil's Modern Masterpiece* (London, 1853) and other pamphlets claiming that tables were turned by diabolic agency. Godfrey's pamphlets and lectures gave rise to a vigorous pamphlet literature, including the Rev. F. Close (incumbent of Cheltenham, Litchfield's home), *The Testers Tested; or Table Moving, Turning, and Talking, Not Diabolical: A review of the publications of the Rev. Messrs. Godfrey, Gillson, Vincent and Dibdin* (London 1853). On Faraday's contribution see Numbers 40 and 41.

APPENDIX II: FRAGMENTS OF AN APOSTLES ESSAY 'WHAT IS THE NATURE OF EVIDENCE OF DESIGN?'

1853[16]

From Campbell and Garnett, *Life of Maxwell*[17]

Design! The very word...disturbs our quiet discussions about *how* things happen with restless questionings about the *why* of them all. We seem to have recklessly abandoned the railroad of phenomonology, and the black rocks of Ontology stiffen their serried brows and frown inevitable destruction.

...The belief in design is a necessary consequence of the Laws of Thought acting on the phenomena of perception.

...The essentials then for true evidence of design are – (1) A phenomenon having significance to us; (2) Two ascertained chains of physical causes contingently connected, and both having the same apparent terminations, viz., the phenomenon itself and some presupposed personality.... If the discovery of a watch awakens my torpid intelligence I perceive a significant end which the watch subserves. It goes, and, considering its locality, it is going well.... My young and growing reason points out two sets of phenomena... (*a*) the elasticity of springs, etc. etc., and (*b*) the astronomical facts which render the mean solar day the unit of civil time combined with those social habits which require a cognisance of the time of day.

...It is the business of science to investigate these causal chains. If they are found not to be independent but to meet in some ascertained point, we must transfer the evidence of design from the ultimate fact to the existence of the chain. Thus, suppose we ascertained that watches are now made by machinery...the machinery including the watch forms one more complicated and therefore more evident instance of design.

...The only subordinate centres of causation which I have seen formally investigated are men and animals; the latter even are often overlooked. But every well-ascertained law points to some central cause, and at once constitutes that centre a *being* in the general sense of the word. Whether that being be *personal* is a question which may be determined by induction. The less difficult question whether the *being* be *intelligent* is more practicable, and should be kept in view in the investigation of organised beings.

The search for such invisible potencies or wisdoms may appear novel and unsanctioned.... For my part I do not think that any speculations about the

---

(16) As given by Campbell, *Life of Maxwell*: 225.   (17) *Life of Maxwell*: 225–7.

personality or intelligence of subordinate agents in creation could ever be perverted into witchcraft or demonolatry.

Why should not the Original Creator have shared the pleasure of His work with His creatures and made the morning stars sing together, etc.?

I suspect that such a hope has prompted many speculations of natural historians, who would be ashamed to put it into words.

...Three fallacies – (1) Putting the final cause in the place of a physical connection, as when Bernoulli saw the propriety of making the curves of isochronous oscillations and of shortest time of descent both cycloids;[18] (2) The erroneous assertion of a physical relation, as when Bacon supplemented the statement of Socrates about the eyebrows by saying that pilosity is incident to moist places;[19] (3) (and worst) applying an argument from final causes to wrongly asserted phenomena: 'Because water is incompressible, it cannot transmit sound, and therefore fishes have no ears.' Every fact here stated is erroneous. ...

Perception is the ultimate consciousness of self and thing together.

If we admit, as we must, that this *ultimate* phenomenon is incapable of further analysis, and that subject and object alone are immediately concerned in it, it follows that the fact is strictly private and incommunicable. One only can know it, therefore two cannot agree in a name for it. And since the fact is simple it cannot be thought of by itself nor *compared alone* with any other *equally simple fact*. We may therefore dismiss all questions about the absolute nature of perception, and all theories of their resemblances and differences. We may next refuse to turn our attention to perception in general, as all perceptions are particular.[20]

---

(18) Johann Bernoulli, 'Curvatura radii in diaphanis non uniformibus...de invenienda *Linea Brachystochrona*', *Acta Eruditorum* (May 1697): 206–11 (= Johann Bernoulli, *Opera Omnia*, 4 vols. (Lausanne/Geneva, 1742), **1**: 187–93), Bernoulli's solution of the path of swiftest fall ('concludo: curvam *Brachystochronam esse Cycloidem vulgarem*'). Huygens had discussed the isochronism of the cycloidal pendulum in his *Horologium Oscillatorium* (Paris, 1673). For contemporary comment see D. F. Gregory, *Examples of the Processes of the Differential and Integral Calculus* (Cambridge, 1841): 135.

(19) Pilosity: hairiness; Francis Bacon, *Advancement of Learning*, Book II, vii, 7 (*OED*); 'that "the hairs about the eye-lids are for the safeguard of the sight", doth not impugn the cause rendered, that "pilosity is incident to orifices of moisture"'.

(20) For an additional fragment see *Life of Maxwell*: 227n.

LETTER TO GEORGE GABRIEL STOKES

3 OCTOBER 1853

From the original in the University Library, Cambridge[1]

Trinity College
Oct 3rd 1853

Dear Sir

What I said about chlorophyll was merely a vague remembrance of one out of a number of facts in Chemistry which Professor Gregory[2] mentioned to me last Winter during a visit to his laboratory. I am very sure that he told me of its isolation but less sure about its colour.[3]

He gave me no reference to any source of information and at that time I had no intention of asking for any.

I have written to him on the subject, but I have not yet received an answer.

With respect to my conjecture about Haidingers brushes, your statement of it is quite accurate and expresses all that deserves the name even of conjecture.[4] In August 1850 I noted down some experiments and deductions from them[5] some of which are the same with those which I have since seen described in the fourth part of Moigno's 'Répertoire d'Optique' particularly the experiment with a plate of quartz.[6]

I have published nothing on the subject. In fact I have had no opportunity of obtaining sufficient foundation for a theory, and a conjecture is best corroborated by the existence of identical or similar conjectures springing up independently.

Yours faithfully
Jas CLERK MAXWELL

---

(1) ULC Add. MSS 7656, M404. Previously published in Larmor, *Correspondence*, 2: 1.

(2) See Number 25 note (2).

(3) On Stokes' interest in the absorption bands displayed by a solution of chlorophyll, see his discussion of the 'solution of leaf-green in alcohol' and of the 'colours of natural bodies' in his paper 'On the change of refrangibility of light', *Phil. Trans.*, **142** (1852): 463–562, esp. 486–92, 527–9 (= *Papers*, 3: 267–409). See also Number 118, esp. note (8). Maxwell made a list of the colour values of plants (ULC Add. MSS 7655, V, b/4); and see Number 58.

(4) On Stokes' interest in Haidinger's brushes see his paper presented to the Edinburgh meeting of the British Association in 1850 'On Haidinger's brushes', *Report of the Twentieth Meeting of the British Association for the Advancement of Science* (London, 1851), part 2: 20–21. Stokes' paper, and a paper presented to the same meeting by Brewster, probably prompted Maxwell's experiments on the phenomenon; see Number 32 notes (5) and (6).   (5) Number 32.

(6) F. N. M. Moigno, *Répertoire d'Optique Moderne*, 4 vols. (Paris, 1847–50), **4**: 1364–5; and see Number 32, 'Experiment VI'.

DRAFT OF SOLUTIONS OF PROBLEMS IN THE
CAMB. & DUBL. MATH. J.[1]

circa OCTOBER 1853[2]

From the original in the University Library, Cambridge[3]

[SOLUTIONS OF PROBLEMS]

[Problem]II. If from a point in the circumference of a vertical circle, two heavy particles be successively projected along the curve, their initial velocities being equal and either in the same or in opposite directions, the subsequent motion will be such that a straight line joining the particles at any instant will touch a fixed circle.[4]

The direct analytical proof would involve the properties of elliptic integrals, but it may be made to depend on the following geometrical theorems.

(1) If from a point in one of two circles a right line be drawn cutting the other, the rectangle contained by the segments so formed is double of the rectangle contained by the perpendicular from the point on the *radical axis* of the two circles,[5] and the line joining their centres.

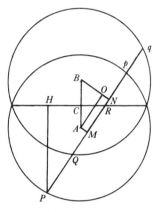

Figure 44,1

In the figure, $PQ \cdot Pq = 2PH \cdot AB$.

Let R be the intersection of Pq with the radical axis CH. Draw AM, BN perpendicular, and AO parallel, to Pq. Then since R is a point in the radical axis –

$$RP \cdot Rp = RQ \cdot Rq$$
$$RP(Pp - RP) = (RP - PQ)(Pq - RP)$$
$$RP \cdot Pp - RP^2 = RP(Pq + PQ) - PQ \cdot Pq - RP^2.$$

---

(1) 'Problems' set (by Maxwell: see notes (4) and (6)) in the *Camb. & Dubl. Math. J.*, **8** (1853): 188; and see [J. C. Maxwell,] 'Solutions to problems', *ibid.*, **9** (1854): 7–11 (= *Scientific Papers*. **1**: 74–9).

(2) A final draft (ULC Add. MSS 7655, V, d/8), corresponding very closely to the published paper, is dated 'Oct 27th 1853'.　　　　(3) ULC Add. MSS 7655, V, d/8.

(4) In a postcard to P. G. Tait of 1 January 1872 (to be reproduced in Volume II) Maxwell states that this 'prop. ... was original to me in 1853 or so'.

(5) In a preliminary draft (ULC Add. MSS 7655, V, d/8) Maxwell adds a note: 'For the definition & properties of the radical axis of two circles see Mulcahy's Modern Geometry Chap. IV'. See John Mulcahy *Principles of Modern Geometry, with numerous applications to Plane and Spherical Figures* (Dublin, 1852): 52; if *A* and *B* are the centres of two circles and *O* a point from which equal tangents are drawn to the two circles, then a line *OP* 'the *radical axis* of the two circles' is drawn perpendicular to the line *AB* joining the centres of the circles.

Taking from these equals $-RP^2$, and observing that
$$Pp = 2PM = 2PN - 2MN$$
and
$$PQ + Pq = 2PN$$
$$\therefore 2RP(PN - MN) = 2RP \cdot PN - PQ \cdot Pq.$$
Hence
$$2RP \cdot MN = PQ \cdot Pq.$$
Now *AOB*, *PHR* are similar triangles & $MN = AO$
$$\therefore AB \cdot HP = RP \cdot MN$$
$$\therefore 2AB \cdot HP = PQ \cdot Pq. \qquad\qquad \text{QED.}$$

Cor. If the line be drawn from P to *touch* the circle, then the square of the tangent $= 2AB \cdot HP$.

Hence if a line be drawn touching one circle and cutting another the squares of the segments of the chord are proportional to the distances of its extremities from the radical axis.

(2) If two straight lines, chords to one circle and tangents to another be made to approach to each other indefinitely, the arcs intercepted by their corresponding extremities will be ultimately proportional to the segments of either chord.

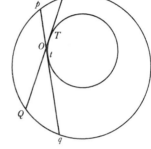

Figure 44,2

In the figure, $Pp : Qq :: PT : QT$.

Join Pq, Qq then if O be the point of intersection of PQ & pq

$$POp, \; qOQ \text{ are similar triangles, and}$$

$$Pp : Qq :: PO : qO.$$

Now when pq is made to approach PQ, Pp & Qq ultimately coincide with the arcs of the circle, t and O coincide with T and PO, qO become respectively equal to PT, QT, therefore ultimately

$$Pq : Qq :: PT : QT. \qquad\qquad \text{QED.}$$

Cor. If particles P, Q be constrained to move at the extremities of the chord PQ, then at any instant the velocity of P is to that of Q as PT to QT and therefore by (1) the squares of the velocities of P and Q are proportional to their respective distances from the radical axis.

Therefore if the plane of the circles be vertical and the radical axis horizontal, and if the velocity of P at any instant be that which would be acquired by falling from the radical axis to P, the velocity of Q will also be that due to its distance from the radical axis.

Hence if the extremity P of the chord be made continually to coincide with the position of a heavy particle projected from q with the velocity due to the distance of q from the radical axis, and if when the other extremity Q of the chord has arrived at q, another particle be projected from q with the same velocity, this second particle will coincide with Q during the whole of the motion.

Therefore the line joining the particles continually touches the second circle, which was to be proved.

Cor (1). If any number of such particles be projected from the same point at equal intervals of time, the lines joining successive particles at any instant will be tangents to the same circle.

Cor (2). If the time of a complete revolution (or oscillation) contain n of these intervals, then the lines joining successive particles at any instant will form a polygon of n sides, and as this instant may be chosen at pleasure, any number of such polygons may be formed.

The following geometrical axiom is therefore true – 'If two circles be such that n lines can be drawn touching one of them, and having their successive intersections, including that of the last and first, in the circumference of the other the result will be the same at whatever point of the first circle we draw the first tangent.'

[Problem]III. A transparent medium is such that the path of a ray of light within it is a given circle, the index of refraction being a function of the distance from a given point in the plane of the circle.

Find the form of this function and shew

(1) That the path of *every other ray* will be circular.

(2) That all the rays proceeding from any point in the medium will meet accurately at another point.

(3) That if rays diverge from any point without the medium and enter it through a spherical surface whose centre is that point, they will be made to converge accurately to a point within the medium. [6]

---

(6) In a letter to James Thomson of 2 September 1872 (to be published in Volume II) Maxwell explained that he had in mind the problem of the path of rays in a medium of continuously variable index of refraction 'when I came to visit your brother Sir William in December 1852 (I think)'. He indicated that he had himself set the problem: 'In 1853 I sent the Cambridge and Dublin M.J. a problem about the path of a ray in a medium...intended to illustrate the fact that the principal focal length of the crystalline lens is very much shorter than anatomists calculate it from the curvature of its surface and the index of refraction of its substance...'.

Lemma I.

Let a transparent medium be so constituted that the refractive index is the same at the same distance from a fixed point, then the path of any ray of light within the medium will be in one plane, and the perpendicular from the fixed point on the tangent to the path of the ray at any point will vary inversely as the refractive index of the medium at that point.

The medium may be considered as composed of a series of concentric spherical shells, the index of refraction being the same throughout each shell but varying from one shell to another.

Let a ray of light pass at the point R from a medium whose refractive index is μ_1 to another for which it is μ_2.

Let the figure represent the section made by a plane passing through C the centre of the shells and RP the tangent to the path of the ray at R in the medium μ_1.

Let RS be a tangent to the path of the ray at R after entering the medium μ_2, & CP, CQ perpendicular to RP & SR.

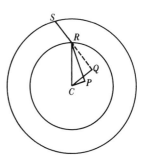

Figure 44,3

Then RS is in the plane of incidence, that is the plane of the paper

$$\text{and } CQ::CP::\sin CRQ:\sin CRP$$

$$::\text{sine of refraction}:\text{sine of incidence}$$

$$::\mu_1:\mu_2.$$

Therefore when the ray passes from one medium to another its plane is still that which passes through C and the original direction, and the perpendicular on the tangent after any number of refractions is inversely proportional to the refractive index at the point of contact.

Lemma II.

If from any fixed point in the plane of a circle, a perpendicular be drawn to a tangent at any point of the circumference the rectangle contained by this perpendicular and the diameter is equal to the square of the line joining the point of contact and the fixed point, together with the rectangle contained by the segments of any chord through the fixed point.

Let APB be the circle, O the fixed point then

$$OY \cdot PR = OP^2 + AO \cdot OB.$$

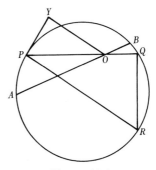

Figure 44,4

Join QR then the triangles OYP, PQR are similar therefore[7]

$$OY \cdot PR = OP \cdot PQ$$
$$= OP^2 + PO \cdot OQ$$
$$\therefore OY \cdot PR = OP^2 + AO \cdot OB.$$

QED.

If we put in this expression $AO \cdot OB = a^2$

$$PO = r \quad OY = p \quad \& \quad PR = 2\rho$$

it becomes

$$2p\rho = r + a$$
$$p = \frac{r^2 + a^2}{2\rho}.$$

To find the law of the refractive index of a medium such that a ray proceeding from A in the direction of AT may describe the circle $APBQ$, μ must be made to vary inversely as p.

Let $AO = r_1$ and let the refractive index of the medium at $A = \mu_1$ then generally $\mu = \dfrac{C}{p}$

$$\mu = \frac{2C\rho}{a^2 + r^2}$$

at A

$$\mu_1 = \frac{2C\rho}{a^2 + r_1^2}.$$

Hence $\mu = \mu_1 \dfrac{a^2 + r_1^2}{a^2 + r^2}$ the law required a result independent of ρ.

Hence the path of any other ray proceeding from A will be a circle for which the value of a^2 is the same.

The value of OB is $\dfrac{a^2}{r_1}$ which is also independent of ρ, so that every ray which proceeds from A must pass through B.

Again if we assume μ_0 as the value of μ when $r = 0$

$$\mu_0 = \mu_1 \frac{a^2 + r_1^2}{a^2}$$
$$\therefore \mu = \mu_0 \frac{a^2}{a^2 + r^2}$$

---

(7) In the final draft (see note (2)) Maxwell states that his geometrical method 'was deduced by analogy from Newton's Principia, lib.I prop.VII'; see note (8). In this proposition – 'A body orbits in the circumference of a circle: the law of centripetal force tending to some point in its circumference is required' – Newton's geometrical demonstration is based on the appeal to the similarity of three triangles. See Isaac Newton, *Philosophiae Naturalis Principia Mathematica* (Cambridge, ₂1713): 42–3; compare the drafts in *The Mathematical Papers of Isaac Newton*, ed. D. T. Whiteside, 8 vols. (Cambridge, 1967–81), **6**: 42–5, 134–5, esp. 43 note (32).

a result independent of r_1. This shows that any point may be taken as the origin of the light instead of A and that the path of every ray will still be circular and will pass through another point on the other side of O.

Next let CP be a ray from C, a point without the medium falling at P on a spherical surface whose centre is C.

Let O be the fixed point in the medium. Join PO & produce to Q till $OQ = \dfrac{a^2}{OP}$. Through P and Q draw a circle touching PC at P, and cutting CO in A & B then this circle is the path of the ray within the medium.

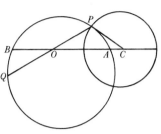

Figure 44,5

It is easy to show that any other circle through A & B will represent the path of another ray cutting the spherical surface at right angles and therefore proceeding from C without the medium.[8]

APPENDIX: A PRELIMINARY DRAFT

1853

From the original in the University Library, Cambridge[9]

[SOLUTION TO PROBLEM II: INTRODUCTION]

It is evident from the propositions in the fourth book of Euclids Elements that in equilateral and equiangular polygons circles can be inscribed and circumscribed and that these circles have the same centre.[10]

---

(8) In the final draft (see note (2)) he adds a supplementary note; as in 'Solutions of problems': 11 (= *Scientific Papers*, **1**: 79).

'(Note to Problem III)

The possibility of the existence of a medium of this kind possessing remarkable optical properties was suggested by the contemplation of the structure of the crystalline lens in fish; and the method of searching for these properties was deduced by analogy from Newton's Principia, lib.I prop.VII.

It would require a more accurate investigation into the law of the refractive index of the different coats of the lens to test its agreement with this supposed medium, which is an optical instrument theoretically perfect for homogeneous light, and might be made achromatic by proper adaptation of the dispersive power of each coat.

On the other hand, we find that the law of the index of refraction which would give a minimum of aberration for a sphere of this kind placed in water, gives results not discordant with facts, so far as they can readily be ascertained'. See note (6).   (9) ULC Add. MSS 7655, V, d/8.

(10) Book IV Props. XIII and XIV of *Elements of Geometry*, ed. John Playfair (Edinburgh, 1795): 122–4.

By drawing consecutive sides chords of the larger circle and tangents to the smaller, we may make a polygon of the same number of sides having one of its angles at any proposed point of the circumference, so that an indefinite number of polygons may be drawn inscribed in the one circle and circumscribing the other.

Again, it may be shewn in the case of a triangle that the distance between the centres of the inscribed & circumscribed circles and their respective radii are connected by an eq^n not involving any other quantities. Hence if this eq^n is satisfied any number of triangles may be drawn inscribed in the one circle & circumscribing the other.

I am not aware that any proof of the following more general theorem has been published.

If a polygon of n sides can be described so that each side may be a tangent to one circle while the intersections of the consecutive sides lie in the circumference of another circle then any number of such polygons may be drawn fulfilling the same conditions.

In the following propositions I have introduced the conception of velocity instead of the notation of the differential calculus. The analytical investigation would involve the properties of elliptic functions, while the mechanical illustration recalls the familiar idea of the simple pendulum.

45

LETTER TO WILLIAM THOMSON

20 FEBRUARY 1854

From the original in the University Library, Cambridge[1]

Trin. Coll.
Feb 20 1854

Dear Thomson

Now that I have entered the unholy estate of bachelorhood[2] I have begun to think of reading. This is very pleasant for some time among books of acknowledged merit wh one has not read but ought to. But we have a strong tendency to return to Physical Subjects and several of us here wish to attack Electricity.

Suppose a man to have a popular knowledge of electrical show experiments and a little antipathy to Murphys Electricity,[3] how ought he to proceed in reading & working so as to get a little insight into the subject wh may be of use in further reading?

If he wished to read Ampère Faraday &c how should they be arranged, and at what stage & in what order might he read your articles in the Cambridge Journal?[4]

If you have in your mind any answer to the above questions, three of us here would be content to look upon an embodiment of it in writing as advice.

I have another question for myself. At Ardmillan while bathing on the rocks you mentioned that Gauss(?) had been investigating the bending of surfaces and had found in particular that the product of the principal radii of curvature at any point is unchanged by bending.[5]

I have no means here of finding the paper from wh you quoted so that I w^{d}

---

(1) ULC Add. MSS 7342, M 87. Previously published in Larmor, 'Origins': 697–8.

(2) Maxwell graduated Second Wrangler (to E. J. Routh) in the Mathematical Tripos in 1854, and in early February was placed equal with Routh in the Smith's Prize examination set by Challis, Stokes and Whewell. See *The Cambridge University Calendar for the Year 1854* (Cambridge, 1854): 372–407, 408–16; Stokes, *Papers*, **5**: 320–2; and [W. Walton and C. F. Mackenzie,] *Solution of the Problems and Riders proposed in the Senate House Examination for 1854* (Cambridge, 1854).

(3) Robert Murphy, *Elementary Principles of the Theories of Electricity, Heat and Molecular Actions. Part I. On Electricity* (Cambridge, 1833). See J. J. Cross, 'Integral theorems in Cambridge mathematical physics, 1830–55', in *Wranglers and Physicists*, ed. P. M. Harman (Manchester, 1985): 112–48, esp. 121–9, for a reassessment of Murphy's mathematical papers.

(4) See Number 51.

(5) Carl Friedrich Gauss, 'Disquisitiones generales circa superficies curvas', *Commentationes Societatis Regiae Scientiarum Göttingensis Recentiores*, **6** (1827): 99–146 (= Gauss, *Werke*, 12 vols. (Göttingen, 1863–1933), **4**: 217–58).

be obliged to you if you could give me some reference to it or even tell me whether he had considered the conditions of bending of a finite portion of a surface in general.

I have been working for some time at the more general problem & have completed the theory for surfaces of revolution and got several results in the case of other surfaces by the consideration of two systems of lines on the surface wh may be called lines of bending.[6]

These being given the effect of bending the surface is reduced to the consideration of one indep^t var^{ble} only.

These lines themselves however are subject to certain conditions that the surface may be bent at all, and to additional conditions that these lines may continue 'lines of bending'.

When the lines of bending are the lines of principal curvature the conditions are very simple & are fulfilled for all surfaces of revol^n.

But some of the operations are long so I am going over them by a process quite different from the first.

By finding what Gauss' results are I may be spared much trouble in pruning my calculations.

Finally I have heard nothing of C. J. Taylor[7] since he went to Glasgow, and w^d gladly receive information about him, Ramsay[8] & the professorship.[9] Are there any *good* classical places about Glasgow or elsewhere, 'where a man might enjoy a comfortable house'.

This is a letter of questions so I go on in the same spirit to the end by enquiring after the prosperity of the College especially No^s 2 and XIII i.e. commend me to the Blackburns[10] & Mrs Thomson.

Yrs truly

J. C. MAXWELL

---

(6) See J. C. Maxwell, 'On the transformation of surfaces by bending', *Trans. Camb. Phil. Soc.*, **9** (1856): 445–70, esp. 449–50 (= *Scientific Papers*, **1**: 80–114, esp. 86–8). See also Numbers 46 and 47.

(7) Charles Johnstone Taylor, Glasgow University 1842, Trinity 1845 (Venn).

(8) William Ramsay, Professor of Humanity at Glasgow 1831–63 (Venn).

(9) Presumably the Professorship of Civil Engineering and Mechanics, held by Lewis Gordon; from January to April 1855 W. J. M. Rankine lectured as Gordon's deputy, being appointed to the professorship in November 1855 (*DNB*).

(10) Hugh Blackburn, Trinity 1840, Professor of Mathematics at Glasgow 1849–79; married to Maxwell's cousin Jemima Wedderburn (Venn).

46

ABSTRACT OF PAPER 'ON THE TRANSFORMATION OF SURFACES BY BENDING'

[13 MARCH 1854][1]

From the *Proceedings of the Cambridge Philosophical Society*[2]

The kind of transformation here considered is that in which a surface changes its form without extension or contraction of any of its parts. Such a process may be called bending or development. The most obvious case is that in which the surface is originally a plane, and becomes, by bending, one of the class called 'developable surfaces'. Surfaces generated by straight lines, which do not ultimately intersect, may also be bent about these straight lines as axes. In this way they may be transformed into surfaces whose generating lines are parallel to a given plane, just as the former class are transformed into planes.

In both these cases, the bending round one straight line of the system is quite independent of that round any other; but in those which follow, the bending at one point influences that at every other point. The case of a surface of revolution bent symmetrically with respect to the axis is taken as an example.

The remainder of the paper contains an elementary investigation of the conditions of bending of a surface of any form.

The surface is considered as the limit of the inscribed polyhedron when the number of the sides is increased and their size diminished indefinitely.

A method is then given by which a polyhedron with triangular facets may be inscribed in any surface; and it is shown, that when a certain condition is fulfilled, the triangles unite in pairs so as to form a polyhedron with quadrilateral facets. The edges of this polyhedron form two intersecting systems of polygons, which become in the limit curves of double curvature; and when the condition referred to is satisfied, the two systems of curves are said to be 'conjugate' to one another.

The solid angle formed by four facets which meet in a point is then considered, and in this way a 'measure of curvature' of the surface at that point is obtained.

It is then shown that if there be two surfaces, one of which has been developed from the other, one, and only one, pair of systems of corresponding lines can be drawn on the two surfaces so as to be conjugate to each other on both surfaces. This pair of systems completely determines the nature of the transformation,

---

(1) The date the paper was read to the Cambridge Philosophical Society.

(2) *Proc. Camb. Phil. Soc.*, **1** (1854): 134–6; and see 'On the transformation of surfaces by bending', *Trans. Camb. Phil. Soc.*, **9** (1856): 445–70 (= *Scientific Papers*, **1**: 80–114).

and is called a double system of 'lines of bending'.[3] By means of these lines the most general cases are reduced to that of the quadrilateral polyhedron. The condition to be fulfilled at every point of the surface during bending is deduced from the consideration of one solid angle of the polyhedron. It is found that the product of the principal radii of curvature is constant.[4]

By considering the angles of the four edges which meet in a point, we obtain certain conditions, which must be satisfied by the lines of bending in order that any bending may be possible. If one of these conditions be satisfied, an infinitesimal amount of bending may take place, after which the system of lines must be altered that the bending may continue. Such lines of bending are in continual motion over the surface during bending, and may be called 'instantaneous lines of bending'. When a second condition is satisfied, a finite amount of bending may take place about the same system of lines. Such a system may be called a 'permanent system of lines of bending'.

Every conception required by the problem is thus rendered perfectly definite and intelligible, and the difficulties of further investigation are entirely analytical. No attempt has been made to overcome these, as the elementary considerations previously employed would soon become too complicated to be of any use.

For the analytical treatment of the subject the reader is referred to the following memoirs:

1 'Disquisitiones generales circa superficies curvas', by M. C. F. Gauss (1827). *Comm. Recentiores Gott.* vol. vi.; and in Monge's 'Application de l'Analyse à la Géométrie', edit. 1850.[5]

2 'Sur un Théorème de M. Gauss, &c.', par J. Liouville. *Liouville's Journal*, 1847.[6]

3 'Démonstration d'un Théorème de M. Gauss', par M. J. Bertrand. *Liouville's Journal*, 1848.

4 'Demonstration d'un Théorème', Note de M. Diguet. *Liouville's Journal*, 1848.

---

(3) See 'On the transformation of surfaces by bending': 450 (= *Scientific Papers*, **1**: 88).

(4) See Number 45 esp. note (5) and Number 47.

(5) Carl Friedrich Gauss, 'Disquisitiones generales circa superficies curvas', *Commentationes Societatis Regiae Scientiarum Göttingensis Recentiores*, **6** (1827): 99–146. Gauss' paper was reprinted under the title 'Recherches générale des surfaces courbes' in Gaspard Monge, *Application d'Analyse à la Géométrie*, ed. J. Liouville (Paris, ₅1850): 505–46.

(6) Joseph Liouville, 'Sur un théorème de M. Gauss, concernant le produit des deux rayons de courbure principaux en chaque point d'une surface', *Journal de Mathématiques Pures et Appliquées*, **12** (1847): 291–304.

5   'Sur le même Théorème', par M. Puiseux. *Liouville's Journal*, 1848.[7]

And two notes appended by M. Liouville to his edition of Monge.[8]

(7) Joseph Bertrand, 'Démonstration d'un théorème de M. Gauss', *Journal de Mathématiques Pures et Appliquées*, **13** (1848): 80–82; 'Note de M. Diguet', *ibid.*: 83–6; V. Puiseux, 'Sur le même théorème', *ibid.*: 87–90.

(8) Joseph Liouville, 'Expressions divers de la distance de deux points infiniment voisins et de la courbure géodésique des lignes sur une surface', and 'Sur le théorème de M. Gauss, concernant le produit des deux rayons de courbure principaux en chaque point d'une surface', in his edition of Monge, *Application d'Analyse à la Géométrie*: 569–76, 583–600.

47

LETTER TO WILLIAM THOMSON

14 MARCH 1854

From the original in the University Library, Cambridge[1]

Trin. Coll.
14th March 1854

Dear Thomson

I have to acknowledge your letter of information, & an Exam^n Paper. I have looked up your references, and have also got a reference to Gauss' Memoir from Stokes, who got it from Cayley.[2] It is in Liouvilles edition of Monges Analyse Appliquée, as an addition to the theory of Surfaces. There is also an account of the researches of Bertrand, &c with further developments of Liouvilles own.[3] Both Gauss' method of referring the surface to a sphere and Bertrand's of describing a small re-entering curve on the surface had occurred to me, & I had worked them out before I saw them elsewhere. But I think the method which I had adopted at first better suited for the investigation of the particular problem of *bending*, as giving it more distinctness. You mentioned an elementary proof of your own which you thought might form a proposition in educational treatises. I sh^d be much obliged if you could send it or mention the principal steps, if they are easily described.

Stokes asked me to fill up a blank in the Phil. Society's proceedings and so I delivered a vivâ voce exposition of bending last night,[4] & I suppose I must proceed to put the subject into a definite form, so any assistance from you wd be acceptable.

As you approve of the reduction of problems to their elements I send you the steps of my analysis of the problem which you will easily follow as they are simple geometrical facts, all well known.

The surface is considered as the limit of the inscribed polygon and the first step is to inscribe a polygon with triangular facets.

(1) This is done by drawing on the surface 2 systems of curves forming quadrilaterals, & joining the diagonals of these by the dotted system, then, drawing plane triangles with same angular points, we have the polyhedron required.

---

(1) ULC Add. MSS 7342, M 88. Previously published in Larmor, 'Origins': 698–701.

(2) Arthur Cayley to Stokes, 2 March 1854, in Larmor, *Correspondence*, **1**: 382.

(3) Carl Friedrich Gauss, 'Recherches générale des surfaces courbes'; and Joseph Liouville, 'Sur le théorème de M. Gauss, concernant le produit des deux rayons de courbure principaux en chaque point d'une surface', in Gaspard Monge, *Application d'Analyse à la Géométrie*, 5e ed. revue...par M. Liouville (Paris, $_5$1850): 505–46, 583–600. See Number 46 notes (5) to (8).

(4) Number 46.

(2) Then it is shewn that if any of these facets be produced it will cut the surface in a curve wh: is ultimately a conic section. I have called this curve the 'Conic of Contact'. It is a bad name, can you devise one instead?[5]

(3) The condition of pairs of triangles forming plane quadrilateral facets is shown to be that lines of the 1st & second system form conjugate diameters of the conic of contact at their inter section. The two systems are then said to be 'conjugate to one another on the surface'.

(4) Normals to four contiguous facets are drawn thro the centre of a sphere whose rad. is unity, their intersections with the sphere form the angles of a spherical quadrilateral whose area is $2\pi - (a+b+c+d)$, a b &c being the plane angles forming a solid angle of the polyhedron & $\therefore$ invariable. But since a & c, b & d are ultimately equal the figure is a small parallelogram and we may find its area as such.

If l be the angle of the planes ab & λ of bc

$$\text{area} = l\lambda \sin a$$

and this is easily shown to be

$$\frac{\text{area of a facet}}{\rho_1 \rho_2}$$

$\rho_1 \rho_2$ being principal radii of curvature. Hence when the polyhedron is bent this is invariable.[6]

(5) When two surfaces are given one of wh: is formed from the other by bending, one & only one pair of systems of lines can be drawn which shall be conjugate to one another in both surfaces.

For let the surfaces touch in corresponding points so as to have corresponding lines in the same direction then the two conics of contact will intersect in 4 p^{ts}

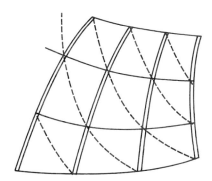

Figure 47,1

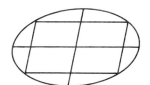

Figure 47,2

Figure 47,3

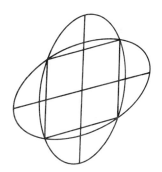

Figure 47,4

---

(5) Maxwell retains the term in his paper 'On the transformation of surfaces by bending', *Trans. Camb. Phil. Soc.*, **9** (1856): 454–5 (= *Scientific Papers*, **1**: 93).

(6) As established by Gauss, 'Disquisitiones generales circa superficies curvas'; see Number 46, esp. note (5). Compare Number 45.

forming angles of a parallelogram. Lines parallel to the sides are the only ones wh: are conjugate diameters to both conics, and 2 systems of curves whose directions at every p^t correspond to these lines are the systems required. Such lines are the '*Lines of Bending*' corresponding to the given change of form of the surface. They are perfectly definite and when drawn the bending about them may be completely expressed in terms of one var^{ble}.

(6) In order that a pair of conjugate systems of lines may be lines of bending at all one condition connecting their directions &c is necessary and that they may *continue* lines of bending 2 conditions are required. Hence the distinction of Instantaneous & Permanent lines of bending the Lines of Bending in 5 correspond to that axis about wh we may suppose a rigid system turned about a point from any one pos^n into any other.[7] Such a motion however may not be possible in every case. The true motion is about a system of Instantaneous lines wh: continually change their pos^n during the transformation.

(7) The principal lines of curvature in surfaces of revolution are shown to be permanent lines of bending and examples are given.

(8) The lines of curvature in surfaces in wh: $(\rho_1 \rho_2)$ is *const* are permanent lines of bending. When $\rho_1 \rho_2$ is positive & $= a^2$ the surface may be transformed into a sphere, radius $= a$. When $\rho_1 \rho_2 = -a^2$ the surface may be transformed in an infinite number of ways into the surface of revolution generated by the equitangential curve.

The latter prop^n is taken from Liouville.[8]

The demonstration is original and of the same kind with the rest of the investigation.

I have also a method of defining surfaces generated by a straight line by 4 indep^t variables.

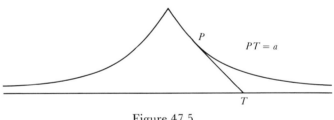

$PT = a$

Figure 47,5

Let the eq^n to the s^t line contain one variable t.

Draw 3 consecutive lines and 2 shortest distances.

(7) The 'lines of bending' define the topology of surfaces; see Maxwell, 'On the transformation of surfaces by bending': 446, 464–5 (= *Scientific Papers*, **1**: 81, 106–7).

(8) Joseph Liouville, 'Expressions divers de la distance de deux points infiniment voisins et de la courbure géodésique des lignes sur une surface', in Monge, *Application d'Analyse à la Géométrie* (Paris, ₅1850): 569–76. Compare Maxwell's discussion in 'On the transformation of surfaces by bending': 452 (= *Scientific Papers*, **1**: 90).

Let dt be the var^n of t

 dz the shortest distance

 $d\phi$ the angle between consecutive lines

 ds the distance between consecutive shortest distances

 $d\psi$ the angle between consec. shortest distances.

Eliminating t, we obtain ϕ, s & ψ in terms of z. Of these $z\,\phi$ & s are invariable but ψ may be altered in any arbitrary manner by bending the surface along straight lines.[9]

So much for Mathematics.

I know one Daltonian[10] very well but I have not seen as much as I expect to see of him yet.

The defect consists in seeing only one gradation of colour in the spectrum. Colour *as perceived by us* is a function of three independent variables at least three are I think sufficient, but time will show if I thrive.[11]

To the Daltonian Colour is a function of two indep^t variables[12] which we

---

(9) Compare Maxwell, 'On the transformation of surfaces by bending': 463–5 (= *Scientific Papers*, **1**: 104–7).

(10) Colour blindness was barely recognised as a defect of vision until John Dalton's description of his own problem in his 'Extraordinary facts relating to the vision of colours: with observations', *Memoirs of the Manchester Literary and Philosophical Society*, **5** (1798): 28–45; see Paul D. Sherman, *Colour Vision in the Nineteenth Century. The Young–Helmholtz–Maxwell Theory* (Bristol, 1981): 117–52. According to Elie Wartmann, 'Mémoire sur le Daltonisme', *Mémoires de la Société de Physique et d'Histoire Naturelle de Genève*, **10** (1843): 273–326, the term 'Daltonism' was current in Geneva. Commenting on Wartmann's paper (in his 'Observations on colour blindness', *Phil. Mag.*, ser. 3, **25** (1844): 134–41) David Brewster objected to the use of 'the offensive name of Daltonism', preferring the term 'colour blindness', which was used in the English translation of Wartmann's paper: 'Memoir on Daltonism [or Colour Blindness]', in *Scientific Memoirs*, ed. Richard Taylor, **4** (London, 1846): 156–87. Brewster's usage was adopted by George Wilson, *Researches on Colour-Blindness* (Edinburgh, 1855): esp. 161–2, where he declares that 'Daltonism *should* signify the Doctrine of Indivisible Chemical Atoms; and a Daltonian, a believer in such'; and see Maxwell's usage in his letter to Wilson of 4 January 1855 (Number 54).

(11) Compare Number 54 for discussion of Thomas Young's theory of colour vision in terms of three primary colours or 'elementary sensations'; and where Maxwell gives a definition of three 'independent variables, of which sensible colour is a function': 'intensity, hue and tint'. In his 'Experiments on colour, as perceived by the eye, with remarks on colour-blindness', *Trans. Roy. Soc. Edinb.*, **21** (1855): 275–98, on 279 (= *Scientific Papers*, **1**: 131) Maxwell shows that these two methods of representing colour 'are capable of exact numerical comparison'. See Number 54 notes (6) and (10).

(12) See Number 54, and 'Experiments on colour, as perceived by the eye': 287–8 (= *Scientific Papers*, **1**: 141–2). For similar arguments compare Thomas Young, *A Course of Lectures on Natural Philosophy and the Mechanical Arts*, 2 vols. (London, 1807), **2**: 315–16; and a letter from John Herschel

may call blue & yellow,[13] so that to him colours are blue or yellow & dark or light. Red & green are between blue & yellow & may be matched in pairs.

But this is only possible under a given kind of incident light for if a red & a green are to him the same by day light, the red will be brightest by candle light.[14] I have enabled such a man to distinguish reds & greens by showing him the opposite effects of red & green glasses upon colours similar to him. A pair of glasses, one red, one green, for the two eyes might become by practice a means of habitually distinguishing colours.[15]

But the subject of colour as perceived by the eye is one which I intend to investigate more fully.[16] I have satisfied myself that by using the aid of others in every experiment, and shutting out obvious sources of error, very exact results may be obtained about the equivalence of two colours as regards the eye. It is possible for people to agree on such points and this is the condition of a science of sensible colour independent of individual peculiarities. Do you know any place short of Munich[17] where tolerable prisms could be got for chromatic purposes, just good enough to show one or two reference lines in the spectrum, and of sufficient size to transmit a pencil of $\frac{3}{4}$ inch square?

Have you seen Hopkins' pamphlet on Public Lectures for Mathematical Men entitled Remarks on Mathematical Studies of Cambridge?[18]

I am glad to hear (from you) that I am to be in Glasgow at Easter. I suppose when people say things they may come true. However I certainly intend to spare some time from home to visit the University of Glasgow.

Yours truly
J. CLERK MAXWELL

---

to Dalton of 20 May 1833, in W. C. Henry, *Memoirs of the Life and Scientific Researches of John Dalton* (London, 1854): 25–7, noted by Maxwell in 'Experiments on colour, as perceived by the eye': 288 (= *Scientific Papers*, **1**: 142). See also Number 179 and Number 64 note (4).

(13) Maxwell here implies the received view, as maintained by J. D. Forbes, 'Hints towards a classification of colours', *Phil. Mag.*, ser. 3, **34** (1849): 161–78, on 172, that the 'primary colours' were 'red, yellow and blue'. Reviewing alternative schemes, Forbes considered Young's suggestion 'that the primary colours are red, green and violet' to be 'a singular opinion'. By January 1855 Maxwell had adopted Young's scheme; see Number 54, esp. note (7).

(14) Compare Dalton's observation of a geranium; 'Extraordinary facts relating to the vision of colours': 29–30.

(15) See Maxwell, 'Experiments on colour, as perceived by the eye': 287–8 (= *Scientific Papers*, **1**: 141); and see Number 59. Compare Wilson's favourable comment in his *Researches on Colour-Blindness*: 173.   (16) See Numbers 54 and 59.

(17) On the manufacture of optical instruments in Munich, pioneered by Joseph Fraunhofer and continued (from 1854) by Carl August Steinheil, see C. Jungnickel and R. McCormmach, *Intellectual Mastery of Nature: Theoretical Physics from Ohm to Einstein. Vol. 1: The Torch of Mathematics 1800–1870* (Chicago/London, 1986): 268–73, 280.

(18) William Hopkins, *Remarks on the Mathematical Teaching of the University of Cambridge* (Cambridge, 1854).

48

FROM A LETTER
TO RICHARD BUCKLEY LITCHFIELD
25 MARCH 1854

From Campbell and Garnett, Life of Maxwell[1]

Trin. Coll.
25 March 1854

I am experiencing the effects of Mill, but I take him slowly. I do not think him the last of his kind.[2] I think more is wanted to bring the connexion of sensation with Science to light, and show what it is not.[3] I have been reading Berkeley on the *Theory of Vision*,[4] and greatly admire it, as I do all his other non-mathematical works; but I was disappointed to find that he had at last fallen into the snare of his own paradoxes, and thought that his discoveries with regard to the senses and their objects would show some fallacy in those branches of high mathematics which he disliked.[5] It is curious to see how speculators are led by their neglect of exact sciences to put themselves in opposition to them where they have not the slightest point of contact with their systems. In the Minute Philosopher there is some very bad Political Economy and much very good thinking on more interesting subjects. Paradox is still sought for and exaggerated. We live in an age of wonder still.

APPENDIX: FROM AN ESSAY FOR THE APOSTLES 'HAS EVERYTHING BEAUTIFUL IN ART ITS ORIGINAL IN NATURE?'[6]
SPRING 1854[7]

From Campbell and Garnett, Life of Maxwell[8]

As the possibility of working out the question within the time forms no part of our specification, we may glance at heights which mock the attenuated triangle

(1) *Life of Maxwell*: 207–8.

(2) John Stuart Mill, *A System of Logic Ratiocinative and Inductive: being a connected view of the principles of evidence and the methods of scientific investigation*, 2 vols. (London, 1843). Compare Number 101.

(3) At about this time Maxwell was preparing an essay for the Apostles concerned with science and visual sensation; see Appendix. On Litchfield's membership of the Apostles see Number 42 note (9).   (4) George Berkeley, *An Essay towards a New Theory of Vision* (Dublin, 1709).

(5) See George Berkeley, *The Analyst, or a Discourse addressed to an Infidel Mathematician* (London, 1734).

(6) Maxwell's comments in this essay suggest his response to Berkeley's discussion of visual

of the mathematician, and throw our pebble into depths which his cord and plummet can never sound.... Nothing beautiful can be produced by Man except by the laws of mind acting in him as those of Nature do without him; and therefore the kind of beauty he can thus evolve must be limited by the very small number of correlative sciences which he has mastered; but as the Theoretic and imaginative faculty is far in advance of Reason, he can apprehend and artistically reproduce natural beauty of a higher order than his science can attain to; and as his Moral powers are capable of a still wider range, he may make his work the embodiment of a still higher beauty, which expresses the glory of nature as the instrument by which our spirits are exercised, delighted, and taught. If there is anything more I desire to say it is that while I confess the vastness of nature and the narrowness of our symbolical sciences, yet I fear not any effect which either Science or Knowledge may have on the beauty of that which is beautiful once and for ever....

All your analysis is cruelly anatomical, and your separated faculties have all the appearance of preparations. You may retain their names for distinctness, but forbear to tear them asunder for lecture-room demonstration.... They separate a faculty by saying it is not intellectual, and then, by reasoning blindfold, every philosopher goes up his own tree, finds a mare's nest and laughs at the eggs, which turn out to be pure intellectual abstractions in spite of every definition....

With respect to beauty of things audible and visible, we have a firm conviction that the pleasure it affords to any being would be of the same kind by whatever organisation he became conscious of it.... Our enjoyment of music is accompanied by an intuitive perception of the relations of sounds, and the agreement of the human race would go far to establish the universality of these conditions of pleasure, though Science had not discovered their physical and numerical significance....

[*Beauty of Form.*] A mathematician might express his admiration of the Ellipse. Ruskin agrees with him.[9]... It is a universal condition of the enjoyable that the mind must believe in the existence of a law, and yet have a mystery to move about in.... All things are full of ellipses – bicentral sources

---

perception (especially on colour and form as the objects of sense) in *The New Theory of Vision*; and to themes (on colour, form, art and nature, and perception) discussed by Ruskin (to whom Maxwell refers *infra*); see John Ruskin, *The Seven Lamps of Architecture* (London, 1849), *The Stones of Venice*, 3 vols. (London, 1851–3), and *Modern Painters*, 2 vols. (London, 1843–6).

(7) As given by Campbell: 'Spring of 1854 – shortly after the Tripos Examination'.

(8) *Life of Maxwell*: 230–2.

(9) Ruskin, *Seven Lamps*, chap. IV, §6; *Modern Painters*, **2**, part 3, I, chap. vi, §11.

of lasting joy, as the wondrous Oken might have said.[10]... [Beauty of form, then, is– 1. Geometrical; 2. Organic; 3.] Rivers and mountains have not even an organic symmetry; the pleasure we derive from their forms is not that of comprehension, but of apprehension of their fitness as the forms of flowing and withstanding matter. When such objects are represented by Art, they acquire an additional beauty as the language of Nature understood by Man, the interpreter, although by no means the emendator, of her expressions.

...The power of Making is man's highest power in connexion with Nature....

[*Beauty of Colour.*] The Science of Colour does, indeed, point out certain arrangements and gradations which follow as necessarily from first principles, as the curves of the second order from their equations. These results of science are, many of them, realised in natural phenomena taking place according to those physical laws of which our mathematical formulæ are symbols; but it is possible that combinations of colours may be imagined or calculated, which no optical phenomenon we are acquainted with could reproduce. Such a result would no more prove the impropriety of the arrangement than ignorance of the planetary orbits kept the Greeks from admiring the Ellipse.

---

(10) Lorenz Oken, exponent of *Naturphilosophie*; see his *Elements of Physiophilosophy*, trans. A. Tulk (London, 1847).

FROM A LETTER TO JANE CAY

3 JUNE 1854

From Campbell and Garnett, *Life of Maxwell*[1]

Trin. Coll.
Whitsun. Eve 1854

I am in great luxury here, having but 2 pups., and able to read the rest of the day, so I have made a big hole in some subjects I wish to know. We have hot weather now, and I am just come from a meeting of subscribers to the Bathing Shed, which we organised into a Swimming Club so as to make it a more sociable affair, instead of mere 'pay your money and use your key'.

A nightingale has taken up his quarters just outside my window, and works away every night. He is at it very fierce now. At night the owls relieve him, softly sighing after their fashion.

I have made an instrument for seeing into the eye through the pupil.[2] The difficulty is to throw the light in at that small hole and look in at the same time; but that difficulty is overcome, and I can see a large part of the back of the eye quite distinctly with the image of the candle on it. People find no inconvenience in being examined, and I have got dogs to sit quite still and keep their eyes steady. Dogs' eyes are very beautiful behind, a copper-coloured ground, with glorious bright patches and networks of blue, yellow, and green, with blood-vessels great and small.

---

(1) *Life of Maxwell*: 208.   (2) An ophthalmoscope; see Numbers 65, 66 and 68.

NOTES ON ELECTRICITY

circa 1854[1]

From the original in the University Library, Cambridge[2]

ELECTRICITY[3]

1 When certain substances are rubbed together and then separated, they acquire the power of attracting each other. We call the agency thus developed Electricity, but we abstain from attributing to it any properties other than the following which are deduced from experiment.

2 Let two pieces of glass be rubbed each upon a piece of resin and separated, then it is found that

1st   The two pieces of glass repel one another.
2nd   Each piece of glass attracts each piece of resin.
3rd   The two pieces of resin repel one another.

The glass and resin have therefore opposite electrical properties which may be called positive and negative. The mode in which the glass is electrified is called positive by consent of men of science but when we wish to avoid ascribing positive existence to one kind of electricity more than the other we speak of them as vitreous and resinous.

3 *Electrical Quantity.* If a small piece of glass and a small piece of resin be electrified by friction with each other so that when separated by unit of distance they attract each other with unit of force then the glass is said to have a unit of positive and the resin a unit of negative electricity.

4 The force between two *small* electrified bodies varies inversely as the square of the distance between them.

5 Let e and e' be the quantities of positive electricity on two small bodies and r the distance between them. They will repel each other with a force equal to

$$\frac{ee'}{r^2}.$$

---

(1) Handwriting and content suggest this date.

(2) ULC Add. MSS 7655, V, c/2.

(3) Compare William Thomson, 'On the mathematical theory of electricity in equilibrium. II. A statement of the principles on which the mathematical theory of electricity is founded', *Camb. & Dubl. Math. J.*, **3** (1848): 131–40, esp. 131–5 (= *Electrostatics and Magnetism*: 42–5), and Maxwell's similar presentation in the *Treatise*, **1**: 30–31 (§27).

On Lines of Force[4]

If a body or a system of bodies is electrified in any manner a small body charged with positive electricity and placed in any position will experience a force urging it in a certain direction. If the small body be now carried slowly along, so as to move always in the direction of the force acting on it, it will describe a line whose direction always coincides with that of the force. This line is called a 'Line of Force'.

If the small body had been charged with negative electricity it would have been urged in the direction exactly opposite with an equal force. It is evident that we may begin a line of force at any point of space and thus describe as many different lines of force as we please.

On Surfaces of Equilibrium[5]

If we describe surfaces perpendicular to the lines of force (which is always possible) then an electrified body will be in equilibrium on each of these surfaces.

Let *ABC* represent a Line of Force. A unit of electricity placed on any part of this line will be acted on by a force in the direction of this line and if allowed to move along the line, work will be done.[6] Let one unit of work be done in moving from *A* to *B*, another in moving from *B* to *C* and so on then if we describe surfaces of equilibrium through *A, B, C* &c it is easy to prove

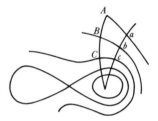

Figure 50,1

(1) that one unit of work will be done by one unit of electricity passing from any one of these surfaces to the next in order at whatever part of the surfaces it takes place.

2nd That the electrical force is always perpendicular to these surfaces and inversely proportional to the distance between them.[7]

---

(4) Michael Faraday, 'Experimental researches in electricity. – Twenty-eighth series. On lines of magnetic force', *Phil. Trans.* **142** (1852): 25–56 (= *Electricity*, **3**: 328–70).

(5) Compare William Thomson, 'Propositions in the theory of attraction', *Camb. Math. J.*, **3** (1842–3): 189–96, 201–6, esp. 190 (= *Electrostatics and Magnetism*: 126); 'surfaces for which the potential is constant are therefore called, by Gauss, *surfaces of equilibrium*'. See C. F. Gauss, 'General propositions relating to attractive and repulsive forces acting in the inverse ratio of the square of the distance', in *Scientific Memoirs*, ed. R. Taylor, **3** (London, 1843): 153–96, esp. 195; and also Hermann Helmholtz, 'On the conservation of force', in *Scientific Memoirs, Natural Philosophy*, ed. J. Tyndall and W. Francis (London, 1853): 114–62, esp. 140. In his letter to Thomson of 13 November 1854 Maxwell uses the term 'equipotential surfaces' (Number 51, esp. note (13)).

(6) Compare Number 51, esp. note (14).   (7) See Number 51 note (3).

Definition of Electrical Tension[8]

The Electrical Tension at any point is the whole force with which electricity from that point would escape into a conductor communicating with the earth.

It is measured by the number of surfaces of equilibrium between the point and the earth.

Definition of Electromotive Force

The Electromotive Force between any two points is the difference of the Electrical Tensions at those points.[9]

When Electricity is in equilibrium on a conductor, the Tension must be the same at every point of it, otherwise electricity would pass from one point to another where the tension was less.

The Tension of any conductor depends partly on the quantity of Electricity it contains, and partly on the quantity of Electricity in neighbouring bodies.

Definition of Electrical Capacity

The Electrical Capacity of a Conductor is the quantity of Electricity it must receive to raise its tension to unity.

(The capacity of a sphere is as the radius.)

---

(8) Subsequent definitions can be contrasted (but compare note (3)) with those given in the *Treatise*, **1**: 47–8 (§§48–50), where Maxwell notes that 'the word Tension has been used by electricians in several vague senses'. On the interpretation of 'tension' as 'potential' see Maxwell, 'On Faraday's lines of force', *Trans. Camb. Phil. Soc.*, **10** (1856): 27–83, esp. 46 (= *Scientific Papers*, **1**: 180–1); and Gustav Kirchhoff, 'Ueber eine Ableitung der Ohm'schen Gesetze, welche sich an die Theorie der Electrostatik anschliesst', *Ann. Phys.*, **78** (1849): 506–13 (= Gustav Kirchhoff, *Gesammelte Abhandlungen* (Leipzig, 1882): 49–55); see Number 84 note (46).

(9) Compare 'On Faraday's lines of force': 47 (= *Scientific Papers*, **1**: 181–2); and Gustav Kirchhoff, 'Ueber die Anwendbarkeit der Formeln für die Intensitäten der galvanischen Ströme in einem Systeme linearer Leiter auf Systeme, die zum Theil nicht aus linearen Leitern bestehen', *Ann. Phys.*, **75** (1848): 189–205 (= *Gesammelte Abhandlungen*: 33–49).

LETTER TO WILLIAM THOMSON

13 NOVEMBER 1854

From the original in the University Library, Cambridge[1]

Trin. Coll.
Nov 13/54

Dear Thomson

I have heard very little of you for some time exept thro' Hopkins & Stokes, but I suppose you are at work in Glasgow as usual. Do you remember a long letter you wrote me about electricity, for wh: I forget if I thanked you?

I soon involved myself in that subject, thinking of every branch of it simultaneously, & have been rewarded of late by. finding the whole mass of confusion beginning to clear up under the influence of a few simple ideas.

As I wish to study the growth of ideas as well as the calculation of forces, and as I suspect from various statements of yours that you must have acquired your views by means of certain conceptions which I have found great help, I will set down for you the confessions of an electrical freshman.

I got up the fundamental principles of electricity of tension easily enough.[2] I was greatly aided by the analogy of the conduction of heat, wh: I believe is your invention at least I never found it elsewhere.[3] Then I tried to make out the theory of attractions of currents but tho' I could see how the effects could

---

(1) ULC Add. MSS 7342, M 89. Previously published in Larmor, 'Origins': 701–5.

(2) That is: 'potential'. See Number 50.

(3) [William Thomson,] 'On the uniform motion of heat in homogeneous solid bodies, and its connection with the mathematical theory of electricity', *Camb. Math. J.*, **3** (1842): 71–84, esp. 73–4 (= *Phil. Mag.*, ser. 4, **7** (1854): 502–15 [June 1854 Supplement]; *Electrostatics and Magnetism*: 1–14). Thomson's analogy is between temperature and potential; fluxes or forces flow continuously across isothermal or equipotential ellipsoidal surfaces. He establishes that: 'the sole condition of equilibrium of electricity, distributed over the surface of a body, is, that it must be so distributed that the attraction on a point at the surface, oppositely electrified, may be perpendicular to the surface. Since, at any of the isothermal surfaces, v [temperature] is constant, it follows that $-dv/dn$, where n is the length of a curve which cuts all the surfaces perpendicularly, measured from a fixed point to the point attracted, is the total attraction on the latter point; and that this attraction is in a tangent to the curve n, or in a normal to the isothermal surface passing through the point... $-dv/dn$ will be the total flux of heat at the variable extremity of n, and the direction of this flux will be along n, or perpendicular to the isothermal surface. Hence, if a surface in an infinite solid be retained at a constant temperature, and if a conducting body, bounded by a similar surface, be electrified, the flux of heat, at any point, in the first case, will be proportional to the attraction on an electrical point, similarly situated, in the second; and the direction of the flux will correspond to that of the attraction.' See note (13); and see M. N. Wise, 'The flow analogy to electricity and magnetism, Part I: William Thomson's reformulation of action at a distance', *Archive for History of Exact Sciences*, **25** (1981): 19–70.

be determined I was not satisfied with the form of the theory which treats of elementary currents & their reciprocal actions, & I did not see how any general theory was to be formed from it. I read Ampère's investigations this term & greatly admired them but thought there was a kind of ostensive demonstration about them wh: must have been got up, after Ampère had convinced himself, in order to suit his views of philosophical inquiry, and as an example of what it ought to be. Yet I believe there is no doubt that Ampère discovered the laws & probably by the method wh: he has given.[4] Now I have heard you speak of 'magnetic lines of force' & Faraday seems to make great use of them,[5] but others seem to prefer the notion of attractions of elements of currents directly. Now I thought that as every current generated magnetic lines & was acted on in a manner determined by the lines thro wh: it passed that something might be done by considering 'magnetic polarization' as a property of a 'magnetic field'[6] or space and developing the geometrical ideas according to this view.

---

(4) André Marie Ampère, *Théorie des Phénomènes Électrodynamiques, uniquement déduite de l'éxpérience* (Paris, 1826); published (with some modifications) as 'Mémoire sur la théorie mathématique des phénomènes électrodynamiques uniquement déduite de l'expérience', *Mémoires de l'Académie Royale des Sciences de l'Institut de France*, **6** (1827): 175–388. Compare Maxwell's remarks on Ampère as the 'Newton of electricity' in the *Treatise*, **2**: 162–3 (§528). On Ampère's method of inductive demonstration see Christine Blondel, *A.-M. Ampère et la Création de l'Électrodynamique (1820–1827)* (Paris, 1982). On Ampère's force law between two infinitesimal line elements carrying currents see Number 66 note (3).

(5) William Thomson, 'On the theory of magnetic induction in crystalline and non-crystalline substances', *Phil. Mag.*, ser. 4, **1** (1851): 177–86 (= *Electrostatics and Magnetism*: 465–80); Michael Faraday, 'Experimental researches in electricity. – Twenty-eighth series. On lines of magnetic force', *Phil. Trans.*, **142** (1852): 25–56; and Faraday, 'On the physical character of the lines of magnetic force', *Phil. Mag.*, ser. 4, **3** (1852): 401–28 (= *Electricity*, **3**: 328–70, 407–37).

(6) Faraday first used the term 'magnetic field' in his 'Experimental researches in electricity. – Twentieth series. On new magnetic actions, and on the magnetic condition of all matter', *Phil. Trans.*, **136** (1846): 21–40, 41–63, on 22–3 (§§2247, 2252) (= *Electricity*, **3**: 29–30). Thomson first used the term 'field of force' in a letter to Faraday which he dated 'Saturday, June 19'; in S. P. Thompson, *The Life of William Thomson, Baron Kelvin of Largs*, 2 vols. (London, 1910), **1**: 214–16. Thomson first used the terms 'field of force' and 'magnetic field' in a published paper in his 'On the theory of magnetic induction', *Phil. Mag.*, ser. 4, **1** (1851): 179 (= *Electrostatics and Magnetism*: 467); and again in two papers 'On certain magnetic curves; with applications to problems in theories of heat, electricity and fluid motion' and 'On the equilibrium of elongated masses of ferromagnetic substance in uniform and varied fields of force', *Report of the Twenty-first Meeting of the British Association for the Advancement of Science; held at Belfast in 1851* (London, 1852), part 2: 18–20 (= *Electrostatics and Magnetism*: 514–17). Maxwell gave the term 'field' its first clear definition, in consonance with previous usage, in his paper 'A dynamical theory of the electromagnetic field', *Phil. Trans.*, **155** (1865): 459–512, on 460 (= *Scientific Papers*, **1**: 527); 'The theory I propose may therefore be called a theory of the *Electromagnetic Field*, because it has to do with the space in the neighbourhood of the electric or magnetic bodies'.

I use the word 'polarization' to express the fact that at a point of space the south pole of a small magnet is attracted in a certain direction with a certain force.

'Polarity' is a property of magnets &c to produce pol[n]. Then come two definitions wh may be modified afterwards.

(1) The pol[n] of a curve in space is the integral of the pol[n] at any point resolved along it mult: into the element of the curve.

(2) The pol[n] of a surface is the integral of the pol[n] resolved in the normal into the element of surface.[7]

The pol[n] of a curve or surface arising from several sources is the algebraic sum of the effects taken separately.

It appears from experiment that the mag: effect of a current forming a helix of small uniform section at any point may be reduced to two forces of opposite kinds directed to the extremities & $\propto \dfrac{1}{d^2}$.[8]

Whence it is easy to find the effect of a small circular current on a point at some distance from it. When forces act centrally according to the law of inverse square the pol[n] of any closed curve is zero & that of any closed surface not passing thro the magnetic substance is zero.[9]

(7) Compare Thomson's distinction between 'solenoidal and lamellar distributions of magnetism' in his paper 'A mathematical theory of magnetism', *Phil. Trans.*, **141** (1851): 243–85, esp. 269–85 (= *Electrostatics and Magnetism*: 341–404, esp. 378–404). Thomson conceives the distribution of magnetisation in two different geometrical modes: a magnet is divided along lines of magnetisation into tubes ('solenoids', a term derived from Ampère; see Number 71 note (29)); or a magnet is divided perpendicular to lines of magnetisation into 'magnetic shells'. 'If a finite magnet of any form be capable of division into an infinite number of solenoids which are either closed or have their ends in the bounding surface, the distribution of magnetism in it is said to be solenoidal. . . . If a finite magnet of any form be capable of division into an infinite number of magnetic shells which are either closed or have their edges in the bounding surface, the distribution of magnetism in it is said to be lamellar. . . .' ('A mathematical theory of magnetism': 270.) Compare Maxwell's discussion of 'magnetic quantity and intensity' in his paper 'On Faraday's lines of force', *Trans. Camb. Phil. Soc.*, **10** (1856): 27–83, esp. 54–5 (= *Scientific Papers*, **1**: 192), a distinction between quantities summed along lines of force (intensity) and across the surface enclosed by lines of force (quantity). This becomes the basis of his distinction between 'magnetic force' (**H**) and 'magnetic induction' (**B**) as 'intensities' and 'quantities' or 'forces' and 'fluxes'; see Maxwell, *Treatise*, **1**: 10–11 (§12); **2**: 31–5 (§§407–13) on 'magnetic solenoids and shells'. See notes (11) and (16) for Thomson's expressions for solenoidal and lamellar distributions of magnetism; and see also Number 71 note (30).

(8) Ampère, 'Mémoire sur la théorie mathématique des phénomènes électrodynamiques', *Mémoires de l'Académie Royale des Sciences*, **6** (1827): 366–71.

(9) Compare Maxwell's discussion of Ampère's laws of electrodynamics in 'On Faraday's lines of force': 55–6 (= *Scientific Papers*, **1**: 193–4).

Now let a number of little circuits be disposed so as to cover a given portion of a surface then the effect on any external point will be the same as that of a uniformly magnetized lamina (see your paper) [10] and if we remove one of the circuits & draw a closed curve so as not to pass thro any one, its pol[n] will be zero. [11] But if we replace the circuit the pol[n] will depend entirely on the strength of the circuit wh: thus is linked into the closed curve & will measure the intensity of the current as may be easily shown.

But in this case we may suppose that all the contiguous currents of the small circuits destroy each other, leaving the current round the boundary. Hence these theorems

(1) The pol[n] of any closed curve is measured by the sum of the intensities of all the currents which pass thro' it.

(2) The pol[n] of any surface, round the boundary of wh: a current passes, is measured by the intensity of that current. [12]

---

(10) Thomson's 'lamellar' distribution of magnetism; see note (7) and his 'A mathematical theory of magnetism': 270n; 'the term *lamellar*...is preferred to "laminated"; since this might be objected to as rather meaning "composed of plates", than *composed of shells*, whether plane or curved, and is besides too much associated with a mechanical structure such as that of slate or mica, to be a convenient term for the magnetic distributions defined in the text'. See also note (11).

(11) If α, β, γ denote the components of the intensity of magnetisation at a point x, y, z, then Thomson shows that the condition that a magnetic distribution is lamellar is

$$\frac{d\beta}{dz} - \frac{d\gamma}{dy} = 0, \quad \frac{d\gamma}{dx} - \frac{d\alpha}{dz} = 0, \quad \frac{d\alpha}{dy} - \frac{d\beta}{dx} = 0;$$

see Thomson, 'A mathematical theory of magnetism': 275. Stokes derived a similar equation for a 'wave of dilatation' in isotropic media in his paper 'On the dynamical theory of diffraction', *Trans. Camb. Phil. Soc.*, **9** (1849): 1–62, esp. 9, 13 (= *Papers*, **2**: 243–328). Maxwell was not familiar with Stokes' results at this time; see 'On Faraday's lines of force': 67 (= *Scientific Papers*, **1**: 209). See also J. J. Cross, 'Integral theorems in Cambridge mathematical physics, 1830–55', in *Wranglers and Physicists. Studies on Cambridge Physics in the Nineteenth Century*, ed. P. M. Harman (Manchester, 1985): 112–48, esp. 138, 142.

(12) These theorems express symmetrical relations between 'polarisation' and electric currents. As stated, the second theorem does not allow for the size and shape of the circuit through which a current passes; see M. N. Wise, 'The mutual embrace of electricity and magnetism', *Science*, **203** (1979): 1310–18. In 'On Faraday's lines of force': 65–6 (= *Scientific Papers*, **1**: 206) Maxwell reformulates these theorems as two laws of electromagnetism ('Law III' and 'Law I'); see Number 87, esp. notes (11) and (19). Maxwell does not state these theorems in an analytic form, but his statement of the theorems suggests the application of 'Stokes' theorem' which transforms line integrals into surface integrals. First stated by Thomson in a letter to Stokes of 2 July 1850 (ULC Add. MSS. 7656, K 39), the theorem was published by Stokes in his Smith's Prize examination of February 1854, where Maxwell was placed equal Smith's Prizeman (see Number 45 note (2)):

'If X, Y, Z be functions of the rectangular co-ordinate x, y, z, dS an element of any limited

On any of the 'equipotential surfaces' (or surfaces perp to the lines of force)[13] draw two systems of curves so as to divide the surface into small elements, the pol^n of each of wh: is the same. Let lines of force move so as to pass thro' these curves so as to generate what I call ('surfaces of no pol^n') then these surfaces will intersect other equipotential surfaces forming elements of the same amount of pol^n as before. Finally let a series of equipotential surfaces be drawn so that the diff. of potential is the same between each.

All space is now cut up into elements. Call the intersections of the surfaces of no pol^n 'lines of pol^n' then

(1) The pol^n of a surface is expressed by the *number* of lines wh cut it.

(2) The pol^n of a line is expressed by the number of equip^1 surfaces wh: it cuts.

Def^n

The pol^n of a portion of space is expressed by the number of parallelopipeds into wh it is thus divided.

---

surface, l, m, n, the cosines of the inclinations of the normal at dS to the axes, ds an element of the bounding line, shew that

$$\iint \left\{ l\left(\frac{dZ}{dy}-\frac{dY}{dz}\right)+m\left(\frac{dX}{dz}-\frac{dZ}{dx}\right)+n\left(\frac{dY}{dx}-\frac{dX}{dy}\right)\right\} dS = \int \left(X\frac{dx}{ds}+Y\frac{dy}{ds}+Z\frac{dz}{ds}\right) ds,$$

the differential coefficients of X, Y, Z being partial, and the single integral being taken all round the perimeter of the surface.' See *The Cambridge University Calendar for the Year 1854* (Cambridge, 1854): 415 (= Stokes, *Papers*, **5**: 320).

(13) Compare Maxwell's use of the term 'surfaces of equilibrium' (Number 50 esp. note (5)). Compare William Thomson, 'On the mathematical theory of electricity in equilibrium. I. On the elementary laws of statical electricity', *Camb. & Dubl. Math. J.*, **1** (1845): 75–95, esp. 85, 89 (= *Electrostatics and Magnetism*: 15–37), where he expresses the analogy between the flux of heat and the flow of electric force by the concept of 'potential'; 'The problem of *distributing sources of heat* ... is mathematically identical with the problem of distributing *electricity in equilibrium*.... In the case of heat, the *permanent temperature* at any point replaces the *potential* at the corresponding point in the electrical system, and consequently the *resultant flux of heat* replaces the *resultant attraction* of the electrified bodies, in direction and magnitude'. The term 'potential', as Thomson indicated, had been introduced by George Green in *An Essay on the Application of Mathematical Analysis to the Theories of Electricity and Magnetism* (Nottingham, 1828) (= *Mathematical Papers of George Green*, ed. N. M. Ferrers (London, 1871): 1–115); and independently by Carl Friedrich Gauss, 'Allgemeine Lehrsätze in Beziehung auf die im verkehrten Verhältnisse des Quadrats der Entfernung wirkenden Anziehungs- und Abstossungs-Kräfte', in *Resultate aus den Beobachtungen des magnetischen Vereins in Jahre 1839*, Herausgegeben von C. F. Gauss und W. Weber (Leipzig, 1840): 1–51, (trans.) 'General propositions relating to attractive and repulsive forces acting in the inverse ratio of the square of the distance', in *Scientific Memoirs*, ed. R. Taylor, **3** (London, 1843): 153–96.

Prop^n

The work done by any given displacement of two circuits may be measured by (1) the change of the pol^n of all space or (2) the increase of the number of lines of pol^n wh pass thro both circuits.

The work done by any displacement of two circuits is measured by the increase of the number of lines of pol^n wh pass thro both circuits.[14]

Theory of Currents

Let u, v, w be the vel^{ies} of a current in xyz that is, the quantity of electricity wh passes across unit section in unit time. XYZ electromotive forces, k coeff^t of resistance p electric tension at any point

$$dp = (X - ku)\, dx + (Y - kv)\, dy + (Z - kw)\, dz$$

also

$$\frac{du}{dx} + \frac{dv}{dy} + \frac{dw}{dz}. \text{[15]}$$

---

(14) Compare 'On Faraday's lines of force': 41–3, 58–9, 61–4 (= *Scientific Papers*, **1**: 173–7, 196–8, 201–4). See William Thomson, 'On the theory of electro-magnetic induction', *Report of the Eighteenth Meeting of the British Association for the Advancement of Science* (London, 1849), part 2: 9–10 (= *Math. & Phys. Papers*, **1**: 91–2); Hermann Helmholtz, *Über die Erhaltung der Kraft, eine physikalische Abhandlung* (Berlin, 1847) (= Helmholtz, *Wissenschaftliche Abhandlungen*, 3 vols. (Leipzig, 1882–95), **1**: 12–68), (trans.) 'On the conservation of force', in *Scientific Memoirs, Natural Philosophy*, ed. J. Tyndall and W. Francis (London, 1853): 114–62, esp. 156–8; and F. E. Neumann, 'Allgemeine Gesetze der inducirten elektrischen Ströme', *Ann. Phys.*, **67** (1846): 31–44, cited by Thomson and Helmholtz. Neumann established that the change in potential caused by the action of two closed currents measures the mechanical work done by the electrodynamic force. The term 'work' was coming into regular use at this time; see Tyndall's rendition of Helmholtz's term 'Arbeitsgrösse' (*Über die Erhaltung der Kraft*: 19) as 'quantity of work' in his translation of 'On the conservation of force': 126. Compare also Thomson's use of the term 'work', in his paper 'On the dynamical theory of heat, with numerical results deduced from Mr Joule's equivalent of a thermal unit and M. Regnault's observations on steam', *Trans. Roy. Soc. Edinb.*, **20** (1851): 261–88; and in *Phil. Mag.*, ser. 4, **4** (1852): 8–21, 105–17, 168–76 (= *Math. & Phys. Papers*, **1**: 174–210).

(15) Read: $\dfrac{du}{dx} + \dfrac{dv}{dy} + \dfrac{dw}{dz} = 0$. In a manuscript note 'To determine the steady motion of a current of electricity in a conductor of any form' (ULC Add. MSS 7655, V, c/3), which was probably written at this time, Maxwell states that: 'The forces parallel to x acting on the electricity within the element dx, dy, dz are

(1)   The electromotive force $= + X\, dx\, dy\, dz$

(2)   The resistance due to the conducting substance $= - ku\, dx\, dy\, dz$

(3)   The difference of electric tensions $= -\dfrac{dp}{dx} dx\, dy\, dz$. Whence we have the equation

$$X - ku - \frac{dp}{dx} = 0.$$

From these notions we get the following which correspond to what we had in magnetism.[16] Surfaces of equal tension cutting lines of electric motion at right angles. Electric motion along any curve made up of sums of electric motions along its elements. Electric motion thro a surface made up of electric motion thro each element. When the surface is closed the whole electric motion is zero.

Electromotive forces

The electromotive force along any line is measured by the number of lines of pol^n wh that line cuts in unit of time. Hence the electromotive force round a given circuit depends on the decrease of the number of lines wh: pass thro it in unit of time, that is, on the decrease of the whole pol^n of any surface bounded by the circuit.

I have applied this to the case of a thin conducting sphere revolving about an axis placed east & west. I find that the currents are circles about an axis perp to axis of rotation & to the line of dip, and that the vel^y varies with the distance

Also
$$Y - kv - \frac{dp}{dy} = 0 \quad Z - kw - \frac{dp}{dz} = 0.$$

The equation for dp is therefore
$$dp = (X\,dx + Y\,dy + Z\,dz) - k(u\,dx + v\,dy + w\,dz)$$

and the equation of continuity is as usual $\dfrac{du}{dx} + \dfrac{dv}{dy} + \dfrac{dw}{dz} = 0$.' See also note (16) for Thomson's comment on the significance of the equation of continuity for the mathematical theory of magnetism. Maxwell made the analogy of the flow of an incompressible fluid in tubes formed by lines of force fundamental to the theory of 'On Faraday's lines of force'. The intensity of electro-motive forces is represented by the velocity of an imaginary fluid in a tube. Maxwell further developed this analogy in May 1855: see Numbers 61, 63 and 66.

(16) Compare Thomson's definition of a 'solenoidal' distribution of magnetism in 'A mathematical theory of magnetism': 273; 'if α, β, γ denote the components of the intensity of magnetization at any internal point (x, y, z), the equation $d\alpha/dx + d\beta/dy + d\gamma/dz = 0$ expresses that the distribution of magnetism is solenoidal'. Thomson notes the 'analogy between the circumstances of this expression and those of the cinematical condition expressed by the "equation of continuity" to which the motion of a homogeneous incompressible fluid is subject. ... Hence the intensity and direction of magnetization, in a solenoid, according to the definition, are subject to the same law as the mean fluid velocity in a tube with an incompressible fluid flowing through it'. Compare Maxwell's appeal to the analogy of an incompressible fluid moving in tubes formed by lines of force: Numbers 83, 84 and 'On Faraday's lines of force': 33–9, 42–3, 46–8, 52–4 (= *Scientific Papers*, **1**: 163–71, 175–7, 181–3, 189–92). Stokes stated an equation similar to Thomson's equation for a solenoidal distribution of magnetism in his paper 'A dynamical theory of diffraction': 9, 13, an equation for a 'wave of distortion' in isotropic media; see note (11). See also a letter from Thomson to Stokes of 25 February 1850 (ULC Add. MSS 7656, K 38).

from this axis. The magnetic effect is uniform within the sphere and the external effect may be simply expressed.[17] I have not yet calculated the effect of the magnetism on itself when the spherical shell is thick but I see how to do it.

Induction in soft Iron &c

In these bodies a polarity is developed in each particle proportional to the pol^n at that point. The result is that the lines of force converge like the lines of heat near good conductors.[18] The perfection of the property is the same as that of a perfect conductor of electricity, for at any point in the interior of perfectly soft iron the resultant pol^n is zero. It is easy to demonstrate the effect of iron ores for electromagnets &c in this way.

I have put the electric state of my mind before you that you may see how I am trying to make everything cohere, perhaps prematurely, & that you may know the kind of answers I want to my enquiries.

(1) Have you published any general theory founded on a theorem you gave in the *Math.* Journ, about the possibility of finding V as to fulfill a condition given in terms of V & α

$$\frac{d}{dx}\left(\alpha^2\frac{dV}{dx}\right)+\frac{d}{dy}\left(\alpha^2\frac{dV}{dy}\right)+\frac{d}{dz}\left(\alpha^2\frac{dV}{dz}\right)=0.\text{[19]}$$

---

(17) Compare Example XII of 'On Faraday's lines of force': 81–3 (= *Scientific Papers*, **1**: 226–9). On the production of electric currents in a metallic globe revolving in a magnetic field see Michael Faraday, 'Experimental researches in electricity. – Second series. Terrestrial magneto-electric induction', *Phil. Trans.*, **122** (1832): 163–94, esp. 168–70 (§§160–9) (= *Electricity*, **1**: 47–50). See Numbers 87 and 91.

(18) Compare Michael Faraday, 'Experimental researches in electricity. – Twenty-sixth series. Magnetic conducting power', *Phil. Trans.*, **141** (1851): 29–84 (= *Electricity*, **3**: 200–73).

(19) William Thomson, 'Theorems with reference to the solution of certain partial differential equations', *Camb. & Dubl. Math. J.*, **3** (1848): 84–7 (= *Electrostatics and Magnetism*: 139–41; *Math. & Phys. Papers*, **1**: 93–6). The equation as stated by Maxwell corresponds to Laplace's equation $\nabla^2 V = 0$ for the potential V of gravitating spheroids; P. S. de Laplace, *Traité de Mécanique Céleste*, 5 vols. (Paris, An VII [1799]–1825), **1**: 136–7 (= *Oeuvres Complètes de Laplace*, 14 vols. (Paris, 1879–1912), **1**: 152–3). Compare Thomson, 'Theorems': 84; 'It is possible to find a function V, of x, y, z, which shall satisfy, for all real values of these variables, the differential equation

$$\frac{d\left(\alpha^2\frac{dV}{dx}\right)}{dx}+\frac{d\left(\alpha^2\frac{dV}{dy}\right)}{dy}+\frac{d\left(\alpha^2\frac{dV}{dz}\right)}{dz}=4\pi\rho$$

α being any real continuous or discontinuous function of x, y, z, and ρ a function which vanishes for all values of x, y, z, exceeding certain finite limits (such as may be represented geometrically by a finite closed surface), within which its value is finite but entirely arbitrary.' Compare Poisson's equation $\nabla^2 V = -4\pi\rho$ for the attraction at an interior point; see S. D. Poisson, 'Remarques sur

(2) Is Weber's theory of the galvanic circuit to be read or can it be got?[20]

(I am looking into Neumann in the Berlin Academy for 1847,[21] also into Ohm's tract.[22] I find some things by Kirchhoff on currents in solid conductors[23] wh: are plain enough.)

(3) Can the case of ordinary electric phenomena be considered as an extreme case of conduction when the tension is enormous & the conductor excessively bad? I have not made it out yet or been able to see why there sh^d be strong attraction along the lines of almost no conduction.

(4) Can you recommend any other places to seek for further information?

This is a long screed of electricity but I find no other man to apply to on the subject so I hope you may not find it difficult to see my drift.

I have made acquaintance with Smith of Peterhouse,[24] but only lately. Stokes has gone to London to lecture for two terms,[25] but will return in May. I have 5 pups at present two of them very good, one in dynamics of particle & the other in rigid dynamics.

I am going to Cheltenham (previous to coming to Scotland) to examine the

---

une équation qui se présente dans la théorie des attractions des sphéroides', *Nouveau Bulletin des Sciences par la Société Philomathique de Paris*, **3** (1813): 388–92, esp. 390. Compare Maxwell's statement of Poisson's equation as Theorem I of 'On Faraday's lines of force': 57 (= *Scientific Papers*, **1**: 195); and see Numbers 71 and 100. The theorems of Laplace and Poisson are discussed in Green's *Essay*: 8, and by Gauss in his 'Allgemeine Lehrsätze'; see note (13). For discussion of the equations in a Cambridge text see J. H. Pratt, *The Mathematical Principles of Mechanical Philosophy, and their application to the Theory of Universal Gravitation* (Cambridge, 1836): 155–6. A question was set on the equations in the 1854 Mathematical Tripos examination; see *The Cambridge University Calendar for the Year 1854*: 404.

(20) Wilhelm Weber, 'Elektrodynamische Maassbestimmungen, über ein allgemeines Grundgesetz der elektrischen Wirkung', *Abhandlungen bei Begründung der Königlichen Sächsischen Gesellschaft der Wissenschaften, Leipzig* (1846): 211–378. See Number 66 note (3).

(21) F. E. Neumann, 'Über ein allgemeines Princip der mathematischen Theorie inducirter elektrischer Ströme', *Physikalische Abhandlungen der Königlichen Akademie der Wissenschaften zu Berlin, Aus dem Jahre 1847* (Berlin, 1849): 1–71. See also Neumann, 'Allgemeine Gesetze der inducirten elektrischen Ströme', *ibid.*, Aus dem Jahre 1845 (Berlin, 1847): 1–87; see note (14).

(22) Georg Simon Ohm, *Die galvanische Kette, mathematisch bearbeitet* (Berlin, 1827), (trans.) 'The galvanic circuit investigated mathematically', *Scientific Memoirs*, ed. R. Taylor, **2** (London, 1841): 401–506.

(23) Gustav Kirchhoff, 'Ueber die Auflösung der Gleichungen, auf welche man bei der Untersuchung der linearen Vertheilung galvanischer Ströme geführt wird', *Ann. Phys.*, **72** (1847): 497–508; 'Ueber die Anwendbarkeit der Formeln für die Intensitäten der galvanischen Ströme in einem Systeme linearer Leiter auf Systeme, die zum Theil nicht aus linearen Leitern bestehen', *ibid.*, **75** (1848): 189–205 (= *Gesammelte Abhandlungen*: 22–33, 33–49).

(24) Charles Abercombie Smith, Peterhouse 1854 from Glasgow University, Second Wrangler and Smith's Prize 1858 (Venn).

(25) Stokes had been appointed Professor of Physics at the Government School of Mines.

College there in Math: so I must be getting up papers and preparing for vivâ voce, but I have not heard the subjects yet.[26] I am glad to hear that Prof Forbes is back to his class. I have not seen him since he has been ill. I am getting up my theory of vision of colours with the help of four Daltonians & a great many ordinary men here. Who are the Daltonians in Edinburgh that Wilson mentioned?[27] I shd like to know a few more, so as to get greater accuracy. Remember me to No 13 & the College generally.

Yours truly
J. C. MAXWELL

(26) See Number 53.

(27) See Number 47 note (10).

FROM A LETTER TO JANE CAY

24 NOVEMBER 1854

From Campbell and Garnett, *Life of Maxwell*[1]

Trin. Coll.
24 Nov./54

I have been very busy of late with various things, and am just beginning to make papers for the examination at Cheltenham, which I have to conduct about the 11th of December.[2] I have also to make papers to polish off my pups. with. I have been spinning colours a great deal, and have got most accurate results, proving that ordinary peoples' eyes are all made alike, though some are better than others, and that other people see two colours instead of three; but all those who do so agree amongst themselves. I have made a triangle of colours by which you may make out everything.

You see that W lies outside the tri-angle B, R, Y, so that White can't be made with Blue, Red, and Yellow; but if you mix blue and yellow you don't get green, but pink – a colour between W and R. Those who see two colours only distinguish blue and yellow, but not red and green: for instance –

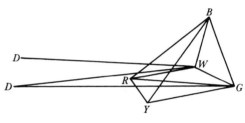

Figure 52,1

6 of blue and 94 of red make a red which looks to them like a gray made of 10 W and 90 Black.

40 of blue and 60 of green make 34 of W and 66 black.[3]

I should like you to find out if the Normans have got Bishop Percy's *Reliques of Ancient Ballad Poetry*,[4] for if they have I would not send them a duplicate; if not I think the book would suit one-half of that family.

If you can find out any people in Edinburgh who do not see colours (I know that Dicksons don't), pray drop a hint that I would like to see them. I have put one here up to a dodge by which he distinguishes colours without fail. I have also constructed a pair of squinting spectacles, and am beginning operations on a squinting man.

---

(1) *Life of Maxwell*: 208–9. (2) See Number 53. (3) See Number 54.

(4) Thomas Percy, *Reliques of Ancient English Poetry*, 3 vols. (London, various editions, 1765–1844).

DRAFT OF A REPORT ON A MATHEMATICS EXAMINATION AT CHELTENHAM COLLEGE[1]

DECEMBER 1854[2]

From the original in the University Library, Cambridge[3]

Gentlemen

I have the honour to report that the mathematical examination with which you have entrusted me is now concluded. I have arranged the names of the boys in each division of the first five classes in order of merit as shewn by the papers which they sent up. In most cases there was no difficulty in determining the order of the boys arrangement, but where the difference between two boys did not appear very decided I have not hesitated to bracket them as equal, and where there happened to be a large difference between the marks I have introduced a line between the names to indicate an hiatus in the scale of merit.

I have now only to make a few remarks on the state of the School in the different departments which I have examined.

In the first division of the first class I have been much struck by the familiarity with the various subjects which some of the higher boys seem to have acquired. The way in which they have written out the papers on Geometrical Curves & Elementary Mechanics is very satisfactory as shewing the aptitude of the boys for those subjects as well as the soundness of the training which they receive at the School.

The introduction of these subjects into the school course seems to me to be extremely useful as an extension of the methods with which they are already acquainted and as preparing them for any mathematial studies which they may afterwards undertake. Even for those whose mathematical reading is to cease when they leave the school the acquisition of sound elementary ideas will be found of the utmost importance as affording the only means of attaining that accuracy of thought and language which the mechanical advancement of the present age so preeminently demands.

With respect to the higher and more abstruse parts of these subjects, although I confess that much ability has been displayed by the boys, still I think that their attention ought not to be prematurely turned to such severe study.

---

(1) The invitation to conduct an examination in mathematics at Cheltenham College undoubtedly arose from Maxwell's friendship with R. B. Litchfield (see Number 42 note (1)). Litchfield's father, Captain Richard Litchfield, was one of the founders of the College, and Hon. Secretary to the Board of Directors 1842–54; see *Cheltenham College Register 1841–1889*, ed. A. A. Hunter (London, 1890): 25.   (2) See Number 52.   (3) ULC Add. MSS 7655, V, k/1a.

Those who do not pursue their mathematical enquiries will find the lower parts of the subject quite sufficient and those who proceed to the Universities will I am sure have no difficulty in mastering them for themselves.

These remarks have special reference to the subject of Trigonometry in which I think more attention ought to be paid to the parts relating to mensuration and practical details.

Deductions from Euclid have been found very useful in schools, as testing and confirming mathematical knowledge. I could have wished that in this Examination, the boys had more generally attempted these, as I feel convinced that many of them have geometrical talent which they might have thus displayed.

I was also disappointed to find so little variation from the precise form in which the demonstrations have been arranged in Euclid. Though I am certain that in most cases the proposition has been understood as well as remembered, a little variation might have increased that certainty.

In Arithmetic several of the boys have done extremely well though in some of those who best understood the principles I observed occasionally fundamental mistakes. These appear to have been due generally to a too entire reliance on mathematical formulæ, the results of which might easily have been tested and found erroneous by the exercise of ordinary thought.

Besides this there have been frequent numerical errors arising sometimes from an impatience and inattention but sometimes also from an entire ignorance of the elementary rules. With these rules the boys ought to be familiar before entering the mathematical classes at all, for it is impossible to make up for elementary deficiencies amid all the various studies of the higher classes.

I would in the last place suggest that the boys ought to be enjoined to depend entirely on their own resources especially during examination. In arithmetic and mathematics there may be different methods but one result. Every boy who has attained that result may judge for himself if it is right. Any information which he may obtain from others can only tend to defeat the purpose of the Examination, and ought to be quite unnecessary to him.

I have the more confidence in reporting these few shortcomings as I see that the mathematical studies of the School are in other respects so healthy and vigorous, that I have no doubt that in a very short time the school would of itself correct these irregularities which must from time to time appear.

I am
Gentlemen
Your Obedient Servent
J. C. MAXWELL

54

LETTER TO GEORGE WILSON

4 JANUARY 1855

From George Wilson, *Researches on Colour-Blindness* (1855)[1]

[ON THE THEORY OF COLOURS IN RELATION TO
COLOUR-BLINDNESS]

Dear Sir,

As you seemed to think that the results which I have obtained in the theory of colours might be of service to you, I have endeavoured to arrange them for you in a more convenient form than that in which I first obtained them. I must premise, that the first distinct statement of the theory of colour which I adopt, is to be found in *Young's Lectures on Natural Philosophy* (p. 345, Kelland's Edition);[2] and the most philosophical inquiry into it which I have seen is that of Helmholtz, which may be found in the Annals of Philosophy for 1852.[3]

It is well known that a ray of light, from any source, may be divided by means of a prism into a number of rays of different refrangibility, forming a series called a spectrum. The intensity of the light is different at different points of this spectrum; and the law of intensity for different refrangibilities differs according to the nature of the incident light. In Sir John F. W. Herschel's *Treatise on Light*, diagrams will be found, each of which represents completely, by means of a curve, the law of the intensity and refrangibility of a ʰeam of solar light after passing through various coloured media.[4]

I have mentioned this mode of defining and registering a beam of light, because it is the perfect expression of what a beam of light is in itself, considered with respect to all its properties as ascertained by the most refined instruments. When a beam of light falls on the human eye, certain sensations are produced,

---

(1) 'On the theory of colours in relation to colour-blindness, by James Clerk Maxwell, Esq., Trinity College, Cambridge, in a letter to Dr G. Wilson', in George Wilson, *Researches on Colour-Blindness. With a Supplement on the Danger attending the Present System of Railway and Marine Coloured Signals* (Edinburgh, 1855): 153–9; and in *Transactions of the Royal Scottish Society of Arts*, **4** (1856): 394–400 (= *Scientific Papers*, **1**: 119–25). On the term 'colour-blindness' see Number 47 note (10).

(2) Thomas Young, *A Course of Lectures on Natural Philosophy and the Mechanical Arts*, ed. P. Kelland, 2 vols. (London, 1845), **1**: 345.

(3) Hermann Helmholtz, 'Ueber die Theorie der zusammengesetzten Farben', *Ann. Phys.*, **87** (1852): 45–66 (= Helmholtz, *Wissenschaftliche Abhandlungen*, 3 vols. (Leipzig, 1882–95), **2**: 3–23); (trans.) 'On the theory of compound colours', *Phil. Mag.*, ser. 4, **4** (1852): 519–34.

(4) J. F. W. Herschel, 'Light', in *Encyclopaedia Metropolitana, or Useful Dictionary of Knowledge. Second Division: Mixed Sciences, Vol. II* (London, 1830): 341–586, esp. 430–38.

from which the possessor of that organ judges of the colour and intensity of the light. Now, though every one experiences these sensations, and though they are the foundation of all the phenomena of sight, yet, on account of their absolute simplicity, they are incapable of analysis, and can never become in themselves objects of thought. If we attempt to discover them, we must do so by artificial means; and our reasonings on them must be guided by some theory.

The most general form in which the existing theory can be stated is this, –

There are certain sensations, finite in number, but infinitely variable in degree, which may be excited by the different kinds of light. The compound sensation resulting from all these is the object of consciousness in a simple act of vision.

It is easy to see that the *number* of these sensations corresponds to what may be called in mathematical language the number of independent variables, of which sensible colour is a function.[5]

This will be readily understood by attending to the following cases:

1 When objects are illuminated by a homogeneous yellow light, the only thing which can be distinguished by the eye is difference of intensity or brightness.

If we take a horizontal line, and colour it black at one end, with increasing degrees of intensity of yellow light towards the other, then every visible object will have a brightness corresponding to some point in this line.

In this case there is nothing to prove the existence of more than one sensation in vision.

In those photographic pictures in which there is only one tint of which the different intensities correspond to the different degrees of illumination of the object we have another illustration of an optical effect depending on one variable only.

2 Now, suppose that different kinds of light are emanating from different sources, but that each of these sources gives out perfectly homogeneous light, then there will be two things on which the nature of each ray will depend: (1) its intensity or brightness; (2) its hue, which may be estimated by its position in the spectrum, and measured by its wave length.

If we take a rectangular plane, and illuminate it with the different kinds of homogeneous light, the intensity at any point being proportional to its horizontal distance along the plane, and its wave length being proportional to its height above the foot of the plane, then the plane will display every possible

---

(5) Compare Hermann Grassmann, 'Zur Theorie der Farbenmischung', *Ann. Phys.*, **89** (1853): 69–84; (trans.) 'On the theory of compound colours', *Phil. Mag.*, ser. 4, **7** (1854): 254–64, esp. 255; 'every impression of colour…may be analysed into three mathematically determinable elements, – *the tint, the intensity of the colour*, and the *intensity of the intermixed white*'.

variety of homogeneous light, and will furnish an instance of an optical effect depending on two variables.

3 Now, let us take the case of nature. We find that colours differ not only in intensity and hue, but also in tint; that is, they are more or less pure. We might arrange the varieties of each colour along a line, which should begin with the homogeneous colour as seen in the spectrum, and pass through all gradations of tint, so as to become continually purer, and terminate in white.

We have, therefore, three elements in our sensation of colour, each of which may vary independently. For distinctness sake I have spoken of intensity, hue, and tint; but if any other three independent qualities had been chosen, the one set might have been expressed in terms of the other, and the results identified.[6]

The theory which I adopt assumes the existence of three elementary sensations, by the combination of which all the actual sensations of colour are produced. It will be shown that it is not necessary to specify any given colours as typical of these sensations. Young has called them red, green, and violet; but any other three colours might have been chosen, provided that white resulted from their combination in proper proportions.[7]

---

(6) Maxwell adapts terminology for the classification of colours from Forbes, Grassmann and Hay. Compare J. D. Forbes, 'Hints towards a classification of colours', *Phil. Mag.*, ser. 3, **34** (1849): 161–78, esp. 175, where the terms '*intensity*', '*quality*' and '*purity*' are used; and Grassmann, 'On the theory of compound colours': 255, where the terms '*intensity*', '*tint*' and '*intensity of the intermixed white*' are cognates for Maxwell's 'intensity', 'hue' and 'tint' (see note (5)). In his paper 'Experiments on colour, as perceived by the eye, with remarks on colour-blindness', *Trans. Roy. Soc. Edinb.*, **21** (1855): 275–98, esp. 279 (= *Scientific Papers*, **1**: 131) Maxwell uses the terms 'shade', 'hue' and 'tint', adapting terms from D. R. Hay, *A Nomenclature of Colours, Hues, Tints and Shades* (Edinburgh, 1845).

(7) Thomas Young's trichromatic theory of vision; see Young, *Course of Lectures on Natural Philosophy*, ed. Kelland, **1**: 344; 'we may consider white light as composed of a mixture of red, green, and violet only'. The painters' triad of primary colorants, red, yellow and blue, was introduced in the seventeenth century, and was assimilated into the scientific tradition by Athanasius Kircher and Robert Boyle, and by Newton. Newton defined simple colours with respect to their refrangibility, as those colours whose rays refracted equally, but he supposed the identity of the mixing rules for lights and pigments; see *The Optical Papers of Isaac Newton. Volume I. The Optical Lectures 1670–1672*, ed. Alan E. Shapiro (Cambridge, 1984): 110–11n, 506n. Although he considered that spectral green was a simple colour, he claimed that green can be compounded of the adjacent spectral colours blue and yellow; see Isaac Newton, *Opticks: Or a Treatise of the Reflections, Refractions, Inflections and Colours of Light* (London, ₃1721): 116 (Book I, Part II, Prop. IV). Reviewing the colours of the Newtonian spectrum, Forbes argued that red, yellow and blue were the 'three primary colours'; 'Hints towards a classification of colours': 162. In 'Ueber die Theorie der zusammengesetzen Farben' (1852) Helmholtz demonstrated that while the mixture of coloured lights is an additive process, pigment mixing is subtractive. In imitating the colours of the spectrum he found 'at least five' spectral colours to be necessary; but he noted that if 'we wish to limit ourselves to three colours, it would be best to choose the three simple ones which admit

Before going farther I would observe, that the important part of the theory is not that three elements enter into our sensation of colour, but that there are only three. Optically, there are as many elements in the composition of a ray of light as there are different kinds of light in its spectrum; and therefore, strictly speaking, its nature depends on an infinite number of independent variables.

I now go on to the geometrical form into which the theory may be thrown. Let it be granted that the three pure sensations correspond to the colours red, green, and violet, and that we can estimate the intensity of each of these sensations numerically.

Let vrg be the angular points of a triangle, and con-ceive the three sensations as having their positions at these points. If we find the numerical measure of the red, green, and violet parts of the sensation of a given colour, and then place weights proportional to these parts at rg and v, and find the centre of gravity of the three weights by the ordinary process, that point will be the position of the given colour, and the numerical measure of its intensity will the the sum of the three primitive sensations.[8]

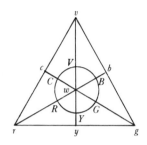

Figure 54,1

In this way, every possible colour may have its position and intensity ascertained; and it is easy to see that when two compound colours are combined, their centre of gravity is the position of the new colour.

The idea of this geometrical method of investigating colours is to be found in Newton's *Opticks* (Book I., Part 2, Prop. 6.),[9] but I am not aware that it has

of the least perfect imitation, namely, red, green, and violet'; Helmholtz, 'On the theory of compound colours': 532–3. See also Young, *Lectures on Natural Philosophy*, **1**: 344n for reference to a work by Chrétien-Ernst Wünsch, *Versuche und Beobachtungen über die Farben des Lichts* (Leipzig, 1792), where Young's selection of red, green and violet as the three primary colours is anticipated. Maxwell noted this work in the 1860s (Notebook 2, Archives, King's College, London). See also Number 47 note (13); Number 64 notes (2), (7) and (9); and Number 72 note (7).

(8) Compare Forbes' colour triangle (the vertices representing red, yellow and blue) in his 'Hints towards a classification of colours': 168. Forbes also described the triangular classification of colours (based on red, yellow and blue) by Tobias Mayer and Johann Heinrich Lambert; see Tobias Mayer, 'De affinitate colorum commentatio', in *Opera Inedita*, ed. G. C. Lichtenberg (Göttingen, 1775): 33–42, and Lichtenberg's commentary, *ibid.*: 93–103; and J. H. Lambert, *Beschreibung einer mit Calauischem Wachse ausgemalten Farbenpyramide* (Berlin, 1772). Compare Thomas Young's colour triangle employing red, green and violet primary colours; Young, *Lectures on Natural Philosophy*, ed. Kelland, **1**: 345, and **2**: Plate XXIX, Fig. 427.

(9) Isaac Newton, *Opticks* (London, ₃1721): 134–7, and Book I, Part II, Table III, Fig. 11. Newton divides the circumference of a circle in a harmonic proportion, each arc representing one of the seven principal colours. He draws a small circle at the centre of gravity of each arc, of an area proportional to the number of rays 'of each colour in the given mixture'. The centre of gravity of all the small circles represents the colour mixture 'accurate enough for practice, though not mathematically accurate'. The centre of the circle represents white: the centre of gravity of all

been ever employed in practice, except in the reduction of the experiments which I have just made. The accuracy of the method depends entirely on the truth of the theory of three sensations, and therefore its success is a testimony in favour of that theory.

Every possible colour must be included within the triangle rgv. White will be found at some point, w, within the triangle. If lines be drawn through w to any point, the colour at that point will vary in hue according to the angular position of the line drawn to w, and the purity of the tint will depend on the length of that line.[10]

Though the homogeneous rays of the prismatic spectrum are absolutely pure in themselves, yet they do not give rise to the 'pure sensations' of which we are speaking. Every ray of the spectrum gives rise to all three sensations, though in different proportions; hence the position of the colours of the spectrum is not at the boundary of the triangle, but in some curve $CRYGBV$ considerably within the triangle. The nature of this curve is not yet determined, but may form the subject of a future investigation.

All natural colours must be within this curve, and all ordinary pigments do in fact lie very much within it. The experiments on the colours of the spectrum which I have made are not brought to the same degree of accuracy as those on coloured papers. I therefore proceed at once to describe the mode of making those experiments which I have found most simple and convenient.

The coloured paper is cut into the form of discs, each with a small hole in the centre, and divided along a radius, so as to admit of several of them being placed on the same axis, so that part of each is exposed. By slipping one disc over another, we can expose any given portion of each colour. These discs are placed on a little top or teetotum, consisting of a flat disc of tin-plate and a vertical axis of ivory. This axis passes through the centre of the discs, and the quantity of each colour exposed is measured by a graduation on the rim of the disc, which is divided into 100 parts.[11]

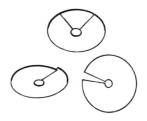

Figure 54,2

seven colours in due proportion. See Number 170. Compare Maxwell, 'Experiments on colour, as perceived by the eye': 279, 298, Fig. 1 ($=$ *Scientific Papers*, **1**: 131, 154).

(10) See Maxwell, 'Experiments on colour, as perceived by the eye': 282 ($=$ *Scientific Papers*, **1**: 135); the qualities of 'hue, tint, and shade [intensity], are represented on the diagram by angular position with respect to w, distance from w, and [a] coefficient; and the relation between the two methods of reducing the elements of colour to three becomes a matter of geometry'. See Number 59.

(11) Compare Forbes, 'Hints towards a classification of colours': 166–7; and see Number 64 esp. note (8). The coloured papers were supplied by D. R. Hay; see 'Experiments on colour, as perceived by the eye': 275 ($=$ *Scientific Papers*, **1**: 126–7).

By spinning the top, each colour is presented to the eye for a time proportional to the angle of the sector exposed, and I have found by independent experiments, that the colour produced by fast spinning is identical with that produced by causing the light of the different colours to fall on the retina at once.

By properly arranging the discs, any given colour may be imitated and afterwards registered by the graduation on the rim of the top. The principal use of the top is to obtain colour-equations. These are got by producing, by two different combinations of colours, the same mixed tint. For this purpose there is another set of discs, half the diameter of the others, which lie above them, and by which the second combination of colours is formed.

The two combinations being close together, may be accurately compared, and when they are made sensibly identical, the proportions of the different colours in each is registered, and the results equated.

These equations in the case of ordinary vision, are always between four colours, not including black.

From them, by a very simple rule, the different colours and compounds have their places assigned on the triangle of colours. The rule for finding the position is this: Assume any three points as the positions of your three standard colours, whatever they are; then form an equation between the three standard colours, the given colour and black, by arranging these colours on the inner and outer circles so as to produce an identity when spun. Bring the given colour to the left-hand side of the equation, and the three standard colours to the right hand, leaving out black, then the position of the given colour is the centre of gravity of three masses, whose weights are as the number of degrees of each of the standard colours, taken positive or negative, as the case may be.[12]

In this way the triangle of colours may be constructed by scale and compass from experiments on ordinary vision. I now proceed to state the results of experiments on Colour-Blind vision.[13]

If we find two combinations of colours which appear identical to a Colour-Blind person, and mark their positions on the triangle of colours, then the straight line passing through these points will pass through all points corresponding to other colours, which, to such a person, appear identical with the first two.

We may in the same way find other lines passing through the series of colours which appear alike to the Colour-Blind. All these lines either pass through one

---

(12) See 'Experiments on colour, as perceived by the eye': 277–82 (= *Scientific Papers*, **1**: 131–5); and Number 59.

(13) Compare 'Experiments on colour, as perceived by the eye': 284–8 (= *Scientific Papers*, **1**: 137–41).

point or are parallel, according to the standard colours which we have assumed, and the other arbitrary assumptions we may have made. Knowing this law of Colour-Blind vision, we may predict any number of equations which will be true for eyes having this defect.

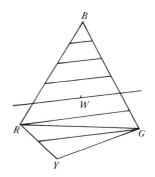

Figure 54,3

The mathematical expression of the difference between Colour-Blind and ordinary vision is, that colour to the former is a function of two independent variables, but to an ordinary eye, of three; and that the relation of the two kinds of vision is not arbitrary, but indicates the absence of a determinate sensation, depending perhaps upon some undiscovered structure or organic arrangement, which forms one-third of the apparatus by which we receive sensations of colour.[14]

Suppose the absent structure to be that which is brought most into play when red light falls on our eyes, then to the Colour-Blind red light will be visible only so far as it affects the other two sensations, say of blue and green. It will, therefore, appear to them much less bright than to us, and will excite a sensation not distinguishable from that of a bluish-green light.[15]

I cannot at present recover the results of all my experiments; but I recollect that the neutral colours for a Colour-Blind person may be produced by combining 6 degrees of ultramarine with 94 of vermilion, or 60 of emerald-green with 40 of ultramarine. The first of these, I suppose to represent to our eyes the kind of red which belongs to the red sensation. It excites the other two sensations, and is, therefore, visible to the Colour-Blind, but it appears very dark to them and of no definite colour. I therefore suspect that one of the three sensations in perfect vision will be found to correspond to a red of the same hue, but of much greater purity of tint. Of the nature of the other two, I can say nothing definite, except that one must correspond to a blue, and the other to a green, verging to yellow.

I hope what I have written may help you in any way in your experiments. I have put down many things simply to indicate a way of thinking about colours which belongs to this theory of a triple sensation. We are indebted to Newton for the original design; to Young for the suggestion of the means of working it out; to Prof. Forbes* for a scientific history of its application to practice; to

* Phil. Mag., 1848.[16]

---

(14) Compare the views of Young and Herschel; see Number 47 note (12); and see 'Experiments on colour, as perceived by the eye': 287–8 (= *Scientific Papers*, **1**: 141–2). See also Wilson, *Researches on Colour-Blindness*: 174–8.

(15) Compare Number 47.   (16) Read: 1849; see note (6).

Helmholtz for a rigorous examination of the facts on which it rests; and to Prof. Grassmann (in the Phil. Mag. for 1852), for an admirable theoretical exposition of the subject.[17] The colours given in Hay's '*Nomenclature of Colours*' are illustrations of a similar theory applied to mixtures of pigments, but the results are often different from those in which the colours are combined by the eye alone.[18] I hope soon to have results with pigments compared with those given by the prismatic spectrum, and then, perhaps, some more definite results may be obtained.[19]

Yours truly,
J. C. MAXWELL

Edinburgh, 4th Jan. 1855

---

(17) See notes (2), (3), (5), (6), (7) and (9).

(18) D. R. Hay, *A Nomenclature of Colours, Hues, Tints and Shades* (Edinburgh/London, 1845).

(19) See Numbers 71 and 102.

DRAFT OF PAPER ON THE PLATOMETER READ TO THE ROYAL SCOTTISH SOCIETY OF ARTS

JANUARY 1855[1]

From the original in the University Library, Cambridge[2]

DESCRIPTION OF A NEW FORM OF THE PLATOMETER AN INSTRUMENT FOR MEASURING THE AREAS OF PLANE FIGURES

The measurement of the area of a plane figure is an operation so frequently occurring in practice that any method by which it may be easily and quickly performed is deserving of attention.

In the case of measurements on a large scale, the method of division into triangles and separate calculation of the areas of each is the only one which is useful, but when we have already mapped the boundary on paper some more expeditious method is to be preferred. One of the simplest of these is that used by some of the earliest of the restorers of mathematical learning & recently recommended by Sir J. F. W. Herschell.[3]

It consists in cutting out a copy of the required figure from a sheet of very uniform paper and comparing its weight with that of a sheet of known size.

But the object of the class of instruments to which I have to direct your attention this evening is to effect the measurement of an area however complicated, by simply moving a tracing point round the boundary of the figure.

The motion of the tracing point is communicated to the instrument and the result is that the area is determined by reading off the number indicated by the graduation.

As many members of the Society have seen Mr Sangs Platometer and heard him explain it they will be the better prepared to follow what I have to say about these instruments in general.[4]

---

(1) The paper was read to the Royal Scottish Society of Arts on 22 January 1855; the published text is dated 30 January 1855. See J. C. Maxwell, 'Description of a new form of the platometer, an instrument for measuring the areas of plane figures drawn on paper', *Transactions of the Royal Scottish Society of Arts*, **4** (1856): 420–8 (= *Scientific Papers*, **1**: 230–7).

(2) ULC Add. MSS 7655, V, d/9.

(3) See the note appended by J. F. W. Herschel, as a member of the Jury for Class X (Philosophical Instruments), to the report on 'Planimeters', in *Exhibition of the Works of Industry of all Nations. Reports by the Juries on the Subjects in the Thirty Classes into which the Exhibition was divided* (London, 1852): 303–4.

(4) John Sang, 'Description of a platometer, an instrument for measuring the areas of figures drawn on paper', *Transactions of the Royal Scottish Society of Arts*, **4** (1852): 119–26 (read 12 January

If we wish to measure anything by means of a hand moving on a dial-plate, we must make the rate of motion of the hand proportional to the rate of increase of the thing measured, whether it be time, as in a clock, distance as in an odometer or superficial measure as in this instance.

Now let it be required to find the area of *OPM*; *OMN* being a fixed horizontal line, *OPQ* a line of any kind, straight, curved, or crooked, and *PM* a vertical line which is capable of moving horizontally from *O* to *M*.

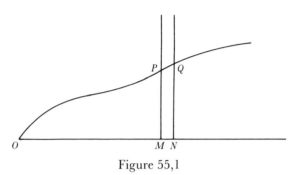

Figure 55,1

Suppose the index which is to measure the area set at zero and *PM* at *O*, then as *PM* moves to the right, the area will increase and the index will move with a velocity depending on the rate of increase of the area.

While the vertical line moves from *M* to *N* the area will have increased by the figure *PMNQ* and therefore supposing *MN* very small the increase of the numerical measure of the area will be the product of *MN* & *PM*.

If we can arrange the parts of the machine so, that while the tracing point moves from *P* to *Q*, the index moves over a distance proportional to *PQMN*, the motion of the index will always indicate the increase of the area, and the final position of the index will indicate the total area after the tracing point has been moved in any manner from one position to another.

The thing to be done, then is to make the velocity of the index proportional to the horizontal velocity of the tracing point, and to its vertical distance above the horizontal line conjointly. For this purpose the horizontal motion of the tracing point *P* must set the index in motion, and the rate of this motion as compared with that of *P* must in some way be made proportional to *PM* the height of *P* above the fixed line.

To make the motion of the index depend on the horizontal part of the motion of *P* is easy, but to make the ratio of these velocities depend on *PM* and vary with the varying height of *P* above the fixed line, requires some contrivance of a nature different from most of those trains of wheelwork usually employed.[5]

1852). See also the Report of a Committee of the Royal Scottish Society of Arts on Sang's platometer, *ibid*: 126–9. In the published paper 'Description of a new form of the platometer': 428 (= *Scientific Papers*, **1**: 237) Maxwell remarks that, 'at the time when I devised the improvements here suggested I had not seen that paper, though I had seen [Sang's] instrument standing at rest in the Crystal Palace [at the Great Exhibition in 1851]'.

(5) The problems of rolling contact, gearing and frictionless motion from toothed wheels are discussed in Robert Willis, *Principles of Mechanism* (Cambridge, 1841). See also Numbers 11, 18 and 79.

It is evident that toothed wheels would not answer, for since *PM* may have an infinite number of values, an infinite number of toothed wheels would be required and these would require to be put in gear & out of gear just as *PM* changed its length.

In all the machines which I have seen, this difficulty is overcome by employing friction instead of teeth to connect the revolving parts of the instrument.

In one of the simplest forms which has been adopted a small vertical disc has its edge resting on the flat surface of a horizontal disc, so that when the horizontal disc revolves in its own plane, it causes the vertical disc to revolve by friction, just as the crown wheel of a watch drives the pinion of the escapement wheel by its teeth.

The rate at which the vertical disc is driven depends on the rate of the horizontal disc and the distance of the point of contact from the vertical axis conjointly.

The rate of the horizontal disc depends on the horizontal velocity of the tracing point, and the vertical disc is made to slide sideways on the horizontal disc so that the distance of the point of contact from the vertical axis is always equal to *PM*.

It is evident that the motion of the vertical disc will be that which is required for the purpose of measuring the area.

Mr Sang's instrument depends on exactly the same principle of having two wheels, one of which slides on the other so as to touch it in points which are at different distances from the axis. In Mr Sang's instrument one of these wheels or rotating parts instead of being a flat surface, is a cone, and the mode in which the cone was made to act the part of a wheel of variable radius by means of the index-wheel sliding along its surface from base to vertex greatly excited my admiration when I saw it at the Great Exhibition of 1851.

One thing however struck me at that time as a possible defect in the instrument. As long as a wheel rolls on a surface straight forward there is no fear of any slipping between the surfaces in contact. But when the wheel, in addition to its forward motion has a sidelong motion impressed on it, so as to be dragged to one side, then slipping actually takes place and we are by no means certain of its direction and amount.

It might appear at first sight that the slipping would be entirely in the sidewards direction and the rolling entirely forwards, so that the rolling would exactly measure the forward part of the motion. In this case the instrument would give perfect results. But it appears on more careful investigation that the smallest irregularities and impediments to the free turning of the wheel, which would have no effect if the rolling had been perfect will produce appreciable errors when the rolling is combined with lateral slipping.

It is but justice to Mr Sang to state that he has himself both minutely investigated the nature of the error thus produced, and devised a method by which its effect may be eliminated at once. In his instrument the whole error arising from this cause must be very small, judging from the specimens of its work which he has given. I have therefore very little to offer as an apology for bringing the subject again before the Society except the natural desire to see ones own device on paper.

I leave it to those who have some experience in the construction of such instruments to say whether it would be worth their while to adopt any part of my contrivance. As I have no intention of making any use of it myself I leave them at liberty either to appropriate it or leave it as I do as a mere mathematical theorem.

The contrivance at present before the Society is intended to show how the relative rate of motion of the tracing point and the index can be varied without the occurrence of slipping between the working parts of the instrument.

Instead of one wheel sliding from one part to another of a revolving disc or cone, I have introduced two spheres which roll on one another without any slipping at the point of contact. These two spheres form the essential part of the instrument, the rest of the machinery being merely a rough indication of the way in which the principle might be applied.[6]

The instrument consists of a frame which is constrained to move up and down the paper parallel to itself.[7]

On this frame is mounted the horizontal axis *AS* which carries a wheel *H* and the hemisphere *KCB*.

The wheel *H* rolls on the plane on which the instrument travels and communicates its motion to the hemisphere which therefore revolves about the axis *BS* with a velocity proportional to that with which the instrument moves backwards or forwards. *GDCSM* is a framework which revolves round a vertical axis thro *S* the centre of the hemisphere and carries with it a ring *EDF* which is capable of turning about a vertical axis thro *D*. This ring supports the

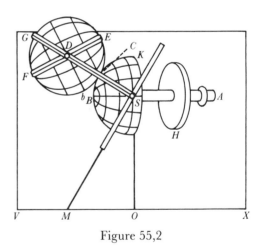

Figure 55,2

---

(6) Maxwell's method of devising an instrument depending on the mutual rolling of two spheres was subsequently developed by James Thomson, 'An integrating machine having a new kinematic principle', *Proc. Roy. Soc.*, **24** (1876): 262–5.

(7) The instrument described was not constructed; compare Numbers 72 and 79.

sphere *ECF* by its poles *EF* exactly as the meridian circle of a globe supports the poles of the globe.

By this arrangement the axis of the sphere *EF* is kept always horizontal while its centre moves in such a way as to be always at the same distance from *S*. This distance is adjusted so as that the two spheres shall always remain in contact at *C*.

Now let us consider the working of the instrument. Suppose the arm *SM* made to coincide with *SO* then *SCD* which is at right angles to *SM* will be in the direction of *AS*. Let the sphere *EF* be placed so that its equator is in contact with the pole of the hemisphere *B*.

Then let the arm *SM* be turned into the position in the figure. The sphere will roll on the hemisphere so that the point of contact will now be at *C* and we shall have the arc *BC* equal to the arc *bC* and the angles *bDC*, *BSC* and *MSO* all equal.

Now let the instrument be moved backwards or forwards so as to turn the wheel *H* and the hemisphere with it then the sphere will be turned about its axis by the action of the hemisphere. The action of the hemisphere on the sphere will correspond to that of a wheel whose radius is the distance of *C* from *BS* on another whose radius is the distance of *C* from *EF* the axis of the sphere.

It is easy to see that the angular velocity of the sphere must be to that of the hemisphere as the sine of the angle *BSC* is to that of *CDE* or what is the same thing the cosine of *CDB*, that is as the tangent of *OSM* to unity that is as *MO* to *OS*.

It appears therefore that the angular velocity of the sphere is always proportional to the product of the motion of the frame multiplied by the distance *MO* and therefore the motion of the sphere measures the area swept out by the line *MO*.

If therefore we make the tracing point move along the edge of the frame *VX* and make the arm *SM* always pass thro the tracing point, the whole motion of the sphere will measure the whole area swept out by the line *OM* as the instrument with its tracing point travels over the area.

56

LETTER TO CECIL JAMES MONRO[1]

7 FEBRUARY 1855

From the original in the Greater London Record Office, London[2]

18 India Street,
Edinburgh Feb 7 '55

Dear Monro

It is a fearful thing to answer when a man tackles you with arguments. I wont argufy at all, leastways not with them as tries to argufix me. I w^d be most happy to give any assistance in my power to the translator of Newton, short of taking the work of his hands. For that I am not prepared.[3] I am prepared to refuse resist & rebel. I w^d as soon think of translating Butlers Analogy[4] for the Mathematical Journal.

I am at present superintending a course of treatment practised on my father for the sake of relieving certain defluxions which take place in his bronchial tubes. These obstructions are now giving way and the medico who is a skillful bellows mender, pronounces the passages nearly clear.

However it will be a week or two before he is on his pins again so w^d you have the goodness to tell Freeman to tell Mrs Jones to tell those whom it may concern that I cannot be up to time at all. I have written to Lee to seek instruction from another source but I doubt whether it w^d be necessary expected conventional proper &c to do so to other pups. Mackenzie doubtless will tell Duncan of Pembroke, Hunter may tell Cormack of Queens and Platt & Vivian will find out of themselves.[5]

---

(1) Trinity 1851, B.A. 1855, 38th Wrangler and 8th Classic, Fellow 1856 (Venn).

(2) Greater London Record Office Acc. 1063/2078.

(3) In reply to Monro's letter of 20 January 1855 (Greater London Record Office Acc. 1063/2095): 'NEWTON MUST BE TRANSLATED, and you are the one to do it'. Monro remarks that Maxwell was sufficiently familiar with Latin 'for practical purposes... [though] it is very true that you don't seem ever to have displayed such acquaintance in your college examinations'. Maxwell had failed to gain a fellowship at Trinity in the previous year; see a letter from J. D. Forbes to William Whewell of 29 December 1854 (Whewell Papers, Trinity College, Cambridge, Add. MS. a.204^{112}), and Whewell's reply of 10 March 1855 (Forbes Papers, St Andrews University Library, 1855/40a) suggesting that 'it would be well that he should attend to his classics more than he has done and give some neatness and polish to his mathematics'. Maxwell was successful at Trinity in October 1855 (see Number 73).

(4) Joseph Butler, *The Analogy of Religion, Natural and Revealed, to the Constitution and Course of Nature* (London, 1736).

(5) William Henry Freeman, Caius 1851; Thomas William Lee, Trinity 1853; Francis Lewis MacKenzie, Trinity 1851; James Duncan, Pembroke 1852; Oswald Hunter, Queens' 1853 (see *Life of Maxwell*: 195); Marcus Tulloch Cormack, Queens' 1852; Francis Thomas Platt, Trinity 1852; Arthur Pendarves Vivian, Trinity 1852 or Richard Glynn Vivian, Trinity 1853 (Venn). Maxwell's mathematics pupils.

I may be up in time to keep the term and so work off a streak of mathematics wh. I begin to yearn after. At present I confine myself to Lucky Nightingales line of business[6] except that I have been writing description of Platometers for measuring plane figures and privately by letter confuting rash mechanics who intrude into things they have not got up and suppose that their devices will act when they cant.

I hope that my absence will not delay the assumption of W. D. Maclagan.[7] He has been here all the vacation if not there and everywhere. The foundation of Ethics, though it may have tickled my core has not germinated at my vertex. Whether it will yet be laid bare either as a paradox or a truism is more than I can tell. Perhaps it may be a pun. What is Horton?[8] I sh^d not like to miss it. It will miss the pruning of the late Sec^y & I dought whether the genial giant[9] will do it as neatly with his stoned club.

Why is it more profitable to put a £5 note in your waistcoat pocket than 5 sovs?

How did Khutor get on in the University scholarship? What of Smiths Prizes?[10]

I was informed by you of Farrar's ordination.[11] Does he stay at Marlboro? J. H. Gedge is now at work near Derby.[12] I have just heard from him.

The doctor was speaking just now of a painful cutaneous affection. It set me thinking of the line of the poet wh seems to indicate a fatal termination if the reading is correct

Die of a *rose* in a rheumatic pain.[13]

How does the rest run?

I have now to do a little cooking & buttling in the shape of toast & beef tea & everfizzing draught. Remember me to Freeman. I hope to see you by the end of term. Does Pomeroy flourish & has he Crimean letters still.

Yrs

J. C. MAXWELL

(6) Florence Nightingale, who organised nursing and hospitals during the Crimean War, then in progress.

(7) William Dalrymple Maclagan, Peterhouse 1852 (Venn). The allusion is to his proposed election into the Society of Apostles, both Maxwell and Monro being members. Maxwell was elected in the winter of 1852–3 (*Life of Maxwell*: 165). In a letter of 7 March [1855] (Greater London Record Office, Acc. 1063/2093) Monro informed Maxwell that 'we have not elected Maclagan'.   (8) Elias Robert Horton, Peterhouse 1852 (Venn).

(9) R. H. Pomeroy (see Number 42 note (3)), so described by Maxwell in a letter to his father of 21 April 1855 (*Life of Maxwell*: 211).

(10) See *The Cambridge University Calendar for the Year 1855* (Cambridge, 1855): 171, the Smith's prizemen being James Savage and Leonard Henry Courtney, both of St John's (Venn). Khutor is unidentified.   (11) Frederic William Farrar, Trinity 1849 (Venn).

(12) Johnson Hall Gedge, Trinity 1849 (Venn); see *Life of Maxwell*: 185.

(13) Pope, 'An Essay on Man', I, 200; 'Die of a rose in aromatic pain'.

57

LETTER TO CECIL JAMES MONRO

19 FEBRUARY 1855

From the original in the Greater London Record Office, London[1]

18 India Street
Edinburgh
Feb 19/55

Dear Monro

I fear you must recommend your Caius brother to look about him for my steps will be no more by the reedy & crooked till Easter term.[2] My fathers recovery is retarded by the frosty weather tho' we have got up an etherial mildness here by means of a good fire & a towel hung up wet at the other end of the room together with an internal *exhibition* of nitric ether.

I wrote to Mackenzie[3] about putting a respectable man in my rooms as a stopper to Cat's Hall men Manns & Boy Joneses but I have not heard of his success.[4] I have no time at present for anything except looking thro novels &c & finding passages wh: will not offend my father & read to him.

He strongly objects to newfangled books and knows the old by heart. But he likes the Essays[5] in intervals of business cause why they have not too many words. The frost here has lasted long & I am beginning to make use of it. I get an uncle to take my place in the afternoon & I rush off to Lochend or Duddington. I have not yet succeeded in skating on one foot for an indefinite time & getting up speed by rising & sinking at the bends of the path but I attribute my failure to want of faith, for I can get up speed for a single bend, only I always slip at a certain critical turning. However I have only been 3 days & I may do it yet. My plans are not fixt but I think it will be some while before my father is on his pins again & when he is I intend to look after him still but do a private streak of work for I will soon be in a too much bottled up condition of mathematics from which even mental collapse would be a relief. I have no intention of doing a Newton or any elegant mathematics. I have a few thoughts on top spinning[6] and sensation generally & a kind of dim outline of Cambridge

---

(1) Greater London Record Office, Acc. 1063/2079. Previously published in *Life of Maxwell*: 209–11.

(2) In reply to Monro's letter of 8 February 1855 (Greater London Record Office, Acc. 1063/2096), requesting Maxwell to tutor his brother Charles Henry Monro, Caius 1853 (Venn).

(3) See Number 56 note (5).

(4) Possibly: Charles Philip Mann, Trinity 1852; Herbert Riversdale Mansel Jones, Trinity 1853 (Venn).

(5) Possibly: F. D. Maurice, *Theological Essays* (Cambridge, 1853); or *Cambridge Essays, contributed by members of the University* (London, 1855).   (6) See Number 58.

palavers tending to shadow forth the influence of mathematical training on opinion & speculation.

I suppose when my father can move I will see him out of this eastern clime & safe located in Gallovidian westnesses, & so be up in Cam. before the beginning of next term.

I sh^d like to know how many kept baccalaurean weeks go to each of these terms, & when they begin & end. Overhaul the calendar & when found make note of.

Is Pomeroy up? or where? This is the [2]^nd time of asking.[7]

Yours truly
J. C. Maxwell

(7) See Number 56.

58

MANUSCRIPT ON THE COMPARISON OF COLOURS USING A SPINNING TOP

27 FEBRUARY 1855

From the original in the University Library, Cambridge[1]

DESCRIPTION OF THE CHROMATIC TEETOTUM AS CONSTRUCTED BY Mr J. M. BRYSON, OPTICIAN EDINBURGH[2]

FEB^y 27th, 1855

The Teetotum consists of a circular plate which is clamped between the two parts of the axis which screw together & hold the disc firm with whatever coloured papers are placed upon it. It may be spun by means of the fingers but if more speed is required it ought to be spun by means of a thread wound round the part of the axis immediately below the disc. This is best done by slipping the knot on the thread behind the little brass pin under the disc and after winding it up placing the axis vertical & so that the two grooves in it rest in the two hooks belonging to the brass handle. When the string is pulled, the hooks keep the axis vertical and thus the teetotum may be spun steadily on the smallest table or tea tray.[3]

The coloured discs which accompany the instrument have each a slit from the circumference to the centre, so that when several discs are placed upon the axis, any required portion of each may be brought into view. They must be placed in the required position over the circular plate, and the upper part of the axis screwed down to keep them in their places. When the teetotum is spun with sufficient velocity, the face of the instrument will appear of a uniform tint resulting from the mixture of the colours placed on it. The proportions of these colours are ascertained by reading off the divisions at the circumference which give the number of parts of each colour in 100 of the resulting tint.

By means of this instrument any natural colour within a very extensive range, may be exactly imitated. For this purpose more than four discs are never required. The best for the purpose are Black, White, Vermillion Ultramarine

---

(1) ULC Add. MSS 7655, V, b/6.

(2) James MacKay Bryson; see D. J. Bryden, *Scottish Scientific Instrument-Makers 1600–1900* (Edinburgh, 1972): 45.

(3) This is the improved colour top as described in 'Experiments on colour, as perceived by the eye', *Trans. Roy. Soc. Edinb.*, **21** (1855): 275–98 (= *Scientific Papers*, **1**: 126–54). See Number 59 and plate III.

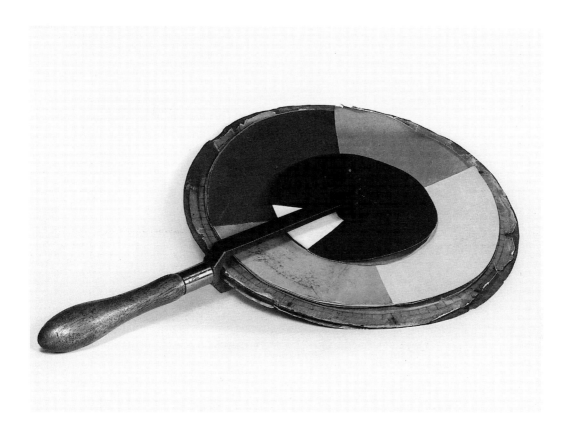

Plate III. Colour top (1855) having two sets of tinted papers arranged in adjustable sectors, enabling colour mixtures to be compared (Number 58).

Emerald Green & Chrome Yellow. It is easy to see whether the given colour inclines to any of the four latter tints or to their intermediates. For instance if the given colour be pink or lilac, red & blue must be taken & combined till a *hue* resembling that to be matched is obtained. But as this will probably be too *rich* a colour it must be mixed white to diminish its purity of tint and with black to lower its intensity or brightness.[4]

In this way the colours of flowers leaves rocks &c may be imitated and their components ascertained by reading off the number of degrees of each colour.[5] It will be found that besides the power of registering all the phenomena of colour the use of this instrument gives a readiness in appreciating the differences of colour in cases where the eye must trust to its own judgement.

But the most accurate method of experimenting is by means of the smaller discs which are to be slipped into one another like the larger ones and placed immediately over them.[6]

It has been said that production of any required tint never requires more than four colours including black. But the same tint may be produced by means of two quite different combinations of colours. For instance any grey may be made by mixing white & black in different proportions and by mixing Vermillion Ultramarine & Emerald Green in proper proportions a perfectly neutral grey may be obtained.

Let these colours be mixed on the outer circle and compared with the grey compounded of black & white on the smaller discs then by carefully altering the proportions the outer and inner circles may be made perfectly identical. This result should be obtained by the unassisted judgement of the eye, the degree of each colour should then be registered in a form like that of a chemical equation stating the equality of the sum of the three sectors of Vermillion Ultramarine & Emerald Green with that of black & white. It will be found that the number of degrees of each colour, as determined by independent trials will not differ by more than two or three divisions, when the same light is used

---

(4) On Maxwell's terms 'hue', 'intensity' and 'tint' see Number 54 esp. note (6).

(5) Maxwell listed the colour values of plants in a manuscript fragment dating from around 1855 (ULC Add. MSS 7655, V, b/4); and see Number 43.

(6) Maxwell had been introduced to experiments on colour mixing by J. D. Forbes in the summer of 1849 (see Number 64). Forbes had observed the hues generated by adjustable coloured sectors fitted to a rapidly spinning disc. Maxwell modified the method, adding the second set of sectors so that colour comparisons could be made. In his letter to Maxwell of 16 May 1855 (*Life of Maxwell*: 214–15; and see Number 64 note (8)), Forbes recollected that: 'My plan was, in fact, the same as yours. ... You will recollect that I had a whirling-machine (made on purpose), in which a number of discs revolved simultaneously with equal velocities. I used black and white on one of these; colours on another. Your teetotum, combining both, I consider preferable for experiments. By the way, I did not get the teetotum you were to leave for me.'

in all the experiments. When different lights are used, or when coloured spectacles are worn, the results are altogether changed.

Different grey colours may be made by substituting any red or yellow for vermillion or making modifications of the blue & green.

We may also produce many of the dull colours such as Brunswick Green, Prussian Blue &c by combinations of the others, which may be easily discovered. It is remarkable that by no combination of Ultramarine & Chrome Yellow can we produce a green by spinning, though one of the greens is a mixture of those colours.

Many persons have a peculiar inability of distinguishing colours. Of these some are very accurate except with respect to certain pairs of colours. Such persons will pronounce a mixture of 90 parts of Vermillion & 10 of Ultramarine a good match for 15 of white and 85 black. The colours must be shown to them while in motion and the effect according to them is perfect identity. Other tests of this defect of vision may easily be applied by means of the teetotum.

But the most interesting result of previous experiments is that the results obtained by different eyes in similar circumstances agree to the most minute accuracy while the same eyes in different lights give different results.

This shows that the human eye is an instrument constructed so as to give uniform results, and is the same in persons of all climates, ages and degrees of cultivation and it is found that even those defects which sometimes occur may be accounted for by supposing a part of the complete organisation wanting. These results however can be completely verified only by a large number of observations. For this purpose standard coloured papers have been prepared, in order to accompany the instrument. Observations with these are capable of comparison, and any chromatic equations which may be obtained and registered should be sent to the address given below for the purpose of comparison with those formerly obtained.

JAMES CLERK MAXWELL
Trinity College, Cambridge

ABSTRACT OF PAPER 'EXPERIMENTS ON COLOUR AS PERCEIVED BY THE EYE, WITH REMARKS ON COLOUR-BLINDNESS'

[19 MARCH 1855][1]

From the *Proceedings of the Royal Society of Edinburgh*[2]

EXPERIMENTS ON COLOUR AS PERCEIVED BY THE EYE, WITH REMARKS ON COLOUR-BLINDNESS

By James Clerk Maxwell, Esq., B.A., Trinity College, Cambridge. Communicated by Professor Gregory.[3]

These experiments were made with the view of ascertaining and registering the judgments of the eye with respect to colours, and then, by a comparison of the results with each other, by means of a graphical construction, testing the accuracy of that theory of the vision of colour which analyses the colour-sensation into three elements, while it recognises no such triple division in the nature of light, before it reaches the eye.

The method of experimenting consisted in placing before the eye of the observer two tints, produced by the rapid rotation of a system of discs of coloured paper, arranged so that the proportions of each of the component colours could be changed at pleasure. The apparatus used was a simple top, consisting of a circular plate on which the coloured discs were placed, and a vertical axis. The discs consisted of paper painted with the unmixed colours used in the arts. Each disc was slit along a radius from centre to circumference, so that several could be interlaced, so as to leave exposed a sector of each. The larger discs, about 3 inches diameter, were first combined and placed on the disc, and the smaller, about $1\frac{3}{4}$ inches diameter above them, so as to leave a broad ring of the larger discs visible.[4]

When the top was spun the observer could compare the resulting tint of the outer and inner circles, and by repeated adjustment, perfect identity of colour could be obtained. The proportions of each colour were then ascertained, by

---

(1) The data the paper was presented to the Royal Society of Edinburgh.

(2) *Proc. Roy. Soc. Edinb.*, **3** (1855) : 299–301.

(3) The memoir was published in *Trans. Roy. Soc. Edinb.*, **21** (1855) : 275–98 (= *Scientific Papers*, **1**, 126–54).

(4) See Number 58 and plate III.

reading off on the circumference of the top, which was divided into 100 parts. As an example, it was found on one occasion, that, –

$$\left.\begin{array}{l} .37 \text{ Vermilion,} \\ +.27 \text{ Ultramarine,} \\ +.36 \text{ Emerald green,} \end{array}\right\} = \left\{\begin{array}{l} .28 \text{ White} \\ +.72 \text{ Black.} \end{array}\right.$$

By experiments on various individuals, it was found (1) that a good eye could be depended upon within two of these divisions, or hundredths, at most; and that by repetition of experiments the *average* result might be made much more accurate.

(2) That the difference of the results of experiments on different individuals was insensible, provided the light used remained the same.

(3) That when different kinds of light were used, or when the resultant tints were examined with coloured glasses, the results were totally changed.

It follows from this that the cause of the equality of the resulting tints is not a true optical identity of the light received by the eye, but must be sought for in the constitution of the sense of sight. The materials for this inquiry are to be found in the equations of colour of which the above is an example, and these are to be viewed in the light of Young's theory of a threefold sensation of colour.[5]

The first consequence of this theory is, that between any *four* colours an equation can be found, and this is confirmed by experiment.

The second is, that from two equations containing different colours a third may be obtained by the ordinary rules, and that this also will agree with experiment. This also was found to be true by experiments at Cambridge which include every combination of five colours.

A graphical method was then described, by which, after fixing arbitrarily the positions of three standard colours, that of any other colour could be obtained by experiments in which it was made to form a neutral gray along with two of the standard colours. In the diagram so formed, the position of any compound tint is the centre of gravity of the colours of which it is composed, their *masses* being determined from the equation, and the resultant *mass* of colour being the sum of the component *masses*. The colour-equations represent the fact that the same tint may be produced by two different combinations.[6] This diagram is similar to those which have been given by Meyer, Hay, and Professor J. D. Forbes, as the results of mixing colours.[7] It is identical with

(5) See Number 54.

(6) See Number 54; and 'Experiments on colour, as perceived by the eye': 279–82 (= *Scientific Papers*, **1**: 131–5).

(7) Compare J. D. Forbes, 'Hints towards a classification of colours', *Phil. Mag.*, ser. 3, **34** (1849): 161–78; and see Number 54 esp. notes (8) and (18).

that proposed by Young, and figured in his *Lectures on Natural Philosophy*.[8] The original conception, however, seems to be due to Newton, who gives the complete theory, with an indication of a construction in his *Optics*.[9]

The success of this method depends entirely on the truth of the supposition that there are three elements of colour as seen by the eye, every ray of the spectrum being capable of exciting all three sensations, though in different proportions. It is at present impossible to define the colours appropriate to these sensations, as they cannot be excited separately. But it appears probable that the phenomena of colour-blindness are due to the absence of one of these elementary sensations, and, if so, a comparison of colour-blind with ordinary vision will show the relation of the absent sensation to those with which we are familiar.[10]

A method was then described, by which one observation by a colour-blind eye was made to determine a certain point representing the absent sensation, which thus appears to be a red approaching to crimson. The results of this hypothesis were calculated in the form of 'equations of colour-blindness' between colours which seem to defective eyes identical. These equations were compared with those previously determined from the testimony of two colour-blind but accurate observers, and found to agree with remarkable precision, rarely differing by more than 0.02 in any colour. The effect of red and green glasses on the colour-blind was then described, and a pair of spectacles having one eye red and the other green was proposed as an assistance to them in detecting doubtful colours.[11]

---

(8) See Number 54 note (8).

(9) See Number 54 note (9).

(10) See Number 47, esp. note (12).

(11) See Number 47, esp. note (15).

FROM A LETTER TO JOHN CLERK MAXWELL

30 APRIL 1855

From Campbell and Garnett, *Life of Maxwell*[1]

Trin. Coll.
Vesp. SS. Philipp. & S. Jac.
1855

I have been working at the motion of fluids, and have got out some results.[2] I am going to show the colour trick at the Philosophical on Monday.[3] Routh has been writing a book about Newton in conjunction with Lord Brougham.[4] Stokes is back again and lecturing as usual.[5]

---

(1) *Life of Maxwell*: 211.

(2) See Number 61.

(3) See Numbers 62 and 65.

(4) Henry Lord Brougham and E. J. Routh, *Analytical View of Sir Isaac Newton's Principia* (London, 1855).

(5) See Number 51 note (25).

MANUSCRIPT ON THE MOTION OF FLUIDS

circa SPRING 1855[1]

From the original in the University Library, Cambridge[2]

ON THE MOTION OF 'FLUIDS'

Definition A 'Fluid' is a collection of distinct particles which are capable of moving independently of each other or according to any assigned law.[3]

NB. Actual fluids are subject to certain mechanical laws like other matter, & have also laws of their own depending on their fluidity but these hypothetical fluids resemble actual fluids in nothing but the complete freedom of their motion, and may therefore be subjected to any arbitrary laws which it is convenient to assume.

The motion of a fluid is to be defined with reference to a fixed origin of coordinates and three rectangular axes fixed in direction. At any given point of space, the fluid will be moving with a velocity v' in a direction making angles $\alpha \beta \gamma$ with the axes of x, y, z. The value of these quantities will depend on the position of the point & on the time & will therefore be functions of $x y z$ & t. Resolving these velocities in the directions of $x y z$ we obtain

$$u = v' \cos \alpha \text{ for velocity parallel to } x$$
$$v = v' \cos \beta \ ----------- \ y$$
$$w = v' \cos \gamma \ ----------- \ z.$$

If ρ be the *density* of the fluid at the point $x y z$, then $\rho u \, dy \, dz, \rho v \, dz \, dx$ & $\rho w \, dx \, dy$ will be the *quantities* of fluid transferred in unit of time across the small planes $dy \, dz, dz \, dx$ & $dx \, dy$ perpendicular to the three axes at the given point. From these expressions we may easily find the whole amount of fluid which enters or departs from a given closed surface.[4]

---

(1) This draft is preliminary to Number 63 (dated 9 May 1855); see also Number 60.

(2) ULC Add. MSS 7655, V, c/4.

(3) Maxwell's undergraduate notes on 'Hydrostatics' (ULC Add. MSS 7655, V, m/8), contain the following definition: 'A fluid is a collection of material particles wh. can be moved among each other by an indefinitely small force'. Compare W. H. Miller, *The Elements of Hydrostatics and Hydrodynamics* (Cambridge, ₄1850): 1.

(4) The derivation of the equation of continuity is drawn from William Thomson, 'Notes on hydrodynamics. I. On the equation of continuity', *Camb. & Dubl. Math. J.*, **2** (1847): 282–6. Maxwell transcribed the first part of this paper ('On the equation of continuity': 282–4) almost verbatim in his undergraduate notes on 'Hydrodynamics' (ULC Add. MSS 7655, V, m/8). The addition of an extract from Thomson's paper constitutes the major difference between Maxwell's notes on hydrodynamics and transcriptions (from Hopkins' manuscripts) by Thomson in 1843 (ULC Add. MSS 7342, PA 16), and earlier by Stokes (ULC Add. MSS 7656, PA 19).

For let δS be the element of surface & $l\,m\,n$ the direction cosines of the normal pointing outwards then the velocity of the fluid in the direction of the normal is $lu+mv+nw$ and the quantity of fluid which enters δS in the time δt is

$$-\rho(lu+mv+nw)\,\delta S\,\delta t$$

and the sum of the values of this expression taken over the whole surface will give the quantity which enters the surface in the given time.

If we consider the expression $-\sum \rho u l \delta S$ we shall see that for the same value of y & z there are two elements of the surface corresponding to the two points in which it is cut by a line parallel to x. The value of ldS' for each of these is $dy\,dz$ so that the expression to be summed becomes

$$-\iint (\rho'u'-\rho u)\,dy\,dz$$

where $\rho'u'$ refers to the greater value of x & ρu to the smaller. Now $\rho'u'-\rho u = \int_x^{x'} \dfrac{d(\rho u)}{dx}$. Substituting this value the expression becomes $-\iiint \dfrac{d}{dx}(\rho u)\,dx\,dy\,dz$, the limits of integration being those of the closed surface. There are two similar expressions corresponding to the other axes so that we have now

$$\sum \rho(lu+mv+nw)\,\delta S\,\delta t = \iiint \left(\frac{d}{dx}(\rho u)+\frac{d}{dy}(\rho v)+\frac{d}{dz}(\rho w)\right)dx\,dy\,dz\,\delta t \quad (1)$$

as the expressions for the quantity of fluid which escapes from the surface. As this is true for every surface it is also true for that of the element $dx\,dy\,dz$ so that

$$\left(\frac{d}{dx}(\rho u)+\frac{d}{dy}(\rho v)+\frac{d}{dz}(\rho w)\right)dx\,dy\,dz\,dt$$

is the quantity of fluid lost out of that element in the time dt. If there is no creation or destruction of fluid then this must be equal to

$$-\frac{d}{dt}(\rho\,dx\,dy\,dz)\,dt$$

and therefore
$$\frac{d}{dx}(\rho u)+\frac{d}{dy}(\rho v)+\frac{d}{dz}(\rho w)+\frac{d\rho}{dt}=0 \quad (2)$$

is the condition of the conservation of the mass of the fluid during the motion.

This equation, which must always be satisfied for actual fluids, is true for the fluids under consideration only when we consider them as endued with the property of permanence. We are at liberty when we please to consider them as produced or destroyed according to any law. In such cases the left hand side of this equation instead of being $=0$ represents the rate of destruction of the fluid at the point $x\,y\,z$ referred to unit volume & unit of time.[5]

---

(5) Compare S. D. Poisson, *Traité de Mécanique*, 2 vols. (Paris, $_2$1833), **2**: 663–92, esp. 676–80.

On a particular case of Fluid Motion

As we are at liberty to assume any laws of motion we please in a hypothetical fluid, let us suppose it regulated entirely by the pressure produced at different points of space.

As the pressure is a function of $x\,y\,z$, this function equated to a constant will give a surface of equal pressure. Let the law of motion of the fluid be this –

The motion of the fluid is normal to the surfaces of equal pressure, and the quantity of fluid traversing any surface is proportional to the rate of variation of pressure along the normal.[6]

Hence it follows that if $a = \rho u \quad b = \rho v$ & $c = \rho w$

$$ak = -\frac{dp}{dx}$$

$$bk = -\frac{dp}{dy}$$

$$ck = -\frac{dp}{dz}$$

where k is the difference of pressure in two planes at unity of distance when unity of mass of fluid per unit of surface is passing from the one to the other in unit of time.

In the most general case k may be a function of the position of the point $x\,y\,z$ but at present we shall consider it constant, and regard the motion of the fluid as steady, that is make p, $a\,b\,c$ &c functions of $x\,y\,z$ without t. Everything then depends on p

$$a = -\frac{1}{k}\frac{dp}{dx} \quad b = -\frac{1}{k}\frac{dp}{dy} \quad c = -\frac{1}{k}\frac{dp}{dz}.$$

Equation (2) becomes

$$\frac{1}{k}\left(\frac{d^2p}{dx^2}+\frac{d^2p}{dy^2}+\frac{d^2p}{dz^2}\right) = P \tag{3}$$

where $P\,dx\,dy\,dz$ is the amount of fluid produced in the element $dx\,dy\,dz$.

---

(6) Compare (11)–(14) of Number 83; and see 'On Faraday's lines of force', *Trans. Camb. Phil. Soc.*, **10** (1856): 27–83, esp. 34–6 (= *Scientific Papers*, **1**: 165–7).

FROM A LETTER TO JOHN CLERK MAXWELL

5 MAY 1855

From Campbell and Garnett, *Life of Maxwell*[1]

[Cambridge]
Saturday 5 May 1855

The Royal Society have been very considerate in sending me my paper on colours[2] just when I wanted it for the Philosophical here. I am to let them see the tricks on Monday evening, and I have been there preparing their experiments in the gas-light.[3] There is to be a meeting in my rooms to-night to discuss Adam Smith's 'Theory of Moral Sentiments', so I must clear up my litter presently.[4] I am working away at electricity again, and have been working my way into the views of heavy German writers.[5] It takes a long time to reduce to order all the notions one gets from these men, but I hope to see my way through the subject and arrive at something intelligible in the way of a theory.

APPENDIX: FRAGMENTS OF AN APOSTLES ESSAY 'IS ETHICAL TRUTH OBTAINABLE FROM AN INDIVIDUAL POINT OF VIEW?'

MAY 1855[6]

From Campbell and Garnett, *Life of Maxwell*[7]

The repeated action of what Smith calls sympathy, calls forth various moral principles, which may be deduced, no doubt, from other theories, as necessary truths, but of which the actual presence is now first accounted for.... Instead of supposing the moral action of the mind to be a speculation on fitness, a calculation of happiness, or an effort towards freedom, he makes it depend on a recognition of our relation to others like ourselves.

---

(1) *Life of Maxwell*: 211–12.

(2) The proofs (ULC Add. MSS 7655, V, b/8) of 'Experiments on colour, as perceived by the eye, with remarks on colour-blindness', *Trans. Roy. Soc. Edinb.*, **21** (1855): 275–98 (= *Scientific Papers*, **1**: 126–54).   (3) See Number 65.

(4) Adam Smith, *The Theory of Moral Sentiments* [1759], new edn, with a biographical and critical memoir of the author by Dugald Stewart (London, 1853). Maxwell had read Smith's *Theory of Moral Sentiments* during his final session at Edinburgh University in 1850; see Number 31.

(5) See Number 66.   (6) As given by Campbell.

(7) *Life of Maxwell*: 234.

MANUSCRIPT ON THE STEADY MOTION OF AN INCOMPRESSIBLE FLUID

9 MAY 1855

From the original in the University Library, Cambridge[1]

On the steady motion of an incompressible fluid, the velocity at every point of which is parallel to the plane of xy and is a function of x & y only.[2]

Let $x\,y\,z$ be the coordinates of a fluid particle $a\,b\,c$ at the time t, $x\,y\,z$ will in general be functions of $a\,b\,c$ & t and so will $\dfrac{dx}{dt}, \dfrac{dy}{dt}, \dfrac{dz}{dt}$, but in this particular case of steady motion $\dfrac{dx}{dt} = u$ $\dfrac{dy}{dt} = v$ & $\dfrac{dz}{dt} = w$ may be expressed as functions of $x\,y\,z$ without t, since these velocities remain constant during the motion.

In the case as stated we have u & v functions of x & y only & $w = 0$.

Since the fluid is incompressible the eq^n of continuity[3] becomes

$$\frac{du}{dx} + \frac{dv}{dy} = 0$$

therefore $u\,dy - v\,dx$ is a perfect differential of a function of x & y. Let us call this function ψ

then

$$\left.\begin{aligned} u &= \frac{d\psi}{dy} \\ v &= -\frac{d\psi}{dx}. \end{aligned}\right\} \tag{1}$$

From these eq^{ns} we obtain

$$u\frac{d\psi}{dx} + v\frac{d\psi}{dy} = 0 \tag{2}$$

which shows that the value of ψ must remain the same for any particle of the fluid during its motion. Hence the eq^n

$$\psi = \text{constant} \tag{3}$$

represents the line described by a fluid particle.[4] If we conceive a series of curves whose equations are of the form (3) to be drawn in the plane of xy, and

---

(1) ULC Add. MSS 7655, V, c/5. Previously published in A. T. Fuller, 'James Clerk Maxwell's Cambridge manuscripts – extracts relating to control and stability – I', *International Journal of Control*, **35** (1982): 792–7.

(2) Compare Stokes, 'On the steady motion of incompressible fluids', *Trans. Camb. Phil. Soc.*, **7** (1842): 439–53 (= *Papers*, **1**: 1–16).   (3) See Number 61 esp. note (4).

(4) On the stream function ψ and stream lines see Horace Lamb, *A Treatise on the Mathematical Theory of the Motion of Fluids* (Cambridge, 1879): 66–8.

a series of cylindric surfaces to be generated by lines parallel to the axis of z and passing through these curves, there will be no motion of the fluid particles across these surfaces. We may therefore treat them as physical surfaces, constraining the motion of the fluid, & modifying the fluid pressures.[5]

Let p be the pressure at any point $x\,y$ of the fluid. Consider the variation of p as we travel round the element $dx\,dy$.

Pressure at $x, y = p$

$$x + dx, y = p + \frac{dp}{dx}\,dx$$

$$(x + dx), y + dy = p + \frac{dp}{dx}\,dx + \left(\frac{dp}{dy} + \frac{d}{dx}\frac{dp}{dy}\,dx\right)dy$$

$$x, y + dy = p + \left(\frac{dp}{dy} + \frac{d}{dx}\frac{dp}{dy}\,dx\right)dy - \frac{d}{dy}\left(\frac{dp}{dx}\,dy\right)dx$$

at $\qquad x, y = p + \left(\frac{d}{dx}\frac{dp}{dy} - \frac{d}{dy}\frac{dp}{dx}\right)dx\,dy.$

Figure 63,1

It appears that the excess of pressure which we thus obtain by going round the element must be balanced by the resistance of the constraining surfaces. The moment of the total force which the surfaces exert on the fluid in the element $dx\,dy$ is

$$\left(\frac{d}{dy}\frac{dp}{dx} - \frac{d}{dx}\frac{dp}{dy}\right)dx\,dy = N\,dx\,dy \text{ suppose.} \tag{4}$$

The ordinary eq^{ns} of fluid motion are[6]

$$\frac{dp}{dx} = \rho\left(X - \frac{d^2 x}{dt^2}\right)\qquad \frac{dp}{dy} = \rho\left(Y - \frac{d^2 y}{dt^2}\right) \tag{5}$$

$$N = \rho\left(\frac{dX}{dy} - \frac{dY}{dx}\right) - \rho\left(\frac{d}{dy}\frac{d^2 x}{dt^2} - \frac{d}{dx}\frac{d^2 y}{dt^2}\right).$$

In all known cases of attractions acting on the fluid elements

$$\frac{dX}{dy} - \frac{dY}{dx} = 0.$$

(5) Compare William Thomson, 'Notes on hydrodynamics. II. On the equation of the bounding surface', *Camb. & Dubl. Math. J.*, **3** (1848): 89–93 (= *Math. & Phys. Papers*, **1**: 83–7). In 'On Faraday's lines of force', *Trans. Camb. Phil. Soc.*, **10** (1856): 27–83, esp. 30–33 (= *Scientific Papers*, **1**: 160–3) Maxwell develops the model of stream lines, supposing an imaginary surface impermeable to the fluid, 'a *tube of fluid motion*'; and he applies the hydrodynamical analogy to Faraday's lines and tubes of force. See Numbers 83 and 84.

(6) X and Y are external forces acting on the fluid at the point x, y, ρ the fluid density. See G. G. Stokes, 'Notes on hydrodynamics. III. On the dynamical equations', *Camb. & Dubl. Math. J.*, **3** (1848): 121–7, esp. 124, equation (1) (= *Papers*, **2**: 1–7).

We may therefore confine our attention to the second term & put

$$N = -\rho\left(\frac{d}{dy}\frac{d^2x}{dt^2} - \frac{d}{dx}\frac{d^2y}{dt^2}\right). \tag{6}$$

We have in steady motion

$$\frac{d^2x}{dt^2} = u\frac{du}{dx} + v\frac{du}{dy}$$

$$= \frac{d\psi}{dy}\frac{d^2\psi}{dx\,dy} - \frac{d\psi}{dx}\frac{d^2\psi}{dy^2}$$

$$\frac{d}{dy}\frac{d^2x}{dt^2} = \frac{d\psi}{dy}\frac{d^3\psi}{dx\,dy^2} - \frac{d\psi}{dx}\frac{d^3\psi}{dy^3}$$

$$\frac{d^2y}{dt^2} = u\frac{dv}{dx} + v\frac{dv}{dy}$$

$$= -\frac{d\psi}{dy}\frac{d^2\psi}{dx^2} + \frac{d\psi}{dx}\frac{d^2\psi}{dx\,dy}$$

$$\frac{d}{dx}\frac{d^2y}{dt^2} = -\frac{d\psi}{dy}\frac{d^3\psi}{dx^3} + \frac{d\psi}{dx}\frac{d^3\psi}{dx^2\,dy}.$$

$$\therefore N = \rho\left\{\frac{d\psi}{dy}\frac{d}{dx}\left(\frac{d^2\psi}{dx^2} + \frac{d^2\psi}{dy^2}\right) - \frac{d\psi}{dx}\frac{d}{dy}\left(\frac{d^2\psi}{dx^2} + \frac{d^2\psi}{dy^2}\right)\right\}$$

$$= \rho\left\{u\frac{d\chi}{dx} + v\frac{d\chi}{dy}\right\} \text{ where } \chi = \frac{d^2\psi}{dx^2} + \frac{d^2\psi}{dy^2}$$

$$= \rho\left(\frac{d\chi}{dt}\right) \text{ where } \frac{d\chi}{dt} \langle\text{means}\rangle \text{ refers [to] the var}^n \text{ of } \chi \text{ as we}$$

follow the particle in its actual motion. (7)

In the case of unconstrained motion we must have $N = 0$

or

$$\frac{d\psi}{dy}\frac{d\chi}{dx} = \frac{d\psi}{dx}\frac{d\chi}{dy} \tag{8}$$

that is χ must be a function of ψ. The general condition of steady motion is therefore –

$$\chi = \frac{d^2\psi}{dx^2} + \frac{d^2\psi}{dy^2} = f(\psi).^{(7)} \tag{A}$$

To determine the stability or instability of this steady motion we must give it a perfectly general derangement & determine whether it will or will not tend to return to its original state.

Let $\psi = F(x, y, h)$ be the function which determines the motion as before, & let the derangement be effected by the variation of h from h to $h + \delta h$ after which

(7) Compare Stokes, 'On the steady motion of incompressible fluids': 445–6.

another form of steady motion is to be considered, differing in any arbitrary manner from the former though in an infinitely small degree.

We have after derangement

$$N = \rho\left(\frac{d\chi}{dt} + \frac{d}{dh}\frac{d\chi}{dt}\,\delta h + \&c\right) \tag{9}$$

& the force exerted by the surfaces in $dx\,dy$ to turn the fluid element round from x to $y = N\,dx\,dy$.

The angle thro which the surfaces are turned by the variation of h

$$\delta\theta = \frac{\dfrac{d^2\psi}{dx\,dh}\dfrac{d\psi}{dy} - \dfrac{d^2\psi}{dy\,dh}\dfrac{d\psi}{dx}}{\left|\dfrac{d\psi}{dx}\right|^2 + \left|\dfrac{d\psi}{dy}\right|^2}\,\delta h$$

$$= \frac{u\dfrac{d}{dx}\dfrac{d\psi}{dh} + v\dfrac{d}{dy}\dfrac{d\psi}{dh}}{u^2 + v^2}\,\delta h$$

$$= \frac{\dfrac{d}{dt}\left(\dfrac{d\psi}{dh}\right)}{u^2 + v^2}\,\delta h \tag{10}$$

$\dfrac{d}{dt}$ being understood as before.

Now if the reaction of the fluid on the surfaces tend to turn them in the same direction as the derangement turns them then the derangement will increase, and the eq^m will be unstable; but if the reaction of the fluid tend to turn the surfaces back again the eq^m will be stable.

If therefore $N\,dx\,dy$ the reaction of the surfaces on the fluid have always the same sign with $\delta\theta$ the eq^m will be stable.

The condition of stability is therefore

$N\,dx\,dy\,\delta\theta$ positive for all forms of the derangement.

$$N\,dx\,dy\,\delta\theta = \rho\,dx\,dy\left(\frac{d\chi}{dt} + \frac{d}{dh}\left(\frac{d\chi}{dt}\right)\delta h\right)\frac{\dfrac{d}{dt}\dfrac{d\psi}{dh}\,\delta h}{u^2 + v^2} \tag{11}$$

here $\rho\,dx\,dy$ is the mass of the element & essentially positive, $u^2 + v^2$ is also essentially positive. We have therefore

$$\frac{d\chi}{dt}\frac{d}{dt}\left(\frac{d\psi}{dh}\right)\delta h + \frac{d}{dh}\left(\frac{d\chi}{dt}\right)\frac{d}{dt}\left(\frac{d\psi}{dh}\right)(\delta h)^2$$

positive for every form of derangement.

That this may be we must have

$$\frac{d\chi}{dt} = 0 \text{ as before} \tag{A}$$

$$\&\ \frac{d}{dh}\frac{d\chi}{dt} \text{ of the same sign with } \frac{d}{dt}\frac{d\psi}{dh} \tag{B}$$

from (A) we obtain as before $\chi = f(\psi)$

$$\& \quad \frac{d}{dh}\left(\frac{d\chi}{dt}\right) = \frac{d}{dt}\left(\frac{d\psi}{dh}\right)$$

because $\chi = f(\psi)$ & ψ = function of h & t & x & y

$$= \frac{d}{dt}f'(\psi)\frac{d\psi}{dh} = f'(\psi)\frac{d}{dt}\frac{d\psi}{dh} \text{ because } \frac{d\psi}{dt} = 0$$

the final condition of stability is therefore

$$f'(\psi)\left(\frac{d}{dt}\frac{d\psi}{dh}\right)^2 \text{ positive}$$

or
$$f'(\psi) \text{ positive}$$

or
$$\frac{d\chi}{d\psi} \text{ positive}$$

or
$$\frac{d}{d\psi}\left(\frac{d^2\psi}{dx^2} + \frac{d^2\psi}{dy^2}\right) = \text{ positive.} \qquad \text{(B)}$$

If $\dfrac{d\chi}{d\psi}$ be negative the motion will be unstable & if it be $= 0$ the investigation here given is not sufficient to determine the stability. The remainder of the process must in that case be considered separately.

JAMES CLERK MAXWELL
May 9th 1855

Conditions of Stability of Steady Motion of
Incompressible Fluid in plane of xy

$$\boxed{\begin{aligned} u &= \frac{d\psi}{dy} \\[2mm] v &= -\frac{d\psi}{dx} \\[2mm] \chi &= \frac{d^2\psi}{dx^2} + \frac{d^2\psi}{dy^2} \end{aligned}}$$

Then $\chi = f(\psi)$

is the condition of steadiness
and $f'(\psi) > 0$
is the condition of stability.

LETTER TO JAMES DAVID FORBES

12 MAY 1855

From the original in the University Library, St Andrews[1]

Trinity College
12th May 1855

Dear Sir[2]

When D[r] Gregory read my paper to the Royal Society[3] I told him of your note with respect to your experiments so he read it aloud after a letter on a similar subject from D[r] G Wilson. Some time afterwards I got proofs of my abstract of contents which I corrected, and it seemed as if D[r] Wilsons letter was to be appended to it.[4] I have since seen the same in the Edinburgh Phil: Mag:

---

(1) St Andrews University Library, Forbes MSS 1855/60.

(2) A reply to a letter from Forbes of 4 May 1855 (*Life of Maxwell*: 213–14), referring to the publication of the abstract of Maxwell's paper 'Experiments on colour, as perceived by the eye, with remarks on colour-blindness', presented to the Royal Society of Edinburgh on 19 March 1855 (Number 59), in the *Edinburgh New Philosophical Journal*, **1** (1855): 359–60, with a note by Forbes, *ibid.*: 362. In his letter of 4 May 1855 Forbes wrote: 'I am informed that my note to you about some of my experiments on colours has been printed in the *Edinburgh Philosophical Journal*. This was by no means what I intended.... What I thought that you might do was to introduce into that part of your paper where you speak of what has been done or written in the subject, mention of the fact that as early as January 18— (I do not at the moment recollect the year I stated to you) I had used the method of rapid motion in blending colours; that I had endeavoured to obtain an equation between certain mixed colours and pure gray; and that I had pointed out before Helmholtz, or I believe any one else, that a mixture of yellow and blue, under these circumstances at least, does not produce green; you yourself being a witness to what I then tried, though I was prevented from resuming the subject by ill health and some experimental occupations (conduction of heat) which I considered more imperative.' (Campbell cut Forbes' letter.) See notes (5) and (9). On Helmholtz's demonstration that spectral indigo blue and yellow (additively) mixed yield a 'pure white', see his 'Ueber die Theorie der zusammengesetzten Farben', *Ann. Phys.*, **87** (1852): 45–66 (= *Wissenschaftliche Abhandlungen*, **2**: 3–23); (trans.) 'On the theory of compound colours', *Phil. Mag.*, ser. 4, **4** (1852): 519–34, esp. 526, 528. Helmholtz demonstrated that while the mixture of coloured lights is an additive process, pigment mixing is subtractive; pigments act as filters to light reflected from interior layers below the surface. In experiments on the mixing of spectral colours Helmholtz found that 'yellow and blue do not furnish green... [which] contradicts in the most decided manner the experience of all painters during the last thousand years'. See Number 54 note (7). Huygens had suggested that 'for the composition of *White*...it may possibly be, that *Yellow* and *Blew* might also be sufficient for that...'; see *Phil. Trans.*, **8** (1673): 6086 (= *Isaac Newton's Papers and Letters on Natural Philosophy*, ed. I. Bernard Cohen (Cambridge, 1958): 136); and see note (6).

(3) See Number 59; on Gregory see Number 25 note (2).

(4) George Wilson, 'Observations on Mr. Maxwell's paper', *Edinburgh New Philosophical Journal*, **1** (1855): 361–2. Wilson's comments were directed to Maxwell's discussion of colour

along with your note as you sent it to me.[5] I have never seen the note itself since, but I have the contents at hand, so that I can refer to them.

I remember seeing several of your experiments in the summer of 1849 both with whirling discs & with the colours of the spectrum. I saw white produced by three colours of the spectrum, the rest being stopped and it was from what I saw then combined with some observations of Newtons about such white light differing from sun light,[6] that turned my attention to this subject. I am

blindness: 'It seems to me exceedingly doubtful whether we sufficiently *fully* define colour-blindness, even in reference to the utilitarian perception of colours, by regarding it as equivalent to the non-perception of red'. See also George Wilson, *Researches on Colour-Blindness* (Edinburgh, 1855): 174–8, for similar comments; and see Numbers 47 and 54.

(5) J. D. Forbes, 'Observations on Mr Maxwell's paper', *Edinburgh New Philosophical Journal*, **1** (1855): 362. Forbes wrote: 'I do not know whether you advert at all to the history of experiments on the mixing of colours, but I may mention that I find by my register, that my chief experiments were made on the 4th and 12th January 1849; and amongst these results I find "yellow 100°, blue 120°, white 140°, produce a quite neutral gray like black 180°, white 180°"'. "Yellow and blue only, equal, produce a yellow grey or citrine – *never green*". [The yellow was gamboge.] On the 1st March 1849, I have the following entry: "Examining the red, yellow, and blue papers by the colours they reflect in a dark room, when a narrow slip of each was strongly illuminated by the sun, and the light examined (not in the plane of reflection) by a prism, the colours appear very complex indeed. Both the red and yellow reflect almost every colour of the spectrum. The blue seems purest but very decidely violet or tinged with red".' See note (9).

(6) See Isaac Newton, *Opticks; Or a Treatise of the Reflections, Refractions, Inflections and Colours of Light* (London, ₃1721): 136–7. In describing his colour-mixing circle (see Number 54 note (9)) Newton observes: 'For I could never yet by mixing only two primary colours produce a perfect white. Whether it may be compounded of a mixture of three taken at equal distances in the circumference I do not know, but of four or five I do not much question but it may. But these are curiosities of little or no moment to the understanding the Phaenomena of nature. For in all whites produced by nature, there uses to be a mixture of all sorts of rays, and by consequence a composition of all Colours'. Compare Maxwell's discussion of the difference between the 'optical constitution, as revealed by the prism' and the '*chromatical* properties' of a mixture of colours, in his paper 'On the theory of compound colours, and the relations of the colours of the spectrum', *Phil. Trans.*, **150** (1860): 57–84, esp. 57–9 (= *Scientific Papers*, **1**: 411–12); 'Newton is always careful, however, not to call any mixture white, unless it agrees with common white light in its optical as well as its chromatical properties, and is a mixture of *all* the homogeneal colours'. Maxwell refers to Newton's '7th and 8th Letters to Oldenburg [relating to the theory of light and colours]', as printed in *Isaaci Newtoni Opera quae exstant omnia*, ed. Samuel Horsley, 5 vols. (London, 1779–85), **4**: 342–52, letters of 23 June and 3 April 1673 in reply to Huygens, printed out of chronological sequence by Oldenburg in *Phil. Trans.*, **8** (1673): 6087–92, 6108–11. Newton observes: 'Concerning the business of colours, when Mr. *Hugens* hath shewn how White may be *produced out of two uncompounded colours, I will tell him, why he can conclude nothing from that*; my meaning was, that such a White, were there any such, would have different properties from the White which I had respect to, when I described my Theory, that is, from the White of the sun's immediate light, of the ordinary objects of our senses, and of all white phenomena that have hitherto fallen under my observation...'; *Opera*, **4**: 343–4; and see *The Correspondence of Isaac Newton*, ed. H. W.

glad to hear that you think blue and yellow were not known to produce pink when combined till you discovered it. I always considered it a new thing while I was with you but I was told of it as an established thing afterwards. I have since found, that most optical writers speak confidently of the same mixture as a fine green. Indeed in your own paper on Classification of Colours you seem to be of that opinion.[7] So that I must consider you as the first person who ever attempted to make green by spinning blue and yellow, in spite of all experiments of the kind formerly described, for if any one had really tried it he would have been struck by the result, and would have made *your* discovery. I think your coloured discs were made with alternate sectors of yellow & blue paper pasted on. I do not recollect how you altered the proportions.[8] In my paper I have not referred to anything of yours except your 'Classification of Colours'. It will be much better now that I have your leave, to refer to your experiments, which were to me the origin of the whole enquiry.[9]

Turnbull (Cambridge, 1959), **1**: 291 (letter of 23 June 1673). For Huygens' claim see note (2); and see also Alan E. Shapiro, 'The evolving structure of Newton's theory of light and color', *Isis*, **71** (1980): 211–35.

(7) J. D. Forbes, 'Hints towards a classification of colours', *Phil. Mag.*, ser. 3, **34** (1849): 161–78, esp. 162, referring to Isaac Newton, *Opticks* (London, ₃1721): 116 (Book I, Part II, Prop. IV); '...the yellow and blue, on either hand, if they are equal in quantity they draw the intermediate green equally towards themselves in Composition...'. Compare the views of Huygens and Helmholtz, cited in note (2).

(8) In his reply dated 16 May 1855 (*Life of Maxwell*: 214–15) Forbes responds: 'My plan was, in fact, the same as yours. I had sectors much larger than I required of each colour, making them overlap, and fixing them down by a screw at the centre, pressing a disc of indiarubber on the discs. When I got the anomalous result of blue and yellow, I got Mr Hay to make a disc of *many* alternating narrow sectors merely to see whether it might be a physiological effect from the imperfect blending of the colours. I still think the experiment ought to be tried *without motion*...'. Compare Forbes, 'Hints towards a classification of colours': 166–7. On D. R. Hay see Number 54 notes (6) and (11).

(9) Maxwell added a supplementary note on Forbes' experiments to the proofs of his paper 'Experiments on colour, as perceived by the eye, with remarks on colour-blindness', *Trans. Roy. Soc. Edinb.*, **21** (1855): 275–98; the passage is at 291–2 (= *Scientific Papers*, **1**: 145–6). The proofs (ULC Add. MSS 7655, V, b/8) are dated May 15, 1855 by the printer, presumably the date of receipt from Maxwell (he had received the proofs by 5 May: see Number 62). The inserted passage, which begins 'It has long been known...', and concludes '...in the manner above referred to', includes the following statement. 'The experiments of Professor J. D. Forbes, which I witnessed in 1849, first encouraged me to think that the laws of this kind of mixture might be discovered by special experiments. After repeating the well-known experiment in which a series of colours representing those of the spectrum are combined to form gray, Professor Forbes endeavoured to form a neutral tint, by the combination of three colours only. For this purpose, he combined the three so-called primary colours, red, blue, and yellow, but the resulting tint could not be rendered neutral by any combination of these colours; and the reason was found to be, that blue and yellow do not make green, but a pinkish tint, when neither prevails in the combination. It was plain, that no addition of *red* to this, could produce a neutral tint. This result of mixing blue and yellow was,

I address to Clifton hoping that your address is known at the Post Office. If not they will send to Edinburgh.

Yours sincerely
JAMES C. MAXWELL

I believe, not previously known. It directly contradicted the received theory of colours, and seemed to be at variance with the fact, that the same blue and yellow paint, when ground together, do make green. Several experiments were proposed by Professor Forbes, in order to eliminate the effect of motion, but he was not then able to undertake them. One of these consisted in viewing alternate stripes of blue and yellow, with a telescope out of focus. I have tried this, and find the resultant tint pink as before.' See note (2) and Number 54 note (7). Compare Maxwell's discussion of the mixture of 'blue and yellow *light*... [establishing that] blue and yellow do *not* make green' in his 'On the theory of compound colours with reference to mixtures of blue and yellow light', *Report of the Twenty-sixth Meeting of the British Association for the Advancement of Science; held at Cheltenham in August 1856* (London, 1857), part 2: 12–13 (= *Scientific Papers*, **1**: 244).

FROM A LETTER TO JOHN CLERK MAXWELL

15 MAY 1855

From Campbell and Garnett, *Life of Maxwell*[1]

Trin. Coll.
15 May 1855

The colour trick came off on Monday, 7th.[2] I had the proof sheets of my paper,[3] and was going to read; but I changed my mind and talked instead, which was more to the purpose. There were sundry men who thought that Blue and Yellow make Green, so I had to undeceive them. I have got Hay's book of colours[4] out of the Univ. Library, and am working through the specimens, matching them with the top. I have a new trick of stretching the string horizontally above the top, so as to touch the upper part of the axis. The motion of the axis sets the string a-vibrating in the same time with the revolutions of the top, and the colours are seen in the haze produced by the vibration. Thomson has been spinning the top, and he finds my diagram of colours agrees with his experiments, but he doubts about browns what is their composition. I have got colcothar brown, and can make white with it, and blue and green; also, by mixing red with a little blue and green and a great deal of black, I can match colcothar exactly.[5]

I have been perfecting my instrument for looking into the eye. Ware[6] has a little beast like old Ask, which sits quite steady and seems to like being looked at, and I have got several men who have large pupils and do not wish to let me look in. I have seen the image of the candle distinctly in all the eyes I have tried, and the veins of the retina were visible in some; but the dogs' eyes showed all the ramifications of veins, with glorious blue and green network, so that you might copy down everything.[7] I have shown lots of men the image in my own eye by shutting off the light till the pupil dilated and then letting it on.

I am reading Electricity and working at Fluid Motion,[8] and have got out the condition of a fluid being able to flow the same way for a length of time and not wriggle about.[9]

---

(1) *Life of Maxwell*: 212–13.

(2) See Number 62; and *Proc. Camb. Phil. Soc.*, **1** (1856): 149, for 7 May 1855: 'Mr Maxwell gave an account of some experiments on the mixture of colours'.

(3) See Number 62 note (2). (4) See Number 54 note (18).

(5) See Number 66. (6) Henry Ware, Trinity 1849 (Venn).

(7) See Numbers 49 and 66. (8) See Numbers 63 and 66.

(9) In a letter of 21 May 1855 John Clerk Maxwell replied: 'Have you put a burn in a fit condition to flow evenly, and not beat on its banks from side to side? That would be the useful practical application'. (*Life of Maxwell*: 213).

LETTER TO WILLIAM THOMSON

15 MAY 1855

From the original in the University Library, Cambridge[1]

Trin. Coll.
May 15/55

Dear Thomson

Many thanks for your list of Electrical matter.[2] I think I can get hold of all you mention. I am reading Weber's Elektrodynamische Maasbestimmungen which I have heard you speak of. I have been examining his mode of connecting electrostatics with electrodynamics, induction &c & I confess I like it not at first.[3] He makes the attraction of two elements of electricity =

$$-\frac{ee'}{r^2}\left(1 - a^2\left|\overline{\frac{dr}{dt}}\right|^2 + b\frac{d^2r}{dt^2}\right)^{[4]}$$

determining a & b from Ampères laws.[5]

---

(1) ULC Add. MSS 7342, M 90. Previously published in Larmor, 'Origins': 705–10.

(2) This letter, like others from Thomson, is not extant.

(3) Wilhelm Weber, 'Elektrodynamische Maassbestimmungen, über ein allgemeines Grundgesetz der elektrischen Wirkung', *Abhandlungen bei Begründung der Königlichen Sächsischen Gesellschaft der Wissenschaften, hrsg. von der Fürstlichen Jablonowskischen Gesellschaft, Leipzig* (1846): 211–378 (= *Wilhelm Weber's Werke*, 6 vols. (Berlin, 1892–4), **3**: 25–214). As Maxwell notes in 'On Faraday's lines of force', *Trans. Camb. Phil. Soc.*, **10** (1856): 27–83, esp. 67 (= *Scientific Papers*, **1**: 208), Weber assumed that currents are electric masses in motion and 'when electricity is moving in a conductor, the velocity of the positive fluid *relatively to the matter of the conductor* is equal and opposite to that of the negative fluid'. The current i is measured by the number of electric masses passing through a cross section of the wire in unit time, and $i = aeu$, where a is a constant, e is the mass of positive electricity per unit length of a conductor, and u is its velocity along the conductor. In Weber's formula for electric action between two moving particles, Coulomb's law for the force between electric charges e and e' at a distance r from each other is modified by terms for the relative velocity and acceleration of the two moving particles, derived from Ampère's formula for the law of force between two infinitesimal line elements carrying currents

$$\frac{ii'\,ds\,ds'}{r^2}(\cos\epsilon - \tfrac{3}{2}\cos\theta\cos\theta'),$$

where i and i' are the intensities of the electric currents, ds and ds' the line elements, r the distance between them, ϵ the angle between the line elements, and θ and θ' the angles between the line elements and the straight line r connecting them. Ampère states the relations

$$\cos\theta = \frac{dr}{ds}, \quad \cos\theta' = -\frac{dr}{ds'}$$

and
$$\cos\epsilon = -r\frac{d^2r}{ds\,ds'} - \frac{dr}{ds}\cdot\frac{dr}{ds'}.$$

But I suppose the rest of his views are founded on experiments which are trustworthy as well as elaborate.

I am trying to construct two theories, mathematically identical, in one of which the elementary conceptions shall be about fluid particles attracting at a distance while in the other nothing (mathematical) is considered but various states of polarization tension &c existing at various parts of space.[6] The result

---

See A. M. Ampère, 'Mémoire sur la théorie mathématique des phénomènes électrodynamiques uniquement déduite de l'expérience', *Mémoires de l'Académie Royale des Science*, **6** (1827): 175–388, esp. 252–3.

Weber states Ampère's law in the form

$$\frac{ii'\,ds\,ds'}{r^2}\left(-\tfrac{1}{2}\frac{dr}{ds}\frac{dr}{ds'}+r\frac{d^2r}{ds\,ds'}\right)$$

and obtains the fundamental electrodynamic law

$$\frac{ee'}{r^2}\left(1-\frac{a^2}{16}\left(\frac{dr}{dt}\right)^2+\frac{a^2}{8}r\frac{d^2r}{dt^2}\right).$$

See also Weber, 'Elektrodynamische Maassbestimmungen', *Ann. Phys.*, **73** (1848): 193–240, esp. 219–30; (trans.) 'On the measurement of electrodynamic forces', *Scientific Memoirs*, ed. R. Taylor, **5** (London, 1852): 489–529. In 'On Faraday's lines of force': 66n (= *Scientific Papers*, **1**: 207n) Maxwell added the note that 'when this was written, I was not aware that part of M. Weber's Memoir is translated in Taylor's *Scientific Memoirs*, Vol. V. Art XIV. The value of his researches, both experimental and theoretical, renders the study of his theory necessary to every electrician'. On Weber see C. Jungnickel and R. McCormmach, *Intellectual Mastery of Nature: Theoretical Physics from Ohm to Einstein. Volume 1: The Torch of Mathematics 1800–1870* (Chicago/London, 1986): 138–48.

(4) Compare Weber's formula: note (3).

(5) Compare Weber's statement of Ampère's law: see note (3). In a subsequent paper, 'Elektrodynamische Maassbestimmungen, insbesondere Widerstandsmessungen', *Abhandlungen der Königlichen Sächsischen Gesellschaft der Wissenschaften, math.-phys. Klasse*, **1** (1852): 199–381, esp. 259–70 (= *Werke*, **3**: 301–471), Weber states his force law in the form

$$\frac{ee'}{r^2}\left(1-\frac{1}{c^2}\left(\frac{dr}{dt}\right)^2+\frac{2r}{c^2}\frac{d^2r}{dt^2}\right)$$

where the constant $c = a/4$ (see note (3)). Weber observes that when the relative velocity dr/dt between the two electrical masses e and e' is constant their relative acceleration $d^2r/dt^2 = 0$, and it follows '*dass c denjenigen constanten Werth der relativen Geschwindigkeit dr/dt bedeutet, bei welchem zwei elektrische Massen gar keine Wirkung auf einander ausüben*' (on 268). Weber interprets the constant c as a limiting velocity of electric masses; and also as a constant whose determination would permit the conversion of measurements of current intensity from electrodynamic to mechanical measures; see Wilhelm Weber and Rudolph Kohlrausch, 'Ueber die Elektricitätsmenge, welche bei galvanischen Strömen durch den Querschnitt der Kette fliesst', *Ann. Phys.*, **99** (1856): 10–25 (= Weber, *Werke*, **3**: 597–608). Weber did not consider the closeness of the measured value of c to the velocity of light to be physically significant; see Number 187 note (15).

(6) See Number 51.

will resemble your analogy of the steady motion of heat.[7] Have you patented that notion with all its applications? for I intend to borrow it for a season, without mentioning anything about heat (except of course historically) but applying it in a somewhat different way to a more general case to which the laws of heat will not apply.[8] By the way do you profess to account for what becomes of the vis viva of heat when it passes thro' a conductor from hot to cold?[9] You must either modify Fourier's laws[10] or give up your theory, at least so it seems to me.

With respect to Colours – I have been thinking of what you say about Brown.[11] I have matched ground coffee tolerably though the surface is bad chocolate cakes & improvised various browns with black, red & a little blue & green. The only brown *paper* I have is Colcothar brown. I find

$$53\ \text{Bk} + \quad 28\ \text{V} \quad + \quad 8\ \text{U} \quad + \quad 11\ \text{EG} \quad = 100\ \text{Colcothar.}$$
Vermillion   Ultramarine   Emerald Green

---

(7) [William Thomson,] 'On the uniform motion of heat in homogeneous solid bodies, and its connection with the mathematical theory of electricity', *Camb. Math. J.*, **3** (1842): 71–84 (= *Electrostatics and Magnetism*: 1–14). Compare Number 84, and Maxwell's discussion in 'On Faraday's lines of force': 28–30 (= *Scientific Papers*, **1**: 156–60).

(8) The analogy of the motion of an 'imaginary' incompressible fluid; see Number 84 and 'On Faraday's lines of force': 30–3 (= *Scientific Papers*, **1**: 160–3).

(9) Thomson had himself raised a similar query: 'When "thermal agency" is thus spent in conducting heat through a solid, what becomes of the mechanical effect which it might produce? Nothing can be lost in the operations of nature – no energy can be destroyed. What effect then is produced in place of the mechanical effect which is lost?'; William Thomson, 'An account of Carnot's theory of the motive power of heat; with numerical results deduced from Regnault's experiments on steam', *Trans. Roy. Soc. Edinb.*, **16** (1849): 541–75, esp. 545n (= *Math. & Phys. Papers*, **1**: 118–19n). Thomson answered his own query by his formulation of the concept of dissipation of energy in his paper 'On the dynamical theory of heat, with numerical results deduced from Mr Joule's equivalent of a thermal unit and M. Regnault's observations on steam', *Trans. Roy. Soc. Edinb.*, **20** (1851): 261–88; and in *Phil. Mag.*, ser. 4, **4** (1852): 8–21, 105–17, 168–76 (= *Math. & Phys. Papers*, **1**: 174–210). See also William Thomson, 'On a universal tendency in nature to the dissipation of mechanical energy', *Proc. Roy. Soc. Edinb.*, **3** (1852): 139–42, esp. 140; and in *Phil. Mag.*, ser. 4, **4** (1852): 304–6 (= *Math. & Phys. Papers*, **1**: 511–14); 'When heat is diffused by *conduction*, there is dissipation of mechanical energy, and perfect *restoration* is impossible'. While there was an 'absolute waste of mechanical energy' in irreversible processes, the heat dissipated in conduction is not destroyed; there 'must be some transformation of energy'. See also C. W. Smith, 'William Thomson and the creation of thermodynamics', *Archive for History of Exact Sciences*, **16** (1976): 231–88. On *vis viva* see note (22).

(10) Joseph Fourier, *Théorie Analytique de la Chaleur* (Paris, 1822). Concerned with Thomson's analogy between the flux of heat and electrostatic force (see Number 51 note (3)), Maxwell probes Thomson's concept of the dissipation of energy.

(11) See also Number 65.

At the same time I made the following equations

$$78 \text{ Colc} + 11 \text{ U} + 11 \text{ EG} = 12 \text{ W}$$
$$36 \text{ V} + 29 \text{ U} + 35 \text{ EG} = 24 \text{ W}.$$

Multiplying the first by 2 & subtracting

$$156 \text{ Colc} = 36 \text{ V} + 7 \text{ U} + 13 \text{ EG}$$

or

$$100 \text{ Colc} = 24 \text{ V} + 5 \text{ U} + 9 \text{ EG}$$

which is a rough approximation to the other & is good considering the light &c & the character of my helper. All my experience of coloured papers has led me to think that it only requires adjustment to make an equation with any four colours & black but in order to preserve the identity the presence of red curtains &c should be avoided & the top viewed at a constant angle (vertically is best).

I have constructed an Eye-Speculum on Helmholtz principle but with convex glasses.[12]

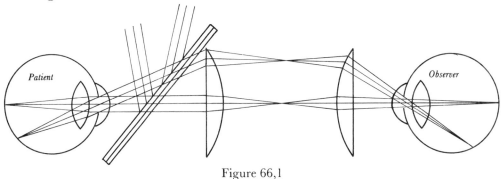

Figure 66,1

The patient looks at the image of the candle as reflected by the oblique glass plates. When he sees it distinctly the rays from the image on his retina come out as they went in and passing thro the plates form an image between the lenses wh. is observed by an eye at an eyehole. The advantage of this arrangement is that the eyehole is conjugate focus to the pupil of the patient & therefore receives all the light which returns thro the pupil. In this way I have seen the image of a candle of a dark brown colour in many men's eyes, & traced some of the blood vessels.

In a dog's eye I saw the brilliant colours of the tapetum with all its reticulations. This is really a beautiful object & by no means difficult to be seen. The dog does not seem to object.

(12) Hermann Helmholtz, *Beschreibung eines Augenspiegels zur Untersuchung der Netzhaut im lebenden Augen* (Berlin, 1851); and 'Ueber eine neue einfachste Form des Augenspiegels', *Archiv für Physiologische Heilkunde*, **2** (1852): 827–52 (= *Wissenschaftliche Abhandlungen*, **2**: 229–60, 261–79). See Numbers 49, 65 and 68.

I have been investigating fluid motion with reference to stability and I have got results when the motion is confined to the plane of xy.[13] I do not know whether the method is new. It only applies to an incompressible fluid moving in a plane.

Stability of Fluid Motion

Let the velocities parallel to x & y at the point (x, y) be

$$u = \frac{d(x)}{dt} \quad v = \frac{d(y)}{dt} \tag{1}$$

then since the fluid is incompressible the eqⁿ of continuity becomes

$$\frac{du}{dx} + \frac{dv}{dy} = 0 \tag{2}$$

∴

$$v\,dx - u\,dy$$

will be a perfect diff¹ of a fⁿ of x & y, say ψ so that

$$u = -\frac{d\psi}{dy}$$

$$v = \frac{d\psi}{dx}. \tag{3}$$

The velocity of the fluid across the curve

$$\psi = \text{const} \tag{4}$$

is

$$u\frac{d\psi}{dx} + v\frac{d\psi}{dy}$$

which vanishes by eqⁿ (3) so that the curves (4) are the paths of the particles of the fluid,[14] & may be considered as ⟨surfaces⟩ partitions in the fluid which constrain & completely determine its motion.[15] We have also

$$\frac{dp}{dx} = \left(X - \frac{d^2(x)}{dt^2} \right) \quad \frac{dp}{dy} = \left(Y - \frac{d^2(y)}{dt^2} \right)^{[16]} \tag{5}$$

where X & Y arise partly from external attractions in wh.

$$X\,dx + Y\,dy + Z\,dz = dV$$

and partly from the ⟨pressure⟩ reaction of the partitions.

---

(13) See Number 63.
(14) On the stream function ψ and stream lines see Number 63 note (4).
(15) See Number 63 note (5).
(16) Compare Number 63, equation (5) and note (6).

By putting
$$\frac{d^2(x)}{dt^2} = u\frac{du}{dx} + v\frac{du}{dy}$$

&
$$\frac{d^2(y)}{dt^2} = u\frac{dv}{dx} + v\frac{dv}{dy}$$

in eq^{ns} (5) and observing that
$$\frac{d}{dy}\frac{dp}{dx} = \frac{d}{dx}\frac{dp}{dy}$$

$$-\frac{dX}{dy} + \frac{dY}{dx} = -\frac{d\psi}{dy}\frac{d}{dx}\left(\frac{d^2\psi}{dx^2} + \frac{d^2\psi}{dy^2}\right) + \frac{d\psi}{dx}\frac{d}{dy}\left(\frac{d^2\psi}{dx^2} + \frac{d^2\psi}{dy^2}\right).$$

Put
$$\chi = \frac{d^2\psi}{dx^2} + \frac{d^2\psi}{dy^2} \quad {}^{(17)}$$

then the twisting force on the element at xy
$$\frac{dY}{dx} - \frac{dX}{dy} = u\frac{d\chi}{dx} + v\frac{d\chi}{dy}$$
$$= \frac{d(\chi)}{dt}$$

where $\dfrac{d(\chi)}{dt}$ refers to the variation of χ as we pass along the path of the particle.

Since the external forces have no power of twisting this force is due entirely to the reaction of the partitions, and therefore when there are no partitions
$$\frac{d(\chi)}{dt} = 0$$

or
$$-\frac{d\psi}{dy}\frac{d\chi}{dx} + \frac{d\psi}{dx}\frac{d\chi}{dy} = 0$$

whence
$$\chi = f(\psi)$$

which is the condition of steady motion as is otherwise known.[18]

If this motion be stable, then if we give the partitions a derangement the twisting force of the fluid on the partitions will be in the opposite direction to the twist they have received.

By making
$$\psi = f^n \text{ of } x, y \;\&\; h$$

we may express any change of the form of ψ by changing the value of h.[19]

Then the twisting force of the fluid on the partitions
$$N = \frac{dX}{dy} - \frac{dY}{dx} = -\left(\frac{d(\chi)}{dt} + \frac{d}{dt}\left(\frac{d\chi}{dh}\right)dh\right)$$

(17) Compare Stokes, 'On the steady motion of incompressible fluids', *Trans. Camb. Phil. Soc.,* **7** (1842): 439–53, esp. 441–5 (= *Papers,* **1**: 1–16).

(18) Stokes, 'On the steady motion of incompressible fluids': 445–6.

(19) See Number 63.

the inclination of partition to axis of x is

$$\theta = \tan^{-1} \frac{\dfrac{d\psi}{dx}}{\dfrac{d\psi}{dy}}$$

$$\therefore d\theta = \frac{\dfrac{d\psi}{dy}\dfrac{d^2\psi}{dx\,dh} - \dfrac{d\psi}{dx}\dfrac{d^2\psi}{dy\,dh}}{\left|\dfrac{d\psi}{dx}\right|^2 + \left|\dfrac{d\psi}{dy}\right|^2}\,dh$$

$$= -\frac{u\dfrac{d}{dx}\left(\dfrac{d\psi}{dh}\right) + v\dfrac{d}{dy}\left(\dfrac{d\psi}{dh}\right)}{u^2 + v^2}\,dh$$

$$= -\frac{\dfrac{d}{dt}\left(\dfrac{d\psi}{dh}\right)}{u^2 + v^2}\,dh$$

the sign of which must be the reverse of N whatever be the values of $\dfrac{d\psi}{dh}$ & of dh that the motion may be stable.

Hence (A)
$$\frac{d\chi}{dt} = 0 \text{ or } \chi = f(\psi)$$

&
$$\therefore \frac{d\chi}{dh} = f'(\psi)\frac{d\psi}{dh}.$$

So that since $\dfrac{d}{dt}\left(\dfrac{d\psi}{dh}\right)$ & $\dfrac{d}{dt}\left(\dfrac{d\chi}{dh}\right)$ must have opposite signs and since we suppose the displacement of the partitions to begin after a certain time

(B) $f'(\psi)$ or $\dfrac{d\chi}{d\psi}$ must be negative for stability.

When $f'(\psi)$ is positive the motion is unstable.

When $f'(\psi) = 0$, χ is constant or 0.

When χ is constant I think eq$^{\rm m}$ is neutral.

When $\chi = 0$ the whole motion is determined by the motion at a limiting curve so that there can be no finite displacement.[20]

---

(20) The condition of irrotational motion of an incompressible fluid in two dimensions; compare Stokes, 'On the steady motion of incompressible fluids': 441.

Example

In a vortex in which the velocity is as the *n*th power of the distance from the centre the value of *n* must be between $+1$ & -1 for stability.

In 3 dimensions this method is complicated, we have

$$u = \frac{d\phi}{dy}\frac{d\phi'}{dz} - \frac{d\phi}{dz}\frac{d\phi'}{dy}$$

$$v = \frac{d\phi}{dz}\frac{d\phi'}{dx} - \frac{d\phi}{dx}\frac{d\phi'}{dz}$$

$$w = \frac{d\phi}{dx}\frac{d\phi'}{dy} - \frac{d\phi}{dy}\frac{d\phi'}{dx} \quad \&\text{c.}$$

Another method[21]

Consider the fluid within the curve

$$u = 0.$$

Its vis viva $= \sum (v^2)$ [22]

$$\iint dx\, dy \left(\overline{\frac{d\psi}{dx}}^2 + \overline{\frac{d\psi}{dy}}^2 \right)$$

between proper limits. Integrating by parts

$$\sum (v^2) = \int dx \frac{d\psi}{dx} \psi_{y_2} - \int dx \frac{d\psi}{dx} \psi_{y_1}$$

$$+ \int dy \frac{d\psi}{dy} \psi_{x_2} - \int dx \frac{d\psi}{dy} \psi_{x_1} - \iint dx\, dy\, \psi \left(\frac{d^2\psi}{dx^2} + \frac{d^2\psi}{dy^2} \right)$$

$$= \int ds\, \psi \frac{d\psi}{du} - \iint dx\, dy\, \psi \left(\frac{d^2\psi}{dx^2} + \frac{d^2\psi}{dy^2} \right)$$

where *ds* is an element of the curve & *u* is measured in direction of normal.

---

(21) Compare Thomson's paper 'Notes on hydrodynamics. V. On the vis-viva of a liquid in motion', *Camb. & Dubl. Math. J.*, **4** (1849): 90–4 (= *Math. & Phys. Papers*, **1**: 107–12), where he had applied energy conditions to the analysis of the motion of an incompressible fluid enclosed within a flexible and extensible envelope. Thomson discussed extremal conditions which determine the bounding surface of the fluid, employing theorems from his papers 'Propositions in the theory of attraction', *Camb. Math. J.*, **3** (1842–3): 189–96, 201–6; and 'Theorems with reference to the solution of certain partial differential equations', *Camb. & Dubl. Math. J.*, **3** (1848): 84–7 (= *Electrostatics and Magnetism*: 126–38, 139–41). In this latter paper Thomson formulates the theorem known as Dirichlet's principle. See also J. J. Cross, 'Integral theorems in Cambridge mathematical physics, 1830–55', and Ole Knudsen, 'Mathematics and reality in William Thomson's electromagnetic theory', in *Wranglers and Physicists*, ed. P. M. Harman (Manchester, 1985): 140–1, 159–60.

(22) The Leibnizian term *vis viva* (mv^2) was in common use. Maxwell follows Thomson's usage: see note (21). On *vis viva* and kinetic energy see Numbers 107 note (77) and 133 note (7).

The first term here refers to the limits & is not altered by any derangement within the limiting curve. The second term is of the familiar form

$$\iint dx\,dy\,V\left(\frac{d^2V}{dx^2}+\frac{d^2V}{dy^2}\right)$$

in which we regard V as the potential &

$$\frac{d^2V}{dx^2}+\frac{d^2V}{dy^2}=4\pi\rho.^{(23)}$$

Now we know that
$$4\pi\iint dx\,dy\,V\rho$$

is a maximum for stable & a minimum for unstable eq^m.[24] In both cases

$$\rho=fV$$

and in the former $f'(V)$ is positive.

Hence
$$\sum(v^2)=\iint dx\,dy\left(\left|\frac{d\psi}{dx}\right|^2+\overline{\left|\frac{d\psi}{dy}\right|^2}\right)$$

is a maximum or a minimum when

$$\chi=\frac{d^2\psi}{dx^2}+\frac{d^2\psi}{dy^2}=f(\psi)$$

maximum when $f'(\psi)$ is negative
minimum when $f'(\psi)$ is positive.

Now if the vis viva be a maximum any derangement will diminish it & the motion will tend to revive again and vice versa.

I hope to hear again of Mrs Thomsons progress. I have good accounts of my father who seems quite recovered.

Yours truly

J. C. MAXWELL

---

(23) See Number 51 note (19). On Poisson's equation for the distribution of electricity in a space of two dimensions see Maxwell, *Treatise*, **1**: 226–7 (§182).

(24) Compare Thomson's discussion in his 'Propositions in the theory of attraction': 202–5; and see Thomson, 'On the vis-viva of a liquid in motion': 94, for the application of the extremal condition to the motion of an incompressible fluid within a flexible envelope. The minimum condition had been discussed by C. F. Gauss, 'Allgemeine Lehrsätze in Beziehung auf die in verkehrten Verhältnisse des Quadrats der Entfernung wirkenden Anziehungs- und Abstossungs-Kräfte', in *Resultate aus der Beobachtungen des magnetischen Vereins in Jahre 1839*, Herausgegeben von C. F. Gauss und W. Weber (Leipzig, 1840): 1–51; (trans.) 'General propositions relating to attractive and repulsive forces acting in the inverse ratio of the square of the distance', *Scientific Memoirs*, ed. R. Taylor, **3** (London, 1843): 153–96, esp. 177–91. See Number 100. See also M. N. Wise, 'The flow analogy to electricity and magnetism, Part I: William Thomson's reformulation of action at a distance', *Archive for History of Exact Sciences*, **25** (1981): 19–70, esp. 56–8.

FROM A LETTER TO JOHN CLERK MAXWELL

23 MAY 1855

From Campbell and Garnett, *Life of Maxwell*[1]

Trin. Coll.
Eve of H.M. Nativity

Wednesday last I went with Hort[2] and Elphinstone[3] to the Ray Club,[4] which met at Kingsley of Sidney's[5] rooms. Kingsley is great in photography and microscopes, and showed photographs of infusoria, very beautiful, also live plants and animals, with oxy-hydrogen microscope.

...I am getting on with my electrical calculations now and then, and working out anything that seems to help the understanding thereof.[6]

---

(1) *Life of Maxwell*: 213.

(2) Fenton John Anthony Hort, Trinity 1845 (Venn). See his letter to Campbell of 4 February 1882, in *Life of Maxwell*: 417–21.

(3) Howard Warburton Elphinstone, Trinity 1850 (Venn).

(4) The 'Society for the Cultivation of Natural History', an informal association of professors, graduates and undergraduates; see Harvey W. Becher, 'Voluntary science in nineteenth-century Cambridge University to the 1850s', *The British Journal for the History of Science*, **19** (1986): 64–5.

(5) William Towler Kingsley, Fellow and Tutor of Sidney Sussex College (Venn).

(6) Compare Number 71. The letter was cut by Campbell.

LETTER TO RICHARD BUCKLEY LITCHFIELD

6 JUNE 1855

From the original in the library of Trinity College, Cambridge[1]

Trinity Coll June 6th, 55

Dear Litchfield

I expect to pass thro London next week & I write to you to know if you could put me up to some of the lodgings about the Temple where I might put myself up for two nights. I intend to come up on Monday morning & do the Crystal Palace trick, do my friends &c on Tuesday & depart on Wednesday evening.

Pomeroy[2] was intending to come, but as his call to town is not till the 20th he is not coming. For me I must be at home before that time else I should have stayed. Pomeroy has been acting conspicuously as the leader of a revolution in the Bathing shed of wh: he was treasurer. The subscribers are now a swimming club with new laws & lots of new members with Cambridge University swimming races going on at present during the examinations.

I suppose you know about Farrar,[3] how he has accepted a mastership at Harrow. There may be some chance of Sale[4] succeeding him at Marlborough.

I have been busy with electrical reading this term & as I had but two pups I got on very well. It is hard work grinding out 'appropriate ideas' as Whewell calls them.[5] However I think they are coming out at last and by dint of knocking them against all the facts and half digested theories afloat I hope to bring them to shape after which I hope to understand something more about inductive philosophy than I do at present.

I have a project of sifting the theory of light & making everything stand upon definite experiments & definite assumptions, so that things may not be supposed to be assumptions when they are either definitions or experiments.

I have been looking into all the dogs eyes here to see the bright coating at the back of the eye, thro an instrument I made to that end.[6] The spectacle is very fine. I remember the appearance of Mungo's eyes at Cheltenham. He would be the dog to sit. Human eyes are very dull and brown as to their retinæ but you can see the image of a candle quite well on it, & sometimes the blood-vessels

---

(1) Trinity College, Cambridge, Add. MS Letters c.l[80]. Previously published (in extract) in *Life of Maxwell*: 215.

(2) Number 42 note (3).

(3) Number 56 note (11).

(4) George Samuel Sale, Trinity 1850 (Venn).

(5) Whewell's term is 'fundamental ideas': see William Whewell, *Philosophy of the Inductive Sciences, founded upon their History*, 2 vols. (London, ₂1847), **1**: 66.

(6) See Numbers 49, 65 and 66.

&c. Do you know of MacLennan?[7] I do not, but I saw a long piece of blank verse in the Guardian, after the Alexander Smith pattern of wh: it w^d have been a good parody had it not been serious.[8] I saw a great deal of him in Edin^h & I rather admired the symptoms.

If you hear of a lodgement just drop me a note here or if you dont leave word with the porter at Temple bar, but a note will be best.

Yours very hot

J. C. MAXWELL

(7) Number 37 note (4).

(8) [Alexander Smith and S. T. Dobell,] *Sonnets on the War* (London, 1855); see the poems on the Crimean War in *The Guardian* (23 May and 30 May 1855): 412, 422. On Maxwell's acquaintance with Smith see Numbers 40 and 119.

LETTER TO CECIL JAMES MONRO

11 JUNE 1855

From the original in the Greater London Record Office, London[1]

9 Lincolns inn New Square
Monday

Dear Monro

I came along here this morning & am put up at Mrs Turner's, 18 Devereux Court Temple, where I am to be found till Wednesday. I am going to the C. Palast[2] today & to various people tomorrow. Will you come in on Wednesday morning & breakfast with Litchfield sometime in the morning. Litchfield goes about $\frac{1}{2}$ past 10 to his business so early is well.

We may then do the Palast as we proposed.

Yours truly
J. C. MAXWELL

NB. Litchfield has no plain envelopes.

---

(1) Greater London Record Office, Acc. 1063/2080.

(2) The Crystal Palace designed by Joseph Paxton and built for the Great Exhibition of 1851 at Hyde Park, London; subsequently re-erected at Sydenham, London. Maxwell had visited the Great Exhibition: see Number 55. The Crystal Palace was used for later exhibitions: on 8 June 1855 *The Times* carried an advertisement on the opening of the raw produce department including mineralogical specimens.

EXTRACT FROM LETTER TO GEORGE WILSON

27 JULY 1855

From George Wilson, *Researches on Colour-Blindness* (1855)[1]

Last year I was observing the spectrum formed by a long slit, using a single prism and the naked eye. I then first observed that a peculiar dark appearance (which I supposed to belong to the blue near the line F) extended only to that part of the spectrum to which the eye was directed; above and below, the blue was much brighter, and by moving the eye, the blue spot ran up and down the blue band of the spectrum as if confined in a groove. I found that none of the other colours became dark in the same way when looked at directly.[2]

Last spring, Professor Forbes kindly allowed me the use of his apparatus for the purpose of repeating this observation. By using a telescope along with the prism, and so obtaining a large field of pure colour, I found that the dark spot could be best seen in the blue, and presented the appearance well known to be due to the yellow spot. It became fainter in the more refrangible rays, but was brightest in the line F, and on to G. Between F and E, I think, it suddenly disappeared, re-appearing as a light spot in the yellow and orange. I am not quite sure, however, that the latter effect was not produced by the former; for I could not attain it without looking first at the blue, and then suddenly at the yellow.

The best way of seeing this is by means of a whirling disc, supported by an axis, so that it may be made to go slow for some time without stopping. By putting on this disc, sectors of ultramarine and chrome yellow in the proportion of 3 to 1, the effect of the change of the light entering the eye from yellow to blue might be observed by slowly turning the disc, so that each colour might be a sensible time before the eye, which was kept steady. Every time the blue came opposite, the dark spot appeared, and seemed to fade away gradually, if the yellow did not come in to relieve the eye from the effect of the blue. By properly adjusting the speed, the dark spot could be made to glimmer in a very conspicuous manner.

The result shows that the yellow spot is *less* sensitive to *blue* rays than other parts of the retina. That it is *more* sensitive to *yellow* rays is less certain. A bright spot is seen when yellow succeeds blue, but I have not produced it by simply admitting yellow light.

---

(1) George Wilson, *Researches on Colour-Blindness* (Edinburgh, 1855): 164–5.

(2) See Maxwell's paper 'On the unequal sensibility of the foramen centrale to light of different colours', *Report of the Twenty-sixth Meeting of the British Association for the Advancement of Science; held at Cheltenham in August 1856* (London, 1857), part 2: 12 (= *Scientific Papers*, **1**: 242).

LETTER TO WILLIAM THOMSON

13 SEPTEMBER 1855

From the original in the University Library, Cambridge[1]

Glenlair
Sept 13th 1855.

Dear Thomson

I hoped to have been in Glasgow by this time, but my departure was postponed, owing to the state of my father's health. I wrote to M[rs] Blackburn[2] to tell what had happened & how things were going on, so I need not repeat particulars. He is now down stairs, and apparently as well as ever, but I think I had better stay here till we see his health reestablished.

If I had seen you in Glasgow, I should have asked you some questions which have been some time in store. Perhaps it is better as it is not to bother you with 'vivâ voce' while you are busy with the wise men,[3] but to write them down for you to answer by post during the lull between the extraordinary & the regular occupation of the College.

I have got a good deal out of you on electrical subjects, both directly & through the printer & publisher[4] & I have also used other helps, and read Faraday's three volumes of researches.[5] My object in doing so was of course to learn what had been done in electrical science, mathematical & experimental, and to try to comprehend the same in a rational manner by the aid of any notions I could screw into my head. In searching for these notions I have come upon some ready made, which I have appropriated.[6] Of these are Faradays theory of polarity which ascribes that property to every portion of the whole sphere of action of the magnetic or electric bodies,[7] also his general notions about 'lines of force' with the 'conducting power' of different media for them.[8]

---

(1) ULC Add. MSS 7342, M 91. Previously published in Larmor, 'Origins': 711–13.

(2) See Number 45 note (10).

(3) The meeting of the British Association at Glasgow: see Number 72.

(4) See Maxwell's letters to Thomson of 13 November 1854 and 15 May 1855 (Numbers 51 and 66).

(5) Michael Faraday, *Experimental Researches in Electricity*, 3 vols. (London, 1839–55).

(6) See Number 84; and Maxwell, 'On Faraday's lines of force', *Trans. Camb. Phil. Soc.*, **10** (1856): 27–83 (= *Scientific Papers*, **1**: 155–229).

(7) Michael Faraday, 'Experimental researches in electricity. – Twenty-eighth series. On lines of magnetic force; their definite character, and their distribution within a magnet and through space', *Phil. Trans.*, **142** (1852): 25–56 (= *Electricity*, **3**: 28–70).

(8) Michael Faraday, 'Experimental researches in electricity. – Twenty-sixth series. Magnetic conducting power', *Phil. Trans.*, **141** (1851): 29–84 (= *Electricity*, **3**: 200–73).

Then comes your allegorical representation of the case of electrified bodies by means of conductors of heat,[9] and your theorem on the eq^n

$$\frac{d}{dx}\left(\alpha^2\frac{dV}{dx}\right) + \frac{d}{dy}\left(\alpha^2\frac{dV}{dy}\right) + \frac{d}{dz}\left(\alpha^2\frac{dV}{dz}\right) = -4\pi\rho.^{[10]}$$

Then Ampères theory of *closed* galvanic circuits,[11] then part of your allegory about incompressible elastic solids[12] & lastly the *method* of the last demonstration in your R.S. paper on Magnetism.[13] I have also been working at Webers theory of Electro Magnetism as a mathematical speculation which I do not believe but which ought to be compared with others and certainly gives many true results at the expense of several startling assumptions.[14]

Now I have been planning and partly executing a system of propositions about lines of force &c which may be *afterwards* applied to Electricity, Heat or Magnetism or Galvanism, but which is in itself a collection of purely geometrical truths embodied in geometrical conceptions of lines, surfaces &c.[15]

---

(9) [William Thomson,] 'On the uniform motion of heat in homogeneous solid bodies, and its connection with the mathematical theory of electricity', *Camb. Math. J.*, **3** (1842): 71–84 (= *Electrostatics and Magnetism*: 1–14). See Number 51 note (3).

(10) William Thomson, 'Theorems with reference to the solution of certain partial differential equations', *Camb. & Dubl. Math. J.*, **3** (1848): 84–7, esp. 84 (= *Electrostatics and Magnetism*: 139–41; *Math. & Phys. Papers*, **1**: 93–6). On the equation for the potential *V* see Number 51 note (19); and compare 'On Faraday's lines of force': 57 (= *Scientific Papers*, **1**: 195).

(11) André Marie Ampère, 'Mémoire sur la théorie mathématique des phénomènes électrodynamiques uniquement déduite de l'expérience', *Mémoires de l'Académie Royale des Sciences*, **6** (1827): 175–88. Compare 'On Faraday's lines of force': 55–6 (= *Scientific Papers*, **1**: 193–4).

(12) William Thomson, 'On a mechanical representation of electric, magnetic, and galvanic forces', *Camb. & Dubl. Math. J.*, **2** (1847): 61–4 (= *Math. & Phys Papers*, **1**: 76–80). Thomson defines the relation between the displacements α, β, γ of a point x, y, z in an incompressible elastic solid and the magnetic force X, Y, Z by equations $X = d\beta/dz - d\gamma/dy$, $Y = d\gamma/dx - d\alpha/dz$, $Z = d\alpha/dy - d\beta/dx$, which represent the rotation of an element of the solid. Compare Maxwell, 'On Faraday's lines of force': 51, 55–65, 67 (= *Scientific Papers*, **1**: 188, 195–205, 209).

(13) William Thomson, 'A mathematical theory of magnetism', *Phil. Trans.*, **141** (1851): 243–85, esp. 283–5 (= *Electrostatics and Magnetism*: 340–404). Given the equation $d\alpha/dx + d\beta/dy + d\gamma/dz = 0$ for a solenoidal distribution of magnetism (see note (29)), where α, β, γ denote the components of the intensity of magnetization at any internal point in a magnet, 'we may assume $\alpha = dH/dy - dG/dz$, $\beta = dF/dz - dH/dx$, $\gamma = dG/dx - dF/dy$ where F, G, H are three functions to a certain extent arbitrary...'. Compare Maxwell, 'On Faraday's lines of force': 59, 67 (= *Scientific Papers*, **1**: 198, 209). See Number 100.

(14) For Maxwell's comments on Weber's theory of electrodynamics see 'On Faraday's lines of force': 66–7 (= *Scientific Papers*, 1: 207–8). On Weber see Number 66 note (3).

(15) See Maxwell's drafts on fluid motion (Numbers 61, 63 and 66); and their application in the drafts of 'On Faraday's lines of force' (Numbers 83 and 84). By 'galvanism' Maxwell means 'electromagnetism'; compare Thomson's usage in his paper 'On a mechanical representation of electric, magnetic, and galvanic forces'.

The first part of my design is to prove by popular, that is not professedly symbolic, reasoning, the most important propositions about V and about the solution of the equation in the last page $\left(\dfrac{d}{dx}\alpha^2\dfrac{dV}{dx}+\&c\right)$ and to trace the lines of force and surfaces of equal V.[16]

The second part would be nothing else than a collection of examples[17] of your 'Electrical Images'[18] worked out by the method of lines of force & equipotential surfaces,[19] for cases of spheres of various conducting powers[20] and for the case of a sphere of a 'magnecrystallic' substance in a uniform field.[21]

The third part would be the theory of the connections of the three divisions of the subject the passage of statical into current electricity and the magnetic properties of closed currents with the laws of induced current.[22]

---

(16) Compare Maxwell's letter to Thomson of 13 November 1854 (Number 51); and see 'On Faraday's lines of force': 27–30 (= *Scientific Papers*, **1**: 155–9).

(17) In the published paper these 'Examples' were placed at the end of the paper; see 'On Faraday's lines of force': 68–83 (= *Scientific Papers*, **1**: 209–29).

(18) William Thomson, 'On the mathematical theory of electricity in equilibrium. III. Geometrical investigations with reference to the distribution of electricity on spherical conductors. IV. Geometrical investigations regarding spherical conductors. V. Effects of electrical influence on internal spherical and on plane conducting surfaces. VI. Geometrical investigations regarding spherical conductors', *Camb. & Dubl. Math. J.*, **3** (1848): 141–8, 266–74; **4** (1849): 276–84; **5** (1850): 1–9 (= *Electrostatics and Magnetism*: 52–85). See also the 'Extrait d'une lettre de M. William Thomson à M. Liouville' [of 8 October 1845], *Journal de Mathématiques Pures et Appliquées*, **10** (1845): 364–7; and, 'Extraits de deux lettres addressées à M. Liouville' [of 26 June 1846 and 16 September 1846], *ibid.*, **12** (1847): 256–64, and Liouville's 'Note au subjet de l'article précédent': 265–90 (= *Electrostatics and Magnetism*: 144–77). Compare Maxwell's definition in the *Treatise*, **1**: 193 (§157); 'An electrical image is an electrified point or system of points on one side of a surface which would produce on the other side of that surface the same electrical action which the actual electrification of that surface really does produce'.

(19) See Maxwell's development of Thomson's method of electrical images in Example I of 'On Faraday's lines of force': 68–70 (= *Scientific Papers*, **1**: 209–12).

(20) See Examples II to VII of 'On Faraday's lines of force': 70–76 (= *Scientific Papers*, **1**: 212–20).

(21) See Example VI of 'On Faraday's lines of force': 74–6 (= *Scientific Papers*, **1**: 217–19). Faraday's discussion of the alignment of crystals of bismuth in a magnetic field led to his suggestion of the '*line or axis of* MAGNECRYSTALLIC *force*' in his Bakerian Lecture, 'Experimental researches in electricity. – Twenty-second series. On the crystalline polarity of bismuth (and other bodies), and on its relation to the magnetic form of force', *Phil. Trans.*, **139** (1849): 1–41, esp. 7 (§2479) (= *Electricity*, **3**: 90). See William Thomson, 'On the theory of magnetic induction in crystalline and non-crystalline substances', *Phil. Mag.*, ser. 4, **1** (1851): 177–86 (= *Electrostatics and Magnetism*: 465–80); and Numbers 104 and 109.

(22) This material is placed in the latter part of Part I of 'On Faraday's lines of force': 42–51 (= *Scientific Papers*, **1**: 175–88). See also Number 84.

I intend next to apply to these facts Faradays notion of an *electrotonic* state.[23] I have worked a good deal of mathematical material out of this vein and I believe I have got hold of several truths which will find a mathematical expression in the electrotonic state.[24]

One thing at least it succeeds in, it reduces to one principle not only the attraction of currents & the induction of currents but also the attraction of electrified bodies without any new assumption.[25]

Now what I have planned out in this way I intend to set about for the sake of acquiring knowledge sufficient to guide me in devising experiments; but supposing that at any time I should be satisfied that the theory would stick together all through, and that I could put it into a form not too abstrusely mathematical, but still exact and leading to numerical results then I would be much assisted by your telling me whether you have not the whole draught of the thing lying in loose papers and neglected only till you have worked out Heat or got a little spare time.

The reasons I have for thinking so are – That you are acquainted with Faradays theory of lines of force & with Ampères laws of currents and of course you must have wished at least to understand Ampère in Faradays sense. You had the advantage of being well acquainted with V and with Green's essay,[26] and you published a fragment of your speculations in the form of an allegory about incompressible elastic solids.[27] In your paper on Electrolysis[28] you state several of the laws of induction of currents which are right as far as they are stated and at the end of your paper of magnetism you have not only stated and applied Ampères properties of currents but used a method in your demonstration about the superficial tangential distribution of magnetism in a

(23) Michael Faraday, *Electricity*, **1**:16 (§60), and **3**:420 (§3269). See Maxwell, 'On Faraday's lines of force': 51–2 (= *Scientific Papers*, **1**: 188–9).

(24) See Part II of 'On Faraday's lines of force': 51–67 (= *Scientific Papers*, **1**: 188–209); and Number 87.

(25) Compare 'On Faraday's lines of force': 63 (= *Scientific Papers*, **1**: 203). See Number 87.

(26) William Thomson, 'On the mathematical theory of electricity in equilibrium. I. On the elementary laws of statical electricity', *Camb. & Dubl. Math. J.*, **1** (1845): 75–95, esp. 76, 94 (= *Electrostatics and Magnetism*: 15–37); and see George Green, *An Essay on the Application of Mathematical Analysis to the Theories of Electricity and Magnetism* (Nottingham, 1828). See Number 51 note (13).

(27) Thomson, 'On a mechanical representation of electric, magnetic, and galvanic forces'. For Thomson's description of his 'mathematical analogy' see his letter to Faraday of 11 June 1847; in S. P. Thompson, *The Life of William Thomson, Baron Kelvin of Largs*, 2 vols. (London, 1910), **1**: 203–4.

(28) William Thomson, 'On the mechanical theory of electrolysis', *Phil. Mag.*, ser. 4, **2** (1851): 429–44 (= *Math. & Phys. Papers*, **1**: 472–89).

solenoidal magnet[29] which seems to me to be part of my results applied to magnetism[30] without acknowledging that you had taken it from a more general theory.

As there can be no doubt that you have the mathematical part of the theory in your desk all that you have to do is to explain your results with reference to electricity. I think that if you were to do so publicly it would introduce a new set of electrical notions into circulation & save much useless speculation.

I do not know the Game laws & Patent laws of science. Perhaps the Association may do something to fix them but I certainly intend to poach among your electrical images, and as for the hints you have dropped about the 'higher electricity', I intend to take them. At the same time if you happen to know where anything on this part of the subject is to be found it would be of great use to me.

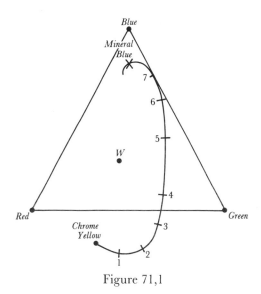

Figure 71,1

---

(29) See Ampère, 'Mémoire sur la théorie mathématique des phénomènes électrodynamiques': 267, for the term 'solénoide électro-dynamique'. Thomson supposes a magnet to be divided along lines of magnetisation into tubes, 'magnetic solenoids'; see Thomson, 'A mathematical theory of magnetism': 269–70. In a solenoidal distribution of magnetism he represents the magnetic force at an internal point by 'the force at a point in an infinitely small crevass tangential to the lines of magnetization' ('A mathematical theory of magnetism': 277). See Number 51 notes (7) and (16).

(30) In the case of a 'lamellar' distribution of magnetism where the magnet is divided into 'magnetic shells' perpendicular to the lines of magnetisation (see Number 51 notes (7), (10), (11)), Thomson shows that the magnetic force 'at a point in an infinitely small crevass perpendicular to the lines of magnetization' differs from the force in a tangential crevass (see note (29)) by a term 'equal to the product of 4π into the intensity of the magnetization'; see Thomson, 'A mathematical theory of magnetism': 275, 277. On Maxwell's reformulation of Thomson's distinction between 'solenoidal' and 'lamellar' distributions of magnetism into a distinction between 'magnetic force' (**H**) and 'magnetic induction' (**B**) as 'intensities' and 'quantities' or 'forces' and 'fluxes', see Number 51 note (7), and 'On Faraday's lines of force': 54–5 (= *Scientific Papers*, **1**: 192). In the *Treatise* Maxwell states 'equations of magnetization', which he writes in vector form as $\mathbf{B} = \mathbf{H} + 4\pi\mathbf{I}$ (where **I** is the 'intensity of magnetization'). See *Treatise*, **2**: 22–4, 228–9 (§§ 397–400, 605).

Since I last wrote to you I have been making mixtures of colours by weight and finding the resulting colour by means of the colour-top. I find that very few mixtures lie in the line joining their components, but that they form a curve generally pretty regular from the one to the other. In the case of Chrome Yellow & Mineral Blue (copper I believe) the curve goes away among the greens.[31]

I have a good many other pairs of colours, but this is the most remarkable.

I am beginning to get results on the colours of the spectrum, but I have great difficulties in getting my apparatus sufficiently steady & accurate & in measuring the diameters of slits.[32]

My father was up to breakfast today but we must take care that he does not do too much. How is Mrs Thomson? I have heard nothing of either of you since Malvern. Dr G. Wilson is to be at Glasgow. I have had several letters from him on Colour blindness.[33]

I shall be here till the 24th Sept. after which address Trin. Coll. If I do not hear from you soon I hope to hear later.

Yours truly

J. C. Maxwell

---

(31) See Maxwell's paper 'On the theory of compound colours with reference to mixtures of blue and yellow light', *Report of the Twenty-sixth Meeting of the British Association for the Advancement of Science; held at Cheltenham in August 1856* (London, 1857), part 2: 12–13 (= *Scientific Papers*, **1**: 243–5); and the draft (Number 102).

(32) See Number 139. Compare Maxwell's description of his portable colour box in letters to Thomson and Stokes of December 1856 and May 1857 (Numbers 109 and 117).

(33) See Number 70.

72

FROM A LETTER TO JOHN CLERK MAXWELL

24 SEPTEMBER 1855

From Campbell and Garnett, *Life of Maxwell*[1]

Holbrooke, by Derby
24 Sept. 1855

We had a paper from Brewster on the theory of three colours in the spectrum,[2] in which he treated Whewell with philosophic pity, commending him to the care of Prof. Wartman of Geneva, who was considered the greatest authority in cases of his kind, cases in fact of colour-blindness.[3] Whewell was in the room, but went out, and avoided the quarrel; and Stokes made a few remarks, stating the case not only clearly but courteously. However, Brewster did not seem to see that Stokes admitted his experiments to be correct, and the newspapers represented Stokes as calling in question the accuracy of the experiments.[4]

I am getting my electrical mathematics into shape, and I see through some parts which were rather hazy before; but I do not find very much time for it at present, because I am reading about heat and fluids, so as not to tell lies in my lectures. I got a note from the Society of Arts about the platometer, awarding thanks, and offering to defray the expenses to the extent of £10, on the machine being produced in working order.[5] When I have arranged it in my head, I intend to write to James Bryson about it.[6]

---

(1) *Life of Maxwell*: 216.

(2) At the meeting of the British Association in Glasgow; see David Brewster, 'On the triple spectrum', *Report of the Twenty-fifth Meeting of the British Association for the Advancement of Science; held at Glasgow in September 1855* (London, 1856), part 2: 7–9.

(3) See the report in *The Athenæum* (6 October 1855): 1156–7. On Wartmann and colour blindness see Number 47 note (10).

(4) Brewster's remarks were made in response to criticisms by Airy and Whewell of his theory that the colours of the spectrum arose from the intermixture, in various proportions, of red, yellow and blue 'primary' colours of light across the spectrum. See Brewster, 'On a new analysis of solar light, indicating three primary colours, forming coincident spectra of equal length', *Trans. Roy. Soc. Edinb.*, **12** (1834): 123–36; G. B. Airy, 'The Astronomer Royal on Sir David Brewster's new analysis of solar light', *Phil. Mag.*, ser. 3, **30** (1847): 73–6. See Number 139 notes (15) and (16).

(5) See Number 55. The Report of a Committee of the Royal Scottish Society of Arts, dated 26 February 1855, read and approved on 12 March 1855, which was sent to Maxwell, stated that: 'the author is entitled to high praise...the instrument if constructed would be found to work well...construction does not appear to involve *any very* considerable difficulty...[and] suggesting that the Society should order the Construction of such an Instrument at an expense not exceeding £20.' (ULC Add. MSS 7655, V, d/9).   (6)   Number 58 note (2).

I got a long letter from Thomson about colours and electricity. He is beginning to believe in my theory about all colours being capable of reference to three standard ones,[7] and he is very glad that I should poach on his electrical preserves.[8]

(7) But compare Thomson's letter to Stokes of 28 January 1856 (ULC Add. MSS 7656, K 89): 'Have you seen Clerk Maxwell's paper in the Trans. R.S.E. on colour as seen by the eye? Are you satisfied with the perfect accuracy of Newton's centre of gravity principle on w^h all theories & nomenclatures on the subject are founded? That is to say do you believe that the whites produced by various combinations, such as two homogeneous colours, three homogeneous colours &^{c} are absolutely indistinguishable from one another & from solar white by the best eye? It will be a corollary from this that every colour (except of course the purples) is identical, in the sensation it produces, with some homogeneous colour mixed with white light. Are you at all satisfied with Young's idea of triplicity in the perceptive organ?' See Numbers 54, 59 and 64. In his reply of 4 February 1856 (ULC Add. MSS 7342, S 387) Stokes stated that 'I have not made any experiments on the mixture of colours, nor attended particularly to the subject.' Compare Stokes' letter to Maxwell of 7 November 1857 (Number 139 note (2)). In questioning Young's theory of three colour sensations, and in his point about obtaining purple, Thomson follows Hermann Helmholtz, 'Ueber die Theorie der zusammengesetzten Farben', *Ann. Phys.*, **87** (1852): 45–66; (trans.) 'On the theory of compound colours', *Phil. Mag.*, ser. 4, **4** (1852): 519–34, esp. 533. Helmholtz argues that five spectral colours are necessary to imitate the colours of the spectrum and hence 'we must also abandon the theory of three primitive colours, which, according to Thomas Young, are three fundamental qualities of sensation'; and states that 'purple-red...cannot be otherwise obtained than from the extreme red and violet, without a loss of brightness'. See Number 54 note (7).

(8) See Maxwell's letter to Thomson of 13 September 1855 (Number 71).

FROM A LETTER TO JOHN CLERK MAXWELL

27 SEPTEMBER 1855

From Campbell and Garnett, *Life of Maxwell*[1]

Trin. Coll.
27 Sept. 1855

It is difficult to keep up one's interest in intellectual matters when friends of the intellectual kind are scarce. However, there are plenty of friends not intellectual, who serve to bring out the active and practical habits of mind, which overly-intellectual people seldom do. Wherefore, if I am to be up this term, I intend to addict myself rather to the working men who are getting up classes, than to pups., who are in the main a vexation. Meanwhile there is the examination to consider.[2]

---

(1) *Life of Maxwell*: 216–17.

(2) On 4 October 1855 John Clerk Maxwell wrote to Robert Cay: 'At this time Jas. is engaged with the exam: for Trin. Coll fellowships the decision will come out on tuesday next' (Peterhouse, Cambridge, Maxwell MSS (5)). This was Maxwell's second – and successful – attempt to gain a Fellowship at Trinity: see Number 56 note (3).

FROM A LETTER TO JOHN CLERK MAXWELL

5 OCTOBER 1855

From Campbell and Garnett, *Life of Maxwell*[1]

Trin. Coll.
5 October 1855

You say Dr. Wilson has sent his book.[2] I will write and thank him. I suppose it is about colour-blindness.[3] I intend to begin Poisson's papers on electricity and magnetism[4] to-morrow. I have got them out of the library; my reading hitherto has been of novels, – *Shirley* and *The Newcomes*, and now *Westward Ho*.[5]

---

(1) *Life of Maxwell*: 217.

(2) George Wilson, *Researches on Colour-Blindness. With a Supplement on the Danger attending the Present System of Railway and Marine Coloured Signals* (Edinburgh, 1855).

(3) Maxwell's letters to Wilson of 4 January 1855 (Number 54) and of 27 July 1855 (Number 70) were appended to Wilson's *Researches on Colour-Blindness*: 153–9, 164–5.

(4) Siméon Denis Poisson, 'Mémoire sur la distribution de l'électricité a la surface des corps conducteurs', *Mémoires de la Classe des Sciences Mathématiques et Physiques de l'Institut de France*, (année 1811): 1–92, 163–274; Poisson, 'Mémoire sur la théorie du magnétisme', *Mémoires de l'Académie Royale des Sciences*, **5** (1826): 247–338; Poisson, 'Second mémoire sur la théorie du magnétisme', *ibid.*, **5** (1826): 488–533; and Poisson, 'Mémoire sur la théorie du magnétisme en mouvement', *ibid.*, **6** (1827): 441–570.

(5) By Charlotte Brontë (published in 1849), William Makepeace Thackeray (1853–5), and Charles Kingsley (1855).

FROM A LETTER TO JOHN CLERK MAXWELL

10 OCTOBER 1855

From Campbell and Garnett, *Life of Maxwell*[1]

Trin. Coll.
10 October 1855

Macmillan[2] proposes to get up a book of optics, with my assistance, and I feel inclined for the job. There is great bother in making a mathematical book, especially on a subject with which you are familiar, for in correcting it you do as you would to pups. – look if the principle and result is right, and forget to look out for small errors in the course of the work. However, I expect the work will be salutary, as involving hard work, and in the end much abuse from coaches and students, and certain no vain fame, except in Macmillan's puffs. But, if I have rightly conceived the plan of an educational book on optics, it will be very different in manner, though not in matter, from those now used.[3]

---

(1) *Life of Maxwell*: 217.
(2) Daniel Macmillan (*DNB*); see Number 40.
(3) Compare *Life of Maxwell*: 204n. See Number 124.

FROM A LETTER TO JOHN CLERK MAXWELL

17 OCTOBER 1855

From Campbell and Garnett, *Life of Maxwell*[1]

Trin. Coll.
17 October 1855

The lectures were settled last Friday. I am to do the upper division of the third year in hydrostatics and optics, and I have most of the exercising of the questionists.[2]

---

(1) *Life of Maxwell*: 219.

(2) The 'appellation of the students during the last six weeks of preparation' for the degree of Bachelor of Arts; *The Cambridge University Calendar for the Year 1850* (Cambridge, 1850): 11.

Maxwell concluded a letter of 19 October 1855 to his aunt Jane Cay (Peterhouse, Cambridge, Maxwell MSS (13)), which is concerned with the administration of property, with similar remarks on his Trinity College duties: 'I have to lecture on Hydrostatics on Monday and to set papers. I may have to lecture the working men too, if any of the other men have to leave off. I am not going to take pups except if old ones come for explanation of difficulties.' See also Number 77 and 82.

FROM A LETTER TO JOHN CLERK MAXWELL

25 OCTOBER 1855

From Campbell and Garnett, *Life of Maxwell*[1]

Trin. Coll.
25 October 1855

I have refused to take pupils this term, as I want to get some time for reading and doing private mathematics, and then I can bestow some time on the men who attend lectures.

I go in bad weather to an institution just opened for sports of all sorts – jumping, vaulting, etc. By a little exercise of the arms every day, one comes to enjoy one's breath, and to sleep much better than if one did nothing but walk on level roads.

---

(1) *Life of Maxwell*: 219.

FROM A LETTER TO JOHN CLERK MAXWELL

1 NOVEMBER 1855

From Campbell and Garnett, *Life of Maxwell*[1]

[Cambridge]
1 November 1855

I have been lecturing two weeks now, and the class seems improving, and they come up and ask questions, which is a good sign.

I have been making curves to show the relations of pressure and volume in gases, and they make the subject easier. I think I told you about the Ray Club.[2] I was elected an associate last Wednesday. ... We had a discussion and an essay by Pomeroy last Saturday about the position of the British nation in India,[3] and sought through ancient and modern history for instances of such a relation between two nations, but found none. We seem to be in the position of having undertaken the management of India at the most critical period, when all the old institutions and religions must break up, and yet it is by no means plain how new civilisation and self-government among people so different from us is to be introduced. One thing is clear, that if we neglect them, or turn them adrift again, or simply make money of them, then we must look to Spain and the Americans for our examples of wicked management and consequent ruin.

---

(1) *Life of Maxwell*: 219–20.

(2) See Number 67 note (4).

(3) A meeting of the Apostles essay club. Pomeroy was to enter the Indian Civil Service: see *Life of Maxwell*: 211, and Number 131 note (4).

FROM A LETTER TO JOHN CLERK MAXWELL

12 NOVEMBER 1855

From Campbell and Garnett, *Life of Maxwell*[1]

Trin. Coll.
12 November 1855

I attended Willis[2] on Mechanism to-day, and I think I will attend his course, which is about the parts of machinery. I was lecturing about the velocity of water escaping from a hole this morning. There was a great noise outside, and we looked out at a magnificent jet from a pipe which had gone wrong in the court. So that I was saved the trouble of making experiments.

I was talking to Willis about the platometer, and he thinks it will work.[3] Instead of toothed wheels to keep the spheres in position always, I think watch-spring bands would be better.

---

(1) *Life of Maxwell*: 220–1.

(2) Robert Willis, Jacksonian Professor of Natural and Experimental Philosophy at Cambridge, 1837–75 (Venn).

(3) See Numbers 55 and 72.

FROM A LETTER TO JOHN CLERK MAXWELL

25 NOVEMBER 1855

From Campbell and Garnett, *Life of Maxwell*[1]

Trin. Coll.
25 November 1855

I have been reading old books of optics, and find many things in them far better than what is new.[2] The foreign mathematicians are discovering for themselves methods which were well known at Cambridge in 1720 but are now forgotten.[3]

I have got a contrivance made for expounding instruments. It is a squared rod, one yard long, on which slide pieces, which will carry lenses. Each piece has a wedge which fixes it tight on the rod, and a saw-shaft, with holes through it, for fastening the pasteboard frame of the lens. By means of this I intend to set up all kinds of models of instruments.

---

(1) *Life of Maxwell*: 221; abridged.

(2) Maxwell is alluding to 'Cotes' theorem', as stated by Robert Smith, *A Compleat System of Optiks*, 2 vols. (Cambridge, 1738), **2**: ('Remarks') 76–8. See Number 91 note (9).

(3) See Number 93 note (5); but compare J. L. Lagrange, 'Sur la théorie des lunettes', *Nouveaux Mémoires de l'Académie Royale des Sciences de Berlin* (année 1778): 162–80, esp. 162, where 'le beau Théorème [de Cotes]' is discussed and compared with the work of Euler.

LETTER TO RICHARD BUCKLEY LITCHFIELD

28 NOVEMBER 1855

From the original in the library of Trinity College, Cambridge[1]

Trin. Coll. Nov 28th 1855.

Dear Litchfield

I was glad to get a letter from you, though it was due to Pomeroys[2] illness of which I have to make you more certain. It was about 4 weeks ago that the thing was coming on but it began to be troublesome about Thursday Nov 8. On Saturday he stayed in rooms & took to bed on Monday night and has been ill since. Humphrey seems to have a very good notion of his ailment when he calls it bilious fever, rather severe but not mixed up with anything else. He has been very weak & sometimes restless but for some time he has been sleeping most of his time and improving in appearance.

M[rs] Pomeroy came up last week & has now established herself in the room, and does everything for him. The nurse is still there but she is getting very sleepy & snores like a hedgehog not but what Pomeroy snores too.

Humphreys notion is that he is going through the regular thing & that we must expect to wait sometime, and that caution should be used to prevent relapses or excitements during recovery.

I have been up and about him to a certain extent but I have let him go on dozing as much as possible. Today he asked about various things & I told him of your enquiries which rather surprised him as he thought no one w[d] know.

M[rs] Pomeroy has not been writing letters as it would only bring down shoals & she has plenty to do.

I am busy with questionists[3] pretty regularly just now, slanging them one after another for the same things. As they have just set upon me for the evening I must stop now & get out some optical things to shew them.

I expect to go north straight after term ends as my father is expecting me.

I have heard nothing of the Cheltenham College.[4] Besant is Moderator[5] and only slowly recovering the use of the side of his face.

Yours

J. C. MAXWELL

---

(1) Trinity College, Cambridge, Add. MS Letters.c.1[81]. An extract is published in *Life of Maxwell*: 221.

(2) See Number 42 note (3).   (3) See Number 76 note (2).   (4) See Number 53.

(5) William Henry Besant, St John's 1846 (Venn), Moderator in the Mathematical Tripos 1856 (*Cambridge University Calendar for the Year 1856*: 117).

FROM A LETTER TO JOHN CLERK MAXWELL

3 DECEMBER 1855

From Campbell and Garnett, *Life of Maxwell*[1]

Trin. Coll.
3 December 1855

I had four questionist papers last week, as my subjects come thick there; so I am full of men looking over papers. I have also to get ready a paper on Faraday's Lines of Force for next Monday.[2]...

Maurice[3] was here from Friday to Monday, inspecting the working men's education. He was at Goodwin's[4] on Friday night, where we met him and the teachers of the Cambridge affair. He talked of the history of the foundation of the old colleges, and how they were mostly intended to counteract the monastic system, and allow of work and study without retirement from the world.

---

(1) *Life of Maxwell*: 221–2; abridged.

(2) Numbers 84, 85.

(3) Frederick Denison Maurice, Trinity 1823 (Venn), a major influence on Cambridge in the 1850s (*DNB*). See *Life of Maxwell*: 171n, 192, 194, 205, 218–19; and Number 42. See also Peter Allen, *The Cambridge Apostles. The Early Years* (Cambridge, 1978); Allen, 'F. D. Maurice and J. M. Ludlow, a reassessment of the leaders of Christian Socialism', *Victorian Studies*, **11** (1968): 461–82; and Sheldon Rothblatt, *The Revolution of the Dons. Cambridge and Society in Victorian England* (London, 1968): 143–51.

(4) Harvey Goodwin, Hulsean Lecturer 1855 (Venn).

THEORY OF THE MOTION OF AN IMPONDERABLE AND INCOMPRESSIBLE FLUID (DRAFT OF 'ON FARADAY'S LINES OF FORCE')[1]

AUTUMN 1855[2]

From the original in the University Library, Cambridge[3]

THEORY OF THE MOTION OF AN IMPONDERABLE AND INCOMPRESSIBLE FLUID THROUGH A RESISTING MEDIUM.[4]

(1)[5] The substance here treated of does not possess any of the properties of ordinary fluids except that of freedom of motion. It is not even a hypothetical fluid which is introduced to explain ⟨real⟩ actual phenomena. It is simply a collection of imaginary properties, which may be employed for establishing certain theorems in pure mathematics in a way more intelligible to ordinary minds than that in which algebraic symbols alone are used.[6] The use of the word 'Fluid' will not lead us into error if we remember that it simply denotes an imaginary substance with the following properties –

(1) The portion of fluid which at any instant occupied a given volume will in any succeeding instant occupy an equal volume.

(2) Any portion of the fluid moving through the resisting medium is opposed by a retarding force proportional to the velocity.

The first of these laws expresses the incompressibility of the fluid, and furnishes us with a convenient measure of its quantity, namely, its volume. The unit of quantity of the fluid will therefore be the unit of volume.

The second law determines the resistance of the medium to the motion of the fluid. Suppose the fluid moving uniformly parallel to a certain direction with a velocity V then if k represents the specific resistance of the medium, there will be a force kV acting on every unit of volume of the fluid in a direction contrary

---

(1) J. Clerk Maxwell, 'On Faraday's lines of force', *Trans. Camb. Phil. Soc.*, **10** (1856): 27–83, esp. 30–42 (= *Scientific Papers*, **1**: 160–75). Compare Numbers 61, 63 and 66.

(2) Compare Numbers 71 and 85.

(3) ULC Add. MSS 7655, V, c/6. Endorsed: 'Imaginary Fluids'.

(4) In 'On Faraday's lines of force' the material in this draft is revised and rearranged into two sections: 'I. Theory of the motion of an incompressible fluid', and 'II. Theory of the uniform motion of an imponderable incompressible fluid through a resisting medium'; see *Trans. Camb. Phil. Soc.*, **10** (1856): 30–33, 33–42 (= *Scientific Papers*, **1**: 160–63, 163–75). These are subdivided into numbered articles, I(1)–(9) and II(10)–(33), to which reference will be made.

(5) Compare I(1) and II(10) of 'On Faraday's lines of force': 30–1, 33–4 (= *Scientific Papers*, **1**: 160, 163–4).   (6) Compare Number 71.

to that of the motion. The effect of this force will be to make the pressure of the fluid on any plane perpendicular to the direction of motion greater than that on a second plane at unity of distance from the first by a quantity $= kV$ on every unit of surface.

More generally if V be the velocity of the fluid at any point, then a force kV due to resistance of the medium will act on the fluid contrary to the direction of motion. This force produces an increase of pressure $= kV$ for every unit of length measured backwards on the line of motion.

Hence if the pressure at every part of the fluid be known we may determine the motion, for if we find that the increase of pressure in passing from one point to another at the small distance h from it is $= p$ then we know that

$$p = -kVh$$

$$\&\ \therefore V = -\frac{p}{kh}$$

which gives the velocity of motion along any line when we know the pressure at different parts of it.

(2)[7] Now suppose the pressure at every point of the fluid to be known, then all the points at which the pressure is equal to a given pressure will be in a certain surface which will be a surface of equal pressure.[8] Suppose a series of such surfaces to be described corresponding to the pressures 0, 1, 2, 3 &c then the number of the surface will indicate the pressure belonging to it and the surface may be referred to as the surface 0, 1, 2 or 3.

If we draw any line in one of these surfaces, the pressure will be the same in every part of it. Hence the velocity of the fluid along this line vanishes, and the motion of the fluid must be entirely normal to the surface of equal pressure.

The lines of motion of the fluid[9] therefore cut the surfaces of equal pressure perpendicularly and the velocity along this line is

$$-\frac{1}{kh}$$

where h is the perpendicular distance between consecutive surfaces.

(3)[10] If we consider the motion of the fluid thro any one of these surfaces we shall see that since the quantity of the fluid is measured by its volume, the

---

(7) Compare II(11)–(12) of 'On Faraday's lines of force': 34 (= *Scientific Papers*, **1**: 164).

(8) Surfaces for which the velocity potential is constant, corresponding to equipotential and isothermal surfaces in the theory of electrostatics and thermal conduction; see Number 84.

(9) See article (3) and note (12).

(10) Compare I(3)–(4) and II(13)–(14) of 'On Faraday's lines of force': 31, 34–5 (= *Scientific Papers*, **1**: 160–1, 165).

quantity passing thro unit of area in unit of time must be measured by the velocity.[11]

If we draw on a surface of equal pressure a line enclosing an area equal to *kh* then unity of volume of fluid will pass through this area in unit of time.

If we now draw through every point of this line a line of motion, that is a line traced by a particle of the fluid in its actual motion,[12] then these lines will form a tubular surface along which the motion of the fluid will take place, no fluid escaping thro the sides because the sides be in the direction of motion.

Hence the amount of fluid which passes thro every section of this tube must be the same, namely unity of volume in unity of time. This must also be true of the sections of the tube formed by the other surfaces of equal pressure and therefore the area of any one of these sections must be represented by *kh* where *h* is the distance of consecutive surfaces at the point where the section is made. (4)[13] To extend this method of conceiving the motion to the whole mass of the fluid, we take any one of the surfaces of equal pressure, and draw upon it a series of curves according to any simple law. We then take any line cutting all these curves and draw another in such a manner that each of the spaces enclosed by these two lines and any two consecutive lines of the first series shall be equal to *kh* where *h* is the distance of two consecutive surfaces of equal pressure at that place. We then go on drawing the second series of curves according to this law and so divide the surface of pressure into little ⟨parallelograms⟩ quadrilaterals the area of which is

$$kh$$

h being the distance from that surface of pressure to the next.

We then draw lines of motion from every point of the first series of curves. These form a series of surfaces across which no motion takes place. By doing the same with respect to the second series of curves we form a second series of surfaces of the same kind which intersect the former and so form tubes in which the fluid travels. The section of each of these tubes made by the first surface of equal pressure is *kh* and therefore by what we have already proved, the section made by any other surface will also be *kh*.

(5) Let us now consider the effect of this construction. We have first a series of surfaces of equal pressure corresponding to the pressures 0, 1, 2 &c.

(11) Compare Number 61.

(12) In 'On Faraday's lines of force': 31 (= *Scientific Papers*, **1**: 160) Maxwell notes that the '*lines of fluid motion*', whose direction always indicates the direction of fluid motion', will 'represent the paths of individual particles of the fluid' (stream lines) only 'if the motion of the fluid be what is called *steady motion*'.

(13) Compare I(6) of 'On Faraday's lines of force': 32 (= *Scientific Papers*, **1**: 161–2).

Besides this we have two series of surfaces which intersect each other, forming tubes of fluid motion each tube delivering unity of volume of fluid in unit of time thro every section.

Each of these tubes is cut at right angles by the surfaces of equal pressure so that every one of these surfaces is divided into quadrilaterals thro each of which unity of volume passes in unity of time.

The whole space occupied by the fluid is therefore divided into small portions by these three systems of surfaces. Each of these is a right prism on the quadrilateral base the height of the prism measured perpendicularly to the surfaces of equal pressure being h and the area of the base $= kh$. The height of every prism is therefore proportional to the area.

In every one of these prisms or cells unity of volume of the fluid passes from a pressure p to a pressure $(p-1)$ in unit of time. The fluid passes through a distance h opposed by a resistance $\dfrac{1}{h}$ at the rate of unity of volume in unit of time. Hence unity of work is done by the fluid on the resisting medium in unit of time in each cell.

(6)[14] Every one of the tubes along which the fluid may be supposed to run delivers through any section unity of volume in unit of time. Hence the amount of fluid which passes across any finite surface in unit of time is measured by the number of tubes which pass through it, those being counted positive which cross the surface in a determined direction. Hence the total efflux from any closed surface will be zero, provided no fluid is produced or destroyed within that surface. In fact it follows from the incompressibility of the fluid that the quantity within the closed surface must remain the same and therefore whatever the influx may be, the efflux must be the same if there be no creation or destruction of the fluid.

(7)[15] As however the properties of the fluid are arbitrary and disposable provided they be consistent with each other we may suppose creation or destruction to go on at a given rate at any given part of the fluid mass.

The rate of production at any place will be measured by the volume produced in unit of time at that place. When the fluid is destroyed, this rate of production will of course be a negative quantity.

If the series of surfaces of equal pressure be given it is easy to determine the centres of production & destruction of the fluid. For by constructing the system of tubes as before we find the total efflux from any closed surface which must be equal to the total volume of fluid produced within it, and by making this closed

---

(14) Compare I(7)–(8) of 'On Faraday's lines of force': 32–3 (= *Scientific Papers*, **1**: 162–3).
(15) Compare II(14) of 'On Faraday's lines of force': 35 (= *Scientific Papers*, **1**: 165).

surface as small as we please we may determine the rate of production in any elementary portion of the fluid.

We have next to show, that if we know the rate of production at every part of the fluid, and also the pressure and the rate at which the fluid enters or departs from every part of the surface which bounds the part of the fluid under consideration, we may determine the pressure at every part of the fluid.

(8)[16] Suppose the pressure at every point of a certain fluid mass known and the velocities and centres of production deduced therefrom and suppose the pressures, velocities & centres of production at corresponding points of a second fluid mass also known, then if in a third fluid mass, the pressure at any point be the sum of the pressures at corresponding points in the two former masses the velocity will be the resultant of the velocities at these two corresponding points and the rate of production at any point will be compounded of the rates in the two former masses at corresponding points.

For the velocity resolved in any given direction is measured by the decrease of pressure in that direction, so that if two systems of pressures be combined, since the rate of decrease of pressure along any line will be the sum of the combined rates, the velocity in the new system resolved along that line will be the sum of the resolved parts in the old systems. The velocity in the new system will therefore be the resultant of the velocities at corresponding points in the two former systems.

The volume of fluid which passes through any given surface in the new system will be the sum of the volumes which pass through the corresponding surfaces in the old systems, because the rate at which the fluid passes thro the surface depends on the resolved part of the velocity normal to the surface, and these resolved parts are simply added together in the new system.

The volume of fluid produced in unit of time within any closed surface in the new system will be the sum of the corresponding quantities in the two old systems, and since the closed surface may be made as small as we please, till it encloses a single element of space, the rate of production at any point of the new system is the sum of the rates of production at corresponding points in the two old systems.

It is obvious that similar reasoning applies to the case in which the pressures in the new system are the differences of pressure in the old ones, regard being had to sign.

(9)[17] If in an infinite mass of fluid ⟨bounded by a closed surface⟩ there be no centres of production or of destruction and if the pressure at an infinite distance

---

(16) Compare I(9) and II(15) of 'On Faraday's lines of force': 33, 35 (= *Scientific Papers*, **1**: 163, 165–6).

(17) Compare II(16) of 'On Faraday's lines of force': 35 (= *Scientific Papers*, **1**: 166).

be zero the pressure at every other point must be ⟨zero⟩ constant and the fluid must be at rest.

For if any pressure exist in any part of space there will be motion from parts of greater pressure to parts of less pressure the lines of motion cutting the surfaces of equal pressure at right angles. Now such a line of motion must begin either at a centre of production or at infinity and terminate either at a centre of destruction or at infinity. But there are no centres of production or destruction, therefore the lines of motion must begin and end at infinity. But the pressure at infinity is everywhere zero, therefore there can be no motion from one point to another since the motion is always from greater pressure to less. Hence there can be no motion whatever and therefore no difference of pressures, and therefore the pressure is zero everywhere.

(10)[18] We are now able to prove the following proposition.

'If the distribution of centres of production & destruction throughout the whole of an infinite fluid mass be given, then only one distribution of pressures can exist in the fluid so as to make the pressure at infinity zero.'

For suppose two different distributions of pressure could exist a third distribution might be formed in which the pressure at any point would be the difference of pressures in the two former cases and the rate of production at any point would be the difference of rates in the two former cases. But since the rates of production are the same in the two former cases the rate of production at every point of the new distribution is zero. Hence by the last proposition since there is no production and since the pressure at infinity is zero the pressure at every point must be zero, and therefore the difference of pressures in the two systems must be zero at every point so that the two distributions must be the same.

(11)[19] Let us next determine the pressures at any point outside a given sphere, the centres of production being uniformly disposed in concentric shells within the sphere, the pressure and infinity being zero. Let the volume of fluid produced in unit of time by the centres within the sphere be P, then a volume of fluid equal to P must pass through every closed surface circumscribing the sphere in the same time.

Suppose this closed surface to be a sphere concentric with the given sphere and to have a radius $= r$, then since everything is symmetrical about the centre of the system, the surfaces of equal pressure will be spheres and the motion of the fluid across each spherical surface will be uniform. If h be the distance of

---

(18) Compare II(17) of 'On Faraday's lines of force': 35–6 (= *Scientific Papers*, **1**: 166–7).

(19) Compare II(18) of 'On Faraday's lines of force': 36 (= *Scientific Papers*, **1**: 167). See 'Section II – Particular Cases'.

two consecutive surfaces of equal pressure then the volume of fluid passing thro' unit of surface in unit of time will be

$$\frac{1}{kh}$$

and since the whole surface of the sphere whose radius $= r$ is

$$4\pi r^2$$

the whole fluid which escapes from the sphere in unit of time will be

$$4\pi \frac{r^2}{kh} = P \quad \text{or} \quad h = \frac{4\pi r^2}{Pk}$$

P being the whole quantity produced within the sphere.

Now if we assume $p = \dfrac{A}{r}$ then $r = \dfrac{A}{p}$.

& when
$$r = r_0 + h \qquad\qquad p = p_0 - 1$$
$$\therefore p + 1 = \frac{A}{r_0 - h} = \frac{A}{r_0}\left(1 + \frac{h}{r_0} + \&c\right).$$

But
$$p_0 = \frac{A}{r_0} \qquad \therefore 1 = \frac{Ah}{r_0^2} + \&c$$

or
$$h = \frac{r_0^2}{A} \text{ approximately.}$$

This equation agrees with that which we have already found if we make $A = \dfrac{Pk}{4\pi}$. The final solution of the question is therefore

$$p = \frac{Pk}{4\pi}\frac{1}{r}$$

which determines a distribution of pressure consistent with the distribution of centres of production & by what we have already proved, no other distribution of pressure can exist. The velocity at any point is in the direction of the radius and equal to $\dfrac{1}{kh} = \dfrac{P}{4\pi}\dfrac{1}{r^2}$.

These results for the space without the sphere are evidently independent of its radius.

(12)[20] It easily follows from the reasoning employed in (9) that if the pressure be uniform at all points of a closed surface and if there be no production of fluid within it, the pressure must be equal to that at the surface throughout and no motion can take place within the surface.

---

(20) Compare II(18) of 'On Faraday's lines of force': 36 ($=$ *Scientific Papers*, **1**: 167).

If therefore the centres of production be uniformly distributed in a spherical shell there must be a uniform pressure within for the pressure on the spherical surface bounding the interior of the shell is uniform by reason of symmetry and there is no production within the shell.

If the shell be infinitely thin with radius r then at the surface forming the shell

$$p = \frac{Pk}{4\pi}\frac{1}{r}.$$

This therefore is the constant value of the pressure within the shell.

(13) To find the pressure at any point of a sphere the centres of production being distributed uniformly through the sphere, so that each unit of volume of the sphere produces P volumes of fluid in unit of time.

Let the radius of the sphere be a then its volume will be $\frac{4}{3}\pi a^3$ and the rate of production $\frac{4}{3}\pi a^3 P$.

Therefore for the space outside the sphere

$$p = \frac{ka^3}{3}P\frac{1}{r}$$

and at the surface

$$p = \frac{ka^2}{3}P.$$

Now suppose the sphere divided into two parts by a spherical surface whose radius is r. We may easily find the pressures produced at this surface, first by the shell which surrounds it and then by the sphere within it.

The shell which surrounds it may be split up into an infinite number of infinitely thin shells producing at the rates $P_1 P_2 P_3$ &c respectively. These will produce at the point within them pressures $\dfrac{k}{4\pi}\dfrac{P_1}{r_1}$ $\dfrac{k}{4\pi}\dfrac{P_2}{r_2}$ respectively.

These pressures will be added together in the actual case so that the sum will be $\dfrac{k}{4\pi}\sum\left(\dfrac{P'}{r}\right)$ where $P' = 4\pi r^2\, \delta r P$ δr being the thickness of each shell. Now the sum of these

$$= kP\sum r\,\delta r = \frac{kP}{2}(a^2 - r^2).$$

The pressure due to the sphere (radius r) at its surface will be

$$\frac{kP}{3}r^2$$

so that the entire pressure at any distance r from the centre is expressed by

$$kP\left(\frac{a^2}{2} - \frac{r^2}{6}\right).$$

(14) [21] If any number of spheres producing respectively at the rates $P_1 P_2$ &c

---

(21) Compare II(19) of 'On Faraday's lines of force': 36 ($=$ *Scientific Papers*, **1**: 167–8).

coexist in the fluid mass, the resulting pressure will be found by adding the pressures which would be produced by each separately. Therefore at any point distant r_1 r_2 &c from the centres of these spheres respectively the resultant pressure is

$$\frac{k}{4\pi}\left(\frac{P_1}{r_1}+\frac{P_2}{r_2}+\&c\right).$$

By supposing these spheres to be reduced in size to any extent while their number is unlimited we may apply this result to the case of any arbitrary distribution of the centres of production and by art (10) the solution thus obtained must be the only possible one.

It appears therefore that when a system of centres of production (or destruction) is given the pressure at any point can be determined by the formula just given, and that when the pressure at every point is given, the centres of production can be found by the method of art: (7) or from the consideration that the rate of production of an element is measured by the number of tubes of fluid motion which originate in it.

(15)[22] To determine the number of cells into which any given space is divided by the tubes of motion and the surfaces of equal pressure.

Consider first a tube originating at a centre of production at the pressure p and terminating in a centre of destruction at the pressure p'. Then all the surfaces of equal pressure between p & p' will intersect the tube and divide it into $(p-p')$ cells. We may therefore conceive the centre of production to contribute p cells to the tube, and the centre of destruction to annihilate them. In fact if the centre of destruction had not existed the tube would have extended to the surface of no pressure and would have contained p cells, but the fluid was destroyed in passing the surface p' & so p' cells were taken away. We may therefore suppose every tube terminating at a pressure p to have p cells and every tube terminating at a pressure p' to have $-p'$ cells and then adding the results we shall obtain the true number of cells.

Now a centre producing at the rate P gives rise to P tubes & if the pressure at the centre be p each tube will (virtually) have p cells. Hence the whole number of cells due to the tube is Pp.

A centre destroying at the rate $-P'$ will form the termination of P' tubes and if the pressure be p' will prevent the formation of $P'p'$ cells, so that the cells due to the centre $-P'$ are $-P'p'$ in number. Hence the whole number of cells in the system is $\sum(Pp)$ where P is the rate of production, and is therefore negative for destruction, and p is the pressure under which the production takes place.

We might determine the number of cells in another way. The work done in

---

(22) Compare II(13) and (29) of 'On Faraday's lines of force': 34, 40–41 (= *Scientific Papers*, **1**: 165, 173).

producing P of fluid in unit of time under pressure p is Pp. Of this $P'p'$ is recovered if P' be destroyed under pressure p'. So that generally the work done in unit of time throughout the system is

$$\sum (Pp).$$

But it has been shown that unity of work is done in each cell therefore this result expresses also the number of cells.[23]

(16) Another expression of the total work done.

Let h be the distance of consecutive surfaces of equal pressure at any point, then the sectional area of the tubes at that point will be kh and the volume of a cell will be kh^2.

Therefore since unity of work is done in a cell whose volume is kh^2 the work done in unity of volume will be $\dfrac{1}{kh^2}$.

Now if v be the velocity at the same point $v = \dfrac{1}{kh}$ so that we have $\dfrac{1}{kh^2} = kv^2$. The quantity of work referred to unity of volume is therefore expressed by kv^2, & the total work done throughout the system is $k \sum (v^2)$ where the summation is taken with respect to every unit of volume.[24]

(17)[25] We have next to show that if we conceive any imaginary surface as fixed in space, and intersecting the moving mass of fluid, we may substitute for the fluid on one side of this surface, a distribution of centres of production over the surface without altering in any way the motion of the fluid on the other side of the surface.

For if we construct the tubes of fluid motion and produce them till they cut the given surface, and place a unit centre of production wherever a tube enters through the surface, and a unit centre of destruction wherever a tube passes out through the surface, then if we render the surfaces impermeable to the fluid these centres of production will replace the remainder of the fluid.

(18) Suppose the given surface to be that of one of the cells into which the fluid is divided, then since unity of volume of fluid enters at one end & escapes at the other in unit of time, we must replace these faces of the cell by a unit centre of production & a unit centre of destruction. When several of these cells are united end to end so as to form a tube, the centres of production & destruction will be united at the junction of every pair of cells so that they will all be neutralized except one centre of production at the origin of the tube and one centre of

---

(23) Compare II(30) of 'On Faraday's lines of force': 41 (= *Scientific Papers*, **1**: 173–4).

(24) Compare II(30) and Theorem III of 'On Faraday's lines of force': 41, 57–8 (= *Scientific Papers*, **1**: 173–4, 196); and see Number 51 note (14).

(25) Compare II(20) of 'On Faraday's lines of force': 36–7 (= *Scientific Papers*, **1**: 168).

destruction at the termination of the tube. The final distribution due to the whole system of cells will be precisely that with which we started.

(19)[26] If a system of centres of production &c be given, and the amount of work done by this system be determined, then any alteration in the motion of the fluid due to the arbitrary introduction of new centres of production &c will increase the amount of work done. For if the work due to the given system by itself be

$$\sum (Pp)$$

& that due to the new system be $\sum (P'p')$, the whole work will be

$$\sum (Pp) + \sum (P'p').$$

Now the work done by any system must be positive for the fluid always moves from places of greater pressure to places of less pressure, therefore the second term of this expression must be positive and therefore $\sum (Pp)$ is the minimum amount of work that can be done by the fluid under the given circumstances.

(20)[27] On the conditions to be satisfied at the bounding surface of two media having different coefficients of resistance.

It is evident that since no fluid is supposed to be produced or destroyed in passing the surface, the velocity normal to the surface will be the same on both sides of it.

If the coefficient for the first medium be k and for the second k' then in the first medium the velocity between two points distant h & differing by unity of pressure will be $\dfrac{1}{kh}$ while in the second it is $\dfrac{1}{k'h}$. If therefore we suppose these two points to lie on the bounding surface we shall have the tangential velocity in the two media inversely as their coefficients of resistance. The direction and velocity of motion of the fluid will therefore be changed in crossing the surface. The law of refraction of the lines of motion may be thus stated

(I) The directions of incidence & refraction lie in the same plane with the normal to the surface.

(II) The tangents of the angles which they make with the normal are inversely proportional to the coefficients of resistance.

(21) On the relations between the rate of production, the resistance & the pressure in an infinite mass of uniform medium.

The expression for the pressure at any point is

$$p = \frac{k}{4\pi}\left(\frac{P_1}{r_1} + \frac{P_2}{r_2} + \&\text{c}\right).$$

---

(26) Compare II(31)–(32) of 'On Faraday's lines of force': 41 (= *Scientific Papers*, **1**: 174).

(27) Compare II(22) of 'On Faraday's lines of force': 37 (= *Scientific Papers*, **1**: 168).

If therefore the rate of production at every centre of production be changed in a given ratio p will change in the same ratio, and if k be varied p will vary with it also, p therefore varies as k & P conjointly.

(22)[28] Let the space within a given closed surface be filled with a medium different from that exterior to it, and let the pressures at any point of this compound system due to a given distribution of centres of production & destruction be known, it is required to determine a distribution of centres which would produce the same pressures in a uniform medium whose coefft of resistance $= 1$.

Let the coefficient of resistance of the exterior medium $= k$ & that of the interior medium $= k'$. Let the external centres of production be $\sum P$ & the internal $\sum (P')$. Now suppose the bounding surface to be made impermeable to the fluid, then in order to maintain the motion of the fluid as before we must place a unit centre of production wherever a tube enters the medium and a unit centre of destruction on the other side of the surface where the tube leaves the other medium. The production at one side of the surface and the destruction at the other are therefore each equal to the amount of fluid which passes through the surface.

Now if we change the resistance of the exterior medium from k to 1 we must change the rate of production at every centre from P to kP in order to maintain the same pressure as before at every point and this must be done not only for the given centres within the medium but for the centres distributed over the surface to account for the fluid that passes through it.

If we next change the resistance of the interior medium from k' to 1 and the rate of production at any point from P' to $k'P'$ the pressure at every point within the surface as well as without will be the same as it was at first. We have therefore the same distribution of pressure as at first, but in a uniform medium. The rate of production of every centre is multiplied by the coefficient of resistance of the medium in which it is and this is true of those distributed over the two sides of the bounding surface. At any point of this surface where one of the unit tubes originally crossed it from the exterior medium to the interior we placed a centre of destruction in the exterior medium and a centre of production in the interior. In the new case, the exterior centre must be multiplied by k & the interior by k' so that the result is, production at the rate of $(k'-k)$. So that to obtain the complete equivalent to the original case we must multiply the rate of production at every centre by k in the exterior medium & by k' in the interior, and also multiply the rate of influx of the original fluid through the bounding surface by $(k'-k)$ to obtain the rate of production at the surface

---

(28) Compare II(23) of 'On Faraday's lines of force': 37 ($=$ *Scientific Papers*, **1**: 168–9).

itself. The whole efflux from the interior medium was originally $= \sum (P')$ so that the whole distribution over the surface is now equal to $(k-k') \sum (P')$ and the efflux from within the surface is now $k' \sum (P')$ so that the total efflux from the interior and the surface together is $k \sum (P')$.

(23) By supposing the coefft of resistance both outside & inside the surface equal to unity while the pressures were maintained as before, we have had to conceive fluid to be actually produced whenever it passes through the bounding surface in one direction & destroyed when it passes in the other direction. But since we have taken the volume of the fluid as the measure of its quantity this apparent creation and destruction may be considered due to a change of density only. Let us try the effect of this supposition making the density ρ & ρ' in the two media while the coefficient of resistance in both is unity. Let Π be the rate of production of the fluid measured by the mass produced in unit of time then

$$\Pi = \rho P.$$

In the original system we have

$$p = \frac{k}{4\pi} \sum \left(\frac{P}{R} \right).$$

In the new

$$p = \frac{1}{4\pi} \sum \left(\frac{\frac{\Pi}{\rho}}{r} \right).$$

These expressions will be identical if we make $\rho = \dfrac{1}{k}$. We may therefore obtain the same distribution of pressures in the new case by supposing the fluid to change its density in crossing the surface from $\dfrac{1}{k}$ to $\dfrac{1}{k'}$. The motion of the fluid measured by the mass will be the same as in the original case, but the velocities will be those in art (22).

(24) Let us now consider the case in art (22) when the second medium is enclosed in a very small space and when it is not very different from the surrounding medium in resisting power. The lines of motion will not be far removed from the position they would have occupied if the second medium had not existed and therefore the apparent effect will be to distribute centres of production & destruction so as to cause a polarity in the direction of motion if the second medium resist more than the first and in the reverse direction if it resist less. This apparent polarity of the second medium is due entirely to the difference in resisting power between it & the first and it would be better to attribute polarity to each medium separately and consider the apparent polarity as due to the difference of the two.

(25) The total amount of work done in the system is always

$$\sum (Pp)$$

but when there are different media the calculation of p becomes complex. In this case the equivalent expression $\sum (kv^2)$ may be of use. For in the case of a small portion of the second medium, the work done in it exceeds that which would be done if it were away by $V(k'-k)\, v^2$ where V is the volume of the portion & v is the velocity at the given point.

(26)[29] When $k = 0$ in the second medium it opposes no resistance to the motion of the fluid and therefore the pressure must be uniform throughout. The bounding surface must therefore be a surface of equal pressure, and the total efflux through this surface must be equal to the quantity of fluid produced within it. If by assuming any arbitrary system of centres of production within the second medium in addition to the given centres in the first medium, we can fulfil these conditions the system of pressures so found for the first medium, together with the constant pressure in the second will be the true and only distribution of pressures in the actual case. For if two such distributions of pressure could be found, then by taking the difference of the two as in art ( )[30] we should find in the second medium, a bounding surface of uniform pressure & as much fluid produced as destroyed within it and therefore no flow across it, and in the outer medium, no flow across the bounding surface & no centres of production or destruction & therefore a uniform pressure & therefore no pressure, since the pressure at infinity is supposed to be zero. Therefore the difference of the two distributions of pressure is zero, that is they are alike.

(27)[31] When $k = \infty$ in the second medium there can be no passage of fluid through it or into it. The bounding surface may therefore be considered impenetrable,[32] the tubes of fluid motion will run along it without cutting it, & the surface of equal pressure will meet it at right angles.

If by assuming any arbitrary distribution of centres within the bounding surface in addition to the given centres in the first medium and by calculating the resulting pressures as in the case of a uniform medium, we can fulfil these conditions, this system of pressures will be the actual one in the first medium.

The proof is the same as before.

---

(29) Compare II(24) of 'On Faraday's lines of force': 38 (= *Scientific Papers*, **1**: 169–70).

(30) Article (15).

(31) Compare II(25) of 'On Faraday's lines of force': 38 (= *Scientific Papers*, **1**: 170).

(32) See articles (2) and (3).

Section II – Particular Cases

(1) Let the centres of production be arranged in concentric shells within a small sphere, then we know by a former demonstration[33] that if the medium outside the sphere be uniform the pressure at any point will be

$$p = \frac{k}{4\pi} \frac{P}{r} + p_0$$

where k is the coeff$^\mathrm{t}$ of resistance, P the rate of production, r the distance of the given point from the centre of the sphere, and p_0 a constant pressure depending on the circumstances of the problem. If it be known that the medium is infinite, and that $p = 0$ at infinity then $p_0 = 0$

$$\& \; p = \frac{k}{4\pi} \frac{P}{r}.$$

(2) Let there be several media arranged in spherical shells concentric with the centres of production and let the outermost extending from radius a_1 to infinity have the coefficient k_1, & let the coefficient be k_2 from a_1 to a_2 & so on then we shall have

$$p = \frac{k_1}{4\pi} \frac{P}{r} \text{ from } \infty \text{ to } a_1$$

$$p = \frac{k_2}{4\pi} \frac{P}{r} + p_0 \text{ from } a_1 \text{ to } a_2.$$

Putting $r = a_1$ we obtain for the pressure at the bounding surface

$$p_1 = \frac{k_1}{4\pi} \frac{P}{a_1} \text{ in the first medium}$$

$$p_1 = \frac{k_2}{4\pi} \frac{P}{a_1} + p_0 \text{ in the second}$$

and the values of p_1 must be equal. Hence

$$p_0 = \frac{P}{4\pi a_1} (k_1 - k_2).$$

Therefore in the second medium

$$p = \frac{P}{4\pi} \left\{ \frac{k_2}{r} + \frac{k_1 - k_2}{a_1} \right\}.$$

In the third

$$p = \frac{P}{4\pi} \left\{ \frac{k_3}{r} + \frac{k_2 - k_3}{a_2} + \frac{k_1 - k_2}{a_1} \right\}.$$

---

(33) See article (11).

If the third medium be the same as the first, $k_3 = k_1$

$$\& \; p = \frac{P}{4\pi}\left\{\frac{k_1}{r} + (k_2 - k_1)\left(\frac{1}{a_2} - \frac{1}{a_1}\right)\right\}.$$

This is the solution of the case where a hollow shell of the second medium is made to surround the centre of production.

(3) Let there be two centres of production or destruction, then the general expression for the pressure is

$$p = \frac{1}{4\pi}\left(\frac{P}{r} + \frac{P'}{r'}\right) + p_0.$$

If P' be negative then there will be a surface for which

$$\frac{r}{r'} = -\frac{P}{P'}$$

and it is easy to show that this surface must be a sphere.

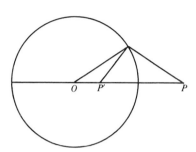

Figure 83,1

For let c be the distance of the centres. Take the line PP' as axis of x & P as origin then

$$r^2 = x^2 + y^2 + z^2 \qquad r'^2 = (x-c)^2 + y^2 + z^2$$

and at the surface required

$$\frac{r^2}{P^2} = \frac{r'^2}{P'^2}$$

$$P'^2(x^2 + y^2 + z^2) = P^2((x-c)^2 + y^2 + z^2)$$

$$(P'^2 - P^2)(x^2 + y^2 + z^2) = P^2(-2cx + c^2)$$

$$\left(x - \frac{P^2}{P^2 - P'^2}c\right)^2 + y^2 + z^2 = \frac{P^2}{P^2 - P'^2}c^2$$

the equation to a sphere of which the radius is $\sqrt{\dfrac{P^2}{P^2 - P'^2}}\,c$ and the distance

from $P = \dfrac{P^2}{P^2 - P'^2}c$ measured in the direction PP' produced.

84

ABSTRACT OF PAPER 'ON FARADAY'S LINES OF FORCE'[1]

DECEMBER 1855[2]

From the original in the University Library, Cambridge[3]

ON FARADAY'S LINES OF FORCE[4]

In studying the Electrical Researches of Professor Faraday[5] every one must have observed, that the conceptions by means of which he interprets the complicated phenomena of electricity and magnetism are not those which are usually employed by other physicists and mathematicians. Instead of directing his attention to the points or centres through which the attractive or repulsive forces are directed and considering the attraction at any other point merely with reference to those centres of force,[6] Faraday treats the distribution of forces in space as the primary phenomenon, and does not insist on any theory as to the nature of the centres of force round which these forces are generally but not always grouped.[7] Although this method of treating the subject has been successfully employed by Faraday himself in his investigations of discovery and has been shown by Professor William Thomson and others to coincide mathematically with the laws of attraction which have been established for certain classes of electrical phenomena[8] an impression still prevails

---

(1) Endorsed: 'On Faraday's Lines of Force. Abstract'. Compare Maxwell's reference to an 'abstract' in his letter to Thomson of 14 February 1856 (Number 91). The manuscript is a summary of Part I of 'On Faraday's lines of force', *Trans. Camb. Phil. Soc.*, **10** (1856): 27–83, esp. 27–51 (= *Scientific Papers*, **1**: 155–88).

(2) Part I of 'On Faraday's lines of force' was read to the Cambridge Philosophical Society on 10 December 1855 (see Number 85).   (3) ULC Add. MSS 7655, V, c/7.

(4) The manuscript is annotated in pencil by William Thomson (see Number 91).

(5) Michael Faraday, *Experimental Researches in Electricity*, 3 vols. (London, 1839–55).

(6) Compare the work of Ampère and Weber; see Number 66 notes (3) and (5).

(7) Michael Faraday, 'Experimental researches in electricity – Twenty-eighth series. On lines of magnetic force', *Phil. Trans.*, **142** (1852): 25–56; Faraday, 'On the physical character of the lines of magnetic force', *Phil. Mag.*, ser. 4, **3** (1852): 401–28 (= *Electricity*, **3**: 328–70, 407–37). Cited by Maxwell, 'On Faraday's lines of force': 29n (= *Scientific Papers*, **1**: 157n).

(8) William Thomson, 'On the mathematical theory of electricity in equilibrium. I. On the elementary laws of statical electricity', *Camb. & Dubl. Math. J.*, **1** (1845): 75–95; reprinted (with additional notes) in *Phil. Mag.*, ser. 4, **8** (1854): 42–62 (= *Electrostatics and Magnetism*: 15–37). The paper is a translation (with considerable additions) of Thomson's paper, prepared at the request of Joseph Liouville, 'Note sur les lois élémentaires de l'électricité statique', *Journal de Mathématiques Pures et Appliquées*, **10** (1845): 209–21. See Ole Knudsen, 'Mathematics and reality in William Thomson's electromagnetic theory', in *Wranglers and Physicists*, ed. P. M. Harman (Manchester,

that there is something vague and unmathematical about the idea of lines of force, and that they will inevitably lead astray any student who is not guided by that familiar acquaintance with the laws of nature which distinguishes the great electrical philosopher. To illustrate the mathematical character and scientific usefulness of lines of force I have developed certain of their properties in a series of propositions which I have proved in full without the introduction of technically mathematical expressions.[9]

The results which I arrive at are not new but I believe that the method by which they are obtained is capable of more extensive applications than those here given. In a future paper I intend to exhibit its application to certain other ideas of Faradays with respect to the mutual action of electrical currents.[10]

The phenomena of statical electricity and magnetism as studied by Coulomb,[11] Poisson[12] Green[13] &c and those of dielectrics, paramagnetics & diamagnetics and crystalline magnetic polarity as investigated by Faraday[14] Plucker[15] and others,[16] have evidently much similarity as to

1985): 149–79, esp. 150–7. In referring to 'others' Maxwell may have been alluding to Green and Gauss; their theorems on the potential function established the basis for Thomson's analogy between Faraday's lines of force and heat flow. See Number 51 note (13); and compare Thomson's reference to the work of Green, Gauss and Chasles in Note III of 'On the elementary laws of statical electricity': 94.   (9) Compare Number 71.

(10) See 'On Faraday's "Electro-tonic state"', Part II of 'On Faraday's lines of force': 51–83 (= *Scientific Papers*, 1: 188–229); and see Number 87.

(11) Charles Augustin Coulomb, 'Sur l'électricité et le magnétisme', *Mémoires de l'Académie Royale des Sciences* (année 1785): 569–77, 578–611, 616–38; (année 1786): 67–77; (année 1787): 421–67; (année 1788): 617–705; (année 1789): 455–505.

(12) S. D. Poisson, 'Mémoire sur la distribution de l'électricité à la surface des corps conducteurs', *Mémoires de la Classes des Sciences Mathématiques et Physiques de l'Institut de France* (année 1811): 1–92, 163–274; Poisson, 'Mémoire sur la théorie du magnétisme', *Mémoires de l'Académie Royale des Sciences*, 5 (1826): 247–338; Poisson, 'Second mémoire sur la théorie du magnétisme', *ibid.*, 5 (1826): 488–533; and Poisson, 'Mémoire sur la théorie du magnétisme en mouvement', *ibid.*, 6 (1827): 441–570.

(13) George Green, *An Essay on the Application of Mathematical Analysis to the Theories of Electricity and Magnetism* (Nottingham, 1828).

(14) Michael Faraday, 'Experimental researches in electricity. – Eleventh series. On induction', *Phil. Trans.*, 128 (1838): 1–40; 'Twenty-sixth series. Magnetic conducting power', *ibid.*, 141 (1851): 29–84; 'Twenty-second series. On the crystalline polarity of bismuth (and other bodies), and on its relation to the magnetic form of force', *ibid.*, 139 (1849): 1–41 (= *Electricity*, 1: 360–416; 3: 200–73, 83–136). On the terms 'paramagnetism' and 'diamagnetism' see note (37).

(15) Julius Plücker, 'Ueber die Abstossung der optischen Axen der Krystalle durch die Pole der Magnete', *Ann. Phys.*, 72 (1847): 315–43, (trans.) 'On the repulsion of the optic axes of crystals by the poles of a magnet', *Scientific Memoirs*, ed. R. Taylor, 5 (London, 1852): 353–75; Plücker, 'Ueber das Verhältnis zwischen Magnetismus und Diamagnetismus', *Ann. Phys.*, 72 (1847): 343–52,

their mathematical expression.[17] If we would abandon physical theory, and use only mathematical formulæ the methods would be identical. But though we might gain in generality of expression by this method we should lose those distinct conceptions which a physical theory presents to the mind and the general laws of the science would be put out of the reach of any but professed mathematicians. There is, however, one method which combines the advantages, while it gets rid of the disadvantages both of premature physical theories and technical mathematical formulæ. I mean the method of Physical Analogy.[18] Of this we have instances in the substitution of numbers for quantities in all calculations, in the use of lines in mechanics to represent forces and velocities, in the partial analogy between the motion of light and that of a particle, and the more complete analogy between the motion of light and that of a vibration in an elastic medium.[19]

The analogy between the formulæ of attractions which vary inversely as the square of the distance and those for the uniform conduction of heat has been pointed out by Professor Thomson (Cam. & Dublin[a] Math. Journ.

(a) {Thomson} ⟨Dublin⟩.

(trans.) 'On the relation of magnetism to diamagnetism', *Scientific Memoirs*, **5**: 376–82; Plücker, 'Ueber Intensitätsbestimmung der magnetischen und diamagnetischen Kräfte', *Ann. Phys.*, **74** (1848): 321–79, (trans.) 'On the determination of the intensity of magnetic and diamagnetic forces', *Scientific Memoirs*, **5**: 713–60; Plücker, 'Ueber die Theorie des Diamagnetismus, die Erklärung des Ueberganges magnetischen Verhaltens in diamagnetische und magnetische Begründung der bei Krystallen beobachten Erscheinungen', *Ann. Phys.*, **86** (1852): 1–34, (trans.) 'On the theory of diamagnetism, the explanation of the transition from the magnetic to the diamagnetic deportment, and the mathematical treatment of the phenomena observed in crystals', *Scientific Memoirs, Natural Philosophy*, ed. J. Tyndall and W. Francis (London, 1853): 332–58. Plücker's experiments on the effects of magnetism on crystals had prompted Faraday's theory of the alignment (axiality) of crystals along lines of magnetic force: see note (14) and Number 71 note (21). Faraday and Plücker had met and corresponded, and Faraday had arranged the translation of Plücker's papers; see *The Selected Correspondence of Michael Faraday*, ed. L. P. Williams, 2 vols. (Cambridge, 1971), **1**: 511–14, 519–21, 524–7, 536–7; **2**: 541–3, 545–8, 550–1, 561–2, 566–8, 574–8.

(16)　Recent papers by Weber and Tyndall; see note (40).

(17)　See William Thomson, 'On the mathematical theory of electricity in equilibrium', *Camb. & Dubl. Math. J.*, **1** (1846): 75–95; **3** (1848): 131–40, 141–8, 266–74; **4** (1849): 276–84; **5** (1850): 1–9; and Thomson, 'A mathematical theory of magnetism', *Phil. Trans.*, **141** (1851): 243–85 (= *Electrostatics and Magnetism*: 15–37, 42–85, and 341–404).

(18)　See Number 88.

(19)　Compare Maxwell, 'On Faraday's lines of force': 27–30 (= *Scientific Papers*, **1**: 155–9).

Vol. III)[b][20] and method of Images, applied by Professor Thomson to the theory of electrified spheres,[21] has been used by Professor Stokes in the case of spheres moving in an incompressible fluid.[22] In fact the mathematical expressions which indicate the laws of attraction are so connected with the laws of the uniform conduction of heat, that if we know the complete solution of any problem in the conduction of heat we have only to substitute centres of attraction for flow of heat at any point, and potential for temperature, and we have the complete solution of a problem in attractions.[23]

Now the laws of the conduction of heat are derived from the supposition that heat is communicated between adjacent parts of the medium with a velocity depending on their difference of temperature, while those of attraction suppose that the action is between bodies at a distance. Here, therefore, we have a mathematical analogy between two sets of phenomena confessedly very different and yet if we bear this difference in mind we may reason with perfect security from one to the other. We should also remember that while the mathematical laws of the conduction of heat derived from the idea of heat as a substance are admitted to be true, the theory of heat has been so modified that we can no longer apply to it the name of substance.[24] It is for this reason that in choosing a concrete form in which to develope the analogy of attraction I have not taken that of heat, as pointed out by Professor Thomson, but at once assumed a purely imaginary fluid as the vehicle of mathematical reasoning.[25]

In this way I have endeavoured to make it plain that I am not attempting to establish any physical theory of a science in which I have not made a single

---

(b) {Thomson} Feby 1842.

(20) [William Thomson,] 'On the uniform motion of heat in homogeneous solid bodies, and its connection with the mathematical theory of electricity', *Camb. Math. J.*, **3** (1842): 71–84; reprinted in *Phil. Mag.*, ser. 4, **7** (1854): 502–15 (= *Electrostatics and Magnetism*: 1–14). See Number 51 note (3).

(21) William Thomson, 'On the mathematical theory of electricity in equilibrium [Parts III to VI]', *Camb. & Dubl. Math. J.*, **3** (1848): 141–8, 266–74; **4** (1849): 276–84; **5** (1850): 1–9 (= *Electrostatics and Magnetism*: 52–85). See Number 71 note (18).

(22) G. G. Stokes, 'On the resistance of a fluid to two oscillating spheres', *Report of the Seventeenth Meeting of the British Association for the Advancement of Science* (London, 1848), part 2: 6; and Stokes, 'On the effect of the internal friction of fluids on the motion of pendulums', *Trans. Camb. Phil. Soc.*, **9** part 2 (1850): 8–106 (= *Papers*, **1**: 230, **3**: 1–141). See also a letter from Thomson to Stokes of 1 February 1848 (ULC Add. MSS 7656, K 23); and Thomson's paper 'On a system of magnetic curves', *Camb. & Dubl. Math. J.*, **2** (1847): 240 (= *Math. & Phys. Papers*, **1**: 81–2).

(23) [William Thomson,] 'On the uniform motion of heat in homogeneous solid bodies, and its connection with the mathematical theory of electricity'; and Thomson, 'On the mathematical theory of electricity in equilibrium. I. On the elementary laws of statical electricity'.

(24) Compare Maxwell, *Treatise*, **1**: 74 (§72).

(25) Compare Maxwell's letter to Thomson of 13 September 1855 (Number 71).

experiment worthy of the name, and that the limit of my design is to show how by a strict application of the ideas and methods of Faraday to the motion of an imaginary fluid, everything relating to that motion may be distinctly represented, and thence to deduce the theory of the attractions of electric and magnetic bodies and of the conduction of electric currents. The theory of electromagnetism including the induction of electric currents, which I have deduced mathematically from certain ideas due to Faraday, I reserve for a future communication. [26]

The following is an abstract of the propositions relating to fluid motion and lines of force.

I
Theory of the steady motion of an incompressible fluid [27]

Definitions – (1). A line of fluid motion is a line so drawn in the fluid that its direction at any point always corresponds with that of fluid motion at that point. [28]

Cor. If a system of such lines be drawn so as to generate a surface in the fluid, then, since the direction of fluid motion always corresponds to that of the generating lines, no fluid can pass through the surface.

Def. (2). If lines of fluid motion be drawn through every point of a given closed curve, the tubular surface so generated is called a tube of fluid motion.

Cor. Since no fluid can pass through the sides of this tube the quantity of fluid which flows through every section will be the same.

Def. (3). When unity of volume of the fluid flows through any section of a tube in unit of time, that tube is called a unit tube.

Scholium – The units of length, time, pressure &c being arbitrary may be taken as small or large as we please so that by properly selecting our units we may always speak of the number of units as if every quantity could be measured by an exact number of [c] units.

Problem I. To define the motion of the fluid by means of a system of unit tubes.

This is done by drawing on any surface cutting the lines of motion two systems of curves which by their intersection form a system of quadrilaterals through each of which unity of fluid passes in unity of time. The unit tubes of

---

(c) {Thomson} such.

(26) See note (10).

(27) Compare 'On Faraday's lines of force': 30–3 (= *Scientific Papers*, **1**: 160–3); and see Number 83.

(28) See Number 83 article (3) and note (12).

which these quadrilaterals are the bases will define the direction and velocity of the motion of the fluid at any point.

Cor. These tubes are either reentrant or have terminations at which the fluid enters and escapes. In the latter case when the fluid enters and leaves the system of tubes under consideration we must suppose it created and annihilated at these points.

Def. (4). The origin of a tube where the fluid is supposed to be created is called a source, and the end of a tube where it is supposed to be annihilated is called a sink. Both however will sometimes be called sources, sources being understood to mean sinks when they have a negative sign.

Def. (5). The rate of production of a source is the volume of fluid created by it in unit of time. This expression is also to be understood negatively of sinks.

Def. (6). When the rate of production is unity the source is called a unit source.

Cor. (1). The origin of every unit tube is a unit source and its end is a unit sink.

Cor. (2). The difference between the number of unit tubes which pass out of a given closed surface and that of those which pass in is equal to the difference of the numbers of unit sources and of unit sinks within the surface.

II
Theory of the uniform motion of an imponderable & incompressible fluid through a resisting medium[29]

Law of resistance – Any portion of the fluid moving through the resisting medium is opposed by a force proportional to the velocity.

The resistance on unity of volume is therefore kv, and this resistance renders necessary a difference of pressure in the different parts of the fluid in order to keep up the motion.

A series of surfaces of equal pressure is drawn in the fluid the difference of pressure between consecutive surfaces being unity.

Having determined the surfaces of equal pressure we can find the direction and velocity of the fluid at any point and so construct a system of unit tubes and mark out the positions of the sources and sinks.

Theorem I. If we know the distribution of pressures and of sources in two different cases, then if in a third case the pressure at any point be the sum or difference of the pressures at corresponding points in the first or second, the

(29) Compare 'On Faraday's lines of force': 33–42 (= *Scientific Papers*, **1**: 163–75); and see Number 83.

distribution of sources in the third case will be the algebraical sum or difference of the distributions in the first & second.

Theorem II. If the pressure at every point of a closed surface be constant, and if no sources exist within it, then the pressure throughout will be the same as that at the surface.

Theorem III. If the distribution of sources within a closed surface be given and the pressures at every point of the surface be known then only one distribution of pressures within the surface can exist. For by Theorem I if two distributions existed we could take their difference which would reduce the case to that of Theorem II.

Theorem IV. The pressure at any point of a medium whose resistance is measured by k due to a source S at a distance r from the point is

$$p = \frac{kS}{4\pi r}. \quad (30)$$

By means of this expression the pressure due to any distribution of sources may be calculated.

Cor. By increasing S in the ratio of k to 1 we may calculate p on the supposition that $k = 1$ without altering the result.

Theorem V. If we conceive an imaginary surface fixed in space and intersecting the lines of motion, we may substitute for the fluid on one side of this surface a distribution of sources on the other side and at the same time render the surface impermeable to the fluid without altering the motion of the remaining fluid.

Theorem VI. – If we know the distribution of pressures at every point of two media of different resisting power this distribution may be reproduced in a medium whose resisting power is unity.

For by Theorem V we may make the bounding surface impermeable with a distribution of sources upon each side of it, and then by Theorem IV we can increase all the sources in each medium in the ratio of the resisting power in that medium to unity and thus obtain a distribution of sources in the first and second media and on the bounding surface which would produce the same pressures if the resistance were everywhere unity.

Cor. (1). If the second medium offers no resistance to the fluid, its bounding surface is a surface of equal pressure.

Cor. (2). If the resistance in the second medium be infinite the lines of fluid motion coincide wholly or partially with its bounding surface.

Methods are given for calculating the resultant pressures in various cases by

---

(30) S is the rate of production by each source.

assuming imaginary distributions of sources in a medium of which the resisting power is unity.

If the resisting power in the given medium varies from point to point then there will be a continuous distribution of imaginary sources through the medium, sources being placed where the fluid passes from places of less to places of greater resistance and sinks when it passes in the opposite direction.

Hitherto we have supposed the resistance of the medium at any point to be the same in whatever direction the fluid passes, but we may conceive a case in which the resistance is different in different directions. In such a case the lines of motion will not in general be perpendicular to the surfaces of equal pressure. There are however three directions at right angles to each other at every point of the medium in which this will be the case and if we refer the points of the medium to coordinate axes parallel to these axes of resistance we shall have

$$\frac{dp}{dx} = k_1 u \qquad \frac{dp}{dy} = k_2 v \qquad \frac{dp}{dz} = k_3 w$$

$k_1 \, k_2 \, k_3$ being three coefficients of resistance and $u \, v \, w$ the resolved parts of the velocity in these three directions.[31]

The methods of theorems I II III may be applied to this and all other cases of resisting media. The following general propositions are true of all cases.

Theorem VII. The surfaces of equal pressure divide the unit tubes into unit cells in each of which unity of fluid passes from pressure p to $(p-1)$ in unit of time and therefore does unity of work in overcoming resistance.

Cor. The whole number of unit cells measures the whole work done by the fluid.[32]

Theorem VIII. The whole number of cells in the fluid is $\sum (Sp)$ where S is the rate of production of each source & p the pressure under which the fluid is produced.

Theorem IX. In a uniform medium if $\sum (S)$, $\sum (S')$ be two systems of sources & p, p' the pressures due to each respectively at any point then

$$\sum (Sp') = \sum (S'p).$$

---

(31) Compare 'On Faraday's lines of force': 39–40 (= *Scientific Papers*, **1**: 171–2), where Maxwell states three '*equations of conduction*' which relate the velocity and the resistance of the medium at a point. These equations are derived from a paper by Stokes on the conduction of heat in crystals, where the flux of heat is expressed in terms of variations of temperature along rectangular axes in a solid; see G. G. Stokes, 'On the conduction of heat in crystals', *Camb. & Dubl. Math. J.*, **6** (1851): 215–38 (= *Papers*, **3**: 203–27). Stokes' method had been discussed in a paper by Thomson, 'On the dynamical theory of heat. Part V. Thermo-electric currents', *Trans. Roy. Soc. Edinb.*, **21** (1854): 123–71, esp. 164–7 (= *Phil. Mag.*, ser. 4, **11** (1856): 214–25, 281–97, 379–88, 433–46; and *Math. & Phys. Papers*, **1**: 232–91).

(32) Compare 'On Faraday's lines of force': 40–1 (= *Scientific Papers*, **1**: 173–4); and see Number 51 note (14).

Theorem X. If a lamina of a uniform medium be bounded by impermeable surfaces at a distance k then the length and breadth of the unit cells will be equal and the lines of motion will be reciprocal to those of equal pressure so that they may be interchanged.

(see Cambridge & Dublin Math. Journ. Vol. III p. 286)[33]

III
Application of the Idea of Lines of Force[34]

(1) *Statical Electricity*

The application of the theory of Potentials to statical electricity is well known. Now it appears from Theorem IV that the expression for the pressure due to a source S in a medium of resistance k is identical with that for the potential due to a quantity e of electricity,[35] provided

$$S = \frac{4\pi e}{k}.$$

Hence all problems in electricity can be represented by the motion of the fluid by putting sources instead of the supposed electric matter and calculating the resultant force in any direction from the rate of decrease of pressure in that direction.

Theory of Dielectrics

The distribution of electricity on conductors is influenced not only by their form & position but by the nature of the intervening medium. According to the theory of Faraday on this subject some substances conduct the lines of inductive action more readily than others.[36] In our imaginary fluid the lines of inductive action are represented by unit tubes and the differences of dielectrics by differences of resisting power. Theorem VI is therefore applicable to such cases, and we can determine the imaginary distribution of electricity over the surface due to the temporary inductive action. When the conduction is perfect or nearly so Cor I becomes applicable and the surface of the conductor is one of equal potential.

(2) *Magnetism*

By considering the system of unit cells (Theorem VII) we arrive at the conception of a system of elements endued with polarity, and by comparing this

---

(33) Read: [William Thomson,] 'Note on orthogonal isothermal surfaces', *Camb. Math. J.*, **3** (1843): 286–8 (= *Math. & Phys. Papers*, **1**: 22–8).

(34) Compare 'On Faraday's lines of force': 42–51 (= *Scientific Papers*, **1**: 175–88).

(35) See 'On Faraday's lines of force': 42–3 (= *Scientific Papers*, **1**: 176).

(36) Michael Faraday, 'On induction' (= *Electricity*, **1**: 360–416).

with the established laws of magnetism we find them to agree. We have only to regard the unit tubes as representing the lines of inductive magnetic action both in direction and in number and the rate of variation of pressure as indicating the resultant force or intensity of magnetism and the methods above given become applicable to the theory of magnetism.

Paramagnetic and Diamagnetic bodies[37]

Faraday conceives of paramagnetic and diamagnetic bodies as conducting the lines of magnetic induction more freely & less freely respectively than the surrounding medium.[38] Admitting this explanation, the method of Theorem VI becomes applicable, as in the theory of dielectrics. If a paramagnetic body be placed in the magnetic field it opposes less resistance to the lines of inductive action than the surrounding medium, and therefore by its presence it diminishes the total resistance by a quantity depending on the square of the number of lines which pass through it. Now the total resistance overcome is $\Big\langle$ expressed by $\sum (Sp)$ and is equal to $\dfrac{4\pi}{k} \sum mp \Big\rangle$ proportional to the total potential of the magnetic system and since the resultant forces act so as to diminish this quantity, they urge the paramagnetic body in the direction in which more lines of force will pass through it, that is from places of weak to places of strong magnetic force.

In diamagnetic bodies the total resistance is increased, and the resultant force is towards places of weak magnetic force.[39] It is plain that we might have obtained similar results by supposing the magnetic force to execute a polarity

(37) Michael Faraday, 'Experimental researches in electricity. – Twenty-fifth series. On the magnetic and diamagnetic condition of bodies', *Phil. Trans.*, **141** (1851): 7–28, esp. 26 (§2790) (= *Electricity*, **3**: 195–6). In introducing a distinction between 'paramagnetic' and 'diamagnetic' bodies Faraday defined his use of these terms: 'in an absolute vacuum or free space, a [para] magnetic body tends from weaker to stronger places of magnetic action, and a diamagnetic body under similar conditions from stronger to weaker places of action' (§2789). The terms 'paramagnetic' and 'diamagnetic' were suggested to Faraday by William Whewell: see their letters in August–September 1850, in *Correspondence of Faraday*, **2**: 586–91.

(38) Michael Faraday, 'Magnetic conducting power' (= *Electricity*, **3**: 200–73). See L. P. Williams, *Michael Faraday* (London, 1965): 408–50; and D. Gooding, 'Final steps to field theory: Faraday's study of magnetic phenomena, 1845–1850', *Historical Studies in the Physical Sciences*, **11** (1981): 231–75, esp. 268–75.

(39) See William Thomson, 'On the forces experienced by small spheres under magnetic influence; and on some of the phenomena presented by diamagnetic substances', *Camb.·& Dubl. Math. J.*, **2** (1847): 230–5 (= *Electrostatics and Magnetism*: 493–9).

in the particles of these bodies in the same or in the opposite direction.[40] We might also explain the phenomena by imaginary magnetic matter disposed on the surface of the bodies.[41]

Magnecrystallic Induction

The theory of Faraday on this subject is that in certain substances the lines of magnetic induction are conducted in certain directions more freely than in

---

(40) Drawing on Ampère's theory that a magnet consists of molecules within each of which circulates a system of electric currents (see A. M. Ampère, 'Mémoire sur la théorie mathématique des phénomènes électrodynamiques uniquement déduite de l'expérience', *Mémoires de l'Académie Royale des Sciences*, **6** (1827): 175–388, esp. 368), Wilhelm Weber argued that paramagnetic and diamagnetic bodies exhibited opposite polarities under the same conditions of induction. He supposed that while permanent electric currents circulated in the molecules of paramagnetic substances, such currents were induced by a magnet in the molecules of diamagnetic substances, these induced currents circulating in the opposite direction to the permanent currents in the molecules of paramagnetic substances. See Wilhelm Weber, 'Ueber die Erregung und Wirkung des Diamagnetismus nach den Gesetzen inducirter Ströme', *Ann. Phys.*, **73** (1848): 241–56 (= *Wilhelm Weber's Werke*, 6 vols. (Berlin, 1892–4), **3**: 255–68), (trans.) 'On the excitation and action of diamagnetism according to the laws of induced currents', *Scientific Memoirs*, ed. R. Taylor, **5** (London, 1852): 477–88; Weber, 'Ueber den Zusammenhang der Lehre vom Diamagnetismus mit der Lehre von dem Magnetismus und der Elektricität', *Ann. Phys.*, **87** (1852): 145–89 (= *Werke*, **3**: 555–90), (trans.) 'On the connexion of diamagnetism with magnetism and electricity', *Scientific Memoirs, Natural Philosophy*, ed. J. Tyndall and W. Francis (London, 1853): 163–99. The issue of diamagnetic polarity was discussed in John Tyndall's Bakerian Lecture, 'On the nature of the force by which bodies are repelled from the poles of a magnet', *Phil. Trans.*, **145** (1855): 1–51, esp. 37. While Tyndall regarded Weber's theory of molecular currents as so 'artificial... that the general conviction of its truth cannot be very strong', he concluded that '*the diamagnetic force is a polar force, the polarity of diamagnetic bodies being opposed to that of paramagnetic ones under the same conditions of excitement*'. Tyndall's conclusions were therefore at variance with Faraday's theory of magnetism, based on the propensity of substances to conduct lines of force, views which Faraday amplified in his paper 'On some points of magnetic philosophy', *Phil. Mag.*, ser. 4, **9** (1855): 81–113 (= *Electricity*, **3**: 528–65). The problems of polarity and Faraday's theory of magnetism were discussed in a series of letters between Tyndall, Faraday, Thomson and Weber published in the *Philosophical Magazine*; see Tyndall, 'On the existence of a magnetic medium in space', *Phil. Mag.*, ser. 4, **9** (1855): 205–9; Faraday, 'Magnetic remarks', *ibid.*: 253–5; Thomson, 'Observations on the "magnetic medium" and on the effects of compression', *ibid.*: 290–3; Weber, 'On the theory of diamagnetism', *ibid.*, **10** (1855): 407–10 (= John Tyndall, *Researches on Diamagnetism and the Magnecrystallic Action, including the question of Diamagnetic Polarity* (London, 1870): 213–29). Tyndall developed his interpretation in his 'Further researches on the polarity of the diamagnetic force', *Phil. Trans.*, **146** (1856): 237–59, to which Maxwell makes reference in 'On Faraday's lines of force': 45n (= *Scientific Papers*, **1**: 180n), where he suggests that Faraday's theory is 'the most precise, and at the same time least theoretic statement'. See also Number 109.

(41) See William Thomson, 'A mathematical theory of magnetism': 250–6.

others.[42] By the method of the imaginary fluid the requirements of the case may be distinctly conceived without any admission implying a physical theory.[43] The case of a crystalline sphere in a uniform field of magnetic force admits of an easy solution which is perfectly accurate,[44] and the lines of force and surfaces of equal pressure may be found by geometrical constructions without calculation both in this case and in those preceding.

(3)

Conduction of current electricity

Here the received theory agrees entirely with that of our imaginary fluid. The laws of conduction, as laid down by Ohm[45] are nothing more than those of our fluid applied to a real phenomenon.[46] In a closed circuit, however, which is the only case in which we can produce a constant current, an electromotive force is necessary to keep up the circulation. The intensity of this electromotive force being represented by F, and the total quantity of electricity which passes any section of the current in unit of time by I, we find, $F = IK$, where K denotes the total resistance of the circuit. As in this case the conducting power of the surrounding medium may be neglected, Cor (2) of Theorem VI becomes applicable.

---

(42) Michael Faraday, 'On the crystalline polarity of bismuth (and other bodies), and on its relation to the magnetic form of force' (= *Electricity*, **3**: 83–136).

(43) Compare 'On Faraday's lines of force': 46 (= *Scientific Papers*, **1**: 180); see Theorem VI *supra* and note (31).

(44) See Example VI of 'On Faraday's lines of force': 74–6 (= *Scientific Papers*, **1**: 217–19), and Numbers 104 and 109.

(45) Georg Simon Ohm, *Die galvanische Kette, mathematisch bearbeitet* (Berlin, 1827), (trans.) 'The galvanic circuit investigated mathematically', *Scientific Memoirs*, ed. R. Taylor, **2** (London, 1841): 401–506.

(46) Compare 'On Faraday's lines of force': 46 (= *Scientific Papers*, **1**: 180); 'This pressure, which is commonly called electrical tension, is found to be physically identical with the *potential* in statical electricity, and thus we have the means of connecting the two sets of phenomena'. On the interpretation of 'tension' as potential see Gustav Kirchhoff, 'Ueber eine Ableitung der Ohm'schen Gesetze, welche sich an die Theorie der Elektrostatik anschliesst', *Ann. Phys.*, **78** (1849): 506–13, (trans.) 'On a deduction of Ohm's laws, in connection with the theory of electrostatics', *Phil. Mag.*, ser. 3, **37** (1850): 463–8. In 'On Faraday's lines of force': 46 Maxwell refers to Émile Verdet's translation of Kirchhoff's paper, 'Démonstration des lois de Ohm fondée principes ordinaires de l'électricité statique; par M. Kirchhoff', *Ann. Chim. Phys.*, ser. 3, **41** (1854): 496–500. See Number 50.

Quantity and Intensity[47]

The quantity of current which passes through any surface is measured by the number of ⟨lines of force⟩ unit tubes which pass through it, and the total quantity in a given current by the total number of unit tubes in it. The intensity of current at any point is the amount of resistance which unity of fluid overcomes in unity of time at that point. The total intensity of a closed current is the amount of resistance overcome by the fluid within it in unit of time.[48]

We may speak in like manner of the quantity of magnetic induction through a given surface, and of the magnetic intensity at a given point or the total intensity of a closed magnetic curve.[49]

4

Action of closed currents at a distance

Ampère has most ably demonstrated the fundamental laws of these currents, on the supposition that the action is between elements of the currents.[50] As, however, all real constant currents are closed, we may confine ourselves to that case. Ampère has shown that the mutual action of two small circuits is similar to that of two small magnets. Now a large circuit may be made up of small ones arranged side by side, in which case the internal currents neutralize one another, and the surrounding current remains equivalent to a magnetic shell having the current for its edge.

It is easy to show that the total intensity of magnetizing force in a closed curve embracing the circuit is measured by the quantity of the current and that the total resistance overcome by the lines of magnetic induction due to any number of closed currents is expressed by the sum of the products of the quantity of each current, into the number of lines of magnetic induction which

---

(47) Compare 'On Faraday's lines of force': 52–5 (= *Scientific Papers*, **1**: 189–92); and see Number 87.

(48) Compare 'On Faraday's lines of force': 47 (= *Scientific Papers*, **1**: 182); and see Number 83.

(49) Compare 'On Faraday's lines of force': 54–5 (= *Scientific Papers*, **1**: 192); and see Number 87. On the distinction between magnetic 'intensity' summed along lines of force and magnetic 'induction' summed across the surface enclosed by lines of force, the basis of Maxwell's distinction between 'magnetic force' (**H**) and 'magnetic induction' (**B**) in the *Treatise*, see Number 51 note (7).

(50) Ampère, 'Mémoire sur la théorie mathématique des phénomènes électrodynamiques uniquement déduite de l'éxpérience'; and see Number 66 note (3).

pass through it.[51] Hence the resultant of the forces which act on any circuit will tend to place it so that as many lines of force as possible pass through it.

5
Induction of closed currents

Helmholtz* has shown the connection between the attraction of currents already established and the induction of new ones,[52] and from his speculations and Faradays own explanations of the phenomena,[53] the resulting law is, that the total electromotive force in a circuit arising from induction is measured by the change in unit of time of the number of lines of magnetic induction which pass through it. This change in the number of lines may arise either from motion of the conductor or from motion of other conductors or magnets, or it may be due to a change in the intensity of other currents or magnets, and even to a change in the intensity of a current in the conductor itself.

The application of ideas due to Faraday to the mathematical investigation of some of these phenomena is reserved for another communication.[54]

JAMES CLERK MAXWELL

* Conservation of Force. Taylors Scientific Memoirs, Feb. 1853.[55]

---

(51) Compare 'On Faraday's lines of force': 49 (= *Scientific Papers*, **1**: 184–5); and see Number 85.

(52) Hermann Helmholtz, *Über die Erhaltung der Kraft, eine physikalische Abhandlung* (Berlin, 1847) (= H. Helmholtz, *Wissenschaftliche Abhandlungen*, 3 vols. (Leipzig, 1882–95), **1**: 12–68), (trans.) 'On the conservation of force', in *Scientific Memoirs, Natural Philosophy*, ed. J. Tyndall and W. Francis (London, 1853): 114–62, esp. 156–60.

(53) Michael Faraday, 'Experimental researches in electricity. – Twenty-eighth series. On lines of magnetic force', *Phil. Trans.*, **142** (1852): 25–56 (= *Electricity*, **3**: 328–70).

(54) See note (10). (55) See note (52).

85

ABSTRACT OF PAPER 'ON FARADAY'S LINES
OF FORCE' (PART I)

[10 DECEMBER 1855][1]

From the *Proceedings of the Cambridge Philosophical Society*[2]

[ON FARADAY'S LINES OF FORCE]

The method pursued in this paper is a modification of that mode of viewing electrical phænomena in relation to the theory of the uniform conduction of heat, which was first pointed out by Professor W. Thomson in the Cambridge and Dublin Mathematical Journal, vol. iii.[3] Instead of using the analogy of heat, a fluid, the properties of which are entirely at our disposal, is assumed as the vehicle of mathematical reasoning. A method is given by which two series of surfaces may be drawn in the fluid so as to define its motion completely. The uniform motion of an imponderable and incompressible fluid permeating a medium, whose resistance is directly as the velocity, is then discussed, and it is shown how a system of surfaces of equal pressure may be drawn, which, with the two former systems of surfaces, divides the medium into cells, in each of which the same amount of work is done in overcoming resistance. It is then shown that if the fluid be supposed to emanate from certain centres, and to be absorbed at others, the position of these centres can be found when the pressure at any point is known; and that when the centres are known, the distribution of pressures may be found. Methods are then given by which the motion of the fluid out of one medium into another, the resistance of which is different, may be conceived and calculated, and the theory of motion in a medium in which the resistance is different in different directions is stated.[4]

The mathematical ideas obtained from the fluid are then applied to various parts of electrical science. It is shown that the expression for the electrical potential at any point is identical with that of the pressure in the fluid, provided that 'sources' of fluid are put instead of positive electrical 'matter', and centres of absorption or 'sinks' for negative 'matter'.[5]

---

(1) The date on which the paper was read to the Cambridge Philosophical Society.

(2) *Proc. Camb. Phil. Soc.*, **1** (1856): 160–2. Part I of 'On Faraday's lines of force' was published in *Trans. Camb. Phil. Soc.*, **10** (1856): 27–51 (= *Scientific Papers*, **1**: 155–88).

(3) Read: *Cambridge Mathematical Journal*; see Thomson's correction appended to Number 84, and Number 84 note (20).

(4) Compare Maxwell, 'On Faraday's lines of force', *Trans. Camb. Phil. Soc.*, **10** (1856): 30–42 (= *Scientific Papers*, **1**: 160–75); see Number 83.　　(5) See Number 84.

The theory of Faraday with respect to the effect of dielectrics in modifying electric induction,[6] is illustrated by the case of different media having different conducting power; and it is shown, that, in order to calculate the effects by the ordinary formulæ of attractions, we must alter in a certain proportion the quantities of electricity within the dielectric, and conceive an imaginary distribution of electricity over the surface which separates it from the surrounding medium.

The theory of magnets and of the phænomena of paramagnetic and diamagnetic bodies is expressed with reference to the 'lines of inductive magnetic action'; and elementary proofs of the tendency of paramagnetic bodies toward places of stronger magnetic action, and of diamagnetic bodies toward places of weaker action, are given. This distinction of paramagnetic and diamagnetic is not here used absolutely, but indicates a greater or less conductivity for the lines of inductive action than that of the surrounding medium.[7]

The magnetic phænomena of crystals are then examined, and referred to unequal magnetic conductivity in different directions;[8] and the case of a crystalline sphere in a uniform field of force is worked out.[9]

The laws of electric conduction, as laid down by Ohm,[10] are shown to agree with those of the imaginary fluid, and definitions of quantity and intensity are given, which will apply to magnetism as well as galvanism.

The theory of the attractions of closed circuits, as established by Ampère,[11] is shown to lead to the following results:

1. The total intensity of the magnetizing force estimated along any closed curve embracing the circuit is a measure of the quantity of the current.

2. The quantity of the current, multiplied by the quantity of inductive magnetic action, from whatever source, which passes through it, gives what may be called the potential of the circuit. The tendency of the resultant forces is to increase this potential.[12]

The theory of Faraday with respect to the induction of currents in closed circuits takes the following form:

When the quantity of inductive magnetic action which passes through a given circuit changes in any way, an electromotive force proportional to the

---

(6) Number 84 note (36).   (7) Number 84 esp. note (40).

(8) Number 84 esp. note (31).

(9) Example VI of 'On Faraday's lines of force': 74–6 (= *Scientific Papers*, **1**: 217–19); and see Numbers 104 and 109.

(10) Number 84 note (45).

(11) Number 84 note (50).

(12) See 'On Faraday's lines of force': 49 (= *Scientific Papers*, **1**: 184–5).

rate of change acts in the circuit, and a current is produced whose quantity is the electromotive force divided by the total resistance of the circuit.[13]

The mathematical discussion of the electro-magnetic laws is reserved for another communication.[14]

---

(13) See 'On Faraday's lines of force': 47 (= *Scientific Papers*, **1**: 181–2).
(14) See Number 87.

FROM A LETTER TO JOHN CLERK MAXWELL

11 DECEMBER 1855

From Campbell and Garnett, *Life of Maxwell*[1]

Trin. Coll.
11 December [1855]

Last night I lectured on Lines of Force at the Philosophical.[2] I put off the second part of it to next term.[3] I have been drawing a lot of lines of force by an easy dodge. I have got to draw them accurately without calculation.[4]

Pomeroy has been improving slowly, but sometimes stopping. He is so big that it requires a great deal to get up his strength again. I saw Dr. Paget[5] at the Philosophical to-day, and he seemed to think him in a fair way to recover.

---

(1) *Life of Maxwell*: 222.

(2) Numbers 84 and 85.

(3) See Number 87.

(4) See J. C. Maxwell, 'On a method of drawing the theoretical forms of Faraday's lines of force without calculation', *Report of the Twenty-sixth Meeting of the British Association for the Advancement of Science; held at Cheltenham in August 1856* (London, 1857), part 2: 12 (= *Scientific Papers*, **1**: 241). See Number 166.

(5) George Edward Paget, Caius 1827, Physician at Addenbrooke's Hospital, Cambridge, 1839–84, President of Cambridge Philosophical Society, 1855, Regius Professor of Physic, 1872–92 (Venn; *DNB*).

ABSTRACT OF PAPER 'ON FARADAY'S LINES OF FORCE' (PART II)

[11 FEBRUARY 1856][1]

From the *Proceedings of the Cambridge Philosophical Society*[2]

[ON FARADAY'S LINES OF FORCE]

This paper was chiefly occupied with the extension of the method of lines of force to the phænomena of electro-magnetism, by means of a mathematical method founded on Faraday's idea of an 'electrotonic state'.[3]

In order to obtain a clear view of the phænomena to be explained, we must begin with some general definitions of *quantity* and *intensity* as applicable to electric currents and to magnetic induction. It was shown in the first part of this paper, that electrical and magnetic phænomena present a mathematical analogy to the case of a fluid whose steady motion is affected by certain moving forces and resistances. (The purely imaginary nature of this fluid has been already insisted on.) Now the amount of fluid passing through any area in unit of time measures the *quantity* of action over this area; and the moving forces which act on any element in order to overcome the resistance, represent the total *intensity* of action within the element.[4]

In electric currents, the *quantity* of the current in any given direction is measured by the quantity of electricity which crosses a unit area perpendicularly to this direction; and the intensity, by the resolved part of the whole electromotive forces acting in that direction.[5] In a closed circuit, whose length $= l$, coefficient of resistance $= k$, and section $= C^2$, if F be the whole electromotive force round the circuit, and I the whole quantity of the current,

$$\frac{I}{C^2}lk = F, \quad I = \frac{C^2 F}{lk}.$$

The laws of Ohm[6] with respect to electric currents were then applied to cases in which the conducting power of the medium is different in different directions. The general equations were given and several cases solved.

In magnetic phænomena, the distinction of quantity and intensity is less

(1) The date the paper was read to the Cambridge Philosophical Society; but see Number 98.

(2) *Proc. Camb. Phil. Soc.*, **1** (1856): 163–6. Part II of 'On Faraday's lines of force' was published in *Trans. Camb. Phil. Soc.*, **10** (1856): 51–83 (= *Scientific Papers*, **1**: 188–229).

(3) Faraday, *Electricity*, **1**: 16 (§60) and **3**: 420 (§3269).   (4) See Number 84.

(5) Compare 'On Faraday's lines of force', *Trans. Camb. Phil. Soc.*, **10** (1856): 52–4 (= *Scientific Papers*, **1**: 189–92).   (6) Number 84 note (45).

obvious, though equally necessary. It is found, that what Faraday calls the quantity of inductive magnetic action over any surface, depends only on the *number* of lines of magnetic force which pass through it, and that the total electromotive effect on a conducting wire will always be the same, provided it moves across the same system of lines, in whatever manner it does so. But though the quantity of magnetic action over a given area depends only on the number of lines which cross it, the intensity depends on the force required to keep up the magnetism at that part of the medium; and this will be measured by the product of the quantity of magnetization, multiplied by the coefficient of resistance to magnetic induction in that direction.[7]

The equations which connect magnetic quantity and intensity are similar in form to those which were given for electric currents, and from them the laws of diamagnetic and magnecrystallic induction may be deduced and reduced to calculation.

We have next to consider the mutual action of magnets and electric currents. It follows from the laws of Ampère, that when a magnetic pole is in presence of a closed electric circuit, their mutual action will be the same as if a magnetized shell of given intensity had been in the place of the circuit and been bounded by it. From this it may be proved, that (1) the *potential* of a magnetized body on an electric circuit is measured by the *number* of lines of magnetic force due to the magnet which pass through the circuit. (2) That the total amount of work done on a unit magnetic pole during its passage round a closed curve embracing the circuit depends only on the *quantity* of the current, and not on the form of the path of the pole, or the nature or form of the conducting wire.

The first of these laws enables us to find the forces acting on an electric circuit in the magnetic field. Give the circuit any displacement, either of translation, rotation, or disfigurement, then the difference of potential before and after displacement will represent the force urging the conductor in the direction of displacement. The force acting on any element of a conductor will be perpendicular to the plane of the current and the lines of magnetic force, and will be measured by the product of the quantities of electric and magnetic action into the sine of the angle between the direction of the electric and magnetic lines of force.[8]

The second law enables us to determine the quantity and direction of the electric currents in any given magnetic field; for, in order to discover the quantity of electricity flowing through any closed curve, we have only to estimate the work done on a magnetic pole in passing round it. This leads to the

---

(7) Compare 'On Faraday's lines of force': 54–5 (= *Scientific Papers*, **1**: 192).

(8) Compare 'On Faraday's lines of force': 49, 55–6, 77 (= *Scientific Papers*, **1**: 184–5, 193–4, 220–1).

following relations between $\alpha_1\,\beta_1\,\gamma_1$, the components of magnetic intensity, and $a_2\,b_2\,c_2$, the resolved parts of the electric current at any point,[9]

$$a_2 = \frac{d\beta_1}{dz} - \frac{d\gamma_1}{dy}, \quad b_2 = \frac{d\gamma_1}{dx} - \frac{d\alpha_1}{dz}, \quad c_2 = \frac{d\alpha_1}{dy} - \frac{d\beta_1}{dx}. \quad (10)$$

In this way the electric currents, if any exist, may be found when we know the magnetic state of the field.[11] When $\alpha_1\,dx + \beta_1\,dy + \gamma_1\,dz$ is a perfect differential, there will be no electric currents.[12]

Since it is the *intensity* of the magnetic action which is immediately connected with the *quantity* of electric currents, it follows that the presence of paramagnetic bodies, like iron, will, by diminishing the total resistance to magnetic induction while the total intensity is constant, increase its quantity. Hence the increase of external effect due to the introduction of a core of soft iron into an electric helix.[13]

From the researches of Faraday into the induction of electric currents by changes in the magnetic field, it appears that a conductor, in cutting the lines of magnetic force, experiences an electromotive force, tending to produce a current perpendicular to the lines of motion and of magnetic force, and depending on the number of lines cut by the conductor in its motion.[14]

It follows that the total electromotive force in a closed circuit is measured by the *rate of change* of the number of lines of magnetic force which pass through it; and it is indifferent whether this change arises from a motion of this circuit, or from any change in the magnetic field itself, due to changes of intensity or position of magnets or electric currents.[15]

This law, though it is sufficiently simple and general to render intelligible all the phænomena of induction in closed circuits, contains the somewhat artificial conception of the number of lines *passing through* the circuit, exerting a physical influence on it.[16] It would be better if we could avoid, in the enunciation of the law, making the electromotive force in a conductor depend upon lines of force external to the conductor. Now the expressions which we obtained for the

---

(9) Maxwell distinguishes magnetic quantities by the suffix (1) and electric quantities by the suffix (2).

(10) See 'On Faraday's lines of force': 56 (= *Scientific Papers*, **1**: 194).

(11) Compare 'Law III' of 'On Faraday's lines of force': 66 (= *Scientific Papers*, **1**: 206); '*The entire magnetic intensity round the boundary of any surface measures the quantity of electric current which passes through that surface*'. See Number 51 esp. note (12).

(12) See 'On Faraday's lines of force': 62–3 (= *Scientific Papers*, **1**: 202–3).

(13) Compare Example IX of 'On Faraday's lines of force': 77–8 (= *Scientific Papers*, **1**: 222).

(14) Michael Faraday, 'Experimental researches in electricity. – Twenty-eighth series. On lines of magnetic force', *Phil. Trans.*, **142** (1852): 25–56 (= *Electricity*, **3**: 328–70).

(15) Compare 'On Faraday's lines of force': 50 (= *Scientific papers*, **1**: 186).

(16) Compare 'On Faraday's lines of force': 63, 66 (= *Scientific Papers*, **1**: 203, 207).

connexion between magnetism and electric currents supply us with the means of making the law of induced currents depend on the state of the conductor itself.

We have seen that from certain expressions for magnetic intensity we could deduce those for the quantity of currents, so that the currents which pass through a given closed curve may be measured by the total magnetic intensity round that curve. Here we have an integration *round the curve itself* instead of one *over the enclosed surface*. In the same way, if we assume the mathematical existence of a state,[17] bearing the same relation to magnetic quantity that magnetic intensity bears to electric quantity, we shall have an expression for the quantity of magnetic induction passing through a closed circuit in terms of quantities depending on the circuit itself, and not on the enclosed space.

Let us therefore assume three functions of xyz, $\alpha_0\,\beta_0\,\gamma_0$,[18] such that $a_1\,b_1\,c_1$ being the resolved parts of magnetic quantity,

$$a_1 = \frac{d\beta_0}{dz} - \frac{d\gamma_0}{dy}, \quad b_1 = \frac{d\gamma_0}{dx} - \frac{d\alpha_0}{dz}, \quad c_1 = \frac{d\alpha_0}{dy} - \frac{d\beta_0}{dx};$$

then it will appear that if we assume $\frac{d\alpha_0}{dt}$, $\frac{d\beta_0}{dt}$, $\frac{d\gamma_0}{dt}$ as the expressions for the electromotive forces at any point in the conductor, the total electromotive force in any circuit will be the same as that expressed by Faraday's law.[19] Now as we know nothing of these inductive effects except in closed circuits, these expressions, which are true for closed currents, cannot be inconsistent with known phænomena, and may possibly be the symbolic representative of a real law of nature. Such a law was suspected by Faraday from the first, although, for want of direct experimental evidence, he abandoned his first conjecture of the existence of a new state or condition of matter. As, however, we have now shown that this state, as described by him (Exp. Res. (60)), has at least a mathematical significance, we shall use it in mathematical investigations, and we shall call the three functions $\alpha_0, \beta_0, \gamma_0$ the *electrotonic functions* (see Faraday's Exp. Res. 60. 231. 242. 1114. 1661. 1729. 3172. 3269.).[20]

---

(17) The electro-tonic state.

(18) The '*components of the Electro-tonic intensity*'; see 'On Faraday's lines of force': 63 (= *Scientific Papers*, **1**: 203).

(19) Compare 'Law I' of 'On Faraday's lines of force': 65 (= *Scientific Papers*, **1**: 206); '*The entire electro-tonic intensity round the boundary of an element of surface measures the quantity of magnetic induction which passes through that surface or, in other words, the number of lines of magnetic force which pass through that surface*'. See Number 51 esp. note (12).

(20) See note (3).

That these functions are otherwise important may be shown from the fact, that we can express the potential of any closed current by the integral

$$\int \left(a_2 \alpha_0 \frac{dx}{ds} + b_2 \beta_0 \frac{dy}{ds} + c_2 \gamma_0 \frac{dz}{ds} \right) ds,$$

and generally that of any system of currents in a conducting mass by the integral

$$\iiint (\alpha_0 a_2 + \beta_0 b_2 + \gamma_0 c_2)\, dx\, dy\, dz.^{(21)}$$

The method of employing these functions is exemplified in the case of a hollow conducting sphere revolving in a uniform magnetic field[22] (see Faraday's Exp. Res. (160)),[23] and in that of a closed wire in the neighbourhood of another in which a variable current is kept up,[24] and several general theorems relating to these functions are proved.

---

(21) See 'On Faraday's lines of force': 61–3 (= *Scientific Papers*, **1**: 201–3); and Number 100.

(22) See Example XII of 'On Faraday's lines of force': 81–3 (= *Scientific Papers*, **1**: 226–9; and Number 91 esp. note (5)).

(23) On the production of electric currents in a metallic globe revolving in a magnetic field see Michael Faraday, 'Experimental researches in electricity. – Second series. Terrestrial magneto-electric induction', *Phil. Trans.*, **122** (1832): 163–94, esp. 168–70 (§§160–9) (= *Electricity*, **1**: 47–50). See Number 51.

(24) See Example XI of 'On Faraday's lines of force': 79–80 (= *Scientific Papers*, **1**: 224–6); and Number 91.

ESSAY FOR THE APOSTLES ON
'ANALOGIES IN NATURE'

FEBRUARY 1856

From Campbell and Garnett, *Life of Maxwell*[1]

ARE THERE REAL ANALOGIES IN NATURE?[2]

In the ancient and religious foundation of Peterhouse there is observed this rule, that whoso makes a pun shall be counted the author of it, but that whoso pretends to find it out shall be counted the publisher of it, and that both shall be fined. Now, as in a pun two truths lie hid under one expression, so in an analogy one truth is discovered under two expressions. Every question concerning analogies is therefore the reciprocal of a question concerning puns, and the solutions can be transposed by reciprocation. But since we are still in doubt as to the legitimacy of reasoning by analogy, and as reasoning even by paradox has been pronounced less heinous than reasoning by puns, we must adopt the direct method with respect to analogy, and then, if necessary, deduce by reciprocation the theory of puns.

That analogies appear to exist is plain in the face of things, for all parables, fables, similes, metaphors, tropes, and figures of speech are analogies, natural or revealed, artificial or concealed. The question is entirely of their reality. Now, no question exists as to the possibility of an analogy without a mind to recognise it – that is rank nonsense. You might as well talk of a demonstration or refutation existing unconditionally. Neither is there any question as to the occurrence of analogies to our minds. They are as plenty as reasons, not to say blackberries. For, not to mention all the things in external nature which men have seen as the projections of things in their own minds, the whole framework

---

(1) *Life of Maxwell*: 235–44; dated by Campbell.

(2) There has been much discussion of Maxwell's ideas on physical analogy: George E. Davie, *The Democratic Intellect. Scotland and her Universities in the Nineteenth Century* (Edinburgh, ₂1964): 138–45, 192–7; Richard Olson, *Scottish Philosophy and British Physics 1750–1880. A Study in the Foundations of the Victorian Scientific Style* (Princeton, 1975): 287–321; D. B. Wilson, 'The educational matrix: physics education at early-Victorian Cambridge, Edinburgh and Glasgow Universities', in P. M. Harman ed., *Wranglers and Physicists. Studies on Cambridge Physics in the Nineteenth Century* (Manchester, 1985): 12–48 esp. 33–44; P. M. Harman, 'Edinburgh philosophy and Cambridge physics: the natural philosophy of James Clerk Maxwell', in *Wranglers and Physicists*: 202–24; Joseph Turner, 'Maxwell on the method of physical analogy', *British Journal for Philosophy of Science*, **6** (1955): 226–38; Robert H. Kargon, 'Model and analogy in Victorian science: Maxwell and the French physicists', *Journal of the History of Ideas*, **30** (1969): 423–36.

of science, up to the very pinnacle of philosophy, seems sometimes a dissected model of nature, and sometimes a natural growth on the inner surface of the mind. Now, if in examining the admitted truths in science and philosophy, we find certain general principles appearing throughout a vast range of subjects, and sometimes re-appearing in some quite distinct part of human knowledge; and if, on turning to the constitution of the intellect itself, we think we can discern there the reason of this uniformity, in the form of a fundamental law of the right action of the intellect, are we to conclude that these various departments of nature in which analogous laws exist, have a real inter-dependence; or that their relation is only apparent and owing to the necessary conditions of human thought?

There is nothing more essential to the right understanding of things than a perception of the relations of *number*. Now the very first notion of number implies a previous act of intelligence. Before we can count any number of things, we must pick them out of the universe, and give each of them a fictitious unity by definition. Until we have done this, the universe of sense is neither one nor many, but indefinite. But yet, do what we will, Nature seems to have a certain horror of partition. Perhaps the most natural thing to count 'one' for is a man or human being, but yet it is very difficult to do so. Some count by heads, others by souls, others by noses; still there is a tendency either to run together into masses or to split up into limbs. The dimmed outlines of phenomenal things all merge into another unless we put on the focussing glass of theory and screw it up sometimes to one pitch of definition, and sometimes to another, so as to see down into different depths through the great millstone of the world.

As for space and time, any man will tell you that 'it is now known and ascertained that they are merely modifications of our own minds'.[3] And yet if we conceive of the mind as absolutely indivisible and capable of only one state at a time, we must admit that these states may be arranged in chronological order, and that this is the only real order of these states. For we have no reason to believe, on the ground of a given succession of simple sensations, that differences in position, as well as in order of occurrence, exist among the causes of these sensations. But yet we are convinced of the co-existence of different objects at the same time, and of the identity of the same object at different times. Now if we admit that we can think of difference independent of sequence, and of sequence without difference, we have admitted enough on which to found the possibility of the ideas of space and time.

But if we come to look more closely into these ideas, as developed in human

---

(3) Compare Hamilton's remarks on 'Kant's doctrine of space and time'; see William Hamilton, *Lectures on Metaphysics and Logic*, ed. H. L. Mansel and J. Veitch, 4 vols. (Edinburgh, 1859–60), **1**: 402–4, and see Number 16.

beings, we find that *their* space has triple extension, but is the same in all directions, without behind or before, whereas time extends only back and forward, and always goes forward.

To inquire why these peculiarities of these fundamental ideas[4] are so would require a most painful if not impossible act of self-excenteration; but to determine whether there is anything in Nature corresponding to them, or whether they are mere projections of our mental machinery on the surface of external things, is absolutely necessary to appease the cravings of intelligence. Now it appears to me that when we say that space has three dimensions, we not only express the impossibility of conceiving a fourth dimension, co-ordinate with the three known ones, but assert the objective truth that points may differ in position by the independent variation of three variables. Here, therefore, we have a *real* analogy between the constitution of the intellect and that of the external world.[5]

With respect to time, it is sometimes assumed that the consecution of ideas is a fact precisely the same kind as the sequence of events in time. But it does not appear that there is any closer connection between these than between mental difference, and difference of position. No doubt it is possible to assign the accurate date of every act of thought, but I doubt whether a chronological table drawn up in this way would coincide with the sequence of ideas of which we are conscious. There is an analogy, but I think not an identity, between these two orders of thoughts and things. Again, if we know what is at any assigned point of space at any assigned instant of time, we may be said to know all the events in Nature. We cannot conceive any other thing which it would be necessary to know; and, in fact, if any other necessary element does exist, it never enters into any phenomenon so as to make it differ from what it would be on the supposition of space and time being the only necessary elements.

We cannot, however, think any set of thoughts without conceiving of them as depending on reasons. These reasons, when spoken of with relation to objects, get the name of *causes*, which are reasons, analogically referred to objects instead of thoughts. When the objects are mechanical, or are considered in a mechanical point of view, the causes are still more strictly defined, and are called *forces*.[6]

---

(4) The term is Whewell's; see William Whewell, *Philosophy of the Inductive Sciences, founded upon their History*, 2 vols. (London, ₂1847), 1: 66. Compare Numbers 68 and 105.

(5) Compare Whewell on the *'Fundamental Antithesis of Philosophy'* between 'Ideas and Senses, Thoughts and Things, Theory and Fact'; Whewell, *Philosophy of the Inductive Sciences*, 2: 647–68, esp. 650.

(6) Compare Whewell's view that the idea of cause construed as force is the 'fundamental idea' of mechanics; see Whewell, *Philosophy of the Inductive Sciences*, 2: 177–254, 473–94; and also Whewell, *An Elementary Treatise on Mechanics* (Cambridge, ₇1848): 1.

Now if we are acquainted not only with the events, but also with the forces, in Nature, we acquire the power of predicting events not previously known.

This conception of cause, we are informed, has been ascertained to be a notion of invariable sequence. No doubt invariable sequence, if observed, would suggest the notion of cause, just as the end of a poker painted red suggests the notion of heat, but although a cause without its invariable effect is absurd, a cause by its apparent frustration only suggests the notion of an equal and opposite cause.

Now the analogy between reasons, causes, forces, principles, and moral rules, is glaring, but dazzling.

A reason or argument is a conductor by which the mind is led from a proposition to a necessary consequence of that proposition. In pure logic reasons must all tend in the same direction. There can be no conflict of reasons. We may lose sight of them or abandon them, but cannot pit them against one another. If our faculties were indefinitely intensified, so that we could see all the consequences of any admission, then all reasons would resolve themselves into one reason, and all demonstrative truth would be one proposition. There would be no room for plurality of reasons, still less for conflict. But when we come to causes of phenomena and not reasons of truths, the conflict of causes, or rather the mutual annihilation of effects, is manifest. Not but what there is a tendency in the human mind to lump up all causes, and give them an aggregate name, or to trace chains of causes up to their knots and asymptotes. Still we see, or seem to see, a plurality of causes at work, and there are some who are content with plurality.

Those who are thus content with plurality delight in the use of the word force as applied to cause. Cause is a metaphysical word implying something unchangeable and always producing its effect. Force on the other hand is a scientific word, signifying something which always meets with opposition, and often with successful opposition, but yet never fails to do what it can in its own favour. Such are the physical forces with which science deals, and their maxim is that might is right, and they call themselves laws of nature. But there are other laws of nature which determine the form and action of organic structure. These are founded on the forces of nature, but they seem to do no work except that of direction. Ought they to be called forces? A force does work in proportion to its strength. These *direct* forces to work after a model. They are *moulds*, not forces. Now since we have here a standard from which deviation may take place, we have, besides the notion of *strength*, which belongs to force, that of *health*, which belongs to organic law. Organic beings are not conscious of organic laws, and it is not the conscious being that takes part in them, but another set of laws now appear in very close connexion with the conscious being. I mean the laws of thought. These may be interfered with by organic laws, or by physical disturbances, and no doubt every such interference is regulated by the laws of the brain and of the connexion between that medulla and the process of thought. But the thing to be observed is, that the laws which regulate the *right* process of the intellect are identical with the most abstract of all laws, those which are found among the relations of necessary truths, and that though these are mixed up with, and modified by, the most complex systems of phenomena in physiology and physics, they must be recognised as supreme among the other laws of thought. And this supremacy does not consist in superior strength, as in physical laws, nor yet, I think, in reproducing a type as in organic laws, but in being right and true; even when other causes have been for a season masters of the brain.

When we consider voluntary actions in general, we think we see causes acting like forces on the willing being. Some of our motions arise from physical necessity, some from irritability or organic excitement, some are performed by our machinery without our knowledge, and some evidently are due to us and our volitions. Of these, again, some are merely a repetition of a customary act, some are due to the attractions of pleasure or the pressure of constrained activity, and a few show some indications of being the results of distinct acts of the will. Here again we have a continuation of the analogy of Cause. Some had supposed that in will they had found the only true cause, and that all physical causes are only apparent. I need not say that this doctrine is exploded.[7]

What we have to observe is, that new elements enter into the nature of these higher causes, for mere abstract reasons are simply absolute; forces are related by their strength; organic laws act towards resemblances to types; animal emotions tend to that which promotes the enjoyment of life; and will is in great measure actually subject to all these, although certain other laws of *right*, which are abstract and demonstrable, like those of reason, are *supreme* among the laws of will.

Now the question of the reality of analogies in nature derives most of its interest from its application to the opinion, that all the phenomena of nature, being varieties of motion, can only differ in complexity, and therefore the only way of studying nature, is to master the fundamental laws of motion first, and then examine what kinds of complication of these laws must be studied in order to obtain true views of the universe. If this theory be true, we must look for indications of these fundamental laws throughout the whole range of science, and not least among those remarkable products of organic life, the results of cerebration (commonly called 'thinking'). In this case, of course, the resemblances between the laws of different classes of phenomena should hardly be called analogies, as they are only transformed identities.

---

(7) Compare Hamilton's account of causality in his *Lectures on Metaphysics*, **2**: 376–413.

If, on the other hand, we start from the study of the laws of thought (the abstract, logical laws, not the *physio*logical), then these apparent analogies become merely repetitions by reflexion of certain necessary modes of action to which our minds are subject. I do not see how, upon either hypothesis, we can account for the existence of one set of laws of which the supremacy is necessary, but to the operation contingent. But we find another set of laws of the same kind, and sometimes coinciding with physical laws, the operation of which is inflexible when once in action, but depends in its beginnings on some act of volition. The theory of the consequences of actions is greatly perplexed by the fact that each act sets in motion many trains of machinery, which react on other agents and come into regions of physical and metaphysical chaos from which it is difficult to disentangle them. But if we could place the telescope of theory in proper adjustment, to see not the physical events which form the subordinate foci of the disturbance propagated through the universe, but the moral foci where the true image of the original act is reproduced, then we shall recognise the fact, that when we clearly see any moral act, then there appears a moral necessity for the trains of consequences of that act, which are spreading through the world to be concentrated on some focus, so as to give a true and complete image of the act in its moral point of view. All that bystanders see, is the physical act, and some of its immediate physical consequences, but as a partial pencil of light, even when not adapted for distinct vision, may enable us to see an *object*, and not merely light, so the partial view we have of any act, though far from perfect, may enable us to see it morally as an act, and not merely physically as an event.

If we think we see in the diverging trains of physical consequences not only a capability of forming a true image of the act, but also of reacting upon the agent, either directly or after a long circuit, then perhaps we have caught the idea of *necessary* retribution, as the legitimate consequence of all moral action.

But as this idea of the *necessary* reaction of the consequences of action is derived only from a few instances, in which we have guessed at such a law among the necessary laws of the universe; and we have a much more distinct idea of *justice*, derived from these laws which we necessarily recognise as supreme, we connect the idea of retribution much more with that of *justice* than with that of *cause and effect*. We therefore regard retribution as the result of *interference* with the mechanical order of things, and intended to vindicate the supremacy of the right order of things, but still we suspect that the two orders of things will eventually dissolve into one.

I have been somewhat diffuse and confused on the subject of moral law, in order to show to what length analogy will carry the speculations of men. Whenever they see a relation between two things they know well, and think they see there must be a similar relation between things less known, they reason from the one to the other. This supposes that although pairs of things may differ widely from each other, the *relation* in the one pair may be the same as that in the other. Now, as in a scientific point of view the *relation* is the most important thing to know, a knowledge of the one thing leads us a long way towards a knowledge of the other. If all that we know is *relation*, and if all the relations of one pair of things correspond to those of another pair, it will be difficult to distinguish the one pair from the other, although not presenting a single point of resemblance, unless we have some difference of relation to something else, whereby to distinguish them.[8] Such mistakes can hardly occur except in mathematical and physical analogies, but if we are going to study the constitution of the individual mental man, and draw all our arguments from the laws of society on the one hand, or those of the nervous tissue on the other, we may chance to convert useful helps into Wills-of-the-wisp.[9] Perhaps the ' book,' as it has been called, of nature is regularly paged; if so, no doubt the introductory parts will explain those that follow, and the methods taught in the first chapters will be taken for granted and used as illustrations in the more advanced parts of the course; but if it is not a ' book' at all, but a *magazine*, nothing is more foolish to suppose that one part can throw light on another.

Perhaps the next most remarkable analogy is between the principle, law, or plan according to which all things are made suitably to what they have to do, and the intention which a man has of making machines which will work. The doctrine of final causes, although productive of barrenness in its exclusive form, has certainly been a great help to enquirers into nature; and if we only maintain the existence of the analogy, and allow observation to determine its form, we cannot be led far from the truth.

There is another analogy which seems to be supplanting the other on its own ground, which lies between the principle, law, or plan according to which the forms of things are made to have a certain community of type, and that which induces human artists to make a set of different things according to varieties of the same model. Here apparently the final cause is analogy or homogeneity, to the exclusion of usefulness.

---

(8) Compare Colin MacLaurin's discussion of mathematics: 'the mathematical sciences treat of the relations of quantities to each other... [hence] we enquire into the relations of things rather than their inward essences. Because we may have a clear conception of that which is the foundation of a relation, without having a perfect or adequate idea of the thing it is attributed to, our ideas of relation are often clearer and more distinct than those of the things to which they belong, and to this fact we may ascribe, in some measure, the peculiar evidence of the mathematics'; MacLaurin, *A Treatise of Fluxions*, 2 vols. (Edinburgh, 1742), **1**: 51–2.

(9) Compare Maxwell's discussion of moral philosophy in May 1855 in his Apostles essay 'Is Ethical Truth obtainable from an Individual Point of View', prompted by Adam Smith's *Theory of Moral Sentiments* (1759); see Number 62.

And last of all we have the secondary forms of crystals bursting in upon us, and sparkling in the rigidity of mathematical necessity and telling us, neither of harmony of design, usefulness or moral significance, – nothing but spherical trigonometry and Napier's analogies.[10] It is because we have blindly excluded the lessons of these angular bodies from the domain of human knowledge that we are still in doubt about the great doctrine that the only laws of matter are those which our minds must fabricate, and the only laws of mind are fabricated for it by matter.[11]

(10) See W. H. Miller, *A Treatise on Crystallography* (Cambridge, 1839): 4. Drawing normals from the faces of crystals to the surface of a sphere, the 'sphere of projection' is divided into a network of spherical triangles; '[crystallographic] calculations will be performed by spherical trigonometry', using the Napier analogies for right-angled spherical triangles. See John Napier, *Mirifici Logarithmorum Canonis descriptio* (Edinburgh, 1614): 30–9.

(11) Compare Whewell on the 'fundamental antithesis of philosophy' and the 'fundamental idea of symmetry of crystals'; *Philosophy of the Inductive Sciences*, **1**: 16–51, **2**: 440–52, 647–68.

FROM A LETTER TO JOHN CLERK MAXWELL

14 FEBRUARY 1856

From Campbell and Garnett, *Life of Maxwell*[1]

Trin.
14 February 1856

Yesterday the Ray Club[2] met at Hort's.[3] I took my great top[4] there and spun it with coloured discs attached to it. I have been planning a form of top, which will have more variety of motion, but I am working out the theory, so that I will wait till I know the necessary dimensions before I settle the plan.[5]

I told Willie (Cay) how I had hung up a bullet by a combination of threads.

I have drawn from theory the curves which it ought to describe, and when I set the bullet a-going over the proper curve, it traces it out over and over again as if it were doing a pre-ordained dance, and kept a steady eye on the line on the paper. I have enlarged my stock of models for solid geometry, made of coloured thread, stretched between two pasteboard ends.

---

(1) *Life of Maxwell*: 249–50.
(2) Number 67 note (4).
(3) Number 67 note (2).
(4) A wooden early version of the 'dynamical top' (Number 116).
(5) See Number 101, and his paper 'On an instrument to illustrate Poinsot's theory of rotation', *Report of the Twenty-sixth Meeting of the British Association for the Advancement of Science; held at Cheltenham in August 1856* (London, 1857), part 2: 27–8 (= *Scientific Papers*, **1**: 246–7).

MANUSCRIPT ON THE THEORY OF OPTICAL INSTRUMENTS[1]

FEBRUARY 1856[2]

From the original in the library of Peterhouse, Cambridge[3]

THEORY OF FOCAL CENTRES & PRINCIPAL FOCI

Definitions

(1) When incident rays, converging to or diverging from a point pass thro an optical instrument and emerge converging to or diverging from some other point, these two points are called the foci respectively of the incident and emergent rays, and the one is said to be the image of the other.

(2) When the emergent rays do not accurately pass through any point the inaccuracy is called aberration. In a perfect instrument there is no aberration.

(3) When the image of a right line perp to the axis becomes curved and when the different parts of it are unequally magnified it is said to be affected with curvature & distortion. In a perfect instrument these errors disappear.

(4) The focus of emergent rays which before incidence are parallel to the axis is called the *emergent principal focus* and the focus of parallel rays coming the other way thro' the instrument is called *incident principal focus*.

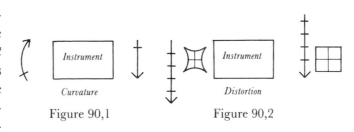

Figure 90,1   Figure 90,2

(5) When the image of an object is erect and equal to itself then the points on the axis in the planes of the object & of the image are called the *incident* & *emergent positive focal centres*. When the image is equal to the object but inverted the positions of the object & image point out the *incident* and *emergent negative focal centres*.

Prop I Given the incident ray, find the emergent ray.

Let $Q_1 a_1 b_1$ be the incident ray, $A_1 A_2$ the incident & emergent positive focal centres $B_1 B_2$ the incident and emergent negative focal centres.

---

(1) Compare Maxwell's paper 'On the elementary theory of optical instruments', *Proc. Camb. Phil. Soc.*, **1** (1856): 173–5 (= *Scientific Papers*, **1**: 238–40), read to the Society on 12 May 1856.

(2) See note (1) and Numbers 91 and 93.

(3) Maxwell MS 48, Peterhouse, Cambridge.

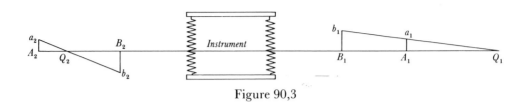

Figure 90,3

Draw $A_1 a_1$, $B_1 b_1$ perp to axis meeting incident ray in $a_1 b_1$.
Draw $A_2 a_2$ parallel & equal to $A_1 a_1$ & $B_2 b_2$ parallel & equal to $B_1 b_1$ but in the opposite direction. Join $b_2 a_2$, $b_2 a_2$ is the emergent ray.

For by def^n of focal centres a_2 must be the image of a_1 & b_2 of b_1 therefore any ray passing through a_1 must pass through a_2 & any ray thro b_1 must pass through b_2. Therefore the ray $a_1 b_1$ must at emergence be $a_2 b_2$.

Cor (1) If the incident ray be parallel the emergent ray will bisect $A_2 B_2$ so that the principal focus is midway between the focal centres.

Cor (2) $Q_1 A_1 : Q_1 B_1 :: Q_2 A_2 : Q_2 B_2$.

Prop II Given the incident focus, find the emergent focus.

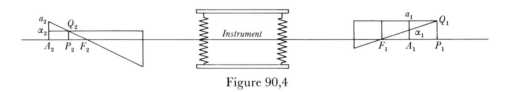

Figure 90,4

Let Q_1 be the incident focus, draw $Q_1 a_1$ parallel to axis & $Q_1 \alpha_1 F_1$ to the principal focus.
Make $A_2 a_2 = A_1 a_1$ & $A_2 \alpha_2 = A_1 \alpha_1$. Join $a_2 F_2$ & draw $\alpha_2 Q_2$ parallel to the axis, then by last prop. the ray $Q_1 a_1$ will emerge as $F_2 a_2$ & $Q_1 \alpha_1$ as $Q_2 \alpha_2$. So that Q_2 must be the image of Q_1.

Now $P_1 Q_1 = A_1 a_1 = A_2 a_2$ & $P_2 Q_2 = A_2 \alpha_2 = A_1 \alpha_1$ & by sim. triangles
$$P_1 F_1 : F_1 A_1 :: P_1 Q_1 : A_1 \alpha_1 \text{ or} :: P_1 Q_1 : P_2 Q_2$$
$$F_2 A_2 : P_2 F_2 :: A_2 a_2 : P_2 Q_2 \text{ or} :: P_1 Q_1 : P_2 Q_2.$$

Therefore
$$\frac{P_1 F_1}{F_1 A_1} = \frac{P_1 Q_1}{P_2 Q_2} = \frac{F_2 A_2}{P_2 F_2}$$

or
$$P_2 F_2 = \frac{F_1 A_1 \cdot F_2 A_2}{P_1 F_1} \qquad P_2 Q_2 = P_1 Q_1 \frac{F_1 A_1}{P_1 F_1}.$$

91

LETTER TO WILLIAM THOMSON

14 FEBRUARY 1856

From the original in the University Library, Cambridge[1]

Trin Coll
14th Feb. 1856

Dear Thomson

I left a paper with you on 'Faradays lines of Force'.[2] If you have done with it & can lay hands on it, I would like to have it, because I have to write up the second part – 'On Faradays Electrotonic State',[3] and there will be a necessity for reference. I think I left an abstract too.[4] Tell me if you have it, but do not seek for it if you do not know where it is.

I was working at the following problems in the vacation.

I A thin hollow shell of conducting matter whose resistance and magnetic coeff[t] are given revolves about any axis in a uniform field of magnetic force. To find the electric currents and the magnetic effects (the magnetism due to the electric currents changes the magnetic field so that the resultant magnetism depends in direction as well as quantity on the velocity of rotation).[5]

II A closed wire of circular section is placed in the neighbourhood of another circuit through which a current whose intensity at any time is known is sent. To find the current at any time in the closed wire, regard being had *to the induction of the wire on itself*.[6]

---

(1) ULC Add. MSS 7342, M 92. Previously published in Larmor, 'Origins': 714–17.

(2) Part I of 'On Faraday's lines of force', *Trans. Camb. Phil. Soc.*, **10** (1856): 27–51 (= *Scientific Papers*, **1**: 155–88).

(3) Part II of 'On Faraday's lines of force': 51–83 (= *Scientific Papers*, **1**: 188–229).

(4) Number 84 is endorsed 'On Faraday's Lines of Force. Abstract', and is annotated by Thomson.

(5) Example XII of 'On Faraday's lines of force': 81–3 (= *Scientific Papers*, **1**: 226–9); compare Michael Faraday, 'Experimental researches in electricity. – Second series. Terrestrial magneto-electric induction', *Phil. Trans.*, **122** (1832): 163–94, esp. 168–70 (§§160–9) (= *Electricity*, **1**: 47–50). See Numbers 51 and 87. See also L. Boltzmann, ed., *Ueber Faraday's Kraftlinien. Von James Clerk Maxwell* (Leipzig, 1895): 122–8. Maxwell refers to an experiment by Foucault on the evolution of heat in a '*gyroscope*' rotating between the poles of an electromagnet; see Léon Foucault, 'De la chaleur produite par l'influence de l'aimant sur les corps en mouvement', *Comptes Rendus*, **41** (1855): 450–2. Referring to James Prescott Joule's paper 'On the calorific effects of magneto-electricity, and on the mechanical value of heat', *Phil. Mag.*, ser. 3, **23** (1843): 263–76, 347–55, 435–43, esp. 273, 353–5, where Joule established that 'heat would be evolved by the rotation of non-magnetic substances' in an electromagnetic field, Thomson commented on 'Foucault's experiment on the heat generated in his revolving mass' in a letter to Stokes of 12 November 1855, declaring that 'I think that Joule in 1843 was rather ahead of France in 1855' (ULC Add. MSS 7656, K 86).

(6) Example XI of 'On Faraday's lines of force': 79–80 (= *Scientific Papers*, **1**: 224–6).

I find that if we know by experiment the potential of the one circuit on the other, supposing each to be traversed by a unit current, and also the potential of each on itself, then we get rid of all the integrations and we have only to consider the peculiar effects due to the difference of the electromotive forces at different parts of the section of the wire. These are found by a very rapid approximation.

III A closed wire is made to rotate about an axis so rapidly that it produces a constant deflection of a needle in its neighbourhood. To find the relation between the velocity &c and the deflection of the needle.

I think something might be made of the following plan of determining the amount of magnetic induction passing through a given closed curve.[7]

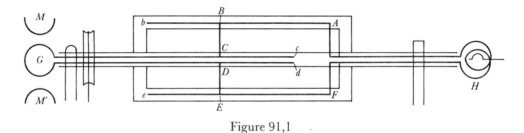

Figure 91,1

MM' are magnetic poles forming a field at G, $cCGDd$ is a wire attached to a horizontal axis and revolving with it. $eEFHABb$ is another wire attached also to the same axis. BC, DE are moveable links connecting one of these wires with the other and so completing the circuit. H is a coil consisting of a small number of turns of thick wire and allowing a small magnetic needle to be introduced suspended by a hook from a fixed rod in the axis produced and protected from wind by a glass globe. Let the horizontal axis be north & south, then the effect of turning the apparatus will be to produce a current alternating in direction as respects the wire but always in the same general direction

Figure 91,2

in space. If the rotation be rapid the deflection of the needle at H will be due to the difference of the inductive magnetic action of the magnets through G and that of the earth through $ABEF$. Now by moving BC, DE we can vary the area $ABEF$ so as to make the whole effect 0. Then the number of the magnet's lines through G will be equal to the number of earths lines through $ABCDEF$.

I have been working at the theory of Optical Instruments and I have reduced the whole theory of Geometrical Foci of compound instruments to a few propositions founded on the following axioms.

(7) See Number 87 esp. note (19).

II A wire of circular section [closed] is placed in the neighbourhood of another circuit through which a current whose intensity, at any time is known is sent. To find the current at any time in the closed wire, regard being had to the induction of the wire on itself

I find that if we know by experiment the potential of the one circuit on the other, supposing each to be traversed by a unit current, and also the potential of each on itself, then we get rid of all the integrals and we have only to consider the peculiar effects due to the difference of the electromotive forces at different parts of the section of the wire. These are found by a very rapid approximation

III A closed wire is made to revolve about an axis so rapidly that it produces a constant deflection of a needle in its neighbourhood. To find the relation between the velocity &c and the deflection of the needle

I think something might be made of the following plan of determining the amount of magnetic induction passing through a given closed curve

MM' are magnetic poles forming a field at G cC GDd is a wire attached to a horizontal axis and revolving with it. eEFHABb is another wire attached also to the same axis. BC, DE are moveable links connecting one of these wires with the other and so completing the circuit H is a coil consisting of a small number of turns of thick wire and allowing a small magnetic needle to be introduced suspended by a hook from a fixed rod in the axis produced and protected from wind by a glass globe. Let the horizontal axis be north & south, then the effect of turning the apparatus will be to produce a current alternating in direction as respects the wire but always in the same direction in space. If the rotation be rapid the deflection of the needle at H will be due to the difference of the inductive magnetic action of the magnets through G and that of the earth through ABEF Now by moving BC, DE we can vary the area ABEF so as to make the whole effect 0 Then the number of the magnets lines through G will be equal to the number of earths lines through ABCDEF.

Plate IV. An experiment to determine the magnetic induction passing through a closed circuit (1856), from a letter to William Thomson (Number 91).

I If a pencil of light falling on an instrument be small, nearly centrical and nearly direct, then if the incident light have a focus the emergent light will have a determinate focus.

II The distances of any two given rays of a pencil from its axis at any section before incidence are proportional to their distances in any other section after emergence.

Definition

If a small cylindrical pencil of light fall directly on an instrument, its form after emergence will be a double cone. The vertex of this cone is the *principal focus* and the two sections which are equal to that of the cylinder give the positions of two *focal centres*. That in which the rays are on the same side of the axis as in the cylinder is called the *positive focal centre*, and the other the *negative focal centre*. Turning the instrument the other way we get another principal focus and two focal centres.

From these axioms, without any assumption of law of refraction, centres of lenses &c we get this construction for any ray.[8]

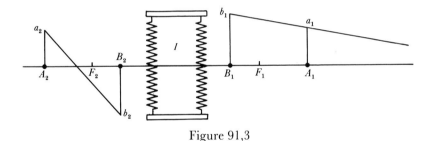

Figure 91,3

Let $A_1 A_2$ be the positive & $B_1 B_2$ the negative focal centres of the instrument *I*. Then if $a_1 b_1$ be any ray whatever cutting the planes of the focal centres in

---

(8) In his papers 'On the elementary theory of optical instruments', *Proc. Camb. Phil. Soc.*, **1** (1856): 173–5 and 'On the general laws of optical instruments', *Quarterly Journal of Pure and Applied Mathematics*, **2** (1858): 233–46 (= *Scientific Papers*, **1**: 238–40, 271–85), Maxwell develops a theory of geometrical optics which separates theorems expressing geometrical relations between object and image and the magnification of an image from the dioptrical properties of lenses. This approach, formulating theorems of geometrical optics independent of the physical process by which images are produced, was subsequently developed (independently) by Ernst Abbe; see Abbe, *Abhandlungen über die Theorie des Mikroskops* in his *Gesammelte Abhandlungen*, **1** (Jena, 1904). See also Max von Rohr, ed., *Die Bilderzeugung in optischen Instrumenten von Standpunkte der geometrischen Optik* (Berlin, 1904); and H. Boegehold, 'Die allgemeinen Gesetze über die Lichtstrahlenbündel und die optische Abbildung', in *Grundzüge der Theorie der optischen Instrumenten nach Abbe*, ed. S. Czapski and O. Eppenstein (Leipzig, ₃1924): 213–33, esp. 215–17.

$a_1 b_1$. Make $A_2 a_2 = A_1 a_1$ on the same side of the axis and $B_2 b_2 = -B_1 b_1$ on the opposite side of the axis. Join $b_2 a_2$ for the direction of the emergent ray.

We have also the following rule for conjugate foci.

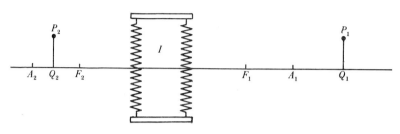

Figure 91,4

Let $A_1 A_2$ be positive focal centres, $F_1 F_2$ principal foci, $P_1 P_2$ conjugate foci for incident and emergent rays.

$$\frac{F_1 Q_1}{F_1 A_1} = \frac{F_2 A_2}{F_2 Q_2} = \frac{P_1 Q_1}{P_2 Q_2}$$

or
$$F_2 Q_2 = \frac{(F_1 A_1)(F_2 A_2)}{F_1 Q_1}, \quad P_2 Q_2 = \frac{(F_1 A_1)(P_1 Q_1)}{F_1 Q_1}$$

so that the position of P_2 may be found from that of P_1 by multiplication & division only.

I have also considered the laws of combination of two instruments and the case in which they form a 'telescope' in which the focal centres &c go off to infinity and instead we have a 'centre' C a 'modulus' n and a 'linear magnifying power' m.

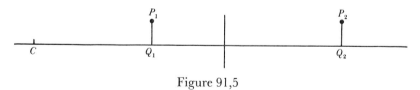

Figure 91,5

We have then

$$CQ_2 = nCQ_1$$
$$Q_2 P_2 = mQ_1 P_1$$

for the law of conjugate foci for a 'telescope'.

I am beginning to apply this method to the theory of primary & secondary foci of compound instruments. I find a great many good things in Smiths 'Opticks'.[9] Do you know where Möbius has put his optical theorems?[10]

I am working at another optical question – the place where a screen must be held so as to give the most distinct image of the edge of anything, the instrument being afflicted with aberration. It is not the 'least circle of aberration' that is a very useless thing, as I find in practice.

Smith (page 86 of Vol 2, art. 526 to 530) gives a good set of illustrations of the principles of the stereoscope.[11]

I have set up a reflecting stereoscope for Solid Geometry.

Two mirrors turning on axes inclined 45° to vertical. Pictures horizontal on table. Spectator looks horizontally.[12]

I hope to hear a good report of the College & all its inmates. Remember me to Rankine.[13]

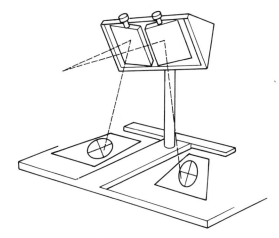

Figure 91,6

Yours truly
JAMES CLERK MAXWELL

(9) See Maxwell's reference to Cotes' theorem in his 'On the general laws of optical instruments': 233–4 (= *Scientific Papers*, **1**: 271), as stated by Robert Smith, *A Compleat System of Optics in Four Books, viz. A Popular, a Mathematical, a Mechanical, and a Philosophical Treatise. To which are added Remarks upon the Whole*, 2 vols. (Cambridge, 1738), **2**: ('Remarks') 76–8. 'To find the apparent magnitude, situation, apparent place and degree of distinctness with which an object is seen through any number of glasses of any sort, at any distances from each other and from the eye and object'. An object at P is viewed by the eye at O through lenses at A, B, C; Ω is the point at which the naked eye would see the object with the same magnitude as which it appears through the lenses; 'The apparent magnitude of the object will bear the same proportion to the true magnitude, as the distance PO bears to the distance $P\Omega$'.

(10) A. F. Möbius, 'Entwickelung der Lehre von dioptrischen Bildern mit Hülfe der Collineationsverwandtschaft', *Berichte über die Verhandlungen (math.-phys. Klasse) der Königlich Sächsischen Gesellschaft der Wissenschaften zu Leipzig*, **7** (1855): 8–32.

(11) Smith, *A Compleat System of Opticks*, **2**: ('Remarks') 86. This reference is given by Charles Wheatstone in his 'Contributions to the physiology of vision – Part the second. On some remarkable, and hitherto unobserved, phenomena of binocular vision', *Phil. Trans.*, **142** (1852): 1–17, on 3n. The paper includes a description of a portable reflecting stereoscope.

(12) Maxwell had first expressed interest in Wheatstone's mirror or reflecting stereoscope (and Brewster's lenticular or refracting stereoscope) in October 1849; see Number 20, esp. note (3).

(13) See Number 45 note (9).

92

FROM A LETTER TO JOHN CLERK MAXWELL

15 FEBRUARY 1856

From Campbell and Garnett, *Life of Maxwell*[1]

Trin. Coll.
15 Feb. 1856

Professor Forbes has written to me to say that the Professorship of Nat. Phil. at Marischal College, Aberdeen, is vacant[2] by the death of Mr. Gray,[3] and

---

(1) *Life of Maxwell*: 250–1.

(2) See a letter from J. D. Forbes to Maxwell dated 13 February 1856 (*Life of Maxwell*: 250). 'You may not perhaps have heard that Mr. Gray, Professor of Natural Philosophy, Marischal College, Aberdeen, is dead. He was a pleasing and energetic person, in the prime of life and health, a few months ago, when I saw him last. I have no idea whether the situation would be any object to you; but I thought I would mention it, as I think it would be a pity were it not filled by a Scotchman, and you are the person who occurs to me as best fitted for it. Do not imagine from my writing that I have the smallest influence in the matter, or interest in it beyond the welfare of the Scottish Universities. It is in the gift of the Crown. The Lord Advocate and Home Secretary are the parties to apply to. I am not acquainted with either. In the Commissioners' Report of 1830 the emoluments are stated at about £350. But they are not always to be depended upon. Another point. I think you ought certainly to be a Fellow of the Royal Society of Edinburgh. I shall be glad to propose you if you wish it.' Having encouraged Maxwell to apply to Marischal College, Forbes exerted his influence to secure the appointment. His testimonial is dated 18 February 1856: 'I have known Mr. Maxwell for a good many years, in fact since he was a youth. ... Since the first day I saw him, his attention appeared to be ardently and incessantly devoted to questions of Mathematics and Natural Philosophy. ... He subsequently attended my Lectures on Natural Philosophy, when he showed an exact acquaintance with many branches of science, and a singular fertility and accuracy in the original application of their principles. ... Mr. Maxwell's peculiar fitness for a Scotch Professorship of Science consists in this – that his scientific character was formed at a very early age under the Scotch system, with which he is therefore thoroughly conversant. ...' (St Andrews University Library, Forbes MSS, Letter Book V, 272; and the printed testimonials, Glasgow University Library, Yl–h.18). Forbes wrote to the Duke of Argyll (at the time Postmaster-General in Palmerston's ministry) on Maxwell's behalf on 7 March 1856 (Letter Book V, 280), pointing to the 'very remarkable coincidence between...[William Hopkins'] view of Maxwell's characteristics & those given by Prof. Stokes & by myself'. In his testimonial Hopkins declared: 'During the last thirty years I have been intimately acquainted with every man of mathematical distinction who has proceeded from the University, and I do not hesitate to assert that I have known no one who, at Mr. Maxwell's age, has possessed the same amount of accurate knowledge in the higher departments of physical science as himself. ... His mind is devoted to the prosecution of scientific studies, and it is impossible that he should not become (if his life be spared) one of the most distinguished men of science in this or any other country. ... It may possibly be thought that the language I have used may be the suggestion of private friendship, rather than that which might be dictated by sound judgment. I can only say, that I leave the justification of those expressions in

he inquires if I would apply for the situation, so I want to know what your notion or plan may be. For my own part, I think the sooner I get into regular work the better, and that the best way of getting into such work is to profess one's readiness by applying for it.

The appointment lies with the Crown – that is, the Lord Advocate and Home Secretary.[4] I suppose the correct thing to do is to send certificates of merit, signed by swells, to one or other of these officers.

I am going to ask about the method of the thing here, and Thacker[5] has promised to get me the College Testimonials.[6] If you see any one in Edinburgh that understands the sort of thing, could you pick up the outline of the process?

In all ordinary affairs political distinctions are supposed to weigh a great deal in Scotland. The English notion is that in pure and even in mixed mathematics politics are of little use, however much a knowledge of these sciences may promote the study of politics. As to Theology, I am not aware that the mathematicians, as a body, are guilty of any heresies, however some of them may have erred. But these are too mysterious subjects to furnish matter for calculation, so I may tell you that the reflecting stereoscope was finished yesterday, and looks well, and that I got a Devil made at the same time, which I play at the Gymnasium for relaxation and breathing time.[7]

Forbes also suggests my joining the Royal Society.[8]

---

their most literal sense to the character which Mr. Maxwell may hereafter establish for himself, whether at Aberdeen or elsewhere.' (Glasgow University Library, Yl–h.18). In 1853 Maxwell's Trinity contemporary W. N. Lawson recorded Hopkins' opinion that Maxwell was 'the most extraordinary man he has met with in the whole range of his experience ... a great genius ...' (*Life of Maxwell*: 133n). For Stokes' testimonial see Number 97 note (4).

(3) David Gray, Professor of Natural Philosophy at Marischal College, died 10 February 1856; see [P. J. Anderson,] *Officers of the Marischal College and University of Aberdeen 1593–1860* (Aberdeen, 1897): 49.

(4) The Lord Advocate (the pre-eminent legal and political functionary in Scotland prior to the appointment of a Secretary of State for Scotland in 1880) was at the time James Moncreiff; see G. W. T. Omond, *The Lord Advocates of Scotland. Second Series. 1834–1880* (London, 1914): 170–90. In his letter to the Duke of Argyll of 7 March 1856 (see note (2)) Forbes stated that Maxwell had sent his testimonials to Sir George Grey, Home Secretary in Palmerston's ministry.

(5) Arthur Thacker, Tutor of Trinity College, Cambridge (Venn); his testimonial is dated 21 February 1856, attesting Maxwell's 'extensive, well-ordered, and mature knowledge of physical science, and [that] he combines with that knowledge a power and fertility of geometrical illustration of which, in my own experience, I have found no parallel ...'. (Glasgow University Library, Yl–h.18).

(6) The testimonial from the Master and Fellows of Trinity College is dated 22 February 1856, and records Maxwell's Cambridge career (Glasgow University Library, Yl–h.18).

(7) The devil on two sticks: see Number 9 note (2).

(8) See Number 115 esp. note (2).

93

LETTER TO GEORGE GABRIEL STOKES

16 FEBRUARY 1856

From the original in the University Library, Cambridge[1]

Trinity College Feb 16, 1856

Dear Sir

Professor Forbes writes to me that the chair of Natural Philosophy at Aberdeen (Marischal College) is vacant and suggests that I should become a candidate.

I have no hesitation in acting on that suggestion at once but I would like to know what sort of men go in, and what chance there is of success, because I have no wish to have what I cannot get.

Now as several of your Cambridge acquaintances have obtained such situations in Scotland & Ireland, I thought you might both know the style of man wanted and whether any such were in the field and might also have learned the outline of the proper method of making application.

I find the presentation belongs to the Crown, that is I suppose equivalent to the fact that the Home Secretary & Lord Advocate are the proper persons to apply to.[2]

I hope to hear the details from someone who knows, meanwhile I have no doubt that anything you could say in my favour would have its weight in a certificate or testimonial, so as soon as I am better acquainted with the matter I will write and ask you for one if it is wanted.

I have been trying to simplify the theory of compound Optical Instruments[3] and I have been reading up all that I can find on that subject. I find a great deal in 'Smiths Opticks' that is good[4] and also investigations by Euler, Lagrange, & Gauss.[5] Do you know of any other researches on this subject?

---

(1) ULC Add. MSS 7656, M 405. Previously published in Larmor, *Correspondence*, **2**: 2–3.

(2) See Number 92.   (3) See Numbers 90 and 91.

(4) 'Cotes' theorem', as stated by Robert Smith, *A Compleat System of Opticks*, 2 vols. (Cambridge, 1738), 2: ('Remarks') 76–8; see Number 91 note (9).

(5) Leonhard Euler, 'Régles générales pour la construction des télescopes et des microscopes de quelques nombres de verres qu'ils soyent composés', *Mémoires de l'Académie Royale des Sciences de Berlin*, **13** (1757): 288–322; Euler, 'Régles générales pour la construction des télescopes et microscopes', *ibid.*, **17** (1761): 201–11; Euler, 'Précis d'une théorie générale de la dioptrique', *Mémoires de l'Académie Royale des Sciences* (année 1765): 555–75; J. L. Lagrange, 'Sur la théorie des lunettes', *Nouveau Mémoires de l'Académie Royale des Sciences de Berlin* (année 1778): 162–80; Lagrange, 'Sur une loi générale d'optique', *ibid.*, (année 1803): 3–12; and Carl Friedrich Gauss, *Dioptrische Untersuchungen* (Göttingen, 1841) (= *Abhandlungen der Mathematische Classe der Königlichen Gesellschaft der Wissenschaften zu Göttingen*, Theil 1, Von den Jahren 1838–1843: 1–34), abstracted by

I have greatly simplified the calculation of the foci of an instrument as far as first approximations by considering the 'principal foci' and 'focal centres' of the instrument. In this way I can begin with two axioms about the *possibility* of conjugate foci, undistorted images and thence immediately deduce the law of conjugate foci and the magnitudes of images in any instrument. I thus render the theory of instruments independent of the particular optical apparatus used in the instrument and make no reference to the propositions about refraction at spherical surfaces.[6]

The geometry is very simple and short and all that is necessary to be remembered is easily expressed in two eq^{ns} – thus –

Let $F_1 F_2$ be the principal foci $A_1 A_2$ focal centres $P_1 P_2$ conjugate foci $P_1 Q_1, P_2 Q_2$ perp. to axis

$$\frac{Q_1 F_1}{F_1 A_1} = \frac{F_2 A_2}{F_2 Q_2} = \frac{P_1 Q_1}{P_2 Q_2}.$$

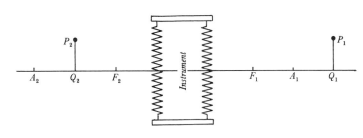

Figure 93,1

There are several other things in Geometrical Optics that I find worth examining.

A modification of the same method seems applicable to the case of small oblique & excentric pencils.

I hope I do not trouble you in the first part of this letter. You must understand that I do not wish to trouble you now but merely to inform you of my design and to intimate that if I went through with it I might find it necessary to write to you for your certificate as Professor.

Yours truly

J. C. MAXWELL

W. H. Miller in *Scientific Memoirs*, ed. R. Taylor, **3** (London, 1843): 490–98. See the sources which Maxwell cites in his paper 'On the general laws of optical instruments', *Quarterly Journal of Pure and Applied Mathematics*, **2** (1858): 233–46, esp. 233 (= *Scientific Papers*, **1**: 271).

(6) See Number 91 note (8).

94

LETTER TO WILLIAM THOMSON

19 FEBRUARY 1856

From the original in the University Library, Cambridge[1]

Trin Coll.
19th Feb 1856

Dear Thomson

Thanks for your letter, written under stress of time, and for the papers on which has come no hurt.[2]

I am glad we are to have some tangible memorial of your experiments[3] for it appeared to me that science would suffer for want of a reporter like the worthies who lived before Agamemnon.[4]

My present business is to trouble you again. I wish you to write out a description of me and sign it that I may send it to the representatives of the Crown and that they may look favourably on my scheme of setting up as professor of nat. phil. at Marischal College Aberdeen.[5] I have written to the Home Sec^y & the Lord Advocate to offer myself and I have said that I would send in Testimonials as soon as my friends could write them out.[6] I had got promise of several but I reckoned on yours too. I trust I was right both in the whole affair and in this particular supposition. I know nothing about the time of election or the kind of candidates but time is always valuable, so the sooner you write the better.

I did not know Mr Gray. Prof. Forbes writes to me that he was a young & healthy man and known to him.[7]

---

(1) ULC Add. MSS 7342, M 94. Previously published in Larmor, 'Origins': 717–18.

(2) See Number 91.

(3) Thomson's Bakerian Lecture (delivered on 28 February 1856) 'On the electrodynamic qualities of metals', *Phil. Trans.*, **146** (1856): 649–751 (= *Math. & Phys. Papers*, **2**: 189–327).

(4) Compare Maxwell's quotation from Horace, *Odes*, IV.ix, 25 in his letter to Thomson of 24 November 1857 (Number 137, esp. note (3)).

(5) Thomson's testimonial is dated 23 February 1856: '... I have been very much struck with the singular acuteness and the great fertility of his mind, and I believe him to possess a power of original research in the most varied fields of investigation, which cannot fail to achieve results wherever it is applied. ... He is possessed of one very available and important qualification for a University Lecturer, – in his extensive and familiar knowledge, not only of modern discovery, but of historical works in science: and ... he possesses to a very remarkable degree ... [the] power of appreciating the value and meaning of a physical hypothesis. ...' (Glasgow University Library, Y1–h.18). John Clerk Maxwell had written to Thomson on 18 February 1856 (ULC Add. MSS 7342, M 93).

(6) See Number 92.

(7) Number 92 note (2).

If Mr Joule[8] can see stereoscopic pictures, putting a division from his nose to between the pictures, he can do more than I can. If he squints with the right eye at the left picture et v.v. he sees everything inside out. Which?

These are two precisely similar curves projections of the helix. If you squint at them straight they will appear plane curves. If you raise the left side of the paper you will see a right handed helix start into solidity and if you raise the right the helix will be left handed (squinting across).

With well drawn curves the effect is wonderful.

I have drawn stereoscopic pictures of the spherical ellipse in all positions of a sphere with two cylinders crossing it (the 'Florentine Enigma')[9] of a knot on a thread and a collection of others all large for the reflecting Stereoscope.

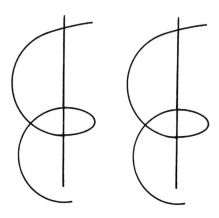

Figure 94,1

Yours

J. C. MAXWELL

---

(8) James Prescott Joule; his letters to Thomson at this time (ULC Add. MSS 7342) make no reference to the topic.

(9) Vincenzo Viviani's anonymously issued challenge problem 'Ænigma Geometricum De Miro Opificio Testudinis Quadrabilis Hemisphæricae', printed in *Phil. Trans.*, **17** (1692/3): 585–6, which drew solutions from Leibniz, Jakob Bernoulli and David Gregory, in response to Viviani's challenge to find a 'quadrable hemispherical tortoise'. The hemispherical surface is pierced by four equal windows at its base cut by cylinders of diameter equal to the hemisphere's radius, so that the surface area of the 'tortoise shell' remaining is the square of the hemisphere's diameter. The 'Florentine enigma' is stated in the following terms by D. F. Gregory in his *Examples of the Processes of the Differential and Integral Calculus* (Cambridge, 1841): 424, 431–2; 'A sphere is cut by a cylinder, the radius of whose base is half that of the sphere, and whose axis bisects the radius of the sphere at right angles...[It is required] to find the area of the intercepted surface of the sphere.... If the sphere be pierced by two equal and similar cylinders, the area of the non-intercepted surface is...twice the square of the diameter of the sphere. This is the celebrated Florentine enigma which was proposed by Vincent Viviani as a challenge to the mathematicians of his day.' For discussion see also F. G. Teixeira, *Traité des Courbes Spéciales Remarquables*, 2 vols. (Coimbra, 1908–9), **2**: 311–20, and D. T. Whiteside's note in *The Mathematical Papers of Isaac Newton*, 8 vols. (Cambridge, 1967–81), **8**: 80–1n.

FROM A LETTER TO JOHN CLERK MAXWELL

20 FEBRUARY 1856

From Campbell and Garnett, *Life of Maxwell*[1]

Trin. Coll,
20 Feb./56

As far as writing (Testimonials) goes, there is a good deal, and if you believe the Testimonials you would think the Government had in their hands the triumph or downfall of education generally, according as they elected one or not.

However, wisdom is of many kinds, and I do not know which dwells with wise counsellors most, whether scientific, practical, political, or ecclesiastical. I hear there are candidates of all kinds relying on the predominance of one or other of these kinds of wisdom in the constitution of the Government.

I had a letter from Dr. Swan of Edinburgh,[2] who is a candidate, asking me for my good opinion, which I gave him, so far as I had one. His printed papers are good, and I hear he is so himself.[3] Maclennan is also a candidate.[4] He has the qualification of making himself understood.

The results of this term are chiefly solid Geometry Lectures, stereoscopic pictures, and optical theorems.[5] My lectures are to be on Rigid Dynamics and Astronomy next term, so I do not expect to be out of work by reason of Aberdeen, and I have plenty to get through in those subjects.

I have been making more stereoscopic curves for my lectures. I intend to select some and draw them very neat the size of the ordinary stereoscopic pictures, and write a description of them, and publish them as mathematical illustrations. I am going to do one now to illustrate the theory of contour lines in maps, and to show how the rivers must run, and where the lines of watershed must be.[6]

---

(1) *Life of Maxwell*: 251–2.

(2) William Swan, subsequently Professor of Natural Philosophy at St Andrews 1859 (*DNB*); and see Number 32 note (3).

(3) William Swan, 'On the gradual production of luminous impressions on the eye, and other phenomena of vision', *Trans. Roy. Soc. Edinb.*, **16** (1849): 581–603, to which Maxwell refers in his paper 'Experiments on colour, as perceived by the eye', *ibid.*, **21** (1855): 275–98, on 288 (= *Scientific Papers*, **1**: 142). See also Swan's papers 'On the total eclipse of the sun, on July 28 1851, observed at Göteborg, with a description of a new position micrometer', *Trans. Roy. Soc. Edinb.*, **20** (1853): 335–46; and 'On the red prominences seen during total eclipses of the sun', *ibid.*: 445–59, 461–73.   (4) Number 37 note (4).   (5) Numbers 91, 93 and 94.

(6) Compare Maxwell's paper 'On hills and dales', *Phil. Mag.*, ser. 4, **40** (1870): 421–7 (= *Scientific Papers*, **2**: 233–40).

LETTER TO WILLIAM THOMSON

22 FEBRUARY 1856

From the original in the University Library, Cambridge[1]

Trin Coll
Feb 22nd 1856

Dear Thomson

I return you Fullers letter; I had written to him myself before I got it, to ask for information, particularly the time of election.[2] He says that Tait is in. He is doing well at Belfast & is resolved to do something in physics. I should not be sorry to see him get some encouragement. I do not think his own college appreciates his mathematical powers. There are two more men I know Swan a teacher in Edin[h] who has done several ingenious things in various subjects, and MacLennan a Trinity man now in Edin[h] who would get on as a lawyer better than as a professor.[3]

Both have asked me for my good will which they have as I do not think either of them men to be avoided and both are very ingenious & MacLennan, though very loose in some parts of science has a great power of expression.

As for what Fuller says about the press I think that if there is a good virulent newspaper editor quarreling with the whole college, the professors may agree to quarrel with him instead of one another.

Now a chronic slanging mania is the normal state of the editor but the utter spoiling of a college. I will therefore give my vote that Carthage should be spared.

Some of my friends are going to act on Lords Ardmillan & Cowan[4] and I hope to organise another line of attack through Sir J. Herschel[5] & Lord Aberdeen.[6]

---

(1) ULC Add. MSS 7342, M 95. Previously published in Larmor, 'Origins': 718–19.

(2) Frederick Fuller, Professor of Mathematics in King's College Aberdeen, formerly Tutor of Peterhouse (Venn). Fuller had been Maxwell's tutor at Peterhouse in 1850, and wrote a testimonial dated 23 February 1856 attesting that he had 'formed a very high estimate of his mathematical abilities...' (Glasgow University Library Y1–h.18). In his letter to Thomson of 18 February 1856 (ULC Add. MSS 7342, M 93) John Clerk Maxwell had requested Thomson's aid in obtaining 'some help from Prof. Fuller'.

(3) See Number 95.

(4) James Craufurd, Lord Ardmillan, Lord of the Court of Session and Lord of Justiciary, Scotland (*DNB*); John Cowan, Lord of Session and Lord of Justiciary (*Modern English Biography*, ed. F. Boase, **1** (Truro, 1892): 735).

(5) See Number 97.

(6) George Hamilton-Gordon, 4th Earl of Aberdeen, Prime Minister 1852–5 (*DNB*).

Swan has sent me his testimonials neatly printed so that I shall be able to give the printer a specimen of how he ought to truss these things up. How came Cayley to be wanting a professorship.[7] I thought he was a great lawyer. One might suppose Adams to be wanting it much more.[8]

I have found a curve in space that has its centre of plane curvature & centre of spherical curvature easily determined. I have made stereoscopic pictures of it and have been able to explain to the 3rd year men the whole theory of normal & osculating planes principal normal, axis of curvature involute in space & so on. The eqns to the curve are

$$x = 4a\cos^3\theta \quad y = 4a\sin^3\theta \quad z = 3c\cos 2\theta.$$

Those of the centre of sphere of curvature

$$x = 4\frac{4a^2+3c^2}{a}\cos^3\theta \quad y = 4\frac{4a^2+3c^2}{a}\sin^3\theta \quad z = 3\frac{4a^2+3c^2}{c}\cos 2\theta$$

and those of the centre of circle of curvature are

$$x = 4a\cos^3\theta + 12\frac{a^2+c^2}{a}\cos\theta\sin^2\theta, \quad y = 4a\sin^3\theta + 12\frac{a^2+c^2}{a}\cos^2\theta\sin^2\theta$$

$$z = 3c\cos 2\theta.$$

I have learnt a great deal about curves by drawing this one.

Here is a stereoscopic view drawn by hand which will show you that in the large curve which

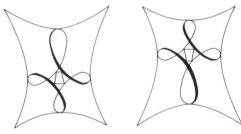

Figure 96,1

is the locus of centres of spheres of contact the upper left & lower right corners are uppermost while the opposite is the case with the original curve which is small . The locus of centre of ⊙[9] of curvature has the

upper right & lower left sweeps uppermost. It is an involute of the large curve.

(7) Arthur Cayley, Trinity 1838, Sadlerian Professor of Pure Mathematics at Cambridge 1863.

(8) John Couch Adams, St John's 1839, Professor of Mathematics at St Andrews 1858, Lowndean Professor of Astronomy and Geometry at Cambridge 1859 (Venn).

(9) Circle.

I hope to hear soon of the Bakerian Lecture.[10] Roscoe[11] told me about it. Do you think my paper on Faradays lines too long for the Phil Mag. I would like to put it in because Faraday reads it and so does Tyndall.[12]

Yours truly

J. C. MAXWELL

(10) See Number 94 note (3).

(11) Henry Enfield Roscoe, subsequently Professor of Chemistry at Owens College, Manchester (*DNB*).

(12) See Number 84 note (40) and Number 33. Compare John Tyndall's letter to Maxwell of 7 November 1857: 'I am very much obliged to you for your kind thoughtfulness in sending me your papers on the Dynamical Top & on the Perception of Colour, as also for your memoir on the Lines of Force received some time ago. I never doubted the possibility of giving Faraday's notions a mathematical form, and you would probably be one of the last to deny the possibility of a totally different imagery by which the phenomena might be represented.' (ULC Add. MSS 7655, II/13; *Life of Maxwell*: 288).

97

LETTER TO GEORGE GABRIEL STOKES

22 FEBRUARY 1856

From the original in the University Library, Cambridge[1]

Trinity College
Feb 22nd 1856

Dear Sir

I have received your two letters. I acted upon your advice with regard to Lord Aberdeen by asking Alex. Herschel to get his father[2] to tell Lord Aberdeen that my case was worth attending to. Two of my friends are writing to two friends of the Ld Advocate.[3] So much for private influence. With respect to testimony, I should think my friends are all saying their utmost in my favour judging by the specimens I have received. I am obliged to you for promising me a testimonial. As I am told it is important to be early I hope you will send it soon.[4]

I have received two applications myself for testimonials.[5] I intend to write out my estimate of the men and leave it to them to consider whether they will use it at once or wait till another time.

I find the nearest thing to my optical propositions in Smiths Optics.[6] You

---

(1) ULC Add. MSS 7656. Previously published in Larmor, *Correspondence*, **2**: 3–4.

(2) Alexander Stewart Herschel, Trinity 1854 (Venn); and Sir John Herschel (Venn; *DNB*), who wrote a testimonial declaring he considered Maxwell's '[paper] "Experiments on Colour, as perceived by the Eye, and Remarks on Colour-Blindness"…as a very able production, and one highly creditable to him as an experimenter and a photologist'. (Glasgow University Library, Yl–h.18). On Herschel's interest in colour blindness see Number 47 note (12).

(3) See Numbers 92, 94 and 96.

(4) Stokes' testimonial is dated 25 February 1856: '…I have often been struck with the singular clearness of his apprehension of physical and geometric subjects, and I have formed the highest opinion of his powers, so that I look on him as destined, should his life and health be prolonged, and should he continue to devote himself to science, to stand in the foremost rank among English scientific men. One thing more is wanted in a teacher, namely, a power of conveying clearly his knowledge to others. That Mr. Maxwell possesses that power I feel satisfied, having once been present when he was giving an account of some of his geometrical researches to the Cambridge Philosophical Society, on which occasion I was struck with the singularly lucid manner of his exposition…' (Glasgow University Library, Yl–h.18). See Number 46, and compare Number 92 note (2). Stokes commended Maxwell's paper 'On the equilibrium of elastic solids' (1850) as 'highly original…involving a profound acquaintance with some of the most abstruse branches of physical investigation…' (see Number 26).

(5) See Numbers 95 and 96.

(6) Robert Smith, *A Compleat System of Opticks*, 2 vols. (Cambridge, 1738), 2: ('Remarks') 76–8; see Number 91 note (9).

will find a good description of a stereoscopic experiment in vol II page 388 and in the 'Remarks' page 108.[7] Miller[8] has lent me a paper by Listing on the Dioptics of the Eye,[9] which looks like good optics as well as anatomy.

I have set up a reflecting stereoscope on a new plan[10] and am using it for lectures on Solid Geometry. I have worked out a curve with the locus of the centre of the circle of curvature & of the sphere of curvature, the curve is

$$x = 4a\cos^3\theta \quad y = 4a\sin^3\theta \quad z = 3a\cos 2\theta.\text{[11]}$$

It is the best curve I know to explain everything on, for all the results have simple expressions.

I read the second part of my paper on Faradays 'lines of force' at the Philosophical some weeks ago. It is an examination of the 'Electrotonic state' and an application of the mathematical expression of it to the solution of problems.[12] I have not got it into shape owing to other things on hand, Aberdeen among the rest, but the first part,[13] which is entirely elementary I consider intelligible without professional mathematics, and I would like to know if a somewhat long investigation of the strict theory of the lines of force would be acceptable in the Phil. Mag. I want to get Tyndall & Faraday to read it. Thomson has read it.[14] It requires attention and imagination but no calculation, at least that is all done by me and put in the form of diagrams of the lines.[15] I have no more time just now.

Yours truly
J. C. MAXWELL

Is it true that Cayley is going in?[16]

---

(7) See Smith's discussion of a binocular telescope in his *Opticks*, **2**: 388 and ('Remarks'): 108.

(8) William Hallowes Miller, Professor of Mineralogy at Cambridge (Venn); see also Number 93 note (5).

(9) Johann Benedict Listing, 'Mathematische Discussion des Ganges der Lichtstrahlen im Auge', in *Handwörterbuch der Physiologie*, ed. Rudolf Wagner, 4 vols. (Braunschweig, 1842–53), **4**: 451–504. Compare Maxwell's reference in 'On the general laws of optical instruments', *Quarterly Journal of Pure and Applied Mathematics*, **2** (1858): 233–46, on 233 (= *Scientific Papers*, **1**: 271).

(10) See Number 91.   (11) Compare Number 96.   (12) Number 87.

(13) See Numbers 84 and 85.   (14) See Numbers 91 and 94.

(15) See Number 86 esp. note (4).   (16) See Number 96.

FROM A LETTER TO JOHN CLERK MAXWELL

12 MARCH [1856]

From Campbell and Garnett, *Life of Maxwell*[1]

Trin. Coll.
12 March

I was at the Working College[2] to-day, working at decimal fractions. We are getting up a preparatory school for biggish boys to get up their preliminaries. We are also agitating in favour of early closing of shops. We have got the whole of the ironmongers, and all the shoemakers but one. The booksellers have done it some time. The Pitt Press keeps late hours, and is to be petitioned to shut up.

I have just written out an abstract of the second part of my paper on Faraday's Lines of Force.[3] I hope soon to write properly the paper of which it is an abstract. It is four weeks since I read it. I have done nothing in that way this term, but am just beginning to feel the electrical state come on again, and I hope to work it up well next term.

---

(1) *Life of Maxwell*: 252–3.
(2) See Number 82.
(3) Number 87.

FROM A LETTER TO LEWIS CAMPBELL

22 APRIL 1856

From Campbell and Garnett, *Life of Maxwell*[1]

Trin. Coll.
22 April 1856

I have two or three stiff bits of work to get through this term here, and I hope to overtake them. When the term is over I must go home and pay diligent attention to everything there, so that I may learn what to do.

The first thing I must do is carry on my father's work of personally super-intending everything at home, and for doing this I have his regular accounts of what used to be done, and the memories of all the people, who tell me everything they know.[2] As for my own pursuits, it was my father's wish, and it is mine, that I should go on with them. We used to settle that what I ought to be engaged in was some occupation of teaching, admitting of long vacations for being at home; and when my father heard of the Aberdeen proposition he very much approved. I have not heard anything very lately, but I believe my name is not yet put out of question in the Ld Advocate's book.[3] If I get back to Glenlair I shall have the mark of my father's work on everything I see. Much of them is still his, and I must be in some degree his steward to take care of them. I trust that the knowledge of his plans may be a guide to me, and never a constraint....

I am getting a new top turned to show my class the motion of bodies of various forms about a fixed point. I expect to get very neat results from it, and agreeing with theory of course.[4]

---

(1) *Life of Maxwell*: 255–6; abridged.

(2) John Clerk Maxwell died on 3 April 1856; see Maxwell's letters to Jane Cay and Mrs Blackburn of that date (*Life of Maxwell*: 253–4).

(3) See Number 92 note (4).

(4) See Number 89 esp. note (5).

LETTER TO WILLIAM THOMSON

25 APRIL 1856

From the original in the University Library, Cambridge[1]

Trinity College
25th April 1856

Dear Thomson

I enclose a note to your brother,[2] for though you told me his address I have forgotten it. Blackburn would tell you of my private affairs since we left Glasgow.[3] My father expected his death continually and the fear of it never alarmed him. It came at last with less outward signs than had been seen on several occasions in which he recovered. He was happy to have got home, and to have accomplished several things he had on hand and though he might have had comfort in his troubles here, he was content to go, though he did not know the time.

I returned here on the 19th as they had got no one to take my lectures. I am getting up the analytical part of my electromagnetic functions.[4]

Have you any propositions on maxima or rather minima[5] of quantities like

$$\iiint (a\alpha + b\beta + c\gamma)\, dx\, dy\, dz$$

through infinity where $a\,b\,c$ are measures of quantity, magnetic or galvanic and $\alpha\,\beta\,\gamma$ of intensity[6] and

$$a = A_1\,\alpha + B_1\,\beta + C_1\,\gamma$$
$$b = A_2\,\alpha + B_2\,\beta + C_2\,\gamma$$
$$c = A_3\,\alpha + B_3\,\beta + C_3\,\gamma.$$

If
$$\alpha = \frac{dp}{dx} \quad \beta = \frac{dp}{dy} \quad \gamma = \frac{dp}{dz}$$

---

(1) ULC Add. MSS 7342, M 96. Previously published in Larmor, 'Origins': 720–2. The letter has a mourning border: see Number 99.

(2) James Thomson, subsequently Professor of Civil Engineering at Queen's College, Belfast.

(3) See Numbers 45 note (10) and 99 note (2).

(4) Compare J. C. Maxwell, 'On Faraday's lines of force', *Trans. Camb. Phil. Soc.*, **10** (1856): 27–83, esp. 57–65 (= *Scientific Papers*, **1**: 195–205).

(5) Maxwell is alluding to Thomson's discussion of extremal conditions in his paper 'Theorems with reference to the solution of certain partial differential equations', *Camb. & Dubl. Math. J.*, **3** (1848): 84–7 (= *Electrostatics and Magnetism*: 139–41); see Number 66 note (21).

(6) The integral measures the energy of an electromagnetic system; see Maxwell, 'On Faraday's lines of force': 61–2 (= *Scientific Papers*, **1**: 201–2), and Number 87. Here a, b, c are the 'quantities' (taken through a surface) of electric current and magnetic induction; and α, β, γ the 'intensities' of electromotive force and magnetic force (taken round its boundary).

and if
$$\frac{da}{dx}+\frac{db}{dy}+\frac{dc}{dz}=-4\pi\rho^{(7)}$$

where ρ is a given f^n of $x\,y\,z$

$$\frac{d}{dx}(A_1\,\alpha+A_2\,\beta+A_3\,\gamma)+\frac{d}{dy}(B_1\,\alpha+B_2\,\beta+B_3\,\gamma)+\frac{d}{dz}(C_1\,\alpha+C_2\,\beta+C_3\,\gamma)=4\pi\rho$$

is the condition of minimum.

If you have any three functions of $x\,y\,z$ such as $a\,b\,c$ and if you make

$$\frac{da}{dx}+\frac{db}{dy}+\frac{dc}{dz}=-4\pi\rho$$

and if you find the value of V

$$V=\iiint\frac{\rho\,dx\,dy\,dz}{\sqrt{\overline{x-x'}|^2+\overline{y-y'}|^2+\overline{z-z'}|^2}}\qquad(8)$$

and make
$$a-\frac{dV}{dx}=a',\;b-\frac{dV}{dy}=b',\;c-\frac{dV}{dz}=c'$$

then
$$\frac{da'}{dx}+\frac{db'}{dy}+\frac{dc'}{dz}=0^{(9)}$$

and therefore by your prop. in theory of magnetism$^{(10)}$ if $a'\,b'\,c'$ be given as

---

(7) ρ is the current density and the '*real magnetic density*' and p the electric and magnetic 'potential or tension'; see 'On Faraday's lines of force': 58, 62 (= *Scientific Papers*, **1**: 197, 202).

(8) The value of V (the potential function) which satisfies Poisson's equation $\nabla^2 V = -4\pi\rho$. See Theorem II of 'On Faraday's lines of force': 57 (= *Scientific Papers*, **1**: 195); and compare Thomson, 'Theorems with reference to the solution of certain partial differential equations': 84. See also George Green, *An Essay on the Application of Mathematical Analysis to the Theories of Electricity and Magnetism* (Nottingham, 1828): 8; and P. S. de Laplace, *Traité de Mécanique Céleste*, 5 vols. (Paris, An VII [1799]–1825), **1**: 136–7.

(9) See Theorem VI of 'On Faraday's lines of force': 61 (= *Scientific Papers*, **1**: 200–1).

(10) William Thomson, 'A mathematical theory of magnetism', *Phil. Trans.*, **141** (1851): 243–85, esp. 283–5 (*Electrostatics and Magnetism*: 401–4). For a solenoidal distribution of magnetism $d\alpha/dx+d\beta/dy+d\gamma/dz=0$, where α, β, γ denote the components of the intensity of magnetisation at any internal point in a magnet, Thomson states $\alpha = dH/dy-dG/dz$, $\beta = dF/dz-dH/dx$, $\gamma = dG/dx-dF/dy$ 'where F, G, H are three functions to a certain extent arbitrary', and he obtains
$$F=\int(\beta\,dz-\gamma\,dy)+\frac{d\psi}{dx}$$
$$G=\int(\gamma\,dx-\alpha\,dz)+\frac{d\psi}{dy}$$
$$H=\int(\alpha\,dy-\beta\,dx)+\frac{d\psi}{dz}$$
where ψ denotes an arbitrary function. Compare Theorem V of 'On Faraday's lines of force': 59–60 (= *Scientific Papers*, **1**: 198–9). See Number 51 note (7) and Number 71 note (29).

continuous functions of xyz and if $\int a'\,dz$ mean the result of integration with respect to z as the quantity stands then if[11]

$$\alpha_0 = \int c\,dy - \int b\,dz$$

$$\beta_0 = \int a\,dz - \int c\,dx$$

$$\gamma_0 = \int b\,dx - \int a\,dy$$

then we shall have[12]

$$a = \frac{dV}{dx} + \frac{d\beta_0}{dz} - \frac{d\gamma_0}{dy}$$

$$b = \frac{dV}{dy} + \frac{d\gamma_0}{dx} - \frac{d\alpha_0}{dz}$$

$$c = \frac{dV}{dz} + \frac{d\alpha_0}{dy} - \frac{d\beta_0}{dx}.$$

Now if we wish to transform

$$\iiint (a\alpha_1 + b\beta_1 + c\gamma_1)\,dx\,dy\,dz$$

where the integration is through infinity and all the quantities vanish there and if $a\,b\,c$ are expressed as above then

$$\iiint \left(\frac{dV}{dx}\alpha_1 + \frac{dV}{dy}\beta_1 + \frac{dV}{dz}\gamma_1 \right) + \left(\frac{d\beta_0}{dz} - \frac{d\gamma_0}{dy} \right)\alpha_1 + \left(\frac{d\gamma_0}{dx} - \frac{d\alpha_0}{dz} \right)\beta_1 + \left(\frac{d\alpha_0}{dy} - \frac{d\beta_0}{dx} \right)\gamma_1$$

$$= -\iiint V\left(\frac{d\alpha_1}{dx} + \frac{d\beta_1}{dy} + \frac{d\gamma_1}{dz} \right) dx\,dy\,dz$$

$$-\iiint \alpha_0\left(\frac{d\beta_1}{dz} - \frac{d\gamma_1}{dy} \right) + \beta_0\left(\frac{d\gamma_1}{dx} - \frac{d\alpha_1}{dz} \right) + \gamma_0\left(\frac{d\alpha_1}{dy} - \frac{d\beta_1}{dx} \right).\text{[13]}$$

Now if we put
$$\iiint \frac{\left(\dfrac{d\alpha_1}{dx} + \dfrac{d\beta_1}{dy} + \dfrac{d\gamma_1}{dz} \right) dx\,dy\,dz}{\sqrt{\overline{x-x'}|^2 + \overline{y-y'}|^2 + \overline{z-z'}|^2}} = p$$

---

(11) $\alpha_0, \beta_0, \gamma_0$ are the components of the electro-tonic intensity; see Number 87.

(12) Compare Theorem VI of 'On Faraday's lines of force': 61 ($=$ *Scientific Papers*, **1**: 200).

(13) Compare Theorem VII of 'On Faraday's lines of force': 61–2 ($=$ *Scientific Papers*, **1**: 201–2). Maxwell uses the suffix (1) to denote magnetic quantities.

then the first term becomes

$$\sum V\left(\frac{d^2p}{dx^2}+\frac{d^2p}{dy^2}+\frac{d^2p}{dz^2}\right)$$

which we know $= \sum p\left(\frac{d^2V}{dx^2}+\frac{d^2V}{dy^2}+\frac{d^2V}{dz^2}\right)$ or $= 4\pi\sum p\rho$ [14]

and if we put

$$\frac{d\beta_1}{dz}-\frac{d\gamma_1}{dy}=a_2 \&\text{c}^{[15]}$$

the second term is $\iiint \alpha_0 a_2 + \beta_0 b_2 + \gamma_0 c_2$.

Now if $a_1\,b_1\,c_1$ be measures of quantity of magnetization ρ is the real magnetic density, and $\alpha_0\,\beta_0\,\gamma_0$ are the electrotonic functions, and $a_2\,b_2\,c_2$ the electric currents if $\alpha_1\,\beta_1\,\gamma_1$ be the intensities of magnetization so that

$$\sum a_1\alpha_1 + b_1\beta_1 + c_1\gamma_1 = -\sum 4\pi p\rho - \sum (\alpha_0 a_2 + \beta_0 b_2 + \gamma_0 c_2).^{[16]}$$

Yours truly
J. C. MAXWELL

I seem to have got the professorship.[17]

---

(14) See Theorems III and VII of 'On Faraday's lines of force': 57–8, 61–2 (= *Scientific Papers*, **1**: 196, 201–2).

(15) Compare 'On Faraday's lines of force': 56 (= *Scientific Papers*, **1**: 194); 'we define the measure of an electric current to be the total intensity of magnetizing force in a closed curve embracing it'; and see Number 87. Maxwell denotes electric quantities by the suffix (2).

(16) Compare Theorem VII of 'On Faraday's lines of force': 61–2 (= *Scientific Papers*, **1**: 201–2); p is the 'magnetic potential or tension'.

(17) At Marischal College, University of Aberdeen; see Numbers 92 to 97.

101

LETTER TO RICHARD BUCKLEY LITCHFIELD

4 JULY 1856

From the original in the library of Trinity College, Cambridge[1]

Glenlair
July 4th 1856

Dear Litchfield

Time is going on & I have to arrange matters in advance.[2] I have been very much engaged here since I came and I expect to have several things to do after a little so that it is well to see that things will fit together. The British Asses meet on the 6th of August and close on the 13th.[3] On the 14th I have to be in Edinburgh at a marriage and thereafter I purpose to go to Aberdeen for a few days and then to Belfast with a cousin of mine whom I have persuaded to go there to learn engineering and his future proprietor has asked him and me to come over & see him.

When that is over I will bring my cousin here to stay as long as is convenient and if you or anyone else such as Lushington[4] should be in Westmoreland &c at that time there might be the more company if you were to turn up.

I spoke to you but doubtfully about any time I might have after Cheltenham.[5] I have certainly no time now & I have much more occupation than I expected such as to examine into the state of two sets of houses & provide wood &c for roofing them and workmen to do it and various things of this kind also to enquire into the merits of the younger clergy and the sentiments of the parish on the subject, for our minister died unexpectedly this week and there are no resident proprietors in the parish except the patron of the living who is a lady of the Romish persuasion who has been for a year in Edinburgh and denies herself to all her friends.

---

(1) Trinity College, Cambridge, Add. MS Letters c.1[83]. Previously published in S. G. Brush, C. W. F. Everitt and E. Garber, *Maxwell on Saturn's Rings* (Cambridge, Mass., 1983): 39–40. The letter has a mourning border: see Number 99.

(2) Maxwell wrote to Litchfield on 4 June 1856 (Trinity College, Add. MS Letters c.1[82]; published in extract in *Life of Maxwell*: 256–7), expressing his feelings on leaving Cambridge for Glenlair after his father's death.

(3) The meeting of the British Association for the Advancement of Science. See Maxwell's papers 'On a method of drawing the theoretical forms of Faraday's lines of force without calculation'; 'On the unequal sensibility of the foramen centrale to light of different colours'; 'On the theory of compound colours, with reference to mixtures of blue and yellow light'; and 'On an instrument to illustrate Poinsot's theory of rotation', *Report of the Twenty-sixth Meeting of the British Association for the Advancement of Science; held at Cheltenham in August 1856* (London, 1857), part 2: 12–13, 27–8 (= *Scientific Papers*, **1**: 241–7).   (4) Vernon Lushington, Trinity 1852 (Venn).

(5) The meeting of the British Association: see note (3).

At the same time I ought to be thankful, and am, that all the people here stick to their duty both in working and agreeing together and in giving advice when wanted, now that they have lost the experience and the wellfitted plans under which they used to act with confidence.

I have got some prisms and opticals from Edinburgh and I am fitting up a compendious colour weaving machine capable of transportation.[6] I have also my top for doing dynamics[7] and several colour-diagrams so that if I come to Cheltenham I shall not be empty handed. At the same time I should like to hear from you soon.

I have been giving a portion of time to Saturns Rings which I find a stiff subject but curious, especially the case of the motion of a fluid ring.[8]

The very forces which would tend to divide the ring into great drops or satellites are made by the motion to keep the fluid in a uniform ring.[9]

I find I get fonder of metaphysics and less of calculation continually and my metaphysics are fast settling into the rigid high style, that is about ten times as far *above* Whewell as Mill is *below* him or Comte or Macaulay *below* Mill using above and below conventionally[10] like *up* & *down* in Bradshaw.[11]

Experiment furnishes us with the values of our arbitrary constants but only suggests the form of the functions. Afterwards, when the form is not only recognised but understood scientifically we find that it rests on precisely the same foundation as Euclid does, that is it is simply the contradiction of an absurdity. Out of which may we all get our legs at last.

Yours truly
JAMES CLERK MAXWELL

---

(6) For his British Association paper 'On the theory of compound colours, with reference to mixtures of blue and yellow light'. See Number 102. On the instrument see Numbers 109 and 117.

(7) For his British Association paper 'On an instrument to illustrate Poinsot's theory of rotation'.

(8) Maxwell's first reference to the subject of the Adams Prize for 1857, on '*The Motions of Saturn's Rings*'; see Number 107 esp. note (1).

(9) See Numbers 107 and 126.

(10) Compare Maxwell's comment in his letter to Litchfield of 25 March 1854 (Number 48).

(11) The allusion is to *Bradshaw's Railway Time Tables*.

102

DRAFT OF PAPER ON COLOURS READ TO THE BRITISH ASSOCIATION IN AUGUST 1856

circa JULY / AUGUST 1856

From the original in the University Library, Cambridge[1]

ON THE THEORY OF COMPOUND COLOURS AND THE EFFECT OF MIXING BLUE & YELLOW LIGHT[2]

The theory of colour as recognized by painters differs in many points from that held by opticians. The painters maintain that all the colours in nature may be imitated by mixtures of three primary colours provided we could discover pigments which would truly represent these primary colours. The optician can exhibit by means of the prism many colours belonging to kinds of light perfectly distinct and incapable of being produced by combination.[3]

This difference of opinion between two classes of observers ought to remind us that the laws of the perception of colour by the eye are probably very different from the law of the separation of the different kinds of light by the prism. The laws of the modifications of light may be considered without reference to the sensations it excites in us. These sensations, however are well worth investigation on their own account, for on their laws depends the whole theory of visible colour and the art of the imitation of colour. The only evidence we can have of these sensations is our own consciousness and though there may be some mechanism connected with them which may be discovered by anatomy, mere anatomy could never demonstrate the sensations themselves.

Now if the laws of the perception of compound colours differ in form from those of the mixture of different kinds of light we shall be able to prove it either by the different appearance of the same kind of light under different circumstances or by the similar appearance of optically different combinations of light under the same circumstances. The effects of what are called subjective or accidental colours are well known and the results of Chevreul[4] show us how

---

(1) ULC Add. MSS 7655, V, b/22(iii).

(2) A draft of 'On the theory of compound colours, with reference to mixtures of blue and yellow light', *Report of the Twenty-sixth Meeting of the British Association for the Advancement of Science; held at Cheltenham in August 1856* (London, 1857), part 2: 12–13 (= *Scientific Papers*, **1**: 243–5).

(3) Compare Number 184.

(4) M. E. Chevreul, *De la Loi du Contraste Simultané des Couleurs et de l'Assortiment des Objets Colorés* (Paris, 1839); (trans. C. Martel) *The Principles of Harmony and Contrast of Colours, and their Applications to the Arts* (London, 1854).

much the judgement of the eye is influenced by circumstances but the apparent identity of optically different kinds of light, although an admitted fact, does not seem to have been much investigated.

By presenting to the eye of an observer two combinations of light optically different and adjusting them till he can detect no difference between them we discover a fact in the theory of sensation – that two different objects can produce the same sensation. In this way we may ascertain completely the conditions under which the sensations of the observer are identical though the things observed are different and we find that the principle of three primary colours though without foundation in physical optics, is completely established as a physiological fact.

It is easy to show that a colour may vary in three different ways for we may make the intensity vary or the hue or the degree of purity of the colour.[5] To show that no other element can enter into our perception of colour requires a rigorous proof for an outline of which I must refer to my paper on Colour in the Transactions of the Royal Society of Edinburgh Vol XXI Part II.[6]

By the experiments there mentioned and by others made since the paper was published I think it has been proved

(1) That in judging of the similarity of two colours the eye can detect a difference of 0.01 in the properties of the colours that compare them.

(2) That the colours apparently identical are optically different as may be shown by viewing them through a coloured glass.

(3) That different observers obtain the same results in these experiments, provided they use the same incident light.

(4) That when the properties of the two component colours are registered and equated together the equation so formed may be treated as an ordinary equation and combined with other equations by the rules of algebra.

(5) That between any four colours and black an equation may be found and thus any colour may be expressed in terms of three standard colours.

The method of observation is by arranging the colours to be mixed as sectors of a circle. When the circle is in rapid rotation the tint appears uniform and the proportions of the colours can be read off on the graduated rim of the circle.[7]

---

(5) Compare Maxwell's terms 'intensity, hue and tint' in his letter to George Wilson of 4 January 1855 (Number 54, esp. note (6)).

(6) 'Experiments on colour, as perceived by the eye, with remarks on colour-blindness', *Trans. Roy. Soc. Edinb.*, **21** (1855): 275–98 (= *Scientific Papers*, **1**: 126–54).

(7) See Numbers 54 and 59.

103

LETTER TO RICHARD BUCKLEY LITCHFIELD

18 AUGUST 1856

From the original in the library of Trinity College, Cambridge[1]

18 India Street
Edinburgh
18ᵗʰ August 1856

Dear Litchfield

I have not had any time to write till now and I doubt whether you will get this in reasonable time, but I report myself accᵈ to order.

I got to Edinʰ on Wednesday morning having had no adventures except one of an unprotected female who made an unsuccessful bolt out of the window because she was afraid of some Wolverhampton revellers who conducted themselves boisterously. So I took measures against a repetition of the fright or the bolting and presently she was reasonable and continued so, yea, and consumed one of the oolites and a biscuit.

At Edinburgh I oriented myself with respect to friends in rural cottages and then drowned care & dust in the firth of Forth off the Chain Pier.

On Thursday I found my man trying to look patient and tranquil but could not prevent him from going to church a quarter before time. I hear that the ladys private bag went astray on the railroad, and I hope she found some substitutes in a bachelors house.

Next day I went out to Penicuik and inspected relatives and had to be photographed four times by reason of the variable lights.

I came round by Dalkeith where other relatives are rusticating in a temporary way.

I found MacLennan at 25 India Street. He has been very busy with his great work on *Law* and has much MS.[2] I think he looks satisfactory he is glad about getting off the trying for the Aberdeen chair without personal bother for certainly he is best here and I think he will *do*.[3]

I go at 8 this morning to Aberdeen and am to be inducted on Tuesday[4] just

---

(1) Trinity College, Cambridge, Add. MSS Letters c.1⁸⁵. The letter has a mourning border: see Number 99.

(2) J. F. McLennan, 'Law', in *Encyclopaedia Britannica* (8ᵗʰ edition), **13** (Edinburgh, 1857): 253–79.

(3) See Numbers 95 and 96.

(4) Maxwell was admitted as a Regent of Marischal College on Tuesday, 19 August 1856; see [P. J. Anderson,] *Officers of the Marischal College and University of Aberdeen 1593–1860* (Aberdeen, 1897): 50.

after you will have concluded the symbolical part of your domestic event.[5] Dr Dyce to whom I am to go just now has his mother with him very ill. I have written to him about it and I hope that I may not come by invitation at a wrong time.

I shall be at home by the end of the month so write.

MacLennan wants a letter though he confesses his debt of letters himself.

Yours truly

JAMES CLERK MAXWELL

(5) The marriage of Litchfield's sister Jane to Henry George Tuke; see [Henrietta Litchfield,] *Richard Buckley Litchfield. A memoir written for his friends by his wife* (Cambridge, 1910): 4n; and *Men-at-the-Bar. A Biographical Handlist*, ed. J. Foster (London, 1885): 474.

104

LETTER TO CECIL JAMES MONRO

14 OCTOBER 1856

From the original in the Greater London Record Office, London[1]

Glenlair
14th Oct 1856

Dear Monro

I saw the Times yesterday so I send you my unnecessary blessing hoping that Elphinstone[2] may find comfort in the rest. As for Farrar he might send me some of his fellowship for I am right busy & will be for a time and will be thankful for the smallest 'favor'. Sale & Hudson I suppose will be the resident luminaries Crompton & Monro will very properly go forth among the lewd men. Is Hardy threatened with instant matrimony? why else did he get his fellowship in such a hurry.[3]

Punctuality you know is the thief of time[4] and Procrastination is the soul of Wit. – That is if they had let him wait a year he would have become witty and not lost time. However I know his pointz better than other men's of that year and he is clever though hardly up to his weight across problems. Dunning[5] I suppose has trotted himself out. I hope he has been well & not lazy and that our revered seniors are not men of a passive imagination who aid their weak judgement with the sough of the Senate House.

During Sept. I had Lushington MacLennan & two cousins 'Cay' here.[6] Now I am writing a solemn address or manifesto to the natural philosophers of the North[7] which I am afraid I must reinforce with coffee & anchovies and a roaring hot fire and spread coat-tails to make it at all natural.

By the way I have proved that if there be nine coefficients of magnetic induction perpetual motion will set in and a small crystalline sphere will inevitably destroy the universe by increasing all velocities till the friction brings

---

(1) Greater London Record Office, Acc. 1063/2081.

(2) Number 67 note (3).

(3) Monro had been elected to a Fellowship at Trinity College, Cambridge, as had Frederic William Farrar, George Samuel Sale, Thomas Percy Hudson, Charles Crompton, and Thomas William Hardy; see *The Times* (11 October 1856). Hardy had matriculated at Trinity in 1852, the others in the years 1849–51 (Venn). Fellows relinquished tenure of their fellowships on marriage; see Numbers 124 and 125.

(4) Edward Young, *Night Thoughts*, I, 393; in *The Poetical Works of Edward Young* (London, 1834): 13.

(5) Joseph William Dunning, Trinity 1851, Fellow 1858 (Venn).

(6) See Numbers 101 and 103.

(7) Number 105.

all nature into a state of incandescence[8] or as Heeley[9] wd say Terrestrial all in Chaos shall exhibit effervescence.

Here is the equation I am come to in Saturns rings.

$$t^4 + (3\omega^2 - L(s^2+2))\, t^2 + 4\sqrt{-1}\,\omega Lst + (2L\omega^2 + 3L)\, s^2 - 2L^2 s^4 = 0$$

a quadratic in t.[10] L is a small quantity relative to ω^2 and s is a whole number not very large. Find the conditions of t being of the forms

$$\sqrt{-1}\, a, \quad \sqrt{-1}\, a + b, \quad \sqrt{-1}\, a - b.$$

Perhaps it might be easier to examine the conditions of

$$x^4 - (3\omega^2 - L(s^2+2))\, x^2 - 4\omega Lsx + (2L\omega^2 + 3L^2)\, s^2 - 2L^2\, s^4$$

having possible roots or roots of the form

$$-a + \sqrt{-1}\, b \text{ or } -a - \sqrt{-1}\, b.$$

If the roots are possible all right. There are two big roots which are *approximately* to 1st power of $\dfrac{L}{\omega^2}$

$$x = +\sqrt{3}\,\omega - \frac{L}{\omega}\frac{8 - 4\sqrt{3}\,s + 2s^2}{3\sqrt{3}}$$

$$x = -\sqrt{3}\,\omega + \frac{L}{\omega}\frac{8 + 4\sqrt{3}\,s + 2s^2}{3\sqrt{3}}.\text{[11]}$$

---

(8) See Example VI, 'On the magnetic phenomena of a sphere cut from a substance whose coefficient of resistance is different in different directions' of 'On Faraday's lines of force', *Trans. Camb. Phil. Soc.*, **10** (1856): 74–6, esp. 76n (= *Scientific Papers*, **1**: 217–19). Maxwell observes that 'no inexhaustible source of work can exist in nature'. To clarify his discussion of magnecrystallic induction he refers to William Thomson's paper 'On the theory of magnetic induction in crystalline and non-crystalline substances', *Phil. Mag.* ser. 4, **1** (1851): 179–86, esp. 186 (= *Electrostatics and Magnetism*: 465–80), where Thomson argues that the number of independent coefficients of magnecrystallic induction must be reduced from nine to six; 'the demonstration being founded on no uncertain or special hypothesis, but on the principle that a sphere of matter of any kind, placed in a uniform field of force, and set to turn round an axis fixed perpendicular to the lines of force, cannot be an inexhaustible source of mechanical effect'. See also Number 109.

(9) Wilfred Lucas Heeley, Trinity 1851 (Venn).

(10) Read: biquadratic. Compare Part II, Proposition IX of the Adams Prize essay (Number 107).

(11) Compare Part II, Proposition X of the Adams Prize essay (Number 107).

These are all right, find me the other two.[12]

What did Galileo say when he had signed his recantation? I have an Encyclopædia which does *not* contain the expression & it forms an appropriate piece of cram.[13]

I go to Edinburgh Thursday & shall be *at Aberdeen* by Monday. Address Mrs Buyers 129 Union Street Aberdeen.

Greet the brethren & may there never be wanting a supply of men for fire screens. As for me I am in Hades & I doubt when I may become angelic O Pie Pie.[14]

---

(12) In response to Maxwell's request, Monro attempted to obtain the two small additional roots, without success; Monro to Maxwell, 4 and 8 November 1856 (Greater London Record Office, Acc. 1063/2097, 2098). Approximating, and writing the equation as

$$-3\omega^2 x^2 - 4\omega Lsx + 2L\omega^2 s^2 = 0$$

the roots are $x = \pm \sqrt{(\frac{2}{3}L)}\, s$. Compare Part II, Proposition X of the Adams Prize essay (Number 107, esp. note (57)).

(13) In a reply of 11 November 1856 (Greater London Record Office, Acc. 1063/2099) Monro provided the required information: '" E pur si muove ". Will that do for you. The rest of the speech is to be found, translated, in Arago's Astronomie populaire vol. 3. What is this appropriate to? Or how is a piece of cram appropriate in the abstract?' Monro's reference is to Dominique Francois Arago, *Astronomie Populaire*, 4 vols. (Paris, 1854–7), **3**: 30; 'On reconte qu'après l'abjuration, Galilée en se relevant dit à demi-voix et en frappant la Terre des pieds, *e pur si muove* (et cependant elle se meut).' Maxwell was in search of a suitable motto for his Adams Prize essay; see Number 107 note (5).

(14) In the parlance of the Apostles – both Maxwell and Monro being members of the Cambridge essay society – a 'brother' is a member of the society, an 'angel' a member released from the obligation to attend meetings.

105

INAUGURAL LECTURE AT MARISCHAL COLLEGE, ABERDEEN, 3 NOVEMBER 1856[1]

OCTOBER 1856[2]

From the original in the University Library, Cambridge[3]

Gentlemen

The work which lies before us this session is the study of Natural Philosophy. We are to be engaged during several months in the investigation of the laws which regulate the motion of matter. When we next assemble in this room we are to banish from our minds every idea except those which necessarily arise from the relations of Space Time and Force. This day is the last on which we shall have time or liberty to deliberate on the arguments for or against this exclusive course of study, for, as soon as we engage in it, the doctrines of the science itself will claim our constant and undivided attention. I would therefore ask you seriously to consider whether you are prepared to devote yourselves during this session to the study of the Physical Sciences, or whether you feel reluctant to leave behind you the humanizing pursuits of Philology and Ethics for a science of brute matter where the language is that of mathematics and the only law is the right of the strongest that might makes right.

In order that you may be able to decide this question for yourself, I intend to set before you in this lecture

(I)   The nature and limits of the sciences which constitute Natural Philosophy as taught in this class.

(II)   The kind of evidence which we have for the truth of the doctrines of these sciences.

(III) The relation of Physical science to those other departments of human knowledge which have their place in our University system.

I What is Natural Philosophy

About the end of the sixteenth century when a Regent in Philosophy began the exposition of the system of the universe his conception of the extent of his subject

---

(1) The *Aberdeen Journal* of 5 November 1856 reported that the university session opened on Monday [3 November]. 'Mr Clerk Maxwell, the recently appointed Professor of Natural Philosophy, delivered his introductory lecture, on the above day, at eleven o'clock, in the presence of the Principal and most of the Professors, Dr Piazzi Smith, Astronomer Royal for Scotland, Rev. Dr McTaggart, Dr Kilgour, and others, besides a very large number of students.' The content of the lecture was summarised, and the report concludes: 'The lecture...was very favourably received.' On Piazzi Smyth see Numbers 115 and 117.   (2) See Number 104.

(3) ULC Add. MSS 7655, V, h/1. Previously published in R. V. Jones, 'James Clerk Maxwell at Aberdeen 1856–1860', *Notes and Records of the Royal Society*, **28** (1973): 69–81.

was definite and satisfactory. The founder of our university had ordained that the knowledge of natural things should be publicly taught, and the Regent simply fulfilled the duties of his office. It forms no part of my duty to describe the system originally taught from this Chair. The progress of discovery threw the wisdom of those days into confusion and accumulated new materials far too various to be classed under one name. Sciences were multiplied and large departments of human knowledge sprung up by the side of the old Natural Philosophy. Thus the new science of Chemistry carried away with it the germs of the sciences of heat and electricity and as these gradually became too important to be treated of as introductory to the chemistry of ponderables they have returned to take possession of the domain of Physics.

It would seem therefore as if the boundaries of our subject were becoming daily more confused and as if the modern student must sacrifice all the higher aims of science to the necessity of extending his acquaintance with the discoveries of the day. And when we remember that our object in this University is not merely to produce philosophers but also men qualified in other ways to serve God both in Church and State it becomes a matter of the greatest importance to decide upon the principle which is to guide us in our study of Nature in this place. Are we to study many things rapidly or a few things calmly? Shall we acquire what information we can, and leave deep thinking to professional men? or shall we employ our limited time on one or two subjects, and be content to be behind the age in general information?

Now I would begin to answer this question by directing your attention to the method of teaching in a University. The same set of students meet at stated hours during the whole session with scarcely a days intermission. Does not this seem as if we had some *one* thing to do requiring steady resolution and repeated efforts. And is not that one thing to learn something of the first principles of the works of nature, the rudiments of the creation so that we may no longer be oppressed with the multitude of wonders, but rather confident that as we have recognized the operation of general principles in some instances, so in due time we shall discover more and more that the whole system of nature is disposed according to a wonderful *order*.

Indeed the very fact that these general principles are to be found shows that it is needless to bewilder our minds with an accumulation of detached facts for in the light of the great laws of nature these facts are no longer detached and bewildering, but necessary consequences of principles which they illustrate.

When we have once made our minds familiar with one or two great physical laws we begin to look upon the Universe as a realization of the highest principles of Order & Beauty and we are prepared to see in Nature not a mere assemblage of wonders to excite our curiosity but a systematic museum designed to introduce us step by step into the fundamental principles which are displayed in the works of Creation.

I would now direct your attention to the limits of the sciences which we are to study.

Natural Philosophy may be divided into two great branches. The mechanical sciences treat of the motion of matter considered simply as matter, and are built upon the fundamental ideas[4] of force and mass without any appeal to experimental measurements.

We have first the theory of the composition and resolution of forces which forms the foundation of statics.

Then we consider the geometrical conditions of motion as exemplified in the motion of a single point or in that of a system of points which preserves the same invariable form or in that of a fluid in which there is perfect freedom of motion in every part.

Lastly we consider the effects of Force in producing or preventing motion in any of these cases. When the motion is prevented by opposing forces we have a case of equilibrium. The science which treats of cases of equilibrium of points and rigid bodies is called *Statics*. That which treats of the equilibrium of fluids is called Hydrostatics.

When the forces produce motion the results are investigated in the science of Dynamics & Hydrodynamics.[5] To these four mechanical sciences we may add two others which have claims to rank beside them although they are founded on assumptions. The first is Theoretical Astronomy which treats of the motions of distant bodies which attract one another according to Newtons law of gravitation. The second is the science of the equilibrium & motion of those hypothetical systems of particles which by virtue of the forces acting among them appear to us as elastic solids and fluids. This science of molecular mechanics rests upon a much less certain basis than that of Astronomy. The superstructure however is an example of strict mechanical reasoning and may be found of use even when the fundamental assumptions have been proved to be erroneous.

We come next to the Physical Sciences in which we build upon facts of observation which we cannot completely explain but which we can mathematically define.

Of these some are so mixed up with mechanics that they are usually treated of as part of pure mechanics. The first is the science of Elastic Bodies treated without any assumptions about molecular action. The most important parts of

---

(4) The term is Whewell's; see William Whewell, *Philosophy of the Inductive Sciences, founded upon their History*, 2 vols (London, ₂1847), **1**: 66.

(5) A division of mechanics commonplace in Cambridge texts; see, for example, J. H. Pratt, *The Mathematical Principles of Mechanical Philosophy, and their application to Elementary Mechanics and Architecture, but chiefly to the Theory of Universal Gravitation* (Cambridge, ₂1845): 3. Compare Number 109 note (14).

this science are the theory of the elasticity and strength of solid materials such as rods and beams and that of the vibrations of elastic bodies which constitutes the physical theory of Sound. The science of Elasticity treats of the effect of Pressure in altering the form and size of bodies. The effects of Heat in producing the same effect are so mixed up with that of pressure that in treating of the theory of gaseous bodies we are compelled to include the phenomena due to changes of temperature as well as those due to changes of Pressure. The practical application of this part of the Science of Heat is to be found in the Theory of the Steam Engine.

The Science of Heat has three main divisions 1st The theory of the expansion and alteration of bodies into which it passes 2nd the theory of the distribution of heat in bodies by *conduction* and 3rd that of the transmission of heat from one place to another by a radiation of the same kind as that of light.

The science of radiations appears to be the same in the case of the heat which we feel and the light which we see. In fact there appears to be no difference between radiant heat and light except that the eye by which we become acquainted with the sensation of light is a much more delicate organ than that of touch although it has a smaller range of sensibility.

Heat and light then being the same thing perceived by two different senses we prefer to use that organ by which we obtain most information. We therefore call our science Optics and begin by stating the properties of a ray of light. From a few simple axioms about reflection and refraction we are able to deduce a systematic science which rivals the mechanical sciences in precision, and has this great advantage as an educational science that the phenomena are easily observed, and that the disturbing effects, which render mechanical experiments fallacious do not here exist.

Last of all we have the Electrical and Magnetic sciences which treat of certain phenomena of attraction, heat, light & chemical action depending on conditions of matter of which we have as yet only a partial and provisional knowledge. An immense mass of facts has been collected and these have been reduced to order as the results of a number of experimental laws, but the form under which these laws are to appear as deduced from central principles is as yet uncertain. The present generation has no right to complain of the great discoveries already made, as if they left no room for further enterprise. They have only given Science a wider boundary, and we have not only to reduce to order the regions that have been conquered, but to keep up constant operations on the frontier, on a continually increasing scale.

When Socrates asked Theætetus what he considered to be the nature of Science, he could get no answer except the names of all the sciences then in existence. In stretching out the limits of Natural Philosophy, we appear to have adopted the same method – of definition by simple enumeration. Perhaps in the beginning of our enquiries there is nothing better than a kind of Index-Knowledge of our subject but still it is right that we should have some *characteristic* of the Physical Sciences to lay hold of, that we may determine what branches of knowledge ought to be recognized here, and what ought to be handed over to some other department of the University.

In the enumeration which we have just made we began with the science of motion and force, a science which differs from pure geometry only by the introduction of the ideas of mass. Now it is this very idea which gives mechanics the character of a physical science. The physical sciences treat of matter as related to force. By this I mean 1st that when we are occupied in defining the *forms* of bodies we are performing the work of a geometer, which, however necessary in mechanical enquiries, is rather a *preparation* for mechanics than mechanics itself.[6]

Again Chemistry is the science which treats of the different *kinds* of matter and their combinations. Here again we are *beyond* the limits of the Physical Sciences for we admit the existence of different kinds of matter whereas in a physical point of view all matter must in itself be the same, and can be modified only by differences of arrangement and motion and by being actuated by different systems of force.

As soon as any chemical phenomenon has been reduced to a change in the arrangement and motion of a material system, produced by forces of a *determinate* nature, the phenomenon will be physically explained.

As long as we have *no* means of conceiving of chemical differences being due to *determinate* physical arrangements of pure matter and its forces, but are obliged to admit differences in *kinds* of matter for which we cannot account mechanically, we must be content to allow the chemists to study their own science from their own point of view.

It appears, therefore, that we may study material objects in a *geometrical* point of view, taking account only of their forms and positions, or we may extend our investigations to their *physical* properties, their mass, velocity and state of aggregation, or we may go still further and consider the chemical qualities of the bodies, that is, the conditions under which they are compounded into entirely new substances.

Thus if we fill a jar with oxygen and hydrogen, the determination of the volume of the gases which the jar will hold is a geometrical question, the specific gravity of the mixture, and its pressure on the sides of the jar, depend on physical considerations, but the result of the introduction of a lighted match depends on the chemical nature of the gases, and the explosion, though a physical effect, cannot be explained without reference to Chemistry.

---

(6) Compare William Whewell, *An Elementary Treatise on Mechanics* (Cambridge, ₇1848) : 2.

Now though the boundary between physics and Chemistry is plain enough in this case, there are two if not three great branches of science which are claimed by both. Electricity and Heat are the chief agents in chemistry and they are revealed to us principally by their chemical actions. Light also is an important agent in chemistry. Now the chemist, accustomed to refer all actions to the properties of the agents which produced them, supposed that in Electricity Heat and Light he had found three *substances* as distinct in their properties as oxygen or iron. The *balance*, however, gives no indication of their presence, and hence, since they could not be estimated by weight, they were classed as *imponderable* substances.

When the universality of the action of the supposed imponderables was more fully recognized, it was seen that they have no analogy to the elementary bodies of chemistry, but are to be considered for the present as the unknown causes of phenomena which are observed in every kind of matter. The laws of heat light and electricity are therefore included in physics, simply because their action extends to all substances whatever be their chemical nature.

II

We come next to consider the evidence upon which the physical sciences rest. The question whether Mechanics has a right to rank as a Pure Science has been often argued. If the Laws of Motion are necessary truths, then the advocates of Mechanics as a Pure Science have gained their point. If on the other hand they are only to be arrived at by experience then mechanics is to be considered as a Mixed science, in which mathematical reasoning is applied to facts, which are admitted as facts simply because they are observed, and not because we know that they could not be otherwise.

You are aware, gentlemen that there is a science which treats of necessary truth, and which carefully distinguishes that which must be from that which only happens to be. The science of Metaphysics is not of modern growth. Philosophers of the greatest eminence have attacked these very questions from the earliest times, and yet it would appear that the one party could never wholly convince the other. One reason was that up to the sixteenth century both parties were wrong, for the laws they contended about were neither necessary truths nor the results of experience, being contrary both to fact and to reason, and after the sixteenth century both parties were on a different ground for now the laws they admitted were at least true if not necessary.

You will perhaps be inclined to think that the root of the matter must lie somewhere among the discoveries of Galileo Kepler or Descartes and that the great practical conclusion is, 'that once we were in error, and now we understand and therefore let us be content'.

But I would have you remember that the men to whom we owe the greatest discoveries in mathematics and physics were metaphysicians. *They* thought it a very important thing to determine the *evidence* on which they built any law. And it is still more remarkable that the greatest and most original metaphysicians have been *nourished* as it were with physical truth. Bacon, though his supply of physical truth was scanty, had his mind fixed upon the discoveries of the future, and he draws both his wisdom and his eloquence from that new era, of which he was himself the prophet.[7] Descartes & Leibnitz need only be named to recall systems of metaphysics which were also systems of science. In fact the things discussed in metaphysics are so intimately connected with the foundations of Natural Philosophy, that we have only to read a few pages of a metaphysical work to ascertain the precise limits of the author's knowledge of physical science.

It is commonly supposed that Scotchmen are born with an instinctive tendency towards metaphysics, and that when they begin to run short of practical arguments, they take refuge in a higher and more impregnable region, where necessary truth reigns in perpetual sterility concealed by the mists of rhetoric, and defended by the thunders of declamation.

Whatever may be the fate of Scotchmen when they run short of matter of discussion, their capacity for physical research ought to ensure them against every danger, for *here* they have an inexhaustible supply of incontrovertible facts, all of which have their bearing on metaphysical principles, and neither rhetoric nor declamation will avail, till *these* metaphysical principles are at peace with *those* incontrovertible facts.

In the course of our studies here you will have abundant matter for the most abstract speculation but you must recollect that in *physical* speculation there must be nothing vague or indistinct. The truths with which we deal are far above the region of mist and storm which conceals them from the popular mind and yet they are solidly built upon the foundations of the world and were established of old, according to number, and measure, and weight.

Nothing that we can say or think here can escape from the ordeal of the measuring rod and the balance. All quantities must be exact quantities, and all laws must be expressed with reference to exact quantities, so that we have a most effectual means of discovering error, and an absolute security against vagueness and ambiguity.

I ought now to tell you what my own opinion is with respect to the necessary truth of physical laws – whether I think them true only so far as experiment can be brought forward to prove them or whether I believe them to be true independent of experiment. On the answer which I give to this question will depend the whole method of treating the foundations of our science.

I have no reason to believe that the human intellect is able to weave a system

(7) Compare Whewell's panegyric on Bacon in his *Philosophy of the Inductive Sciences*, 2: 229.

of physics out of its own resources without experimental labour. Whenever the attempt has been made, it has resulted in an unnatural and self-contradictory mass of rubbish. In fact unless we have something before us to theorize upon, we immediately lose ourselves in that misty region from which I have just warned you.

As long as a confused idea of the nature of force prevailed, and when men talked of the force of a cannonball the force of gunpowder and the force of so many pounds, without seeing that different names were required for each, of course no conclusions could be arrived at as to the laws of force.

If we found a person calling a right angle, a square and a cube by the same name and unable to distinguish them in words, we could not expect him to have true ideas of these geometrical entities, but we know that as soon as men have learned the true meaning of terms in geometry they are able to deduce all the propositions in mathematics without possibility of error.

In the same way I maintain that as soon as we clearly understand what motion is, and that force is that which alters motion we are able to prove all the laws relating to force.[8] The laws relating to the higher departments of physics can hardly be said to be so easily proved. We know mathematically what motion is but we do not know in the same way what temperature is. But we can reason mathematically about temperature by adopting a definition and reasoning on principles for some of which we see the ultimate necessity while for others we can only say that it appears so by experiment.

But as Physical Science advances we see more and more that the laws of nature are *not* mere arbitrary and unconnected decisions of Omnipotence, but that they are essential parts of one universal system in which infinite Power serves only to reveal unsearchable Wisdom and eternal Truth.

When we examine the truths of science and find that we can not only say

---

(8) Compare the comment on the laws of motion, transcribed by Maxwell from Hopkins' notes, in his Cambridge notebook 'Statics Dynamics' (ULC Add. MSS 7655, V, m/10, ff. 62–3): 'The Laws of Motion...are frequently considered as having been established by experiments. It will be more correct however to consider them rather as suggested than established by any direct experiments made for the express purpose of verifying them. They must be considered as established by that inductive power of reasoning by which in fact the truth of known laws of Nature is confirmed. In this case the process is this: the fundamental laws are assumed and the motion of a material point or System of points is calculated with this assumption. The results of such Calculations are then compared with the actually observed motion of the system. It is in the perfect accordance of these results of calculation and observation that the ultimate proof of all the elementary principles assumed consists.' The same 'General remark' on the laws of motion was transcribed in 1843 by William Thomson from Hopkins' notes on 'Dynamics' (ULC Add. MSS 7342, PA11, ff. 10–11); and the same passage was transcribed by Stokes (ULC Add. MSS 7656, PA6). Compare William Whewell, 'On the nature of the truth of the laws of motion', *Trans. Camb. Phil. Soc.*, **5** (1834): 149–72; and *Philosophy of the Inductive Sciences*, **2**: 245–54.

'This *is* so' but 'This *must* be so, for otherwise it would not be consistent with the first principles of truth' – or even when we can only say 'This *ought* to be so according to the analogy of nature' we should think what a great thing we are saying, when we pronounce a sentence on the laws of creation, and say they are true, or right, when judged by the principles of reason. Is it not wonderful that man's reason should be made a judge over Gods works, and should measure, and weigh, and calculate, and say at last 'I understand I have discovered – It is right and true'. Man has indeed but little knowledge of the *simplest* of Gods creatures, the nature of a drop of water has in it mysteries within mysteries utterly unknown to us at present, but what we do know we know *distinctly*; and we see before us distinct physical truths to be discovered and we are confident that these mysteries are an inheritance of knowledge, not revealed at once, lest we should become proud in knowledge, and despise patient inquiry, but so arranged that, as each new truth is unravelled it becomes a clear, well-established addition to science, quite free from the mystery which must still remain, to show that every atom of creation is *unfathomable* in its perfection. While we look down with awe into these unsearchable depths and treasure up with care what with our little line and plummet we can reach, we ought to admire the wisdom of Him who has so arranged these mysteries that we find first that which we can *understand* at first and the rest in order so that it is possible for us to have an ever increasing stock of *known* truth concerning things whose nature is absolutely incomprehensible.

These reflections may appear premature to some of you. We have not yet become familiar with these truths, and the work which we have to go through may seem of a nature foreign to such contemplations. But as the work is before us, I have taken this opportunity of pointing out some of those trains of thought which may serve to lead the mind at times from the matter we have to deal with, to subjects with which as *students of this class* we are supposed to have nothing to do, but with which as *human beings* we have a great deal to do and I have rather directed your attention to these at present for the more material advantages of the physical sciences are dilated upon by all who write in these times and they will be more particularly treated of as we advance.

I have also thought it unnecessary to tell you that the study of the world in which we live is our obvious duty as a condition of our fulfilling the original command '*to subdue the earth and have dominion over the creatures*'. Those who have raised objections to the engrossing pursuit of physical science have done so on the ground of the supposed effects of exact science in making the mind unfitted to receive truths which it cannot comprehend. I have endeavoured to show that it is the peculiar function of physical science to lead us to the confines of the incomprehensible, and to bid us behold and receive it in faith, till such time as the mystery shall open. I trust also to be able to point out the fallacy of reasoning upon the supposed existence of laws of which we have no distinct idea, and assuming that the higher laws which we do *not* yet know are capable of being stated in the same forms as the lower ones which we *do* know. When vague ideas are put into the forms of mathematical or physical arguments we can expect nothing but vague conclusions, and great discredit to the mathematics or physics so desecrated. Vague ideas may give picturesqueness to a declamation, but we must be very cautious of them when they are disguised in the form of an exact science.

We have next to consider the relation of Physics to the other studies of the University.

The old University course consisted of seven fundamental studies. Of these the first three formed what was called the Trivium. First came Grammar teaching us words with their relation to one another. Then Logic treating of the relation of words to their meaning, and the laws of valid reasoning. Lastly Rhetoric illustrated the relation of words to living men, and laid down the laws of persuasion. The student was now furnished with his stock in trade and might amass knowledge for himself by the only known method – disputation.

But if he aspired to academic perfection, there were yet other four things to be learnt – the Quadrivium. Mathematics discussed necessary truth with regard to measurable quantity, and Metaphysics extended its authority over all other species of necessary truth.

Natural Philosophy took charge of the external world while Moral Philosophy made the student acquainted with a more extensive range of phenomena within himself.

This classification may be considered as dividing education into the acquisition of language and the study of philosophy. The first thing to be done was to make the student speak and think as a man ought, and then they gave him something to speak and think about before they turned him loose on the world, with his new powers of thought and speech. In the old time the student had very few pieces of genuine fact given him, but he had to conform to College rules, to talk such Latin as was possible to him, and to dispute a great deal according to the true maxims of Grammar Logic and Rhetoric.

Now a days we have too much to teach and too little time to teach it. It is understood that the essence of a liberal education consists in a familiarity with all the departments of existing knowledge. We have no time for academic disputation or any other exercise in the shape of intellectual drill. The modern mind is content with a due supply of rations and considers discipline and a uniform system of drill as needless formalities.

When the demand alters the supply must be altered to suit it so that the system of education must change with the times.

I am happy to see that language is still considered the primary element in the education of all classes. The right use of speech is the first thing to be taught. We must first speak grammar and then speak the truth.

How far these objects are promoted by what is called a Classical Education does not come into our consideration here. We need only observe that the studies commonly pursued under that name have always been of the severe order and therefore are still cultivated under the sanction of authority. In a University we may regard the study of languages as forming a preparation for the cultivation of Philology 'the love of words' whether English or foreign.

I have spoken of these sciences of human speech as a preparation for that system of science which begins with arithmetic and embraces mathematics and physics and endeavours to understand the world we live in and to express its laws in accurate and appropriate language.

We have in these sciences a different kind of discipline from that developed by philology. We have less of the necessities of human society and its intercourse, and more of the necessity of material laws, that is, the laws of nature, or in other words that portion of the Divine Order which relates to things without life.

Now if this portion of truth be conversant with subjects less noble than that which relates to the language and actions of man as a spiritual being, it has this to recommend it, that the laws we have to do with are better understood and more perfectly expressed than those which relate to higher things.

The expression of these laws depends in a great measure upon the power which we have of referring all physical events to changes which we can estimate with regard to their quantity. Now whatever can be referred to quantity comes under the power of mathematics and in this way mathematics becomes the instrument of the physical sciences and physical laws and physical arguments are reduced to symbolic notation and algebraic calculation.

I ought now to state my opinion as to the necessity of mathematics for the study of Natural Philosophy. I need say no more than that I hold that Natural Philosophy is, and ought to be, Mathematics, that is, the science in which laws relating to Quantity are treated according to the principles of accurate reasoning.

It by no means follows that those who have not found themselves at home in the study of Pure Mathematics will find themselves still more out of their element in this class. For there are many men indifferent to abstract truth, and indeed hardly capable of appreciating it *as* truth, who, when the same truth is clothed in a material form, are not only eager to make themselves familiar with it, but perceive it in its true relation to the principles on which it rests.

Many, who care not for algebra, delight in geometry, and those who are unfamiliar with calculation will yet reason rightly about forces when they meet with them in an actual experiment.

I might point out some of the most remarkable specimens of physical reasoning of the most intricate and yet most accurate kind in Faradays Electrical Researches,[9] a work in which there is hardly an allusion to technical mathematics. But while I believe that every human mind is capable of understanding something of the elementary laws of the universe and that in some cases these laws are seen most clearly when embodied in material forms, still I would recommend the study of pure mathematics as a most valuable aid to that of Natural Philosophy and I would also point out the fact that the greatest advances in mathematics have been due to enquirers into physical laws.

To those who have profited by their mathematical studies I hope to give abundant opportunity of exercising their powers, but they must for their own sake as well as that of others go through the examination of physical laws apart from symbolic language.

I have in the last place to speak of the relation of Natural Philosophy to those sciences which belong to the more advanced part of the University course. I need say nothing of the importance of a knowledge of physical action to the student of chemistry, and I need hardly remind the anatomist that the *body* is subject to the elementary laws of matter.

I have already said enough about the regions of obscure mental science which can only be explored by the light of physical discoveries. The student of Law will find a respite from his professional studies in contemplating a state of things where every law carries itself into execution, and where the success of every enterprise is determined by the precision with which the Laws of Nature are complied with. Those who intend to pursue the study of Theology will also find the benefit of a careful and reverent study of the order of Creation. They will learn that though the world we live in, being made by God, displays His power and goodness even to the careless observer, yet that it conceals far more than it displays, and yields its deepest meaning only to patient thought.

They will learn that the human mind cannot rest satisfied with the mere *phenomena* which it contemplates, but is constrained to seek for the principles embodied in the phenomena, and that these elementary principles compel us to admit that the laws of matter and the laws of mind are derived from the same source, the source of all wisdom and truth. It is right that we should sometimes think of these things for such meditation is the *salt* of our worldly life. My duty however is to give you the requisite foundation and to allow your thoughts to arrange themselves freely. It is best that every man should be settled in his own mind, and not be led into other mens ways of thinking under the pretence of studying science. By a careful & diligent study of natural laws I trust that we shall at least escape the dangers of vague and desultory modes of thought and

(9) Michael Faraday, *Experimental Researches in Electricity*, 3 vols. (London, 1839–55).

acquire a habit of healthy and vigorous thinking which will enable us to recognize error in all the popular forms in which it appears and to seize and hold fast truth whether it be old or new.

Our constant attention to the foundation of our beliefs will also be of service in showing us the limits of our knowledge and of our reasoning powers, and when either the one or the other is assumed to be greater than it is we shall be able to detect the assumption which underlies all the logic & rhetoric which seems at first so formidable, and by disowning the cunningly devised and plausible wisdom of men we shall enter into possession of that secret wisdom of which we know only this that it is not our own.

106

LECTURE ON THE PROPERTIES OF BODIES TO THE NATURAL PHILOSOPHY CLASS AT MARISCHAL COLLEGE, ABERDEEN

NOVEMBER 1856[1]

From the original in the University Library, Cambridge[2]

PROPERTIES OF BODIES

In the successful study of Nature there is one fundamental rule to be observed – We must begin with those phenomena which are the most abstract simple and general, and make ourselves familiar with these before we proceed to those which are concrete complex and particular. In arranging a course of scientific study this is the essential principle of our progress for the abstract laws relating to motion and force must be understood before we can apply them to the explanation of the properties of concrete bodies and in all cases it is only by a recognition of general principles that we can properly understand the particular instances of their application.

But although this is the necessary arrangement of our physical studies considered in a scientific point of view, it is not that which presents itself most readily to the mind. We wish to have some knowledge of the ultimate results of nature before we begin to study her secret processes and we suppose that it is more natural to enquire what bodies are, than to study their motions.

Now the result of the direct inquiry into the nature of bodies has been only endless disputation without prospect of agreement among the different schools of philosophy; while the study of the motions of bodies has led to the establishment of certain fundamental principles which are the foundation of all the physical sciences and to which we must look for all that we are ever to know of the constitution of bodies.

The Book of Nature is so written in the world that we must understand the first chapter before we can make out the meaning of the second, and yet the leaves are so scattered that we have ourselves to pick out and arrange what appear to be the earlier pages. The part which we have to study is one of the introductory chapters and we must avoid as much as possible the confusion

---

(1) Handwriting indicates *circa* 1856; and content suggests that this lecture was written for the Natural Philosophy Class as an introduction to the scope of physical science. In 1848–9 J. D. Forbes commenced his Edinburgh natural philosophy course (with Maxwell attending) with 16 lectures 'Introduction and properties of bodies'; see Number 16, note (3). In 1847–8 Forbes also commenced with this topic (with Maxwell attending): see Number 8.

(2) ULC Add. MSS 7655, V, g/4.

arising from the intrusion of stray leaves from the chapters of chemistry or physiology with which we have as yet nothing to do.

There is one part of a book however in which we may lawfully mix up all the subjects to which the book is devoted, and when we are about to begin the study of the book, especially if it is a very large work, it may be advisable to turn first to that part.

In France, the table of contents is generally placed at the end of the work. In this country it follows the preface and we read it naturally before the first chapter. Now you may consider our present enquiry into the properties of bodies as mere preliminary matter, serving only to indicate or point out what we are afterwards to study more thoroughly. We are in fact glancing over the index or table of contents which according to our national custom I have placed at the beginning.

And I think it is not only admissible but even necessary that those who are about to enter upon a serious course of study should have the opportunity of seeing the field of their labour afar off that they may judge for themselves whether the work to which they are invited be worthy of their efforts.

In the Physical Sciences we consider the mutual actions of bodies in so far as these actions are or may be exhibited by all bodies or in other words the general properties of bodies. We leave the consideration of the special properties exhibited by particular substances to Chemistry, the science of the specific actions of different kinds of matter.

We have therefore in the first place to make a general classification of the physical properties of bodies, beginning with the most abstract and simple and proceeding to the more concrete and complex in their order.

In this arrangement we must be guided by this principle alone and must take no account either of the mode in which we become aware of these properties, whether by touch, sight or any other sense. We have nothing to do with questions of Physiology and the Philosophy of the Senses. Neither must we alter our arrangement in favour of a chronological order, by stating those properties first which have been longest known and admitted. This would be more a matter of history than of science. Our business is to form our system according to the best of our present knowledge, neither disguising what we know to be true nor insisting upon what we only conjecture.

We therefore divide the properties of bodies scientifically into Four great divisions.

I Geometrical properties, relating to Space, such as the position size and form of bodies.

II Mechanical properties, relating to Force, such as Mass or quantity of matter, density gravity.

III Structural properties relating to the state of aggregation or consistency of bodies, such as Solidity Fluidity Elasticity Tenacity.

IV Physical properties relating to the behaviour of bodies with respect to heat, light, electricity &c. The classification of these properties must for the present depend rather on the progress of discovery than on true scientific principles, for while new properties are coming to light every day we cannot maintain that we know the arrangement of the whole which would necessarily include those which we have yet to discover. We are under no such difficulty with respect to the geometrical properties, for our conception of space is complete and satisfactory so far as it goes. We know that space has three dimensions, that one point of space has the same properties as any other, and so on.

I Position

Now the first essential property of all matter is that it exists in space. Every portion of matter must therefore have its own *position* in space.

The *position* of a body at a given time is therefore the first thing we have to consider and it is the leading geometrical property of matter that it has position or locality.

Extension
Figure

Next to this comes the subordinate principle that every system of matter occupying space takes up a definite portion of it and this portion must have a definite extent and boundary. A material system therefore must have Extension and Figure.

Divisibility

Again since each part of the system occupies a definite portion of space, we may confine our attention to certain parts of the system by defining the portion of space which they occupy. We may therefore conceive of the *divisibility* of the material system. This property does not imply any actual divisibility which depends on the structure of the system and the nature of our tools. Here we only divide bodies in imagination in order to confine our attention to a convenient portion for our analytical reasoning.

There is another geometrical property belonging to this class which was considered of great importance by the metaphysical philosophers. I mean the so called property of the Impenetrability of Matter or the impossibility of two material things occupying the same space. I cannot at present explain the reasons which may be brought against the necessary truth of this dogma.[3]

I trust, however to convince you in due time that the abhorrence of Nature

---

(3) A philosophical issue that received some discussion in Maxwell's writings. See his essay on 'Atom', in *Encyclopaedia Britannica* (9th edition), **3** (Edinburgh, 1875): 36–49, esp. 37 (= *Scientific Papers*, **2**: 448), where he refers to 'the vulgar opinion that two bodies cannot co-exist in the same place. This opinion is deduced from our experience of the behaviour of bodies of sensible size, but we have no experimental evidence that two atoms may not sometimes coincide.' Compare Whewell's dismissal of the assumption that atoms possess properties of hardness and solidity as 'an incongruous and untenable appendage, to the Newtonian view of the Atomic Theory'; see his *Philosophy of the Inductive Sciences*, 2 vols. (Cambridge, 1847), **1**: 432. On Maxwell's critique of Newton's 'analogy of nature' (the 'third rule of philosophising'), see P. M. Harman, *Metaphysics and Natural Philosophy* (Brighton, 1982): 140–5.

for a superplenum or two things in the same place only differs from her abhorrence for a vacuum or a place with nothing in it, by the fact that the forces
in the one case are somewhat stronger than in the other, being in the one case
the enormous repulsion of the two bodies when coming into contact and in the
other the very moderate pressure of the atmosphere.

These are matters however which we cannot discuss at present. I proceed to
the second leading property of matter with respect to Space.

That a given portion of matter may at successive instants of time occupy
different portions of space but that this alteration of position must be continuous that is it must proceed gradually.

It is absurd to explain by words what is meant by the position of a body in
space. It is equally absurd to try to explain that alteration of position which we
call motion. Every one has an idea of motion and however vague it may be it
is capable of being rendered definite by scientific treatment, and careful study.

We have now arrived at the conception of the Mobility of Matter and we   Mobility
must now conceive bodies as changing their position while they retain their
identity.

Subordinate to this idea are the conceptions of the Velocity and direction of
this motion with the changes and accelerations which may take place either in
the velocity or the direction.

The theory of Motion becomes very important when the material systems
considered are very complex. We have then to devote our special attention to
the motion of several parts and the science by which our investigations proceed
is called Kinetics or sometimes Kinematics, that is the science of Motion.

II The Mechanical Properties relate to the action of Force. Force is that
which produces and alters Motion. Now the same force when applied to
different bodies will not produce the same velocity. The velocity produced will   Mass
depend upon the magnitude of the body as well as on the force. Now since we
are considering Matter only with reference to its relation to Force & Motion
we define equal quantities of Matter to be those which are acted on equally by
equal forces. The quantity of Matter estimated in this way is called its Mass.
We may estimate the mass of any body by the number of times it contains the
mass of a standard pound.

Under the head of Mass we may class the so called property of Inertia or the   Inertia
resistance which a body offers to a forcible change of motion.

Density is the relation of Mass to Extension or Volume it may be measured   Density
by the number of pounds in a cubic foot.

The second class of Mechanical Properties depends on attraction from without. The chief of these is the attraction of gravitation which on the surface of
our earth is called *weight*. It is a remarkable fact for which we cannot yet account   Weight
but which is perhaps the most important in the whole range of Physics, that the

attraction of gravitation depends only on the mass of a body and not at all on the kind of matter of which it is composed. The weight of a body is not like its mass invariable but changes as the body moves into different regions at different distances from the centre of the earth.

III The Structural properties are those which determine the consistence of the body which may be in the first place Solid, liquid or gaseous.

A Solid body has some determinate form which cannot be altered without the application of force.

Solids are classified according to the relation between the force and the change of form which it produces. When a great force produces a small alteration the solid is called a hard solid and in the imaginary case in which no alteration of form is produced the solid is termed Rigid.

When a small force produces a great alteration the solid is said to be soft and in this property we may have an approximation to the fluid state.

When the solid recovers its form when the force is removed it is said to be elastic. When its form is permanently altered it is said to be plastic.

When a small force will break the body it is said to be brittle. When it is not easily broken it is tough. The properties of bodies with respect to toughness & plasticity are tenacity flexibility ductility malleability   [4] cleavage.

A Liquid body has no permanent form. It has however a certain volume. When put into a vessel it covers the lower part accurately and the upper surface of the liquid becomes horizontal. The reason of this is that the very smallest force is sufficient to make the liquid alter its form and it therefore assumes that form which is assigned it by the forces which act on it.

The best test of a liquid is that when at rest it has a perfectly level surface. There are some liquids, such as treacle tar &c which must be left for a long time before they become perfectly level. They are then said to be viscous liquids. Viscosity is a property of all liquids although some are much more viscous than others. Those which like ether or chloroform are not perceptibly viscous are termed mobile liquids. All liquids may be compressed by the application of force. This property of compressibility varies in different liquids but it is always very small, that is a great force produces but a small compression.

The third state of bodies is called the gaseous state. Both liquids & gases are termed fluids but there is this distinction between them, that a gas always expands till it fills the vessel which contains it so that a vessel cannot for any length of time be half full of a gas, neither will a gas remain in a vessel without a cover, whereas a liquid will remain at the bottom of a vessel and will present a distinct upper surface even when there is no pressure upon it.

The most remarkable law relating to gases is that discovered by Boyle &

---

(4) Space in the MS.

Mariotte and it may be stated thus. When a gas is submitted to various pressures the volume which it occupies is inversely as the pressure. When a gas obeys this law accurately it is said to be a *perfect* gas. When the law of pressure and volume is different the gas is said to be imperfect. This distinction between perfect & imperfect gases depends entirely upon the result of experiments with respect to the behaviour of the gas under different pressures.

We come in the last place to the Physical properties of bodies, relating to their behaviour with respect to Heat Light Electricity Magnetism.

And first with respect to Heat.

Bodies are altered by heat in three ways.

1st Their temperature is raised.

2nd Their volume is increased & their form altered.

3rd Their state is changed from solid to liquid & from liquid to gaseous.

The measurement of these changes and the laws which connect them with the amount of heat entering the body form important branches of the science of heat and the numbers which enter into the experimental results indicate properties of the bodies with respect to heat into which we cannot now enter. When we come to the theory of the Steam Engine we shall find the importance of these determinations.

Heat passes through bodies in two ways.

1st By conduction when it passes from a hotter part of a body to a colder part in contact with it.

2nd By radiation when it passes through the body like a sunbeam without warming it.

The quality of bodies which conduct heat is called conductivity that of bodies which allow heat to radiate through them is called diathermancy or transparency for heat.

Radiant Heat and Light obey the same laws as they are two different aspects of the same thing,[5] so that transparency for heat and transparency for light differ no more than transparency for red light differs from transparency for blue light.

Until we have acquired some of the preliminary notions relating to Electricity & Magnetism it would be absurd to give you names of properties of bodies which can only be understood in connexion with the science of those agents.

---

(5) Compare Forbes' discussion of the analogies between light and radiant heat; see J. D. Forbes, 'On the refraction and polarization of heat', *Trans. Roy. Soc. Edinb.*, **13** (1836): 131–68, and Forbes, 'Note respecting the undulatory theory of heat and on the circular polarization of heat by total reflexion', *Phil. Mag.*, ser. 3, **8** (1836): 246–9.

107

ADAMS PRIZE ESSAY 'ON THE STABILITY OF THE MOTION OF SATURN'S RINGS'[1]

OCTOBER–DECEMBER 1856[2]

From the original in the University Library, Cambridge[3]

ON THE STABILITY OF THE MOTION OF SATURNS RINGS[4]

'*E pur si muove.*'[5]

There are some questions in Astronomy to which we are attracted rather on account of their peculiarity than from any direct advantage which their solution would afford to mankind. The theory of the Moons inequalities, though

---

(1) The subject for the Adams Prize for 1857 was announced in March 1855: '*The Motions of Saturn's Rings.*/** The Problem may be treated on the supposition that the system of Rings is exactly or very approximately concentric with Saturn and symmetrically disposed about the Planes of his Equator, and different hypotheses may be made respecting the physical constitution of the Rings. It may be supposed (1) that they are rigid: (2) that they are fluid, or in part aeriform: (3) that they consist of masses of matter not mutually coherent. The question will be considered to be answered by ascertaining on these hypotheses severally, whether the conditions of mechanical stability are satisfied by the mutual attractions and motions of the Planet and the Rings./ It is desirable that an attempt should also be made to determine on which of the above hypotheses the appearances both of the bright Rings and the recently discovered dark Ring may be most satisfactorily explained; and to indicate any causes to which a change of form, such as is supposed from a comparison of modern with the earlier observations to have taken place, may be attributed.' See Cambridge Observatory Archives, Library of the Institute of Astronomy, Cambridge; and James Clerk Maxwell, *On the Stability of the Motion of Saturn's Rings. An Essay which obtained the Adams Prize for the Year 1856, in the University of Cambridge* (Cambridge, 1859): iv (= *Scientific Papers*, **1**: 288). The Adams Prize was however to be 'adjudged in 1857', as stated in the 'Examiners' Notice'; see also James Challis' 'Book of Minutes relating to the Adams Prize, kept by the Plumian Professor' (Cambridge Observatory Archives). The dark 'obscure ring' interior to the two bright rings A and B had been first observed by G. P. Bond in 1850; see 'Inner ring of Saturn', *Monthly Notices of the Royal Astronomical Society*, **11** (1851): 20–7. Otto Struve had recently claimed that the inner bright ring was approaching the planet; see Otto Struve, 'Sur les dimensions des anneaux de Saturne', *Mémoires de l'Académie Impériale des Sciences de Saint-Petersbourg*, ser. 6, **5** (1853): 439–75, esp. 473 (= *Recueil de Mémoires présentès à l'Académie des Sciences par les Astronomes de Poulkova*, **1** (1853): 349–85, esp. 383); '1) Le bord intérieur des anneaux s'approche continuellement du globe de la planète; 2) Le rapprochement du bord intérieur est combiné avec un accroissement de la largeur totale des anneaux; 3) Dans l'intervalle entre les observations de J. D. Cassini et de W. Herschel, la largeur de l'anneau *B* a augmenté en plus forte raison que celle de l'anneau *A*.' Struve's claim led Challis to propose the subject for the Adams Prize, as he explained in a letter to William Thomson of 28 February 1855 (ULC Add. MSS 7342, C 76A).

(2) Compare Maxwell's letter to Monro of 14 October 1856 (Number 104). The official date for receipt of the essays, as stated on the Examiners' Notice, was 16 December 1856; and on 31

interesting to all physicists in its first stages, has been pursued into such intricacies of calculation as can be followed up only by those who make the improvement of the Lunar Tables the object of their lives. The value of the labours of these men is recognised in some measure by all who know the importance of such tables in Practical Astronomy and Navigation. The methods by which the results are obtained are admitted to be sound, and we leave to professed astronomers the labour and the merit of developing these results. The questions which are suggested by the appearance of Saturns Rings cannot, in the present state [of][(a)] Astronomy, call forth so great an amount of labour among mathematicians. I am not aware that any practical use has been made of Saturns Rings, either in Practical Astronomy or in Navigation. They are too distant, and too insignificant in mass, to produce any appreciable effect on the motions of other parts of the Solar System; and for this very reason it is difficult to determine those elements of their motion which we obtain so accurately in the case of bodies of greater mechanical importance.

But, when we contemplate the Rings from a purely scientific point of view, they become the most remarkable objects in the heavens, except, perhaps, those still less *useful* bodies – the spiral nebulæ.

(a) {Challis} of.

December 1856 Challis informed Thomson that 'I received Dec. 17 from the Vice-Chancellor one exercise for the Adams Prize' (ULC Add. MSS 7342, C 76D), Maxwell's essay apparently being the only one submitted.

(3) ULC Add. MSS 7655, V a/8. Previously published in A. T. Fuller, 'James Clerk Maxwell's Cambridge manuscripts: extracts relating to control and stability – II and III', *International Journal of Control*, **36** (1982): 547–74; *ibid.*, **37** (1983): 1197–238.

(4) The Examiners' Notice required that 'any Candidate is at liberty to send in his Essay *printed* or *lithographed*'; Maxwell's essay is the autograph manuscript, annotated in pencil by Challis and Thomson (though not by Stephen Parkinson, St John's College, the third examiner). Informing Thomson on 31 December 1856 that 'I shall soon have completed my examination of it', Challis clearly was the first to read the essay. He subsequently acknowledged its receipt back from Thomson on 28 April 1857 (ULC Add. MSS 7342, C 76E). On 30 May 1857 Challis recorded: 'The Adams Prize to be awarded in 1857 was this day adjudged to James Clerk Maxwell B.A. of Trinity College' ('Book of Minutes relating to the Adams Prize'). See also note (38).

(5) 'And yet it does move.' To preserve anonymity, candidates were required to prefix a motto to their essay, which was 'to be accompanied by a paper sealed up, with the same motto on the outside; which paper is to enclose another, folded up, having the Candidate's name and College written within' ('Examiners' Notice', Cambridge Observatory Archives). As his motto Maxwell used words attributed to Galileo after disavowing Copernicus – see his letter to Monro of 14 October 1856 (Number 104, esp. note (13)) – probably because Galileo was the first to observe that Saturn has a complex form. See P. S. de Laplace, *Traité de Mécanique Céleste*, 5 vols. (Paris, An VII [1799]–1825), **5**: 288–91, esp. 288; 'Galilée observa le premier l'anneaux de Saturne. Il le vit, dans ses faibles lunettes, sous la forme de le deux corps lumineux contingus à deux parties opposées de la surface de la planète.'

We cannot bring our minds to rest after we have actually seen that great arch swung over the equator of the planet without any material connection.[6] We cannot simply admit that such is the case, and describe it as an observed fact of Nature, not admitting or requiring explanation. We must either explain its motion on mechanical principles, or admit that, in the Saturnian realms, there can be motion regulated by laws which we are unable to explain. We must account for the rings remaining suspended above the planet, and remaining concentric with Saturn, and in his equatoreal plane, for the flattened figure of the section of each ring, and for whatever other phenomena the telescope may reveal.

Our curiosity with respect to these questions is rather stimulated than appeased by the investigations of Laplace. That great mathematician, though occupied with many questions which more imperiously demanded his attention, has devoted several chapters, in various parts of his great work, to points connected with the Saturnian System.[7]

He has investigated the formula for the attractions of a ring of small section on a point very near it, (Mec. Cel. Liv III, Chap. VI)[8] and from this he deduces the equation from which the ratio of the breadth to the thickness of each ring is to be found,[9]

$$e = \frac{R^3}{3a^3}\frac{\rho}{\rho'} = \frac{\lambda(\lambda-1)}{(\lambda+1)(3\lambda^2+1)}$$

where R is the radius of Saturn & ρ his density a the radius of the ring and ρ' its density, and λ the ratio of the breadth of the ring to its thickness. The equation

---

(6) Compare John F. W. Herschel, *Outlines of Astronomy* (London, ₄1851): 314–22.

(7) P. S. de Laplace, *Traité de Mécanique Céleste*, **2**: 155–66, 373–82, and **5**: 288–91 (= *Oeuvres Complètes de Laplace*, 14 vols. (Paris, 1878–1912), **2**: 166–77, 393–402, and **5**: 319–23). Laplace's 'Mémoire sur la théorie de l'anneau de Saturne', *Mémoires de l'Académie Royale des Sciences* (année 1787): 249–67, was revised as Livre III, Chapitre VI 'De la figure de l'anneau de Saturne' of the *Traité de Mécanique Céleste*, **2**: 155–66.

(8) Laplace, *Traité de Mécanique Céleste*, **2**: 155–66.

(9) Supposing that the ring is a homogeneous fluid mass and that its figure is elliptical, Laplace obtains the equations

$$g = \frac{S}{a^3}, \qquad \frac{\dfrac{4\pi\lambda}{\lambda+1}+\dfrac{S}{a^3}}{\dfrac{4\pi}{\lambda+1}-\dfrac{3S}{a^3}} = \lambda^2,$$

where g is the centrifugal force arising from the rotatory motion at unit distance from the axis of rotation, S the mass of Saturn, a the distance between the centre of the ring and that of Saturn, and λ the ratio of the breadth of the ring to its thickness. The first equation determines the rotation of the ring, the second the ellipticity of the generating figure. Putting $e = S/4\pi a^3$, the second equation gives

$$e = \frac{\lambda(\lambda-1)}{(\lambda+1)(3\lambda^2+1)}.$$

for determining λ has one negative root, which must be rejected, and two roots which are positive while $e < 0.0543$, and impossible when e has a greater value. At the critical value of e $\lambda = 2.594$ nearly.

The fact that λ is impossible when e is above this value shows that the ring cannot hold together if the ratio of the density of the planet to that of the ring exceeds a certain value. This value is estimated by Laplace at 1.3, assuming $a = 2R$.[10]

We may easily see the physical interpretation of this result, by observing that the forces which act on the ring may be reduced to –

(1) The attraction of Saturn, varying inversely as the square of the distance from his centre.

(2) The centrifugal force, acting outwards, and varying directly as the distance.

(3) The attraction of the ring itself, depending on its form and density, and directed roughly speaking towards the centre of its section.

The first of these forces must balance the second somewhere near the mean distance of the ring, and beyond this distance their resultant will be outwards, within this distance it will act inwards.

If the attraction of the ring itself is not sufficient to counterbalance these residual forces, the outer and inner portions of the ring will separate, and the ring will be destroyed, and it appears from Laplace's result that this will be the case if the density of the ring is less than $\frac{10}{13}$ of that of the planet.

This result appears to bear upon the question of the material of the rings, and to overturn[b] any theory which supposes them entirely made of an aeriform substance. Another result[c] of Laplace's theory is, that there are probably many independent rings moving with different velocities. The phenomena of dark lines observed on the surface of the rings and concentric with them appears favourable to this deduction.

Laplace has also shown (Liv V Chap. III) that on account of the oblate-

---

(b) {Thomson} Not so, because if the rings were fluid the parts at different distances from the axis would revolve with different angular velocities.[11]

(c) {Challis} The existence of several independent rings, if established by observation, would not be inconsistent with the theory.

If R is the radius of Saturn and ρ its density (the density of the ring being taken as unity) then $S = \frac{4}{3}\pi\rho R^3$ and hence $e = \dfrac{\rho R^3}{3a^3}$. See Laplace, *Traité de Mécanique Céleste*, **2**: 161–2.

(10) The limiting value of $e = 0.0543026$, and as $e = \rho R^3/3a^3$, hence the maximum value of ρ is $0.1629078\, a^3/R^3$. Laplace supposes that for the inner ring the ratio $a/R = 2$; then the value of ρ will be $\frac{13}{10}$ (Laplace, *Traité de Mécanique Céleste*, **2**: 162).

(11) Maxwell responded to Thomson's comment in the revise, *Saturn's Rings*: 3 (= *Scientific Papers*, **1**: 293); 'This condition applies to all rings whether broad or narrow, of which the parts are separable, and of which the outer and inner parts revolve with the same angular velocity.'

ness of the figure of Saturn the planes of the rings will follow that of Saturn's equator through every change of its position due to the disturbing action of other heavenly bodies.[12]

Besides this, he proves most distinctly (Liv III Chap VI) that a solid uniform ring cannot possibly revolve about a central body in a permanent manner, for the slightest displacement of the relative positions of the centre of the ring & that of the planet, would originate a motion, which would never be checked, and would inevitably precipitate the ring upon the planet, not necessarily by breaking the ring, but by the ring and planet coming into internal contact.[13]

He therefore infers that the rings are irregular solids whose centres of gravity do not coincide with their centres of figure.[14] Perhaps we may be allowed to express his conclusion under this form. 'If the rings were solid and uniform their motion would be unstable, and they would be destroyed. But they are not destroyed therefore they are either not uniform or not solid.'

I have not discovered either in the works of Laplace or in those of more recent mathematicians, any investigation of the motion of a ring either not uniform or not solid.[15] So that in the present state of mechanical science, we are unable to decide whether an irregular solid ring, or a fluid or soft ring, can revolve permanently about a central body; and the Saturnian system still remains an unregarded witness in heaven to some necessary development of the laws of the universe.

We know, since it has been demonstrated by Laplace, that a solid uniform ring cannot fulfil a necessary condition of permanence. We propose in this essay to determine the amount and nature of the irregularity which would be required to make a permanent rotation possible. We shall find that the stability of the motion of the ring would be ensured by loading one side of the ring more than the other but when we come to estimate the necessary amount of this inequality we shall find that it must be so enormous as to be quite inconsistent with the observed appearance of the rings.

We are therefore constrained to abandon this theory of irregular solid rings,

---

(12) Laplace, 'Des mouvements des anneaux de Saturne, autour de leurs centres de gravité', Livre V, Chapitre III of the *Traité de Mécanique Céleste*, **2**: 373–82, esp. 381.

(13) Laplace, *Traité de Mécanique Céleste*, **2**: 163–5.

(14) Laplace, *Traité de Mécanique Céleste*, **2**: 165.

(15) Maxwell appended a note on work by G. P. Bond and Benjamin Peirce to the published essay on *Saturn's Rings*: 3–4n (= *Scientific Papers*, **1**: 294n). See his letter to Thomson of 1 August 1857 (Number 126, esp. notes (3), (4) and (5)).

and to consider the case of a ring, the parts of which are not rigidly connected, as in the case of a ring of disconnected satellites, or a fluid ring.

There is now no danger of the whole ring or any part of it being precipitated on the body of the planet. Every particle of the ring is now to be regarded as a satellite of Saturn, disturbed by the presence of a ring of satellites at the same mean distance from the planet. The mutual action of the parts of the ring will be so small, compared with the attraction of the planet, that no part of the ring can ever cease to move round Saturn as a satellite.

But the question before us now is altogether different from that relating to the solid ring. We have now to take account of variations in the form and arrangement of the parts of the ring, as well as its motion as a whole, and we have as yet no security that these variations may not accumulate, till the ring entirely loses its original form, and collapses into one or more satellites, circulating round Saturn. In fact, such a result is one of the leading doctrines of the 'nebular theory' of the formation of planetary systems,[16] and we are familiar with the actual breaking up of fluid rings[d] in the beautiful experiments of M. Plateau.[17]

(d) {Thomson} under the action of 'capillary' force.[18]

(16) According to Laplace's 'nebular' cosmogony, planets were formed by the condensation of gaseous matter surrounding the primeval sun, the particles of vapour coalescing into rings and finally forming planets. Similarly, he suggested that 'les satellites et les anneaux de Saturne out été formé par les zones que l'atmosphère de la planète a successivement abandonnées à mésure qu'elle s'est resserrée en se refroidissant', *Traité de Mécanique Céleste*, **5**: 291. Laplace referred to observations by William Herschel of the periods of rotation of Saturn and the ring in 10 hours 16′ 0″.4 and 10 hours 32′ 15″.4, respectively; see Herschel, 'On the satellites of the planet Saturn, and the rotation of its ring on an axis', *Phil. Trans.*, **80** (1790): 427–95, esp. 479; Herschel, 'On the rotation of the planet Saturn upon its axis', *ibid.*, **84** (1794): 48–66, esp. 66; and claimed that the faster rotation of the planet 'comme cela doit être suivant l'hypothèse que j'ai proposée sur la formation des planètes, des satellites et des anneaux'.

(17) Joseph Plateau, 'Sur les phénomenès qui présente une masse liquide libre et soustraite à l'action de la pesanteur', *Mémoires de l'Académie Royale des Sciences de Bruxelles*, **16** (1843), (trans. with the assistance of James Challis) as 'On the phenomena presented by a free liquid mass withdrawn from the action of gravity', *Scientific Memoirs*, ed. R. Taylor, **4** (London, 1846): 16–43. Plateau had carried out experiments on the effect of rotation on a sphere of oil immersed in a mixture of alcohol and water; 'under the influence of a sufficient centrifugal force' the sphere '*is transformed into a perfectly regular ring*'. He noted that the 'heavens exhibit to us also a body of a form analogous to our liquid ring…Saturn's ring'; and observed that the rupturing of the fluid ring presented 'an image in miniature of the formation of the planets, according to the hypothesis of Laplace, by the rupture of the cosmical rings attributable to the condensation of the solar atmosphere'. See Plateau, 'On the phenomena presented by a free liquid mass': 27–8, 35–6. See also Numbers 113 and 128.

(18) Compare *Saturn's Rings*: 4 (= *Scientific Papers*, **1**: 295), where Maxwell adds Thomson's modification.

If we now investigate the case of a fluid ring, originally uniform, but afterwards slightly disturbed, so that the section of the ring is greater at some points of its circumference than at others, we shall find that the resultant attraction is always from thinner to thicker parts of the ring, so that, on a statical theory, the remaining fluid would rush towards the thickened parts, and the whole ring would collapse into satellites, of which the original accumulations of fluid would be the nuclei.

It is only when we treat the question dynamically that we are able to understand the possibility of the stable motion of a fluid ring. We must reserve the complete discussion of the motion for analytical investigation, but in order to give a first impression of the nature of the compensating motion, we may here attempt to show how the same tangential force, which would, of itself, break up the ring by increasing its irregularities, becomes converted by the motion of the ring into a cause which actually tends to diminish the irregularities and to reduce the ring to uniformity.

Let us fix our attention upon a part of the ring where the section is increasing in the direction of the motion of the ring, so that the parts of the ring immediately in front of the part considered are thicker than those immediately behind it.

The result will be that the fluid will be attracted in a forward direction – towards the thicker parts of the ring, which would tend to increase the irregularity, if the ring were at rest.

But this attraction forwards will increase the velocity of the fluid in the part considered, so that its path will be rendered less concave towards the planet, and its distance from the planet will therefore increase.

The increased velocity due to the tangential attraction is expended in overcoming the attraction of the planet, while the distance from the planet is increased, so that at this increased distance the actual linear velocity, and *à fortiori* the angular velocity about Saturn, is less than it was originally.

The angular velocity being now less than the mean angular velocity of the ring, the fluid in question, instead of advancing on the ring will lag behind it, and will go to thicken the thinner part of the ring which lies behind it.

If the element of the ring had been situated so that the thicker parts lay behind it, the tangential force would have been backwards, the velocity would have diminished,[e] the element would have gone nearer to Saturn, the angular velocity would have increased, and the element would have overtaken the thinner parts of the ring.

In this way there is a continual tendency towards those parts of the ring

---

(e) {Thomson} increased. A retarding tangential force causes the velocity to be increased, and an accelerating tangential force causes it to be diminished, as is clearly stated in the preceding page.[19]

(19) Thomson has misunderstood the argument: Maxwell is now considering the opposite

where the fluid is deficient. In reality, the effect of any disturbance of the ring is such as to produce four sets of waves in the fluid in the plane of the ring besides two sets of waves oscillating perpendicularly to the plane of the ring. The consideration of the laws of the first four of these systems of waves forms the essential part of our theory of a fluid ring, but in this introduction I have thought it sufficient to show, how a force, which, statically, would cause the destruction of the ring, becomes, dynamically, the condition of its permanence.[20]

As the astronomical value[21] of Saturns Rings is very slight compared with the scientific interest of the question of their stability, I have considered their motion rather as an illustration of general principles than as a subject for elaborate calculation, and therefore I have confined myself to those parts of the subject which bear upon the question of the permanence of a given possible form of motion.

There is a very general and very important problem in Dynamics, the solution of which would contain all the results of this essay and a great deal more. It is this –

'Having found a particular solution of the equations of motion of any material system, to determine whether a slight disturbance of the motion would cause a small periodic variation, or a total derangement of the motion, as given by the particular solution.'

The question may be made to depend upon the conditions of a maximum or a minimum of a function of several variables, but the theory of the tests for distinguishing maxima from minima by the Calculus of Variations become[f] so intricate when applied to functions of several variables, that I think it doubtful whether the physical or the abstract problem will be first solved.[22]

(f) {Challis} becomes.

case. The paragraphs *supra* are cancelled in pencil: compare the abbreviated discussion of this point in *Saturn's Rings*: 4 (= *Scientific Papers* 1: 295).

(20) See Part II, Proposition X; and Maxwell's letter to Thomson of 1 August 1857 (Number 126).

(21) Thomson queries this phrase; compare Maxwell, *Saturn's Rings*: 5 (= *Scientific Papers*, 1: 295), where he uses the phrase 'scientific interest'.

(22) Compare J. L. Lagrange, *Méchanique Analitique* (Paris, 1788): 36–44, 241–62. He obtains the energy equation $\sum mv^2 = C - 2\phi$, where each mass is represented by m and its velocity by v, C is a constant and the potential energy function ϕ depends on the coordinates and determines the condition of equilibrium. Lagrange observes that 'si cette fonction est un *minimum* l'équilibre aura de la stabilité; en sorte que le système étant d'abord supposé dans l'état d'équilibre, & venant ensuite à être tant soit peut déplacé de cet état, il tendra de lui-même à s'y remettre, en faisant des oscillations infiniment petites'; *Méchanique Analitique*: 38. See also P. G. Lejeune Dirichlet, 'Über die Stabilität des Gleichgewichts', *Journal für die reine und angewandte Mathematik*, 22 (1846): 85–8, (trans.) 'Note sur la stabilité de l'équilibre', *Journal de Mathématiques Pures et Appliquées*, 12 (1847):

Part I
On the Motion of a Rigid Body of any form about a Sphere

We confine our attention for the present to the motion in the plane of reference, as the instability, if it exists, relates to displacements in this plane.

Let S be the centre of gravity of the sphere, which we may call Saturn; and R that of the Rigid body, which we may call the Ring. Join RS, and divide it in G, so that

$$SG : GR :: R : S$$

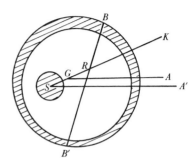

Figure 107,1

R & S being the masses of the ring & of Saturn respectively. Then G will be the centre of gravity of the system, and its position will be independent of the mutual action of the parts of the system.

Assume G as the point to which the motions of the system are to be referred. Draw GA in a direction fixed in space. Let $AGR = \theta$ and $SR = r$ then

$$GR = \frac{S}{S+R}r \text{ and } GS = \frac{R}{S+R}r$$

so that the positions of S & R are determined. Let BRB' be a straight line through R, fixed with respect to the substance of the ring, and let $BRK = \phi$. This determines the angular position of the ring, so that from the values of r, θ & ϕ the configuration of the system may be deduced, as far as relates to the plane of reference.

We have next to determine the forces which act between the ring and the sphere, and this we shall do by means of the *potential*, V, due to the ring.[23]

The value of V for any point of space S depends on its position relatively to the ring, and it is found from the eq^{n} $V = \sum \left(\dfrac{dm}{r'}\right)$ where dm is an element of mass of the ring and r' is the distance of that element from the given point, and the summation is extended over every element of mass belonging to the ring.

474–8 (= Lagrange, *Mécanique Analytique*, ed. Joseph Bertrand, 2 vols. (Paris, ₃1853–5), **1**: 399–401). Dirichlet rigorised Lagrange's theorem on the minimum energy condition for the point of equilibrium of a conservative system. See also the discussion by S. D. Poisson, *Traité de Mécanique*, 2 vols. (Paris, ₂1833), **1**: 690–4.

(23) Compare Laplace, *Traité de Mécanique Céleste*, **1**: 136–7. See Number 51 notes (13) and (19).

V will then depend entirely upon the position of the point S relatively to the ring; and may be expressed as a function of r, the distance of S from R, the centre of gravity of the ring, and ϕ the angle which the line SR makes with the line RB, fixed in the ring. A particle P placed at S will, by the theory of potentials, experience a moving force $= P\dfrac{dV}{dr}$ in the direction which tends to increase r, and $P\dfrac{1}{r}\dfrac{dV}{d\phi}$ in a tangential direction tending to increase ϕ.

Now we know that the attraction of a sphere is the same as that of a particle of equal mass placed at its centre. The forces acting between the sphere and the ring are therefore $S\dfrac{dV}{dr}$, tending to increase r, and a tangential force $S\dfrac{1}{r}\dfrac{dV}{d\phi}$, tending to increase ϕ.

Since this tangential force is applied at the centre of the sphere, in order to estimate its effect on the ring, we must resolve it into a tangential force $S\dfrac{1}{r}\dfrac{dV}{d\phi}$ acting at R, and a couple $= S\dfrac{dV}{d\phi}$ tending to increase ϕ.

We are able now to form the equations of motion for the Planet and the Ring –

For the Planet

$$S\frac{d}{dt}\left(\frac{R^2}{(S+R)^2}r^2\frac{d\theta}{dt}\right) = -\frac{R}{S+R}S\frac{dV}{d\phi} \tag{1}$$

$$S\frac{d^2}{dt^2}\left(\frac{R}{S+R}r\right) - S\frac{R}{S+R}r\left|\frac{\overline{d\theta}}{dt}\right|^2 = S\frac{dV}{dr} \tag{2}$$

For the Centre of Gravity of the Ring

$$R\frac{d}{dt}\left(\frac{S^2}{(R+S)^2}r^2\frac{d\theta}{dt}\right) = -\frac{S}{R+S}S\frac{dV}{d\phi} \tag{3}$$

$$R\frac{d^2}{dt^2}\left(\frac{S}{R+S}r\right) - R\frac{S}{R+S}r\left|\frac{\overline{d\theta}}{dt}\right|^2 = S\frac{dV}{dr} \tag{4}$$

For the rotation of the Ring about its centre of gravity

$$Rk^2\frac{d^2}{dt^2}(\theta+\phi) = S\frac{dV}{d\phi}.^{(24)} \tag{5}$$

Equations (3) & (4) are identical with (1) & (2), because the orbit of the centre of gravity of the ring must be similar to that of the centre of gravity of the planet.

---

(24) k is the radius of gyration of the ring about its centre of gravity G.

Equations (3) (4) & (5) are the necessary and sufficient data for determining the motion of the Ring. Clearing them of fractions, and performing the differentiations, they may be written –

$$R\left(2r\frac{dr}{dt}\frac{d\theta}{dt}+r^2\frac{d^2\theta}{dt^2}\right)+(R+S)\frac{dV}{d\phi}=0 \tag{6}$$

$$R\left(\frac{d^2r}{dt^2}-r\overline{\left.\frac{d\theta}{dt}\right|}^2\right)-(R+S)\frac{dV}{dr}=0 \tag{7}$$

$$Rk^2\left(\frac{d^2\theta}{dt^2}+\frac{d^2\phi}{dt^2}\right)-S\frac{dV}{d\phi}=0. \tag{8}$$

Problem I To find the conditions under which a uniform motion of the Ring is possible.

By a uniform motion is here meant a motion of uniform rotation, during which the distances of the parts of the ring from the planet do not change, so that the centre of gravity of the planet has an invariable position with respect to the ring.

In this case r and ϕ are constant and therefore V and its differential coefficients are given. Equation (7) becomes –

$$Rr\overline{\left.\frac{d\theta}{dt}\right|}^2+(R+S)\frac{dV}{dr}=0$$

which shows that the angular velocity ω is constant, and that

$$\omega^2=-\frac{R+S}{Rr}\frac{dV}{dr}. \tag{9}$$

Hence $\dfrac{d^2\theta}{dt^2}=0$ and therefore by equation 8

$$\frac{dV}{d\phi}=0. \tag{10}$$

Equations (9) & (10) give the conditions under which the uniform motion is possible, and if they were perfectly fulfilled, the uniform motion would go on for ever if not disturbed. But it does not follow that if these conditions were *nearly* fulfilled, or that if when accurately adjusted they were *slightly* disturbed, the motion would go on for ever *nearly* uniform.

The effect of the disturbance might either be to produce a periodic variation in the elements of the motion, the amplitude of the variation being small, or it might produce a displacement which would increase indefinitely and derange the system altogether.

In the one case the motion would be dynamically stable and in the other it would be dynamically unstable. The investigation of these displacements while still very small will form the next subject of inquiry.

Problem II To find the Equations of the motion when slightly disturbed.

Let $r = r_0$ $\theta = \omega t$ $\phi = \phi_0$ in the case of uniform motion and let

$$r = r_0 + r_1$$
$$\theta = \omega t + \theta_1$$
$$\phi = \phi_0 + \phi_1$$

when the motion is slightly disturbed
where $r_1 \theta_1$ & ϕ_1 are to be treated as small quantities of the first order. We may expand $\dfrac{dV}{dr}$ and $\dfrac{dV}{d\phi}$ by Taylors Theorem.[25]

$$\frac{dV}{dr} = \frac{dV}{dr} + \frac{d^2V}{dr^2}r_1 + \frac{d^2V}{dr\,d\phi}\phi_1$$
$$\frac{dV}{d\phi} = \frac{dV}{d\phi} + \frac{d^2V}{dr\,d\phi}r_1 + \frac{d^2V}{d\phi^2}\phi_1$$

where the values of the differential coefficients on the right hand side of the equations are those in which r_0 stands for r and ϕ_0 for ϕ.

Calling $\left(\dfrac{d^2V}{dr^2}\right)_0 = L$ $\left(\dfrac{d^2V}{dr\,d\phi}\right)_0 = M$ & $\left(\dfrac{d^2V}{d\phi^2}\right)_0 = N$ and remembering that $\left(\dfrac{dV}{dr}\right)_0 = -\dfrac{Rr_0}{R+S}\omega^2$ & $\left(\dfrac{dV}{d\phi}\right)_0 = 0$ we may write these equations

$$\frac{dV}{dr} = -\frac{R}{R+S}r_0\,\omega^2 + Lr_1 + M\phi_1$$
$$\frac{dV}{d\phi} = Mr_1 + N\phi_1.$$

Substituting these values in equations (6) (7) (8) and retaining all small quantities of the first order, while omitting their powers and products, we have the following system of linear equations in $r_1 \theta_1 \phi_1$

$$R\left(2r_0\omega\frac{dr_1}{dt} + r_0^2\frac{d^2\theta_1}{dt^2}\right) + (R+S)(Mr_1 + N\phi_1) = 0 \tag{11}$$

$$R\left(\frac{d^2r_1}{dt^2} - \omega^2r_1 - 2r_0\omega\frac{d\theta_1}{dt}\right) - (R+S)(Lr_1 + M\phi_1) = 0 \tag{12}$$

$$Rk^2\left(\frac{d^2\theta_1}{dt^2} + \frac{d^2\phi_1}{dt^2}\right) - S(Mr_1 + N\phi_1) = 0. \tag{13}$$

---

(25) Methods familiar to a Cambridge 'wrangler'; Maxwell's notes (from Hopkins) on 'Taylors theorem' are in his notebook 'Differential & Integral Calculus' (ULC Add. MSS 7655, V, m/7, ff. 1–24). Compare also the contemporary texts by D. F. Gregory, *Examples of the Processes of the Differential and Integral Calculus* (Cambridge, 1841): 52–6, and by Augustus de Morgan, *The Differential and Integral Calculus* (London, 1842): 70–6.

Problem III To reduce the three simultaneous equations of motion to the form of a pure linear equation.

Let us write n instead of the symbol $\dfrac{d}{dt}$ then arranging the equations in terms of $r_1\,\theta_1\,\phi_1$ they may be written

$$\{2Rr_0\,\omega n + (R+S)\,M\}\,r_1 + (Rr_0^2\,n^2)\,\theta_1 + (R+S)\,\mathcal{N}\phi_1 = 0 \qquad (14)$$

$$\{Rn^2 - R\omega^2 - (R+S)\,L\}\,r_1 - (2Rr_0\,\omega n)\,\theta_1 - (R+S)\,M\phi_1 = 0 \qquad (15)$$

$$-(SM)\,r_1 + (Rk^2n^2)\,\theta_1 + (Rk^2n^2 - SN)\,\phi_1 = 0. \qquad (16)$$

Here we have three equations to determine three quantities $r_1\,\theta_1\,\phi_1$ but it is plain that these three quantities will vanish together and leave the following relation among the coefficients –

$$\left.\begin{aligned}
&-\{2Rr_0\,\omega n + \overline{R+S}\,M\}\{2Rr_0\,\omega n\}\{Rk^2n^2 - SN\}\\
&+\{Rn^2 - R\omega^2 - \overline{R+S}\,L\}\{Rk^2n^2\}\{(R+S)\,\mathcal{N}\}\\
&+(SM)\,(Rr_0^2\,n^2)\,(\overline{R+S}\,M) - (SM)\,(2Rr_0\,\omega n)\,(\overline{R+S}\,\mathcal{N})\\
&+\{2Rr_0\,\omega n + \overline{R+S}\,M\}\{Rk^2n^2\}\{\overline{R+S}\,M\}\\
&-\{Rn^2 - R\omega^2 - \overline{R+S}\,L\}\{Rr_0^2\,n^2\}\{Rk^2n^2 - SN\}
\end{aligned}\right\} = 0. \qquad (17)$$

Note. The last factor of the first & last term is $\{Rk^2n^2 - SN\}$.[26]

By multiplying up and arranging by powers of n and dividing by $-Rn^2$ we get

$$\left.\begin{aligned}
&Rr_0^2\,k^2n^4 + \{3R^2r_0^2\,k^2\omega^2 - R(R+S)\,Lr_0^2\,k^2 - R(\overline{R+S}\,k^2 + Sr_0^2)\,\mathcal{N}\}\,n^2\\
&+ R(\overline{R+S}\,k^2 - 3Sr_0^2)\,\mathcal{N}[\omega^2]^{(a)} + (R+S)\,(\overline{R+S}\,k^2 + SR_0^2)\,(LN - M^2)
\end{aligned}\right\} = 0. \qquad (18)$$

Here we have a biquadratic equation in n, which may be treated as a quadratic in n^2 it being remembered that n stands for the operation $\dfrac{d}{dt}$. We proceed to the discussion of the equation, writing it more shortly –

$$An^4 + Bn^2 + C = 0. \qquad (19)$$

Problem IV To determine whether the motion of the Ring is stable or unstable from the relations of the coefficients A, B, C.

The equations to determine $r_1\,\theta_1$ & ϕ_1 are all of the form –

$$A\frac{d^4u}{dt^4} + B\frac{d^2u}{dt^2} + Cu = 0 \qquad (20)$$

and if n_1 be one of the four roots of (19) then $u = C_1\,\epsilon^{n_1 t}$ will be one of the four terms of the solution.

(a) {Challis} ω^2.

(26) Maxwell has corrected these terms, and adds the note for clarification.

Problem III To reduce the three simultaneous equations of motion to the form of a pure linear equation.

Let us write n instead of the symbol $\frac{d}{dt}$ then arranging the equations in terms of r, θ, φ, they may be written

$$\{2\,R\,r_0\,\omega\,n +(R+S)M\}r_i + (R\,r_0^2\,n^2)\theta_i +(R+S)N\varphi_i = 0 \quad (14)$$

$$\{R\,n^2 - R\,\omega^2 -(R+S)L\}r_i - (2\,R\,r_0\,\omega\,n)\theta_i - (R+S)M\varphi_i = 0 \quad (15)$$

$$-(S\,M)\,r_i \qquad\qquad + (R\,k^2\,n^2)\theta_i + (R\,k^2\,n^2 - SN)\varphi_i = 0 \quad (16)$$

Here we have three equations to determine three quantities r, θ, φ, but it is plain that these three quantities will vanish together and leave the following relation among the coefficients —

$$\left.\begin{aligned}
&- \{2\,R\,r_0\,\omega\,n + \overline{R+S}\,M\}\{2\,R\,r_0\,\omega\,n\}\{\overline{R+S}\,N\}\{R\,k^2\,n^2 - SN\}\\
&+ \{R\,n^2 - R\,\omega^2 - \overline{R+S}\,L\}\{R\,k^2\,n^2\}\{\overline{R+S}\,N\}\\
&+ (S\,M)(R\,r_0^2\,n^2)(\overline{R+S}\,M) - (S\,M)(2\,R\,r_0\,\omega\,n)(\overline{R+S}\,N)\\
&+ \{2\,R\,r_0\,\omega\,n + \overline{R+S}\,M\}\{R\,k^2\,n^2\}\{\overline{R+S}\,M\}\\
&- \{R\,n^2 - R\,\omega^2 - \overline{R+S}\,L\}\{R\,r_0^2\,n^2\}\{R\,k^2\,n^2 - SN\}
\end{aligned}\right\} = 0 \quad (17)$$

x Note. the last factor of the first & last term is $\{R\,k^2\,n^2 - SN\}$

Plate V. On Saturn's rings: To establish the equation of motion of a rigid ring, from the 1856 Adams Prize essay (Number 107).

(1) If n_1 be positive, this term would indicate a displacement which must increase indefinitely so as to destroy the arrangement of the system.

(2) If n_1 be negative the disturbance which it indicates will gradually die away.

(3) If n_1 be a pure impossible quantity of the form $a_1 \sqrt{-1}$ then there will be a term in the solution of the form $- C \cos(a_1 t + \alpha_1)$, and this would indicate a periodic variation, whose amplitude is C and period $\dfrac{2\pi}{a_1}$.

(4) If n_1 be the sum of two terms, of which one is positive and the other impossible, as thus –

$$n_1 = b_1 + \sqrt{-1}\, a_1$$

there will be a term in the solution of the form

$$C\epsilon^{b_1 t} \cos(a_1 t + \alpha)$$

which indicates a periodic disturbance, whose amplitude continually increases till it destroys the system.

(5) If n_1 be the sum of a negative quantity and an impossible one as $n_1 = -b_1 + \sqrt{-1}\, a_1$ the corresponding term is of the form

$$C\epsilon^{-b_1 t} \cos(a_1 t + \alpha)$$

which indicates a periodic disturbance, continually dying away.

It is manifest that the first and fourth cases are inconsistent with the permanent motion of the system.

Now since the given equation contains only even powers of n it must have pairs of equal and opposite roots, so that every root coming under the second or fifth cases implies the existence of another root belonging to the first or fourth.

Now if the root exists, some disturbance may occur to produce the corresponding derangement, so that the system is not safe unless the roots of the first and fourth kind are altogether excluded. This cannot be done without excluding also those of the second and fifth, so that to ensure stability all the roots must be of the third kind, that is, pure impossible quantities.

That this may be the case both values of n^2 must be negative, and the condition of this is

1st   That A, B & C should be of the same sign
2nd   that $B^2 > 4AC$.

When these conditions are fulfilled a periodic disturbance is possible. When they are not both fulfilled the motion cannot be permanent.

(b)*Problem V To find the mass, the centre of gravity, the radius of gyration, and the variations of the potential near the centre of a circular ring of small but variable section.*

Let a be the radius of the ring, and let θ be the circular measure of the angle subtended at the centre, between the line through the centre of gravity and the line through a given point of the ring. Then, if μ be the mass of unit of length of the ring near the given point, μ will be a periodic function of θ and may be expanded in the series –

$$\mu = A_0 + A_1 \cos \theta + A_2 \cos 2(\theta - \alpha_2) + A_3 \cos 3(\theta - \alpha_3) + \&\text{c}.$$

(That the value of μ may be thus expanded is a result of a theorem in Fourier's 'Traité de Chaleur'.)[27]

(1) The mass of the ring is given by the equation

$$R = \int_0^{2\pi} \mu a \, d\theta = 2\pi a A_0.$$

(2) To find the moment of the ring about the axis of y we have

$$Rr_0 = \int_0^{2\pi} \mu a^2 \cos \theta \, d\theta = \pi a^2 A_1$$

therefore
$$r_0 = \frac{a A_1}{2 A_0}.$$

[(3)](c) The radius of gyration about the centre is evidently $= a$ and to find that about the centre of gravity we have the equation $a^2 = k^2 + r_0^2$

therefore $k^2 = a^2 - r_0^2 = a^2 \left(1 - \dfrac{A_1^2}{4 A_0^2}\right)$.

(4) The potential at any point is found by dividing the mass of each element by its distance from the given point and integrating over the whole mass.

Let the given point be near the centre of the ring and let its position be defined by the coordinates r' and ψ, of which r' is small compared with a. The distance ρ between this point and a point on the ring is

$\rho = a \sqrt{1 - 2\dfrac{r'}{a} \cos(\psi - \theta) + \dfrac{r'^2}{a^2}}$ whence we obtain by the binomial theorem

$$\frac{1}{\rho} = \frac{1}{a} \left\{ 1 + \frac{r'}{a} \cos(\psi - \theta) + \tfrac{1}{4}\frac{r'^2}{a^2} + \tfrac{3}{4}\frac{r'^2}{a^2} \cos 2(\psi - \theta) + \&\text{c} \right\}.$$

(b) {Maxwell} See alterations.[28]   (c) {Maxwell} Omit.

(27) Joseph Fourier, *Théorie Analytique de la Chaleur* (Paris, 1822): 250–61 (§§ 231–5).

(28) Compare the revision in *Saturn's Rings*: 11–14 (= *Scientific Papers*, 1: 302–6).

The other terms contain powers of $\dfrac{r'}{a}$ higher than the second. We have now to determine the value of the integral –

$$V = \int_0^{2\pi} \frac{\mu}{\rho}\, a\, d\theta$$

and in multiplying the terms of μ by those of $\dfrac{1}{\rho}$ we need retain only those which contain constant terms, for all those which contain sines or cosines of multiples of $(\psi - \theta)$ will vanish when taken between limits.

[d] In this way we find

$$V = 2\pi\left\{ A_0 + \tfrac{1}{2}A_1\frac{r'}{a}\cos\psi + A_0\frac{r'^2}{4a^2} + \tfrac{3}{8}A_2\frac{r'^2}{a^2}\cos 2(\alpha-\psi) + \&c \right\}.$$

The other terms contain powers of $\dfrac{r'}{a}$ higher than the second. We have now obtained the value of V near the centre of the ring in terms of r' & ψ. In order to express it in terms of r_1 and ϕ_1 as we have assumed in the former investigations we must put

$$r'\cos\psi = -r^{(29)}$$
$$r'\sin\psi = -r_0\phi_1.$$

$$V = 2\pi\left\{ A_0 - \tfrac{1}{2}A_1\frac{r_1}{a} + \left(\tfrac{1}{4}A_0 + \tfrac{3}{8}A_2\cos 2\alpha\right)\frac{r_1^2}{a^2} \right.$$
$$\left. + \tfrac{3}{4}A_2\frac{r_0\sin 2\alpha}{a^2}r_1\phi_1 + \left(\tfrac{1}{4}A_0 - \tfrac{3}{8}A_2\cos 2\alpha\right)\frac{r_0^2}{a^2}\phi_1^2 \right\}.$$

When $r_1 = 0$ & $\phi_1 = 0$ we have

$$\frac{dV}{dr_1} = -\pi A_1\frac{1}{a} = -\frac{Rr_0}{R+S}\omega^2 \quad \text{[e]}$$

$$\frac{dV}{d\phi_1} = 0$$

$$\frac{d^2V}{dr_1^2} = \pi\{A_0 + \tfrac{3}{2}A_2\cos 2\alpha\}\frac{1}{a^2} = L$$

$$\frac{d^2V}{dr_1\,d\phi_1} = \pi\{\tfrac{3}{2}A_2\sin 2\alpha\}\frac{r_0}{a^2} = M$$

$$\frac{d^2V}{d\phi_1^2} = \pi\{A_0 - \tfrac{3}{2}A_2\cos 2\alpha\}\frac{r_0^2}{a^2} = N.$$

(d) {Maxwell} Omit.   (e) {Challis} which shews that if $A_1 = 0$ $r_0 = 0$.

(29) Compare *Saturn's Rings*: 12 ($= Scientific Papers$, **1**: 304), where he corrects the equation to read
$$r'\cos\psi = -r_1 + \tfrac{1}{2}r_0\phi_1^2.$$

See his letter to Challis of 24 November 1857 (Number 138); and Problem VI and note (31).

(f) *Problem VI   To determine the values of the coefficients A, B, C of the equation* (19).

The quantities which enter into these coefficients are $R\,S\,k^2\,r_0\,\omega^2\,L\,M\,N$. Of these we must observe that S is very large compared with R and therefore those terms are most important which contain the highest powers of S.

We have found

$$R = 2\pi a A_0 \quad S = a^3 \omega^2 \text{ nearly}$$

$$k^2 = a^2\left(1 - \frac{A_1^2}{4A_0^2}\right)$$

$$r_0 = a\,\frac{A_1}{2A_0}$$

$$L = \pi\{A_0 + \tfrac{3}{2}A_2 \cos 2\alpha\}\frac{1}{a^2}$$

$$M = \pi\{\tfrac{3}{2}A_2 \sin 2\alpha\}\frac{1}{a}\frac{A_1}{2A_0}$$

$$N = \pi\{A_0 - \tfrac{3}{2}A_2 \cos 2\alpha\}\frac{A_1^2}{4A_0^2}. \quad \text{(g)}$$

$$A = R^2 r_0^2 k^2 = \pi^2 a^6 A_1^2\left(1 - \frac{A_1^2}{4A_0^2}\right)$$

$$B = 3R^2 r_0^2 k^2\omega^2 - RS(Lr_0^2 k^2 + N(r_0^2 + k^2)) + \text{smaller terms}$$

$$= 3\pi^2 a^6 A_1^2\left(1 - \frac{A_1^2}{4A_0^2}\right)\omega^2 - 2\pi^2 a^6 \omega^2 A_0\left\{A_0\left(\frac{A_1^2}{4A_0^2}\overline{1 - \frac{A_1^2}{4A_0^2}}\right) + A_0\frac{A_1^2}{4A_0^2}\right\}$$

$$+ 2\pi^2 a^6 \omega^2 A_0 \tfrac{3}{2}\cos 2\alpha\left\{-\frac{A_1^4}{16A_0^4}\right\}$$

$$B = \pi^2 a^6 A_1^2\,\omega^2\left\{2 - \tfrac{5}{8}\frac{A_1^2}{A_0^2} + \tfrac{3}{16}\frac{A_1^2 A_2}{A_0^3}\cos 2\alpha\right\}$$

$$C = RS(k^2 - 3r_0^2)\,N + S^2(r_0^2 + k^2)\,(LN - M^2)$$

$$= 2\pi^2 a^6 \omega^2\left(1 - \frac{A_1^2}{A_0^2}\right)\{A_0 - \tfrac{3}{2}A_2 \cos 2\alpha\}\frac{A_1^2}{4A_0^2}$$

$$+ \pi^2 a^6 \omega^4(A_0^2 - \tfrac{9}{4}A_2^2)\frac{A_1^2}{4A_0^2}.$$

(h) Putting $\dfrac{A_1}{A_0} = x$ & $\dfrac{A_2}{A_0} = y$ and dividing out $\pi^2 a^6 A_0^2 x^2$ (i)

$$A = 1 - \tfrac{1}{4}x^2$$

$$B = \omega^2\{2 - \tfrac{5}{8}x^2 + \tfrac{3}{16}x^2 y \cos 2\alpha\}$$

$$C = \omega^4\{\tfrac{3}{4} - \tfrac{3}{4}y \cos 2\alpha - \tfrac{1}{2}x^2 - \tfrac{9}{16}y^2 + \tfrac{3}{4}x^2 y \cos 2\alpha\}.$$

(f) {Maxwell} omit.[30]
(g) {Thomson} Wrong.

(h) {Maxwell} omit.
(i) {Challis} This factor is significant.

Here if x & $y = 0$ $B^2 > AC$ and the motion[j] is unstable.[31] If $y = 0$ $\dfrac{1}{x^2}$ must

exceed $\dfrac{\sqrt{58}-1}{8}$ [k] this gives $x = 1.1$ nearly which would make the section on

one side of the ring negative.[32] By trying various values of x and y we may show that *small* values of these quantities will not render the motion permanent. Large values of these quantities would imply observable differences in the different parts of the ring. These are small by observation. Theory gives them large.[l]

[m]The result of the theory of a rigid ring shows not only that a perfectly uniform ring cannot revolve permanently about the planet, but that the irregularity of a permanently revolving ring must be a very observable quantity.

As there is no appearance about the rings justifying a belief in so great an irregularity the theory of the solidity of the rings becomes very improbable.

When we come to consider the additional difficulty of the tendency of all the

---

(j) {Challis} $B^2 > 4AC$. If x and $y = 0$, $B = 2\omega^2$ $A = 1$ $C = \frac{3}{4}\omega^4$. Hence $B^2 > 4AC$ and the motion is *stable*.[33]

(k) {Challis} not correct see page opposite. The explanation of this difficulty is to be derived from the factor $\pi^2 a^6 A_0^2 x^2$ which corresponds to a singular point. If $x = 0$, the equation (19) is satisfied, $r_0 = 0$, and the centres of gravity of Saturn & the Ring coincide. The motion is possible but not stable. If x do not vanish & $y = 0$, in order that the roots may be both negative x^2 must be less than $\frac{3}{2}$ otherwise C becomes negative. A, B, and C

being of the same sign, both roots are negative if $\dfrac{1}{x^2}$ be greater than $\dfrac{\sqrt{8}-1}{8}$ or x less than 2.1. From $x = 0$ to $x = \sqrt{\frac{3}{2}}$ there would appear to be different degrees of stability, vanishing at these limits. Laplace's Theory implies that stability is possible for small values of x.[34]

(l) {Challis} The above requires reconsideration.

(m) {Maxwell} take in.[35]

(30) Compare Maxwell's revisions of Problem VI in *Saturn's Rings*: 14–17 (= *Scientific Papers*, **1**: 306–10); and see his letter to Thomson of 24 August 1857 (Number 128).

(31) Compare Problem IV where Maxwell established the condition $B^2 > 4AC$ for the stability of a ring. The case he is considering here is that of a uniform solid ring (where $x = 0$ and $y = 0$); as Challis points out, for these values of x and y the values of the coefficients A, B and C are $A = 1$, $B = 2\omega^2$, $C = \frac{3}{4}\omega^4$. For a uniform solid ring $B^2 > 4AC$, implying stability. However, Maxwell knew that Laplace had demonstrated that a uniform solid ring would be *unstable*; see the introduction to the Essay. Compare Maxwell's letter to Challis of 24 November 1857 (Number 138) acknowledging his mistake, and perceiving that his error lay in establishing the equations for the potential of the ring in Problem V (4); see note (29). He consequently obtained incorrect values for the coefficients A, B and C. See also his letter to Thomson of 24 August 1857 (Number 128); and *Saturn's Rings*: 15 (= *Scientific Papers*, **1**: 307), where he establishes, with Laplace, 'the instability of a uniform ring'.

(32) Maxwell reconstructs this argument for the case of a ring thicker on one side than the other in his letter to Thomson of 24 August 1857 (Number 128, esp. note (6)); and see *Saturn's Rings*: 15 (= *Scientific Papers*, **1**: 307).   (33) See note (31).

(34) Using Maxwell's values for the coefficients A, B, and C, Challis attempts to correct his argument. But see notes (29), (31) and (32), and Maxwell's letter to Challis of 24 November 1857 (Number 138).

(35) The concluding paragraphs of Part I of the Essay are deleted from *Saturn's Rings*.

fluid and loose parts of the ring to accumulate at the thicker parts, and thus to weaken and at length destroy the thinner parts, we have another powerful argument against solidity.

And when we consider the immense size of the rings, and their comparative thinness, the absurdity of treating them as rigid bodies becomes self evident. An iron ring of such a size would comport itself as a viscous fluid, and we have no reason to believe these rings to be strengthened with any material unknown on this earth.[n]

Part II[a]
On the Motion of a Fluid Ring, or a Ring consisting of loose materials

In the case of the Rigid Ring we took advantage of the principle that the mutual actions of the parts of any system form at all times a system of forces in equilibrium, and we therefore took no account of the attraction between one part of the ring and any other part since no motion could result from this kind of action.

But when we regard the different parts of the Ring as capable of independent motion, we must take account of the attraction on each portion of the ring as affected by the irregularities of the other parts, and therefore we must begin by investigating the statical part of the problem, in order to determine the forces that act on any portion of the ring, as depending on the instantaneous condition of the rest of the ring.

The calculation of these attractions involves mathematical difficulties which it would be worthwhile to undertake if we were actually settled on the ring. I have only attempted to show what must be the general result of these attractions. This I have done by general reasoning in the first instance, as appropriate to an essay like the present, which aims rather at being intelligible than at improving the tables, or developing those elliptic integrals which we meet with in this subject, whenever we attempt any valuation of the attractions.

If the method adopted in the first part of these enquiries should not appear satisfactory I would refer the reader to a note at the end, where the same result is arrived at by a more mathematical and less fertile method of enquiry, and I may also state that I have suppressed many calculations of these attractions, because I think that the reading of them would be waste of time after the results have been deduced from the preliminary propositions. The methods appro-

---

(n) {Challis} True.   (a) {Maxwell} omit all the rest.[36]

(36) Part II of the Essay is substantially revised in *Saturn's Rings*. See Maxwell's letters to Thomson of 14 November 1857 and to Challis of 24 November 1857 (Numbers 134 and 138).

priate to these calculations may, perhaps, be of use to mathematicians, but they have no special claims to the attention of speculative astronomers.

Prop I The whole attraction between similar material systems, similarly situated varies as the square of the density, and the fourth power of the linear dimensions.

For the whole attraction is directly as the product of the masses, and inversely as the distance squared. Now each mass varies as the density & the cube of the linear dimensions, & the distance varies as the linear dimension so that the total attraction $\propto \dfrac{D^3\rho \cdot D^3\rho}{D^2} \propto \rho^2 D^4.$ [37]

Prop II The whole fluid pressure in a section of a fluid system acted on by its own gravitation is as the square of the density, & the fourth power of the linear dimensions.

For at similarly situated points the resultant attractions are as the density and the first power of the dimension, and the pressure is as the density, the attraction, and the depth, conjointly, and the surface on which it acts is as the square of the linear dimensions, therefore the whole pressure

$$P \propto \rho D \cdot \rho D \cdot D^2 \propto \rho^2 D^4.$$

Prop III In an elongated body, the attraction between the two parts separated by a transverse section is greater than the whole fluid pressure[b] on that section.

For the fluid, if left to itself would fall into a spherical form, and it would do so by the attraction across the section overcoming the fluid pressure.

Prop IV In an elongated body, both the attraction[c] across the section and the fluid pressure on it are principally affected by changes of form occurring near the section and the results of changes of form occurring at a distance from the section may be neglected.

For all effects of gravitation diminish with the square of the distance, so that changes[d] occurring at points removed from the section by many of its diameters will be much smaller than those occurring near the section.

(b) {Challis} Under what conditions of (equilibrium impossible) rest or motion?

(c) {Thomson} may be easily proved to be totally independent of the prolongations of the body on each side

beyond distances infinitely large compared with the lineal transverse dimensions.

(d) {Challis} not conclusive: but the proposition is an obvious consequence of the remark added above.[38]

(37) Linear dimension is D, density ρ.

(38) Although Challis read the Essay first – see his letter to Thomson of 31 December 1856 cited in note (4) – he apparently reviewed it in the light of Thomson's comments.

Prop V The whole action between the parts of a fluid filament or very elongated body may be roughly estimated as a kind of *tension* varying as the square of the density and the square of the sectional area.

For both the attraction and the fluid pressure vary according to this law,[e] but the attraction is the greatest[39] therefore we may consider the resultant effect as a *tension*.

Prop VI In a filament of varying section, the longitudinal force on any portion is the difference of the tensions at its extremities.

Prop VII In a curved filament, the force along the radius of curvature acting on any element $= T\dfrac{ds}{R}$, where T is the tension, ds the element & R the radius of curvature.

Prop VIII To express the disturbances of a fluid filament originally circular in terms of the original position of each element of the filament.

Let S be the centre, & SA a fixed line.
Let SB be a radius, revolving with the mean velocity of the ring; and let $ASB = \omega t$.
Let Q' be an element of the fluid in its actual position, and let Q be the position it would have had if it had moved uniformly with the mean velocity ω & had not been disturbed, then BSQ is a constant angle $= s$ and the value of s enables us to identify any element of the ring. Let the original radius $SQ = a$.

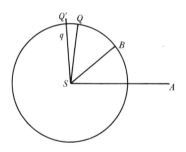

Figure 107,2

Now Q' is displaced in three ways generally
1st   by change of distance from S through a space $qQ' = a\rho$[40]
2nd   by change of angular position through a space $Qq = a\sigma$
3rd   perpendicular to the plane of the circle through a space $= a\zeta$.

In its disturbed position the radius vector of Q'

$$SQ' = a(1+\rho).$$

The true anomaly $ASQ' = \theta$

$$\theta = \omega t + s + \sigma.$$

(e) {Challis} by Prop[s] I & II.

(39) From Proposition III.

(40) Here ρ denotes the ratio of the change of distance of an element of the fluid through a space qQ' along the radius vector to the radius of the ring.

The displacement from the plane of reference $= z$

$$z = a\zeta.$$

The section of the filament varies inversely as $r\dfrac{d\theta}{ds}$ [41]

or as $c^2 \left\{ a(1+\rho)\left(1+\dfrac{d\sigma}{ds}\right)\right\}^{-1}$

where c^2 is the mean section. The Tension therefore

$$= L\frac{c^4}{a}\left(1-2\rho-2\frac{d\sigma}{ds}\right).$$

The length of an element [42]

$$= a\left(1+\rho+\frac{d\sigma}{ds}\right)ds.$$

The difference of tensions of the extremities of this element

$$= L\frac{c^4}{a}\left(-2\frac{d\rho}{ds}-2\frac{d^2\sigma}{ds^2}\right)ds$$

which is the force urging the element forward along the direction of the filament. [43]

The curvature $= \dfrac{1}{a}\left(1-\rho-\dfrac{d^2\rho}{ds^2}\right) = \dfrac{1}{R}$ in plane of reference therefore the force normal to the filament

$$P = T\frac{ds}{R}\,^{[44]} = \frac{Lc^4}{a}\left(1-2\rho-2\frac{d\sigma}{ds}\right)a\left(1+\rho+\frac{d\sigma}{ds}\right)ds\frac{1}{a}\left(1-\rho-\frac{d^2\rho}{ds^2}\right)$$

$$= \frac{Lc^4}{a}\left(1-2\rho-\frac{d\sigma}{ds}-\frac{d^2\rho}{ds^2}\right)ds.$$

Since this force does not act exactly towards the centre S, but forward, on account of the angle $\dfrac{d\rho}{ds}$ between the radius vector and the normal, we must add a quantity $= \dfrac{Lc^4}{a}\dfrac{d\rho}{ds}\,^{[45]}$ to the longitudinal force to get the force

---

(41) Where $r = a(1+\rho)$; and see note (42).

(42) In Proposition VII Maxwell denotes the length of an element of a fluid filament by ds; note his different usage here, where $s = $ angle BSQ.

(43) From Proposition VI.

(44) From Proposition VII; the force normal to the filament $P = T\,ds/R$, where ds is the length of an element of a fluid filament. Note Maxwell's subsequent usage, where $s = $ angle BSQ.

(45) Read: $\dfrac{Lc^4}{a}\dfrac{d\rho}{ds}ds.$

' perpendicular to the radius vector. We now have in the radius vector

$$P = \frac{Lc^4}{a}\left(1 - 2\rho - \frac{d\sigma}{ds} - \frac{d^2\rho}{ds^2}\right)ds.$$

Perpendicular to the radius vector

$$Q = \frac{Lc^4}{a}\left(-\frac{d\rho}{ds} - 2\frac{d^2\sigma}{ds^2}\right)ds.$$

Perpendicular to plane of reference

$$S = \frac{Lc^4}{a} \cdot \frac{d^2\zeta}{ds^2}\,ds. \text{(46)}$$

Prop IX To form the equations of motion.
    We have as usual[47]

$$r\overline{\frac{d\theta}{dt}}\bigg|^2 - \frac{d^2r}{dt^2} = \frac{P}{dm}$$

$$r\frac{d^2\theta}{dt^2} + 2\frac{dr}{dt}\frac{d\theta}{dt} = \frac{Q}{dm}$$

$$\frac{d^2z}{dt^2} = \frac{S}{dm}.$$

Substituting the values of r & θ from the last proposition

$$a(1+\rho)\left(\omega + \frac{d\sigma}{dt}\right)^2 - a\frac{d^2\rho}{dt^2} = \frac{P}{dm}$$

$$a(1+\rho)\left(\frac{d^2\sigma}{dt^2}\right) + 2a\frac{d\rho}{dt}\left(\omega + \frac{d\sigma}{dt}\right) = \frac{Q}{dm}$$

$$a\frac{d^2\zeta}{dt^2} = \frac{S}{dm}$$

or omitting powers and products of small quantities

$$a\left(\omega^2 + \omega^2\rho + 2\omega\frac{d\sigma}{dt} - \frac{d^2\rho}{dt^2}\right) = \frac{P}{dm}$$

$$a\left(2\omega\frac{d\rho}{dt} + \frac{d^2\sigma}{dt^2}\right) = \frac{Q}{dm}$$

$$a\frac{d^2\zeta}{dt^2} = \frac{S}{dm}.$$

If we put $\dfrac{Lc^4}{a^2}\dfrac{ds}{dm} = K$ then we must have besides the attractions of the parts of the ring, that of the planet which must be $= a\omega^2(1 - 2\rho)$ in the radius vector

---

(46) Note the new meanings for the symbols Q and S.
(47) *dm* is the mass of an element of the ring.

and $-a^2\zeta$ perpendicular to the plane of reference. Remembering that K is small compared with ω^2 we may write our equations

$$\omega^2\rho + 2\omega\frac{d\sigma}{dt} - \frac{d^2\rho}{dt^2} + 2\omega^2\rho + K\left(2\rho + \frac{d\sigma}{ds} + \frac{d^2\rho}{ds^2}\right) = 0^{(f)}$$

$$2\omega\frac{d\rho}{dt} + \frac{d^2\sigma}{dt^2} + K\left(\frac{d\rho}{ds} + 2\frac{d^2\sigma}{ds^2}\right) = 0$$

$$\frac{d^2\zeta}{dt^2} + \omega^2\zeta - K\frac{d^2\zeta}{ds^2} = 0.$$

The last of these equations depends on ζ only but the first and second contain both ρ & σ.

In order to simplify these equations let us write n, for the operation $\dfrac{d}{dt}$ and m, for $\dfrac{d}{ds}$. The first & second may then be written

$$\left(3\omega^2 + 2K + Km_{,}^2 - n_{,}^2\right)\rho + \left(Km_{,} + 2\omega n_{,}\right)\sigma = 0$$

$$\left(Km_{,} + 2\omega n_{,}\right)\rho + \left(2Km_{,}^2 + n_{,}^2\right)\sigma = 0.$$

By performing the operation $(2Km_{,}^2 + n_{,}^2)$ on the first of these, and the operation $(Km_{,} + 2\omega n_{,})$ on the second and subtracting

$$\left\{\left(3\omega^2 + 2K + Km_{,}^2 - n_{,}^2\right)\left(2Km_{,}^2 + n_{,}^2\right) - \left(Km_{,} + 2\omega n_{,}\right)^2\right\}\rho = 0$$

$$\left\{\left(6\omega^2 + 4K^{(g)} + 2Km_{,}^2\right)Km_{,}^2 - 4\omega Km_{,}n_{,} + \left(-\omega^2 + 2K - Km_{,}^2\right)n_{,}^2 - n_{,}^4\right\}\rho = 0.$$

If we had eliminated ρ we should have had the very same equation in σ.

We may therefore leave out of account the symbols ρ & σ, and attend only to the equation in m & n

$$n^4 + \left(\omega^2 + (m^2 - 2)K\right)n^2 + 4\omega Kmn - \left(3\omega^2 + (m^2 + 2)^{(h)}K\right)2Km^2 = 0$$

where n denotes the operation $\dfrac{d}{dt}$ & $m\dfrac{d}{ds}$.[49] The equation is therefore a partial differential equation of the fourth order.

The equation in ζ is of the second order

$$n^2 - Km^2 + \omega^2 = 0.$$

(f) {Challis} The term $-Ka$ is omitted. The reason for the omission ought to be distinctly stated.[48]

(g) {Challis} This coefficient should be 3.

(h) {Challis} $\frac{3}{2}$.

(48) In calculating the force exerted by the planet in the plane of reference as $a\omega^2(1-2\rho)$, Maxwell ignores the effect of the attraction of the ring in opposing the centrifugal force due to the rotation of the ring. This gives rise to an additional term $-Ka$ in the equation for the attraction of the planet in the plane of reference: $a(\omega^2 - K - 2\rho\omega^2)$. This additional term $-Ka$ cancels the missing term $-Ka$ in the equation to which Challis refers. See Fuller, 'Maxwell's Cambridge manuscripts – III': 1205–6.

(49) Read: $n_{,}$ and $m_{,}$. In Propositions X and XI there is again some inconsistency.

Prop X To solve the equations of motion.

A partial differential equation requires the introduction of arbitrary functions in its solution. The first of these equations would involve 4 of these, and the second two.

But the circumstances of the case point out a special method of investigation.

The fluid ring forms a simple closed curve, and therefore the values of σ & ρ as depending on s must recur every time that s increases by 2π.

σ & ρ are therefore *periodic functions* of s and their period is $2\pi s$.[50]

Now Fourier has shown[51] that any periodic function whose period is $2\pi s$ can be expanded as the sum of a series of circular functions whose periods are submultiples of $2\pi s$ and whose general term

$$\text{is } \rho = A \cos m(s+\alpha)^{\text{(i)}}$$

where m is necessarily a whole number.

Now in all such cases we may arrive at solutions by making $\dfrac{d}{ds} = \sqrt{-1}\, m$.

The equation of the last proposition then becomes for any given value of m (which must be integral) a biquadratic in n wanting the second term. There will therefore be four roots of n

$$\text{of the form } a + \sqrt{-1}\, b$$

and the complete solution will be

$$\rho = A\epsilon^{at} \cos\left(bt + mt^{\text{(j)}} + B\right)$$

where A & B are constants.

For stability it is necessary that a should vanish or be negative in every possible case but since in our biquadratic the second term is wanting, if it is negative in one case it will be positive in another, and therefore stability requires that it should vanish so that all the roots must be of the form

$$\sqrt{-1}\, b$$

or pure imaginary quantities.

We have therefore to solve the equation on the supposition that m is a pure imaginary quantity and to find the condition that n may be so too.[52] If we multiply both n, & m[53] by $\sqrt{-1}$ whenever they occur, the problem will be changed into the more familiar one of determining whether all the roots are possible the terms of the equation being all real

$$n = \sqrt{-1}\, n, \ \& \ m = \sqrt{-1}\, m_{,}.$$

(i) {Thomson} not the m of the preceding page.[54]   (j) {Challis} *ms.*

(50) Read: 2π.

(51) Joseph Fourier, *Théorie Analytique de la Chaleur*: 250–3 (§231).

(52) Read: $m_{,}, n_{,}$.   (53) Read: $m_{,}$.   (54) See note (49).

The equations which we have now to solve are

$$n^4 - (\omega^2 - (m^2+2)\,K)\,n^2 - 4\omega Kmn + (3\omega^2 + (2-m^2)^{(k)}\,K)\,2Km^2 = 0$$
$$n'^2 - Km'^2 - \omega^2 = 0.$$

We have to determine n on the supposition of m being an integral number.[55]

In the second equation which determines the motion perpendicular to the plane of reference it is manifest that n[56] is always possible & therefore the equilibrium must always be stable with respect to this kind of motion. The question of stability is therefore entirely dependent on the nature of the first equation, which relates to the motion in the plane of reference.

Let $\ n^4 - \omega^2 n^2 + (m^2+2)\,Kn^2 - 4\omega Kmn + (3\omega^2 + (2-m^2)^{(l)}\,K)\,2Km^2 = U.$
It is evident that since K is a small quantity compared with ω^2 the value of U will depend principally on the first two terms. It would appear therefore that n will have one value nearly $= \omega$ another nearly equal to $-\omega$ and two small values of the order $\sqrt{K}$.[57]

U is positive when $n = \pm\infty$ and also when $n = 0$ and it may be made negative by putting $n = \pm\dfrac{\omega}{\sqrt{2}}$ provided mK be not very large.

K is necessarily small and it does not appear that m has any reason to be immoderately large.[m] If m were very large the ring would be like a circular saw,[58] and our former calculations would be of no use. We therefore assume that m is a moderately large number, and that K is very small compared with

---

(k) {Challis} $(\tfrac{3}{2} - m^2)$.

(l) {Challis} $(\tfrac{3}{2} - m^2)$.

(m) {Thomson} Not exactly so, as the amplitude of the vibrations might be small in proportion to $1/m$. If the ring (supposed of infinitely small lineal dimensions in section) is not stable for displacements represented by harmonic curves with ever so great values of m, it is not stable at all.

(55) Each value of n represents the angular velocity (relative to the velocity of the ring) with which a system of m waves would travel round the ring.

(56) Read: n'.

(57) In the equation

$$n^4 - [\omega^2 - (m^2+2)\,K]\,n^2 - 4\omega Kmn$$
$$+ [3\omega^2 + (2-m^2)\,K]\,2Km^2 = 0$$

as K is a very small quantity compared with ω^2 the equation becomes (approximately)

$$n^4 - \omega^2 n^2 = 0.$$

The two principal roots approximate to $n = \pm\omega$.

To evaluate the two small roots, the equation may be written approximately as

$$-\omega^2 n^2 - 4\omega Kmn + 6\omega^2 Km^2 = 0.$$

If K is a very small quantity compared with ω^2, the roots are $n = \pm\sqrt{(6K)}\,m$, as Maxwell states *infra*. See Fuller, 'Maxwell's Cambridge manuscripts – III': 1208. See Maxwell's letter to Monro of 14 October 1856 (Number 104).

(58) See Maxwell's letter to Thomson of 24 August 1857 (Number 128).

ω^2, which is certainly true, and then we may assert that our equation has four real roots.[n]

The possibility of the ring being stable is now established. We have next to calculate the precise nature of the disturbances, and their mode of propagation, and first to find the values of n.

If we put $\qquad n = \omega$

$$U = \omega^2 K(7m^2 - 4m + 2) + 2K^2 m^2 (2 - m^2)^{[o]}$$

$$\frac{dU}{dn} = 2\omega^3 + K(m^2 + 2)\,\omega^{[p]} - 4Km\omega.$$

Hence the corrected value of n is

$$n = \omega - \frac{K}{2\omega}(7m^2 - 4m + 2) + \text{terms in } K^2.$$

If we had assumed $-\omega$ as the approximate root we should have found for the negative root

$$n = -\omega + \frac{K}{2\omega}(7m^2 + 4m + 2).$$

Besides these roots there are two others which we may estimate roughly at

$$+\sqrt{6K}\,m \ \& \ -\sqrt{6K}\,m.$$

Taking the first we have

$$U = 36K^2 m^4 - 6Km^2 + (m^2 + 2)\,6K\omega^2 m^2$$
$$- 4\sqrt{6}\omega^2 K^{\frac{3}{2}} m + 6K\omega^4 m^2 + (2 - m^2)\,2K^2 m^2.$$

The terms in K are equal and opposite and the lowest term is $-4\sqrt{6}\omega^2 K^{\frac{3}{2}} m$.

$\dfrac{dU}{dn} = -2\omega^3 \sqrt{6K}\,m$ with terms containing higher powers of K. We have therefore a second approximation

$$n = \sqrt{6K}\,\omega m - 2\frac{K}{\omega}.$$

The other values of n are nearly

$$+\sqrt{6K}\,m \ \& \ -\sqrt{6K}\,m.^{(59)}$$

(n) {Thomson} That this equation has not four real roots when mK is very large seems to show that a nearly linear ring (i.e. one of which the transverse section is infinitely small in all its dimensions) would not in fact be stable, and if created so, and of detached masses, it would spread itself out into something like one of Saturn's actual rings.[60]

(o) {Challis} $(\frac{3}{2} - m^2)$.

(p) {Challis} $2K(m^2 + 2)\,\omega$.

(59) See note (57).

(60) See Maxwell's letter to Thomson of 1 August 1857 (Number 126).

Taking the first, and arranging according to powers of K

$$U = -6Km^2\omega^2 + 6Km^2\omega^2 - 4\sqrt{6K}\,K\omega m^2 + \&c$$

$$\frac{dU}{dn} = -2\sqrt{6K}\,m\omega^2 + \&c.$$

We have therefore as a second approximation

$$n = \sqrt{6K}\,m - \frac{2K}{\omega}m.$$

Approximation to the other root would give

$$n = -\sqrt{6K}\,m - \frac{2K}{\omega}m.$$

We have therefore the four roots

$$n_1 = \omega - \frac{K}{2\omega}(7m^2 - 4m + 2) + \&c$$

$$n_2 = \sqrt{6K}\,m - \frac{2K}{\omega}m + \&c$$

$$n_3 = -\sqrt{6K}\,m - \frac{2K}{\omega}m + \&c$$

$$n_4 = -\omega + \frac{K}{2\omega}(7m^2 + 4m + 2) + \&c$$

arranged according to powers of $\dfrac{K}{\omega^2}$.

The value of ρ as given by the equation[q][61] is therefore

$$\rho = A_1 \cos(n_1 t + ms + \alpha_1) + A_2 \cos(n_2 t + ms + \alpha_2)$$
$$+ A_3 \cos(n_3 t + ms + \alpha_3) + A_4 \cos(n_4 t + ms + \alpha_4)$$
$$+ \text{other sets, of four terms each, for every integral value of } m.$$

Let $\rho = A \cos(nt + ms + \alpha)$ be a term of the value of ρ. Let us find the corresponding term in the value of σ

$$\sigma = B \sin(nt + ms + \beta)$$

where B & β are to be determined.

The equation for the action of the tangential force is[62]

$$2\omega \frac{d\rho}{dt} + \frac{d^2\sigma}{dt^2} + K\left(\frac{d\rho}{ds} + 2\frac{d^2\sigma}{ds^2}\right) = 0$$

$$\left.\begin{array}{l} \text{or } -2\omega nA \sin(nt + ms + \alpha) - n^2 B \sin(nt + ms + \beta) \\ - K(mA \sin(nt + ms + \alpha) + {}^{(r)}m^2 B \sin(nt + ms + \beta)) \end{array}\right\} = 0$$

(q) {Challis} of 8 dimensions.　　　　　　(r) {Challis} 2.

(61) The equation $\{(6\omega^2 + 4K + 2Km_r^2)\,Km_r^2 - 4\omega Km_r n_r + (-\omega^2 + 2K - Km_r^2)\,n_r^2 - n_r^4\}\rho = 0$; see Proposition IX.　　　　　　(62) See Proposition IX.

which can only be satisfied for all values of s & t by making

$$\alpha = \beta \text{ and } B = -A\frac{2\omega n + Km}{n^2 + 2Km^2}.$$

We may therefore write the value of σ

$$\sigma = -A\frac{2\omega n_1 + Km}{n_1^2 + 2Km^2}\sin(n_1 t + ms + \alpha_1)$$

-3 other terms of the same set besides as many sets
of four terms as there are integral values of m.

It appears therefore that for each value of m there are values of ρ and σ containing eight arbitrary constants: $A_1\,\alpha_1\,A_2\,\alpha_2\,A_3\,\alpha_3\,A_4\,\alpha_4$. These must be determined from the position and velocity of the ring at a given time, say the time $t = 0$.

Both the position and the velocity must be periodic functions whose period is $m(2\pi s)$ [63] where m is integral. When $t = 0$ therefore let

$$\rho = C_1\cos(ms + \gamma_1)$$
$$\sigma = C_2\cos(ms + \gamma_2)$$
$$\frac{d\rho}{dt} = C_3\sin(ms + \gamma_3)$$
$$\frac{d\sigma}{dt} = C_4\cos(ms + \gamma_4).$$

These conditions contain eight arbitrary constants which are connected with those of the equations for ρ & σ

$$A_1\cos\alpha_1 + A_2\cos\alpha_2 + A_3\cos\alpha_3 + A_4\cos\alpha_4 = C_1\cos\gamma_1$$
$$n_1 A_1\cos\alpha_1 + n_2 A_2\cos\alpha_2 + n_3 A_3\cos\alpha_3 + n_4 A_4\cos\alpha_4 = C_3\cos\gamma_3$$
$$\frac{2\omega n_1 + Km}{n_1^2 + 2Km^2}A_1\cos\alpha + \&c = C_2\cos\gamma_2$$
$$n_1\frac{2\omega n_1 + Km}{n_1^2 + 2Km^2}A_1\cos\alpha + \&c = C_4\cos\gamma_4. \text{[64]}$$

From these four simple equations the values of $A_1\cos\alpha_1$ &c may be determined in terms of $C_1\cos\gamma_1$ &c and similarly those of $A_1\sin\alpha_1$ in terms of $C_1\sin\gamma_1$ &c so that the eight constants may be completely determined from the circumstances of the motion at the origin of the time.

It may now be seen that, if the ring receive any disturbance, the subsequent motion may be analyzed into 4 waves, travelling along the ring with different velocities.

---

(63) Read: $m(2\pi)$. (64) Read: α_1.

The velocity of the first is nearly $= \omega$ relatively to the ring and therefore nearly $= 2\omega$ in space.

That of the second is nearly $-\omega$ in relation to the ring, or nearly stationary in space. Its actual velocity in space is nearly $+\dfrac{K}{2\omega} \left(7m^2 + 4m + 2\right)$.

The velocities of the third and fourth waves are nearly $\pm \sqrt{6K}\, m$ in relation to the ring.

If the disturbance communicated to the ring is expressed by a single circular function of ms, these four waves will express the whole motion, but if the original disturbance depend upon several terms, in which m has different integral values, each of them will produce four independent waves, whose velocities will be slightly different from each other in so far as they depend on the values of m.

To understand the effect of the disturbance on the motion of a given particle of the ring, let us consider *one* value of m, and *one* of the four waves belonging to it.

The values of ρ and σ are given by the equations

$$\rho = A \cos (nt + ms + \alpha)$$

$$\sigma = -A \frac{2\omega n + Km}{n^2 + 2Km^2} \sin (nt + ms + \alpha).$$

By making s constant we find the conditions of motion of an individual element of the ring. It oscillates or rather revolves about its mean position in an ellipse of which the major axis is in the direction of a tangent to the ring and the proportion of the axes $= \dfrac{2\omega n + Km}{n^2 + 2Km^2}$.

Now when $n = \pm\omega$ nearly this quantity $= 2$ nearly, and when $n = \pm \sqrt{6K}\, m$ nearly it is $\dfrac{\sqrt{6}}{4} \dfrac{\omega}{m\sqrt{K}}$ nearly.

The direction of motion in this ellipse is always opposite to the motion of the wave relatively to the ring[s] that is when the wave travels along the ring forwards the particles move in ellipses in the direction contrary to that of the ring but when the wave lags behind the ring and travels *in antecedentia*, the particles move in the *same* direction. In the case of the two waves which move rapidly in the ring the axes are as 1 to 2, but in the slow waves the tangential motion is far greater than the radial.

It will be seen that these results are in accordance with the general description of the motion in the introduction to this essay.

(s) {Challis} Do not the above values of ρ and σ prove that the motion in the ellipse is in the *same* direction whether n be positive or negative? {Thomson} Yes. {Maxwell} Always opposite to ring. J. C. M.

Prop XI To determine the effect due to internal friction in the matter of the ring.

When a filament of a plastic solid or a viscous fluid is altered in length, besides the forces of attraction and elasticity which may be called into play, there is an additional force, which always acts against the actual motion or change of form. This is the resistance arising from internal friction. We have nothing here to do with the explanation of the actual internal motions. If we can state the result, we can apply it to our purpose. The general law of internal friction is, that in the distortion of an element of the substance, a force is called into play, which resists the distortion and is proportional to the velocity of the distortion.

The particular kind of distortion which we have to deal with is that which occurs when an elongated filament is made longer or shorter. The result on the whole must be of the same *kind* with the result for each element, that is, there must be a force resisting the change of length, proportional to the rate of that change. The mathematical expression for this is that, if $ds\left(1+\dfrac{d\sigma}{ds}\right)^{(65)}$ be the length of an element and if it be undergoing elongation at the rate $\dfrac{d^2\sigma}{ds\,dt}$ the tension at that point will be increased by a quantity

$$k\,\frac{d^2\sigma}{ds\,dt}$$

where k is a coefficient depending on the internal friction, and on the particular form & diameter of the filament.

It is easy to see that this is the only considerable effect due to the friction, for the effects due to change of curvature will be very small compared with those due to change of length. We have therefore only to investigate the effect of this increase of tension on the radial and tangential forces, and to modify the equations accordingly.

Let $t = k\dfrac{d^2\sigma}{ds\,dt}$

then the tangential force on an element acting forwards is

$$k\,\frac{d^3\sigma}{ds^2\,dt} = km^2n\sigma^{(66)}$$

and the radial force acting inwards is

$$k\,\frac{d^2\sigma}{ds\,dt} = kmn\sigma.^{(67)}$$

---

(65) Read: $ds\left(1+\dfrac{d\sigma}{ds}\right)a$; compare Proposition VIII. (66) Read: $km^2_, n, \sigma\,ds$.

(67) From Proposition VII. Read: $km, n, \sigma\,ds$.

The equations for the motion in plane of reference become[68]

$$(3\omega^2 + 2K + Km^2 - n^2)\,\rho + (Km + 2\omega n - kmn)\,\sigma = 0$$
$$(Km + 2\omega n)\,\rho + (2Km^2 + n^2 - km^2n)\,\sigma = 0.$$

Eliminating as before we get for our general equation what we had before[69] with the addition of the terms

$$-km^2n^3 - 2k\omega mn^2 + \{3\omega^2 + K(1+m^2)\}\,km^2n\,{}^{[70]}$$

all involving k. When we multiply n & m[71] each by $\sqrt{-1}$ whenever it occurs these terms become

$$-\sqrt{-1}\,k\{m^2n^3 - 2\omega mn^2 + (3\omega^2 + K(1-m^2))\,m^2n\}.$$

We must consider k as a very small quantity compared with ω and proceed to find the effect of these additional terms in approximating to the values of n in the four cases determined before.

(1) When n is put equal to $+\omega$ for a first approximation we shall have an additional term

$$-\sqrt{-1}\,k(2m^2 - m).$$

To get rid of the impossible quantity we must return to the exponential form $\epsilon^{k(2m^2-m)}$[72] which will form part of the solution – thus

$$\rho_1 = A\epsilon^{-k(2m^2-m)t}\cos{(n_1 t + ms + \alpha)}\,{}^{[73]}$$

an expression which indicates a continual decrease in the amplitude of the wave, due to the distruction of *vis viva*[74] by the internal friction.

(2) In the case of the second wave in which n nearly $= -\omega$ the index of the exponential will be $-k(2m^2 + m)\,t$.

(3) In the case of the two slow waves, the index of the exponential is the same nearly

$$= -\tfrac{3}{2}km^2t.$$

These results show us that all four waves will continually diminish, so that in equal times they are diminished in equal ratios. The rate of diminution depends on m^2 so that those waves will last longest which are the longest, and those which have many alternations in the circumference of the circle will be rapidly extinguished. In treating of the four roots of the general equation, we remarked that we had nothing to do with very large values of m. This investigation shows what will happen to those waves for which the value of m is large. They will be

---

(68) See Proposition IX. Read: $m_{,}$ and $n_{,}$.

(69) See Proposition IX; the general equation in $n_{,}$ and $m_{,}$.

(70) Read: $m_{,}$ and $n_{,}$.

(71) Read: $n_{,}$ and $m_{,}$.   (72) Read: $\epsilon^{-k(2m^2-m)}$.

(73) Read: A_1 and α_1.   (74) See note (77).

rapidly extinguished by internal friction, and hence we need have no fear of their causing the destruction of the ring.

[Prop] XII We come next to consider the effect produced by an external disturbing force, due either to the planet's irregularities, to the attraction of satellites, or to the effect of the irregularities in the other rings.

All disturbing forces of this kind may be expressed in series of which the general term is

$$A \cos (vt + \mu s + \alpha)$$

where v is an angular velocity and μ a whole number.

Let us confine our attention to that part of the radial force which is expressed by

$$P_1 \cos (vt + \mu s) + P_2 \sin (vt + \mu s)$$

acting inwards, and the corresponding part of the tangential force

$$Q_1 \cos (vt + \mu s) + Q_2 \sin (vt + \mu s) \text{ forwards.}$$

The equations of motion become

$$\left(3\omega^2 - \frac{d^2}{dt^2} + K\left(2 + \frac{d^2}{ds^2}\right)\right)\rho + \left(2\omega \frac{d}{dt} + K\frac{d}{ds}\right)\sigma = P_1 \cos (vt + \mu s) + P_2 \sin (vt + \mu s)$$

$$\left(2\omega \frac{d}{dt} + K\frac{d}{ds}\right)\rho + \left(\frac{d^2}{dt^2} + 2K\frac{d^2}{ds^2}\right)\sigma = Q_1 \cos (vt + \mu s) + Q_2 \sin (vt + \mu s).$$

If we can find a single solution of these equations, we can render it perfectly general by adding to the values of ρ & σ so obtained those which we formerly arrived at by making the second member of these equations vanish. For, if a value satisfy the equation it is manifestly capable of doing so if we add a quantity which makes that part of the equation into which it enters disappear so that its introduction does not affect the value of the entire equation.

Now the part which immediately depends on the disturbing terms must have the form

$$\rho = A_1 \cos (vt + \mu s) + A_2 \sin (vt + \mu s)$$
$$\sigma = B_1 \cos (vt + \mu s) + B_2 \sin (vt + \mu s).$$

Substituting these values in the equations and collecting by themselves the terms involving $\cos (vt + \mu s)$ & $\sin (vt + \mu s)$ we obtain four equations for $A_1 \, A_2 \, B_1 \, B_2$

$$(3\omega^2 + v^2 + K(2 - \mu^2)) A_1 + (2\omega v + K\mu) B_2 = P_1$$
$$(3\omega^2 + v^2 + K(2 - \mu^2)) A_2 - (2\omega v + K\mu) B_1 = P_2$$
$$(2\omega v + K\mu) A_2 - (v^2 + 2K\mu^2) B_1 = Q_1$$
$$-(2\omega v + K\mu) A_1 - (v^2 + 2K\mu^2) B_2 = Q_2.$$

If we make

$$U = v^4 - (\omega^2 - (\mu^2 + 2))^{(75)} v^2 - 4\omega K\mu v + (3\omega^2 + (2-\mu^2) K) 2K\mu^2$$

$$A_1 = \frac{P_1(v^2 + 2K\mu^2) + Q_2(2\omega v + K\mu)}{U}$$

$$A_2 = \frac{P_2(v^2 + 2K\mu^2) - Q_1(2\omega v + K\mu)}{U}$$

$$B_1 = \frac{P_2(2\omega v + K\mu) - Q_1(3\omega^2 + v^2 + K(2-\mu^2))}{U}$$

$$B_2 = \frac{-P_2(2\omega v + K\mu) - Q_1(3\omega^2 + v^2 + K(2-\mu^2))}{U}.$$

We have now obtained the coefficients which determine the disturbance due to the given cause. It will be observed that U occurs in the denominator of each, and therefore if U were to vanish for the given values of μ & v there might be a catastrophe. But it will be remembered, that $U = 0$ is the equation by which we calculate the relations of m & n in the disturbed ring, and that it contains ω the angular velocity of that ring. U cannot vanish, therefore unless μ & v are connected as m & n are.

Now suppose the disturbance due to waves in *another* ring for which the velocity is ω'. Let the equation belonging to this ring be $U' = 0$ & let m' & n' be the values on which the waves depend. Then m' is the same as μ, and $n' + \omega' = v + \omega^{(t)}$ and therefore

$$v = n' + \omega' - \omega.$$

If for n' we substitute ω', the first approximate value, we shall have $v = 2\omega' - \omega$. If we make n' small we find $v = \omega' - \omega$ nearly and if we make $n' = \omega'$ $v = -\omega$ as a first approximation.

This last value of v is the only one which may happen to coincide with a value of n, and so render $U = 0$ but we shall find that in this case the second terms of the approximations depend on the different values of ω in the two rings, and therefore U will not vanish.

Prop XIII To determine the effect which long continued disturbances will have on a system of rings.

(t) {Thomson} Why? {Maxwell} Because n' is the relative angular velocity of the disturbing wave in its ring. v is the relative angular velocity of the disturbing force relative to the disturbed ring and the actual velocities of the disturbing wave and disturbing force must be equal or $n' + \omega' = v + \omega$.

(75) Read: $(\mu^2 + 2) K$.

We may take advantage of two general principles in Dynamics. The first is the principle of the Conservation of Angular Momenta.[76]

Let dm be the mass of a ring, ω its mean angular velocity and r its radius. Then its angular momentum $= \omega r^2\, dm$ and the principle of the equality of action and reaction shows us that whatever be the mutual actions of the rings of the system

$$\sum (\omega r^2\, dm) = A \quad \text{a constant.}$$

Half the vis viva[77] of the system is

$$\sum (\tfrac{1}{2}\omega^2 r^2\, dm) = V.$$

And the potential energy[78] of the system due to Saturn's attraction is

$$-\sum \left(\frac{S}{r}\, dm\right) = P.$$

If the whole motion of the system were angular rotation, and if no loss of power took place on account of internal friction we should have

$$V + P = \text{const.}^{[79]}$$

But if there be loss of power by internal friction then $V + P$ will continually diminish.

---

(76) Compare R. B. Hayward, 'On a direct method of estimating velocities, accelerations, and all similar quantities with respect to axes, moveable in any manner in space, with applications', *Trans. Camb. Phil. Soc.*, **10** (1856): 1–20, esp. 7–10. Hayward's paper, read on 25 February 1856, discusses 'The principle of the *Conservation of Momentum* ... as applied to angular momentum'. For Maxwell's application of '[Hayward's] use of the mechanical conception of Angular Momentum', and his statement of 'the permanence of the *original angular momentum* [of a rotating body] in direction and magnitude', see his paper 'On a dynamical top, for exhibiting the phenomena of the motion of a system of invariable form about a fixed point', *Trans. Roy. Soc. Edinb.*, **21** (1857): 559–70, esp. 560–2 (= *Scientific Papers*, **1**: 249–51). See also S. D. Poisson, *Traité de Mécanique*, **2**: 452–4; and J. H. Pratt, *The Mathematical Principles of Mechanical Philosophy* (Cambridge, ₂1845): 442–3, on the principle of moment of momentum.

(77) That is 'kinetic energy'. The Leibnizian term *vis viva* (mv^2) was in common use. In his *Über die Erhaltung der Kraft* (Berlin, 1847): 9, Hermann Helmholtz had noted the significance of the quantity $\tfrac{1}{2}mv^2$ as a measure of the energy of a body. See also note (79) and Number 133 note (7).

(78) For the term 'potential energy', see W. J. M. Rankine, 'On the general law of the transformation of energy', *Phil. Mag.*, ser. 4, **5** (1853): 106–17, esp. 106. See Number 133 note (11).

(79) The principle of the conservation of energy. See Helmholtz's statement in his *Über die Erhaltung der Kraft*: 7–20; (trans.) 'On the conservation of force', in *Scientific Memoirs, Natural Philosophy*, ed. J. Tyndall and W. Francis (London, 1853): 114–62, esp. 118–26. For general statements of energy principles in mechanics see William Thomson, 'On the mechanical antecedents of motion, heat and light', *Report of the Twenty-fourth Meeting of the British Association for the Advancement of Science* (London, 1855), part 2: 59–63 (= *Math. & Phys. Papers*, **2**: 34–40); and W. J. M. Rankine, 'Outlines of the science of energetics', *Edinburgh New Philosophical Journal*, **2** (1855): 120–41.

We may simplify these expressions by putting

$$\omega = \sqrt{(S)}\, r^{-\frac{3}{2}}$$

since the angular vel^{y} of a ring is that of a satellite at the same distance. We then have

$$\sqrt{(S)} \sum (r^{\frac{1}{2}} dm) = A \tag{1}$$

$$-\tfrac{1}{2} S \sum (r^{-1}\, dm) = V + P. \tag{2}$$

It is evident that the more the rings spread out from one another (under the condition (1)) the less will be the value of $(V+P)$.[u]

In fact if we put $r^{1/2} = s$

and make

$$s_{,} = \frac{\sum s\, dm}{\sum dm}$$

then if

$$s = s_{,} + s'$$

$$\sum (s^{-2}\, dm) = \sum dm \left(s_{,}^{-2} - 2\frac{s'}{s_{,}^{3}} + 3\frac{s'^{2}}{s_{,}^{4}} - \&c \right)$$

$$= M s_{,}^{-2} + \frac{3}{s_{,}^{4}} \sum (s'^{2}\, dm) + \&c.\,^{\text{[v]}}$$

$$V + P = -\tfrac{1}{2} S \left\{ M\frac{1}{s_{,}^{2}} + \frac{3}{s_{,}^{4}} \sum (s'^{2}\, dm) + \&c \right\}$$

$s_{,}$ is an unchangeable quantity but s' is only subject to the restriction $\sum (s'\, dm) = 0$.

By making the outer rings wider & the inner rings narrower we increase $\sum (s'^{2}\, dm)$ and it must be by this means that $V+P$ is diminished.

We therefore conclude that the ultimate effect of internal friction is to make the outer rings extend farther from the planet and the inner rings come nearer to it.[w]

(u) {Thomson} This investigation shows satisfactorily the chief effect of fluid friction not referred to in what precedes – namely that of resisting the greater angular motion of the inner parts and less angular motion of the outer parts which the conditions of steady motion require for a ring either of disjointed particles or of fluid.

(v) {Challis} $\because \sum \sigma'\, dm = 0$.

(w) {Thomson} As observation shows no such effect on the outer parts, and seems to show such an effect on the inner parts,[80] we infer that internal friction is not, but that resistance from other matter than that of the rings probably is the cause of the changing appearances, if *resistance* at all is the cause.

(80) See Otto Struve, 'Sur les dimensions des anneaux de Saturne', *Recueil de Mémoires*: 382; see note (1). Compare the abstract of Struve's paper in the *Monthly Notices of the Royal Astronomical Society*, **13** (1852): 22–4, esp. 23–4; 'By a comparison of the micrometrical measures of Huyghens, Cassini, Bradley, Herschel, W. Struve, Encke and Galle, with the corresponding measures executed by himself, *he found that the inner edge of the interior bright ring is gradually approaching the body of the planet while at the same time the total breadth of the two bright rings is constantly increasing ... [and] that during the interval which elapsed between the observations of J. D. Cassini and those of Sir William Herschel, the breadth of the inner ring had increased in a more rapid ratio than that of the outer ring'.*

Conclusion

The preceding investigation may be considered as a mere indication of the probable form which a theory of a system of fluid rings would assume, if seriously worked out by mathematical astronomers.

The difficulty of making observations, and the purely speculative interest of the subject, will probably prevent much of the time and labour of working astronomers from being directed to the formation of theories which we have no immediate prospect of verifying.

Considerations of the same kind have induced me to exhibit the results of my enquiries in as compact a form as possible avoiding as much as I could digressions relating to the calculation of attractions, which, however necessary in numerical calculations, are useless when we do not possess the data from which to determine the interpretations of our formulae.

This is the reason why all that relates to the attractions of the parts of the ring has been thrown into the form of general reasoning, and why our attention has been so much confined to the discussion of the biquadratic equation in $\frac{d}{ds}$ & $\frac{d}{dt}$.

We may finish this part of the subject by repeating the principal results at which we have arrived.

(1) The resultant action on any portion of a thin filament of fluid, due to the attractions of the other parts and the pressures which they exert upon it, may be represented as a *tension*, similar to the tension of a string, & varying inversely[81] as the square of the sectional area.

(2) The effects of this tension on an element of the ring depend partly on the difference in the *magnitude* of the tension at the extremities of the element, due to the *varying section* of the ring, and partly on the difference of *direction* of the tension at the extremities of the element arising from the *curvature* of the ring.

The longitudinal force acts from the thinner towards the thicker parts of the ring, and would cause the ring to break up into satellites, if dynamical reasons did not interfere.

The normal force acts towards the centre of curvature and would act, in conjunction with Saturn, to precipitate the ring on the planet, if it were not for the rotation of the ring.

(3) The values of these forces being found and added to the attraction of Saturn in the equations of motion, we omit powers and products of small disturbances, and obtain three linear partial differential equations of the second order from which to determine the displacements in the radius vector, the tangent, and the normal to the plane of the ring.

---

(81) Read: directly; compare Proposition V.

As the equation of motion perpendicular to the plane of the ring is of such a form that no doubt can arise as to the stability of the ring with respect to this kind of motion, we confine our attention to the disturbances in the plane of the ring.

By eliminating either ρ or σ from these equations we obtain an equation containing neither, and entirely made up of symbols of operation. This is the biquadratic in m & n.[82]

(4) We then make use of Fouriers theorem to show that m must be of the form $\sqrt{-1}\, m'$ where m' is a whole number, and we show also that the stability of the ring depends on n being of the form $\sqrt{-1}\, n'$ where n' is real.[83]

We therefore multiply n & m[84] by $\sqrt{-1}$ wherever they occur, and then treat the resulting equation as if we were seeking for the possible roots of n, where m is an integer.

The method consists essentially in reducing the motion of the ring to that of a system of waves, travelling round the ring with a certain velocity. A complex disturbance may always be reduced to a combination of a number of such systems of waves, travelling with different velocities. In each system of waves there must be a definite number of waves in the circumference of the ring. We call this number m, put it in the equation, and deduce four values of n.

Each of these values of n represents an angular velocity *relative to the ring itself* with which a system of m waves *might* travel round the ring. One of these values of n is nearly $= \omega$ which indicates a wave moving in the same direction as the ring so as to move round Saturn twice as fast as the ring does.

Another value of n is nearly $= -\omega$ indicating a wave, the position of which in space is nearly stationary. It does, however, travel slowly in the same direction as the ring.

Two other values of n are small and indicate two waves, one of which travels a little faster, & the other a little slower than the ring itself.

(5) With respect to the nature of the oscillations of the particles of the ring about their mean positions, we find that they move in ellipses, of which the shorter axis is directed towards Saturn. In the first and second waves the proportion of the axes is as 1 to 2 in the third and fourth it is as $4m\sqrt{K}$ to $\sqrt{6}\,\omega$ nearly. In[a] the first and third waves the *direction* of motion in this ellipse is opposite to the angular motion of the ring. In the second and fourth it is in the same direction as that of the ring.[85]

(6) With respect to the generality of our solution, we remark, that the original

---

(a) {Thomson} all.[86]

(82) Read: m, and n, for m and n.

(83) Read: m and n for m' and n'; and m, and n, for m and n.

(84) Read: m, and n, for m and n.

(85) Compare the conclusion of Part II, Proposition X; and the annotations by Challis, Thomson and Maxwell.   (86) See note (85).

displacement and velocity of any point in the ring may be completely expressed by a series of periodic functions of s whose periods are submultiples of 2π. Selecting those whose period is $\dfrac{2\pi}{m}$ we find that we have eight things to express.

To express the displacement in the direction of the radius, we have to find the maximum displacement, and the angular position of that maximum, so that this displacement cannot be expressed without two constants being given.

The tangential displacement requires two more, and the original velocities in the radial and tangential directions each require two, so that eight independent constants enter into the determination of the original state of the ring.

Now the solution which we have found requires exactly eight arbitrary constants to render it definite, namely the amplitude and phase of each of the four waves, so that we may be certain that we have obtained the *complete solution* of the equation.

(7) We then examined the effect of internal friction on the motion of these waves, and we found that the result would be that the amplitude of the waves would diminish in geometrical progression while the time increase in arithmetical progression, and that the rate of diminution depended on m^2 so that short waves would diminish most rapidly.[87]

(8) In the next place we endeavoured to estimate the effect of a disturbance arising from external circumstances such as the attraction due to a satellite or to the irregularities of a neighbouring ring. Our result was that these disturbances depended for their amount on the angular velocity of the disturbing cause.

If the angular velocity (in space) is nearly that of one of the four natural waves of the ring, the amplitude of the disturbance will increase to a great extent, but if the satellite or the waves of a neighbouring ring move with a velocity corresponding to none of the four waves of the ring, the disturbance will be confined within smaller limits. We have come to the conclusion that this will be the case, and that no disturbance can rise to a large value in the case of a number of rings at different distances, with satellites beyond them all.

(9) The result of a long-continued series of disturbances among the rings, continually deadened and obliterated by internal friction, was shown to be, that the exterior rings would recede from the planet and the interior ones approach towards his surface. This perhaps is the only one of our results which has been observed, or believed to have been observed.[b][88]

(b) {Thomson} Does observation indicate that the outer rings recede?[89]

(87) See Proposition XI.

(88) Otto Struve, 'Sur les dimensions des anneaux de Saturne', *Recueil de Mémoires*: 383; see notes (1) and (80). See also Number 126.

To ascertain the existence of this continual source of decay we must compare observations of the ring made at different epochs. Now it is admitted that new portions of the ring have been seen within the old ones but this might be due to the improvement of telescopes, not to the decay of the ring. It is to be hoped, that observations made with the actual telescopes used by the old astronomers may do something towards clearing up this point, and that new observations with the best instruments may be so carefully made, that a comparatively short time will render any change perceptible.[90]

Note
On the mutual attractions of a ring of Satellites

The notation is the same as that of the previous propositions.
(1) To find the distance PQ between two particles after displacement.

Let the normal and tangential displacements of P be ρ_0 and σ_0, those of Q ρ_1 and σ_1.

Then if PQ be originally $2a\sin\dfrac{s_1-s_0}{2}$ after displacement

$$PQ = 2a\sin\frac{s_1-s_0}{2}\left\{1+\frac{\rho_0+\rho_1}{2}+\frac{\rho_1-\rho_0}{2}\cot\frac{s_1-s_0}{2}\right\}.$$

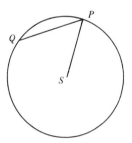

Figure 107,3

(2) To find the alteration of the angle SPQ, its sine and cosine.

Before displacement $SPQ = \dfrac{\pi}{2}-\dfrac{s_1-s_0}{2}$

and the alteration $= +\dfrac{\rho_1-\rho_0{}^{(c)}}{2\sin\dfrac{s_1-s_0}{2}}-\dfrac{\sigma_1-\sigma_0}{2}$

(c) {Maxwell} $\cos\dfrac{s_1-s_0}{2}$.

(89) For Maxwell's response see his letter to Thomson of 1 August 1857 (Number 126); see also note (80).

(90) Compare the conclusion of *Saturn's Rings*: 68 (= *Scientific Papers*, **1**: 374); 'If the changes already suspected should be confirmed by repeated observations with the same instruments, it will be worth while to investigate more carefully whether Saturn's rings are permanent or transitory elements of the Solar System, and whether in that part of the heavens we see celestial immutability, or terrestrial corruption and generation, and the old order giving place to new before our own eyes'. Compare the line 'The old order changeth yielding place to new' in Tennyson's 'Morte D'Arthur'; see Alfred Tennyson, *Poems* (London, 1857): 200. On celestial mutability, compare the Galilean motto of the Essay.

$$\sin SPQ = \cos \frac{s_1 - s_0}{2}\left\{1 + \frac{\rho_1 - \rho_0}{2}^{(d)} - \frac{\sigma_1 - \sigma_0}{2}\tan\frac{s_1 - s_0}{2}\right\}$$

$$\cos SPQ = \sin\frac{s_1 - s_0}{2}\left\{1 - \frac{\rho_1 - \rho_0}{2}\cot^2\frac{s_1 - s_0}{2}^{(e)} + \frac{\sigma_1 - \sigma_0}{2}\cot\frac{s_1 - s_0}{2}\right\}.$$

(3) The attraction between P & Q estimated along PS

$$= \frac{P \cdot Q}{(PQ)^2}\cos SPQ$$

$$= \frac{P \cdot Q}{4a^2 \sin\frac{s_1 - s_0}{2}}\left\{1 - \frac{\rho_1 - \rho_0}{2}\cot^2\frac{s_1 - s_0}{2}^{(f)} - \frac{\sigma_1 - \sigma_0}{2}\cot\frac{s_1 - s_0}{2} - (\rho_0 + \rho_1)\right\}.$$

Now suppose ρ_1 & σ_1 to be expanded in the form

$$\rho_1 = \rho_0 + \frac{d\rho}{ds}(s_1 - s_0) + \frac{d^2\rho}{ds^2}\frac{(s_1 - s_0)^2}{2} + \&c$$

$$\sigma_1 = \sigma_0 + \frac{d\sigma}{ds}(s_1 - s_0) + \frac{d^2\sigma}{ds^2}\frac{(s_1 - s_0)^2}{2} + \&c$$

& the values of $\sin\frac{s_1 - s_0}{2}$ & $\cot\frac{s_1 - s_0}{2}$ near P written

$$\frac{s_1 - s_0}{2} \qquad \frac{2}{s_1 - s_0}$$

the attraction between P & Q resolved in the direction PS will be

$$\frac{P \cdot Q}{4a^2}\frac{2}{s_1 - s_0}\left\{1 - \frac{d\rho}{ds}\frac{2}{s_1 - s_0} - \frac{d^2\rho}{ds^2} - \frac{d\sigma}{ds} - 2\rho_0\right\}$$

together with terms depending on $s_1 - s_0$, which may be neglected compared with these. In summing up all terms of this kind, it is plain that for every positive value of $s_1 - s_0$ there will be an equal negative one, so that the second term will disappear, and if we put $\sum\frac{P \cdot Q}{a^2}\frac{1}{s_1 - s_0} = L$ the whole attraction in direction of radius, due to the other parts of the ring will be, approximately,

$$P = L\left(1 - \frac{d^2\rho}{ds^2} - \frac{d\sigma}{ds} - 2\rho\right).$$

(d) {Maxwell} $\cos\dfrac{s_1 - s_0}{2}$.

(e) {Maxwell} $\left\langle\cot^2\dfrac{s_1 - s_0}{2}\right\rangle\dfrac{\cos\frac{1}{2}(s_1 - s_0)}{\sin^2\frac{1}{2}(s_1 - s_0)}$.

(f) {Maxwell} $\left\langle\cot^2\dfrac{s_1 - s_0}{2}\right\rangle\dfrac{\cos\frac{1}{2}(s_1 - s_0)}{\sin^2\frac{1}{2}(s_1 - s_0)}$.

(4) The attraction between P & Q resolved perpendicular to PS is

$$= \frac{P \cdot Q}{PQ^2} \sin SPQ^{(91)}$$

$$= \frac{P \cdot Q \cos \frac{s_1 - s_2}{2}^{(92)}}{4a^2 \sin^2 \frac{s_1 - s_2}{2}} \left\{ 1 - 2\rho_0 - \frac{\rho_1 - \rho_2}{2}^{(g)} \right.$$

$$\left. - (\sigma_1 - \sigma_0) \cot \frac{s_1 - s_0}{2} - \frac{\sigma_1 - \sigma_0}{2} \tan \frac{s_1 - s_0}{2} \right\}$$

treating this expression in the same way as that for the radial force we find for the value of the tangential force acting forwards

$$Q = L \left\{ -2 \frac{d^2\sigma}{ds^2} - \frac{d\rho}{ds} \right\}.$$

The agreement of these expressions with those which we had before obtained in a very different way (see Prop VIII)[93] shows that a ring of discrete satellites will conduct itself similarly to a fluid ring. The calculation of the value of L involves elliptic integrals, even when rendered as simple as possible. There is no use, however, of determining L, as long as we are sure that the forces arising from the mutual action of the parts of the ring are small compared with the attraction of Saturn.

(g) {Maxwell} $-(\rho_0 + \rho_1) + \dfrac{\rho_1 - \rho_0}{2 \cos \frac{s_1 - s_0}{2}}.$

(91) For PQ^2 read: $(PQ)^2$.   (92) For s_2 read: s_0.

(93) Compare these expressions for the forces on a satellite in the radius vector (P) and perpendicular to the radius vector (Q), with the expressions obtained in Proposition VIII for the forces on an element of a fluid ring in the radius vector (P) and perpendicular to the radius vector (Q).

108

LETTER TO PETER GUTHRIE TAIT

3 DECEMBER 1856

From the original in the University Library, Aberdeen[1]

129 Union Street
Aberdeen
Dec 3rd 1856

Dear Tait

I send you this by my cousin William Dyce Cay about whom I think I have spoken to you before and if not I give you leave to make use of your private judgement. He is with James Thomson C.E. office 16 Donegall Place.[2] He has plenty to do so do not consider yourself as bound to keep him in countenance. My reason for writing in this way is to project two stones at one bird so as to have the satisfaction of making an Irish Cannon & paying a double debt.

So much for explanation.

As to myself I should feel deeply gratified in perusing anything that might fall from your pen illustrative of the present state and occupation of the Mathematical Chair of your College.

I may say that the last vestige of the chair of Natural Philosophy here disappeared some time ago, and that my only visible means of support consists in two hooks which I caused to be inserted in the roof.

However the work here is much more to my taste than lecturing the second lot of men at Trinity. Even the first lot, though they could do a great deal more than we can here had their time occupied with their coaches, and could not work much for lecture.

I have one man here who has answered all my weekly questions (à la Forbes) and there are several who do well, though accuracy & precision in answers is not their strong point.

I had an hours talk with Fuller yesterday.[3] His son his wife & himself have been ill successively and now his cook takes her turn.

We have unusually early & hard & long frost & deep snow all last week & this. I had a note from MacLennan at Inverness busy at 'Law' '*Encyc. Brit.*' article approved by editors coming out in Jan^y.[4] Do you remember Forbes Catenaries.[5] I have set one up ⌀⌀⌀⌀ cylinders and square bits with a groove to fit.

---

(1) Aberdeen University Library, MS 980/1. The letter has a mourning border: see Number 99.

(2) See Number 100 note (2). Tait was Professor of Mathematics at Queen's College, Belfast (Venn).   (3) Number 96 note (2).   (4) Number 103 note (2).

(5) See Number 15. Some notes on the 'Catenary' (ULC Add. MSS 7655, V, i/10) written *circa* 1856, were clearly prepared for class use.

It makes a capital arch which will come down with a stamp on the floor.

I have also set up an adjustible false balance and a pendulum in two planes, and I am getting a man here to make a brass top with my last improvements in gyroscopy.[8] See Athenæum. Brit. Ass. Cheltenham.[7] If he does it well I shall send a circular to the natural philosophers.[8] Faradays lines of Force is in a state of proof.[9] I shall send you a copy.

Dr Maclure[10] sends you his love. I have seen a good deal of him since I came here, and having had the advantage of previous acquaintance I was ready to appreciate his kind offices and adapt my receptivity to his deliverances. Tomorrow I attend the class of Practical Religion. Prof. Pirie the greatest wag in College.[11] He studies variety and the oldest statements become startling when uttered by him. He told us that he knew for certain that the age of the boys that entered the College was increasing every year & he calculated (wrong) the present value of future life as a deferred annuity.

Yours

J. C. MAXWELL

---

(6) Charles Ramage of the Aberdeen firm of instrument makers Smith and Ramage; see D. J. Bryden, *Scottish Scientific Instrument-Makers 1600–1900* (Edinburgh, 1972): 31, 56.

(7) *The Athenæum* (23 August 1856): 105, describing Maxwell's paper 'On an instrument to illustrate Poinsot's theory of rotation', *Report of the Twenty-sixth Meeting of the British Association for the Advancement of Science; held at Cheltenham in August 1856* (London, 1857), part 2: 27–8 (= *Scientific Papers*, **1**: 246–7).

(8) See Number 116; and his paper 'On a dynamical top', *Trans. Roy. Soc. Edinb.*, **21** (1857): 559–70 (= *Scientific Papers*, **1**: 248–62).

(9) Maxwell, 'On Faraday's lines of force', *Trans. Camb. Phil. Soc.*, **10** (1856): 27–83 (= *Scientific Papers*, **1**: 155–229).

(10) Robert Maclure, Professor of Humanity at Marischal College; see [P. J. Anderson,] *Officers of the Marischal College and University of Aberdeen 1593–1860* (Aberdeen, 1897): 63.

(11) William Robinson Pirie, Professor of Divinity at Marischal College; see *Officers of the Marischal College*: 52.

109

LETTER TO WILLIAM THOMSON

17 DECEMBER 1856

From the original in the University Library, Cambridge[1]

129 Union Street
Aberdeen Dec 17 1856

Dear Thomson

I do not know what special subject you are busy with now. My special study is Elementary Mechanics and just at present parabolic motion. I am glad to find that the students are better pleased with Dynamics than with Statics. I was afraid that the new ideas would be more difficult than those of Couples or Friction. However they did a paper on definitions and rectilinear motion far better than they did on friction and I had some very sensible difficulties sent up today about the measure of force by momentum produced and about the dimensions of the quantities in $S = \frac{1}{2}\frac{F}{M}t^2$.

I want to hear from you about your method of making one man look over another's exercise. It would do them good but I do not yet see how to get my men into the way of it. I would like to see what sort of subjects you give & to know how you arrange the critics.

At present the hour from 9 to 10 is supposed to be oral exam[n] and 11 to 12 lecture but I find it best to do both at both hours and examine without warning, for pure examination is tiresome for those who are not examined and pure lecturing encourages passivity in passive men, not to say talking and note writing among the oblique minded. On Tuesdays I dictate 10 questions (with short answers) to be answered in writing on the spot. On Wednesday I explain them and hang up a list of the numbers each man has answered. They do not admit of half answers.

I also give out exercises to be done at home, of a more difficult kind, but these are voluntary, and confer no distinction only I correct them and explain them in the class, and to such men as wish explanations.

Volunteers of the fourth year come two days a week.[2] We do Newton I II III.[3] They have diff. Calc. afterwards from D[r] Cruickshank.[4]

---

(1) ULC Add. MSS 7342, M 97. Previously published in Larmor, 'Origins': 722–5. The letter has a mourning border: see Number 99.

(2) In Maxwell's second session (1857–8) at Marischal College he established an advanced class: see Numbers 136, 142, 144 and 146.

(3) As a text he probably used William Whewell's edition of *Newton's Principia. Book I. Sections I. II. II. In the original Latin; with explanatory notes and references* (London, 1846); four copies are preserved in Maxwell's library (Cavendish Laboratory, Cambridge). See also the commentary by

Most of them had Whewell's Mechanics, so I have continued it this year, but couples & friction are wanting, and the proof of composition of forces is not well done[5] even admitting all the axioms.[6] It is better done in Poinsôts Statique from the lever.[7] But men *must* learn the 'transmission of force' proof. I have set it up in a concrete form and I think it is understood now. But next year I shall take Phears book.[8] Do you think it good? I would rather have every man his own mechanician but that is not easy. I dont intend to have a text book on hydrostatics. Optics I must consider.

I have arranged to have Geometrical Conics taught in time for Projectiles but I must inculcate limits and increments myself. I have not used the words yet, but some of us understand the things and I shall venture on the words before long.

I should much like to hear from you something about your class and methods.

With respect to other things. I constructed in July my SHORT instrument for exhibiting compound colours with a prism.[9]

The rays go twice through the prism and so what would need a box with two chambers four feet long at an awkward angle is done in a rectangular box 2 feet × 7 inch × 4 inch. Also the entrance of light is close to

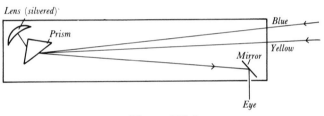

Figure 109,1

eye hole and the slits can be arranged by the observer. Bryson[10] had it in his shop for a week and got it up as I think pretty well by heart, but I dont suppose

Percival Frost, *Newton's Principia. Sections I., II., III., with Notes and Illustrations* (Cambridge, 1854). See Numbers 136 and 144.

(4) John Cruickshank, Professor of Mathematics at Marischal College; see [P. J. Anderson,] *Officers of the Marischal College and University of Aberdeen* (Aberdeen, 1897): 54.

(5) William Whewell, *An Elementary Treatise on Mechanics* (Cambridge, ₇1848): 30–40.

(6) Whewell expanded his treatment of the composition of forces in his *Analytical Statics. A Supplement to the Fourth Edition of An Elementary Treatise on Mechanics* (Cambridge, 1833): 1–26.

(7) Louis Poinsot, *Éléments de Statique* (Paris, ₈1842): 15–47; (trans. T. Sutton) *Elements of Statics* (Cambridge, 1847): 9–22.

(8) J. B. Phear, *Elementary Mechanics* (London, 1850).

(9) Maxwell, 'On the theory of compound colours with reference to mixtures of blue and yellow light', *Report of the Twenty-sixth Meeting of the British Association for the Advancement of Science; held at Cheltenham in August 1856* (London, 1857), part 2: 12–13 (= *Scientific Papers*, **1**: 243–5). See Numbers 102 and 117.

(10) James M. Bryson, the Edinburgh instrument maker; see Number 58 note (2).

he quite knows the adjustments. The focussing is very simple = turning the prism on its axis.

Smith & Ramage here[11] are constructing an improved edition of my dynamical top. See report of British Association 56.[12]

I have a better notion of the necessary dimensions and proportions and of the most convenient arrangement of the screws.[13]

There are six horizontal & 3 vertical screws on the ring, which give the 9 adjustments 3 for centre of gravity 6 for magnitude & pos^n of principal axes. But for large changes I have two more adjustments – the axis screws up and down, and a bob screws on the axis. I have found a blacksmith who made a capital balance for showing stability &c and false arms, adjustible, but very strong, and fit for the class to handle. He has also made me a pendulum on gymbals which goes well and long, and can be used as a Katers pendulum.[14]

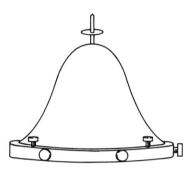

Figure 109,2

I have constructed a catenary 3 feet long of bits $\frac{3}{4}$ inch diam^r which will stand as an arch of any form, flat or lofty.[15] I am also getting a machine for throwing bullets by means of a weight which I hope will be a trustworthy engine.

So much for the concrete. Here is a piece of geometry. Do you know if it is old.

Def. If from a fixed point in a plane we draw lines to every point in a plane curve and cut off from each radius a part inversely proportional to

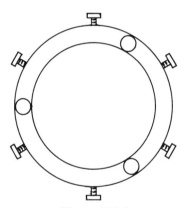

Figure 109,3

---

(11) See Number 108 note (6).   (12) See Number 108 note (7).

(13) See Figure 6 of Maxwell's paper 'On a dynamical top, for exhibiting the phenomena of the motion of a system of invariable form about a fixed point, with some suggestions as to the Earth's motion', *Trans. Roy. Soc. Edinb.*, **21** (1857): 559–70, facing 570 (= *Scientific Papers*, **1**: 262). See Number 116, and plate VI.

(14) Henry Kater, 'An account of experiments for determining the length of the pendulum vibrating seconds in the latitude of London', *Phil. Trans.*, **108** (1818): 33–102. In his Cambridge notebook 'Lunar Theory & Rigid Dynamics' (ULC Add. MSS 7655, V, m/9, f.71) Maxwell had noted: 'For Katers Pendulum see Pratt'; see John Henry Pratt, *The Mathematical Principles of Mechanical Philosophy, and their application to Elementary Mechanics and Architecture, but chiefly to the Theory of Universal Gravitation* (Cambridge, ₂1845); 403–5.   (15) See Numbers 15 and 108.

that radius, the new points form a curve and the transformation is called 'Inversion'. See Mulcahy Salmon &c.[16]

Prop[n]. If the operation of inversion be performed any number of times on a plane curve, using different fixed points each time, one inversion can always be determined which shall be equivalent to the result of the whole.

Proof. The operation of inversion if performed on a stereographic projection of the globe is equivalent to changing the pole of projection to the latitude and longitude of the fixed point as given by the map. Therefore after any number of inversions the pole of projections will have a certain latitude & longitude and the scale of the map will be determined by the moduli.

But we can by a proper fixed point and modulus produce a map with given pole and scale by a single inversion.   Q.E.D.

Another in particle Dynamics.

A particle is projected along the internal surface of a smooth sphere and attracted towards a fixed plane outside the sphere with a force varying as inverse cube of dist. The velocity of proj[n] is that due to ∞. Show that the path of the point is a circle whose plane passes through the *pole* of the given plane.

My paper on Faradays Lines has stuck somewhere & I am trying to find out where the latter part of the MS. is. I have proofs of all the first part.[17] Stokes had looked over it and had pointed out some awkward blunders, a page of 'divide' instead of 'multiply'. Did Blackburn[18] tell you about the 9 coeff[ts] of magnetic conductivity allowing perpetual motion?[19]

Tyndalls paper on Diamagnetism is satisfactory[20] and Faradays 30th Series[21] confirms my results about magnecrystallic action in different surrounding media.[22] Has your lecture been printed?[23] I have been looking at

(16) John Mulcahy, *Principles of Modern Geometry, with numerous applications to Plane and Spherical Figures* (Dublin, 1852): 167–70; and George Salmon, *A Treatise on the Higher Plane Curves: intended as a sequel to A Treatise on Conic Sections* (Dublin, 1852): 306–7.

(17) Maxwell, 'On Faraday's lines of force' [Part I], *Trans. Camb. Phil. Soc.*, **10** (1856): 27–51 (= *Scientific Papers*, **1**: 155–88).

(18) Number 45 note (10).

(19) See Number 104 note (8); and see Example VI of 'On Faraday's lines of force': 74–6 (= *Scientific Papers*, **1**: 217–19).

(20) John Tyndall, 'Further researches on the polarity of the diamagnetic force', *Phil. Trans.*, **146** (1856): 237–59, to which Maxwell makes reference on 'On Faraday's lines of force': 45n (= *Scientific Papers*, **1**: 180n). See Number 84 note (40).

(21) Michael Faraday, 'Experimental researches in electricity. – Thirtieth series. Constancy of differential magnecrystallic force in different media', *Phil. Trans.*, **146** (1856): 159–80.

(22) See Maxwell, 'On Faraday's lines of force': 45–6 (= *Scientific Papers*, **1**: 180), where he explains magnecrystallic induction in terms of the alignment of crystals in a magnetic field.

Rankine's Thermodynamic diagrams.[24] It is a great relief after the Umbral notation.[25] I expect a few papers on the Oscan or Ligurian[26] notation soon.

I am glad to hear better accounts of M[rs] Thomson. I owe M[rs] Wedderburn a letter, I hope young Hugh is better. I hear from W. Cay, at Belfast who seems taking root there. I made a splendid vortex lately quite smooth and 7 inches deep in the middle.

W. A. Porter[27] writes from Lincolns Inn that he wonders he was ever anything else than a young lawyer. Such is life out of Peterhouse.

Yours truly
J. C. Maxwell

Compare John Tyndall, 'On the relation of diamagnetic polarity to magnecrystallic action', *Phil. Mag.*, ser. 4, **11** (1856): 125–37, where magnecrystallic induction is explained in terms of the theory of induced polarity. See Number 84.

(23) William Thomson's Bakerian Lecture 'On the electrodynamic qualities of metals', *Phil. Trans.*, **146** (1856): 649–751 (= *Math. & Phys. Papers*, **2**: 189–327).

(24) W. J. M. Rankine, 'On the geometrical representation of the expansive action of heat, and the theory of thermo-dynamic engines', *Phil. Trans.*, **144** (1854): 115–76, esp. 116. Noting that James Watt had 'contrived the well-known Steam-Engine Indicator' to represent the expansive action of heat, and declaring that 'the principles of the expansive action of heat are capable of being presented to the mind more clearly by the aid of diagrams of energy than by means of words and algebraical symbols alone', Rankine develops the graphical pressure–volume representation of the Carnot cycle introduced by Émile Clapeyron, 'Mémoire sur la puissance motrice de la chaleur', *Journal de l'École Polytechnique*, **14**, cahier 23 (1834): 153–90. Rankine's 'diagrams of energy' give a graphical representation of the Carnot cycle by '*Isothermal Curves, and Curves of No Transmission of Heat*'; he subsequently introduced the term 'Adiabatic curves': see Rankine, 'On the thermo-dynamic theory of steam-engines with dry saturated steam, and its application to practice', *Phil. Trans.*, **149** (1859): 177–92, esp. 180.

(25) W. J. M. Rankine, 'On axes of elasticity and crystalline forms', *Phil. Trans.*, **146** (1856): 261–85, esp. 284–5 for Rankine's note 'On Sylvestrian umbrae', the application of Sylvester's 'umbral notation' for determinants to the expression of coefficients of elasticity. See James Joseph Sylvester, 'On the principles of the calculus of forms', *Camb. & Dubl. Math. J.*, **7** (1852): 52–97, 197–217; *ibid.*, **8** (1853): 256–96; and Sylvester, 'On a theory of syzygetic relations of two rational integral functions, comprising an application to the theory of Sturm's functions, and that of the greatest algebraical common measure', *Phil. Trans.*, **143** (1853): 407–548. On reviewing Rankine's paper for the Royal Society Thomson had remarked that: 'I am sorry that I have not been able to enter on Rankine's paper so fully as I would have liked, and I cannot therefore say I have verified or even fully understood all his investigations. So far as I can see, they appear exceedingly good, and I believe almost entirely new. I do not know enough of Sylvester's 'umbral' notation to be able to judge whether the analytical treatment is perfectly satisfactory' (Thomson to Stokes, 12 January 185[6], Royal Society, *Referees' Reports*, **3**: 229).

(26) The Oscans and the Ligurians: ancient peoples of Campania and Cisalpine Gaul.

(27) William Archer Porter, Trinity 1845, Peterhouse 1845 (Venn).

LETTER TO WILLIAM THOMSON

18 DECEMBER 1856

From the original in the University Library, Cambridge[1]

129 Union Street
18 Dec 1856

Dear Thomson

Your letter & mine crossed. With respect to the history of the optical effects of compression[2] my belief is, that Seebeck & Brewster discovered independently about 1814 the effects of heat[3] in 'developing the depolarizing structure in glass'.[4]

Brewster about 1815 discovered the effects of compression, dilatation & induration on glass gum jelly wax – and resin &c.[5] I do not know whether Seebeck had found the effects of a permanent kind in unannealed glass.[6]

---

(1) ULC Add. MSS 7342, M 98. Previously published in Larmor, 'Origins': 725–7. The letter has a mourning border: see Number 99.

(2) Compare William Thomson, 'On the thermo-elastic and thermo-magnetic properties of matter', *Quarterly Journal of Pure and Applied Mathematics*, **1** (1855): 57–77; Thomson, 'Elements of a mathematical theory of elasticity', *Phil. Trans.*, **146** (1856): 481–98 (= *Math. & Phys. Papers*, **1**: 291–313; **3**: 84–112); and Thomson, 'Dynamical illustrations of the magnetic and heliocoidal rotatory effects of transparent bodies on polarized light', *Proc. Roy. Soc.*, **8** (1856): 150–58 (= *Phil. Mag.*, ser. 4, **13** (1857): 198–204). On induced double refraction in strained glass see Number 26, a draft of 'On the equilibrium of elastic solids', *Trans. Roy. Soc. Edinb.*, **20** (1850): 87–120 (= *Scientific Papers*, **1**: 30–73).

(3) T. J. Seebeck, 'Von den entoptischen Farbenfiguren und den Bedingungen ihrer Bilder in Gläsern', *Journal für Chemie und Physik*, **12** (1814): 1–16; and David Brewster, 'Results of some recent experiments on the properties impressed upon light by the action of glass raised to different temperatures, and cooled under different circumstances', *Phil. Trans.*, **104** (1814): 436–9. See John Herschel, 'Light', *Encyclopædia Metropolitana, or Useful Dictionary of Knowledge. Second Division: Mixed Sciences*, Vol. II (London, 1830): 341–586, esp 562.

(4) See Brewster, 'Results of some recent experiments...': 436; Brewster, 'Experiments on the depolarisation of light', *Phil. Trans.*, **105** (1815): 29–53, esp. 31–2; and Brewster, 'On the production of regular double refraction in the molecules of bodies by simple pressure; with observations on the origin of the doubly-refracting structure', *ibid.*, **120** (1830): 87–95, esp. 87.

(5) David Brewster, 'On the effects of simple pressure in producing that species of crystallization which forms two oppositely polarized images, and exhibits the complementary colours by polarized light', *Phil. Trans.*, **105** (1815): 60–4; and 'On the communication of the structure of doubly-refracting crystals to glass, muriate of soda, fluorspar and other substances, by mechanical compression and dilatation', *ibid.*, **106** (1816): 156–78.

(6) On Seebeck's work on the effects of pressure in inducing polarizing effects, see a 'Note sur le développement des forces polarisantes par la pression. (Extrait de quelques lettres de MM. Brewster et Seebeck à M. Biot)', *Bulletin des Sciences par la Société Philomathique de Paris* (1816): 49–51.

Fresnel made the first experiment to demonstrate the separation of the pencils by compressed glass.[7]

If, in 1815, Brewster considered depolarizing structure equivalent to double refraction then he is the discoverer of both. Fresnel certainly is the first to prove to the eye that there is double refraction and I do not think Brewster in 1815 considered the two things identical.[8]

Therefore as Brewster discovered the phenomenon and Fresnel developed it you had better say

Thus Sir D. Brewster discovered that mechanical stress induces temporarily in transparent solids directional properties with respect to polarized light, and Fresnel has identified these properties with the double refraction of crystals.

With respect to the effects of heat Brewster seems to consider them due to caloric and Herschel seems to be the first to reduce them to cases of mechanical stress. See Encyc. Met. 'Light'.[9]

Herschel's explanations are *very good*. He however connects the optical properties with pressure or force rather than with compression or disfigurement and therefore holds that in all cases, including jelly gutta percha wax & resin &c there must be actual pressure.

In my paper on elasticity I produced instances (not new phenomena) to show that in these cases the pressure could not exist as the effects were in the same direction over the whole body & therefore could not be balanced. At the same time I stated that gutta percha when heated to a certain point goes back to its original form.[10]

Wertheim Ann de Chimie Jan 1854? has the best experiments I have seen on

---

(7) A. J. Fresnel, 'Note sur la double réfraction du verre comprimé', *Ann. Chim. Phys.*, **20** (1822): 376–83; and see Herschel, 'Light': 567; 'M. Fresnel has succeeded in rendering sensible the bifurcation of the pencils produced by glass subjected to pressure by an ingenious combination of prisms having their refracting angles turned opposite ways, and of which the alternate ones are compressed in planes at right angles.'

(8) In his treatise on 'Light': 562, Herschel had noted that Brewster had remarked on the analogy between double refraction and the optical properties induced by the mechanical compression of glass; but that Fresnel had 'verified by direct experiment... that a peculiar species of double refraction is thus produced'. See note (7).

(9) Herschel, 'Light': 562–8; 'we have little hesitation in regarding the inequality of temperature as merely the remote, and the mechanical tension or condensation of the medium as the proximate cause of the phenomena in question, and are very little disposed to call in the agency of a peculiar crystallizing fluid...'. See also Number 26 note (78).

(10) See Maxwell's discussion in Case I of 'On the equilibrium of elastic solids': 97–8 (= *Scientific Papers*, **1**: 43–4); and the draft, Number 26 §6.

compressed glass.[11] He maintains & I think with truth that the effects are due to the '*strains*' not to the '*stresses*'.[12] As far as I know glass is not capable of maintaining strain without stress at least I cannot cut a piece out of an unannealed plate which has the strain all in one direction. When it is cut out the strain disappears for want of stress.

If you examine any plane section of a piece of unannealed glass you will find that $\sum dSp = 0$ where p is the stress perp. to the element dS. That is the whole pressure = whole tension in the section.

You should not make me a partaker in Sir D. B.'s experiments. The *phenomena* are all due to him (except gutta percha of which I do not know the optical history)[13] and they date from 1815 & so on.

The reduction to double refraction is Fresnels.

That of Heat to mechanical action – Herschel.

Proof that this action is not stress   Maxwell.

Proof that it is strain with numerical data for various kinds of glass – Wertheim.

In connection with this see your brothers papers on strained wires after the stress has been removed.[14]

Here is my present notion about plasticity of homogeneous amorphous solids.

Let $\alpha\,\beta\,\gamma$ be the 3 principal strains at any point $P\,Q\,R$ the principal stresses connected with $\alpha\,\beta\,\gamma$ by symmetrical linear equations the same for all axes. Then the whole work done by $P\,Q\,R$ in developing $\alpha\,\beta\,\gamma$ may be written

$$U = A(\alpha^2 + \beta^2 + \gamma^2) + B(\beta\gamma + \gamma\alpha + \alpha\beta)$$

(11) Guillaume Wertheim, 'Sur la double réfraction temporairement produite dans les corps isotropes, et sur la relation entre l'élasticité mécanique et entre l'élasticité optique', *Ann. Chim. Phys.*, ser. 3, **40** (1854): 156–221. See Number 26 note (48).

(12) Compare Thomson, 'Elements of a mathematical theory of elasticity': 481; 'Def. 1. A stress is an equilibrating application of force to a body. ... Def. 2. A strain is any definite alteration of form or dimensions experienced by a solid.' Thomson had derived these terms from Rankine (*ibid.*: 481n) modifying Rankine's usage. See W. J. M. Rankine, 'On axes of elasticity and crystalline forms', *Phil. Trans.*, **146** (1856): 261–85, esp. 262; 'In this paper the word '*Strain*' will be used to denote the change of volume and figure constituting the deviation of a molecule of a solid from that condition which it preserves when free from the action of external forces; and the word '*Stress*' will be used to denote the force, or combination of forces, which such a molecule exerts in tending to recover its free condition, and which for a state of equilibrium, is equal and opposite to the combination of external forces applied to it.'

(13) Compare Maxwell, 'On the equilibrium of elastic solids': 97n (= *Scientific Papers*, **1**: 43n).

(14) James Thomson, 'On the strengths of materials, as influenced by the existence or non-existence of certain mutual strains among the particles composing them', *Camb. & Dubl. Math. J.*, **3** (1848): 252–8; J. Thomson, 'On the elasticity and strength of spiral springs, and of bars subjected to torsion', *ibid.*: 258–66.

where A & B are coeffts, the nature of which is foreign to our inquiry. Now we may put

$$U = U_1 + U_2$$

where U_1 is due to a symmetrical compression $(\alpha_1 = \beta_1 = \gamma_1)$ and U_2 to distortion without compression $(\alpha_2 + \beta_2 + \gamma_2 = 0)$

$$\& \ \alpha = \alpha_1 + \alpha_2, \quad \beta = \beta_1 + \beta_2, \quad \gamma = \gamma_1 + \gamma_2.$$

It follows that

$$U_1 = \tfrac{1}{3}(A+B)(\alpha+\beta+\gamma)^2$$

$$U_2 = \frac{2A-B}{3}(\alpha^2 + \beta^2 + \gamma^2 - (\beta\gamma + \gamma\alpha + \alpha\beta)).$$

Now my *opinion* is, that these two parts may be considered as independent U_1 being the work done in condensation and U_2 that done in distortion. Now I would use the old word 'Resilience' to denote the work necessary to be done on a body to overcome its elastic forces.[15]

The cubical resilience R_1 is a measure of the work necessary to be expended in compression in order to increase the density permanently. This *must* increase rapidly as the body is condensed, whether it be wood or lead or iron.

The resilience of rigidity R_2 (which is the converse of plasticity) is the work required to be expended in pure distortion in order to produce a permanent change of form in the element. I have strong reasons for believing that when

$$\alpha^2 + \beta^2 + \gamma^2 - \beta\gamma - \gamma\alpha - \alpha\beta$$

reaches a certain limit $= R_2$ then the element will begin to give way. If the body be tough the disfigurement will go on till this function U_2 (which truly represents the work which the element *would do* in recovering its form) has diminished to R by an alteration of the *permanent dimensions*.

Now let $a\ b\ c$ be the *very small* permanent alterations due to the fact that $U_2 > R_2$ for an instant. Whenever $U_2 = R_2$ the element has as much work done to it as it can bear. Any more work done to the element will be consumed in permanent alterations.

Therefore if $U_2 = R_2$, and in the next instant U be increased, dU must be lost in some way.

My rough notion on this subject is that

$$a = \frac{dU}{U}\alpha, \quad b = \frac{dU}{U}\beta, \quad c = \frac{dU}{U}\gamma$$

the new values of $\alpha\ \beta\ \gamma$ will be

$$\alpha' = \alpha - a \quad \beta' = \beta - b \quad \gamma' = \gamma - c.$$

---

(15) See James Thomson, 'On the elasticity and strength of spiral springs': 263.

This is the first time that I have put pen to paper on this subject. I have never seen any investigation of the question, 'Given the mechanical strain in 3 directions on an element, when will it give way?' I think this notion will bear working out into a mathemat theory of plasticity when I have time to be compared with experiment when I know the right experiments to make.

Condition of not yielding

$$\alpha^2 + \beta^2 + \gamma^2 - \beta\gamma - \gamma\alpha - \alpha\beta < R_2.$$

I have not had time to read your magnecrystalline spheres, but I shall tomorrow.[16]

I have been proving the 'laws of areas' experimentally in the case of an ellipse performed by a funnel of sand hanging by a string. I draw conjugate diameters $AC\,BD$ and collect the sand from AD & BC into one scale of a balance and that from AB & CD in the other and show that they very nearly balance when all goes well. The drawing conj. diam^r is the chief difficulty when the ellipse is like this, owing to precession of apses.

Figure 110,1

Figure 110,2

Where is Helmholtz on the Eye to be found?[17]

I have contrived a very adjustable pseudo scope[18] wh: I am getting made by degrees.[19]

Yours

J. C. MAXWELL

---

(16) [William Thomson,] 'Sui fenomini magneto-cristallini, Lettera del Prof. W. Thomson al Prof. Matteucci', *Il Nuovo Cimento*, **4** (1856): 192–8.   (17) See Number 111 note (5).

(18) The term 'pseudoscope' for a stereoscope which 'conveys to the mind false perceptions of all external objects', was coined by Charles Wheatstone in his 'Contributions to the physiology of vision – Part the second. On some remarkable, and hitherto unobserved, phenomena of binocular vision', *Phil. Trans.*, **142** (1852): 1–17, on 11. Wheatstone used the term ironically, in reference to Brewster's opinions on the invention of the stereoscope, in a letter to *The Times* (15 November 1856). Brewster had stated his view that Wheatstone was not the true inventor of the stereoscope, asserting the importance of his lenticular stereoscope, in *The Stereoscope; its History, Theory, and Construction, with its Application to the Fine and Useful Arts, and to Education* (London, 1856). Brewster's claims were dismissed in W. B. Carpenter's anonymous review 'Binocular vision', *Edinburgh Review*, **108** (1858): 437–73.

(19) See Number 91. On 5 January 1857 the Edinburgh instrument maker J. M. Bryson wrote to Maxwell about the manufacture of a stereoscope: 'Would you be good enough to inform me if the Stereoscope you mentioned to me when you were here was your contrivance, as I intend fitting up several and calling it yours...' (ULC Add. MSS 7655, II/4).

LETTER TO GEORGE GABRIEL STOKES

27 JANUARY 1857

From the original in the University Library, Cambridge[1]

129 Union Street
Aberdeen
27th Jan 1857

My dear Stokes

Although I am not likely to be much in London, I am fully sensible of the honour and advantages of belonging to the Royal Society and therefore I am much flattered with your remembrance of me in connection with that body. I do not know whether your office precludes you from proposing a candidate.[2] If not, then in the event of my coming forward I should be glad of your paternity.[3]

With respect to the composition I have no immediate prospect of anything in the way of a paper for you. I have made preparations for an analysis of the colours of the spectrum with respect to the sensations they produce, but it is doubtful whether anything presentable will be produced this summer.

I constructed an instrument 2 feet long which acts very well as a portable instrument of illustration[4] but I must make great progress before I come into the field with such men as Helmholtz. Have you seen his physiology of the eye in Karstens Encyc. ?[5] I hope he will give his last words on colour & sensation in the next fasciculus.

So my conclusion is that I am desirous of joining but not yet. I would rather wait a year if you think the same as I do.

My fingers are frozen and the post is ready.

Yours truly
J. C. MAXWELL

---

(1) ULC Add. MSS 7656, M 407. Previously published in Larmor, *Correspondence*, **2**: 4–5. The letter has a mourning border: see Number 99.

(2) Stokes served as Secretary of the Royal Society from 1854–85 (*DNB*).

(3) See Maxwell's letter to Stokes of 7 May 1860 (Number 178) requesting that his name be placed on the list of candidates for election.   (4) See Numbers 109 and 117.

(5) Hermann Helmholtz, *Handbuch der physiologischen Optik*, in *Encyklopädie der Physik*, Hrsg. G. Karsten, **9** (Leipzig, 1856). Maxwell wrote notes on 'Physiologische Optik von H. Helmholtz Karsten's Allgemeine Encyclopädie der Physik' (ULC Add. MSS 7655, V, b/10).

FROM A LETTER TO LEWIS CAMPBELL

6 FEBRUARY 1857

From Campbell and Garnett, *Life of Maxwell*[1]

129 Union Street
Aberdeen
6 February 1857

But as far as I can learn I have not been misunderstood in anything, and no one has heard a single oracle from my lips. Of course I do not mean that my class do not mistake my meaning sometimes. That is found out and remedied day by day. I speak of professors, ministers, doctors, advocates, matrons, maidens, and phenomenal existences (Chimeræ bombylantes in vacuo).[2] We are through mechanics. I had an ex^n on bookwork on 24th Jan. I got answers to all the questions and riders, though no one floored them all right. I have now to be brewing experiments on Heat, as well as determining the form of doctrine to be presented to the finite capacities of my men.

---

(1) *Life of Maxwell*: 265–6; abridged.

(2) A misquotation from Rabelais, *Pantagruel*, Book II, chap. 7; 'Quaestio subtilissima, utrum Chimæra in vacuo bombinans possit comedere secundas intentiones.' (The most subtle question, whether a chimera bombinating in the void can devour secondary intentions); *The Works of Francis Rabelais Translated from the French*, by Sir Thomas Urquhart and Motteux, new edn, 2 vols. (London, 1849), **1**: 325.

LETTER TO PETER GUTHRIE TAIT

15 FEBRUARY 1857

From the original in the University Library, Aberdeen[1]

129 Union Street
Aberdeen Feb 15/57

Dear Tait

An ancient writer hath well observed that a discreet question lieth half way to the answer. You at least are a notable example of 'Discretion in asking'. Punctuality we know is the thief of time[2] as my late hours testify. Procrastination however is the soul of wit judging from the time you took to answer my letter. For myself, brevity is the best policy, seeing that to answer a man who has his questions in a bag, honesty forbids. But though decency dreads cold water a burnt child may take the first plunge, so I send you divers answers that you may find content at the bottom of the well.

(1) With respect to Brazier he is doing the whole of the duties of Dr Clark[3] and is a favorite with old & young.

(2) Fox hunting has been hardly if at all influenced in America by the introduction of Iodide of Potassium.

(3) Welsh slate litharge plate glass and black putty constitute the medium in which Professor Fuller[4] labours for the preservation of his fellow creatures who may be stranded on the coast.

(4) A strong toddy has been brewed in the Natural Phil. Class M.C. in which a sphere of olive oil has remained suspended since 3rd Feb. save when it is made to revolve about an axis after the example of M. Plateau whose experiments are here verified.[5]

(5) An exam^n on bookwork Mechanics was held 24th Jan. The *mean proficiency* was approximately that of the middle classes the whole being divided into ten classes by the system of equal temperament.

(6) The existence of native teachers of hydrostatics has been proved by the

---

(1) Aberdeen University Library, MS 980/2. The letter has a mourning border: see Number 99.

(2) Edward Young, *Night Thoughts*; see Number 104 note (4).

(3) Thomas Clark, Professor of Chemistry at Marischal College; and James Smith Brazier, Fordyce Lecturer on Agriculture; see [P. J. Anderson,] *Officers of the Marischal College and University of Aberdeen 1593–1860* (Aberdeen, 1897): 58–9, 73.

(4) Number 96 note (2).

(5) Joseph Plateau, 'Sur les phénomènes qui présente une masse liquide libre et soustraite à l'action de la pesanteur', *Mémoires de l'Académie Royals des Sciences de Bruxelles*, **16** (1843). See Number 107, esp. note (17).

appearance of peculiar views of density and specific gravity. These views have been thrown into strong perspective by the exhibition of copious definitions and a drastic course of written examinations.

(7) The theory of the expansion of bodies by heat change of state and measure of 'quantity of heat' are now before us. We have had quantitative experiments on the heat required to melt ice & boil water the calculations being done by the men on the spot and the experiments not cooked in any way. The results are not farther from the truth than class experiments without corrections generally are. We get quantities like 960 & 1024 instead of 990.[6]

(8) The volunteers who are from the higher year have done Newtons 3 sections[7] and heard general talk on the Lunar Theory and the rest of Principia and are now meditating on the Greenwich Transit Circle.[8]

(9) The analogical argument on Faradays Lines of Force is in the Press & will shortly be published.[9]

(10) Mess.rs Smith & Ramage of Aberdeen are engaged on a brass top with adjustible moments of inertia whereby the theory of rotation becomes violently intelligible.[10]

The cousin précieux has my profound regard. I have followed (humbly) his track, having been out always 2.ce a week sometimes 4.⁰. Why *other people* should acquire *solitary* smoking habits for your sake I do not see. Here we have very little smoking. Hob à Nob is far more in esteem. Our other developments are curious & interesting but must be asked after by personal application or (if by letter) free of mystery.

Yours
J. C. MAXWELL

---

(6) Maxwell's values are in (British) thermal units Fahrenheit. Compare the values for the latent heat of vaporisation of steam cited by Henri Victor Regnault, 'Sur les chaleurs latentes de la vapeur aqueuse à saturation sous diverses pressions', *Mémoires de l'Académie Royale des Sciences de l'Institut de France*, **21** (1847): 635–728, esp. 635–54.

(7) See Number 109 note (3).

(8) See [G. B. Airy,] 'Appendix I. Description of the transit circle of the Royal Observatory, Greenwich', in *Astronomical and Magnetical and Meteorological Observations made at the Royal Observatory, Greenwich, in the Year 1852* (London, 1854). In notes on 'The Transit' in his Cambridge notebook on 'Astronomy' (ULC Add. MSS 7655, V, m/5), Maxwell refers to 'Hymers'; see John Hymers, *The Elements of the Theory of Astronomy* (Cambridge, ₂1840): 61–4. See Numbers 140 and 142.

(9) See Number 108 note (9).

(10) See Number 108 note (6).

FROM A LETTER TO JANE CAY

27 FEBRUARY 1857

From Campbell and Garnett, *Life of Maxwell*[1]

129 Union Street
Aberdeen
27 February 1857

We have been at the theory of Heat and the Steam Engine this month, and on Monday we begin Optics. I have a volunteer class who have been thro' astronomy, and we are now at high Optics. Tuesday week I give a lecture to operatives, etc., on the Eye. I have just been getting cods' and bullocks' eyes, to refresh my memory and practise dissection. The size of the cod and the ox eye is nearly the same. As this was our last day of fluids, I finished off with a splendid fountain in the sunlight. We were not very wet.

---

(1) *Life of Maxwell*: 264; abridged.

LETTER TO JAMES DAVID FORBES

30 MARCH 1857

From the original in the University Library, St Andrews[1]

129 Union Street
Aberdeen 30th March 1857

Dear Sir

Although I had the honour of being admitted into the Royal Society of Edinburgh some time ago[2] I have not been able to be at a meeting. Now I hope to be in Edinburgh some time after the 13th April, and I should like to know whether the 20th is the night of meeting and also whether I might introduce to the notice of the Society a brass spinning top of refined construction for the illustration of the laws of Rotation and (I have reason to hope) the demonstration of the Earths Rotation on the Gyroscope principle.[3]

I have just succeeded in getting the essentials of the instrument through the hands of the workman and it works well.[4]

It is merely a refinement of a wooden one which I showed at Cheltenham[5] but this one can do most of the Gyroscope experiments as well as its own, differing from Foucault's Gyroscope[6] principally in being *supported* on one end of the axis exactly at the centre of gravity of the system and only *guided* at the other end by a separately balanced swing-frame.

However I intend to bring it to Edinburgh and to let Prof. Smyth[7] and you see it at any rate. If it should be acceptable to the Society I will write a careful explanation nearly free from technical mathematics.[8]

Our Session closes on the 3rd and our graduation is on the 7th. All my class examinations are over. We have had Statics & Dynamics without the calculus

---

(1) University Library, St Andrews, Forbes MSS 1857/40.

(2) On 1 December 1856; see *Proc. Roy. Soc. Edinb.*, **3** (1856): 413.

(3) Number 116.

(4) Charles Ramage; see Number 108 note (6).

(5) See Number 108 note (7).

(6) Léon Foucault, 'Sur une nouvelle démonstration expérimentale du mouvement de la terre, fondée sur la fixité du plan de rotation', *Comptes Rendus*, **35** (1852): 421–4; Foucault, 'Sur les phénomènes d'orientation des corps tournant entraînés, par un axe fixe à la surface de la terre', *ibid.*: 424–7.

(7) Charles Piazzi Smyth, Astronomer Royal for Scotland, Regius Professor of Practical Astronomy in the University of Edinburgh (*DNB*). See Number 105 note (1).

(8) In his reply of 31 March 1857 (*Life of Maxwell*: 266–7) Forbes remarked that 'I shall like much to see your Top, of which I read the account in the *Athenæum*'; see Number 108 note (7); see also his letter of 13 April 1857 (Number 116 note (1)).

but including Catenaries and Compound Pendulums, Hydrostatics and theory of Heat (not radiant) and geometrical Optics.[9]

I have had good fortune in a good class but I am afraid that the class next year will be probably very stupid and better at appearing attentive than at being so.

I have had an advanced class on Newtons Principia, Practical Astronomy (Observatory) and higher Optics,[10] and I hope next year with some of my present students to be able to do more, and take in a little Magnetism & Electricity too.[11]

I have had a year's experience of the advice you gave me in February 1856.[12] I have reason to thank you for it. I do not think I should have done anything in the matter if you had not written to me. And now I am quite sure that I am better here than at Cambridge. I have had no trouble whatever in any matter either College or out of College. Of course I must expect the ordinary accidents and misfortunes here as elsewhere but I think that my present confidence in being able to accomplish my plans is a sign that I am put to my appointed business, and I hope that what non-professional people say about monotony may be proved untrue. I never heard working men complain of having to repeat the same things again unless it appeared that the labour was useless.

I hope to see you when I come to Edinburgh and to find you well.

Yours truly

J. C. MAXWELL

---

(9) This account of the Natural Philosophy class for 1856–57 accords closely with Maxwell's course in the sessions 1857–58, 1858–59 and 1859–60; see Number 132 esp. note (6).

(10) Compare Numbers 109 and 113.   (11) See Numbers 136 and 144.

(12) See Number 92 note (2).

116

ABSTRACT OF PAPER 'ON A DYNAMICAL TOP'

[20 APRIL 1857]

From the *Proceedings of the Royal Society of Edinburgh*[1]

ON A DYNAMICAL TOP, FOR EXHIBITING THE PHENOMENA OF THE MOTION OF A SYSTEM OF INVARIABLE FORM ABOUT A FIXED POINT; WITH SOME SUGGESTIONS AS TO THE EARTH'S MOTION

By Professor Clerk Maxwell

The top is an instrument similar to that exhibited by the author at the meeting of the British Association in 1856.[2] It differs from it in being of smaller size and entirely of brass, except the ends of the axle; and in having six horizontal adjusting screws and three vertical ones, instead of four of each kind.

It consists of a hollow cone, with a heavy ring round the base, and an axle, terminating in a steel point, screwing through the vertex. In the ring are the nine adjusting screws, and on the axle is a heavy bob, which may be fixed at any height.[3]

By means of these adjustments the centre of gravity of the whole is made to coincide with the steel point, and the axle of the top is made one of the principal axes of the *central ellipsoid*.

The whole theory of the spinning of such a system about its centre of gravity depends on the form of Poinsot's ellipsoid corresponding to the particular arrangement of the screws. The top is intended to exhibit those cases in which the three axes of this ellipsoid are nearly equal. In these cases the instantaneous axis is never far from the normal to the invariable plane, which we may call the invariable axis. This axis is fixed in space, but not in the body; for it describes, with respect to the body, a cone of the second order, whose axis is either the greatest or the least of the principal axes of inertia.[4]

---

(1) *Proc. Roy. Soc. Edinb.*, **3** (1857): 503–4. The paper was read on 20 April 1857; see a letter from J. D. Forbes to Maxwell of 13 April 1857: 'I am afraid that we can probably allow you but a short time for expounding it – say 20 minutes' (ULC Add. MSS 7655, II/7). The paper was published in *Trans. Roy. Soc. Edinb.*, **21** (1857): 559–70 (= *Scientific Papers*, **1**: 248–62).

(2) See Number 108 note (7).

(3) See Number 109, esp. note (13). See plate VI.

(4) Louis Poinsot, *Théorie Nouvelle de la Rotation des Corps* (Paris, 1834). See Poinsot's memoir 'Théorie nouvelle de la rotation des corps', *Journal de Mathématiques Pures et Appliquées*, **16** (1851): 9–129, 289–336, esp. 24–6, 85, 102; Poinsot's memoir was reprinted as *Théorie Nouvelle de la Rotation des Corps* (Paris, 1851): see esp. 16–18, 77, 94; and republished in another edition, *ibid.* (Paris,

To observe the path of the invariable axis in the rapidly revolving body, we must have the means of recognizing the part of the body through which it passes at any time. For this purpose a disc of card is placed near the upper end of the axle. The four quadrants of this disc are painted red, yellow, green, and blue, and various other marks are added; so that by observing the colour of the spot which appears the centre of motion, and the diameter of the coloured spot, the position of the invariable axis in the body at any instant may be known, and its path traced out.

This path is a conic section, whose centre is in the principal axis. If that axis be the greatest or least, it is an ellipse with its major axis parallel to the mean axis. If the axle of the top be the mean axis, the path is an hyperbola as projected on the disc.

When the axle is the axis of greatest inertia, the direction of motion in the ellipse is the same as the direction of rotation. When it is the axis of least inertia these directions are opposite. All these results may be deduced from Poinsot's theory, and verified by means of the coloured disc.

The theory of precession may be illustrated by this top in the way pointed out by Mr Elliot,[5] by bringing the centre of gravity to a point a little below or above the point of support.

The theory and experiments with the top suggest the question – Does the earth revolve *accurately* about a principal axis?[6] If not, then a change of the

---

1852): esp. 16, 62, 75. In Poinsot's theory the rotation of a body about an axis which varies in its position round a fixed point is represented ('image sensible de cette rotation') by a cone which rolls without sliding on the surface of another cone, whose vertex coincides with this point; the 'axe instantané' of rotation is the line of contact of the two cones. The rotation of a body is represented by the 'ellipsoide central' which has its axes coincident with the principal axes of inertia of the body. The 'invariable plane' is the plane of the angular momentum; and Maxwell terms the 'axis of angular momentum' through the centre of inertia the 'invariable axis'. When the ellipsoid rolls on the invariable plane its point of contact traces a curve on the surface of the ellipsoid (which Poinsot terms a 'polhodie') and a curve on the invariable plane (a 'herpolhodie'). On these curves see F. Gomes Teixeira, *Traité des Courbes Spéciales Remarquables*, 2 vols. (Coimbra, 1908–9), **2**: 467–76. See Number 174 note (4). G. G. Stokes' 'Note on the axis of instantaneous rotation', *Camb. & Dubl. Math. J.*, **3** (1848): 128–30 is quoted in Maxwell's notes on the 'Motion of a Rigid System' in his undergraduate notebook on 'Lunar Theory & Rigid Dynamics' (ULC Add. MSS 7655, V, m/9, f.112); and is cited in a manuscript fragment (ULC Add. MSS 7655, V, c/9), probably part of a draft of 'On a dynamical top'.

(5) James Elliott, 'A description of certain mechanical illustrations of the planetary motions, accompanied by theoretical investigations relating to them, and in particular, a new explanation of the stability of equilibrium of Saturn's rings', *Transactions of the Royal Scottish Society of Arts*, **4** (1856): 318–44 (= *Edinburgh New Philosophical Journal*, **1** (1855): 310–35). See Number 128 for Maxwell's discussion of Elliott's paper with reference to the rings of Saturn.

(6) If the earth's instantaneous axis of rotation does not coincide with its principal axis of greatest moment of inertia. The problem had been discussed by Leonhard Euler, *Theoria Motus Corporum Solidorum seu Rigidorum* (Rostock/Greifswald, 1765): 293–6, 369–75. The question of the

Plate VI. Dynamical top (1857) to illustrate the motions of a spinning body; the disc has four coloured quadrants, indicating the position of the axis about which the top is revolving (Number 116). The lower end of the axle is a steel point which runs in an agate cup set in the top of the pillar.

position of the axis will take place, not in space, but with respect to the earth, so that the apparent positions of stars with respect to the pole will remain the same, but the latitude of every place will undergo a periodic variation, whose period is about 325 days. To detect this variation, the observations of Polaris with the Greenwich transit circle for four years have been examined. There appeared some doubtful indications of a variation not exceeding half a second.[7] A more extensive investigation would be required to determine accurately the period, and the epoch of maximum latitude at a given observatory,[8] which must depend on the longitude of the station, as the pole of the 'invariable' axis travels round the mean axis from west to east.

---

earth's axis of rotation and the change in latitude is discussed by S. D. Poisson, *Traité de Mécanique*, 2 vols. (Paris, ₂1833), **2**: 194–5. See also Number 185.

(7) Maxwell made notes on observations of Polaris with the Greenwich transit circle for the years 1851–4 (ULC Add. MSS 7655, V, a/3); and on observations for 1852 on the *verso* of a letter from Alexander Jardine Lizars (Professor of Anatomy at Marischal College) dated 9 February 1857 (ULC Add. MSS 7655, II/5). See especially 'Apparent right ascensions of Polaris and δ Ursae Minoris...', in *Astronomical and Magnetical and Meteorological Observations made at the Royal Observatory Greenwich for the Year 1852* (London, 1854): 122. In his paper 'On a dynamical top' Maxwell drew upon the exposition of the conservation of angular momentum in a paper by R. B. Hayward, 'On a direct method of estimating velocities, accelerations, and all similar quantities with respect to axes, moveable in any manner in space, with applications', *Trans. Camb. Phil. Soc.*, **10** (1856): 1–20. See 'On a dynamical top': 560–62 (= *Scientific Papers*, **1**: 249–51); and also Number 107 note (76). In response to Maxwell's paper, Hayward wrote to him on 14 and 26 November 1857, drawing his attention to a paper by C. A. F. Peters, 'Recherches sur la parallax des étoiles fixes', *Recueil de Mémoires présentès à l'Académie des Sciences par les Astronomes de Poulkova* **1** (1853): 1–180. In the letter of 14 November 1857 Hayward wrote that: 'Peters has used the observations of Polaris at Poulkova, & finds an angular radius of 0″.079 (with a probable error of 0″.017) for the circle described on the surface of the earth by the pole of the inst.ˢ axis. The period which he adopts, is however somewhat different from that which you assign, being 303.867 mean solar days.' See Peters, 'Recherches sur la parallax des étoiles fixes': 146–7. Hayward reported that Albert Marth (his colleague at Durham) 'attests that the Greenwich observations of Polaris cannot be depended on for indicating such minute variations of latitude' (ULC Add. MSS 7655, II/15). For Marth's critical review of Greenwich observations, see Albert Marth, 'On the polar distances of the Greenwich transit circle', *Astronomische Nachtrichten*, **53** (1860): col. 177–230. Writing again on 26 November 1857 Hayward observed that: 'Peters expresses a doubt whether after all the motion of the pole indicated by his results is real... with so small a radius for the path of the pole as 0″.079, it would seem almost hopeless to detect with certainty any inequality due to a difference in the equatoreal axes' (ULC Add. MSS 7655, II/16). See also Numbers 136 and 138. The question of the periodic variation of the earth's instantaneous axis is discussed by William Thomson and P. G. Tait in their *Treatise on Natural Philosophy*, new edn, Part I (Cambridge, 1879): 82–5 (§ 108 and footnote). A periodic motion of the earth's axis (an oscillation with a period of 14 months) was subsequently detected by S. C. Chandler, 'On the variation of latitude', *Astronomical Journal*, **11** (1891): 59–61, 65–70, 75–9, 83–6.

(8) The last part of the sentence was added in proof to the MS of the abstract (ULC Add. MSS 7655, V, a/4).

LETTER TO GEORGE GABRIEL STOKES

8 MAY 1857

From the original in the University Library, Cambridge[1]

Glenlair
Springholm
Dumfries
8th May 1857

Is it nitrate of strontian[2] which becomes
doubly absorbent when crystallized out
of logwood infusion?[3]

Dear Stokes

In conversing with Prof. Smyth of Edinburgh[4] about his excursion to Teneriffe and his experiments on the solar spectrum[5] it occurred to me that the method which I arranged last year for exhibiting the mixture of the colours of the spectrum[6] might be made useful in experiments on the spectrum by travellers and out-of-door observers.

My portable instrument consists of a wood box $24 \times 6 \times 4$ inch inside measure. The light enters at slits at the end of the box and falls on a prism at the other end. After refraction it is reflected by a concave mirror back through the same faces of the prism, and finally forms a pure spectrum near the place of entrance.

In my arrangement the light enters at the side and is reflected by a small plane mirror the prism is equilateral & of glass, and the mirror is a lens silvered at the back of about 1 foot focal length by *reflexion*. For the refrangible light, the light should enter directly at the end of the

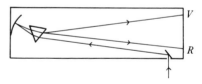

Figure 117,1

---

(1) ULC Add. MSS 7656, M 408. Previously published in Larmor, *Correspondence*, **2**: 5–7.

(2) Strontium nitrate; see Leopold Gmelin, *Hand-Book of Chemistry*, trans. Henry Watts, **3** (London, 1849): 179.

(3) See George Gabriel Stokes, 'On the change of refrangibility of light', *Phil. Trans.*, **142** (1852): 463–562 (= *Papers*, **3**: 267–409), on the absorption of light.

(4) Charles Piazzi Smyth; see Number 115 note (7). Maxwell may have met Smyth at the Royal Society of Edinburgh on 20 April 1857; see Number 116 note (1). See also Number 168 note (2).

(5) Charles Piazzi Smyth, 'Astronomical experiment on the Peak of Teneriffe', *Phil. Trans.*, **148** (1858): 465–534, esp. 503–6.   (6) See Numbers 102 and 109 esp. note (9).

box, the prism should be quartz, and the mirror metallic with a tool for repolishing.

I found this arrangement give a very large and distinct spectrum. The prism should be cleaned well and the mirror adjusted to make the reflexion in the plane of refraction. Then the focussing is done by turning the prism on its axis, which gives a very long range of focal lengths, and different parts of the spectrum are examined by turning the mirror & prism, (which are on the same frame) together.

There should be a finder outside the box in the fashion of an alidade[7] or better a spectacle glass at one end and a mark for the sun's image at the other.

Fluorescent[8] substances may be viewed by looking down from the side of the box and photographic materials might be slid into the same place.

The main advantage is the getting rid of the long rays inclined at an angle and the boxes to hold them. They are all doubled up and put in one box. I am going to try a new plan for comparative trials of different mixtures of light.

The various light is to be admitted by slits as in other experiments and sent through two different angles of the same prism and reflected by two mirrors separately. Thus an eye looking into a mirror M will see the two sides of the prism coloured in two different ways, and will give judgement on the difference.

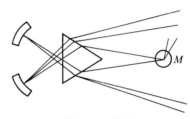

Figure 117,2

Can you give me references to the experiments of Foucault Fizeau & others if any. 1st on the velocity of light in air[9] 2nd comparative velocity in air & water,[10] and 3rd comparative velocity in water running opposite ways.[11]

Moignos Répertoire & his 'Cosmos' are the only authorities I know of.[12]

---

(7) The index of a graduated circle (*OED*).

(8) For Stokes' introduction of the term '*fluorescence*, from fluor-spar' see his 'On the change of refrangibility of light': 479n.

(9) Hippolyte Louis Fizeau, 'Sur une expérience relative à la vitesse de propagation de la lumière', *Comptes Rendus*, **29** (1849): 90–2.

(10) Leon Foucault, 'Méthode générale pour mésurer la vitesse de la lumière dans l'air et les milieux transparents. Vitesses relatives de la lumière dans l'air et dans l'eau', *Comptes Rendus*, **30** (1850): 551–60; and Foucault, 'Sur les vitesses relatives de la lumière dans l'air et dans l'eau', *Ann. Chim. Phys.*, ser. 3, **41** (1854): 129–64.

(11) H. L. Fizeau, 'Sur les hypothèses relatives a l'éther lumineux, et sur une expérience qui paraît démontrer que le mouvement des corps change la vitesse avec laquelle la lumière se propage dans leur intérieur', *Comptes Rendus*, **33** (1851): 349–55.

(12) F. N. M. Moigno, *Répertoire d'Optique Moderne*, 4 vols. (Paris, 1847–50), **3**: 1159–69; and *Cosmos*, **2**: 545–52.

I have had a dynamical top of brass made at Aberdeen[13] and have been simplifying the theory of the motion of the 'invariable axis' (normal to invariable plane) in the body. The extremity of this axis describes spherical ellipses about the greatest or least principal axes and the areas (projected on coordinate planes) described about these axes are proportional to the number of revolutions of the body about that axis (not to the absolute time). I have made some very delicate experiments with the top, to exhibit these movements, and it appears to be very little disturbed by accidental causes even when the difference of principal axes is less than $\frac{1}{300}$. I have also succeeded in setting the axis on the path leading *to* the mean axis so that the axis *approached* the mean axis for a considerable time before it left it for ever.[14]

Yours truly

J. C. Maxwell

(13) See Number 109, esp. note (13).

(14) Maxwell, 'On a dynamical top', *Trans. Roy. Soc. Edinb.*, **21** (1857): 559–70 (= *Scientific Papers*, **1**: 248–62); and Number 116, esp. note (4).

118

LETTER TO CECIL JAMES MONRO

20 MAY 1857

From the original in the Greater London Record Office, London[1]

Glenlair
Springholm
Dumfries.
20th May 1857

Dear Monro

That question of yours about physiognomic ethnology has turned up every time I intended to write to you.[2] By lapse of time its relapses are less painful & now I trust I may be able to get through without searching of heart (that heart by which I did not get the name of the fateful seven). I wont bother you with loose shots at answers to your other questions which probably both of us have forgot. I may do so next session. I went to Old Aberdeen for Fourier which was needed thereto but I have forgotten what was to be discovered out of him.[3]

The session went off smoothly enough. I had seen all the beginning of optics & worked off the experimental part up to Fraunhofers lines[4] which were glorious to see with a water prism I have set up in the form of a cubical box, 5 inch side. The only things not generally done that I attempted last session were the undulatory medium made of bullets for advanced class, and Plateau's experiments on a sphere of oil in a mixture of spirit & water of exactly its own density.[5]

I succeeded very well with heat. The experiments on latent heat came out very accurate. That was my part & the class could explain and work out the results better than I expected.

---

(1) Greater London Record Office, Acc. 1063/2082 (extract in *Life of Maxwell*: 267–8).

(2) In a letter of 15 February 1857 (Greater London Record Office, Acc. 1063/2101) Monro had asked Maxwell about a 'French author you once spoke of who goes about twigging the countenances of men & telling Kelt & Kymry, Hoch-Deutsch & Platt-Deutsch, by their teeth, noses & hair'. See Maxwell's reference to Johann Caspar Lavater in his letter to R. B. Litchfield of 7 February 1858 (Number 144, esp. note (11)).

(3) In his letter of 15 February 1857 (note (2)) Monro had asked a question about harmonic overtones: 'They who deal in instruments of strings say that if you strike a certain note you hear certain others above. Is that because of the further terms in a Fourier's integral, or because a sympathetic vibration is excited in certain other of the strings of the same instrument?' (see *Life of Maxwell*: 266).

(4) See Number 139 note (8); and compare William N. Griffin, *A Treatise on Optics* (Cambridge, 1838): 90–3, to which Maxwell made reference in his Cambridge notebook on 'Hydrostatics, Hydrodynamics & Optics' (ULC Add. MSS 7655, V, m/8).

(5) See Number 113, esp. note (5).

Next year I intend to mix exp. phys. with mechanics devoting Tuesday and THURSDAY (what would Stokes say) to the science of experimenting accurately.

I got a glorified top made at Aberdeen. I think you saw the wooden type at Cambridge.[6] I have made it the occasion of a short screed on rotation coming out in the Roy. Soc Edinburgh presently.[7]

Last week I brewed chlorophyll (as the chemists word it) a green liquor which turns the invisible light red. My pot of all the winter spinach that remained was portentous so I exhibited the optical effects which were allowed to be worth the potful.[8]

My last grind was the reduction of equations of colour which I made last year. The result was eminently satisfactory.

But I have most of my time occupied with other guess affairs. The most notable at present is a mill race, which I am getting to run downhill which it didnt use to do so the water did in spite of it. Now I am to pay for an extensive system of drains in the same latitudes which would be drenched by the mill water so as I know what the occupants of the soil would make of hydraulics I have undertaken the job, in self defence.

I heard various things from Puller which oughtnt to prevent you from giving me any of your own news. As for Hort I got his tokens a few days back. Is Elphinstone at Cambridge. I heard he was forsaking mercies as is the manner of some.[9]

I must write to Litchfield presently. Is Speer about London?[10]

I shall be at Cambridge 7th July I suppose for M.A. I have no other engagement, so if you should have opportunity to come so far north I should be glad to hear when it is to be. We have better arrangements now for letters and live stock than we used to. Lushington is the only brother who has been there, except Hugh Blackburn who is an old angel.[11]

---

(6) See Number 89.   (7) Number 116.

(8) A phenomenon described by David Brewster in his papers 'On the colours of natural bodies', *Trans. Roy. Soc. Edinb.* **12** (1834): 538–45, and 'On the decomposition and dispersion of light within solid and fluid bodies', *Trans. Roy. Soc. Edinb.*, **16** (1846): 111–21, esp. 111, where he describes the solution of the colouring matter of laurel leaves in alcohol: 'At first its colour is a bright green, afterwards changing into a fine olive colour; but in all its stages it disperses light of a *brilliant blood red colour*, which forms a striking contrast to the transmitted tint. After a long exposure to light, the transmitted tint almost wholly disappears, while the dispersed light retains its red colour.' The phenomenon had subsequently been discussed by George Gabriel Stokes in his major paper 'On the change of refrangibility of light', *Phil. Trans.*, **142** (1852): 463–562, esp. 486–92, 527–9 (= *Papers*, **3**: 267–409); and see Maxwell's reference to chlorophyll in his letter to Stokes of 3 October 1853 (Number 43).

(9) Charles Puller, Trinity 1852 (Venn); and for Hort and Elphinstone see Number 67 notes (2) and (3). Members of the Apostles.   (10) Wilfred Dakin Speer, Trinity 1852 (Venn).

(11) Vernon Lushington (Number 101 note (4)) and Hugh Blackburn (Number 45 note (10)), both members of the Apostles. On 'brother' and 'angel' see Number 104 note (14).

Talking of angels I saw at Glasgow one of whose existence I had but a speculative faith. From all my experience of Prof. Lushington I supposed him a single man in spite of In Memoriam (and a rising family in England).[12] It happened that M^rs L was in Glasgow on a sort of visit of a day or two to Prof. L and I never was more reminded of our angels than by her conversation which was as if old Stephen & Williams &c had got Luard & Watson set to historical sketches &c.[13]

We had a pour of rain all day which was all in good time. Things have been very late up to this time but everything is going ahead now. As for young beasts they are multiplying satisfactorily.

I got the Pomeroy MS up to Madras.[14] Do you know if the thing is to go on or if it is exceptional – part of nautical life? This was a wet day & I have been grinding at many things and lately during this letter at a Vortical theory of magnetism & electricity[15] which is very crude but has some merits, so I spin & spin.

Yours truly

J. C. MAXWELL

Ill be sure of your initials.[16]

---

(12) Edmund Law Lushington, whose marriage to the poet's sister Cecilia is described by Tennyson in 'In Memoriam' (1850), Professor of Greek at Glasgow 1838–75 (Venn).

(13) James Fitzjames Stephen, Trinity 1846; John Daniel Williams, Trinity 1847; Henry Richards Luard, Trinity 1843, Fellow 1849; Henry William Watson, Trinity 1846, Fellow 1851 (Venn).   (14) See Number 78 note (3).

(15) See Maxwell's letter to Faraday of 9 November 1857 (Number 133) where he refers to William Thomson's paper 'Dynamical illustrations of the magnetic and heliocoidal rotatory effects of transparent bodies on polarized light', *Proc. Roy. Soc.*, **8** (1856): 150–8 (= *Phil. Mag.*, ser. 4, **13** (1857): 198–204). Compare Maxwell's letters to Thomson of 24 November 1857 and 30 January 1858 (Numbers 137 and 143); and Lewis Campbell's recollection of a discussion on this topic (possibly in 1857), *Life of Maxwell*: 199n. On Maxwell's vortical theory of magnetism and electricity see Parts I and II of his paper 'On physical lines of force', *Phil. Mag.*, ser. 4, **21** (1861): 161–75, 281–91, 338–48 (= *Scientific Papers*, **1**: 451–88). A manuscript fragment 'On the Stability of the Steady Motion of an Incompressible fluid, the motion being entirely parallel to one plane' (ULC Add. 7655, V, c/5) may possibly have been written around this time. He proceeds as in his manuscript on fluid motion of 9 May 1855 (Number 63), and states a condition of the steady motion of a fluid in terms of the stream function ψ; 'curved channels' between stream lines for which ψ = constant 'constrain the motion of the fluid'. Considering the effect of forces applied to the fluid he obtains an equation for the forces in terms of a function χ, which is defined as in the manuscript of 9 May 1855. He concludes: '... $\chi = \dfrac{d^2\psi}{dx^2} + \dfrac{d^2\psi}{dy^2}$ & $\dfrac{d\chi}{dt}$ is the rate of variation of χ as a particle passes along its course. In fact χ is an expression for the velocity of rotation of each element of the fluid & $\dfrac{d\chi}{dt}$ is the change of rotatory motion produced by the action of the fixed channels.'

(16) In a letter of 14 November 1856 (Greater London Record Office, Acc. 1063/2100) Monro had observed: 'You write the initials in the wrong order'.

119

LETTER TO RICHARD BUCKLEY LITCHFIELD

29 MAY 1857

From the original in the library of Trinity College, Cambridge[1]

Glenlair
Springholm
Dumfries
29 May 1857

Dear Litchfield

I should like to hear about London & the Working Men.[2] I have heard nothing of you for long except that you had caught an elemental Hamburger[3] and then I had the symbols of Laurence Harrison.[4] By the way where are his souls that he looks after now? what quarter of England?

I heard from Monro about Pomeroy.[5] My perusal of the Journal had rather 'imprepared me' for such intelligence. It remains to be seen whether the Heeley plan is really any better.[6] I have Indian friends who have tried a good many different sol^{ns} of the problem, but they have all been acclimatized (perhaps too much) first.

It is with a profound feeling of pity that I write to a denizen of Hare Court after participating in the blessings of Heaven the whole of this splendid day. We had just enough of cloud to prevent scorching and the grass seemed to like to grow just as much as the beasts to eat it.

Sowing turnips is the home work with the usual accompaniments, but I have to be looking after the building of houses & dykes and the cutting of drains and water courses by reason of a new lease of a farm commencing.

I have not had a mathematical idea for about a fortnight when I wrote them all away to Prof. Thomson,[7] and I have not got an answer yet with fresh ones.

But I believe there is a department of mind conducted independent of

---

(1) Trinity College, Cambridge, Add. MS Letters c.1^{86}. Published in extract in *Life of Maxwell*: 268–9.

(2) On Litchfield's work with the London Working Men's College, established by F. D. Maurice, see [Henrietta Litchfield,] *Richard Buckley Litchfield. A memoir written for his friends by his wife* (Cambridge, 1910). See also Number 82; and Maxwell's references in his letters to his lectures for artisans.

(3) A young German described by C. J. Monro in a letter to Maxwell of 15 February 1857 (Greater London Record Office, Acc. 1063/2101).

(4) Laurence Harrison, University College, Oxford, 1850, curate of Leckhampton, Gloucestershire 1856–9; see Joseph Foster, *Alumni Oxonienses: the Members of the University of Oxford, 1715–1886*, 4 vols. (Oxford, 1888), 2: 617.

(5) See Number 42 note (3); and Numbers 118 and 124.

(6) See Number 104 note (9).   (7) This letter is not extant.

consciousness where things are fermented and decocted, so that when they are run off, they come clear.

By the way I found it useful at Aberdeen to tell the students what parts of the subject they were *not* to remember, but to get up and forget at once as being rudimentary notions necessary to development but requiring to be sloughed off before maturity.

I have no one with me but the domestics and dog. The valley seems deserted of its gentry but we have one gentleman from Dumfriesshire who is living in a hired house and building with great magnificence an Episcopal chapel in Castle Douglas, at his own expense. His own house is 20 miles off a capital place, and this is perhaps the least Episcopal part of Scotland by reason of the memory of the dragoons. Only one old family in the Stewartry is of that persuasion and most of the persecutors families are now presbyterian & whig so that the congregation is but feeble.

It is very different at Aberdeen where the Presbyterians persecuted far more than the Prelatists, so there I actually found a true Jacobite[8] (female, I could not undertake to produce a male specimen) and there are three distinct Episcopal religions in Aberdeen all pretty lively.

Can you tell me what the illustrated Tennyson is like.[9] I shan't see it till I go to Edinburgh. I dont mean are the prints the best possible or impervious to green spectacles, but are they nice diagrams as such things go. I should like to know before long about it, and whether the characters are of the Adamic type and in reasonable condition or PreRaphaelitic in all but colour and symbolizing everything except the 'archetypal skeleton' and the 'Nature of Limbs'.

I shall be in Cambridge probably at 7 July after which I may go to the far north near Skye and pick up an Aunt to stay here in autumn. I hope you will arrange yourself so as to do your Galloway this year at least.

I told you of McLennan having been here last Sept.[10] He seems getting on. He had 6 fees between passing in Jany and April 1 and he was rapidly acquiring the confidence of the thieves & murderers of Inverness. Alexander Smith has got new lodgings & has accomplished his marriage in Skye.[11] Do you see anything of Isaac Taylor?[12] Let me hear of your movements for the summer. How are the Tukes.[13] Have the parents remained at Cheltenham or do they

(8) Supporter of the Stuarts after the revolution of 1688.

(9) Alfred Tennyson, *Poems* (London, 1857), illustrated by J. E. Millais, Holman Hunt, Dante Gabriel Rossetti (and others). See Number 107 note (90).

(10) See Number 104.   (11) See Numbers 40 and 68.

(12) Isaac Taylor, Trinity 1849 (Venn); see *Life of Maxwell*: 176–7, 180–1.

(13) Litchfield's sister and brother-in-law; see Number 103 note (5).

gather into a sort of covey when the young birds pair off. Do you hear of R. E. Richards of Corfe Castle.[14] What a different lot of people I am among at Aberdeen. Variety is charming but the best of it is that the small talk of the one set would be babble nothing but babble to the other.

Yours truly
J. C. MAXWELL

(14) Robert Edward Richards, Trinity 1850 (Venn).

120

DRAFT ON THE PRINCIPLES OF MECHANICS

circa 1857[1]

From the original in the University Library, Cambridge[2]

[MECHANICAL PRINCIPLES][3]

Mechanics is the science which treats of matter considered with reference to its states of rest and motion.

We take for granted everything which is proved in geometry with respect to the properties of space, and the modes of defining the position of a point, the length of a line, the area of a surface and the volume of a solid figure.

We add to these geometrical conceptions the idea of time and thus we arrive at the idea of motion. For if we conceive that the same thing may exist in space for a certain portion of time, say one minute and if its position at the end of that time be different from what it was at the beginning, it must have moved during that time.

It may seem a very simple thing to say that when a body is first in one place and then in another it must have moved in the interval, but we shall find that it is necessary to be even more precise in order to have a practical knowledge of motion sufficient for our purposes. In all our reasoning we must be careful to admit nothing vague or indistinct.

We have a distinct idea of a mathematical point. However difficult it may be to represent it, we know how to define it by its position. Now suppose this point not only to have position in space but also duration in time that is let the same point continue to exist and let us put a mark upon it so that we may know it again and let us call it P. By giving it a name we distinguish it from the mere position it occupies and we are able to trace it from one position to another. Suppose now that at two different times we find it in two different positions in space, then we know that it must have travelled from the one to the other along a path and that every portion however small of this path was traversed by the p^t P during a definite portion of time. We have now the idea of the motion of a point. It travels along a path, straight or curved so that at successive instants of time it occupies successive positions on the path, and if we name any instant of time we could also point out the corresponding position on the path of motion.

A familiar instance of this is a railway timetable. The positions on the path

---

(1) See Maxwell's letter to Monro of 5 June 1857 (Number 124).

(2) ULC Add. MSS 7655, V, e/17(ii).

(3) The manuscript is endorsed 'Mechanical Principles'.

of motion are indicated by the names of the stations, and the corresponding times are printed opposite so that when we wish to go to any station we can find when the train ought to arrive. The Nautical Almanack gives the positions of the Sun & Moon at a great many specified times and it is easy to find from these the position of the Sun or Moon at any intermediate time.

Now that we have acquired the ideas of position & motion we may observe –
1st That a point can have only one position at any one time but that this position may change at any given rate.
2nd This change of position is called motion and the rate of change is called velocity.
3rd The motion of a point takes place along a line straight or curved, which is called the path of the point.
4th Since the position of a point is always definite the change of position must be definite, that is to say a point can only move in one way at one time, so that both the direction of motion and its velocity must be definite. In fact if a body could move either with two velocities or with two directions of motion, it would be immediately in two places at once.
5th If neither the velocity nor the direction of motion change the point will move uniformly along a straight line so as to describe equal distances in equal times.
6th If the motion be not in a straight line or not uniform the direction or the velocity must have altered.
7th This alteration must be definite in magnitude & direction for if it had two values at once the point would immediately have two different velocities or directions.

Def. Force is any cause which produces, or tends to produce change in the state of a body's rest or motion.

In so far as it actually produces a change it is called efficient force. When no change is produced it is because the forces balance and destroy one another.

The effect of an efficient force in changing motion depends upon the mass on which it acts. If however the forces destroy one another no effect will be produced and the body if originally at rest will remain so whatever be its mass. In fact if the forces be balanced, we may leave them out of account altogether in considering the material system.

The science of Statics forms the first part of Mechanics. It treats of bodies at rest.

Now if a body be at rest the forces acting on it must balance or destroy one another, for if any force remain it would set the body in motion.

Statics therefore treats of systems of forces in equilibrium that is of forces which destroy each other and therefore produce no motion.

We may therefore treat Statics as the science of pure Force, leaving out of account the motion which would be produced if the forces were not balanced.

The science of motion as produced by force is called Dynamics.[4] We have seen that the motion & the [...]

[Momentum]

The absolute position of a point in space being determined by its distances from three planes of reference, these distances are called its coordinates.

The rates of increase of three coordinates determine the components of the absolute velocity in the three directions perpendicular to the planes.

The *relative position* of a point B with respect to A is defined by the difference between the corresponding coordinates of A & B.

The relative velocity of B with respect to A is found by subtracting the component velocities of A from those of B (the velocity of A relative to B is equal and opposite to that of B with respect to A).

The absolute momentum of any mass is defined to be the product of the number denoting the mass into the number denoting its velocity.

Thus if mass be measured in terms of the standard pound and velocity by feet per second a mass of 4 lbs moving with a velocity of 5 feet per second will have a momentum represented by $20 \dfrac{\text{lb ft}}{\text{second}}$.

The relative momentum of B with respect to A is measured by the mass of B multiplied by the relative velocity.

Moving force is measured by the momentum it produces in unit of time and its direction is that of the momentum generated.

General Principle of Action & Reaction

Moving force acts in every case *between* two portions of matter, in such a way that the moving force of A on B is equal and opposite to that of B on A, and in the line AB.

Cor (1). If only one portion of matter be in existence its momentum must be invariable.

Cor (2). The sum of the components of the momenta of the bodies of a system is not altered by their mutual actions.

Def^{n}. The sum of the linear momenta of the parts of a system estimated in any direction is called the momentum of the system in that direction.

---

(4) A commonplace distinction in Cambridge texts; compare William Whewell, *An Elementary Treatise on Mechanics* (Cambridge, ₇1848): 5.

Def[n]. The linear momentum of a portion of a body into the perpendicular between the direction of motion and a given axis is called the angular momentum of the portion of the body round that axis.

The sum of such quantities is the angular momentum of the system about that axis.

Prop. The mutual action of two portions of a system does not affect the sum of their angular momenta about any axis. For the linear momenta produced are equal and directly opposite, and the perpendicular from the axis is the same for both.

Hence the angular momentum of a system is not altered by internal forces.

121

NOTES ON SPACE AND POSITION

circa 1857[1]

From the originals in the University Library, Cambridge[2]

On Absolute Space

There are certain geometrical truths connected with Form and Dimension which lead us to the idea of Absolute Space.

And first, to render distinct the idea of space as independent of matter, conceive a cubical apartment of which the walls are material but impenetrable and immoveable. Then though it is impossible to ascertain whether anything is inside the cube or not we can take measurements outside and determine completely the values of the cube or the amount of space which it contains and we are sure that though all the matter inside the cube were to be annihilated the volume within its sides would remain the same, and so would all the distances measured through the cube so that all the geometrical properties of the space within the cube remain whatever else is conceived to be removed.

Next, to show that the relations of bodies to space are independent of the consciousness of the observer, we may consider the case of two observers measuring two different objects with the same rule. If a third observer compares the two objects, the consistence of his result with that of the former observers depends upon a law which includes in its operation the two objects and the rule as well as the three observers.

On the Relative Positions of a System of Objects

The methods of Geometry afford the means of defining the position of any given body with reference to others. The method which is best adapted for our purposes is that in which the distances of a point from each of three coordinate planes at right angles to each other are employed to define the position of the point.

In laying down these coordinate planes we must first fix on a point of space through which they are to pass, then we must choose any three directions at right angles to each other and finally draw the three planes through the given point perpendicular to the three directions.

But we have still something more to do before we can define numerically the

---

(1) See Maxwell's letter to Monro of 5 June 1857 (Number 124).
(2) ULC Add. MSS 7655, V, e/16.

position of a point with reference to these planes. We must fix upon a unit of length and our choice here again is perfectly arbitrary. As soon as we have agreed on these planes of reference and unit of length the position of every point in space can be defined.

The relative coordinates of any point A with respect to B are the differences of the coordinates A and B in the three directions respectively

$$x_A - x_B \quad y_A - y_B \quad z_A - z_B.$$

But we may conceive of this process being begun by a different set of observers in a different part of space and using of course different coordinate planes and a different unit of length. If these observers have the means of communication with the first set, a little application of geometrical principles will suffice to harmonise all their observations, but whether communication be effected or not, it is always possible, in the nature of things, that the one set of observations should be compared with the other. We thus arrive at a mathematical conception of the unity of all space. We cannot conceive the selection of the coordinate planes and the unit of length as otherwise than perfectly arbitrary – it may be to suit the existing position and dimensions of some system of matter in which we are interested.

Time

The epoch of reckoning and the unit of time are arbitrary but any two different systems of chronology may be compared.

On Absolute and Relative Motion

A Body which changes its position with respect to the system of coordinates is said to move relatively to that system and the extent of its motion is measured by the changes of the three coordinates respectively. These are called the resolved displacements of the body in the three directions.

122

DRAFT ON THE SCIENCE OF MECHANICS

circa 1857[1]

From the original in the University Library, Cambridge[2]

Sketch of an Introduction to the Higher Parts of Mechanics, being the Science of Energy as it relates to the Motion and Rest of a Single Particle, of Two Particles or of a system of Particles, this system being either of invariable form, elastic or fluid.

(1) In mechanics we trace the connexion between the motions of material things and the forces which produce them. Those properties of different kinds of matter which are not capable of being defined as relations of motion or force are not treated of in mechanics, but whenever any phenomenon is explained by determinate motions caused by determinate forces, that explanation becomes part of mechanical science.

(2) An Exact Science is one the facts of which can be stated with numerical accuracy.

(3) A Mathematical science is one in which these numerical statements may be deduced from one another by mathematical laws.

(4) A Pure Mathematical science is one in which these laws are deduced from axioms whose evidence is perfect without quantitative experiment.

Thus the science which records the direction and force of the wind is as much an exact science as that which ascertains the direction of the planets, although no mathematical ⟨laws⟩ relations have been discovered among the observations of the wind like those which regulate the positions of the stars.

(5) To bring any phenomenon under the power of exact science we must analyze it into elements capable of observation and measurement.

(6) In order to measure anything, we must be able to conceive it as a quantity fix upon a qu[antity] of the same kind with the qu[antity] to be measured, as a standard of reference. We must then discover the ratio between the given quantity and the standard quantity and since these are of the same kind the ratio will be a number.

Hence every exact statement of quantity consists of a number and a unit as 'ten minutes' 'four pence' 'three score of sheep'.

All quantities of the same kind may be expressed in terms of the same unit, thus a mile, a stadium or a parasang[3] may be expressed in feet a bushel, a

---

(1) See Maxwell's letter to Monro of 5 June 1857 (Number 124).

(2) ULC Add. MSS 7655, V, e/16.

(3) Ancient Greek/Roman and Persian measures of length.

gallon, a litre in cubic feet, but we cannot express any quantity in terms of any unit of a different kind.

But units of different kinds may be connected by mathematical relations, for instance a square foot may be constructed when we know the linear foot by Euc 1.46.[4]

(4) Book I, Prop. XLVI of *Elements of Geometry*, ed. J. Playfair (Edinburgh, 1795): 45; 'To describe a square upon a given straight line'.

123

DRAFT ON THE MEASUREMENT OF QUANTITIES

circa 1857[1]

From the original in the University Library, Cambridge[2]

ON THE MEASUREMENT OF QUANTITIES

Every object in nature has a determinate place, and is moving in a certain direction with a determinate velocity. The quantity of matter contained in that object is determinate and the mode in which it is arranged at any given instant is definite. Whatever may happen to that portion of matter in the course of its existence the result can only be some definite though new arrangement of things. At this instant the condition of every particle of the universe might be ascertained with absolute accuracy – and the result would be a mathematical definition of the rudiments of the world as they now exist.

In order to make this survey of the universe we should require, first a fixed point in space to reckon from, with fixed directions to lay off our principal lines, and also a measuring line of a fixed length to compare all other lines with. But before we proceed any farther we find neither a fixed point in space, nor fixed divisions to begin our survey nor any fixed standard of length to appeal to. All things are in motion all distances are altering, our own bodies are changing. Our whole system of measurement seems to be without foundation either in absolute position or in scale, and in order to recover our belief in exact quantity, we must enquire into the necessary conditions of what can and what cannot be known about it. It will be seen that those questions to which no answer can be returned, and those which do not affect any phenomenon in Nature are and must be the very same set of questions.

The questions with which we shall be occupied relate entirely to matter and motion and are all capable of being verified by experiment, but in examining such questions we are exercising faculties of our mind conversant with metaphysical truth over a far wider range than can be found in pure mathematics and to depths far more profound than in any other merely intellectual subject. It is worth observing how much pure philosophy has borrowed from these fundamental ideas in physical science and how all 'appropriate ideas' in physics have been the result of much philosophical speculation.[3]

---

(1) See Maxwell's letter to Monro of 5 June 1857 (Number 124).

(2) ULC Add. MSS 7655, V, e/16.

(3) Whewell's term is 'fundamental ideas'; see William Whewell, *Philosophy of the Inductive Sciences, founded upon their History*, 2 vols. (London, ₂1847), **1**: 66. Compare Numbers 68 and 105. In his paper 'On a dynamical top', *Trans. Roy. Soc. Edinb.*, **21** (1857): 559–70, esp. 560–2 (= *Scientific Papers*, **1**: 250–2) Maxwell refers to the 'appropriate ideas' of Poinsot and Hayward on the theory of rotation; see Number 107 note (76) and Number 116 notes (4) and (7).

We have seen that our first great difficulty is a standard of measure and a starting point from which to begin. The ancient geometers evaded this difficulty by using as much as possible the ratios of two things to be compared, both of which were supposed to be in view. In this way each investigator drew his own figure to his own scale and yet arrived at the same ratios because everything was proportional. This method is very convenient and is used in most mathematical works of the geometrical cast, including Newtons Principia.

In order to compare things measured by different persons it is necessary to assume a standard of measure. Some one must indicate a particular specimen of the thing to be measured as the standard of that quantity and every one else must agree to measure by that standard or by copies of it. This method introduces experiment into mathematics, and greatly disturbs the easy elegance of the adherents of proportions who are not accustomed to the apparatus of the market place; but those who make up their minds to study Nature with measuring rod time-piece and weights will find that these arbitrary and perhaps inaccurate standards are intended to represent something uniform and independent of any individual man, which depends on an ancient decree and is preserved by the power of Nature but which neither a new decree nor new actions of Nature could restore if it were destroyed.

Definitions & Axioms on the Measurement of Quantity

Quantities are of the same kind which differ only in magnitude.

The unit of any kind of quantity is a certain quantity of the same kind fixed and agreed upon as the standard to which all other quantities of that kind shall be compared.

The principal units used in ordinary life have proper names such as a foot, an inch, a crown, an hour.

The name of any quantity consists of two parts, the number of units and the name of the unit as ten inches.

In calculations, arithmetical operations may be performed upon the numerical part of a quantity as upon any other number. The significance of the result, however, will depend upon the nature of the unit to which the quantity belongs. The result of many operations may appear under such a complicated form that we may be unable to recognise whether the number we have found belongs to a unit of length or of weight or of velocity. When we use the ordinary figures of arithmetic, after separating them from the units to which they owe their significance they cease to retain any mark of their origin but if we take care to use only algebraical symbols, the result of every operation contains the traces of all that have gone before and we may at once discover whether the final result be a quantity of the kind sought for.

Now since the value of a quantity of one kind has often to be deduced from the values of given quantities of other kinds, it becomes necessary to know the relations between units of quantity of different kinds and to be able to interpret relations between quantities of an artificial or impossible character.

On the Numerical part of a Quantity

The number standing before the name of the unit of quantity indicates the number of times that unit is contained in the specified quantity. The distinction of whole numbers, fractions and incommensurable numbers is entirely out of place in physical measurements. All ratios arising from any operations may be regarded as numbers, but those which are connected with particular operations may have significance as relating to these operations which they would lose if transferred to others.

Thus the ratio of one line to another differs in nothing from the ratio of any other pair of homogeneous quantities, but if one line be the arc of a circle and the other the radius, then the ratio of the two has an important significance as the *circular measure* of an angle and any expression into which this ratio enters will probably have some intimate relation to that angle. The nature of such a ratio is the same in every respect whether it denote an angle or the sine of an angle or a mechanical advantage.

Numerical quantities therefore enter into our expressions as ratios of two things of the same kind. All angles are numerical both plane and solid, all their trigonometrical functions are numerical all logarithms are numerical and every logarithm must be that of a number. The index of any exponential is necessarily numerical.

All numerical quantities, angles & logarithms are capable of exact definition without reference to any external standard unit. The definitions of these quantities rest on their own foundation and the equality of Euclids right angle to that of Newton does not depend on the preservation of a series of copies of an original right angle but each geometer constructs his own.

The other quantities which we meet in mechanics cannot be specified without reference to some standard unit external to the mind. The result of our analysis is that three such units are necessary, before we can measure all dynamical quantities, but that from these three all other units may be deduced. In choosing three fundamental units we may take any three independent of each other, but it is most to our purpose to take the three simplest namely Length, Time and Mass. The unit of Length may be a foot, a mile or the radius of the earths orbit.[4] The unit of Time may be a second a day or a year. The unit of mass may be a grain, a pound or the mass of the Sun, or any other known portion of matter.

When these three units have been fixed on, all other units can be compared with one another by reference to these. The units employed in various branches of art or science may have been fixed without reference to any fundamental unit, but in mathematical calculation all quantities are referred to a system of units derived from the three primary units and are easily reduced to any other set of units afterwards.

We shall use the symbols l, t, m to denote the length, time and mass which are arbitrarily assumed as the units. l is actually a line, not a number representing a line, for the number denoting the length of the line l is 1 since l is a unit of length. In the same way t is a certain time and m is a certain portion of matter.

We proceed to enumerate the principal units derived from the unit of space. 1st The unit of area may be of any magnitude, as an acre but in calculation we assume the area contained by the square of the unit of length as the unit of area. The magnitude of this unit varies of course as the second power of the unit of length.

2nd In like manner the cube whose side is the unit of length is assumed as the unit of volume and its magnitude varies as the third power of the unit of length.

Any other measures such as bushels, hogsheads must be reduced to the assumed unit.

3rd A plane angle is measured by the length of an arc of a circle of radius unity subtended by the angle placed at its centre. The unit of angular measure is therefore that angle which subtends an arc equal to the radius. The measure of an angle is not altered by changing the unit of length because it is a ratio or number and can be defined independently of any arbitrary units.

(4) A solid angle is measured by the area of a spherical surface whose centre is that angle, whose radius is unity and which is cut off from the sphere by the boundary of the solid angle.

The numerical value of a solid angle is not altered by changing the unit of length for it is the ratio of one surface to another namely the surface cut off by the angle and the square of the unit of length, and a solid angle is therefore to be regarded as itself a numerical quantity.

---

(4) Compare the recollection by David Gill of a lecture by Maxwell at Marischal College in 1859, suggesting the wave-length of the D line of sodium vapour as an invariable standard of length; David Gill, 'President's address', *Report of the Seventy-seventh Meeting of the British Association* (London, 1908): 3–26, esp. 4. Compare Maxwell's remarks on spectra as standards in his 1873 British Association lecture on 'Molecules', *Nature*, **8** (1873): 437–41; and *Phil. Mag.*, ser. 4, **46** (1873): 453–69 (= *Scientific Papers*, **2**: 361–78, on 376).

(5) We have often to speak of the curvature of a curved line. The unit of curvature is that of a circle whose radius is unity and since the curvature of a circle is inversely as its radius, the unit of curvature varies inversely as the unit of length.

There are many other quantities connected with the properties of pure space but these may serve as examples of the different ways in which the units of such quantities may be derived from the unit of length.

When we conceive of matter as existing in space and occupying a greater or less volume the quantity of matter being the same we form the conception of density. The mathematical measure of density is the number of units of mass of the substance contained in unit of volume, and the unit of density is therefore that of a substance of which unit of mass is contained in unit of volume.

The unit of density therefore varies directly as the unit of mass and inversely as the unit of volume.

If we assume the density of a particular substance, as water at standard temperature and pressure, for the unit of density, all other densities may be referred to this without any reference to the units of mass and volume. The ratio of the density of a substance to that of water is commonly called the specific gravity of the substance. The numerical value of the specific gravity of a substance depending on the properties of pure water is the same among all the nations who agree to use this natural standard. We must remember however that it is still a material standard and is incapable of pure definition as a standard pound.

When we conceive of matter changing its position in space we arrive at the conception of motion. To render this conception distinct we must identify a particular point in the moving body and trace its successive positions between the beginning and the end of the motion considered. We find by reflexion, that it must pass through every point in the path between these two extremities, that at any given instant it occupies one of these points and is travelling with a certain velocity in a definite direction – that of a tangent to the path at that point. We have therefore to consider the velocity, the direction and the mass moved. The velocity is compared with a unit velocity which is that of a body moving through one unit of length in a unit of time.

124

LETTER TO CECIL JAMES MONRO

5 JUNE 1857

From the original in the Greater London Record Office, London[1]

Glenlair
June 5th 1857

Dear Monro

Thanks for two letters about your usual proportion especially when one is political and both utter news to me.[2]

I certainly did expect that Pomeroy[3] would marry some one or be married himself some of these days but I did not think he would have chosen a voyage around the Cape as an opportunity for making selection.

But there are some chemicals of great affinity which require in the solid form minute tituration and long continued contiguity or no combination will take place.

I have not seen article seven but I agree with your dissent from it entirely.[4] On the vested interest principle I think the men who intended to keep their fellowships by celibacy and ordination and got them on that footing should not be allowed to desert the virgin choir or neglect the priestly office but on those principles should be allowed to live out their days provided the whole amount of souls cured annually does not amount to £20 in the Kings Book.

But my doctrine is that the various grades of College officers should be set on such a basis that although chance lecturers might be sometimes chosen from among fresh fellows who are going away soon, the reliable assistant tutors and those that have a plain calling that way should after a few years be elected permanent officers of the college and be tutors and deans in their time and seniors also, with leave to marry or rather never prohibited or asked any questions on that head and with leave to retire after so many years service as seniors.

As for the men of the world we should have a limited term of existence, and that independent of marriage or 'parsonage'.[5]

---

(1) Greater London Record Office, Acc. 1063/2083 (extract in *Life of Maxwell*: 269–70).

(2) In reply to a letter from Monro of 2 June 1857 (Greater London Record Office, Acc. 1063/2102) and another letter (not extant).

(3) Number 42 note (3).

(4) In his letter of 2 June 1857 Monro described 'article seven' of a report by a Trinity College committee on the reform of the College Statutes: 'no one should lose by the changes who was a fellow when they were made, & placing a slight & rather arbitrary restriction on the extent to which any one sh[d] gain by them'. Monro records his dissent: 'I think no one ought to gain or lose, or be any way affected by them who was a fellow when they are made'. The committee reported to the Trinity Governing Body on 9 June 1857; see D. A. Winstanley, *Early Victorian Cambridge* (Cambridge, 1940): 343–6.   (5) See Number 125.

What day is the dinner? I think of going to Edinburgh on the 15th June and attending a marriage and seeing friends for a while, then doing London and Cambridge till Commencement. I have some books to seek at Cambridge in Libraries. Then I may take Manchester on my way home. In the beginning of August or so I may run up to Moidart where Hugh Blackburn lives and fetch an aunt of mine back here to stay till the end of the long and make tea for anyone who may come, you if you like. I hope you understand that this is not Aberdeen being much nearer Carlisle than any other large town and that we are in view of Skiddaw & Saddleback and also the isle of Man. The 9 pm mail NW[rn] from London brings you to Dumfries about $7\frac{1}{2}$ a.m. and you arrive at Springholm about 12 miles from Dumfries by coach and can breakfast here or in Dumfries as your maw directs.

I saw a paragraph about the Female Artists exhib[n] and that M[rs] Hugh Blackburn[6] had her Phaethon there. It is long since I saw it in 48 and I did not admire it then. Have you seen it? She has done a very small picture of a haystack making somewhat preraphaelite in pose but graceful withal and such that the Moidart natives know every lass on the stack whether seen behind or before. It was at the Edinburgh Academy of P[rs].

I wanted to hear from Litchfield about the illustrious Tennyson[7] but I am likely to get an answer from you before he replies, so I want you to look at him and say what moral lies in being etched or grained or tooled as he is. I wish to know rather early if the diagrams are such as one may look at once and again and then when one pleases or whether they are fated to hasten the oblivion of the Laureate for his hundred years of cocoon dormancy.

Come. Care & Pleasure Hope & Pain
& bring the fated fairy Prints.

I can't see them in this location so I send to you.

I have been practising my hand of write at exposition of fundamentals. I have done a screed of introduction to Optics[8] and am at a sort of general summary of mechanical principles, doctrines relating to absolute and relative motion, analysis of the doctrine of Force into the smallest number of independent truths. Theory of Angular Momentum & Couples of Work done & Vis Viva, of Actual & Potential Energy, with continual jaw on the doctrine of measurement by units all through.[9]

Yours truly
J. C. MAXWELL

I expect my disquisition on tops from printer you shall have one.[10]

---

(6) Maxwell's cousin; see Number 45 note (10).   (7) See Number 119, esp. note (9).

(8) See Number 75; and *Life of Maxwell*: 204n. The manuscript is no longer extant.

(9) Numbers 120 to 123.   (10) See Number 116 note (1).

LETTER TO JOHN WESTLAKE[1]

9 JUNE 1857

From the original in the University Library, Cambridge[2]

Glenlair
9 June 1857

John Westlake Esq.[re]
Trinity College,
Cambridge.

Dear Sir[3]

I hereby authorise you to append my name to the Memorial on Collegiate Celibacy to which you act as Secretary.[4]

JAMES CLERK MAXWELL
M.A. Fellow of Trinity College

---

(1) Fellow of Trinity 1851–60 (Venn).   (2) ULC Add. MSS 725 f. 185.

(3) In reply to a printed circular letter from Westlake of January 1857, urging that: 'Is it not desirable to retain the most able and best qualified men to conduct the work of Education at Cambridge?... The concession of marriage to Fellows, under judicious regulations, would not discourage literary or scientific merit...'; and asking: 'If the object of the Committee should meet with your approbation, I am requested to offer the enclosed copy of the Memorial for your signature,' (ULC Add. MSS 725 ff. 1–2). See note (4).

(4) 'Memorial to the Cambridge University Commissioners, on the Celibacy of Fellows of Colleges', presented 13 June 1857 (ULC Add. MSS 725 ff. 7–15, esp. f. 8^v); 'Among the many questions which will be submitted to you for deliberation and decision, we feel that none is more important in all its bearings, whether academical or social, than that affecting the restriction upon marriage, to which the tenure of Fellowships is now subject. We therefore venture to solicit your earnest consideration of the subject, and to express the hope, that you will give your sanction to such changes in the present "conditions of the tenure of Fellowships", as are calculated to augment the influence and extend the utility of the Colleges and University of Cambridge'. On collegiate and university reform at Cambridge in the 1850s see D. A. Winstanley, *Early Victorian Cambridge* (Cambridge, 1940).

LETTER TO WILLIAM THOMSON

1 AUGUST 1857

From the original in the University Library, Glasgow[1]

Glenlair
Springholm, Dumfries
1 August 1857

Dear Thomson

I have been brewing at Saturns Rings with infusion of your letters[2] for a month during most of which I have been on the move but I hope to explain myself now. I have had talks with Challis and hunts in the University Library and sights of diagrams.[3] As for the rigid ring I ought first to speak of Prof. Peirce. He communicated a large mathematical paper to the American Academy on the Constitution of Ring but up to the present year he has no intention of publishing it. There is a 'popular' abstract of it in Goulds Astron journ June 16, 1851.[4] His result is about 20 fluid rings unable to preserve themselves but guarded by the satellites in some unexplained way.

To understand him it is best to refer to a paper 'On the Adams Prize Problem of 1856' in Goulds journal Sept 5, 1855.[5]

---

(1) Glasgow University Library, Kelvin Papers, M 7. Previously published in S. G. Brush, C. W. F. Everitt, and E. Garber, *Maxwell on Saturn's Rings* (Cambridge, Mass./London, 1983): 41–4.

(2) Following the award of the Adams Prize on 30 May 1857; see Number 107 note (4).

(3) See James Clerk Maxwell, *On the Stability of the Motion of Saturn's Rings* (Cambridge, 1859): 3–4n (= *Scientific Papers*, **1**: 294n), referring to papers by G. P. Bond and B. Peirce; see notes (4) and (5). Maxwell wrote a list of references to works on the rings of Saturn on the *verso* of a letter from Andrew Milligan (of the Castle Douglas branch of the Bank of Scotland) dated 4 July 1857 (ULC Add. MSS 7655, II/8): 'Receuil de Mémoires Astronomique Poulkova par W. Struve/Sur les dimensions des Anneaux de Saturne Par M. Otto Struve 14 November 51/Phil. Trans 1791 on Saturns ring Herschel/1805/Galileo/Scheiner/Hevelius/Riccioli/Huyghens 1659 Systema Saturnium Annulo cingitur tenui plano nusquam cohaerente ad eclipticum inclinati/breadth between ring & S. equal or greater than that of Ring, ring to Saturn as 9 to 4. Cassini J. D. obs 1691/Newton princip. vol III p. 7 Horsley on Pound⁵ obs./Lassel March 11 1853 Mem. Ast. soc./Lardner Uranography of Saturn do. do./Ueber der Ring des Saturn 8 Nov Von J. F. Encke, Berlin acad. 1838/Peirce on the Adams Prize Problem for 1856. Benjamin Peirce LL.D. Perkin prof of Ast. & Math. in Harvard University Goulds Ast. journ. Sept. 5 1855./Peirce on the constitution of Saturns Ring June 16 1851./G. P. Bond on Rings of Saturn 2 May 51. Disappearance of Rings 7 Jan 1850 read to Am. Acad. 15 Ap. 51.' The classical references are cited in the paper by Otto Struve: see *infra* and note (15).

(4) Benjamin Peirce, 'On the constitution of Saturn's rings'. *Astronomical Journal*, **2** (1851): 17–19; see also G. P. Bond, 'On the rings of Saturn', *ibid.*: 5–8, 9–10.

(5) Benjamin Peirce, 'On the Adams Prize-problem for 1856', *Astronomical Journal*, **4** (1855): 110–12.

The number of rings is found on the equilibrium theory by conceiving the condition of the surface of a section of each ring.

Then the continuity of each ring is disposed of by asserting the tendency of the fluid pressures to equalize the section all round. Now the fluid pressures are those due to the attraction of the ring on itself towards the middle of its section and these are exceedingly small compared with the longitudinal attractions. In fact it is impossible for an elongated fluid mass to be stable without a solid core, much denser than the fluid. (The condition is that the potential at the surface of the fluid, due to the fluid and core, together, shall be less as the fluid is thicker.)

It is only dynamically that stability can exist, and it appears from my solution that if there had been statical stability it would have been dynamically unstable.[6]

Peirce also tacitly assumes that the character of the motion of each ring is 'steady' and deduces from that the equality of attractions on all sides of the planet, and therefore, he says, if the centre of the planet were in motion with respect to the ring it would continue to move till it met the ring as the attractions of the parts of the ring always balance by reason of the nearer and faster moving parts being thinner.

But he forgets that *during* these changes this law of thickness is so much modified that the reasoning fails, and besides there is nothing but a mathematical assumption to preserve the 'steady motion' of the ring all the time. With such an assumption he finds it easy to convert the ring into a comet at aphelion by becoming more and more elliptic.

Now for the body guard of Satellites.

It is necessary that the centres of Saturn & the Rings should keep together. According to Peirce there is no reason why they should as far as they are concerned. But the Satellites act on both and pretty much the same way so that they cause them to perform pretty nearly the same paths and *so* to keep pretty nearly together.

I hope to state the argument right for it is greatly abridged in the journal.

His plan of drawing the potential surfaces for the solid ring is very good. I intend to do it actually on paper as I have done electric lines.[7] But I strongly

---

(6) See Maxwell's comments in the introduction, and Part II, Proposition X, of the Adams Prize Essay 'On the Stability of the Motion of Saturn's Rings' (Number 107); compare *Saturn's Rings*: 4 (= *Scientific Papers*, **1**: 295).

(7) Maxwell, 'On the method of drawing the theoretical forms of Faraday's lines of force without calculation', *Report of the Twenty-sixth Meeting of the British Association for the Advancement of Science; held at Cheltenham in August 1856* (London, 1857), part 2: 12 (= *Scientific Papers*, **1**: 241). See also Number 166.

suspect the dynamics of his reasoning on the stability. The motion (in the plane of the ring) depends on 3 variables, 2 for the centre of gravity and one for rotation. All these are interdependent so that an oscillation in one affects the rest. Now the man who without previous analysis in the ordinary way, can so conceive the motion as to predict the whole effects of a disturbance, and then describe in ordinary language the nature of his reasoning, has no business to contribute to mathematical journals and such like. He ought to sit down at once and write off a new Principia on Spiral Nebulae. When I have carefully worked out the motion I hope to state its nature and give a rough description of the sequence of events but that is not done yet.

I must get up the question put by you and Challis about the stability of a ring nearly uniform. That is also not done nor do I relish it.[8]

Now for the streams of particles or of fluid. The equations will do if mK is not too large. Now K is inconceivably small in fact the smaller the better, m is the number of undulations round the ring.[9] If m be very great the supposition, that the undulations were long with respect to the breadth of the ring will not be true, and the assumed value of K will be too large.

The calculation in this case is too difficult for me but I think that mK cannot exceed a certain value which depends on the section and density of the ring, so that a broad ring would stand where a thin one would snap. If you admit the fluid rings with internal friction in each ring, these short waves disappear most rapidly and are therefore less to be dreaded.[10]

If you take the discrete particle theory with a linear ring of them, still K decreases as m increases.

Also consider the smallness and regularity of all disturbance by satellites as long as Saturn is so large and they are so small.

What shall we say to a great stratum of rubbish jostling and jumbling round Saturn without hope of rest or agreement in itself, till it falls piecemeal and grinds a fiery ring round Saturns equator, leaving a wide tract of lava with dust and blocks on each side and the western side of every hill battered with hot rocks.

First then there is no way of making a visible thing without refraction except of discrete parts with interstices. That is the cheapest way of attenuating comets and other heavenly bodies. Gas is quite a mistake. The bigger the bits the more

---

(8) See Challis' annotations to Part I, Problem VI of 'On the Stability of the Motion of Saturn's Rings' (Number 107); and Maxwell's letters to Thomson of 24 August 1857 (Number 128) and to Challis of 24 November 1857 (Number 138). Compare *Saturn's Rings*: 15–17 (= *Scientific Papers*, **1**: 307–10).

(9) See Thomson's annotations to Part II, Proposition X of Number 107; for the definiton of K see Proposition IX.   (10) See Part II, Proposition XI of Number 107.

transparent at the density. Nebulae go naturally in bits, why not rings. Now the edge of Saturn is seen quite undisturbed by the dark ring.

Next what will be the effect of this stratum not disposed in thin rings as before but spread over an annular surface of great breadth and moving with various velocities.

The calculation of the undulations due to attraction is far beyond me.[11] We have two dimensions for one and all the spreading effects of the different velocities.

What will be the effect of collisions.[12]

1st They destroy vis viva but not angular momentum so that if the orbits remain circular they spread the ring wider on the whole tending to make a division in it.

2nd Suppose that disturbance produces ellipticity in an orbit of a particle so that it meets a particle belonging to a wider orbit. The first particle at the most distant part of its course travels slower than the second so that the first is kicked forward and the second backward. The mean distance of the inner particle is increased and that of the outer diminished by this collision and both orbits become less excentric.

This is the opposite effect, tending to concentrate the ring again.

It will be seen that this effect tells most on the inner and outer edges of the whole system, so that these will be held in while the rings generally are spreading. Now the obscure ring[13] is sensibly thicker at its inner edge. As for the outer rings they are far too massy in proportion to the inner ones for any visible result to take place on them. The general result is therefore rather a spreading of the mean parts of the rings towards the outer and inner edges, than an enlargement of the whole ring, but in the inside the spreading effect may be so much the stronger as the stuff spread is thinner.[14]

The best historical treatise is in Recueil de Mémoires Astronomique de Poulkova – Sur les dimensions des Anneaux de Saturne Par M. Otto Struve 14 Nov 1851.[15]

---

(11) Compare *Saturn's Rings*: 46–51 (= *Scientific Papers*, **1**: 346–52).

(12) See Part II, Proposition XIII of Number 107; and compare *Saturn's Rings*: 51–3 (= *Scientific Papers*, **1**: 352–4).

(13) First observed by G. P. Bond on 15 November 1850; see 'Inner ring of Saturn', *Monthly Notices of the Royal Astronomical Society*, **11** (1851): 20–27.

(14) Compare Thomson's queries on Maxwell's discussion of this point in Part II, Proposition XIII and the Conclusion of 'On the Stability of the Motion of Saturn's Rings' (Number 107, esp. notes (80) and (90)).

(15) Otto Struve, 'Sur les dimensions des anneaux de Saturne', *Recueil des Mémoires présentès à l'Académie des Sciences par les Astronomes de Poulkova*, **1** (1853): 349–85. See note (3) and Number 107 note (1).

The encroachment of the inner ring seems very certain and not due to the improvement of telescopes.[16]

On the whole I should not recommend any one to feu[17] a building stance on any of the Rings without security that his parallelograms may not be spun out into spirals of unknown extent in a few hours. Arctic ice packs are most secure in comparison. As for the men of Saturn I should recommend them to go by tunnel when they cross the 'line'.

Yours truly

J. C. MAXWELL

(16) Struve, 'Sur les dimensions des anneaux de Saturne': 385; see Number 107 notes (1) and (80) for Struve's review of recorded observations of Saturn's rings since the seventeenth century.

(17) Compare John Austin, *Lectures on Jurisprudence or the Philosophy of Positive Law*, rev. and ed. Robert Campbell, 2 vols. (London, ₃1869), **2**: 879n; 'The *feu-contract* is in the nature of a perpetual lease, and is in Scotland the usual mode of letting ground for building purposes'.

FROM A LETTER TO LEWIS CAMPBELL

7 AUGUST 1857

From Campbell and Garnett, *Life of Maxwell*[1]

Glenlair
7 August 1857

I have succeeded in establishing the existence of an error in my Saturnian mazes, but I have not detected it yet.[2] I have finished the first part of the Réligion Naturelle.[3] I am not a follower of those who believe they know what perfection must imply, and then make a deity to that pattern; but it is very well put, and carries one through, though if the book belongs to this age at all, it is eminently unlike most books of this century in England. But I only know one other book of French argument on the positive (not positiviste) side, and that also worked by 'demonstration'. My notion is, that reason, taste, and conscience are the judges of all knowledge, pleasure, and action, and that they are the exponents not of a code, but of the unwritten law, which they reveal as they judge by it *in presence of the facts*. The facts must be witnessed to by the senses, and cross-examined by the intellect, and not unless everything is properly put on record and proved as fact, will any question of law be resolved at headquarters.

---

(1) *Life of Maxwell*: 272–3; abridged.   (2) See Number 128.
(3) Jules Simon, *La Religion Naturelle* (Paris, 1856/₃1857); (trans.) J. W. Cole, ed. J. B. Marsden, *Natural Religion* (London, 1857).

LETTER TO WILLIAM THOMSON

24 AUGUST 1857

From the original in the University Library, Glasgow[1]

Glenlair

24th August 1857

Dear Thomson

Since I last wrote to you I have wrought at the theory of a solid ring. I had made a mistake in the evaluation of the quantities $\dfrac{d^2V}{dr^2}, \dfrac{d^2V}{d\phi^2}$.[2] I ought to have tested them by the condition

$$\frac{d^2V}{dr^2} + \frac{1}{r}\frac{dV}{dr} + \frac{1}{r^2}\frac{d^2V}{d\phi^2} + \frac{d^2V}{dz^2} = 0$$

where $\dfrac{d^2V}{dz^2}$ refers to a direction perp. to plane of r & ϕ. In this way I find

$$\frac{d^2V}{dr^2} = \frac{R}{2a^2}(1+g) \left| \frac{d^2V}{dr\,d\phi} = \frac{R}{2a^2}fh \right| \frac{d^2V}{d\phi^2} = \frac{R}{2a}f^2(3-g)$$

where f is the ratio of the distance of centre of gravity of ring from centre to radius and g & h depend on the form of the ring.[3] The density to unit of length being of the form

$$\mu = \frac{R}{2\pi a}\{1 + 2f\cos\theta + \tfrac{2}{3}g\cos 2\theta + \tfrac{2}{3}h\sin 2\theta + \&c\}.$$

The general differential equation of motion becomes

$$(1-f^2)\frac{n^4}{\omega^4} + (1 - \tfrac{5}{2}f^2 + \tfrac{1}{2}f^2g)\frac{n^2}{\omega^2} + \tfrac{9}{4} - 6f^2 - \tfrac{1}{4}(g^2+h^2) + 2f^2g = 0$$

$$A\frac{n^4}{\omega^4} + B\frac{n^2}{\omega^2} + C = 0^{(4)}$$

---

(1) Glasgow University Library, Kelvin Papers, M 8. Previously published in S. G. Brush, C. W. F. Everitt, and E. Garber, *Maxwell on Saturn's Rings* (Cambridge, Mass./London, 1983): 44–8.

(2) *V* is the potential of the ring, and the variables r, ϕ and θ determine the configuration of the ring: r is the distance between the centre of gravity of Saturn and the centre of gravity of the ring; θ the angle between the line r and a fixed line in the plane of the ring; ϕ the angle between the line r and a line fixed with respect to the ring. See Part I of 'On the Stability of the Motion of Saturn's Rings' (Number 107).

(3) *R* is the mass, *a* the radius of the ring. See Part I, Problem V (4) of 'On the Stability of the Motion of Saturn's Rings' (Number 107, esp. notes (29) and (31)) for Maxwell's original argument. See also Maxwell's letter to Challis of 24 November 1857 (Number 138); and Maxwell, *On the Stability of the Motion of Saturn's Rings* (Cambridge, 1859): 12–14, esp. equation (24) (= *Scientific Papers*, **1**: 303–6).

(4) In the equations of motion n stands for the operation d/dt, ω for the angular velocity of the ring; and for the coefficients A, B, C compare Part I, Problem III of Number 107. Compare equations (28) and (18) of *Saturn's Rings*: 9, 15 (= *Scientific Papers*, **1**: 301, 307).

and this equation must give two real negative values of n^2 which implies that A B & C must have the same sign and that $B^2 > 4AC$.[5]

Now if we put $g = 0, h = 0$ the first condition gives $f^2 < 3.75$ and the second $f^2 > 37445$. But in order that the section of the ring may be real throughout (no *negative mass* in it) we must have $f < \frac{1}{2}$ so that we are blocked out.[6]

Now try a uniform ring, mass $= Q$ with a heavy particle P at a point in its circumference.[7]

$$P + Q = R \quad f = \frac{P}{R} \quad g = \frac{3P}{R} = 3f.$$

The first condition gives $f < .8279$
The second $\qquad\qquad f > .815865.$

Between these values there is stability and there is nothing to hinder the mass of P to be to that of Q as 82 to 18.

Now take $f = 82$[8] and we find

$$\sqrt{-1}\,\frac{n}{\omega} = \pm .5916 \text{ or } \pm 3076$$

indicating periods of 1.69 or 3.25 revolutions.
Working out the whole motion we find[9]

$$\frac{r_1}{a} = A \sin (.59\theta_0 - \alpha) + B \sin (.3\theta_0 - \beta)$$
$$\theta_1 = 3.21 A \cos (.59\theta_0 - \alpha) + 5.72 B \cos (.3\theta_0 - \beta)$$
$$\phi_1 = -1.2 A \cos (.59\theta_0 - \alpha) - .79 B \cos (.3\theta_0 - \beta).$$

Here are 3 variables $r_1\, \theta_1\, \phi_1$ expressed with 4 arbitraries $A\, B\, \alpha\, \beta$. Since $r_1\, \theta_1$ ϕ_1 & their diffls $\dfrac{dr_1}{dt}\, \dfrac{d\theta_1}{dt}\, \dfrac{d\phi_1}{dt}$ are all disposable we should expect 6 arbitraries. But two of them are swallowed up by assuming r_0 & θ_0 as the mean values of r & θ, and we prove the thing analytically by searching for dropped roots in the equations.

It is easy to show that for points near C the centre of the ring in this case the attractive force is directed always towards a point O, further from C than R the centre of gravity. Now how is the ring to revolve around S (Saturn) if the force acts at a point beyond the centre of gravity?

(5) See Part I, Problem IV of Number 107.

(6) The case of a ring thicker at one side than the other; see *Saturn's Rings*: 15 ($= Scientific Papers$, **1**: 307), where he corrects these values; 'The condition of stability is therefore that f^2 should lie between .37445 and .375...'. Compare Part I, Problem VI of Number 107; and see Number 138.

(7) Compare *Saturn's Rings*: 15–17 ($= Scientific Papers$, **1**: 307–10); and see Number 138.

(8) Read: $f = .82$.

(9) Putting $\theta_0 = \omega t$; compare Part I, Problem II of Number 107.

Suppose the line *RO* in advance of *SR* the force in *OS* tends to increase the rotation of the ring and so to increase *KRO* the angle of libration.

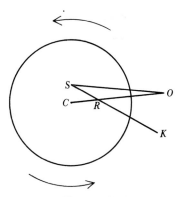

But in order to estimate the whole effect we must consider the force resolved into a couple acting as aforesaid and radial & tangential forces on *R* of which the tangential will be negative and by its *indirect* effect will *increase* the angular velocity of *SR*.

Whether *SR* will *overtake RO* must depend on numerical considerations. In that case there is *stability and periodic inequality.*

Figure 128,1

I have tested the whole of this part well and found it all right. Next comes the fluid filament. Now if you or anybody else can tell me what is the attraction towards the centre upon a particle in the transverse section of a ring whose mean radius is *a*, radius of section $b \left(\dfrac{b}{a} \text{ small} \right)$ and density ρ I will be greatly obliged for I am very bad at such calculations.

Or if a solid ring of these dimensions were cut in two and the parts placed in contact what would be the whole pressure between them?

I am more entangled with questions of this kind than with anything else.

Then here is another nice problem a right cylinder of radius $= b$ has electricity (or anything else) spread over its surface according to the law

$$\rho = A \cos \frac{z}{2\pi C}$$

find the potential at any point.
Or solve the eq^n

$$\frac{d^2 u}{dr^2} + \frac{1}{r} \frac{du}{dr} - n^2 u = 0.$$

To return to visibles & tangibles –

Let there be an infinitely long cylinder of any substance, and let there be spread upon it a coating of another substance to the thickness

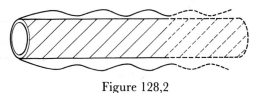

Figure 128,2

$T = A + B \cos \dfrac{z}{2\pi C}$ where *A* and *B* are very small, so as to be in waves of thickness & thinness along the cylinder then conceive this coating melted suddenly. Will the irregularities increase or diminish owing to the attractions of the mass?

If C, the wave length is *small* compared with the diameter of the cylinder they will diminish. If it is *large* they will increase. What is the critical value of C?

If the force were *capillary tension* on the surface it is easy to show that $C = 2\pi b =$ circumference of cylinder at critical period, and Plateaus experiments show that $\frac{c}{b}$ is between 3 & 3.5.[10] But the same problem in *attraction* is beyond my calculating power.

My theory of *long waves* is all right.[11] The longitudinal force is towards the thicker parts and the transverse force is similar to that due to a tension increasing with the thickness.

But if the waves are *short* the tendency is toward the thinner parts, while the transverse force is still of opposite sign to the transverse displacement but depending more on *displacement* from the mean position than on *curvature* at the point.

The case is like that of long and short waves of light. The attractions are not confined to contiguous particles so that though long waves may be treated so, short ones cannot, so that the theory of long waves is not applicable to them.

I intend to apply what I know of this to the fluid ring as I go along. The last sentence applies equally to the hailstorm ring,[12] and my scoffing allusion to the *circular saw* was on account of my not having anything ready at the time to explain it completely.[13] I intend also to accumulate notions about the hailstorm ring and to give the explanation of Mr Elliot's experiment of the iron ring spinning round a magnet.[14] That is a problem in 3 dimensions and it depends

(10) Joseph Plateau, 'Recherches expérimentales et théoriques sur les figures d'équilibre d'une masse liquide sans pesanteur', *Mémoires de l'Academie Royale des Sciences de Bruxelles*, **23** (1849); (trans.) 'Experimental and theoretical researches on the figures of equilibrium of a liquid mass withdrawn from the action of gravity', *Scientific Memoirs*, ed. R. Taylor, **5** (London, 1852): 584–620, 621–712, esp. 673; 'when a liquid cylinder is formed between two solid bases, if the proportion of its length to its diameter exceeds a certain limit, the exact value of which is comprised between 3 and 3.6, the cylinder constitutes an unstable figure of equilibrium'. See also Number 107 note (17).

(11) See Part II, Proposition X of Number 107; but compare his letters to Thomson and Challis of 14 and 24 November 1857 (Numbers 134 and 138).

(12) See Numbers 129, 134 and 138; and *Saturn's Rings*: 38–40 (= *Scientific Papers*, **1**: 336–8).

(13) See Part II, Proposition X of Number 107.

(14) As an illustration of Saturn's ring rotating round the planet; see James Elliott, 'A description of certain mechanical illustrations of the planetary motions, accompanied by theoretical investigations relating to them, and in particular, a new explanation of the stability of equilibrium of Saturn's rings', *Transactions of the Royal Scottish Society of Arts*, **4** (1856): 318–44, esp. 335–44 (= *Edinburgh New Philosophical Journal*, **1** (1855): 310–35, esp. 327–35). See Number 116 for Maxwell's discussion of Elliott's paper with reference to the illustration of precession by his dynamical top.

on the plane of the ring having to turn when its centre moves, owing to there being a fixed point connected with the ring, out of its own plane.

Next Monday my address will be at

$$\left\{\begin{array}{l} \text{John MacCunn Esq}^{\text{re}[15]} \\ \qquad \text{Ardhallow} \\ \qquad \text{Dunoon.} \end{array}\right.$$

After Monday 7th Sept. I shall be at Roshven, Strontian,[16] for a while after which I hope to be here again with Mrs Wedderburn for the rest of my time.

Yours truly

J. C. MAXWELL

---

(15) The brother-in-law of Katherine Mary Dewar, the future Mrs Maxwell; see Number 146 note (7).

(16) With his cousin Jemima (Wedderburn) Blackburn; see Number 45 note (10).

FROM A LETTER TO LEWIS CAMPBELL

28 AUGUST 1857

From Campbell and Garnett, *Life of Maxwell*[1]

Glenlair
28 August 1857

I have been battering away at Saturn, returning to the charge every now and then. I have effected several breaches in the solid ring, and now I am splash into the fluid one, amid a clash of symbols truly astounding. When I reappear it will be in the dusky ring, which is something like the state of the air supposing the siege of Sebastopol[2] conducted from a forest of guns 100 miles one way, and 30,000 miles the other, and the shot never to stop, but go spinning away round a circle, radius 170,000 miles.

---

(1) *Life of Maxwell*: 278.
(2) During the Crimean War, in 1854–5; for the 'fall of Sebastopol' see *The Times* (11 September 1855).

FROM A LETTER TO LEWIS CAMPBELL

4 SEPTEMBER 1857

From Campbell and Garnett, *Life of Maxwell*[1]

Glenlair
4 Sept. 1857

The road along Loch Eck is the most glorious for shape and colour of hills and rocks that I have seen anywhere, specially on a fine calm day, with clouds as well as sun, and with large patches of withered bracken mixed with green on the less steep parts of the hills. Then the crushing and doubling up of the strata, and the slicing and cracking of the already doubled up strata, quite without respect to previous torment, gives a notion of active force, as well as passive, even to ungeological minds. . . . I shall continue my road with my aunt to wait upon the faithful Tobs, and realise Saturn's Rings,[2] and probably feed a few natives of the valley with the produce of its soil. I was writing great screeds of letters to Professor Thomson about those Rings, and lo! he was a-laying of the telegraph which was to go to America, and bringing his obtrusive science to bear upon the engineers, so that they broke the cable with not following (it appears) his advice.[3]

---

(1) *Life of Maxwell*: 278–80; abridged.
(2) See Number 131.
(3) See Number 134.

LETTER TO RICHARD BUCKLEY LITCHFIELD

15 OCTOBER 1857

From the original in the library of Trinity College, Cambridge[1]

(What is M^{rs} P^s address in Cheltenham?)[2]

Glenlair
15th Oct 1857

Dear Litchfield

I enclose what Sale[3] sent me.[4] We are all losing our friends. My Aunt Mrs Wedderburn who is with me has for the last 3 months been gradually learning the details of the murders of her cousin John Wedderburn and his wife & child. Even now there are manifestly contradictory reports by Englishmen of what they have seen or done. Her own son John & wife are at Moulton where they have had to disarm the troops and to despatch 9 of those who remained with them & professed friendship but were traitors.

She is 68, and after four years of constant pain of body and torture of mind she has completely recovered health and strength and comfort. It is fearful to think what may happen to one through some bodily infirmity, so that while appearing not unwell there may be a dull nameless pressure on the mind relieved only by bodily pains or unreasonable mental distress of a more acute kind, but all this is away since this time last year and now she can bear anything and talk about anything, even the darkness that is past.

I was glad Sale sent me the letter. Remember that besides all the danger and distance from friends there was the fever, of which he had already long experience and which in such time he knew to be as inconvenient to his friends as himself. But it is no use saying this and that. Some men redeem their characters by their deeds and we praise them. Those that merely show their character by their deeds should be remembered, not praised and a complete true man will live longest in the memory and I cannot but think will be less changed in reality than one who has doubtfully struggled with duplicity in his constitution and has walked with hesitation though along a good path. I know

---

(1) Trinity College, Cambridge, Add. MS Letters c.1^{88}. Published (in extract) in *Life of Maxwell*: 282–3; and in S. G. Brush, C. W. F. Everitt and E. Garber, *Maxwell on Saturn's Rings* (Cambridge, Mass./London, 1983): 50–1.

(2) R. H. Pomeroy's mother; see Number 42 note (3).   (3) Number 68 note (4).

(4) Presumably the letter from Pomeroy to his mother of 28 July 1857 (from India), which Maxwell transcribed (ULC Add. MSS 7655, II/9; *Life of Maxwell*: 285–7). Pomeroy died during the Indian Mutiny; see Maxwell's letters to Litchfield of 23 September 1857 (Trinity College, Add. MS Letters c.1^{87}; *Life of Maxwell*: 280–2) and 25 October 1857 (*Life of Maxwell*: 284–5); and his letter to Jane Cay of 28 September 1857 (*Life of Maxwell*: 282).

that both do deny and renounce themselves in favour of duty & truth as they come to see them, and as they come to see how goodness having the knowledge of evil has passed through sorrow to the highest state of all, they accept it as a token that they have found their true head and leader, and so with their eyes on him they complete the process called the knowledge of Good and Evil which they commenced so early and so ignorantly.

Now what the 'completion of the process' is, I cannot conceive, but I feel the difference of Good & Evil in some degree and I can conceive the perception of that difference to grow by contemplating the Good till the confusion of the two becomes an impossibility. Then comes the mystery. I have memory and a history or I am nothing at all. That memory and history contains evil which I renounce and must still maintain that *I* was evil. But it contains the image of absolute good and the fight for it and the consciousness that all this is right.

So there the matter lies, a problem certain of solution.

I am & have been very busy here. I leave this on 19th. Address Lauriston Lodge, Edinburgh. After 26th – Marischal College Aberdeen.

I am grinding hard at Saturn and have picked many holes in him and am fitting him up new and true. I am sure of most of him now and have got over some stumbling blocks wh: kept me niggling at calculations 2 years.[5]

I am to have some artisans as weekly students this winter.

I was with Wigram M. P. Camb.[6] lately and discussed University matters a little. His wisdom was not equalled by his information but the sense was good.

It is awfully late and Toby[7] always makes a row when he goes to bed so I must shut you up as I have more to write.

I am so glad you are glad after you have seen Farrar[8] that I will not even demand a local habitation or a name for the twin soul.

But Lewis Campbell (a much older friend) I believe to be nearly as glorious since January and my people had an actual visé the old ones and the youngest ones conjointly & severally & they are all gune[9] grad votes down to a young lady of 6 years who has had letters from several young men already & ought to know all about it.

We are still in our domestic troubles.[10] The sister of the little girl that died is just breathing for the last few days and there are bad symptoms but not hopeless. A very little one is very ill and several of other of our people have been ill but in a slighter way than in that family. So we wait.

Yours truly
J. C. MAXWELL

---

(5) See Numbers 126 and 128.

(6) Loftus Tottenham Wigram, Trinity 1821, MP for Cambridge University 1850–59 (Venn).

(7) A dog.   (8) Number 56 note (11).   (9) Possibly: γῠνή (wife).

(10) Described in Maxwell's letter to Litchfield of 23 September 1857 (see note (4)).

132

INTRODUCTORY LECTURE TO THIRD-YEAR STUDENTS AT MARISCHAL COLLEGE, ABERDEEN[1]

EARLY NOVEMBER 1857[2]

From the original in the University Library, Cambridge[3]

Gentlemen[4]

At the close of last session I addressed a few words to those of you who were then present in which I attempted to point out the general character of the studies which are to occupy you during the third session of your academical course. I now wish to address you on the same subject in a more practical manner, as we are now to contemplate these studies from a nearer point of view, intending as we do to present them vigorously throughout the session. According to the system of study pursued in this University each year is devoted to the special cultivation of some one or more of those faculties which are given us to win our livelihood to help our fellow men to do our duty, and to worship God, in short to equip us completely for the business of life. In the first year you study languages, as the means whereby all truth must be expressed before it can pass from one man to another. In the second year your attention is directed to the various natural objects in the midst of which we are placed. You are taught to distinguish them, to name them, and to discern the resemblances and differences between them, and you are also led to see the beauty and usefulness of the various parts of creation. But while you cultivate the faculty of observation you have also been occupied with mathematical studies in which geometrical diagrams and algebraic calculations are employed not so much to enable you to answer hard questions as to cultivate the inward eye of reason and to point out the existence of truths resting on a foundation higher than that of the senses.

You are now entering the third session of study in which your senses are still to be cultivated by the habit of accurate observation of facts while you are to extend your knowledge of the truths of Reason far beyond the limits of algebra and geometry to the whole science of force & motion where cause and effect first become the subjects of exact reasoning.

You now study phenomena not merely to recognize them when you meet with them or to name them or classify them. You now study every thing in

---

(1) The manuscript is endorsed 'Introductory 1857'.
(2) The session at Marischal College commenced in November; see Number 105 note (1).
(3) ULC Add. MSS 7655, V, h/2.
(4) Compare the inaugural lecture delivered in November 1856 (Number 105).

relation to the cause that produced it and you are not to rest satisfied till you can not only explain the cause of an event but can even predict exactly what would happen under circumstances which have never yet occurred. It is by a familiarity with such reasonings and predictions that you begin to understand the position of man as the appointed lord over the works of Creation and to comprehend the fundamental principles on which his dominion depends which are these – To know, to submit to, and to fulfil, the laws which the Author of the Universe has appointed. Attend to these laws and keep them, you succeed, break them, you fail and can do nothing.

It is by searching out and applying these laws to our human wants that we have been enabled to accomplish all the boasts of the civilisation of the present day, and it is by steady pursuit of the same course that we expect our posterity to put those boasts to shame with far greater achievements, and to stop at no limit, till the whole earth be subdued, and every mans reason enlightened with physical truth.

Now gentlemen you must remember that although this department of study is that which properly belongs to us this year and though the certainty, accuracy and universality of physical truth affords matter worthy of your utmost efforts, your education is not complete without a systematic study of other truths of reason with which we have little or nothing to do this year. After the exercise of your minds in the three different directions of which I have spoken you are prepared to make your minds themselves the object of your study and to trace the steps by which you were enabled to acquire your information and to establish your reasonings. You will then be introduced to the discussion of the laws of thought which is Logic, to the first principles of human knowledge or Metaphysics and to the consideration of human action regarded as right or wrong, which is Moral Philosophy.

I have said thus much as to the studies of the fourth year in order to show you that the studies you pursue are arranged in a regularly ascending scale. You learn first how to express yourselves and that you may learn it more thoroughly you are made to transform expressions from one language to another. You are next taught to observe with your senses and to know the names of things. You also cultivate your reason commencing with the most abstract truths and gradually comprehending a large number of the principles of the material world. You then turn your thoughts to the nature of Man and are introduced to the conception of right and wrong as a philosophical truth.

We must now return to the study of Natural Philosophy which is that which is to occupy us in future. There are three distinct parts or branches of every system of education whatever be the subject taught. The first is discipline or the regulation of our faculties, a process in which the student and teacher must combine to repress everything which disturbs the mind or leads it away from the course it is desired to follow. This part of education includes the eradication or weeding out of all errors and prejudices and the cultivation of habits of strict accuracy of thought as well as of calculation. The proper effect of the discipline of study is to produce that well ordered steady frame of mind and manners which belongs to educated men and by which they are distinguished from the undisciplined and ill regulated in mind whatever may be their rank in life or the amount of facts which they may know.

The second part of education is the cultivation and nursing into vigour of those faculties which discipline has regulated. In this process far more is done by the student than by the teacher for the only method ever discovered for gaining any kind of strength or vigour is exercise. All the passive merits of quietness and attention will do nothing towards making you master of your subject unless you summon up courage and attack the works of activity which are proposed to you. By overcoming difficulties and doing work you acquire a feeling of confidence not only in the performance of work of a similar kind in future, but even in matters of a kind different from those to which you have been accustomed, so that the man, who, in his youth, has trained himself to overcome difficulties and always to do the work that lies before him becomes in after life an able man, or in other words a business-like man a character highly deserving of praise and sure of getting it.

The third part of education is the acquisition of truth. Here the teacher comes more prominently forward as the student looks to him for a supply of the truths he seeks. But even here it is doubtful whether the services of the teacher are more important than the efforts of the student, for although truth be precious the search after truth and success in that search is more precious still, and it may even be said that truth acquired without search is not truth at all, for though it may be true it is believed as an opinion merely and not as a firm conviction. This is more especially the case in Natural Philosophy where the educational importance of the doctrines taught arises not so much from their admitted truth as from their being arrived at by a strict process of reasoning so that though the teacher may act as guide it is the student who conquers the truth for himself.

This third part of education has also its special use in the formation of a manly character, for the man that has been accustomed to contemplate truth of any kind, physical or moral, as truth and not as mere opinion, has already approached to that noble frame of mind which is averse to quarrel about accidental differences in unessential matters but firm and unyielding whenever truth and duty are concerned.

I have endeavoured to set before you a few of the advantages you may gain for yourselves by a systematic course of liberal education. You will have many opportunities of judging of the material or worldly advantages which have accrued to men of science and to the world by means of physical study as we proceed with our course but at present I wish to direct your attention rather to the outside view of the subject and to point out the part which the physical sciences enact in the history of the world. And first I would remind you of a set of men of whom we know very little – The most ancient philosophers of Ionia and the Greek states of Italy. From them is derived that long continued race of speculators who kept alive the zeal for wisdom among the ablest young men of Greece till Plato and Aristotle became the chiefs of all ancient philosophy.

All that we know of these men is that they attempted to explain the system of Nature. Some speculated on the material of which all things were made, others on their forms, some puzzled themselves with the conception of motion and the wisest, seeing the harmony of the universe and the arithmetical accuracy of all natural quantities likened the world to music and spoke of Intelligence as Number. But however far sighted and ingenious the ancient philosophers might be, their labours can teach us only the workings of the human intellect craving for the materials of science.

We study their writings as the memorials of the most subtile thinkers that the world has seen, but for all our actual knowledge we must appeal to Nature herself. The ancient philosophy after producing the science of geometry recoiled from the study of Nature and finished its course by exploring all that could then be discovered of the constitution of Man and the nature of Virtue. In every one of these enquiries it reached a limit beyond which mere speculation has never passed and human science went no further till after centuries of ingenious attempts to fabricate a philosophy out of nothing, patient experiment effected what unaided reason could never do, and many secrets were unveiled and many rebellious elements of the world subdued till Reason led by Experiment rose to a higher level than Reason disdaining Experiment could ever reach.

But there has always existed and there still exists a kind of opposition Philosophy not so much resting on experience as wilfully opposed to the less obvious dictates of Reason.

Among the Greeks there was a school of acute and eloquent men who maintained, and with justice, that all our conceptions of magnitude or intensity are merely relative to us, that they depend on that to which we have been accustomed.

A locust, for instance, is a large insect compared with bees and ants but a small animal compared with camels and horses. What would be considered a cold wind in summer would be felt warm in winter, and what is delightful to a healthy man is disgusting to one who is ill.

By reasoning from examples of this kind they endeavoured to shew that no statement could be true or false except in so far as it agreed or disagreed with the feelings of the person making that statement. Truth they said is that which each man knoweth or believeth to be true. So that if a man sincerely believes a statement to be true which to another man appears not to be true, then it is true for the first man and untrue for the second.

I cannot either set forth these arguments in their complete form or give you the reasons which have been brought forward on the other side. It is true that all our feelings have reference to our own present state and, so far, we measure the Universe by ourselves and our own standards. But the modern philosophy, confessing the impossibility of pronouncing anything absolutely great or small confines itself entirely to expressing the relations between the things themselves.[5] Though we cannot say whether twenty yards is a great or a small distance we know that forty yards is twice that distance and so we compare things not with our own feelings but with each other. In the same way we do not trust to our own sensations to determine whether anything is hot or cold, but we use a thermometer, a mere machine, which always gives the same indications under the same circumstances, and is not subject to the causes which modify human determinations on such matters.

Man is indeed the measurer of the Universe but he does not use himself as his standard of measure nor his own variable feelings as an indication of real existence. We have a certain knowledge of Nature because we have used material measures to measure material things, and one of the principal duties before me is to show you how all distances in earth and heaven may be compared with the length of a foot rule and all forces with the weight of a pound, how colour heat and electricity may be measured and calculated and how errors of observation depending on unknown causes, and existing only as defects, are expressible by laws of their own.

This habit of regarding all material things as capable of exact measurement will grow stronger as our acquaintance with Nature advances till we fully recognise that every law of Nature must express some relation before the magnitudes of measurable quantities, so that number measure and weight are the elements of all knowledge of the material world.

In lecturing and examining on Natural Philosophy I propose continually to set before you this exactness in measureable quantity which belongs to our science, and therefore I shall devote much time to the illustration of the science of measurement, and ask many questions about it. In most of our experiments there will be some result which ought to agree with calculation. In all such cases we must make the comparison. Experiments performed before a class cannot be so accurately prepared or observed as those which are made in

---

(5) Compare Maxwell's discussion of scientific knowledge as relational in his 1856 essay 'Are there Real Analogies in Nature?' (Number 88, esp. note (8)).

private, but it is better that you should sometimes see the failure of an experiment than that you should think that rough experiments can give accurate results, or that scientific investigations are conducted in the rude way adapted for exhibition to a class.

If by following out this course I can impress on your minds the conviction that we live in a world where exact principles rule supreme, one great object of your coming here will be gained, and having once seen that there is such a thing as truth in material Nature you will be the more ready to seek for it in the nature of Man and not to rest satisfied with mere opinion in any department of knowledge.

During the session I hope to give you some idea of the doctrines of Statics and Dynamics, of the laws of fluids, of the properties of Heat and of the elementary parts of Optics. We shall find these subjects amply sufficient for our study during the session.[6] The higher parts of Astronomy, the whole of Electricity and Magnetism and the higher parts of the theory of sound and light all require a familiarity with physical reasoning which is hardly to be looked for in the first year, and I have thought it advisable not to attempt these at first.

There will be written examinations once a week on the subjects of previous lectures. There will also be several general examinations in writing on the subjects of our study. The weekly examinations will differ from the general ones in requiring rather attention to the lectures than mathematical skill, and cultivating brevity of expression rather than fullness of description. The general prizes will depend on both sets of exam[ns]. Besides these exercises there will be others to be done at home. These will be partly mathematical, and partly descriptive. There will be a special prize for proficiency in these.

---

(6) This outline of proposed topics accords with his practice; see his notebooks on 'Examinations in Natural Philosophy' and 'Exercises in Natural Philosophy Course' (ULC Add. MSS 7655, V, k/3, 4) for the sessions 1857–8, 1858–9, and 1859–60. He had covered the same topics in his first session; see his letter to Forbes of 30 March 1857 (Number 115).

133

LETTER TO MICHAEL FARADAY

9 NOVEMBER 1857

From the original in the library of the Institution of Electrical Engineers, London[1]

129 Union Street
Aberdeen
9th Nov[r] 1857

Dear Sir

I have to acknowledge receipt[2] of your papers on the Relations of Gold &c to Light[3] and on the Conservation of Force.[4] Last spring you were so kind as to send me a copy of the latter paper and to ask what I thought of it.[5] That question silenced me at that time, but I have since heard and read various opinions on the subject which render it both easy and right for me to say what I think.[6] And first I pass over some who have never understood the known doctrine of conservation of force and who suppose it to have something to do

---

(1) Institution of Electrical Engineers, London, Special Collection MSS 2. Previously published in *Life of Maxwell* (2nd edn): 202–4.

(2) See a letter from Faraday to Maxwell of 7 November 1857 (ULC Add. MSS 7655, II/11; in *Life of Maxwell*: 288).

(3) Michael Faraday, 'Experimental relations of gold (and other metals) to light', *Phil. Trans.*, **147** (1857): 145–82 (= Michael Faraday, *Experimental Researches in Chemistry and Physics* (London, 1859): 391–443).

(4) Michael Faraday, 'On the conservation of force', *Proceedings of the Royal Institution of Great Britain*, **2**(1857): 352–65; and in *Phil. Mag.*, ser. 4, **13** (1857): 225–39 (= *Chemistry and Physics*: 443–60).

(5) In a letter to Maxwell of 25 March 1857 (ULC Add. MSS 7655, II/6; in *Life of Maxwell*: 519–20; and in *The Selected Correspondence of Michael Faraday*, ed. L. P. Williams, 2 vols. (Cambridge, 1971), **2**: 864). Faraday acknowledged receipt of Maxwell's paper 'On Faraday's lines of force', *Trans. Camb. Phil. Soc.*, **10** (1856): 27–83 (= *Scientific Papers*, **1**: 155–229). He remarked that 'I was at first almost frightened when I saw such mathematical force made to bear upon the subject and then wondered to see that the subject stood it so well. I send by this post another paper ['On the conservation of force'] to you. I wonder what you say to it. I hope however that bold as the thoughts may be you may perhaps find reason to bear with them'. Faraday went on to say that he hoped to undertake 'experiments on the time of magnetic action', speculating that the 'time must probably be short as the time of light'; see Maxwell's letter to Faraday of 19 October 1861 (Number 187, esp. note (15)).

(6) Faraday's 'On the conservation of force' had been critically reviewed in *The Athenæum* (28 March 1857): 397–9. Writing to Maxwell on 31 March 1857 J. D. Forbes asked: 'Have you observed in that same flippant paper for last Saturday an attack upon Faraday (as it seems to me) of a most presumptuous and ignorant kind? Though by no means as yet a convert to the views which Faraday maintains, yet I have so far a general appreciation of them as to believe that this conceited mathematician (some fifteenth Cambridge wrangler, I guess) is ignorant altogether of what Faraday wishes to prove' (*Life of Maxwell*: 266–7).

with the equality of action & reaction. Now first I am sorry that we do not keep our words for distinct things more distinct and speak of the 'Conservation of Work or of Energy' as applied to the relations between the amount of 'vis viva' and of 'tension' in the world;[7] and of the 'Duality of Force' as referring to the equality of action and reaction.

Energy is the power a thing has of doing work arising either from its own motion or from the 'tension' subsisting between it and other things.[8]

Force is the tendency of a body to pass from one place to another and depends upon the amount of change of 'tension' which that passage would produce.[9]

---

(7) The terms '*vis viva*' and 'tension' were used by John Tyndall (as renditions of Helmholtz's terms *lebendige Kraft* and *Spannkraft*), in his trans. of Hermann Helmholtz, *Über die Erhaltung der Kraft* (Berlin, 1847) (= *Wissenschaftliche Abhandlungen*, **1**: 12–68), 'On the conservation of force', in *Scientific Memoirs, Natural Philosophy*, ed. J. Tyndall and W. Francis (London, 1853): 114–62, esp. 122. The expression 'the law of the conservation of energy' was first used by W. J. M. Rankine, 'On the general law of the transformation of energy', *Phil. Mag.*, ser. 4, **5** (1853): 106–17, esp. 106. On concepts of 'work' and 'energy' see Rankine, 'Outlines of the science of energetics', *Edinburgh New Philosophical Journal*, **2** (1855): 120–41, esp. 129; 'The term "*energy*" comprehends every state of a substance which constitutes a capacity for performing work'. On Helmholtz's paper see P. M. Heimann, 'Helmholtz and Kant: the metaphysical foundations of *Über die Erhaltung der Kraft*', *Studies in History and Philosophy of Science*, **5** (1974): 206–38, especially on the relation between Helmholtz's *lebendige Kraft* and *Spannkraft* and 'kinetic' and 'potential' energy.

(8) See William Thomson, 'An account of Carnot's theory of the motive power of heat; with numerical results, deduced from Regnault's experiments on steam', *Trans. Roy. Soc. Edinb.*, **16** (1849): 541–74, esp. 545n; 'Nothing can be lost in the operations of nature; no energy can be destroyed' (= *Math. & Phys. Papers*, **1**: 118n). See also Rankine, 'On the general law of the transformation of energy': 106; 'In this investigation the term *energy* is used to comprehend every affection of substances which constitutes or is commensurable with a power of producing change in opposition to resistance, and includes ordinary motion and mechanical power, chemical action, heat, light, electricity, magnetism, and all other powers, known or unknown, which are convertible or commensurable with these'.

(9) In his reply to Maxwell's letter, dated 13 November 1857 (ULC Add. MSS 7655, II/14; in *Life of Maxwell*: 288–90; in H. Bence Jones, *The Life and Letters of Faraday*, 2 vols. (London, 1870), **2**: 390–3; and in *Correspondence of Faraday*, **2**: 884–5), Faraday stated that: 'I perceive that I do not use the word "force" as you define it "the tendency of a body to pass from one place to another". What I mean by the word is the *source* or *sources* of all possible actions of the particles or materials of the universe: these being often called the *powers* of nature, when spoken of in respect of the different manners in which their effects are shewn'. In an addendum (dated June 1858) to his paper 'On the conservation of force', *Phil. Mag.*, ser. 4, **17** (1859): 166–9 (= *Chemistry and Physics*: 460–3), Faraday states that some comments on his original paper – in *Phil. Mag.*, ser. 4, **13** (1857): 225–39 – had led him to 'think that I have not stated the matter with sufficient precision.... What I mean by the word "force", is the *cause* of a physical action; the source or sources of all possible changes amongst the particles or materials of the universe'. In response to Faraday's addendum Rankine published two notes 'On the conservation of energy', *Phil. Mag.*, ser. 4, **17** (1859): 250–3, 347–8, esp. 251, emphasising that: 'Inasmuch as the word "force" has for a long time been used to denote a *tendency* of a pair of bodies to change of relative motion, a kind of magnitude to which

Now as far as I know you are the first person in whom the idea of bodies acting at a distance by throwing the surrounding medium into a state of constraint has arisen, as a principle to be actually believed in. We have had streams of hooks and eyes flying around magnets, and even pictures of them so beset, but nothing is clearer than your description of all sources of force keeping up a state of energy in all that surrounds them, which state by its increase or diminution measures the work done by any change in the system. You seem to see the lines of force curving round obstacles and driving plump at conductors and swerving towards certain directions in crystals, and carrying with them everywhere the same amount of attractive power spread wider or denser as the lines widen or contract.[10]

You have also seen that the great mystery is, not how like bodies repel and unlike attract but how like bodies attract (by gravit[at]ion). But if you can get over that difficulty, either by making gravity the residual of the two electricities or by simply admitting it, then your lines of force can 'weave a web across the sky' and lead the stars in their courses without any necessarily immediate connection with the objects of their attraction.

The lines of Force from the Sun spread out from him and when they come near a planet *curve out from it* so that every planet diverts a number depending on its mass from their course and substituting a system of its own so as to become something like a comet, *if lines of force were visible.*

The lines of the planet are separated from those of the Sun by the dotted lines. Now conceive every one of these lines (which never interfere but proceed from sun & planet to infinity) to have a *pushing* force, instead of a *pulling* one and then sun and planet will be pushed together with a force which comes out as it ought proportional to the product of the masses & the inverse square of the distance.

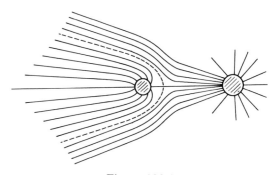

Figure 133,1

The difference between this case and that of the dipolar forces is, that instead of each body catching the lines of force from the rest all the lines keep as clear of other bodies as they can and go off to the infinite sphere against which I have supposed them to push.

no law of conservation applies, the use of the phrase "conservation of force" gives rise to misapprehensions and groundless disputes, by implying a different meaning of the word "force".' See P. M. Harman, *Energy, Force, and Matter. The Conceptual Development of Nineteenth-century Physics* (Cambridge, 1982): 33–44, 58–64.   (10) See Number 84.

Here then we have conservation of energy (actual & potential)[11] as every student of dynamics learns, and besides this we have conservation of 'lines of force' as to their *number* and total strength for *every* body always sends out a number proportional to its own mass, and the pushing effect of each is the same.

All that is altered when bodies approach is the *direction* in which these lines push. When the bodies are distant the distribution of lines near each is little disturbed. When they approach, the lines march round from between them, and come to push behind each so that their resultant action is to bring the bodies together, with a *resultant* force increasing as they approach.

Now the mode of looking at Nature which belongs to those who can see the lines of force deals very little with 'resultant forces' but with a network of lines of action of which these are the final results, so that I for my part cannot realise your dissatisfaction with the law of gravitation provided you conceive it according to your own principles.[12] It may seem very different when stated by the believers in 'forces at a distance', but there can be only differences in form and conception not in quantity or mechanical effect between them and those who trace force by its lines. But when we face the great questions about gravitation Does it require time? Is it polar to the 'outside of the universe' or

---

(11) See Rankine, 'On the general law of the transformation of energy': 106; '*Actual energy* is a measurable, transferable, and transformable affection of a substance, the presence of which causes the substance to change its state in one or more respects; by the occurrence of which changes actual energy diappears, and is replaced by *Potential energy*, which is measured by the amount of a change in the condition of a substance, and that of the tendency or force whereby that change is produced (or, what is the same thing, of the resistance overcome in producing it), taken jointly.' On the relation between Rankine's 'actual' and 'potential' energy and 'kinetic' and 'potential' energy – the term 'kinetic energy' being introduced by W. Thomson and P. G. Tait, *A Treatise on Natural Philosophy* (Oxford, 1867): 195 – see K. Hutchison, 'W. J. M. Rankine and the rise of thermodynamics', *British Journal for the History of Science*, **14** (1981): 1–26.

(12) In his letter of 13 November 1857 (see note (9)) Faraday remarked in reply that 'your words…give me great comfort'. In 'On the conservation of force', *Phil. Mag.*, ser. 4, **13** (1857): 228–9 (= *Chemistry and Physics*: 446–7) Faraday had discussed gravity in terms of the '*creation* of power' and the '*annihilation* of force' when bodies approached and receded, concluding that 'the idea of gravity appears to me to ignore entirely the principle of the conservation of force'. See note (9). Compare Michael Faraday, 'On some points of magnetic philosophy', *Proceedings of the Royal Institution of Great Britain*, **2** (1855): 6–13, esp. 12–13 (= *Electricity*, **3**: 566–74); 'The power is always existing around the sun and through infinite space, whether secondary bodies be there to be acted on by gravitation or not…this case of a constant necessary condition to action in space…I can conceive, consistently, as I think, with the conservation of force'. In his letter of 13 November Faraday states: 'I have nothing to say against the law of action of gravity. It is against the law which measures its total strength as an inherent force that I venture to oppose my opinion. … The idea that we may possibly have to connect *repulsion* with the lines of gravitation force (which is going far beyond anything my mind would venture on at present except in private cogitation) shows how far we may have to depart from the view I oppose.'

to anything? Has it any reference to electricity? or does it stand on the very foundation of matter – mass or inertia? then we feel the need of tests, whether they be comets or nebulae or laboratory experiments or bold questions as to the truth of received opinions.[13]

I have now merely tried to show you why I do not think gravitation a dangerous subject to apply your methods to, and that it may be possible to throw light on it also by the *embodiment* of the same ideas which are expressed *mathematically* in the functions of Laplace[14] and of Sir W. R. Hamilton[15] in Planetary Theory.

But there are questions relating to the connexion between magneto electricity and certain mechanical effects which seem to me opening up quite a new road to the establishment of principles in electricity and a possible confirmation of the physical nature of magnetic lines of force. Professor W. Thomson seems to have some new lights on this subject.[16]

Yours sincerely

JAMES CLERK MAXWELL

Prof^r Faraday
&c &c

(13) In his letter of 13 November 1857 (see note (9)) Faraday observes: 'When a mathematician engaged in investigating physical actions and results has arrived at his conclusions, may they not be expressed in common language as fully, clearly, and definitely as in mathematical formulae?...I have always found that you could convey to me a perfectly clear idea of your conclusions...so clear in character that I can think and work from them.'

(14) Laplace's equations for the potential; P. S. de Laplace, *Traité de Mécanique Céleste*, 5 vols. (Paris, An VII [1799]–1825), **1**: 136–7. See Theorems I and II in Maxwell, 'On Faraday's lines of force': 57 (= *Scientific Papers*, **1**: 195).

(15) Hamilton's 'characteristic function' of motion of a dynamical system: the application of the varying action principle to the determination of the orbits and perturbations of the planets; see W. R. Hamilton, 'On a general method in dynamics, by which the study of the motions of all free systems of attracting or repelling points is reduced to the search and differentiation of one central relation or characteristic function', *Phil. Trans.*, **124** (1834): 247–308; and Hamilton, 'Second essay on a general method in dynamics', *ibid.*, **125** (1835): 95–144.

(16) William Thomson 'Dynamical illustrations of the magnetic and heliocoidal rotatory effects of transparent bodies on polarized light', *Proc. Roy. Soc.*, **8** (1856): 150–8 (= *Phil. Mag.*, ser. 4, **13** (1857): 198–204). See Number 137 note (5). Thomson argued that the phenomenon of the rotation of the plane of polarisation of linearly polarised light by a magnetic field, discovered by Faraday, could be explained by a vortical theory of magnetism; see Michael Faraday, 'On the magnetization of light and the illumination of magnetic lines of force', *Phil. Trans.*, **136** (1846): 1–20 (= *Electricity*, **3**: 1–26). See also Maxwell's letters to Monro of 20 May 1857 (Number 118), and to Thomson of 24 November 1857, 30 January 1858, and 10 December 1861 (Numbers 137, 143 and 189).

134

LETTER TO WILLIAM THOMSON

14 NOVEMBER 1857

From the original in the University Library, Glasgow[1]

129 Union Street,
Aberdeen
14 Nov 1857

Dear Thomson,

I suppose, after a busy vacation, you are very busy in the beginning of the session. Nevertheless you must allow me to report progress on Saturns rings, so that if you have any remarks to make on my results, I may have the benefit of them. I have already reported on the rigid ring. I got your investigation of it,[2] but I have been entirely occupied with the non-solid ring since so that I have only *read* it, not *worked* it yet. In the first place, then, I have abolished my off hand theory of the attractions of a thin fluid filament affected by waves, *long* compared with the diameter of the filament, and this because the *short* waves are the only dangerous ones.[3]

I have therefore *begun* with the theory of a ring consisting of μ satellites affected with undulations of the form, $A \cos m\theta$ normal, $B \sin m\theta$ tangential, or rather putting ρ & σ for radial & tangential displacements & s for θ[4]

$$\rho = A \cos(ms + nt + \alpha)$$
$$\sigma = B \sin(ms + nt + \alpha)$$

I get a biquadratic for n whose four values are possible, provided Saturn be large enough.[5] The force on which the value of n chiefly depends is the tangential force arising from the attraction of the satellites towards the crowded parts of the ring. This must never exceed $\frac{1}{14}\omega^2$.[6] It is greatest when successive

---

(1) Glasgow University Library, Kelvin Papers, M 9. Previously published in S. G. Brush, C. W. F. Everitt, and E. Garber, *Maxwell on Saturn's Rings* (Cambridge, Mass./London, 1983): 54–7.

(2) See William Thomson, 'On the stability of the steady motion of a rigid body about a fixed centre of force', printed as an Appendix to Maxwell's memoir *On the Stability of the Motion of Saturn's Rings* (Cambridge, 1859): 69–71 (= *Scientific Papers*, **1**: 374–6).

(3) Compare Part II, Proposition X of 'On the Stability of the Motion of Saturn's Rings' (Number 107); and his letter to Thomson of 24 August 1857 (Number 128); and see his letter to Challis of 24 November 1857 (Number 138).

(4) See *Saturn's Rings*: 18–19 (= *Scientific Papers*, **1**: 311–12); s and θ determine the angular position of an element of the ring. Compare Part II, Prop. VIII of Number 107.

(5) Compare *Saturn's Rings*: 18–23 (= *Scientific Papers*, **1**: 311–17); n is the angular velocity of a system of m waves in the ring.

(6) See *Saturn's Rings*: 34–5 (= *Scientific Papers*, **1**: 330–32); ω is the angular velocity of the ring.

particles are in opposite phases, that is, when $m = \frac{1}{2}\mu$. In order that there may be stability in this case $S > .4352\mu^2 R$ where S is Saturn R the mass of ring, μ the number of satellites. If $\mu = 100$, $S > 4352R$.[7]

If the mass of the ring were too great, then it would run into lumps of satellites, till the number of satellites were reduced to what the Planet could govern.

I have traced the relative & absolute motion of each satellite the instantaneous form of the ring, and the mode of propagation of the ring, and the mode of propagation of the waves both in the case of stability[8] and in the cases where the ring must break up.[9] When the tangential force is *too great* the effect is that all the waves increase in amplitude till the ring gives way, (Case I) but when the tangential force is negative, that is when it tends *from* crowded portions there is *no propagation* but the irregularities increase without limit in the parts of the ring where they originated (Case II). This effect is very remarkable as showing the destructive effect of an apparently conservative force.[10]

I then apply my method to a liquid ring very much flattened in section. I find that the tangential force is positive for long waves and has a maximum when the wave length is 10.353 times the thickness. It is 0 when the wave length is 5.471 times the thickness and is negative for shorter waves.[11]

That Case I of instability may be avoided the density of Saturn must be at least 42.5 times that of the ring.[12] Now that the outer and inner parts should have the same angular velocity Laplace shows that Saturn must not be more than 1.3 times as dense as the ring.[13] Besides, the *very* short waves having a negative tangential effect (from thick to thin) will inevitably produce the case No 2 of instability so that the liquid continuous ring is doomed.[14]

If a ring were made of a cloud of aerolites the condition is that the density of Saturn should be more than 334 times that of the ring.[15]

I am just entering on the case of many rings of satellites mutually disturbing

---

(7) See *Saturn's Rings*: 24–5 (= *Scientific Papers*, **1**: 318–19); and Number 138.

(8) See *Saturn's Rings*: 26–30 (= *Scientific Papers*, **1**: 321–5); and Number 138.

(9) See *Saturn's Rings*: 34–7 (= *Scientific Papers*, **1**: 330–5).

(10) Compare *Saturn's Rings*: 36 (= *Scientific Papers*, **1**: 333); and Part II, Proposition X of Number 107.

(11) See *Saturn's Rings*: 40–4 (= *Scientific Papers*, **1**: 338–44).

(12) See *Saturn's Rings*: 45 (= *Scientific Papers*, **1**: 344–5).

(13) P. S. de Laplace, *Traité de Mécanique Céleste*, 5 vols. (Paris, An VII [1799]–1825), **2**: 162; see Number 107 note (10).

(14) See *Saturn's Rings*: 45 (= *Scientific Papers*, **1**: 344–5).

(15) Compare *Saturn's Rings*: 38–40 (= *Scientific Papers*, **1**: 336–8); and his letter to Challis of 24 November 1857 (Number 138). On the 'hailstorm' ring see also Numbers 128 and 129.

one another which I think I can attack.[16] The general case of a fortuitous concourse of atoms each having its own orbit & excentricity is a subject above my powers at present, but if you can give me any hint as to the point of attack I will go at it.[17]

Mean while I am busy putting what I have into intelligible form by means of a good deal of verbal explanation and tracing out of the mechanical phenomena.

I also wish to hear from you what you think a good galvanic battery for class purposes [is] and in particular about the expense, convenience or suitableness of the plan of that you mention in your Bakerian lecture.[18] My object is chiefly quantity, to do magnetic experiments well, and to be of use if I set up a great magnet for diamagnetism.

I was talking to Prof. Forbes about the Atlantic Telegraph[19] and heard from him that the slipping of the cable down the incline in the direction of its length had given rise to great waste of cable. In fact the lateral resistance of the water must convert the line of cable into an inclined plane down which it slides and lies in folds below. I hear of buoys &c but would not *kites* be better as thus – flat pieces of iron or wood fastened at intervals so.

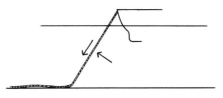

Figure 134,1

The kite to be attached by two ropes so that the upper one keeps the plane of the kite right, while the lower pulls the rope forwards so as to keep it stretched the whole way as it gradually sinks down to the bottom. The size & frequency of kites to depend on the diameter & weight of the rope and the rate at which it is to be payed out.

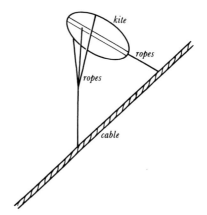

Figure 134,2

---

(16) See *Saturn's Rings*: 45–54 (= *Scientific Papers*, **1**: 345–56).

(17) See *Saturn's Rings*: 53 (= *Scientific Papers*, **1**: 354); 'When we come to deal with collisions among bodies of unknown number, size, and shape, we can no longer trace the mathematical laws of their motion with any distinctness'.

(18) William Thomson, 'On the electro-dynamic qualities of metals', *Phil. Trans.*, **146** (1856): 649–751, esp. 658–62 (= *Math. & Phys. Papers*, **2**: 189–327).

(19) On Thomson's involvement with the Atlantic telegraph project see S. P. Thompson, *The Life of William Thomson, Baron Kelvin of Largs*, 2 vols. (London, 1910), **1**: 325–96.

Besides keeping the lower part tight the kites would relieve the tension at the surface which is what the public dread, as it really broke the rope, but that was because too much was being spent and the break had been applied too strong.

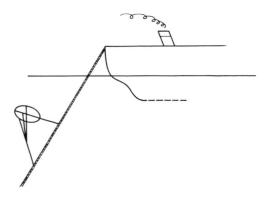

Figure 134,3

My class is small this year. I have viva voce'd them all and will begin written questioning on Tuesday. The second year (my old students) have been grinding in summer and have done well in exam[n]. A good many of them come to me as an advanced class.[20]

I have had correspondence with Faraday on the 'Lines of Force' as applied to Gravitation.[21] What a painful amount of modesty he has when he talks about things which may possibly be of a mathematical cast.[22] If I were free to think about these things I think I could put any part of the theory of Gravitation into Faraday's language which I think a most powerful one instead of a mere complication of simple ideas, as some conceive it.

I read your paper on electric qualities of metals[23] with the more pleasure as I knew how you had divided the labour and exercise with your students. I think many people have failed to follow some of the theoretical arguments, for want of faith in reason. I think you might use a little rhetoric with effect, for the sake of such philosophers as the dry proof spirit repels.

I have not heard of your brother James lately. Is he Godwin's successor yet?[24]

I hope M[rs] Thomson is able to thrive in Glasgow after her vacation. I met an aunt of hers M[rs] Graham at Dunoon.

Yours truly
J. C. MAXWELL

(20) See Numbers 136 and 144.

(21) Number 133. (22) See Number 133 notes (5), (12) and (13).

(23) Thomson, 'On the electro-dynamic qualities of metals'.

(24) John Godwin; see *The Belfast Queen's College Calendar, 1851* (Belfast, n.d.): 71; the post of Professor of Civil Engineering, held by James Thomson from 1857–73 (*DNB*).

FROM A LETTER TO HENRY RICHMOND DROOP[1]

14 NOVEMBER 1857

From Campbell and Garnett, *Life of Maxwell*[2]

129 Union Street
Aberdeen
14 November 1857

I am very busy with Saturn on the top of my regular work.[3] He is all remodelled and recast, but I have more to do to him yet, for I wish to redeem the character of mathematicians, and make it intelligible.[4] I have a large advanced class for Newton, physical astronomy, the electric sciences, and high optics. What is your department by the way?

I have also a mechanics' class in the evening, once a week, on mechanical principles, such as doctrine of lever work done by machines, etc. So I have 15 hours a week, which is a deal of talking straight forward.

I am getting several tops (like the one I had at Cambridge) made here for various parties who teach rigid dynamics.

---

(1) Trinity 1849, Fellow 1855 (Venn).
(2) *Life of Maxwell*: 291.
(3) See Numbers 134 and 138.
(4) Compare Faraday's letter to Maxwell of 13 November 1857 (Number 133 note (13)).

LETTER TO PETER GUTHRIE TAIT

21 NOVEMBER 1857

From the original in the University Library, Aberdeen[1]

129 Union Street
Aberdeen Nov 21 1857

Dear Tait

I feel great satisfaction in being able to appease your anxiety and that of Mrs Tait by so small an enclosure as that which I now send you. Account its rectangular simplicity a feeble emblem on the insufficiency of all invented methods of congratulation.[2]

I have but one paper bearing on Electricity that on Faradays Lines of Force which you have.[3] MacLennan[4] must have had a confused or perhaps (mentally) intoxicated memory when he conceived of two. The thing was read in two parts but published together.[5]

I shall be glad to hear of Ozone having its hash settled.[6] I shall tell Fuller[7] about it as far as I know, and inform him of what he knows, namely, what thus came of W. A. Porters sister.[8]

Hayward of John's now of Durham writes to me about the Earths axis & the variation in Latitude.[9] Peters of Königsberg has worked up the Poulkova obs^{ns} of Polaris & finds a var^{n} of 0″.079 (probable error 0″.017) and period = 303.867 mean solar days.[10] I am getting him a top constructed here.

My regular class is small this year as I expected from the character of the

---

(1) Aberdeen University Library, MS 980/3.   (2) See note (8).

(3) J. C. Maxwell, 'On Faraday's lines of force', *Trans. Camb. Phil. Soc.*, **10** (1856): 27–83 (= *Scientific Papers*, **1**: 155–229).

(4) Number 37 note (4).   (5) Numbers 85 and 87.

(6) Thomas Andrews and P. G. Tait, 'Note on the density of ozone', *Proc. Roy. Soc.*, **8** (1857): 498–500; *idem*, 'Second note on ozone', *ibid.*, **9** (1859): 606–8; and *idem*, 'On the volumetric relations of ozone, and the action of the electrical discharge on oxygen and other gases', *Phil. Trans.*, **150** (1860): 113–31.

(7) Number 96 note (2).

(8) See Number 109 note (27). Tait married Porter's sister on 13 October 1857; see C. G. Knott, *Life and Scientific Work of Peter Guthrie Tait* (Cambridge, 1911): 14.

(9) Robert Baldwin Hayward, St John's 1846 (Venn); a letter of 14 November 1857 (ULC Add. MSS 7655, II/15); see Number 116 note (7). The variation in latitude if the earth did not rotate accurately about its principal axis is discussed by Maxwell in his paper 'On a dynamical top', *Trans. Roy. Soc. Edinb.*, **21** (1857): 559–70, esp. 568–70 (= *Scientific Papers*, **1**:'259–61); see Number 116.

(10) C. A. F. Peters, 'Recherches sur la parallax des étoiles fixes', *Recueil de Mémoires présentès à l'Académie des Sciences par les Astronomes de Poulkova*, **1** (1853): 1–180, esp. 146–7. See Number 116 note (7).

men in the year below last year. My advanced class (an institution of mine) is large and evidently zealous. With them I expect to do Newton 1.2.3 with a sketch of Physical Astronomy, Magnetism and Electricity in the frosty weather and Undulatory when the Sun appears in Spring.[11]

My ordinary men are of course at Statics and are likely to do Dynamics Hydrostatics Heat and common Optics.[12] I send you the paper which was done by last years men *after* the vacation when they pass some of their degree subjects. 4 men got full marks and only 4 were under half and 2 of these were plucked. Of course some did not go in as they have another chance if they only wish to pass.

I am still grinding at Saturns Rings. I have shown that any solid ring must be horribly disfigured in order to go at all and in fact dismissed it.[13]

A liquid ring will either fly asunder from inequalities of centrifugal force or break up into drops by longitudinal forces. *But* the drops so formed may continue as a ring of satellites and will be *spaced out* regularly of themselves provided they be not too large or too numerous for Saturn to govern.[14]

I have found the numerical conditions of the amount of satellites which Saturn can maintain as a ring, how many can rally round him and how many must form coalitions among themselves.

I am now busy with two rings of satellites with different velocities disturbing one another.[15]

I am becoming skilful in the conduct of students' teas and breakfasts. I almost prefer to get together a rough lot and set them agoing. I find the uproarious and idle ones are the best at a tea.

I have not heard of the fate of James Thomsons application for the Engineering chair.[16] I must write to Cay[17] soon, but I have written to very few since I came here for I have only Friday & Saturday evenings without some regular work to do.

I have received the cards of M^r & Mrs Stewart. MacLennan was a day with me on his way south. Your hand of write is very different from what it was at Cambridge but the direction is as it used to [be]. Well Time tries all. 'We take no note of Time, but bye it slaps'. *Youngs Night Thoughts*.[18]

J. C. M.

---

(11) On the advanced class see also Number 144. In the previous year he had taught Newton to a voluntary group; see Number 109 esp. note (3).

(12) See Number 132 note (6).   (13) See Numbers 128 and 138.

(14) See Numbers 134 and 138; and Maxwell, *Saturn's Rings*: 45 (= *Scientific Papers*, **1**: 344–5).

(15) See Numbers 134 and 138.   (16) See Number 134 note (24).

(17) See Number 108.

(18) Compare Edward Young, *Night Thoughts*, I, 55; 'We take no note of time/But from its loss'. See *The Poetical Works of Edward Young*, **1** (London, 1834): 3.

137

LETTER TO WILLIAM THOMSON

24 NOVEMBER 1857

From the original in the University Library, Glasgow[1]

129 Union Street
Nov 24th 1857

Dear Thomson

I have been hard at work all last week at class work and correspondence, and I have done no more *writing* to Saturn yet. As to the strip of paper[2] I suppose many must have dropped since the invention of paper, but fruitlessly, quia carent vate sacro.[3] But I will ask you another thing. What is the history of your Boomer with one side white and the other black?[4] for he that boometh with a boomer may let fly a flying chariot of paper, and both are explained the same way.

The conditions of the phenomenon are

1st a motion of the centre of gravity of the strip relatively to the surrounding medium.

2nd An excess of pressure on the first surface of the strip over that behind.

---

(1) Glasgow University Library, Kelvin Papers, M 10. Previously published in A. T. Fuller, 'James Clerk Maxwell's Glasgow manuscripts: extracts relating to control and stability', *International Journal of Control*, **43** (1986): 1593–5.

(2) J. C. Maxwell, 'On a particular case of the descent of a heavy body in a resisting medium', *Camb. & Dubl. Math. J.*, **9** (1854): 145–8 (= *Scientific Papers*, **1**: 115–18); see Number 38. Thomson referred to this paper, in relation to his current interest in hydrodynamical ether models which would generate rotatory motion, in a letter to Stokes of 23 May 1857 (ULC Add. MSS 7656, K 97): 'I have been [in] great difficulties about hydrodynamics for some time. I thought I had made out that a mote in a perfect (non frictional) liquid, would if set into in any state of motion, end with its centre of gravity at rest and all the energy of the given motion transformed into energy of rotation of the mote & of corresponding motion of the surrounding liquid. I am shaken however from this comfortable theory, and I fear it comes to nothing. Take a rectangular slip of paper & let it fall from a height, with its length nearly horizontal, & you will see a curious illustration. (Clerk Maxwell showed me this a long time ago.) I thought first, and am now again nearly convinced, that the generation of rotatory motion here is inexplicable without taking into account the viscosity of air. If the theory I supposed I had made out had held, the same kind of phenomenon would take place in a perfect liquid.'

(3) Horace, *Odes*, Book IV. ix, 28; 'because they lack a sacred bard'. The full quotation is: 'vixere fortes ante Agamemnona/multi; sed omnes illacrimabiles/urgentur ignotique longa/nocte, carent quia vate sacro'; ['many heroes lived before Agamemnon; but all are overwhelmed in unending night, unwept, unknown, because they lack a sacred bard']; see *Horace, the Odes and Epodes*, with an English translation by C. E. Bennett (London/New York, 1914): 320–1. See also Maxwell's letter to Thomson of 19 February 1856 (Number 94).

(4) Perhaps a wooden ruler (possibly from 'boomerang').

3ʳᵈ This excess greatest at the foremost edge of the strip and near it.

4ᵗʰ These two last effects increasing with the relative velocity of the strip and the medium.

The results are

I A permanent rotation of the strip indifferently either way round.

II A force at right angles to the relative motion acting on the strip in a line turned round from the direction of motion 90° in the direction of rotation.

In May you thought that these effects would take place in an incompressible fluid without friction, and now you think that opinion a delusion, because if all motions at any instant were reversed all would go back as it came.[5]

---

(5) See note (2). Thomson's speculations, pursued in letters to Stokes in 1857, were developed from his suggestion that the magneto-optical effect discovered by Faraday (see Number 133 note (16)) could be explained by a vortical motion in the ether; see William Thomson, 'Dynamical illustrations of the magnetic and heliocoidal rotatory effects of transparent bodies on polarized light', *Proc. Roy. Soc.*, **8** (1856): 150–8 (= *Phil. Mag.*, ser. 4, **13** (1857): 198–204). Thomson drew on the concept of molecular motion which he had outlined in his paper 'On the dynamical theory of heat', *Trans. Roy. Soc. Edinb.*, **20** (1851): 261–88 (= *Math. & Phys. Papers*, **1**: 174–210), and on Rankine's theory of heat as a vortical motion of ethereal atmospheres surrounding molecular nuclei; see W. J. M. Rankine, 'On the centrifugal theory of elasticity and its connection with the theory of heat', *Trans. Roy. Soc. Edinb.*, **20** (1853): 425–40, esp. 428. In a letter to Stokes of 17 June 1857 Thomson elaborated on his discussion of hydrodynamics and the rotation of 'motes' in a perfect fluid: 'an infinite number of such motes all rotating with great angular velocities will repel one another & keep up the kind of stability & relative stiffness required for luminiferous vibrations. If there be a preponderance of axes set in one direction, undulations among the system would have Faraday's rotatory property when the planes of the waves are perp. to this direction...' (ULC Add. MSS 7656, K 98). In a letter of 20 December 1857 Thomson informed Stokes that: 'Now I think hydrodynamics is to be the root of all physical science, and is at present second to none in the beauty of its mathematics' (K 101). In a letter of 23 December 1857 he declared that: 'What I am most anxious to make out however is the mutual action of motes, separated by a perfect liquid. A molecular theory on this foundation would include the generation of heat in liquids by stirring, or in solids by electric conduction; magnetic force; Faraday's rotation of the plane of polarisation under magnetic influence, as most elementary deductions. It might include a great deal more, such as the elasticity of gases & solids, chemical affinity, thermo electricity, &c &c' (K 102). In these two letters Thomson indicated the scope of his current speculations, which sheds some light on Maxwell's remarks. In the letter of 20 December he observes that 'You first told me that if the bounding surface of a perfect liquid originally at rest, be moved in any way (including of course change of shape) and be brought again to rest in its original position, or in any changed shape & position, the whole liquid will come to rest at the same time' (K 101). Breaking off to catch the post, he continued in his letter of 23 December: 'That principle, in the hydrodynamics of a "perfect liquid", which I first learned from you, is something that I have always valued as one of the great things of science, simple as it is, and I now see more than ever its importance. One conclusion from it is that instability, or a tendency to run to eddies, or any kind of dissipation of energy, is impossible in a perfect liquid (a fluid with neither viscosity nor incompressibility) ... I have been trying to make out whether a slight degree of compressibility can possibly give rise to instability & eddies,

Now I cannot see why, if you could gather up all the scattered motions in the fluid, and reverse them *accurately*, the strip should not fly up again. All that you need is to catch all the eddies, and reverse them not approximately, but accurately.

If you pour a perfect fluid from any height into a perfectly hard or perfectly elastic basin its motion will break up into eddies innumerable forming on the whole one large eddy in the basin depending on the total moments of momenta for the mass.

If after a given time say 1 hour you reverse every motion of every particle, the eddies will all unwind themselves till at the end of another hour there is a great commotion in the basin, and the water flies up in a fountain to the vessel above. But all this depends on the *exact* reversal for the motions are *unstable* and an approximate reversal would only produce *a new set of eddies multiplying by division*.

Now I do not see why the unstable motion of a perfect fluid should not produce eddies which can never be gathered up again except by miracle, and therefore, since the diminution of pressure at the back of a body depends on the formation and dissipation of eddies by unstable motion, I do not see why it makes much difference whether these eddies are soon converted into heat, or remain in the fluid in a state of subdivision which is as nearly that of molecular vortices as any finite motion can be.

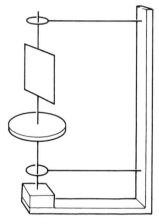

This is all general talk. You may have something more accurate to guide you. I have not.

By the way, a neat form of the experiment is a flat board connected with a fly wheel and set a spinning about a diameter in a strong wind. The board changes periodically the velocity of the wind near it and so fulfils the conditions of permanent rotation.

Figure 137,1

---

but I cannot as yet fully convince myself that I am right in thinking not' (K 102). See also G. G. Stokes, 'Notes on hydrodynamics. IV. Demonstration of a fundamental theorem', *Camb. & Dubl. Math. J.*, **3** (1848): 209–19 (= *Papers*, **2**: 36–47), and William Thomson, 'V. On the vis-viva of a liquid in motion', *Camb. & Dubl. Math. J.*, **4** (1849): 90–4 (= *Math. & Phys. Papers*, **1**: 107–12). In a notebook entry on 6 January 1858 (ULC Add. MSS 7342, NB 35) Thomson speculated that 'it might be possible to conceive that all the phenomena of matter might be explained by the consequences of contractility in a universal fluid constituting the material world'. Because of inherent conceptual difficulties he suggested the theory 'as a temporary mechanical illustration of some of the agencies hitherto looked upon as among the most inscrutable phenomena of inorganic physics'; see Ole Knudsen, 'From Lord Kelvin's notebook: ether speculations', *Centaurus*, **16** (1972): 41–53, esp. 47–50.

I am improving the theory of forced waves[6] in rings of satellites, so as to attack the question of the mutual disturbance of two rings of which the distance is small compared with their radius but great compared with the distance of consecutive satellites of the same ring.[7]

Yours truly

J. C. MAXWELL

(6) By 'forced waves' Maxwell denotes waves 'due to external disturbance, and following different laws from the natural waves of the ring'; see Maxwell, *On the Stability of the Motion of Saturn's Rings* (Cambridge, 1859): 30–4, esp. 33 (= *Scientific Papers*, **1**: 326–30, esp. 329).

(7) See Numbers 134 and 138; and *Saturn's Rings*: 45–54 (= *Scientific Papers*, **1**: 345–56).

LETTER TO JAMES CHALLIS

24 NOVEMBER 1857

From the original in the library of the Institute of Astronomy, Cambridge[1]

129 Union Street
Aberdeen. 24 Nov 1857

Dear Sir

In reply to your letter[2] I shall report on Saturns Rings.

I have been steadily employed on them with one month's interval since Aug. 1st,[3] stopping a week at the end of each hard pull.

1st I have gone over the rigid ring, and find all correct up to the valuation of the attractions, which is wrong, and, indeed, violates Laplace's equation.[4] Putting it right, I find the necessary inequalities of the ring to be very great, that if the section is $(A_1 + A_2 \cos \theta)$, $A_2 > A_1$ and there must be *negative mass* somewhere, which is absurd,[5] and that in another simple case, when the ring is uniform, with a heavy sphere stuck on like the stone on a ring, the stone must be to the ring as 82 to 18 nearly. The limits of stability are .8159 & .8279 which run it very close.[6]

I have worked out the motion when the satellite is exactly .82 of the whole. It is compounded of uniform angular motion ω with two librations performed respectively in 1.69 and 3.25 revolutions of the ring. I have examined the radial and tangential and rotatory parts of these librations, and given a detailed mechanical explanation of them.[7]

2nd I have abolished my rough theory of long waves in a thin fluid filament,

---

(1) Cambridge Observatory Archives, Letter 1857/53 (Institute of Astronomy, Cambridge University). Previously published in A. T. Fuller, 'James Clerk Maxwell's Cambridge manuscripts: extracts relating to control and stability – III', *International Journal of Control*, **37** (1983): 1231–3.

(2) Not extant.

(3) See Maxwell's letter to Thomson of 1 August 1857 (Number 126).

(4) See Challis's comments on Part I, Problem VI of Maxwell's Adams Prize essay 'On the Stability of the Motion of Saturn's Rings' (Number 107); see Number 107 notes (29) and (31). See also Maxwell's letter to Thomson of 24 August 1857 (Number 128). Laplace had shown that a uniform solid ring would be unstable; see the introduction to the Adams Prize essay.

(5) See Challis's comments on Maxwell's treatment of a solid ring thicker at one side than the other in Part I, Problem VI of the Adams Prize essay (Number 107). Compare Maxwell's letter to Thomson of 24 August 1857 (Number 128); and see Maxwell, *On the Stability of the Motion of Saturn's Rings* (Cambridge, 1859): 15 (= *Scientific Papers*, **1**: 307).

(6) See Number 128; and *Saturn's Rings*: 15–16 (= *Scientific Papers*, **1**: 307–8).

(7) See Number 128; and *Saturn's Rings*: 15–17 (= *Scientific Papers*, **1**: 307–10).

and substituted a calculation of the attractions in a ring composed of μ equal satellites, and affected with m waves of radial and tangential displacement.[8]

From these I have deduced the equations of motion of such a set of waves, which are reduced to a biquadratic in n which must have 4 real roots if the motion is stable.[9]

I then find in what cases imaginary roots are to be dreaded, and I find that the *tangential* force is the destructive one, and reaches its maximum when $m = \frac{1}{2}\mu$, but that the mass of the planet may in every case be made large enough to keep any number of satellites at peace with one another. The condition of a stable ring is

$$S > .4352\mu^2 R$$

where S is Saturn, R the whole mass of the Ring, and μ the number of satellites into which it is divided, so that if $S = 4352R$ there must not be more than 100 satellites in the ring.[10]

Supposing this condition satisfied I examine the motion of stable waves, and find 4 for each value of m. I trace the librations of each satellite about its mean position, and its absolute orbit round Saturn, the instantaneous form of the ring, and the rate of propagation of that form.[11]

I then describe the general method of deducing the whole motion of the ring from its condition at a given epoch, the analysis into waves by Fouriers theorem, the solutions of the equation in n, and the evaluation of all the constants.[12]

Then I trace the steps of the destruction of the ring by motions depending on imaginary values of n. When the tangential forces are too great, n is of the form $p + \sqrt{-1}q$ and the waves increase in height till the ring breaks. When the tangential force is *negative* (producing statical stability) the irregularities remain where they were, and increase to destruction without being propagated from satellite to satellite.[13]

Then I consider *forced waves* due to external disturbing forces, with the usual results.[14]

Then I extend the method to a ring of unequal satellites,[15] and to a ring of

(8) Compare Part II, Proposition X of Number 107; and see Number 134.

(9) See *Saturn's Rings*: 22–3 (= *Scientific Papers*, **1**: 316–17); n is the angular velocity of a system of m waves in the ring.

(10) See Number 134; and *Saturn's Rings*: 24–5 (= *Scientific Papers*, **1**: 318–19).

(11) See *Saturn's Rings*: 25–30 (= *Scientific Papers*, **1**: 319–25).

(12) See *Saturn's Rings*: 18–24 (= *Scientific Papers*, **1**: 311–18).

(13) See Number 134; and *Saturn's Rings*: 34–7 (= *Scientific Papers*, **1**: 330–5).

(14) See *Saturn's Rings*: 30–4 (= *Scientific Papers*, **1**: 326–30); and Number 137 note (6).

(15) See *Saturn's Rings*: 37–8 (= *Scientific Papers*, **1**: 335).

incompressible liquid with very elliptical section. This last case leads to the result that a liquid ring would come to sure destruction from the short waves in it and that it would break into drops.[16]

I then conceive a cloud of brickbats, subject only to gravitation and producing no mutual pressure revolving as a ring, and I find that to prevent agglomeration of particles the mean density of the cloud must be less than $\frac{1}{330}$ of that of Saturn.[17] But such a ring could not revolve with uniform angular velocity (See Laplace)[18] so we are driven to a plurality of rings with independent angular velocities.

I have ventured to attack this problem by supposing the distance between consecutive rings large compared with that of the satellites of the same ring, and then finding the laws of the mutual disturbances of any two such rings.[19]

This is the point at which I am at present. I did the attractions of the liquid flat ring since my work here began and I have not yet had spare power to go on, especially as the subject requires a good deal of mental preparation before I sit down to work it on paper.

However I know how the natural waves of one ring will produce forced waves in the other, and I have to investigate whether there may be any waves generated which are not natural in either ring, or whether any peculiar *reactions* take place, owing to the difference of velocities of natural waves in the two rings.[20]

I have taken pains always to have a distinct conception of every motion and of every step of proof, and have endeavoured to express these conceptions as I went along, so that however unreal my different sorts of rings may be, I have some confidence that their motions are rightly described, and would work if set a-going.

I have also conceived some terrestrial experiments, to illustrate the oppositions of statical and dynamical stability.[21]

---

(16)　See Numbers 134 and 136; and *Saturn's Rings*: 40–45 (= *Scientific Papers*, **1**: 338–45).

(17)　See Number 134; and *Saturn's Rings*: 38–40 (= *Scientific Papers*, **1**: 336–8).

(18)　P. S. de Laplace, *Traité de Mécanique Céleste*, 5 vols. (Paris, An VI [1799]–1825), **2**: 162. See Number 134; and Number 107 note (10).

(19)　See *Saturn's Rings*: 45–54 (= *Scientific Papers*, **1**: 345–56); and his letter to Challis of 21 January 1859 (Number 155).

(20)　Compare *Saturn's Rings*: 45–6 (= *Scientific Papers*, **1**: 345–6); and Number 155.

(21)　Compare *Saturn's Rings*: 62 (= *Scientific Papers*, **1**: 366). The equilibrium of each satellite is unstable with respect to tangential displacements; but as the tangential attraction is dependent on the rate of rotation of the ring 'this very force which tends towards destruction may become the condition of the preservation of the ring'. For Maxwell's description of his model illustrating the movements of the satellites forming a discontinuous ring, see his letter to Thomson of 30 January 1858 (Number 143); and *Saturn's Rings*: 59–62 (= *Scientific Papers*, **1**: 363–6).

With respect to the variation in Latitude, Hayward writes[22] that Peters of Königsberg ('Recherches sur la parallaxe' §87) has found it $= 0''.087$ at Poulkowa, with probable error $0''.017$ and his period is 303.867 days.[23] We have no observatory here, so Peters is unknown.

Yours truly

(a)

J. C. MAXWELL

(a) {Challis} I wrote to Prof^r. Maxwell Jan. 18. 1859.[24]

(22) See Number 136 note (8).

(23) C. A. F. Peters, 'Recherches sur la parallaxe des étoiles fixes', *Recueil de Mémoires présentès à l'Académie des Sciences par les Astronomes de Poulkova*, **1** (1853): 1–180, esp. 146–7. See Number 116 note (7). The figures given by Peters, and by Hayward in his letter of 14 November 1857 (ULC Add. MSS 7655, II/15), were a variation in latitude of $0''.079$ with a probable error of $0''.017$. Maxwell gives the correct figures in his letter to Tait of 21 November 1857 (Number 136).

(24) Not extant; but see Number 155.

LETTER TO JAMES DAVID FORBES

26 NOVEMBER 1857

From the original in the University Library, St Andrews[1]

129 Union Street
Aberdeen Nov 26, 1857

Dear Sir

I am very sorry that you did not find my short note on the Theory of Colours[2] sufficiently developed. I did not expect that many people would take much interest in the subject at present, and the reason of my publishing these numerical results, is merely to indicate to competent persons, that the methods of observation are trustworthy, and that equations can be found, by the eye alone, to satisfy the requirements of theory.

The scientific truth brought forward is, that on a particular theory, 6 equations are equivalent to two only, and this is shown to be true in fact.[3] This somewhat abstract fact is not calculated to enlighten the minds of most readers of short articles in the Phil. Mag. who expect something more definite, especially as, in the theory of colour, they have always been accustomed to regard it as settled that red blue and yellow ARE THE primary colours.

My theory only professes to prove that there are three, and only three, and does not define them. Besides, my equations are not between the 'simple colours', which people think they know when they see them, but between the colours of certain pigments, which have no abstract title to have their properties investigated first, so that people do not see the use of knowing the exact relation between 5 colours which are merely paints out of Lithgow and Purdies shop.

---

(1) University Library, St Andrews, Forbes MSS 1857/99.

(2) James Clerk Maxwell, 'Account of experiments on the perception of colour', *Phil. Mag.*, ser. 4, **14** (1857): 40–7 (= *Scientific Papers*, **1**: 263–70). Compare a letter from G. G. Stokes to Maxwell, of 7 November 1857 (ULC Add. MSS 7655, II/12); 'I have just received your papers on a dynamical top &c, & the account of experiments on the perception of colour. The latter, which I missed seeing at the time when it was first published, I have just read with great interest. The results afford most remarkable and important evidence in favour of the theory of 3 primary colour perceptions, a theory which you and you alone so far as I know have established on an exact numerical basis.' (= *Life of Maxwell*: 287–8).

(3) Using coloured discs cut from papers painted with 'Vermilion, Ultramarine, Emerald-green, Snow-white, Ivory-black, and Pale Chrome-yellow', by leaving out each colour in turn he obtained colour equations between five colours, which could be reduced to two equations; 'Experiments on the perception of colour': 42–3 (= *Scientific Papers*, **1**: 265–7).

To explain, that it is the agreement of the equations, and not their separate importance, that constitutes the value of the experiments is a much more difficult thing, and one which I have every intention of doing at greater length, when I have perfected my experiments on the spectrum. I do not like to speak about experiments that should be made, till I know whether they will work. I have had some experience of spectrum mixing, as I have made 3 different sets of apparatus which all succeeded partially. I began in 1852 with one 3 feet long and a water prism. In 1855 I made one 7 feet long with a glass prism and arrangements for seeing two mixtures of pure colours spread uniformly over two contiguous fields.[4]

In 1856 I made a reflecting portable apparatus for showing the phenomena roughly to strangers.[5] The reason why all these things lie bye is that I have a piece of labour to do which will require a new apparatus (on the old plan) and a great deal of very troublesome observations. Hitherto I have seen a few mixtures in juxtaposition in a way that I do not think has been tried before, and which gives better results, and requires less light, than Helmholtzs methods,[6] but the results have all been rough compared with Helmholtz and have only verified his, and paved the way for a new labour which I shall describe to you, that you may see that there is a good deal to do before I am satisfied.[7]

Fraunhofers lines are the land marks of the spectrum,[8] and his values of the corresponding wave lengths enable us to interpolate the wave-lengths of any ray

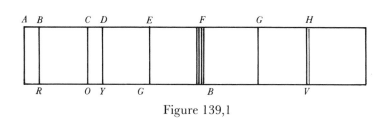

Figure 139,1

---

(4) See Number 71. Compare Maxwell, 'Experiments on colour, as perceived by the eye, with remarks on colour blindness', *Trans. Roy. Soc. Edinb.*, **21** (1855): 275–98, esp. 290; and Maxwell, 'On the theory of compound colours, and the relations of the colours of the spectrum', *Phil. Trans.*, **150** (1860): 57–84, esp. 61 (= *Scientific Papers*, **1**: 144, 415).

(5) See Numbers 109 and 117.

(6) Hermann Helmholtz, 'Ueber die Theorie der zusammengesetzten Farben', *Ann. Phys.*, **87** (1852): 45–66, (trans.) 'On the theory of compound colours', *Phil. Mag.*, ser. 4, **4** (1852): 519–34; and 'Ueber die Zusammentsetzung von Spectralfarben', *Ann. Phys.*, **94** (1855): 1–28 (= *Wissenschaftliche Abhandlungen*, 3 vols. (Leipzig, 1882–95), **2**: 3–23, 45–70).

(7) See Numbers 163 and 170.

(8) Joseph Fraunhofer, 'Bestimmung des Brechungs- und Farben-Zerstreuungs-Vermögens verschiedener Glasarten, in bezug auf die Vervollkommnung achromatischer Fernröhre', *Annalen der Physik*, **56** (1817): 264–313, and Table IV.

when we know its position among the dark lines, and so we can refer any spectrum to a standard spectrum such as the diffraction spectrum.[9]

Neglecting the effects of absorption, the intensity of the light between given lines in different prisms will vary inversely as the distance of the lines and the whole amount of light in the spectrum between given lines will be known when the lines are known.

Helmholtz has found, and I have verified it that there are two points in the spectrum where the light rapidly changes its properties as related to Colour and the eye, while nothing extraordinary takes place with regard to the wavelength. These are on the yellow and the blue borders of the green.[10]

These parts of the spectrum must be avoided in choosing standard colours. When I made my experiments I took my standard rays out of the middles of the three divisions of the spectrum where the change of colour is slow one in the red, one in green the other in blue.[11] The amount of each kind of light was measured (relatively) by the breadth of the slit through which the light came. These measures must be reduced to an absolute scale by the adoption of units of colour.

This being done every other part of the spectrum may be referred to these three. For if we take any other ray, (say violet) we can combine it with two of the standard rays (say red and green) so as to produce white, and by measuring the *quantities* of the given ray the two standards and the resulting white we can determine the value of the ray in terms of the three primaries.[12]

Or – we can combine the violet ray with the standard green, and the standard red and blue rays together in certain proportions till they produce the same result and thence deduce the value of the violet.

I have tried both methods with some success but I have reason to prefer the first, because the resultant tint is always white.

It is curious in the second method to take a prism and analyse two apparently similar colours and find one composed of violet with a very little green and the other of blue with a very little red.

In this way, having fixed upon certain units for the measurement of standard

(9) See Number 163; and Maxwell, 'On the theory of compound colours': 67–9 (= *Scientific Papers*, **1**: 423–5). Compare Helmholtz, 'On the theory of compound colours': esp. 522–3, for the use of Fraunhofer lines as reference marks. See also Helmholtz, 'Ueber die Zusammensetzung von Spectralfarben': 15–16.

(10) Helmholtz, 'Ueber die Zusammensetzung von Spectralfarben': 15–17, and Table I, Fig. 3.

(11) See Maxwell, 'On the theory of compound colours': 68–9 (= *Scientific Papers*, **1**: 424–5); and Number 163.

(12) See Maxwell, 'On the theory of compound colours': 69–71, 73 (= *Scientific Papers*, **1**: 426–8, 430–1).

red (R) blue (B) and Green (G) light the description of any other light would be

$$X \text{ (unit of the given light)} = rR + gG + bB. \text{[13]}$$

Perhaps the best units for the purpose would be derived from the wave measurements thus –

Divide the spectrum into divisions such that the wavelengths at successive divisions differ by equal quantities. Each of these divisions contains unit of its own light when unity of solar light falls on the prism.[14]

After finding the laws of composition of colours when the light is solar, we pass to light from other sources or to solar light altered by absorption. If the Undulatory theory be true, the only effect on each simple ray is to alter its intensity relatively to others without changing its quality.

So that we have only to find each coefficient of absorption for the different rays which is a matter of pure physical optics and then the composition of the *colours* of these altered rays proceeds as before. This is the crucial test of Sir D. Brewsters suspicion of the alterable character of a ray in transitu.[15] His theory or the Undulatory must fall.[16]

You see I have a good deal to begin on. When I do, I shall try to explain, but at present to get the scientific world to see the difference between a theory of colour and a theory of light might be possible but not prudent, without a good show of very scientific-looking experiments to back it. Coloured papers and

---

(13) Maxwell, 'On the theory of compound colours': 69–71 (= *Scientific Papers*, **1**: 426–8); and see Number 170.

(14) Compare Maxwell, 'On the theory of compound colours': 68–9 (= *Scientific Papers*, **1**: 424–5), and see Numbers 163 and 170.

(15) David Brewster, 'On a new analysis of solar light, indicating three primary colours, forming coincident spectra of equal length', *Trans. Roy. Soc. Edinb.*, **12** (1834): 123–36. See Maxwell, 'On the theory of compound colours': 59 (= *Scientific Papers*, **1**: 413); 'Brewster... regards the actual colours of the spectrum as arising from the intermixture, in various proportions, of three primary kinds of light, red, yellow, and blue,... [he] employed coloured media, which, according to him, absorb the three elements of a single prismatic colour in different degrees'. Brewster's experiments on the selective absorption of light were effectively challenged, and his conclusions refuted, by Hermann Helmholtz, 'Ueber Herrn D. Brewster's neue Analyse des Sonnenlichts', *Ann. Phys.*, **86** (1852): 501–23 (= *Wissenschaftliche Abhandlungen*, **2**: 24–44); (trans.) 'On Sir David Brewster's new analysis of solar light', *Phil. Mag.*, ser. 4, **4** (1852): 401–16. See also Number 180; and compare Number 72. On Brewster's theory see P. D. Sherman, *Colour Vision in the Nineteenth Century* (Bristol, 1981): 20–59.

(16) David Brewster, 'Observations on the absorption of specific rays, in reference to the undulatory theory of light', *Phil. Mag.*, ser. 3, **2** (1833): 360–3. Brewster claimed that the selective absorption of light 'militated strongly against the undulatory theory' of the propagation of light in a luminiferous ether; drawing the analogy with the propagation of sound, he questioned 'how any medium could transmit two sounds of nearly adjacent pitches, and yet obstruct a sound of an intermediate pitch'. See Number 184.

spinning tops though capable of far greater accuracy than most spectrum experiments convey no absolute facts about definite colours.

I have really nothing for the Royal Society for I am very busy at Saturn, and my class. If I get Saturn well in hand, it will give me great pleasure to draw up a readable report on him for the Royal Society it being understood that it is a report of a paper otherwise published.[17]

I have been doing a great deal in making what I hope are intelligible expositions (apart from the calculations) first of the methods of investigation and their results, and then of the principles which might be seen to regulate the particular motion, whether it were necessary to go into calculation or not.

M[r] Ramage is getting very expert in the making of Dynamical Tops.[18] Yours will be ready before long.[19]

I hear that Peters of Konigsberg 'Recherche sur la Parallax des Etoiles Fixes' §87 has found the periodic variation of the latitude at Poulkova 0″.087 either way with period = 303.867 mean solar days. I have not been in position to see his book yet.[20]

My class is small but I am working it hard and they seem to be obeying the helm. They had a bad reputation for stupidity this year.

The advanced students form a large class and I hope to make it an institution of the College that men should make some effort the second year to learn their Newton besides the other matters we hope to learn.

Yours truly

J. C. MAXWELL

---

(17) J. C. Maxwell, 'On the theories of the constitution of Saturn's rings', *Proc. Roy. Soc. Edinb.*, **4** (1859): 99–101 (= *Scientific Papers*, **1**: 286–7). See Number 149.

(18) See Number 108 note (6) and Number 116.

(19) In a letter to Maxwell of 14 September 1857 (ULC Add. MSS 7655, II/10) Forbes had requested: 'I should like if, at your *entire convenience* you would get me a Dynamical Top made for my Class on the pattern of your own.' Forbes acknowledged receipt in a letter to Maxwell of 16 March 1858 (*Life of Maxwell*: 308).

(20) See Number 116 note (7), and Maxwell's letters to Tait and Challis of 21 and 24 November 1857 (Numbers 136 and 138).

140

LETTER TO CECIL JAMES MONRO

26 NOVEMBER 1857

From the original in the Greater London Record Office, London[1]

121 Union Street
Aberdeen Nov 26, 1857

Dear Monro

The enclosed letters came from Mrs Pomeroy to me. I think they are one of them at least yours. I doubt of the smaller one but you will know. They seem dropping in still on poor Mrs Pomeroy. Even ordinary returned letters are strange things to read again as if you had been talking on when everybody had gone away.

I got your letter of the 6th.[2] I have been grinding so hard ever since I came here that I have left many letters unanswered. When I have time I shall write to you and meanwhile only thank you for your letters.

I am at full College work again. A small class with a bad name for stupidity so there was the more field for exciting them to more activity. So I have got into regular ways, and have every man viva voced once a week and the whole class examined in writing on Tuesdays and roundly and sharply abused on Wednesday morning, and lots of exercises, which I find it advantageous to brew myself overnight.

Public Opinion here says that what our Colleges want is inferior professors and more of them for the money. Such men says P.O. would devote their attention more to what would pay and would pay more deference to the authority of the local press than superior or better paid men.

Therefore although every individual but one who came before the Commission was privately convinced that the best thing in itself w^{d} be to fuse our two institutions into one with one staff of teachers yet they all agreed that the public opinion of the whole was the opposite of the private opinion of each and that more harm than good would result from adopting the course which seemed good to the members, but not to the body, of the public.

So the battle rages hot between Union (of Universities only) and Fusion (of classes and professors). Almost all the prof^{rs} in Arts are fusionists and all the country south of the Dee together with England & the rest of Europe & the World but Aberdeen (on the platform) is Unionist and nothing else is listened

---

(1) Greater London Record Office, Acc. 1063/2084. Previously published in *Life of Maxwell*: 291–2.

(2) Greater London Record Office, Acc. 1063/2103; containing information about R. H. Pomeroy; see Number 131, note (4).

to at public meetings though perhaps a majority of those present might be fusionists.[3]

Such is Public Opinion but we are all quiet again now & I am working at various high matters for I have a very good class for Physical Astronomy, Electricity & Undulations &c., and I want to do them justice.

I have had a lot of correspondence about Saturn's Rings, Electric Telegraphs, Tops & Colours. I am making a Collision of Bodies machine & a model of Airys Transit Circle (with lenses)[4] and I am having students teas when I can. Also a class of operatives on Monday evening who do better exercises than the University men about false balances, Quantity of Work &c.

Yours truly
J. C. MAXWELL

---

(3) Commissioners were appointed to administer the Act 21 & 22 Vict. c. 83, which enacted the union of King's and Marischal Colleges to form the University of Aberdeen. See Numbers 142 and 174 esp. note (6).

(4) See Number 113 note (8) and Number 142 note (8).

FROM A LETTER TO JANE CAY

28 NOVEMBER 1857

From Campbell and Garnett, *Life of Maxwell*[1]

129 Union Street
Aberdeen
28 Nov. 1857

I had a letter from Willy to-day about jet pumps to be made for real drains, but not saying anything about the Professorship of Engineering.[2]

I have been pretty steady at work since I came. The class is small and not bright, but I am going to give them plenty to do from the first, and I find it a good plan. I have a large attendance of my old pupils, who go on with the higher subjects. This is not part of the College course, so they come merely from choice, and I have begun with the least amusing part of what I intend to give them. Many had been reading in summer, for they did very good papers for me on the old subjects at the beginning of the month. Most of my spare time I have been doing Saturn's Rings, which is getting on now, but lately I have had a great many long letters to write, – some to Glenlair, some to private friends, and some all about science...I have had letters from Thomson and Challis about Saturn – from Hayward of Durham University about the brass top, of which he wants one. He says that the Earth has been really found to change its axis regularly in the way I supposed.[3] Faraday has also been writing about his own subjects.[4] I have had also to write Forbes a long report on colours,[5] so that for every note I have got I have had to write a couple of sheets in reply, and reporting progress takes a deal of writing and spelling....

I have had two students' teas, at which I am becoming expert. I have also indulged in long walks, and have seen more of the country. The evenings are beautiful at this season. There have been some very fine waves on the cliffs south of the Dee.

---

(1) *Life of Maxwell*: 292–3; the cuts are Campbell's.
(2) See Number 134 note (24).
(3) See Number 116 note (7); and Numbers 136 and 138.
(4) See Number 133 notes (9), (12) and (13).
(5) Number 139.

FROM A LETTER TO LEWIS CAMPBELL

22 DECEMBER 1857

From Campbell and Garnett, *Life of Maxwell*[1]

129 Union Street
Aberdeen
22 Dec. 1857

I have been reading Butler's *Analogy*[2] again, specially with reference to obscurities in style and language, and also to distinguish the merits of the man, and what habits of thought they depended on. Also Herschel's Essays, of which read that on Kosmos,[3] and Froude's History.[4] One night I read 160 pages of Buckle's *History of Civilisation*[5] – a bumptious book, strong positivism, emancipation from exploded notions, and that style of thing, but a great deal of actually original matter, the true result of fertile study, and not mere brain-spinning. The style is not refined, but it is clear, and avoids fine writing. Froude is very good that way, though you can see the sort of pleasure that a University man takes in actually realising what he has talked over at Hall about showing what England was in the middle ages, and transfusing himself, style and all, thereinto, that his friends may see. A solitary student never does that sort of thing, nor can he appreciate the graces of imitation. I wish Froude would state whether he translates, and from what language, in each document.

I am still at Saturn's Rings. At present two rings of satellites are disturbing one another. I have devised a machine to exhibit the motions of the satellites in a disturbed ring;[6] and Ramage is making it, for the edification of sensible image worshippers.[7] He has made four new dynamical tops, for various seats of learning.

I have set up a model of Airy's Transit Circle, and described it to my

---

(1) *Life of Maxwell*: 294–6; extracted from a letter on Campbell's ordination (*Life of Maxwell*: 293–7). The letter includes a parody of Byron's 'The Bride of Abydos', beginning: 'Know ye the Hall where the birch and the myrtle/Are emblems of things half profane, half divine', describing the 'Murtle Lecture' at Marischal College.

(2) Joseph Butler, *The Analogy of Religion, Natural and Revealed, to the Constitution and Course of Nature* (London, 1736).

(3) J. F. W. Herschel, 'Humboldt's Kosmos', in *Essays from the Edinburgh and Quarterly Reviews* (London, 1857): 257–64.

(4) John Anthony Froude, *History of England from the Fall of Wolsey to the Death of Elizabeth*, 2 vols. (London, 1856).

(5) Henry Thomas Buckle, *History of Civilization in England*, **1** (London, 1857).

(6) See Maxwell's letter to William Thomson of 30 January 1858 (Number 143); and Number 162. See plate VII.   (7) See Number 108 note (6).

advanced class to-day.[8] That institution is working well, with a steady attendance of fourteen, who have come of their own accord to do subjects not required by the College, *and the dryest first.*

To the present time we have been on Newton's *Principia* (that is, Sects. i. ii. iii., as they are, and a general view of the Lunar Theory, and of the improvements and discoveries founded on such inquiries). Now we go on to Magnetism, which I have not before attempted to explain.[9]

The other class is at two subjects at once. Theoretical and mathematical mechanics is the regular subject, but two days a week we have been doing principles of mechanism, and I think the thing will work well. We now go on to Friction, Elasticity, and Breakage, considered as subjects for experiment, and as we go on we shall take up other experimental subjects germane to the regular course.[10] I am happy in the knowledge of a good tinsmith, in addition to a smith, an optician, and a carpenter. The tinsmith made the Transit Circle.

College Fusion is holding up its head again under the fostering care of Dr. David Brown (father to Alexander of Queen's).[11] Know all men I am a Fusionist, and thereby an enemy of all the respectable citizens who are Unionists (that is, unite the three learned faculties, and leave double chairs in Arts). But there is no use writing out their theory to you. They want inferior men for professors – men who will find it their interest to teach what will pay to small classes, and who will be more under the influence of parents and the local press than more learned or better paid men would be in a larger college.[12]

---

(8) See Number 113 note (8). For a recollection (in Maxwell's advanced class of 1859–60) see David Gill, *A History and Description of the Royal Observatory, Cape of Good Hope* (London, 1913): xxxi; and see George Forbes, *David Gill: Man and Astronomer* (London, 1916): 12–13.

(9) Compare Number 132.   (10) Compare Number 132.

(11) David Brown, Professor of Divinity in the Free Church College, Aberdeen; Alexander Brown, Glasgow University 1852, Queen's College, Oxford 1856; see W. Innes Addison, *The Matriculation Albums of the University of Glasgow from 1728 to 1858* (Glasgow, 1913): 500.

(12) See Number 140 and compare Number 174.

143

LETTER TO WILLIAM THOMSON

30 JANUARY 1858

From the original in the University Library, Glasgow[1]

129 Union Street
Aberdeen Jan 30, 1858

Dear Thomson

I have never thanked you for your propositions on the rotation of a body in a perfect fluid.[2] I hope to be able to give a proper attention to them in the vacation as I am tolerably busy at present. The *rings* have been stationary since New Year. I have had a model of my theory of the wave in a ring of satellites made by Ramage.[3] It is now set up and nearly ready.[4]

It consists of two wheels on the parallel parts of a cranked axle. There are 36 little cranks ranged round the circumference which all move parallel to the axle-crank and to each other. Each carries a satellite placed excentrically on the part of the crank which sticks through the wheel, so that when the axle turns and the wheel moves, every

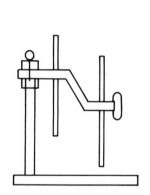
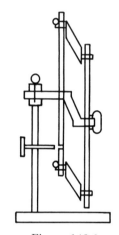

Figure 143,1　　　　Figure 143,2

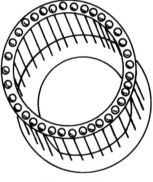
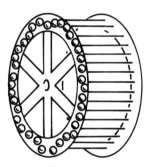

Figure 143,3　　　　Figure 143,4

---

(1) University Library, Glasgow, Kelvin Papers, M 11. Previously published in Brush, Everitt, and Garber, *Maxwell on Saturn's Rings*: 60–3, where the date is transcribed as 1859; but see *ibid*: 63 note (a).

(2) See Maxwell's letter to Thomson of 24 November 1857 (Number 137).

(3) See Number 108 note (6).

(4) See figs. 7 and 8 of Maxwell, *On the Stability of the Motion of Saturn's Rings* (Cambridge, 1859) (= *Scientific Papers*, **1**: at 376). The model is preserved in the Cavendish Laboratory, Cambridge; see plate VII.

Plate VII. Model (1858) illustrating the movements of the satellites constituting the rings of Saturn (Number 143).

satellite describes a small circle. The position of the satellite in this circle is arbitrary and its arm may be adjusted on the crank at pleasure, so that by properly arranging these arms the ring of satellites may be thrown into waves of any length which travel round the ring, the wheel being fixed.

Fixing the crank and moving the wheel we exhibit the absolute motion of the satellites in the '1st' & 'fourth' waves of my essay.[5] The others are less interesting and cannot be well exhibited in any way.

I have been lecturing on statical electricity to the second year, and next week I shall have half a dozen to study 'electrical images'[6] over a cup of tea. I begin current electricity on Tuesday. Next year I must set up Daniell[7] properly.

I have not had time to think it all over but there seems to be something in this which follows –

Suppose magnetism to consist in the revolution or rotation of any material thing, in the same or the opposite direction to the current of positive electricity which is equivalent to the magnet.[8]

Let an unmagnetized steel bar be fixed to an axis passing perp. thro its centre of gravity and let it be mounted in a brass frame so adjusted that the moments of inertia in the plane perp. to the axis are nearly equal but yet that round the steel bar decidedly the least.[9]

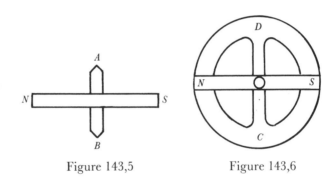

Figure 143,5   Figure 143,6

---

(5) See Maxwell, *Saturn's Rings*: 59–61, and figs. 8 and 12 (= *Scientific Papers*, **1**: 363–5, 376). Compare Part II, Proposition X of Number 107.

(6) William Thomson, 'On the mathematical theory of electricity in equilibrium', *Camb. & Dubl. Math. J.*, **3** (1848): 141–8, 266–74; *ibid.*, **4** (1849): 276–84; *ibid.*, **5** (1850): 1–9 (= *Electrostatics and Magnetism*: 52–85). See Number 71 note (18).

(7) The electric battery which maintained a constant current invented by John Frederic Daniell; see his papers 'On voltaic combinations', *Phil. Trans.*, **126** (1836): 107–24; 'Additional observations on voltaic combinations', *ibid.*: 125–9; and 'Further observations on voltaic combinations', *ibid.*, **127** (1837): 141–60.

(8) Compare J. C. Maxwell, 'On physical lines of force. Part I. The theory of molecular vortices applied to magnetic phenomena. Part II. The theory of molecular vortices applied to electric currents', *Phil. Mag.*, ser. 4, **21** (1861): 161–75, 281–91, 338–48 (= *Scientific Papers*, **1**: 451–88). On a vortical theory of magnetism see Number 118; and see Number 133 for reference to Thomson's paper 'Dynamical illustrations of the magnetic and heliocoidal rotatory effects of transparent bodies on polarized light', *Proc. Roy. Soc.*, **8** (1856): 150–8 (= *Phil. Mag.*, ser. 4, **13** (1857): 198–204).

(9) See Numbers 187, 189 and 190; and Maxwell's comment in 'On physical lines of force': 345n (= *Scientific Papers*, **1**: 485–6n) on an experiment to determine the angular momenta of the

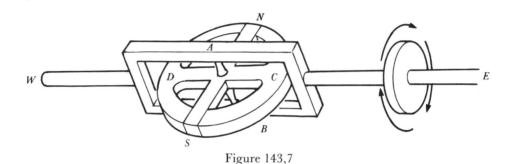

Figure 143,7

Let this apparatus be placed in a revolving frame, so that the axle of the magnet is perp. to that of the frame. If the axle of the frame be placed magnetic east and west, the axle of the magnet AB will revolve in the magnetic meridian. Now, when all is balanced, let the steel bar be made a magnet (NS). The Earth will set it perp. to the axle $E.W.$ Now let it be set in rotation about EW. Because the moment of inertia about NS is less than that about the perp. axis the motion will be stable (by centrifugal force) when NS is perp. to EW. But the brass frame may be so adjusted that this stability is very small, so that a slight force will produce a deviation of sensible magnitude.

Let the axle EW revolve in the direction of the arrows then if any thing in the magnet is in a state of rotation, the axis of its rotation will tend to be in the same direction as EW so that if N be the end which *points to the North* and if the rotation be in the direction of the *positive* current, N will go towards W and from E.

Now let everything be the same and let the magnet be reversed by re-magnetizing it without taking it out, then it ought to deviate the other way. Also the deviation ought to be reversed by reversing the direction of rotation of EW.

There are various ways of observing the deviation, the best is by marks at C and D.

I have not been able to detect any absurdity about this experiment and I think this plan gets rid of all induced currents and makes the two experiments the same every way except the direction of magnetization.

I have just heard of Slesser being Senior Wrangler.[10] I suppose Smith will be 2nd.[11] I got a letter from Hopkins,[12] praising Smith highly. I am glad to learn (from Mr Hardy Robinson) that Mrs Thomson is better this winter.

---

magnetic vortices by the free rotation of a magnet. Compare Maxwell, *Treatise*, **2**: 202–4 (§575) for a description of the apparatus; and see Number 187 note (20) and plate X.

   (10) George Middleton Slesser, Queens' 1854, from King's College, Aberdeen (Venn).

   (11) Charles Abercrombie Smith, Glasgow University 1848, Peterhouse 1854 (Venn).

   (12) William Hopkins; see Number 37 note (2).

If you can spare time I should like to hear your opinion on the desirableness of teaching the whole of Statics before the theory of Motion. This year I introduced practical Kinetics in the form of toothed wheels cranks, Hooks joint &c, which we studied with respect to their *motion* only, bringing in the forces in a different set of lectures. I devoted a little time to more theoretical matters such as the motion of the nail of a wheel tracing a cycloid, an ellipse traced by epicycloidal motion &c.

I am beginning to think that the best way is to drive Statics and Kinetics abreast and so prepare for Dynamics.

I intend to lay more carefully the foundations of the theory of absolute and relative 1st position, 2nd velocity, 3rd acceleration. I find that the systematic study of the theory of units and measurement is practically useful and intelligible to the students, and I have had but few bad errors of this kind.[13]

Of course I must teach Statics in the usual way, because it is good training and because that is what men will be examined on afterwards.

By the way what do you think of a board of Examiners for the Scotch Colleges to ensure a uniform standard of excellence. It would certainly prevent any ambitious deviations from the usual methods on the part of individual professors, for the students would study the examiners text book and not listen to the professor. Now I had a respectable voluntary audience after the hour yesterday to go through the convertible pendulum, and you have experimentalists and essayists. At Cambridge such things are almost crushed by the enlightened Senate House. How much more in Scotland by partially informed examiners.

Yours truly
J. C. MAXWELL

---

(13) See Numbers 120 to 123.

144

LETTER TO RICHARD BUCKLEY LITCHFIELD

7 FEBRUARY 1858

From the original in the library of Trinity College, Cambridge[1]

129 Union Street
Aberdeen 7 Feb 1858

Dear Litchfield

It is a long time since I have heard of you, I very much wish to hear how you are and what you are doing or what is doing around you. When I last wrote I was on my way here. Since then I have been at work, Statics and Dynamics; two days a week being devoted to Principles of Mechanism, and afterwards to Friction Elasticity and Strength of Materials, and also Clocks & Watches when we come to the pendulum. We have just begun hydrostatics. I have found a better text book for hydrostatics than I had thought for the run of them are so bad both Cambridge and other ones – Galbraith & Haughtons Manual of Hydrostatics Longmans 2/–. There are also manuals of Mechanics & Optics of the same set.[2] There is no humbug in them, and many practical matters are introduced instead of mere intricacies. The only defect is a somewhat ostentatious resignation of the demonstrations of certain truths, and a leaning upon feigned experiments instead of them. But this is exactly the place where the students trust most to the professor so that I care less about it. I shall adopt the Optics which have no such defect and possibly the Mechanics next year.

My students of last year to the number of about 14 form a voluntary class and continue their studies. We went through Newton I II III and took a rough view of Lunar Theory, and of the present state of Astronomy. Then we have taken up Magnetism & Electricity static & current, and now we are at Electromagnetism & Ampères Laws.[3] I intend to make Faradays book[4] the backbone of all the rest, as he himself is the nucleus of everything electric since 1830.

So much for Class work. Saturns Rings are going on still but this month I am clearing out some spare time to work them in. I have got up a model to show the

---

(1) Trinity College, Cambridge, Add. MS Letters c.1⁸⁹. Previously published in S. G. Brush, C. W. F. Everitt, and E. Garber, *Maxwell on Saturn's Rings* (Cambridge, Mass./London, 1983): 63–5.

(2) J. A. Galbraith and S. Haughton, *Manual of Hydrostatics* (London, 1854): *idem, Manual of Mechanics* (London, ₄1856); *idem, Manual of Optics* (London, 1854).

(3) Compare Number 136. On Newton see Number 109 note (3); and on Ampère see Number 66 note (3) and Number 84.

(4) Michael Faraday, *Experimental Researches in Electricity*, 3 vols. (London, 1839–55).

motions of a ring of satellites,[5] a very neat piece of work, by Ramage the maker of the 'top'.[6]

For other things – I have not much time in winter for improving my mind. I have read Froudes History,[7] Aurora Leigh[8] & Hopkins' Essay on Geology,[9] also Herschel's collected Essays[10] which I like much also Lavaters life and Physiognomy[11] which has introduced me to him pleasantly though verbosely. I like the man very much quite apart from his conclusions and dogmas. They are only results and far inferior to methods. But many of them are true if properly understood and applied, and I suppose the rest are worth respect as the statements of a truth telling man.

Well work is good and reading is good but friends are better. I have but a finite number of friends and they are dropping off one here one there. A few live and flourish. Let it be long. And let us work while it is day, for the night is coming, and work by day leads to rest by night.

How is C. J. Monro? Do you ever hear of Mrs Pomeroy? What of Farrar?[12]

I suppose you know that Slesser the Senior Wrangler is of Kings College Aberdeen and that Smith the 2nd is of Glasgow.[13] Smith will do something yet, better than 2nd Wrangler. Another Aberdonian (Kings), Stirling, was 1st math. in the May at Trinity.[14]

I think one of my men is inclined that way. I am not encouraging him to it as he is very ambitious, but if he determines on it he will do well, if he keeps out of a small college. He is far too versatile to be trusted where boating billiards beer &c are more immediate paths to distinction than the pursuit of wisdom either mathematical, classical or social. If he could, I think he would naturally seek those whom he could respect as his superiors in something or other, and so avoid being the cock of a dunghill.

Yours truly
JAMES CLERK MAXWELL

---

(5) See Number 143. (6) Number 108 note (6). (7) Number 142 note (4).

(8) Elizabeth Barrett Browning, *Aurora Leigh* (London, 1857).

(9) William Hopkins, 'Geology', in *Cambridge Essays, contributed by Members of the University* (London, 1857): 172–240.

(10) J. F. W. Herschel, *Essays from the Edinburgh and Quarterly Reviews* (London, 1857).

(11) *The Life of Johann Kaspar Lavater* (London, 1845); and Johann Caspar Lavater, *Physiognomische Fragmente*, 4 vols. (Leipzig, 1775–8); (trans.) John Caspar Lavater, *Essays on Physiognomy, designed to promote the Knowledge and Love of Mankind*, 3 vols. (London, 1789–98). See Number 118. (12) Number 56 note (11).

(13) See Number 143 notes (10) and (11). (14) James Stirling, Trinity 1856 (Venn).

FROM A LETTER TO LEWIS CAMPBELL

17 FEBRUARY 1858

From Campbell and Garnett, *Life of Maxwell*[1]

129 Union Street
Aberdeen
17 February 1858

I have not been reading much of late. I have been hard at mathematics. In fact I set myself a great arithmetical job of calculating the tangential action of two rings of satellites, and I am near through with it now.[2] I have got a very neat model of my theoretical ring, a credit to Aberdeen workmen. Here is a diagram, but the thing is complex and difficult to draw:

Two wheels turning on parallel parts of a cranked axle; thirty-six little cranks of same length between corresponding points of the circumferences; each carries a little ivory satellite.[3]

---

(1) *Life of Maxwell*: 302–3.
(2) See Maxwell, *Saturn's Rings*: 45–51 (= *Scientific Papers*, **1**: 345–52). See Numbers 134, 138 and 155.
(3) See Number 143.

LETTER TO CECIL JAMES MONRO

27 FEBRUARY 1858

From the original in the Greater London Record Office, London[1]

129 Union Street
Aberdeen 27th Feb 1858

Dear Monro

I send you a copy of a certain lucubration on Optical Instruments[2] written at Elphinstones[3] instigation. The last two props are new since I last put anything of the sort together and I think the proof of the impossibility of an opt. inst. perfect for more than one distance is an achievement by low mathematics of what has not been tried by high.[4] However it is very clumsy still & I have had no time to sweeten its outline. I wrote to Litchfield some weeks ago but have not heard of him so I enclose you his copy, case he may not be in town. In fact I have written to Elphinstone Wilkinson & Droop without success and to Hopkins Miller Lightfoot and other hard wrought men with success.[5] The rule is doubtful but I know you write.

I have got a pamphlet on the College questions from our defunct friend Roby very like his style of thinking. I wish he would have written essays.[6]

I have taken up the Celibate question in a practical point of view and I hope that before the summer is well begun two members of that profession will have passed into a holier estate.[7] The 'party' who is to accompany me through that change is daughter of Principal Dewar of this College and University. Of course I am not going to give you a description nor to tell you who she is like because of course I never saw anyone the least like her even to look at. Perhaps the best

---

(1) Greater London Record Office, Acc. 1063/2085.

(2) J. C. Maxwell, 'On the general laws of optical instruments', *Quarterly Journal of Pure and Applied Mathematics*, **2** (1858): 232–46 (= *Scientific Papers*, **1**: 271–85).

(3) Number 67 note (3).

(4) Compare Maxwell's paper 'On the elementary theory of optical instruments', *Proc. Camb. Phil. Soc.*, **1** (1857): 173–5 (= *Scientific Papers*, **1**: 238–40); and Numbers 91, 92, and 93.

(5) Michael Marlow Umfreville Wilkinson, Trinity 1850; Joseph Barber Lightfoot, Trinity 1847 (Venn); and see Numbers 37 note (2), 93 note (5), and 135 note (1).

(6) Henry John Roby, St John's 1849 (Venn); elected to the Apostles in 1855, and resigned from the Society. See Henry John Roby, *Remarks on College Reform* (Cambridge, 1858); and compare Number 124. On his election to the Apostles see Monro's letter to Maxwell of 7 March [1855] (Greater London Record Office, Acc. 1063/2093).

(7) See Maxwell's letter to Jane Cay of 18 February 1858 (*Life of Maxwell*: 303–4) announcing that 'I am going to have a wife'; Katherine Mary Dewar, daughter of the Rev. Dr Daniel Dewar, Principal of Marischal College; see [P. J. Anderson,] *Officers of the Marischal College and University of Aberdeen 1593–1860* (Aberdeen, 1897): 30.

plan will be to wait till we are regularly settled and then pay us a visit. We are not going to make pilgrimages this year but to get well rooted at home first.

I am quite overflowed with congratulation letters from my people, but they are all based on my own statements, for none of them know her or can get any information unless they take my word, which they do. But if the high & lofty theories given by 'authors' are vulgarly supposed far above the actual and practicable, I can just reverse the sentence, and assure you that though the sentiments of some authors are a great deal loftier than those of others, we have discovered a more excellent way and without ever having uttered a grand sentiment on either hand we manage to be united in all things, high & deep, and all between, so that working together and thinking together, we shall both be free.

This rests on only about a years experience and that not all very profound, but I have tried all my life to understand people and it will be to little purpose if I fail in understanding her. But I do, and I am understood, which nails it QED. So we hope to domesticate in Galloway this summer. We have not much to alter there, so it is quite possible that after a while we may see a friend or two. But we must see.

Just finishing Hydrostatics, and going on to Optics. The advanced men have been attending well and have done Newton & Lunar roughly, Magnetism, Static & Current Electricity, Electro-magnetism Induction of currents & Diamagnetism. After a lecture on Telegraphs we go on to Physical Optics with them.[8]

I have been giving the 'mechanics' a good deal about Heat, especially Latent ditto to explain natural phenom. and on Monday we go on to Steam Engine.

What sort of faculty is Geometrical Instinct. I see it advertised, but I do not twig what it can be.[9] Have you heard anything of Mrs Pomeroy?

Yours truly

J. C. Maxwell

---

(8) See Numbers 136 and 144.

(9) Cecil James Monro, 'An example of the instinct of constructive geometry', *Quarterly Journal of Pure and Applied Mathematics*, **2** (1858): 225–9, esp. 225; 'There is a class of optical illusions, as they may be called, a readiness in the eye to detect any simple geometrical relations which may exist among an assemblage of points presented to it.' See Number 147.

LETTER TO RICHARD BUCKLEY LITCHFIELD

5 MARCH 1858

From the original in the library of Trinity College, Cambridge[1]

129 Union Street
Aberdeen 5th March 1858

Dear Litchfield

I was very glad to get your letter and to enter into correspondence with you again. My 'lines' are so pleasant to me that I think that everybody ought to come to me to catch the infection of happiness. This College work is what I and my father looked forward to for long and I find we were both quite right, that it was the thing for me to do. And with respect to the particular College I think we have more discipline and more liberty and therefore more power of useful work than anywhere else. It is a great thing to be the acknowledged 'regent' of ones class for a year so as to have them to oneself except in mathematics and what additional classes they take. Then the next year I get those that choose to come, which makes a select class for the higher branches. They have all great power of work.

In Aberdeen I have met with great kindness from all sects of people and you know of my greatest achievement in the way of discovery namely the method of converting friendship & esteem into something far better. We are following up that discovery and making more of it every day, getting deeper and deeper into the mysteries of personality so as to know that we ourselves are united and not merely attracted by qualities or virtues either bodily or mental. Don't suppose we talk metaphysics. Her father and mother will certainly become mine for I respect them as such already yet I know that she is well prepared to follow the general law on that subject and to leave them for me, and surely nothing else could ever come between us.

You will easily see that my 'confession of faith' must be liable to the objection that Satan made against Job's piety. One thing I would have you know, that I feel as free from compulsion to any form of compromised faith, as I did before I had any one to take care of, for I think we both believe too much to be easily brought into bondage to any set of opinions.

With respect to the 'material sciences', they appear to me to be the appointed road to all *scientific* truth, whether metaphysical, mental or social. The knowledge which exists on these subjects derives a great part of its value from

---

(1) Trinity College, Cambridge, Add. MS Letters c.1[90]. Published in extract in *Life of Maxwell*: 304–7.

ideas suggested by analogies from the material sciences, and the remaining part, though valuable and important to mankind is not *scientific* but aphoristic.

The chief *philosophical* value of physics is that it gives the mind something distinct to lay hold of, which if you dont, Nature at once tells you you are wrong. Now every stage of this conquest of truth leaves a more or less presentable trace on the memory so that materials are furnished here more than anywhere else for the investigation of *the* great question 'How does Knowledge come'.

I have observed that the practical cultivators of science (eg Sir J. Herschel, Faraday, Ampère, Oersted, Newton, Young) although differing excessively in turn of mind have all a distinctness and a freedom from the tyranny of words in dealing with questions of Order, Law, &c, which pure speculators and literary men never attain.[2]

Now I am going to put down something on my own authority which you must not take for more than it is worth. There are certain men who write books who assume that whatever things are orderly certain and capable of being accurately predicted by men of experience, belong to one category, and whatever things are the result of conscious action, whatever are capricious, contingent and cannot be foreseen belong to another category.

All the time I have lived and thought I have seen more and more reason to disagree with this opinion and to hold that all want of order caprice and unaccountableness results from interference with liberty, which would, if unimpeded, result in order, certainty and trustworthiness (certainty of success of predicting).

Remember I do not say that caprice and disorder are not the result of free will (so called) only I say that there is a liberty which is not disorder and that this is by no means less free than the other, but more.

In the next place there are various states of mind and schools of philosophy corresponding to various stages in the evolution of the idea of liberty.

In one phase, human actions are the resultant (by par^{m} of forces) of the various attractions of surrounding things, modified in some degree by internal states regarding which all that is to be said is that they are subjectively capricious, objectively the 'Result of Law', that is, the wilfulness of our wills feels to us like liberty, being in reality necessity.

In another phase, the wilfulness is seen to be anything but free will, since it is merely a submission to the strongest attraction after the fashion of material things. So some say that a mans will is the root of all evil in him and that he should mortify it out till nothing of himself remains and the man and his

---

(2) Compare Maxwell's comments in Numbers 105 and 183.

selfishness disappear together. So said Gotama Buddha (see Max Muller)[3] and many Christians have said and thought nearly the same thing.

Nevertheless there is another phase still in which there appears a possibility of the exact contrary to the first state, namely an abandonment of wilfulness without extinction of will but rather by means of a great development of will whereby instead of being consciously free and really in subjection to unknown laws, it becomes consciously acting by law, and really free from the interference of unrecognised laws.

There is a screed of metaphysics. I dont suppose that is what you wanted. I have no nostrum that is exactly what you want. Every man must brew his own, or at least fill his own glass for himself, but I greatly desire to hear more from you just to get into rapport.

As to the Roman Catholic question, it is another piece of the doctrine of Liberty. People get tired of being able to do as they like and having to choose their own steps and so they put themselves under holy men who, no doubt are really wiser than themselves. But it is not only wrong but impossible to transfer either will or responsibility to another, and after the formulæ have been gone through the patient has just as much responsibility as before, and feels it, too. But it is a sad thing for any one to lose sight of their work and to have to seek some conventional, arbitrary treadmill occupation prescribed by sanitary jailers.

[4][...] each others affairs, and though I have a great respect for my married friends, I greatly doubt the possibility of their knowing each other at all up to the mark, for we find it hard work, though we set to it in earnest every day all by ourselves, and have really no temptation to any other kind of amusement.

With respect to the Class, I send you the paper they did last week. 5 floored it approximately 2 first rate.

I got half a dozen correct answers to questions on the effects of mixtures of ice and steam in various proportions, and on the effects of heating and cooling on the thrust of iron beams (numerical). From the higher class I have an essay on Vision (Construction of Eye, spectacles, stereoscopes &c) so the work done is equivalent to the work spent.

[...] begin with a friend or two here at home [...] and run the gauntlet with the entire clan as we shall have to do in Edinburgh some time in Autumn, before next session.

Give my blessing to Monro & Lushington.[5]

I knew there was a low cunning faculty for the secretion of Algebra but that

---

(3) Max Muller, *Buddhism and Buddhist Pilgrims. A Review of M. Stanislas Julien's " Voyages des pèlerins bouddhistes"* (London/Edinburgh, 1857) : 16–17.

(4) The manuscript is defaced.   (5) Number 101 note (4).

constructive geometry should be the result of an instinct[6] seemed an insinuation against the humanity of Monge himself.[7]

Yours truly
J. C. MAXWELL

(6) See Number 146 esp. note (9).
(7) The allusion is to Gaspard Monge's *Géométrie Descriptive*, new edn (Paris, 1811).

FROM A LETTER TO LEWIS CAMPBELL

15 MARCH 1858

From Campbell and Garnett, *Life of Maxwell*[1]

Aberdeen
15 March 1858

When we had done with the eclipse to-day,[2] the next calculation was about the conjunction. The rough approximations bring it out early in June....

The first part of May I will be busy at home. The second part I may go to Cambridge, to London, to Brighton, as may be devised. After which we concentrate our two selves at Aberdeen by the principle of concerted tactics. This done, we steal a march, and throw our forces into the happy valley, which we shall occupy without fear, and we only wait your signals to be ready to welcome reinforcements from Brighton.... Good night. – Your affectionate friend.

J. C. MAXWELL

N.B. – We are going to do optical experiments together in summer.[3] I am getting two prisms, and our eyes are so good as to see the *spot* on the sun to-day without a telescope.

---

(1) *Life of Maxwell*: 307.

(2) The total eclipse of the sun on 15 March 1858; see the reports on the eclipse as seen in London, Doncaster, Leeds and Liverpool in *The Times* (16 March 1858). Compare Forbes' letter to Maxwell of 16 March 1858: 'We saw the Eclipse very badly, and it seems that in England it was no better' (*Life of Maxwell*: 308).

(3) See Numbers 149 and 152.

LETTER TO CECIL JAMES MONRO

29 APRIL 1858

From the original in the Greater London Record Office, London[1]

Glenlair
Springholm
Dumfries
29th April 1858

Dear Monro

I shall be in London probably on the evening of the 12th May, and may stay the rest of that week. Will the house of Harris be open for my reception? Otherwise I shall betake myself to the relatives in the West, but on the whole I should prefer the sign of the spectacles when I am in London, and visit my relatives under the ancestral roof and not in hired houses, especially as the said roof is being widened this year.

So if N° 52 is able to contain me pray write to me here.

I shall be here till the 6th. I go that evening to London and on to Milford near Lymington Hantz, Revd Lewis Campbell, Vicar, stay Sunday with him and on Monday to Brighton, where dwells one Mrs Andrews, widow, and F. P. Andrews, spinster an infant. On Tuesday 11th the infant is to be made a Vicaress or 'Vixter' as Trench[2] would have it and then there will be no need of me as I shall have executed the function of paranymph for the 3rd and positively the last time. So I shall tread North again, and do a short visit to London and see Litchfield & you and Whitt[3] at least. Try and find out the addresses of any friends of mine that you know and be sure to remember the Brethren that are scattered abroad.

As to myself I shall not make long stay but go to Edinburgh and thence to Aberdeen or Perth as the case may be. When the month of May is out my business will be transacted and we are to be married on the 2nd June,[4] and go straight home. On the 5th we begin our duties here with receiving Lewis Campbell & wife and we purpose to be here a good while if nothing interferes.

---

(1) Greater London Record Office, Acc. 1063/2086.

(2) See Richard Chenevix Trench, *English: Past and Present* (London, 1855): 110; noting the growing infrequency in the use of 'the female termination which we employ in certain words, such as from "heir" "heiress"...'. Compare Maxwell's letter to Campbell of 17 October 1855 and the fragments of his Apostles essay on the 'Modern vocabulary of the English language' (*Life of Maxwell*: 218, 235).   (3) John Whitt, Trinity 1850 (Venn).

(4) At Aberdeen. See Maxwell's letters to Jane Cay of 24 May 1858 (Autographen-Sammlung Geigy-Hagenbuch Nr.2168, Universitätsbibliothek, Basel) and to Elizabeth Cay of 27 May 1858 (Trinity College, Cambridge, Add. MS b.52¹).

Give Litchfield my love. I hope not to miss seeing him in London at least, and he must certainly not forget his procrastinated determination of coming to Scotland.

I have been studying Lear. Have you been to the Princesses?

I displayed my model of Saturns Ring at the Edinburgh Royal Society on the 19th.[5] The anatomists seemed to take most interest in the construction of it. We are going to do some experiments on colour this summer if my prisms turn out well.[6] I have got a beautiful set of slits made by Ramage,[7] to let in the different pencils of light at the proper places and of the proper breadths.[8]

Yours truly

J. C. MAXWELL

(5) J. C. Maxwell, 'On theories of the constitution of Saturn's rings', *Proc. Roy. Soc. Edinb.*, **4** (1858): 99–101 (= *Scientific Papers*, **1**: 286–7). See a letter from J. D. Forbes to Maxwell of 16 March 1858 (*Life of Maxwell*: 307–8): 'Your notice of Saturn will be very acceptable. But it should not run too great a length, as the 19th [April] will probably be our last meeting, and is always a crowded billet'. See plate VII.

(6) But see Number 152.   (7) Number 108 note (6).   (8) Number 154.

150

LETTER TO JAMES DAVID FORBES

3 MAY 1858

From the original in the University Library, St Andrews[1]

Glenlair
May 3rd 1858

Dear Sir

I have read your remarks on Ice[2] and have thought a little about the gradual liquefaction of ice, though I have not yet seen Persons own statement[3] nor anything else about it.

Conceive a block of ice below 32 placed in a vessel of water. The water will soon be reduced to the temperature 32, that being by definition the temperature of thawed ice that is, water which has just been melted out of ice. Now heat will be conducted into the body of the ice. What will be the law of conduction. Suppose the conductivity of ice to be the same in all states, then heat always passes from warm to cold and the gain or loss of heat by any portion can be found.

The effect of heat on a particle of ice is partly to warm it and partly to change its state, the more change of state, the less change of temperature and for equal changes of temperature, more heat is required. Now if the change of state takes place through a finite range of temperature all that is required to adapt the theory of conduction to this case is to suppose the 'capacity for heat' very great between the limits of the transformation.

Now if the water be kept above 32 it will communicate heat to the fresh melted ice and so on, the effect of which will be to drive back the surface of the ice, at a certain rate, all the heat being absorbed as it enters the successive strata of the half melted ice.

If the ice be kept below the temperature of hard ice and the water at 32° the surface will march forward into the water.

But if both ice and water are left together for a sufficient time the debatable portion will become wider, till the whole is reduced to the same condition of half melted ice.

If there is very little water the result will be solid. If very little ice it will be

---

(1) University Library, St Andrews, Forbes MSS 1858/51.

(2) J. D. Forbes, 'On some properties of ice near its melting point', *Proc. Roy. Soc. Edinb.*, **4** (1858): 103–6. The paper was presented on 19 April 1858; see Number 149, esp. note (5). See Forbes' letter to Faraday of 14 August 1858, in *The Selected Correspondence of Michael Faraday*, ed. L. P. Williams, 2 vols. (Cambridge, 1971), **2**: 910–11.

(3) C. C. Person, 'Sur la chaleur latente de fusion de la glace', *Comptes Rendus*, **30** (1850): 526–8. To explain the freezing and liquefaction of ice Person claimed that ice absorbs latent heat at a temperature below its freezing point.

liquid. The first is the case of the experiments cited. The second is the case of a thin piece of ice in a vessel of water at 32°.

So that the half melted portion is sure to increase unless the ice is cooled or the water warmed continually.

Now what is the theory of freezing, for there it is the water that is cooled, and often below 32° and then starts into ice very nearly at 32°.

This is a case of unstable equilibrium apparently, and the ice so formed has not, I think, very good surfaces.

In ponds, the ice is cooled and thickens at the bottom, and the faster this goes on the more distinct will be the surface. The surface of a crystal of ice in air under 32° is of course perfect.

I have perhaps made rather a confused statement just now because I am only putting my mind in order at present. I had settled it as far as to see that you 'had reason' but I thought you would like to see the results of the admission in a crude form as I have not had time to digest them. I shall read about it in course of time and next winter try some experiments on freezing in the interior of a vessel of water, if that is possible by any means.[4]

I remain
Yours truly
J. C. MAXWELL

---

(4) Compare Maxwell's discussion of 'regelation' in his *Theory of Heat* (London, 1871): 176.

LETTER TO CECIL JAMES MONRO

24 JULY 1858

From the original in the Greater London Record Office, London[1]

Glenlair
July 24th 1858

Dear Monro

The books came all right by carrier & do credit to the packer being all perfect. Though massive they have some chance of being read. We are no great students at present, preferring various passive enjoyments resulting from the elemental influences of sun, wind & streams.

This week I have begun to make a small hole into Saturn who has slept on his voluminous ring 5 months.

I had a letter from Farrar lately. He seems like the ostrich to have run his head into Smith's Dictionary as a last resource. I think his labours in that labyrinth have wrought him relief.[2]

The Thames seems very bad at present.

Stir up Litchfield to write me a letter if you cant yourself for we are hardly able here to conceive of dwellers in the Village or to exchange words with them that will be understood. There is nothing like out of door life and quiet for imparting a wholesome conception of things as they are to the much revolving brain.

Will you attempt to convey the thanks of my wife & myself to Litchfield & Lushington and we are both glad when we hear of any of our friends by letter as our present condition is not favourable to much active correspondence with the outer world.

Elphinstone tells me that the Master[3] (whose cards we received lately) is flourishing. I hope he is better advised than his enemy Sir D Brewster.[4] It is well I am not a Literary Man. Such men are being divided from their wives by the dozen, and who knows how far the Law may extend.

I remain
Yours truly
J. C. MAXWELL

---

(1) Greater London Record Office, Acc. 1063/2087.

(2) Alluding to F. W. Farrar's philological interests; and to William Smith's *Latin–English Dictionary* (London, 1855) and *A Smaller Latin–English Dictionary* (London, 1855).

(3) William Whewell, Master of Trinity College, Cambridge.

(4) Maxwell is alluding to Brewster's long series of controversies with Whewell. See David Brewster, 'Whewell's History of the Inductive Sciences', *Edinburgh Review*, **66** (1837): 110–51; 'Whewell's Philosophy of the Inductive Sciences', *ibid.*, **74** (1842): 265–306; and his review of Whewell's anonymous essay *Of the Plurality of Worlds* (London, 1853) in *North British Review*, **21** (1854): 1–44, expanded into Brewster's *More Worlds than One* (London, 1854). See also Number 72 for Maxwell's comments on Brewster's defence of his theory of the spectrum against Whewell.

LETTER TO GEORGE GABRIEL STOKES

7 SEPTEMBER 1858

From the original in the University Library, Cambridge[1]

$$\left\{ \begin{array}{l} \text{Glenlair} \\ \text{Springholm} \\ \text{Dumfries} \end{array} \right.$$

Sept 7th 1858

Dear Stokes

I am just finishing my essay on Saturns Rings,[2] and there is a problem in it about a set of fluid rings revolving each with its proper velocity with friction against its neighbours.[3] Now I am in want of the *coefficient* of *internal friction* in water and in air, in order to condescend to numbers. I have your paper on pendulums but it is locked up at Aberdeen so I have written to you rather than wait till I go.[4]

It is the coefft *B* of your paper on Elasticity & fluid friction,[5] that is the number of units of force acting tangentially on unit of area sliding with unit of relative velocity past a fluid plane at unit of distance[6] stating units of force & length.[7]

---

(1) ULC Add. MSS 7656, M 409. Previously published in Larmor, *Correspondence*, 2: 7–8.

(2) See Numbers 153 and 155.

(3) J. C. Maxwell, *On the Stability of the Motion of Saturn's Rings* (Cambridge, 1859): 53–4 (= *Scientific Papers*, 1: 354–6).

(4) George Gabriel Stokes, 'On the effect of the internal friction of fluids on the motion of pendulums', *Trans. Camb. Phil. Soc.*, 9, part 2 (1851): [8]–[106] (= *Papers*, 3: 1–136).

(5) See G. G. Stokes, 'On the theories of the internal friction of fluids in motion, and of the equilibrium and motion of elastic solids', *Trans. Camb. Phil. Soc.*, 8 (1845): 287–319 (= *Papers*, 1: 75–129). The coefficient *B* is the coefficient of 'cubical compressibility'; in Stokes' equations for the equilibrium of elastic solids *B* corresponds to the coefficient of viscosity μ in the 'Navier–Stokes' equations for fluid motion; see Stokes, 'On the theories of the internal friction of fluids in motion': 288, 297, 311; see equations (12) and (30). See Number 26 notes (11) and (94). See also G. G. Stokes, 'Report on recent researches in hydrodynamics', *Report of the Sixteenth Meeting of the British Association...in 1846* (London, 1847): 1–20 (= *Papers*, 1: 157–87).

(6) See Stokes, 'On the effect of the internal friction of fluids on the motion of pendulums': 16–17. 'The motion [of the fluid] will evidently consist of a sort of continuous sliding'; if the fluid be supposed as moving in planes parallel to the plane of xy, the motion taking place with a velocity v in a direction parallel to the axis of y, then 'the total pressure referred to a unit of surface is compounded of a normal pressure corresponding to the density, and a tangential pressure expressed by $\mu\, dv/dz$, which tends to reduce the relative motion'. The constant μ varies with the density ρ, so $\mu = \mu'\rho$. 'The constant μ' may conveniently be called the *index of friction* of the fluid, whether liquid or gas, to which it relates.'

(7) See Stokes, 'On the effect of the internal friction of fluids on the motion of pendulums': 65, 81. Stokes gives values of $\sqrt{\mu'} = 0.116$ for air and $\sqrt{\mu'} = 0.0564$ for water; where '$\sqrt{\mu'}$ expresses

I think you have got it somewhere at hand. I have not.

It is not long since I began to do mathematics after a somewhat long rest. I have set up a neat model for showing the disturbances among a set of satellites forming a ring.[8] It was made by Ramage of Aberdeen who made my dynamical top.

I have done no optics this summer but have been vegetating in the country with success.

Yours truly

JAMES CLERK MAXWELL

a length divided by the square root of a time', units of length and time being an inch and a second. Maxwell uses these values in *Saturn's Rings*: 54. Maxwell uses this result in discussing the viscosity of gases; see his letter to Stokes of 30 May 1859 (Number 157). See also Horace Lamb, *A Treatise on the Mathematical Theory of the Motion of Fluids* (Cambridge, 1879): 219.

(8) See Number 143.

LETTER TO [ALEXANDER MACMILLAN] [1]

17 SEPTEMBER 1858

From the original in the Smithsonian Institution, Washington DC. [2]

Glenlair
17th Sept 1858

Dear Sir

The figures may all be gathered into a lithographic plate at the end, only numbers must be put to them, and I will make references at the proper places when I get proof. [3]

I do not think it likely that we shall be in Cambridge for a good while.

Yours truly
J. C. MAXWELL

---

(1) Daniel Macmillan died in 1857; his brother Alexander was his partner in the publishers Macmillan & Co. (*DNB*).

(2) Dibner Library, Smithsonian Institution Libraries, Washington DC.

(3) James Clerk Maxwell, *On the Stability of the Motion of Saturn's Rings* (Macmillan: Cambridge, 1859) (= *Scientific Papers*, **1**: 288–376).

154

DESCRIPTION OF AN INSTRUMENT FOR COMPARING MIXTURES OF THE COLOURS OF THE SPECTRUM

circa 1858[1]

From the original in the University Library, Cambridge[2]

DESCRIPTION OF AN INSTRUMENT FOR COMBINING THE PURE RAYS OF THE PRISMATIC SPECTRUM[3]

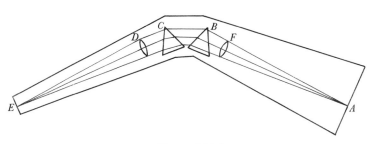

Figure 154,1

The framework of the instrument consists of a wooden base on which the optical apparatus is fixed. This consists of one or two flint-glass prisms *BC*, two lenses *D* & *F* & a slit to let in the light at *A* and another at which the eye is placed at *E*. The whole is covered with a lid of thick pasteboard to keep out stray light.

(A) The slit *A* is of a pecular construction to be afterwards explained.
(F) The Lens *F* is placed so that *FA* is its principal focal distance.

Rays from *A* therefore after passing thro *F* become parallel.

(B) They then pass thro the prisms *B* & *C* which are
(C) placed in the position of least deviation & then
(D) thro the lens (*D*) so as to be brought to a focus
(E) again at the slit *E*, *E* being the principal focus of *D*.

---

(1) See Number 149.
(2) ULC Add. MSS 7655, V, b/22(i).
(3) Compare '§V. Description of an instrument for making definite mixtures of the colours of the spectrum' of Maxwell's paper 'On the theory of compound colours, and the relations of the colours of the spectrum', *Phil. Trans.*, **150** (1860): 57–84, esp. 65–7, and figs. 1 and 2 opposite 84 (= *Scientific Papers*, **1**: 420–2, 444). The instrument as finally perfected enabled mixtures of spectral colours to be directly compared with white light, the white light being kept separate by means of a partition within the apparatus. Compare the shortened portable version of the instrument: Figure 168, 1.

In this way all rays of a certain refrangibility which proceed from *A* are brought to a focus at *E*. Those of greater and less refrangibility are in like manner brought to foci near *E* & form a pure spectrum on the screen at *E* but do not pass thro it.

An eye placed at *E* will therefore see the field of view uniformly illuminated by one kind of light only.

If we now change the position of the slit *A* we shall have another pure spectrum at *E* with the colours shifted in position owing to the alteration of the source of the incident light. A different colour will therefore fall on the eye-slit at *E*.

If both slits at *A* be now opened, both colours will be seen by the eye at *E* uniformly illuminating the field of view.

In this way any number of colours may be mixed by opening various slits near *A*.

On the construction of the part *A*

$A_1 A$ is a frame in which the apparatus works.

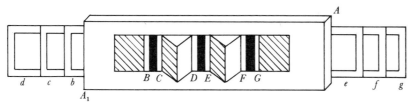

Figure 154,2

BC, *DE*, *FG* are three pairs of knife edges which form the moveable slits. The ends of these slide in two grooves and are moved by means of the handles *b c d e f g* which project at the sides of the frame. The handles are of thin metal or cardboard of the form ▭ .

The upper & lower branches support the knife edge and move it parallel to itself along the grooves. The handles themselves slide in a groove in the frame parallel to that in which the knife edges slide and the knife edges are connected with them so as to work freely in their own groove without interference.

The spaces between the knife-edges *CD*, & *EF* are filled up by folding screens of which the edges are fastened to the knife edges while the joint opens and shuts so as to allow of the variation of the distance of the slits.

The spaces beyond *B* & *G* are filled up with flat screens.

In this way the positions and breadths of the three slits may be varied at pleasure within certain limits, and thus various mixtures of light may be produced.

There should be a scale of inches and parts both above & below the knife edges for the purpose of ascertaining their positions and on one of these the colours & principal lines of the spectrum should be indicated.

This must be done by turning the instrument so as to let the light in at *E* while the eye observes the pure spectrum in the air at *A*, and marks the lines on the scale.

In order to compare the effects of two mixtures there should be another frame like *A* placed above it at as short a distance as possible. This frame need not have more than two slits.

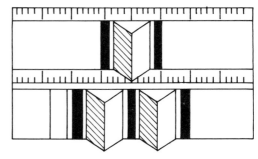

Figure 154,3

On the lens at *F*

This should consist of two similar lenses joined as here represented. The distance *AB* between the centres must be equal to the distance between the middles of the upper and under slits, and the lenses ought to be large enough to have a considerable joining. The joint *CD* must be ground smooth and cemented so as to appear as a fine line only.

The action of this part of the instrument may be understood more easily if we omit the prisms and suppose everything else the same.

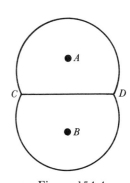

Figure 154,4

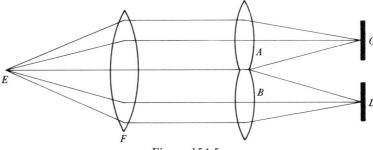

Figure 154,5

The above is a vertical section. *C* & *D* are the middles of the upper & lower slits. Rays from these points after falling on the two lenses at *A* & *B* become all parallel & pass on to *F* so as to be all brought to a single focus at *E*. The whole distance between *F* & *AB* must not be too great to admit of *A B* being distinctly visible by an eye at *E*.

An eye placed at *E* will see the two tints due to the two arrangements of the slits above & below the line of junction of the lenses *A B*, and in this way the results may be compared and equated. J. C. MAXWELL

LETTER TO JAMES CHALLIS

21 JANUARY 1859

From the original in the Library of the Institute of Astronomy, Cambridge[1]

Marischal College
Aberdeen Jan 21 1859

Dear Sir

I have received yours of the 18th.[2]

I sent the whole essay to press on 10th Sept, and it is now all printed and will, I suppose, be ready very soon.[3]

I am aware that I have to send copies to the University Library, the Library of St John's College and the examiners for the prize. May I ask if there were any examiners besides yourself & Prof. W. Thomson for although I have not heard of any other there may have been and I do not think the names of the examiners were published in the Calendar.[4]

I was prevented from working during the latter half of last session, but in the summer I got the theory of the mutual perturbations of the rings put into shape; and this is the most important part of the whole, as it shows that a system of numerous concentric rings, each stable in itself, is liable to perturbations which increase *slowly* without limit, till certain pairs of the rings are thrown into confusion, after which a new temporary arrangement takes place, till other disturbances have grown to a destructive height & so on.[5]

As the last sheet is now corrected and all the others are worked off I hope that publication and distribution will take place very soon.

I am Dear Sir
Yours truly
J. CLERK MAXWELL

Professor Challis[a]

(a) {Challis} Ans[d] Jan 25.[6]

(1) Cambridge Observatory Archives, Letter 1859/4 (Institute of Astronomy, Cambridge University). Previously published in A. T. Fuller, 'James Clerk Maxwell's Cambridge manuscripts: extracts relating to control and stability – III', *International Journal of Control*, **37** (1983): 1233–4.　　　　(2) Not extant.　　　　(3) See Number 153.

(4) The third examiner was Stephen Parkinson, St John's (Venn). The names of the examiners were given in the published 'Advertisement' for the Adams Prize of 23 March 1855; see Number 107 note (1), and compare J. C. Maxwell, *On the Stability of the Motion of Saturn's Rings* (Cambridge, 1859): iv (= *Scientific Papers*, **1**: 288). Only the terms of the prize, not its subject, were published in *The Cambridge University Calendar for the Year 1856* (Cambridge, 1856): 179–80.

(5) See Maxwell, *Saturn's Rings*: 45–54, 65–6 (= *Scientific Papers*, **1**: 345–56, 371); and Maxwell's letter to Challis of 24 November 1857 (Number 138).　　　　(6) Not extant.

LETTER TO PETER GUTHRIE TAIT

9 MARCH 1859

From the original in the University Library, Cambridge[1]

Marischal College
Aberdeen March 9 1859

Dear Tait

Many thanks for the felicitations of the 26th Ult. and may you long be preserved from the claw of the cat, the iodide of cyanogen and the puncture of the scalpel.

The damage of the brass top is £3..3, complete with handle to spin it and prevent drilling of the finger.[2] The patterns are all in existence.[3] The Trinity lecturers broke the cup of theirs by bumping it down inconsiderately. They have got a new cup and are mounted again. All the experiments succeed best at a moderate velocity such as can be got up by finger & thumb twirling.

For a great spin the balancing should be done carefully so that the standpunkt is the true centre of gravity, otherwise the punkt will stagger in its cup and will deteriorate it, and also much Vis viva[4] will leak out through the pillar. Care must also be taken that the knuckles or other mutilable portions of the spectator are not brought into collision with the milled heads. After making the standpunkt the centre of gravity make the axle a permanent axis by means of the vertical screws then lap on string between the card C and the bob B.

Take the handle in one hand and the string in the other & give a graduated pull, strong without rage, and you will have a spin about the vertical axle. Then with your finger on the point move it a little to one side and give it a good jerk to one side, (cuff-on-the-head style) and it will start off on an inst[s] axis and so on.

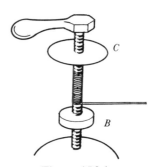

Figure 156,1

---

(1) ULC Add. MSS 7655, I, b/1.

(2) See Number 116 and plate VI.

(3) See figures 1 to 5 of Maxwell's paper 'On a dynamical top', *Trans. Roy. Soc. Edinb.*, **21** (1857): 557–70, opposite 570 (= *Scientific Papers*, **1**: 262).

(4) See Number 133 note (7).

If you make a reticulation of hexagons and colour them red blue & green as thus ——

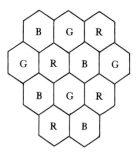

Figure 156,2

then if the 'invariable' axis[5] coincides with a red spot there will be a blue-green circle round it.

If with a green, a purple. If with a blue, a brown-yellow from the vermillion & emerald green.[6]

Fuller Brazier & Maclure are in complete preservation the latter much excited.[7]

I have got a good specimen of colour blindness in my class. I experimented on him yesterday. He is absent today I hope not in fear of the top.[8]

I expect dull weather in a week so I have got through the optical experiments in the sun so as to prevent disappointed men.

Winter I suppose will begin here soon. We have had neither frost nor snow since November and for the last 3 weeks most splendid warm sunny weather.

Yours truly

J. CLERK MAXWELL

---

(5) See Number 116 note (4).

(6) Compare Maxwell, 'On a dynamical top': 566, 570, where he described a disc with four quadrants (vermilion, chrome yellow, emerald green, and ultramarine), that 'combine into a grayish tint' when the top spins axially, but 'burst into brilliant colours when the axis is disturbed'.

(7) See Numbers 96 note (2), 113 note (3), and 108 note (10).

(8) See Numbers 58 and 59, and plate III.

157

LETTER TO GEORGE GABRIEL STOKES

30 MAY 1859

From the original in the University Library, Cambridge[1]

Glenlair
Springholm
Dumfries
30 May 1859

Dear Stokes

I saw in the Philosophical Magazine of February/59 a paper by Clausius on the 'mean length of path of a particle of air or gas between consecutive collisions'[2] on the hypothesis of the elasticity of gas being due to the velocity of its particles and of their paths being rectilinear except when they come into close proximity to each other, which event may be called a collision.[3]

The result arrived at by Clausius is that of N particles Ne^{-n} reach a distance greater than nl where l is the 'mean path' and that

$$\frac{1}{l} = \tfrac{4}{3}\pi s^2 N$$

where s is the radius of the 'sphere of action' of a particle,[4] and N the number of particles in unit of volume.

Note I find $\qquad\qquad \dfrac{1}{l} = \sqrt{2}\pi s^2 N.$[5]

---

(1) ULC Add. MSS 7656, M 410. Previously published in Larmor, *Correspondence*, **2**: 8–11.

(2) Rudolf Clausius, 'On the mean length of the paths described by the separate molecules of gaseous bodies on the occurrence of molecular motion: together with some other remarks upon the mechanical theory of heat', *Phil. Mag.*, ser. 4, **17** (1859): 81–91; trans. of Clausius, 'Ueber die mittlere Länge der Wege, welche bei der Molecularbewegung gasförmiger Körper von den einzelnen Molecülen zurückgelegt werden; nebst einigen anderen Bemerkungen über die mechanische Wärmetheorie', *Ann. Phys.*, **105** (1858): 239–58.

(3) Clausius ignores the effect of 'the forces of chemical affinity', and the attractive force between molecules; he considers only collisions, the case of '*impact*' when 'a rebounding of the molecules takes place'; Clausius, 'On the mean length of the paths': 82–3.

(4) Clausius defines the 'mean length of path' between impacts: '*how far on an average can the molecule move, before its centre of gravity comes into the sphere of action of another molecule*'. By the term 'sphere of action [Wirkungssphäre] of the molecule' he supposes 'a sphere of radius ρ, described around a molecule and having its centre of gravity for a centre'; Clausius, 'On the mean length of the paths': 83–4; 'Ueber die mittlere Länge der Wege': 243.

(5) See Clausius, 'On the mean length of the paths': 88; and J. C. Maxwell, 'Illustrations of the dynamical theory of gases. Part I. On the motions and collisions of perfectly elastic spheres', *Phil. Mag.*, ser. 4, **19** (1860): 19–32, esp. 28 (= *Scientific Papers*, **1**: 386–7). For Clausius' response to Maxwell's correction see his paper 'On the dynamical theory of gases', *Phil. Mag.*, ser. 4, **19** (1860): 434–6; and see W. D. Niven's comment in *Scientific Papers*, **1**: 387n.

As we know nothing about either *s* or *N* I thought that it might be worth while examining the hypothesis of free particles acting by impact and comparing it with phenomena which seem to depend on this 'mean path'. I have therefore begun at the beginning and drawn up the theory of the motions and collisions of free particles acting only by impact, applying it to internal friction of gases diffusion of gases and conduction of heat through a gas (without radiation). Here is the theory of gaseous friction with its results.

Divide the gas into layers on each side of the plane in which you measure the friction, and suppose the tangential motion uniform in each layer but varying from one to another.[6] Then particles from a layer on one side of the plane will always be darting about and some of them will strike the particles

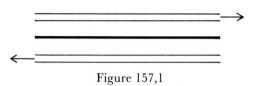

Figure 157,1

belonging to a layer on the other side of the plane, moving with a different mean velocity of translation.

Now though the velocities of these particles are very great and in all directions, their mean velocity, that of their centre of gravity, is that of the layer from which they started so that the other layer will receive so many particles per second, having a different velocity from its own.

Taking the action of all the layers on one side on all the layers on the other side I find for the force on unit of area[7]

$$F = \tfrac{1}{3} M N l v \frac{du}{dz}$$

where M = mass of a particle, N no of particles in unit of volume l = mean

---

(6) Compare Maxwell's letter to Stokes of 7 September 1858 (Number 152, esp. note (6)), on the application of Stokes' theory of viscosity to the motion of Saturn's rings. See G. G. Stokes, 'On the effect of the internal friction of fluids on the motion of pendulums', *Trans. Camb. Phil. Soc.*, **9** part 2 (1851): [8]–[106], esp. 16–17. Compare Maxwell, 'Illustrations of the dynamical theory of gases. Part I': 30–2 (= *Scientific Papers*, **1**: 390–1).

(7) Compare Maxwell, 'Illustrations of the dynamical theory of gases. Part I': 30 (= *Scientific Papers*, **1**: 389), where he obtains the result $p = \tfrac{1}{3} N M v^2$, where p is the pressure on unit of area. See R. Clausius, 'Ueber die Art der Bewegung welche wir Wärme nennen', *Ann. Phys.* **100** (1857): 353–80, esp. 374–5; (trans.) 'On the kind of motion which we call heat', *Phil. Mag.*, ser. 4, **14** (1857): 108–27, esp. 122–3. Compare also 'Problems' and 'Solutions of Problems' in the *Camb. & Dubl. Math. J.*, **8** (1853): 96, 270–2; '8(a). If an infinite number of perfectly elastic material points equally distributed through a hollow sphere, be set in motion each with any velocity, shew that the resulting continuous pressure (referred to unit of area) on the internal surface is equal to one-third of the *vis viva* of the particles divided by the volume of the sphere. – *St Peter's College Examination Papers*, June 1852. (b) Prove the same proposition for a hollow space of any form.' See C. Truesdell, 'Early kinetic theories of gases' *Archive for History of Exact Sciences*, **15** (1975): 1–66, esp. 19.

path, and v = mean molecular velocity and $\frac{du}{dz}$ velocity of slipping. Now you put

$$F = \mu \frac{du}{dz} \,^{(8)}$$

$$\therefore \mu = MNlv = \rho lv.^{(9)}$$

Now $$v = \sqrt{\frac{8k}{\pi}} \text{ and if we take}$$

$$\sqrt{k} = 930 \text{ feet per second}^{(10)} \; v = 1505 \text{ feet per second,}$$

---

(8) F is the 'tangential force which constitutes the internal friction of the gas'; μ the 'coefficient of internal friction'; see Maxwell, 'Illustrations of the dynamical theory of gases. Part I': 31 (= *Scientific Papers*, **1**: 390–1). Compare Stokes 'On the effect of the internal friction of fluids on the motion of pendulums': 16–17, and Number 152 note (6).

(9) Read: $\mu = \frac{1}{3}MNlv = \frac{1}{3}\rho lv$. See Maxwell, 'Illustrations of the dynamical theory of gases. Part I': 31 (= *Scientific Papers*, **1**: 391).

(10) See Maxwell, 'Illustrations of the dynamical theory of gases. Part I': 23, 30, 32 (= *Scientific Papers*, **1**: 381, 389, 391). k is a constant determined by the relation between the pressure and density of a gas $p = k\rho$, 'which is Boyle and Mariotte's law'. The velocity of sound is given by $\sqrt{\left(\gamma\frac{p}{\rho}\right)}$, where γ (the symbol being introduced by Poisson) is the ratio of the specific heat at constant pressure to that at constant volume; see W. J. M. Rankine, 'On Laplace's theory of sound', *Phil. Mag.*, ser. 4., **1** (1851): 225–7. See also P. S. de Laplace, *Traité de Mécanique Céleste*, 5 vols. (Paris, An VII [1799]–1825), **5**: 119–44, esp. 123; S. D. Poisson, *Traité de Mécanique*, 2 vols. (Paris, ₂1833), **2**: 693–720, esp. 714–15; and W. J. M. Rankine, 'On the mechanical action of heat, especially in gases and vapours', *Trans. Roy. Soc. Edinb.*, **20** (1850): 147–90, esp. 167, where Rankine gives a value of $\gamma = 1.401$ for air. Compare [W. J. M. Rankine,] 'Heat, theory of the mechanical action of, or thermodynamics', in J. P. Nichol, *A Cyclopaedia of the Physical Sciences* (London, 1857): 338–54, esp. 345; and W. J. M. Rankine, *A Manual of Applied Mechanics* (London/Glasgow, 1858): 563, where he gives $\gamma = 1.408$. Maxwell gives $\gamma = 1.408$ in Number 181 and in 'Illustrations of the dynamical theory of gases. Part III. On the collision of perfectly elastic bodies of any form', *Phil. Mag.*, ser. 4, **20** (1860): 33–7, esp. 36 (= *Scientific Papers*, **1**: 409). From a value for the velocity of sound in air, hence $\sqrt{k}$. In 'Illustrations of the dynamical theory of gases. Part I'; 32, Maxwell states that '$\sqrt{k} = 930$ feet per second for air at 60°'. Rankine gives the velocity of sound in air as 1092 ft/sec in dry air at 32° Fahr; and correcting for temperature as $1092\sqrt{\left(\frac{T+461°.2}{493°.2}\right)}$, where T is the temperature in degrees Fahrenheit; see Rankine, *Applied Mechanics*: 563–4. Maxwell possessed a presentation copy of Rankine's *Applied Mechanics* (Cavendish Laboratory, Cambridge). In a notebook entry (the notebook is inscribed 'J. Clerk Maxwell/King's College/London', so the entry presumably dates from after October 1860) Maxwell noted that:

'Vel. of sound in pure dry air at 32° = 1090.2 = V

Experimental – MM. Bravais & Martins

$$= 1090.5$$

$$- \text{Moll and Van Beek} = 1090.1$$

and if $\sqrt{\dfrac{\mu}{\rho}} = .116$ in inches & seconds[11] we find

$$l = \tfrac{1}{447000} \text{ of an inch}$$

and the number of collisions per second

$$8,077,000000,$$

for each particle.

The rate of diffusion of gases depends on more particles passing a given plane in one direction than another, owing to one gas being denser towards the one side and the other towards the other.

It appears that l should be the same for all pure gases at the same pressure and temperature but I have found only a very few experiments by Prof Graham on diffusion through measurable apertures, and these seem to give values of l much larger than that derived from friction.[12]

I should think it would be very difficult to make experiments on the conductivity of a gas so as to eliminate the effect of radiation from the sides of the containing vessel so that this would not give so good a method of determining l.

I do not know how far such speculations may be found to agree with facts,

---

at any temp. $T, v = V\sqrt{\dfrac{461.2 + T}{493.2}}$.' (Notebook 1, King's College, London). See Auguste Bravais and Charles Martins, 'Expériences relatives à la vitesse du son dans l'atmosphère', *Comptes Rendus*, **19** (1844): 1164–74; Gerard Moll and Albert Van Beek, 'An account of experiments on the velocity of sound, made in Holland', *Phil. Trans.*, **114** (1824): 424–56. These references and values are given by Rankine in his paper 'On the mechanical action of heat', *Trans. Roy. Soc. Edinb.*, **20** (1853): 565–89, on 589; and in his *A Manual of the Steam Engine and other Prime Movers* (London/Glasgow, 1859): 322.

(11) See Stokes, 'On the effect of the internal friction of fluids': 17, 65. Stokes writes the 'index of friction' as $\sqrt{\dfrac{\mu}{\rho}} = 0.116$ for air (in inches and seconds). See Number 152 notes (6) and (7).

(12) See Maxwell, 'Illustrations of the dynamical theory of gases. Part II. On the process of diffusion of two or more kinds of moving particles among one another', *Phil. Mag.*, ser. 4, **20** (1860): 21–33, esp. 31 (= *Scientific Papers*, **1**: 403). Using data from experiments by Thomas Graham, 'A short account of experimental researches on the diffusion of gases through each other, and their separation by mechanical means', *Quarterly Journal of Science*, **28** (1829): 74–83, he calculates a value for the mean free path, $l = 1/389000$ inch. He states that he obtained Graham's experimental data on gaseous diffusion as 'quoted by Herapath from Brande's *Quarterly Journal of Science*, Vol. XVIII p. 76'; see John Herapath, *Mathematical Physics; or the Mathematical Principles of Natural Philosophy: with a Development of the Causes of Heat, Gaseous Elasticity, Gravitation, and other Great Phenomena of Nature*, 2 vols. (London, 1847), **2**: 24–5. See also Maxwell's notes on 'Thomas Graham on Diffusion of Gases, Brandes Journal 1829 pt. 2 p. 74' (ULC Add. MSS 7655, V, f/7); in E. Garber, S. G. Brush and C. W. F. Everitt, *Maxwell on Molecules and Gases* (Cambridge, Mass., 1986): 284–5.

even if they do not it is well to know that Clausius' (or rather Herapath's)[13] theory is wrong and at any rate as I found myself able and willing to deduce the laws of motion of systems of particles acting on each other only by impact, I have done so as an exercise in mechanics. Now do you think there is any so complete a refutation of this theory of gases as would make it absurd to investigate it further so as to found arguments upon measurements of strictly 'molecular' quantities before we know whether there be any molecules? One curious result is that μ is independent of the density

$$\text{for } \mu = MNlv = \frac{Mv}{\sqrt{2\pi s^2}}.\text{[14]}$$

This is certainly very unexpected, that the friction should be as great in a rare as in a dense gas. The reason is, that in the rare gas the mean path is greater, so that frictional action extends to greater distances.

Have you the means of refuting this result of the hypothesis?[15]

Of course my particles have not all the same velocity, but the velocities are distributed according to the same formula as the errors are distributed in the theory of 'least squares'.[16]

If two sets of particles act on each other the mean vis viva of a particle will become the same for both, which implies, that equal volumes of gases at same press. & temp. have the same number of particles, that is, are chemical equivalents. This is one satisfactory result at least.[17]

---

(13) Herapath, *Mathematical Physics*, *passim*; and Herapath, 'On the physical properties of gases', *Annals of Philosophy*, **8** (1816): 56–60.

(14) Read: $\mu = \frac{1}{3}MNlv = \frac{1}{3}\dfrac{Mv}{\sqrt{2\pi s^2}}$.

(15) On the independence of μ and the density of a gas compare Maxwell's comment in 'Illustrations of the dynamical theory of gases. Part I': 32 (= *Scientific Papers*, **1**: 391), that 'the only experiment I have met with on the subject does not seem to confirm it'. See Stokes, 'On the effect of the internal friction of fluids on the motion of pendulums': 9, 16, where he refers to an experiment by Edward Sabine, 'On the reduction to a vacuum of the vibrations of an invariable pendulum', *Phil. Trans.*, **119** (1829): 207–39. See Number 193.

(16) See Maxwell, 'Illustrations of the dynamical theory of gases. Part I': 22–3 (= *Scientific Papers*, **1**: 380–1). On Maxwell's early interest in the calculus of probabilities see Number 31 esp. note (11); and P. M. Heimann, 'Molecular forces, statistical representation and Maxwell's demon', *Studies in History and Philosophy of Science*, **1** (1970): 189–211, esp. 190–1; Elizabeth Garber, 'Aspects of the introduction of probability into physics', *Centaurus*, **17** (1972): 11–39, esp. 19–29; and Theodore M. Porter, 'A statistical survey of gases: Maxwell's social physics', *Historical Studies in the Physical Sciences*, **12** (1981): 77–116.

(17) See Maxwell, 'Illustrations of the dynamical theory of gases. Part I': 30 (= *Scientific Papers*, **1**: 390). On 'Avogadro's hypothesis' see the 'Lettera al Prof. Stanislao Cannizzaro al Prof. S. di Luca; sunto di un corso di filosofia chimica', *Il Nuovo Cimento*, **7** (1858): 321–66, esp. 321–3, where Cannizzaro makes reference to the work of Clausius.

I have been rather diffuse on gases but I have taken to the subject for mathematical work lately and I am getting fond of it and require to be snubbed a little by experiments and I have only a few of Prof Grahams, quoted by Herapath,[18] on diffusion so that I am tolerably high minded still.

I had a colour-blind student last season who was a good student & a prizeman so I got pretty good observations out of him. Any four colours can be arranged three against one or two against two so as to form an equation to a colour-blind eye. I took 6 colours Red Blue Green Yellow White Black, and formed the 15 sets of four of them and tried 14 of these by the help of my student, making no suggestions of course to him, after I had instructed him in the things to be observed.

I then found the most probable values of the 3 equations, which ought to contain the whole 15, and then deducing the 15 equations from the 3 I found them very near the observed ones, the average error being 2 hundredths of the circle on each colour observed, and much less where the colours were very decided to a colour blind eye. I hope to see him again next year.[19]

I suppose we shall hear something of the theory of absorption of light in crystals from you in due time.[20] I hope Cambridge takes an interest in light in May.[21]

Yours truly

JAMES CLERK MAXWELL

---

(18) See note (12).

(19) See Maxwell's letters to Stokes of 25 February, 12 March, 23 April and 7 May 1860 (Numbers 175, 176, 177 and 178).

(20) See Maxwell's letter to Stokes of 8 May 1857 (Number 117); and G. G. Stokes, 'On the change of refrangibility of light', *Phil. Trans.*, **142** (1852): 463–562 (= *Papers*, **3**: 267–409).

(21) Stokes gave his lectures on 'Hydrostatics, Pneumatics, and Optics' in the Easter term (April–May) at Cambridge. Maxwell had attended these lectures in 1853; see Number 39 and Stokes' fee notebook, ULC Add. MSS 7656, NB1.

LETTER TO GEORGE BIDDELL AIRY

5 JULY 1859

From the original in the Royal Greenwich Observatory Archive[1]

Glenlair
Springholm
Dumfries.
5 July 1859

Sir

The Cambridge Philosophical Society have arranged so as to bind up my essay on Saturn along with their Transactions.[2]

I shall desire Messrs MacMillan to send a copy to the Royal Observatory.[3]

Your abstract of the results[4] may perhaps lead to some observations serving to put more of the hypothetical constitutions of the Rings out of the field.

Mr Ramage,[5] of the firm of Smith & Ramage, Regents Quay, Aberdeen is I believe a son of the telescope maker of whom I have heard.[6]

I do not know if he is the eldest son.

When I left Aberdeen in April he was in very bad health & I do not know whether he is still alive.

The address I have given will find him.

I am Sir
Your faithful servant
J. CLERK MAXWELL

G. B. Airy Esqre
Astronomer Royal
Greenwich

---

(1) Royal Greenwich Observatory Archive, ULC, Airy Papers 6/378, 76R–V.

(2) The essay on *Saturn's Rings*, published by Macmillans, is not included in the *Transactions*. On 14 March 1859 the Council of the Society had 'ordered that Mr. Maxwell's Adams Prize Essay form part of the transactions of the Society.' The minute was confirmed on 2 May 1859. (Council Minute Book (1853–71), Scientific Periodicals Library, Cambridge.)

(3) Maxwell's letter was written in reply to a letter from Airy of 30 June 1859 (Airy Papers 6/378, 75R–V), inquiring 'whether the Essay will be permanently preserved in the Cambridge Philosophical Transactions or elsewhere'. In a note of 1 March 1859 (Airy Papers 6/379, 209), Airy thanked Maxwell for having a copy of *Saturn's Rings* sent by Macmillans.

(4) George Biddell Airy, 'On the stability of the motion of Saturn's rings', *Monthly Notices of the Royal Astronomical Society*, **19** (10 June 1859): 297–304. Airy concluded his review with the observation that 'the essay which we have abstracted is one of the most remarkable contributions to mechanical astronomy that has appeared for many years'. See also Number 167 note (2).

(5) Charles Ramage; see Number 143 on his construction of the model of Saturn's rings.

(6) John Ramage; see D. J. Bryden, *Scottish Scientific Instrument-Makers 1600–1900* (Edinburgh, 1972): 31, 55.

159

LETTER TO JAMES DAVID FORBES

15 JULY 1859

From the original in the University Library, St Andrews[1]

Glenlair
Springholm
Dumfries
1859 July 15

My dear Sir

I am sorry that I do not know of any one likely to answer as your assistant.[2] Of my own students there are two that I think competent, but I think you might find better; I had better men last session, but they are too young yet. The Cambridge Scotchmen who would do are mostly Thomsons pupils, e.g. C. A. Smith[3] and W Jack of Peterhouse.[4] There is also Slesser of Queen's from Kings College Aberdeen[5] whom I am sorry I don't know yet. I should think any of these would be competent if willing to come, and I consider a session so spent would not be thrown away and that a man would return to Cambridge with a much more entertaining manner of treating his subjects than prevails among those who study by book and not by sight.[6]

I shall write to one or other of my Aberdeen men if you think it worth while. One is much occupied (or should be) with preparation for Indian examination. The other has a bursary for two years after degree on condition of study. He is a fair mathematician and very laborious and indefatigable, but he has chosen classical subjects for his special reading for the bursary.

---

(1) St Andrews University Library, Forbes MSS 1859/91.

(2) At Edinburgh University.

(3) Number 143 note (11).

(4) William Jack, Glasgow University 1848, Peterhouse 1855 (Venn).

(5) Number 143 note (10).

(6) Forbes replied on 16 July 1859 (ULC Add. MSS 7655, II/17), reporting that Jack and Smith had already declined, but inquiring about Slesser. He added in conclusion that: 'I am very glad to hear by private channel that your paper on Colour is *proposed* for a Royal Medal in the R.S. London.' Forbes' informant may have been William Whewell who, with Stokes, had nominated Maxwell for a Royal Medal at a meeting of the Council of the Royal Society on 30 June 1859. See *Minutes of Council of the Royal Society from December 16th 1858 to December 16th 1869*, 3 (London, 1870): 27; 'On the motion of Professor Stokes, seconded by the Rev. Dr Whewell, – Resolved, – That Professor J. C. Maxwell be placed on the list of Candidates for a Royal Medal, for his Mathematical Theory of the Composition of Colours, verified by quantitative experiments, and for his Memoirs on Mathematical and Physical subjects.' The medal was awarded to Arthur Cayley. Maxwell was again nominated for the Royal and Rumford medals in 1860, and was awarded the Rumford Medal; see Number 176 note (6).

I hope the Aberdeen meeting of the Association will succeed so far north and amid so much granite.[7]

The personal arrangements of the Commissioners are not yet known. We must see what effect may be produced by the exertions of the Town and College for the preservation of the Arts Classes.[8]

Yours sincerely
JAMES CLERK MAXWELL

[7] The meeting of the British Association for the Advancement of Science at Aberdeen in September 1859; see Numbers 160 to 162.

[8] On the union of King's and Marischal Colleges see Numbers 140, 142 and 174.

160

BRITISH ASSOCIATION PAPER ON THE THEORY OF GASES

[SEPTEMBER 1859]

From the *Report of the British Association for 1859*[1]

ON THE DYNAMICAL THEORY OF GASES[2]

The phenomena of the expansion of gases by heat, and their compression by pressure, have been explained by Joule,[3] [Clausius],[4] Herapath,[5] &c., by the theory of their particles being in a state of rapid motion, the velocity depending on the temperature. These particles must not only strike against the sides of the vessel, but against each other, and the calculation of their motions is therefore complicated. The author has established the following results: 1. The velocities of the particles are not uniform, but vary so that they deviate from the mean value by a law well known in the 'method of least squares'. 2. Two different sets of particles will distribute their velocities, so that their *vires vivæ* will be equal; and this leads to the chemical law, that the equivalents of gases are proportional to their specific gravities. 3. From Prof. Stokes's experiments on friction in air,[6] it appears that the distance travelled by a particle between consecutive collisions is about $\frac{1}{447,000}$ of an inch, the mean velocity being about 1505 feet per second; and therefore each particle makes 8,077,200,000 collisions per second. 4. The laws of the diffusion of gases, as established by the Master of the Mint,[7] are deduced from this theory, and the

---

(1) *Report of the Twenty-ninth Meeting of the British Association for the Advancement of Science; held at Aberdeen in September 1859* (London, 1860), part 2: 9.

(2) Compare J. C. Maxwell, 'Illustrations of the dynamical theory of gases. Part I. On the motions and collisions of perfectly elastic spheres', *Phil. Mag.*, ser. 4, **19** (1860): 19–32 (= *Scientific Papers*, **1**: 377–91); and Number 157.

(3) J. P. Joule, 'Some remarks on heat and on the constitution of elastic fluids', *Memoirs of the Manchester Literary and Philosophical Society*, **9** (1851): 107–14; reprinted (with added notes) in *Phil. Mag.*, ser. 4, **14** (1857): 211–16.

(4) R. Clausius, 'Ueber die Art der Bewegung welche wir Wärme nennen', *Ann. Phys.*, **100** (1857): 353–80; Clausius, 'Ueber die mittlere Länge der Wege, welche bei der Molecularbewegung gasförmiger Körper von den einzelnen Molecülen zurückgelegt werden', *ibid.*, **105** (1858): 239–58. There is a misprint in the original which reads: Claussens.

(5) John Herapath, *Mathematical Physics*, 2 vols. (London, 1847).

(6) G. G. Stokes, 'On the effect of the internal friction of fluids on the motion of pendulums', *Trans. Camb. Phil. Soc.*, **9** part 2 (1851): [8]–[106], esp. 17, 65. See Numbers 152 note (7) and 157.

(7) Thomas Graham, 'A short account of experimental researches on the diffusion of gases through each other, and their separation by mechanical means', *Quarterly Journal of Science*, **28** (1829): 74–83. See Number 157 note (12).

absolute rate of diffusion through an opening can be calculated. The author intends to apply his mathematical methods to the explanation on this hypothesis of the propagation of sound,[8] and expects some light on the mysterious question of the absolute number of such particles in a given mass.[9]

(8) Maxwell did not discuss the propagation of sound in Parts II and III of 'Illustrations of the dynamical theory of gases', *Phil. Mag.*, ser. 4, **20** (1860); 21–37 (= *Scientific Papers*, **1**: 392–409); but see J. J. Waterston, 'On the theory of sound', *Phil. Mag.*, ser. 4, **16** (1858): 481–95.

(9) Not discussed in Parts II and III of 'Illustrations of the dynamical theory of gases'; but see J. C. Maxwell, 'On Loschmidt's experiments on diffusion in relation to the kinetic theory of gases', *Nature*, **8** (1873): 298–300 (= *Scientific Papers*, **2**: 343–50).

BRITISH ASSOCIATION PAPER ON THE COLOURS OF THE SPECTRUM

[SEPTEMBER 1859]

From the *Report of the British Association for 1859*[1]

ON THE MIXTURE OF THE COLOURS OF THE SPECTRUM

The author described his apparatus for obtaining a uniform field, illuminated by the light of any one or more definite portions of the spectrum, and comparing this mixture with a field of white light in contact with it.[2] The experiments consisted in obtaining perfect equality between a combination of three definite portions of the spectrum and this white light. The relations of these portions are then ascertained by mathematical treatment of the equations so obtained, and it results that Newton's 'circle of colours'[3] is found to be really two sides of a triangle; red, yellow-green, and blue being the angular points, and yellow being on the side between red and green. The extreme red and violet form small portions of the third side of which the middle part representing purple is wanting in the spectrum.[4]

The peculiar dimness of the spectrum near the line F, as described to the Section in 1856,[5] was further investigated, and shown to be more marked to the author's eyesight than to that of others.[6] It results from this that a mixture may be formed, which appears green to one eye and red to another, and this was found experimentally true.

These results are only part of a complete investigation of the colours of the spectrum, of which the experimental portion is considerably advanced and will shortly be published.[7]

---

(1) *Report of the Twenty-ninth Meeting of the British Association for the Advancement of Science; held at Aberdeen in September 1859* (London, 1860), part 2: 15.

(2) See Numbers 154 and 170.

(3) See Number 54 note (9) and Number 170.   (4) See Number 163.

(5) J. C. Maxwell, 'On the unequal sensibility of the foramen centrale to light of different colours', *Report of the Twenty-sixth Meeting of the British Association for the Advancement of Science; held at Cheltenham in August 1856* (London, 1857), part 2: 12 (= *Scientific Papers*, **1**: 242).

(6) See Number 170.   (7) See Number 170.

BRITISH ASSOCIATION PAPER ON THE MODEL TO ILLUSTRATE THE MOTIONS OF SATURN'S RINGS

[SEPTEMBER 1859]

From the *Report of the British Association for 1859*[1]

ON AN INSTRUMENT FOR EXHIBITING THE MOTIONS OF SATURN'S RINGS

The author exhibited an instrument made by Messrs. Smith and Ramage of Aberdeen, to exhibit the motion of a ring of satellites about a central body, as investigated in his 'Essay on the Motion of Saturn's Ring'.[2] It is there shown that a solid or fluid ring will be broken up, and that the fragments will continue in the form of a ring if certain conditions are fulfilled. The instrument exhibits the motion of these fragments as deduced from the mathematical theory.

---

(1) *Report of the Twenty-ninth Meeting of the British Association for the Advancement of Science; held at Aberdeen in September 1859* (London, 1860), part 2: 62.

(2) See Number 143, esp. note (4), and plate VII.

LETTER TO GEORGE GABRIEL STOKES

8 OCTOBER 1859

From the original in the University Library, Cambridge[1]

Glenlair
Springholm
Dumfries
1859 Oct. 8

Dear Stokes

I received your letter some days ago. I intend to arrange my propositions about the motions of elastic spheres in a manner independent of the speculations about gases, and I shall probably send them to the Phil. Magazine which publishes a good deal about the dynamical theories of matter & heat.[2] I have not done much to it since I wrote to you,[3] as all my optical observations must be done when I have sun and that is getting lower every day. But I think I have made as much optical hay as will make a small bundle for you as Secretary R. S. some time in winter if you can tell me under what forms it should be sent in. I would call it on the Relations of the Colours of the Spectrum as seen by the human eye.[4]

As I have only lately reduced some of my observations I cannot say what more results I may come to but I shall tell you the present state of affairs that you may judge whether it is better to shew it up at once or to keep it a year till the experiments are more numerically formidable.[5]

My experiments have been of this nature. I mix 3 colours of the spectrum so as to produce a white equal to a given white, thus

$$18.4 \ (24) + 30.7 \ (44) + 30 \ (68) = W$$

where the numbers in brackets denote the position of the colour in the spectrum and the coeff^t denotes the quantity of that colour.

red green blue
(24) (44) & (68) are the 3 standards of colour which I have taken.[6]

---

(1) ULC Add. MSS 7656, M 411. Previously published in Larmor, *Correspondence*, **2**: 11–14.

(2) J. C. Maxwell, 'Illustrations of the dynamical theory of gases. Part I. On the motions and collisions of perfectly elastic spheres', *Phil. Mag.*, ser. 4, **19** (1860): 19–32 (= *Scientific Papers*, **1**: 377–91); see Number 160.· (3) Number 157.

(4) J. Clerk Maxwell, 'On the theory of compound colours, and the relations of the colours of the spectrum', *Phil. Trans.*, **150** (1860): 57–84 (= *Scientific Papers*, **1**: 410–44); see Number 170.

(5) See Maxwell's subsequent letters to Stokes: Numbers 169, 172, 175, 176, 177, 178, and 180.

(6) See Maxwell, 'On the theory of compound colours': 68–9 (= *Scientific Papers*, **1**: 424–5); and Number 139. The numbers in brackets are the positions on Maxwell's scale; see Number 154.

Then I take any other colour say (28) of my scale and mix with 44 & 68 to produce the same white in quality and intensity as before. I get –

$$16 \ (28) + 25.2 \ (44) + 29.7 \ (68) = W.$$

Subtracting one from the other

$$16 \ (28) = 18.4 \ (24) + 5.5 \ (44) + 0.3 \ (68)$$

which gives the relation of (28) to the 3 standard colours and from this I can place it in Newtons diagram when I have fixed the three standards any where. In this way I have laid down 16 points of the spectrum with great care from observations by myself and two others and 15 intermediate points from observations by myself, and I find,

1st the spectrum forms a curve which is not a circle as Newton imagined[7] but two sides of a triangle with doubtful portions at each end of the third side. This shows that all the colours can be made of three, which are marked (24) (46) & (64). These being the purest colours in the spectrum and therefore in nature may be reckoned standard colours although they may not coincide with those assumed at first to guide the experiments.[8]

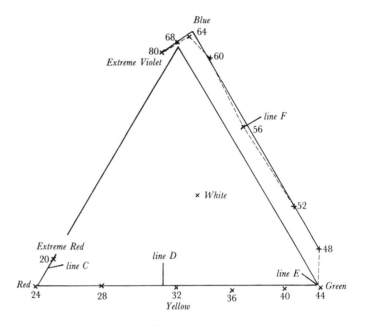

Figure 163,1

(The position of these colours is different for different observers, and I find *real differences* of large amount between the results of different eyes in certain parts of the spectrum. The figure is from the results of a very good observer.)

---

(7) Isaac Newton, *Opticks* (London, ₃1721): 134–7, and Book I, Part II, Table III, Fig. 11; see Number 170, and Number 54 note (9).

(8) See Maxwell, 'On the theory of compound colours': 74 and Fig. 4 opposite 84 (= *Scientific Papers*, **1**: 431, 444).

I mix 3 colours of the spectrum so as to produce a white equal to a given white, thus

$$18.4 (24) + 30.7 (44) + 30 (68) = W$$

where the numbers in brackets denote the position of the colour in the spectrum and the coefft denotes the quantity of that colour. (24) (44) & (68) are the 3 standards of colour which I have taken. Then I take any other colour say (28) of my scale and mix with 44 & 68 to produce the same white in quality and intensity as before. I get —

$$16 (28) + 25.2 (44) + 29.7 (68) = W$$

Subtracting one from the other

$$16 (28) = 18.4 (24) + 5.5 (44) + 0.3 (68)$$

which gives the relation of (28) to the 3 standard colours and from this I can place it in Newton's diagram when I have fixed the three standards any where. In this way I have laid down 16 points of the spectrum with great care from observation by myself and two others and 15 intermediate points from observation by myself, and I find

1st the spectrum forms a curve which is not a circle as Newton imagined but two sides of a triangle with doubtful portions at each end of the third side. This shows that all the colours can be made of three, which are marked (24) (46) & (64). These being the purest colours in the spectrum and therefore in nature may be reckoned standard colours

although they may not coincide with those assumed at first to guide the experiments.

(The position of these colours is different for different observers and I find real differences of large amount between the results of different eyes in certain parts of the spectrum. The figure is from the results of a very good observer.)

I determined the wave length in air for the different divisions of the spectrum by observing the bands seen by analysing light reflected from a thin plate of air giving a retardation of 41 undulations for the line D and comparing with Fraunhofers measures. I thus found for the different divisions

	Wave length
20	2450
24 to	2328
44	1951 46 — 1924
64	1721
68	1688

and so on

Plate VIII. Chart of the spectrum as a colour triangle (1859) from a letter to George Gabriel Stokes (Number 163).

I determined the wave length in air for the different divisions of the spectrum by observing the bands seen by analysing light reflected from a thin plate of air giving a retardation of 41 undulations for the line *D* and comparing with Fraunhofers measures.[9]

I thus found for the different divisions

		Wave length
20	——	2450
24	——	2328
44	&c	1951
∟46	——	1924˩
64		1721
68		1688

and so on.[10]

I propose to give a short geometrical theory of Newtons diagram of colours and to lay down the spectrum upon it and so to establish a method of ascertaining the colour produced by any mixture of any number of colours in the spectrum.[11] I have come on great differences between different observers but they all agree in having 3 primary colours, though they differ in defining them.

---

(9) Compare Numbers 139 and 170. To calibrate the scale of the colour box Maxwell correlated the position of Fraunhofer's lines with the interference bands produced by a thin stratum of air enclosed by two plane surfaces of glass kept apart by two parallel strips of gold leaf. In Maxwell's experiments 'the line *D* corresponded with the seventh dark band' and he found that the *D* line corresponded to a retardation (the path difference between the rays) of 41 wavelengths. See Maxwell, 'On the theory of compound colours': 67–8 (= *Scientific Papers*, **1**: 423–4). For Fraunhofer's lines, see his paper 'Bestimmung des Brechungs- und Farben-Zerstreuungs-Vermögens verschiedener Glasarten, in bezug auf die Vervollkommnung achromatischer Fernröhre', *Annalen der Physik*, **56** (1817): 264–313, and Table IV. On the use of Fraunhofer's lines as reference marks see Hermann Helmholtz, 'Ueber die Theorie der zusammengesetzten Farben', *Ann. Phys.*, **87** (1852): 45–66.

(10) See Maxwell, 'On the theory of compound colours': 68–9 (= *Scientific Papers*, **1**: 424–5). Maxwell gives the wavelengths in millionths of a Paris inch; compare Hermann Helmholtz, 'Ueber die Zusammensetzung von Spectralfarben', *Ann. Phys.*, **94** (1855): 1–28, esp. 15–16, for a similar table of wavelengths (in the same units) corresponding to different colours, and a list of values of Fraunhofer's lines for comparison.

(11) See Maxwell, 'On the theory of compound colours': 62–4 (= *Scientific Papers*, **1**: 416–19), where he develops Grassmann's representation of colour combinations; 'if colours are represented in quantity and quality by the magnitude and direction of straight lines, the rule for the composition of colours is identical with that for the composition of forces in mechanics'. Compare Hermann Grassmann, 'Zur Theorie der Farbenmischung', *Ann. Phys.*, **89** (1853): 69–84; (trans.) 'On the theory of compound colours', *Phil. Mag.*, ser. 4, **7** (1854): 254–64. To 'form a new geometrical conception by the aid of solid geometry' Maxwell suggests a three-dimensional representation of colours as the vector sum of colour combinations. See Number 170.

I can make a colour which, placed beside white, looks green to one and red to another, the reason being, that the space just before the line *F* is dim to all but much dimmer to one than another, so that if I mix this with red it will be neutral red & green according to the degree of obscurity of this part of the spectrum.[12] If you think any of these results proper aliment for the Royal Society pray tell me how I may administer it in due form, as I do not purpose to be in London.

The subject is one about which we have had conversation before, so that I have not thought it necessary to do more than indicate new results as you know by what method they must be obtained and how they are expressed in Newtons method.

I remain
Yours truly
J. Clerk Maxwell

After the 25th Oct address Aberdeen.

(12) On the insensitivity of Maxwell's eyes to greenish-blue light near the *F* line, see Number 170.

LETTER TO JAMES DAVID FORBES

8 NOVEMBER 1859

From the original in the University Library, St Andrews[1]

2 Union Place
Aberdeen
1859 Nov 8

Dear Sir

I was very glad to hear today of your appointment as Principal of S. S. Salvator & Leonard's College.

When you go to St Andrews, I suppose your Chair in Edinburgh will become vacant. Do you think that if I were to become a candidate I would have any chance of getting that appointment, and if so, could you give me any assistance or advice as to how I should bring my claim under the notice of the Curators, and how soon my application to them should be made.

If you approve of my being a candidate, I should feel obliged if you would give me a testimonial for the present occasion, for though you were kind enough to give a very favourable opinion of me when I was a candidate for Aberdeen,[2] you have much more interest in Edinburgh, and understand better than any one can do what are the requirements of the office; so that your influence on this occasion will be very important to me.[3]

Yours very truly
JAMES CLERK MAXWELL

Professor J D Forbes
3 Park Place
Edinburgh

---

(1) St Andrews University Library, Forbes MSS 1859/163.

(2) Number 92 note (2).

(3) In a letter of 10 November 1859 (Forbes MSS Letter Book VI/124) Forbes states that he could have 'added little' to his 1856 testimonial for Marischal College. In a letter of 6 August 1860 (Letter Book VI/292), after Maxwell had failed to be appointed to the Edinburgh Chair, but had been appointed Professor of Natural Philosophy at King's College, London (see Number 183, note (3)), Forbes explained that 'during the vacancy I was not consulted on the subject by *any one* of the Curators'.

LETTER TO PETER GUTHRIE TAIT

11 NOVEMBER 1859

From the original in the University Library, Cambridge[1]

Marischal College
Aberdeen
1859 Nov 11

Dear Tait[2]

Let a unit of north magnetism be placed at the origin of coordinates then the potential function anywhere is

$$-\frac{1}{r} = u \quad \text{(suppose).}$$

If this unit be moved through a distance $= \delta x$ in direction x from the origin the effect on the potential at a point P will be the same as that of a reverse & equal motion of P, the result is therefore that u becomes $u - \frac{du}{dx}\delta x$ and similarly for y & z.

Let a small magnet be placed at the origin in a direction whose direction-cosines are a, b, c whose length (small) is s and which has $+m$ units of magnetism at the $+$ end and $-m$ at the other.

The potential due to the $+$ end is

$$mu - \tfrac{1}{2}mas\frac{du}{dx} - \tfrac{1}{2}mbs\frac{du}{dy} - \tfrac{1}{2}mcs\frac{du}{dz}.$$

That due to the $-$ end is

$$-mu - \tfrac{1}{2}mas\frac{du}{dx} - \tfrac{1}{2}mbs\frac{du}{dy} - \tfrac{1}{2}mcs\frac{du}{dz}$$

therefore the potential function due to the small magnet anywhere is

$$-ms\left(a\frac{du}{dx} + b\frac{du}{dy} + c\frac{du}{dz}\right) = V \text{ suppose.}$$

The *potential* i.e. the total work done in bringing m' of north magnetism from

---

(1) ULC Add. MSS 7655, I, b/2.

(2) The letter was presumably occasioned by a query from Tait. On 18 March 1859 Tait had informed Sir William Rowan Hamilton that 'I have at last attacked the subject of Potentials which was the cause of my recent (and, this time, successful so far) attempt at the study of Quaternions, and I think I have got the method of applying the calculus to the matter'; quoted by C. G. Knott, *Life and Scientific Work of Peter Guthrie Tait* (Cambridge, 1911): 133. See P. G. Tait, 'Quaternion investigations connected with electrodynamics and magnetism', *Quarterly Journal of Pure and Applied Mathematics*, **3** (1860): 331–42; and Tait, 'Quaternion investigation of the potential of a closed circuit', *ibid.*, **4** (1861): 143–4.

∞ to the point P is $m'V$. That of bringing a small magnet $+m'-m'$ length s' from ∞ to P and turning it so that its direction cosines are a' b' c' is

$$m's'\left(a'\frac{dV}{dx}+b'\frac{dV}{dy}+c'\frac{dV}{dz}\right)=W.$$

The forces acting on the centre of gravity of this magnet will be

$$X=\frac{dW}{dx}\quad Y=\frac{dW}{dy}\quad Z=\frac{dW}{dz}.$$

Let ω_1 ω_2 ω_3 be three small angles of rotation about the axes of $x\,y$ & z then the three couples will be

$$L=\frac{dW}{\omega_1}\ \&\mathrm{c}$$

now
$$da'=c'\omega_2-b'\omega_3$$
$$db'=a'\omega_3-c'\omega_1$$
$$dc'=b'\omega_1-a'\omega_2.$$

$$\therefore L=m's'\left(b'\frac{dV}{dz}-c'\frac{dV}{dy}\right)$$

$$M=m's'\left(c'\frac{dV}{dx}-a'\frac{dV}{dz}\right)$$

$$N=m's'\left(a'\frac{dV}{dy}-b'\frac{dV}{dx}\right).$$

Let p stand for $\dfrac{d}{dx}$ q for $\dfrac{d}{dy}$ r for $\dfrac{d}{dz}$

then
$$V=-(ap+bq+cr)\,msu$$
$$W=-(ap+bq+cr)(a'p+b'q+c'r)\,mm'ss'u$$
$$X=pW\quad Y=qW\quad z=rW$$
$$L=-(b'r-c'q)(ap+bq+cr)\,mm'ss'u$$
$$M=-(c'p-a'r)(ap+bq+cr)\,mm'ss'u$$
$$N=-(a'q-b'p)(ap+bq+cr)\,mm'ss'u.$$

We know that $(p^2+q^2+r^2)\,u=0$.
Same is true for V & W.

If we put
$$m'a'=k'\frac{dV}{dx}$$
$$m'b'=k'\frac{dV}{dy}$$
$$m'c'=k'\frac{dV}{dz}$$

we have the case of a ball of matter having magnetism induced by the first magnet k' being a const. of induction for the ball then ·

$$W = s'k'\left(\overline{\frac{dV}{dx}}\Big|^2 + \overline{\frac{dV}{dy}}\Big|^2 + \overline{\frac{dV}{dz}}\Big|^2\right)$$

and

$$L = M = \mathcal{N} = 0$$

if the 2nd magnet is free to rotate and permanent then

$$\frac{1}{a}\frac{dV}{dx} = \frac{1}{b}\frac{dV}{dy} = \frac{1}{c}\frac{dV}{dz}$$

but $a^2 + b^2 + c^2 = 1$

$$\therefore W = m's'\sqrt{\overline{\frac{dV}{dx}}\Big|^2 + \overline{\frac{dV}{dy}}\Big|^2 + \overline{\frac{dV}{dz}}\Big|^2}.$$

Yours truly
J. C. MAXWELL

166

LETTER TO MICHAEL FARADAY

30 NOVEMBER 1859

From the original in the library of the Institution of Electrical Engineers, London[1]

Marischal College
Aberdeen
1859 Nov 30

Dear Sir

I am a candidate for the Chair of Natural Philosophy in the University of Edinburgh, which will soon be vacant by the appointment of Professor J. D. Forbes to St Andrews.[2] If you should be able, from your knowledge of the attention which I have paid to science, to recommend me to the notice of the Curators, it would be greatly in my favour and I should be much indebted to you for such a certificate.[3]

I was sorry that I had so little time in September that I could not write out an explanation of the figures of lines of force which I sent you, but Professor W. Thomson to whom I lent them, seems to have indicated all that was necessary, and most of them can be recognised from their resemblance to the curves made with Iron filings.[4]

The only thing to be observed is, that these curves are due to the action either of long wires perpendicular to the paper or of elongated magnetic poles such as the edge of a long ribbon of steel magnetized transversely. By considering infinitely long currents or magnetic poles perpendicular to the paper, we obtain systems of curves far more easily traced than in any other case, while their general appearance is similar to those produced in the ordinary experiments.

All the diagrams have two sets of lines at right angles to each other and the width between the two sets of lines is the same so that the reticulation is nearly

---

(1) Institution of Electrical Engineers, London, Special Collection MSS 3. Previously published in *Life of Maxwell* (2nd edn): 226–7.

(2) See Number 164.

(3) Faraday declined to write testimonials; see his letter to John Tyndall of 1 August 1851, in *The Selected Correspondence of Michael Faraday*, ed. L. P. Williams, 2 vols. (Cambridge, 1971), 2: 641–2.

(4) Faraday had received figures of lines of force, which he assumed had been sent by Thomson. In a letter to Faraday of 31 October Thomson explained that: 'The diagrams which came to you by post were from Professor Clerk Maxwell. He wished to speak of them to you at Aberdeen [at the British Association meeting in September 1859] but I suppose did not find an opportunity in the closely packed time'. For the Faraday–Thomson correspondence, 28 October to 7 November 1859, see *Correspondence of Faraday*, 2: 932–6, esp. 933.

square. If one system belongs to poles, the other belongs to currents, so that if the meaning of one be known, that of the other may be deduced from it.[5]

I remain
Yours truly
JAMES CLERK MAXWELL

Professor Faraday

---

(5) Compare *Treatise*, **1**: 147–9 (§123); and see Number 86 esp. note (4).

LETTER TO GEORGE BIDDELL AIRY

30 NOVEMBER 1859

From the original in the Royal Greenwich Observatory Archive[1]

Marischal College
Aberdeen
1859 Nov 30

Dear Sir

I am a candidate for the Chair of Natural Philosophy in the University of Edinburgh. If you should be able to give me a certificate of fitness for that office, judging from what you know of me, I should be greatly obliged to you.[2]

I remain
Dear Sir
Yours truly
JAMES CLERK MAXWELL

---

(1) Royal Greenwich Observatory Archive, ULC, Airy Papers 6/378, 77R.

(2) In reply Airy enclosed a testimonial, dated 2 December 1859 (Airy Papers 6/378, 78R–V), stating that the 'Essay on the Stability of Saturn's Rings...is one of the most remarkable applications of mathematics to physics that I have ever seen'; see also Glasgow University Library, Y1–h.18. Compare Number 158 note (4).

LETTER TO WILLIAM THOMSON

8 DECEMBER 1859

From the original in the University Library, Glasgow[1]

Marischal College
Aberdeen Dec 8, 1859

Dear Thomson

I shall send you tomorrow 3 copies of testimonials,[2] which if you can use to profit, you will oblige me. I have sent to the Town Council and have written to most of the Professors. Prof. Balfour[3] has given me some information as to the proceedings and time of electing Curators.[4]

The University Court elects on Monday & the Town Council on Tuesday. I intend to go up on Tuesday evening, and call on the Curators on Wednesday. I have not heard anything of candidates more than those you told me except that Professors David Thomson and Fuller of Kings College are Candidates.[5]

Would you recommend me to get more certificates from parents with respect to my teaching? I have got some already but I could easily get more from respectable people, including the Provost and some magistrates if it would be of use.[6] I have printed what I had to save time and I did not think it worth

---

(1) Glasgow University Library, Kelvin Papers, M 12.

(2) One set is preserved in Glasgow University Library, Y1–h.18. This includes the testimonials submitted in 1856 to Marischal College (from Thomson, Forbes, Stokes, Hopkins and others: see Numbers 92 note (2), 94 note (5) and 97 note (4)); and new testimonials from colleagues at Marischal College (including John Cruickshank, Professor of Mathematics and Thomas Clark, Professor of Chemistry: see Numbers 109 note (4) and 113 note (3)), attesting to his success as a lecturer, and from G. B. Airy, Astronomer Royal (see Number 167 note (2)) and Charles Piazzi Smyth, Astronomer Royal for Scotland (see Numbers 105 note (1) and 117). Smyth declared that he 'had the pleasure of hearing Mr. J. Clerk Maxwell's first scientific memoir' (Number 1) and could 'testify to all those high philosophical qualities, of which he early gave such rich promise, having been refined and extraordinarily developed, by increase of years and by the continual efforts of a soul that lives for the promotion of knowledge...'. See Introduction notes (88) and (123).

(3) John Hutton Balfour, Professor of Botany at Edinburgh; see *Edinburgh University Calendar for Session 1859–60* (Edinburgh, 1859): 5.

(4) Under the recently reformed statutes of the University, the 'power of appointing to the office of Principal, and to all professorships in the University hitherto held by the Town-Council [including that of Natural Philosophy], shall be transferred to seven Curators, four to be nominated by the Town-Council, and three by the University Court'; see *Edinburgh University Calendar for Session 1858–59* (Edinburgh, 1858): 53. See Number 174.

(5) See Number 174 notes (9) and (11).

(6) Maxwell obtained additional testimonials from other colleagues and a late Lord Provost of Aberdeen (ULC Add. MSS 7655, V, i/12). See Introduction note (123).

while to make a great volume but rather to pick out one or two witnesses of good repute who would draw up a clear statement of their opinion.

My colour blind student of last session seems inclined to speculation and has been perplexing his mind with images on the retina and the sensorium. The latter are gradually being resolved, and he may come to a good understanding of the matter presently.

Here is my shortened colour mixer.[7] Rays from *RGB* fall through two prisms at *P* are reflected from concave mirror at *S* and up through the prisms again to M_3 where they pass to the eye tube.

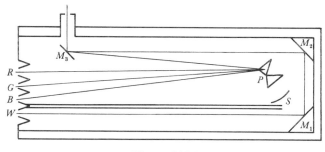

Figure 168,1

At the same time white light from *W* is reflected at M_1 M_2 and M_3 to the eye.[8] Both are from the same source and vary together in intensity so that 3 coloured lights may be mixed & compared with white of intensity moderated by 2 reflexions.

I hope to get it made portable and get good subjects to look through it.

Yours truly

J. C. MAXWELL

(7) Compare the apparatus constructed in 1856: see Figure 109,1. Unlike the earlier instrument, the new portable instrument enabled mixtures of spectral colours to be directly compared with white light. See also Number 139.

(8) In the instrument as finally perfected Maxwell introduced a lens between mirrors M_1 and M_2. See Maxwell, 'On the theory of compound colours', *Phil. Trans.*, **150** (1860): 57–84, esp. 78 and fig. 8 opposite 84 (= *Scientific Papers*, **1**: 437, 444); see plate IX.

LETTER TO GEORGE GABRIEL STOKES

24 DECEMBER 1859

From the original in the University Library, Cambridge[1]

Marischal College
Aberdeen 1859 Dec 24

Dear Stokes

I have addressed to you as Secretary of the Royal Society an abstract of my paper on the Colours of the Spectrum.[2] I have the paper itself in a presentable form and can send it also if required, but I do not wish it to be considered as finished yet, for I am getting more apparatus and expect to do more before long. If my present paper could be considered as Part I of a more complete one it would answer best for me.[3] I suppose the Society will send back the MS. of the paper, if I send it, so that I may refer to it in preparing a second part, for some of the tables in it are the result of calculations which I do not wish to go over again without good cause and I have not yet taken copies.

An optician here is making me a reflecting colour apparatus with which I intend to get colour-blind equations among the prismatic colours which will determine accurately the missing colour.[4]

I intend to produce whites with one fixed colour and one variable one and so find the neutral point of the spectrum and the two points of greatest intensity of colour as seen by my observer. If afterwards Mr Pole[5] should wish to try experiments I have no doubt the instrument could be sent up to him, as I am making it strong, so as to be portable,[6] it is

$$3 \text{ feet } 6 \times 9\tfrac{1}{2} \text{ in.} \times 5 \text{ in. in dimensions.}$$

I should think a quartz prism of 35° or so with one side ground convex, radius 20 or 30 inches and well silvered would make a good instrument for analysing the more refrangible rays.

I remain
Yours truly
J. C. MAXWELL

---

(1) ULC Add. MSS 7656, M 412. Previously published in Larmor, *Correspondence*, **2**: 14–15.

(2) Number 170.　　　(3) See Number 179.　　　(4) See Numbers 175 to 179.

(5) Compare William Pole, 'On colour blindness', *Phil. Trans.*, **149** (1859): 323–39. Pole wrote to Maxwell on 4 June 1860 (ULC Add. MSS 7655, II/19) enclosing this paper.

(6) See Numbers 168 and 182 and plate IX.

ABSTRACT OF PAPER 'ON THE THEORY OF COMPOUND COLOURS, AND THE RELATIONS OF THE COLOURS OF THE SPECTRUM'

[27 DECEMBER 1859]

From the *Proceedings of the Royal Society*[1]

ON THE THEORY OF COMPOUND COLOURS, AND THE RELATIONS OF THE COLOURS OF THE SPECTRUM[2]

By J. Clerk Maxwell, Esq., Professor of Natural Philosophy, Marischal College and University, Aberdeen. Communicated by Professor Stokes, Sec. R.S.

Received December 27, 1859

(Abstract)

Newton (in his 'Optics,' Book I. part ii. prop. 6)[3] has indicated a method of exhibiting the relations of colour, and of calculating the effects of any mixture of colours. He conceives the colours of the spectrum arranged in the circumference of a circle, and the circle so painted that every radius exhibits a gradation of colour, from some pure colour of the spectrum at the circumference, to neutral tint at the centre. The resultant of any mixture of colours is then found by placing at the points corresponding to these colours, weights proportional to their intensities; then the resultant colour will be found at the centre of gravity, and its intensity will be the sum of the intensities of the components.

From the mathematical development of the theory of Newton's diagram it appears that if the positions of any three colours be assumed on the diagram, and certain intensities of these adopted as units, then the position of every other colour may be laid down from its observed relation to these three. Hence Newton's assumption that the colours of the spectrum are disposed in a certain manner in the circumference of a circle, unless confirmed by experiment, must be regarded as merely a rough conjecture, intended as an illustration of his

---

(1) *Proc. Roy. Soc.*, **10** (1860): 404–9. Read 22 March 1860.

(2) Published in *Phil. Trans.*, **150** (1860): 57–78 (= *Scientific Papers*, **1**: 410–36). The paper was refereed by John Tyndall, who recommended its publication in the *Phil. Trans.* in a brief letter to Stokes of 10 May 1860 (Royal Society, *Referees' Reports*, **4**: 181).

(3) Isaac Newton, *Opticks* (London, ₃1721): 134–7, and Book I, part II, Table III, Fig. 1. See Number 54 esp. note (9).

method, but not asserted as mathematically exact. From the results of the present investigation, it appears that the colours of the spectrum, as laid down according to Newton's method from actual observation, lie, not in the circumference of a circle, but in the periphery of a triangle, showing that all the colours of the spectrum may be *chromatically* represented by three, which form the angles of this triangle.

Wave-length in millionths of Paris inch.

Scarlet . . . . 2328, about one-third from line C to D.,

Green . . . . 1914, about one-quarter from E to F.,

Blue . . . . 1717, about half-way from F to G.

The theory of three primary colours has been often proposed as an interpretation of the phenomena of compound colours, but the relation of these colours to the colours of the spectrum does not seem to have been distinctly understood till Dr. Young (Lectures on Natural Philosophy, Kelland's edition, p. 345) enunciated his theory of three primary sensations of colour which are excited in different proportions when different kinds of light enter the organ of vision.[4] According to this theory, the threefold character of colour, as perceived by us, is due, not to a threefold composition of light, but to the constitution of the visual apparatus which renders it capable of being affected in three different ways, the relative amount of each sensation being determined by the nature of the incident light. If we could exhibit three colours corresponding to the three primary sensations, each colour exciting one and one only of these sensations, then since all other colours whatever must excite more than one primary sensation, they must find their places in Newton's diagram within the triangle of which the three primary colours are the angles.

Hence if Young's theory is true, the complete diagram of all colour, as perceived by the human eye, will have the form of a triangle.

The colours corresponding to the pure rays of the spectrum must all lie within this triangle, and all colours in nature, being mixtures of these, must lie within the line formed by the spectrum. If therefore any colours of the spectrum correspond to the three pure primary sensations, they will be found at the angles of the triangle, and all the other colours will lie within the triangle.

The other colours of the spectrum, though excited by uncompounded light, are compound colours; because the light, though simple, has the power of exciting two or more colour-sensations in different proportions, as, for instance, a blue-green ray, though not compounded of blue rays and green rays, produces a sensation compounded of those of blue and green.

The three colours found by experiment to form the three angles of the

---

(4) Thomas Young, *A Course of Lectures on Natural Philosophy and the Mechanical Arts*, ed. P. Kelland, 2 vols. (London, 1845), **1**: 345. See Number 54 note (8).

triangle formed by the spectrum on Newton's diagram, *may* correspond to the three primary sensations.

A different geometrical representation of the relations of colour may be thus described. Take any point not in the plane of Newton's diagram, draw a line from this point as origin through the point representing a given colour on the plane, and produce them so that the length of the line may be to the part cut off by the plane as the intensity of the given colour is to that of the corresponding point on Newton's diagram. In this way any colour may be represented by a line drawn from the origin whose direction indicates the quality of the colour, and whose length depends upon its intensity. The resultant of two colours is represented by the diagonal of the parallelogram formed on the lines representing the colours (see Prof. Grassmann in Phil. Mag. April 1854).[5]

Taking three lines drawn from the origin through the points of the diagram corresponding to the three primaries as the axes of coordinates, we may express any colour as the resultant of definite quantities of each of the three primaries, and the three elements of colour will then be represented by the three dimensions of space.

The experiments, the results of which are now before the Society, were undertaken in order to ascertain the exact relations of the colours of the spectrum as seen by a normal eye, and to lay down these relations on Newton's diagram. The method consisted in selecting three colours from the spectrum, and mixing these in such proportions as to be identical in colour and brightness with a constant white light. Having assumed three standard colours, and found the quantity of each required to produce the given white, we then find the quantities of two of these combined with a fourth colour which will produce the same white. We thus obtain a relation between the three standards and the fourth colour, which enables us to lay down its position in Newton's diagram with reference to the three standards.

Any three sufficiently different colours may be chosen as standards, and any three points may be assumed as their positions on the diagram. The resulting diagram of relations of colour will differ according to the way in which we begin; but as every colour-diagram is a perspective projection of any other, it is easy to compare diagrams obtained by two different methods.[6]

The instrument employed in these experiments consisted of a dark chamber about 5 feet long, 9 inches broad, and 4 deep, joined to another 2 feet long at an angle of about 100°. If light is admitted at a narrow slit at the end of the shorter chamber, it falls on a lens and is refracted through two prisms in succession, so as to form a pure spectrum at the end of the long chamber. Here there is placed

---

(5) Hermann Grassmann, 'Zur Theorie der Farbenmischung', *Ann. Phys.*, **89** (1853): 69–84; (trans.) 'On the theory of compound colours', *Phil. Mag.*, ser. 4, **7** (1854): 254–64. See Number 163 note (11).   (6) See Number 163.

an apparatus consisting of three moveable slits, which can be altered in breadth and position, the position being read off on a graduated scale, and the breadth ascertained by inserting a fine graduated wedge into the slit till it touches both sides.[7]

When white light is admitted at the shorter end, light of three different kinds is refracted to these three slits. When white light is admitted at the three slits, light of these three kinds in combination is seen by an eye placed at the slit in the shorter arm of the instrument. By altering the three slits, the colour of this compound light may be changed at pleasure.

The white light employed was that of a sheet of white paper, placed on a board, and illuminated by the sun's light in the open air; the instrument being in a room, and the light moderated where the observer sits.

Another portion of the same white light goes down a separate compartment of the instrument, and is reflected at a surface of blackened glass, so as to be seen by the observer in *immediate contact* with the compound light which enters the slits and is refracted by the prisms.

Each experiment consists in altering the breadth of the slits till the two lights seen by the observer agree both in colour and brightness, the eye being allowed time to rest before making any final decision. In this way the relative places of sixteen kinds of light were found by two observers. Both agree in finding the positions of the colours to lie very close to two sides of a triangle, the extreme colours of the spectrum forming doubtful fragments of the third side. They differ, however, in the intensity with which certain colours affect them, especially the greenish blue near the line F, which to one observer is remarkably feeble, both when seen singly, and when part of a mixture; while to the other, though less intense than the colours in the neighbourhood, it is still sufficiently powerful to act its part in combinations. One result of this is, that a combination of this colour with red may be made, which appears red to the first observer and green to the second, though both have normal eyes as far as ordinary colours are concerned; and this blindness of the first has reference only to rays of a definite refrangibility, other rays near them, though similar in colour, not being deficient in intensity.[8] For an account of this peculiarity of the author's eye, see the Report of the British Association for 1856, p. 12.[9]

---

(7) See Number 154.

(8) See Number 163, and 'On the theory of compound colours', *Phil. Trans.*, **150** (1860): 75–7 (= *Scientific Papers*, **1**: 433–5); he supposes 'the yellow spot at the foramen centrale of Soemmering' to absorb rays between the E and F lines of the spectrum.

(9) J. C. Maxwell, 'On the unequal sensibility of the foramen centrale to light of different colours', *Report of the Twenty-sixth Meeting of the British Association* (London, 1857), part 2: 12 (= *Scientific Papers*, **1**: 242). The second observer was his wife; see Number 180 note (10).

By the operator attending to the proper illumination of the paper by the sun, and the observer taking care of his eyes, and completing an observation only when they are fresh, very good results can be obtained. The compound colour is then seen in contact with the white reflected light, and is not distinguishable from it, either in hue or brilliancy; and the average difference of the observed breadth of a slit from the mean of the observations does not exceed $\frac{1}{30}$ of the breadth of the slit if the observer is careful. It is found, however, that the errors in the value of the sum of the three slits are greater than they would have been by theory, if the errors of each were independent; and if the sums and differences of the breadth of two slits be taken, the errors of the sums are always found greater than those of the differences. This indicates that the human eye has a more accurate perception of differences of hue than of differences of illumination.

Having ascertained the chromatic relations between sixteen colours selected from the spectrum, the next step is to ascertain the positions of these colours with reference to Fraunhofer's lines.[10] This is done by admitting light into the shorter arm of the instrument through the slit which forms the eyehole in the former experiments. A pure spectrum is then seen at the other end, and the position of the fixed lines read off on the graduated scale. In order to determine the wave-lengths of each kind of light, the incident light was first reflected from a stratum of air too thick to exhibit the colours of Newton's rings. The spectrum then exhibited a series of dark bands, at intervals increasing from the red to the violet. The wave-lengths corresponding to these form a series of submultiples of the retardation; and by counting the bands between two of the fixed lines, whose wave-lengths have been determined by Fraunhofer, the wave-lengths corresponding to all the bands may be calculated; and as there are a great number of bands, the wave-lengths become known at a great many different points.

In this way the wave-lengths of the colours compared may be ascertained, and the results obtained by one observer rendered comparable with those obtained by another, with different apparatus. A portable apparatus, similar to one exhibited to the British Association in 1856,[11] is now being constructed in order to obtain observations made by eyes of different qualities, especially those whose vision is dichromic.[12]

---

(10) See Number 163 note (9).

(11) See Numbers 109 and 117; and compare Number 168.

(12) The term 'dichromic' had been used by John Herschel; see Number 179 esp. notes (6) and (7).

171

CANCELLED PASSAGES IN THE MANUSCRIPT OF 'ON THE THEORY OF COMPOUND COLOURS, AND THE RELATIONS OF THE COLOURS OF THE SPECTRUM'[1]

DECEMBER 1859[2]

From the original in the library of the Royal Society, London[3]

[Passage at the end of § VII of the paper][4]

The average error from the means in this set of equations was 1.16. See note.†

The equation between the standard colours is

$$18.6 \ (24) + 31.4 \ (44) + 30.5 \ (68) = W.*{}^{(5)}$$

By eliminating W from each of the former equations by means of this last we obtain equations involving each of the 14 selected colours of the spectrum along with the three standard colours and by transposing the selected colour to one side of the equation we obtain its value in terms of the three standards. If any of the terms of these equations come out negative, we must interpret it by transposing it to the other side. It is physically impossible to produce a negative colour on one side of the field, but by putting the positive colour on the other side we may make the tints of both sides equivalent.

† *Turn the page.*

Note.[6] The equation between the three standards was deduced from 20 observations. The average error (that is the sum of all the errors from the mean value divided by the number of observations) was

$$.54 \ \text{in} \ (24) \quad 1.22 \ \text{in} \ (44) \quad \text{and} \ 1.15 \ \text{in} \ (68).$$

---

(1) *Phil. Trans.*, **150** (1860): 57–84 (= *Scientific Papers*, **1**: 410–44).

(2) The manuscript is marked: 'Rec^d. Jan 5th 1860'; and see Numbers 169 and 172.

(3) Royal Society, PT.60.4. The manuscript is titled: 'On the Theory of Compound Colours, and the ⟨Chromatic⟩ Relations of the Colours of the Spectrum'.

(4) Compare 'On the theory of compound colours': 71–3 (= *Scientific Papers*, **1**: 428–30). The passage is cancelled in pencil.

(5) The asterisk indicates that the colour equation involves the three standard colours; the numbers in brackets denote the positions (corresponding to these colours) on the scale of Maxwell's colour box. See Numbers 154 and 163.

(6) Maxwell's appended 'Note' is on the verso.

In order to determine whether the eye is more sensitive to variations in brightness or to variations in colour I have determined also the average error in the sum of the three terms and also that of their differences.

Average error of the sum of the three terms   = 2.67

Average error of the hue between red & green = 0.86

———————————————————— green & blue = 0.99

———————————————————— blue & red   = 0.85.

If the errors in each of the three colours were independent of each other then the error of the sum would be the square root of the squares of the errors of each colour, that is, 1.76. Since the true error is 2.67 it follows that the errors are not independent but that they generally are in the same direction in each colour. Again if the errors in red and green were independent the error of (*red−green*) should be the same as that of (*red+green*) and both should be 1.33 whereas that of red + green is 1.57 and that of (*red−green*) is only 0.86 showing that red and green vary together and that a variation in hue between red and green is more easily detected than a variation of brightness due to an increase or diminution of both simultaneously.

The errors in the determination of the other colours are

.74 for red   1.55 for green and 1.34 for blue.[a]

(a) {Stokes} All wrong.

LETTER TO GEORGE GABRIEL STOKES

3 JANUARY 1860

From the original in the University Library, Cambridge[1]

Marischal College
Aberdeen 1860 Jan 3

Dear Stokes

I have sent you my paper on the colours of the Spectrum, which is as complete as I hope to make it.[2] If I do anything more worth publishing it will be upon dichromatic vision,[3] with reference to the spectrum, and I shall not have light for it before March, even if I get my instrument into good order, and find a convenient time for the observer & myself.[4]

I have written out the theory of Newtons diagram, and its relation to the method of representing colours by lines, drawn from a point in various directions. I have described the method of observation and its results as given by two observers, and explained what I believe to be the reason of the differences between them, which are very far beyond any errors of observations, but appear to me to arise from unequal absorption of certain rays by media which are more absorptive in some eyes than others, and not from any radical difference in the sensations excited by the different rays.[5]

Though I expect to get better observations by taking more pains, I believe the results which I send you to be not far from the truth, and certainly I think I have evidence that the spectrum on Newtons diagram is a triangular figure, with one side broken out.[6] So that if the R.S. is willing, I am ready to put my name to the paper and have it printed. I have copied out all that I shall need in the way of tables &c so I shall not want it back. I have addressed it to Cambridge as I do not know whether the Royal Society meets about this time.

I remain
Yours truly
J. CLERK MAXWELL

---

(1) ULC Add. MSS 7656, M 413. Previously published in Larmor, *Correspondence*, **2**: 15.

(2) See Number 171. The first part of the paper 'On the theory of compound colours, and the relations of the colours of the spectrum', *Phil. Trans.*, **150** (1860): 57–84, esp. 57–78 (= *Scientific Papers*, **1**: 410–36).

(3) The term 'dichromic' (sensitivity to only two primary colour sensations) for colour blindness had been used by John Herschel; see Number 179 esp. notes (6) and (7).

(4) See Numbers 175 to 178.

(5) On the absorptive power of the yellow spot see Number 170, esp. note (8).

(6) See Number 163; and 'On the theory of compound colours': 74 (= *Scientific Papers*, **1**: 431).

FROM A LETTER TO LEWIS CAMPBELL

5 JANUARY 1860

From Campbell and Garnett, *Life of Maxwell*[1]

Marischal College
Aberdeen
5 January 1860

I have been publishing my views about Elastic Spheres in the *Phil. Mag.* for Jany.,[2] and am going to go on with it as I get the prop[ns.] written out.[3] I have also sent my experiments on Colours to the Royal Society of London,[4] so I have two sets of irons in the fire, besides class work. I hope you get on with Plato, and that your pupils are all Theætetuses, and that wisdom soaks like oil into their inwards.[5] There is a man here who is striving after a general theory of things, but he has great difficulty in so churning his thoughts as to coagulate and solidify the vague and nebulous notions which wander in his head. He has been applying to me very steadily whenever he can pounce on me, and I have prescribed for him as I best could, and I hope his abstract of his general theory of things will be palatable to the readers of the *British Ass. Reports* for 1859.[6]

(1) *Life of Maxwell*: 328.

(2) J. C. Maxwell, 'Illustrations of the dynamical theory of gases. Part I', *Phil. Mag.*, ser. 4, **19** (1860): 19–32 (= *Scientific Papers*, **1**: 377–91). See Numbers 157 and 160.

(3) Maxwell, 'Illustrations of the dynamical theory of gases. Parts II and III', *Phil. Mag.*, ser. 4, **20** (1860): 21–37 (= *Scientific Papers*, **1**: 392–409); and see Number 181.

(4) See Number 172.

(5) Lewis Campbell, *The Theætetus of Plato, with a revised text and English notes* (Oxford, 1861).

(6) J. S. Stuart Glennie, 'A proposal of a general mechanical theory of physics', *Report of the Twenty-ninth Meeting of the British Association for the Advancement of Science; held at Aberdeen in September 1859* (London, 1860), part 2: 58–9; and see also Glennie, 'The principles of the science of motion', *Phil. Mag.*, ser. 4, **21** (1861): 41–55; and Glennie, 'The principles of the science of energetics', *ibid.*, **21** (1861): 274–81, 350–8; **22** (1861): 62–4.

LETTER TO CECIL JAMES MONRO

24 JANUARY 1860

From the original in the Greater London Record Office, London[1]

Marischal College
Aberdeen 1860 Jan 24

Dear Monro

Communications are to be found from the Revd. A. Thacker in the Philosophical Magazine for 1851 August, October & November giving the motion of a Free pendulum swinging elliptically (wh: is the general case) and finding the motion of the major axis both when the Earth is fixed & when it moves.[2] If I mistake not, the same is done better & shorter in Haywards Kineticdynamical paper, Cambridge Phil Trans. 1856 or 7.[3] There are many others in the Comptes Rendus Phil Mag &c but these are free from nonsense. With respect to your old letter[4] I dont think it was the parental trick but rather the lazy-idle-boy-trick I was playing. I hope you will gather a legal tone from an approx^n to the remains of the Forum.

If you should have no headache & should write to me the best thing you could write wd be a report on Litchfield Farrar Butler Elphinstone &c &c. Lushington wrote to me in Nov^r & I to him.[5]

As to Nat Phil I am to be turned out by the Commissioners (that is if their

---

(1) Greater London Record Office, Acc. 1063/2088.

(2) Arthur Thacker, 'Pendulum experiments: formula for calculating the apsidal motion', *Phil. Mag.*, ser. 4, **2** (1851): 159–60; Thacker, 'On the motion of a free pendulum', *ibid.*: 275–8; Thacker, 'Formulae connected with the motion of a free pendulum', *ibid.*: 412–13. On Thacker's pendulum experiments witnessed by Maxwell at Trinity College in 1851 see *Life of Maxwell*: 153–4; and on Thacker see Number 92 note (5).

(3) R. B. Hayward, 'On a direct method of estimating velocities, accelerations, and all similar quantities with respect to axes, moveable in any manner in space, with applications', *Trans. Camb. Phil. Soc.*, **10** (1856): 1–20. See Numbers 107 note (76) and 116 note (7).

(4) See a letter from Monro of 18 January 1859 (Greater London Record Office, Acc. 1063/2104); 'At this moment however I write on a particular occasion viz. for Cohen. He has been meditating about gyrations and polhodies & such, and wants to know, from some one who can tell him, what literature there is on the subject. Sylvester told him of some very good German book about the gyroscope, but could not remember the name; do you know it? I told him at the same time that you had written something about rotation but I did not know where it was to be found.' The reference is to Arthur Cohen, Magdalene 1849 (Venn); see Arthur Cohen, 'On the differential coefficients and determinants of lines and their application to analytical mechanics', *Phil. Trans.*, **152** (1862); 469–510, esp. 492–510. Monro is alluding to Maxwell, 'On a dynamical top', *Trans. Roy. Soc. Edinb.*, **21** (1857): 559–70 (= *Scientific Papers*, **1**: 248–62). On 'polhodies' see Number 116 note (4).

(5) Henry Montague Butler, Trinity 1850 (Venn); and see Numbers 56 note (11), 67 note (3), 101 note (4).

ordinances are to be carried)[6] but there may be an opposition, as last year, by the citizens and various bodies interested in the preservation of the College and it may be again successful.[7]

I have a much larger class[8] than the Kings Coll Prof[r] who supercedes me[9] & there are no signs of decrepitude either in me or the college and I can stand a trial any day.

However I am a candidate for the Nat Phil in Edinburgh[10] which is promotion in any case and I suppose it will be decided in March or April.

The candidates are Tait of Peterhouse Routh & Fuller 3 peterhousers,[11] the judges are 7 men, ranging from M[r] W E Gladstone to gentlemen of the Town Council of Edin[h].[12] The weapons are testimonials common gossip & diplomacy and the situation is not yet vacant.[13]

---

(6) On the fusion of King's and Marischal Colleges see Numbers 140 and 142. See 'Copies of two ordinances made by the Commissioners under the Act 21 and 22 Vict. c. 83' (which enacted the union of the two Colleges to form the University of Aberdeen), dated 10 January 1860, *Parliamentary Papers*, LIII (1860): 617–32; 'The two Professorships of Natural Philosophy in the said existing Colleges shall be conjoined, and shall be the Professorship of Natural Philosophy in the University of Aberdeen, which shall be held by David Thomson, Master of Arts, now Professor of Natural Philosophy in King's College... [and] compensation shall be made, in terms of the said Act, to...James Clerk Maxwell, Master of Arts.'

(7) Unavailing efforts were made to repeal the Act uniting the two Colleges; see Peter John Anderson, ed. *Fasti Academiae Mariscallanae Aberdonensis. Selections from the Records of the Marischal College and University MDXCIII–MDCCCLX*, 3 vols. (Aberdeen, 1889–98), 1: 547.

(8) Maxwell's statement of the current attendance at his class is supported by the testimonial (for Edinburgh) from the Principal of Marischal College, Daniel Dewar (his father-in-law): 'I feel it to be my duty, as Principal of this College, to state, in proof of his popularity and efficiency as a Lecturer on Natural Philosophy, that his Class is much larger this Session than it has been for many years in Marischal College, and considerably larger than the Class of Natural Philosophy in King's College. The Natural Philosophy Class in Marischal College is the only Class in the Arts that is larger than the corresponding Class in Arts in King's College.' (ULC Add. MSS 7655, V, i/12). Compare 'Copies of the Report...in relation to the Universities and Colleges of Aberdeen', Universities (Scotland), *Parliamentary Papers*, LIII (1860): 597–616, where the average number of students in each session over the previous seven years is given as King's 49, Marischal 44.

(9) David Thomson, Professor of Natural Philosophy and Sub-Principal of King's College, Aberdeen (Venn); see note (6).   (10) See Numbers 164, 166, 167 and 168.

(11) P. G. Tait, Peterhouse 1848, Senior Wrangler 1852, Professor of Mathematics at Queen's College, Belfast, the successful candidate; Edward John Routh, 1850, Senior Wrangler 1854, equal Smith's Prize (with Maxwell) 1854; Frederick Fuller, 1838, Professor of Mathematics at King's College, Aberdeen (Venn).

(12) The Curators: see Number 168 note (4), Gladstone was, at the time, Chancellor of the Exchequer and Lord Rector of the University of Edinburgh; see John Morley, *The Life of William Ewart Gladstone*, 3 vols. (London, 1903), 1, 626, 634.

(13) On Tait's appointment see *Life of Maxwell*: 258–60, 277; and C. G. Knott, *Life and Scientific Work of Peter Guthrie Tait* (Cambridge, 1911): 16.

I am brewing more elastic spheres and have proved a prop. about how they bob about in a plug of plaster of Paris.[14]

On the whole we are going on briskly all the merrier for having something to fight for and all the fiercer for having our natural enemies set over us to exterminate us after the fashion of Henricus Octavus and our other founders & confounders.

I date from Marischal College but the safest address is Glenlair

Springholm
Dumfries

as the College is abolished by the Commissioners.

However, I shall do Nat: Phil: whether in College or out of it and I may perhaps publish more if I teach less.

Mixing colours of spectrum is my summer recreation, and I seem already to have seen something in that line, & am getting up an apparatus for it like a bagatelle board and a Newtonian telescope with a general Post office receiving box in the front and 2 prisms and a concave speculum behind.[15]

Yours truly
J. C. Maxwell

---

(14) Proposition XIX of Part II of Maxwell's paper 'Illustrations of the dynamical theory of gases', *Phil. Mag.*, ser. 4, **20** (1860): 27–9 (= *Scientific Papers*, **1**: 398–400).

(15) See Number 168.

175

LETTER TO GEORGE GABRIEL STOKES

25 FEBRUARY 1860

From the original in the University Library, Cambridge[1]

Marischal College
Aberdeen 1860 Feb 25

Dear Stokes

I have received yours of yesterday and am much honoured by the Council of the Royal Society in being appointed Bakerian Lecturer.[2]

I am not aware of the constitution or laws of the Bakerian Lectureship but from what I have read under the title of Bakerian Lectures I suppose that the Lecturer reads in the ordinary way the paper which the Council have honoured with the title of Lecture, and that his duties are then over.

I cannot leave Aberdeen till April, but after that I shall be able to go to London.

I should myself prefer the 19th April if that would suit the Society.

The instrument I used in summer is not at all portable but I have now got a reflecting one

$$3^{ft} 6^{in} \times 11^{in} \times 4$$

which is warranted to be safe when turned upside down and moderately jolted.[3] I have not had much sun but I find that a certain uniform cloud which we have sometimes is not bad for observations and I have had one set of observations from a colour blind gentleman who spent a day in Aberdeen and two sets from a colour blind student from whom I expect about 3 sets a week till the degree examinations interfere with optics.

They are both perfect cases like Mr Pole's[4] and the student is a good observer. Both agree that the neutral point of the spectrum is near F and the student sees a dark cloud in the axis of vision when he looks at this light just as I do myself.

All less refrangible light is yellow, all more refrangible blue, and by equating white to selected pairs of colours from each side of the spectrum I propose to determine the brightness and blueness and yellowness of every point of the spectrum, in terms of two selected colours. Before April I shall have probably

---

(1) ULC Add. MSS 7656, M 414. Previously published in Larmor, *Correspondence*, **2**: 16–17.

(2) See *Minutes of Council of the Royal Society from December 16th 1858 to December 16th 1869*, **3** (London, 1870): 52. At a meeting of the Council on 23 February 1860 it was 'Resolved, – That James Pettigrew, Esq. be appointed Croonian Lecturer, and Professor Clerk Maxwell Bakerian Lecturer for the present year'. Maxwell did not deliver the Bakerian Lecture in 1860; see Number 176 note (6).

(3) See Numbers 168, 169 and 182.　　　　(4) See Number 169 note (5).

ascertained the accuracy of the observations and got some normal observations with the same instrument to compare with them.[5]

If I should be able to arrive at results and to communicate them to the Society before I go up, would it be lawful to append them (with proper date) to the former paper?

I suppose I had better prepare one or two diagrams and bring the instrument with me.

If I have mistaken what I have to do pray let me know, and if the 19th April will not do, then any day after that will do for me.[a]

I remain
Yours very truly
J. C. MAXWELL

(a) {Stokes} Ans[d]. March 2. Fixed provisionally on April 19. Asked him to send diagrams to be drawn if he wanted them – and give a viva voce exposition – said I was sure the electric lamp of the R.S. w[d]. be at his service.

(5) See Number 179.

176

LETTER TO GEORGE GABRIEL STOKES

12 MARCH 1860

From the original in the University Library, Cambridge[1]

Marischal College
Aberdeen 1860 March 12

Dear Stokes

If it is found necessary for the convenience of anyone to have the day changed from the 19th April,[2] I am willing to come on the 26th, but I cannot go to London at all at any later time. I should prefer the 19th to the 26th if it should turn out equally convenient for Dr Sharpey[3] to have his lecture at a different time, but if the 19th secures advantages to the Croonian Lecture[4] which are not to be had afterwards (such as the presence of a man like Kölliker)[5] then do not scruple to change my day to the 26th as my reasons are all private. After the 26th I should prefer to have it read in absence by deputy, for I do not think I can go up after that time.

What are the rules about one who is not a fellow reading a paper? I see the Bakerian lecturer is called one of the fellows in the foundation, but I suppose from what you say that an extraneous person may deliver it.[6]

I have seen a large diagram of the spectrum with the fixed lines in various lecture rooms. Do you think I could get one in London to point to during lecture in order to indicate the colours to be mixed?

---

(1) ULC Add. MSS 7656, M 415. Previously published in Larmor, *Correspondence*, **2**: 17.

(2) See Number 175.

(3) William Sharpey, Secretary of the Royal Society.

(4) The Croonian Lecture was read on 19 April 1860 by James Pettigrew; *Proc. Roy. Soc.*, **10** (1860): 433–44. See Number 175 note (2).

(5) Alfred Kölliker, elected a Foreign Member of the Royal Society on 24 May 1860; *Proc. Roy. Soc.*, **10** (1860): 473.

(6) See *Minutes of Council of the Royal Society from December 16th 1858 to December 16th 1869*, **3** (London, 1870): 59; at a meeting of the Council on 26 April 1860 it was 'Resolved, – That Mr Fairbairn, FRS be appointed Bakerian Lecturer for the present year instead of Professor J. C. Maxwell, who it appears is not eligible, not being a Fellow of the Society'. The Bakerian Lecture was read on 10 May 1860 by William Fairbairn; *Proc. Roy. Soc.*, **10** (1860): 469–73. Maxwell had already in 1859 been nominated (by Stokes and Whewell) for a Royal Medal for his 'Mathematical Theory of the Composition of Colours, verified by quantitative experiments, and for his Memoirs on Mathematical and Physical Subjects' (*Minutes*, **3**: 27); and was nominated again on 24 May 1860 for the Royal Medal (by the Rev. John Barlow and Prof. W. H. Miller), and for the Rumford Medal (by Stokes and Miller) for his 'Researches on the Composition of Colours, and other Optical Papers' (*ibid*.: 62–3). He was awarded the Rumford Medal by the Council on 1 November 1860 (*ibid*.: 72; and *Proc. Roy. Soc.*, **11** (1860): 19–21). Maxwell was elected FRS on 2 May 1861; see *Minutes*, **3**: 85, and *Proc. Roy. Soc.*, **11** (1861): 193.

I hope to have the triangle of colour ready on a large scale in time for the lecture.

My colour blind student has been ill some time and I have got no more observations.

I remain
Yours truly
J. CLERK MAXWELL

I have had a letter from a colour-blind man in Manchester and expect to hear more of his case in answer to my reply.

177

LETTER TO GEORGE GABRIEL STOKES

23 APRIL 1860

From the original in the University Library, Cambridge[1]

Address to   ⎰ Care of J. MacCunn Esq[re]
30th April   ⎱   Ardhallow
           Dunoon
           23 April 1860

Dear Stokes

I had hopes when I last wrote to you that I should have got a considerable number of colour blind observations of the spectrum from a last-year's student. Unfortunately he fell ill just when the bright weather came and only recovered at the end of the session so I got his observations for two days only, but as these seem pretty consistent perhaps they might be worth reducing and publishing till some one finds a better opportunity to observe. The results are very similar to those of another gentleman whom I tried on a cloudy day. Here are the results.[2]

Taking a green at the line E and a blue $\frac{2}{3}$ from F towards G as arbitrary standard colours,[3] the first appears 'yellow' the second 'blue' to the colour blind. The red end of the spectrum is visible as far to them as to us, using a blue glass to cut off the bright part of the spectrum but the colour is 'yellow' apparently of the same kind as that produced by a very narrow aperture at E in the green. In order to apply my ordinary method we must have enough of the colour to produce the standard white when combined with blue, and this cannot be done without taking all the light from the end of the spectrum to the line D.

Now this quantity of light which to our eyes is much brighter than the standard white, requires an addition of blue to make it equal to white to the colour blind showing that the red end is very feeble in its effects upon them.

The quality of the colour of this end of the spectrum, as shown by the quantity of blue required to produce the standard white seems to be nearly uniform from the extremity as far as a point half way between E & F and is what the colour blind call 'yellow'. The quantity or intensity as shown by the amount of

---

(1) ULC Add. MSS 7656, M 416. Previously published in Larmor, *Correspondence*, **2**: 18–19.

(2) See Number 179; and the 'Postscript' to Maxwell's paper 'On the theory of compound colours, and the relations of the colours of the spectrum', *Phil. Trans.*, **150** (1860); 78–84 (= *Scientific Papers*, **1**: 436–44).

(3) For the standard colours and the correlation with Fraunhofer's lines see Number 170.

the given colour required to neutralise the blue is very small up to D is greatest at $\frac{2}{3}$ from D to E and then diminishes.

This part of the spectrum therefore is nearly uniform in colour and any part may be taken as a representative of 'yellow' by a colour blind person.

From $\frac{1}{2}$ between E & F to $\frac{1}{4}$ from F towards G the colour is to them a mixture of blue and yellow, being neutral close to F. Near F the intensity is feeble so that three times the breadth of slit is required to produce white when compared with the breadth at the standards. Also in viewing a uniform field of this colour the 'yellow spot' of the retina is seen dark on a light ground.[4]

From the point $\frac{1}{4}$ from F to G to the end of the spectrum the colour seems quite pure blue increasing in intensity to $\frac{2}{3}$ from F to G and then diminishing but being visible as far as we see it and apparently as bright.

Comparing my observations with these I find

	Red	Green	Blue	
J.C.M.	$22.6 +$	$26.0 +$	$37.4 =$	W
J.S.[5]		33.7	$33.1 =$	W

$$D = 22.6 \ (R) - 7.7 \ (G) + 4.3 \ (B).\ [6]$$

The colour which if it could be exhibited we would see and they would not is a red with a little blue, and purged of that green (or 'yellow') which renders it still visible to them.

I send you a plan of the spectrum according to my student, showing the intensity of blue & of yellow at different points of the spectrum.[7] I find my new instrument is not altered by being carried about so that if I find any more subjects I can compare them wherever they are.

I remain

Yours truly

J. Clerk Maxwell

After the 30th April my address will be Glenlair by Dumfries.

---

(4) See Number 170 note (8).

(5) James Simpson; see Number 178.

(6) In the portable colour box the 'three standard colours' are 'referred to as (104), (88), and (68), these being the positions of the red, green, and blue on the scale of the new instrument'; Maxwell, 'On the theory of compound colours': 79 (= *Scientific Papers*, **1**: 438). Compare Number 163.

(7) Compare Fig. 9 in Maxwell's paper 'On the theory of compound colours': opposite 84 (= *Scientific Papers*, **1**: 444).

The curve on the right represents the amount of the 'blue' element and that on the left the 'yellow' element of colour in different parts of the spectrum as seen by the colour blind. The dotted line is the sum of the elements.

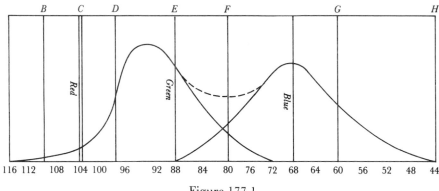

Figure 177,1

178

LETTER TO GEORGE GABRIEL STOKES

7 MAY 1860

From the original in the University Library, Cambridge[1]

Glenlair
Springholm
Dumfries
May 7 1860

Dear Stokes

I sent you this morning an account of observations by Mr James Simpson, on the colours of the spectrum and also a comparison of his observations on coloured papers with mine.[2] It is not very easy to find colourblind persons who know what is required to make good observations. If Mr Simpson had not been in ill health, and by no means strong when he observed the spectrum I would have been able to send you colourblind equations better than most ordinary ones. However they are really very good and certainly correct in the main features. I still believe that the great difficulty in comparing eyes of either the dichromic or the trichromic[3] type lies in the absorptive power of the media of the eye, especially the yellow spot[4] the seat of 'Haidingers brushes'.[5]

Observations of colours near the line F are extremely unsatisfactory as the brightness and even colour of the field vary with the direction of the axis of the eye.[6]

It is very remarkable that this peculiarity should be confined to a small portion of the spectrum while neighbouring parts, *very similar in colour*, show nothing of the kind. I expect that if I could get enough of the yellow material of the spot to try it, it would cut a dark band out of the spectrum near F, and that in a very marked manner.[7] Do you know whether there is any other more abundant yellow substance of the same composition in the animal kingdom? for if it is peculiar to a small spot on the thinnest part of the human retina, my conscience recoils from the amount of slaughter necessary to verify my conjecture.

---

(1) ULC Add. MSS 7656, M 417. Previously published in Larmor, *Correspondence*, **2**: 20–1.

(2) The 'Postscript' to Maxwell's paper 'On the theory of compound colours, and the relations of the colours of the spectrum', *Phil. Trans.*, **150** (1860): 78–84 (= *Scientific Papers*, **1**: 436–44) is endorsed 'Received May 8, – Read May 24, 1860'. See Number 179.

(3) On the terms 'dichromic' (for colour-blind vision) and 'trichromic' (for normal vision) see Number 179, esp. notes (6) and (7).

(4) See Number 170 note (8).   (5) See Number 32 esp. note (2).

(6) Compare 'On the theory of compound colours': 76 (= *Scientific Papers*, **1**: 434).

(7) Compare Numbers 163 and 170.

With respect to comparisons of brightness of dissimilar colours. I find all my observers decline to say which of two very different colours is brightest. I have made attempts to compare white with pure colours and the result is that if I make red (e.g.) and white equal in brightness (that is = in impressing my eye with a sense of something being there) I find that the red requires green and blue to be added to it to neutralise it and still is not brighter than the white with which it was compared thus if

$$R = W$$
$$\& \ R + G + B < W$$

something must be wrong. I think positive colours have more justice done them by the eye than white light.

I was some time ago doubtful about presenting myself as a candidate to the R. S.[8] If you could get me placed as a candidate for admission according to the rules of the society I should be much obliged to you[9] in addition to the thanks I owe you for your advice as to my paper and other optical cooperation.

Yours truly
J. CLERK MAXWELL

(8) See Maxwell's letter to Stokes of 27 January 1857 (Number 111).

(9) Maxwell was elected a Fellow of the Royal Society on 2 May 1861; *Proc. Roy. Soc.*, **11** (1861): 193.

179

ABSTRACT OF THE POSTSCRIPT TO PAPER 'ON COMPOUND COLOURS'

[8 MAY 1860]

From the *Proceedings of the Royal Society*[1]

POSTSCRIPT TO A PAPER 'ON COMPOUND COLOURS, AND ON THE RELATIONS OF THE COLOURS OF THE SPECTRUM'[2]

By J. Clerk Maxwell, Esq. Communicated by Professor Stokes, Sec. R.S.

Received May 8, 1860

(Abstract)
Account of Experiments on the Spectrum as seen by the Colour-blind

The instrument used in these observations was similar to that already described. By reflecting the light back through the prisms by means of a concave mirror, the instrument is rendered much shorter and more portable, while the definition of the spectrum is rather improved.[3] The experiments were made by two colour-blind observers, one of whom, however, did not obtain sunlight at the time of observation. The other obtained results, both with cloud-light and sun-light, in the way already described. It appears from these observations –

I. That any two colours of the spectrum, on opposite sides of the line '*F*', may be combined in such proportions as to form white.

II. That all the colours on the more refrangible side of F appear to the colour-blind 'blue', and all those on the less refrangible side appear to them of another colour, which they generally speak of as 'yellow', though the green at E appears to them as good a representative of that colour as any other part of the spectrum.

III. That the parts of the spectrum from A to E differ only in intensity, and not in colour; the light being too faint for good experiments between A and D, but not distinguishable in colour from E reduced to the same intensity. The *maximum* is about $\frac{2}{3}$ from D towards E.

---

(1) *Proc. Roy. Soc.*, **10** (1860): 484–6. Read 24 May 1860.

(2) *Phil. Trans.*, **150** (1860): 78–84 (= *Scientific Papers*, **1**: 436–44). The 'Postscript' was refereed by John Tyndall, who recommended publication in a brief letter to Stokes of 24 May 1860 (Royal Society, *Referees' Reports*, **4**: 180).   (3) See Numbers 168, 169 and 182.

IV. Between E and F the colour appears to vary from the pure 'yellow' of E to a 'neutral tint' near F, which cannot be distinguished from white when looked at steadily.

V. At F the blue and the 'yellow' element of colour are in equilibrium, and at this part of the spectrum the same blindness of the central spot of the eye is found in the colour-blind that has been already observed in the normal eye, so that the brightness of the spectrum appears decidedly less at F than on either side of that line; and when a large portion of the retina is illuminated with the light of this part of the spectrum, the *limbus luteus* appears as a dark spot, moving with the movements of the eye.[4] The observer has not yet been able to distinguish Haidinger's 'brushes' while observing polarized light of this colour, in which they are very conspicuous to the author.[5]

VI. Between F and a point $\frac{1}{3}$ from F towards G, the colour appears to vary from the neutral tint to pure blue, while the brightness increases, and reaches a maximum at $\frac{3}{5}$ from F towards G, and then diminishes towards the more refrangible end of the spectrum, the purity of the colour being apparently the same throughout.

VII. The theory of colour-blind vision being '*dichromic*',[6] is confirmed by these experiments, the results of which agree with those obtained already by normal or '*trichromic*' eyes, if we suppose the 'red' element of colour eliminated, and the 'green' and 'blue' elements left as they were,[7] so that the '*red-making rays*', though dimly visible to the dichromic eye, excite the sensation not of red but of green, or as they call it, 'yellow'.

VIII. The extreme red ray of the spectrum appears to be a sufficiently good representative of the defective element in the colour-blind. When the ordinary eye receives this ray, it experiences the sensation of which the dichromic eye is incapable; and when the dichromic eye receives it, the luminous effect is probably of the same kind as that observed by Helmholtz in the ultra-violet part of the spectrum – a sensibility to light, without much appreciation of colour.[8]

A set of observations of coloured papers by the same dichromic observer was

---

(4) See Numbers 163, 170 and 177.

(5) See Number 178; on 'Haidinger's brushes' see Number 32 note (2).

(6) The term used by John Herschel in his letter to Dalton of 20 May 1833; in W. C. Henry, *Memoirs of the Life and Scientific Researches of John Dalton* (London 1854): 26.

(7) Compare Numbers 54 and 59, and Herschel's comment to Dalton that 'we have three primary sensations when you have only two', in Henry, *Memoirs*: 26.

(8) Hermann Helmholtz, 'Ueber die Zusammensetzung von Spectralfarben', *Ann. Phys.*, **94** (1855): 1–28, esp. 18–23.

then compared with a set of observations of the same papers by the author, and it was found –

1. That the colour-blind observations were consistent among themselves, on the hypothesis of *two* elements of colour.

2. That the colour-blind observations were consistent with the author's observations, on the hypothesis that the two elements of colour in dichromic vision are identical with two of the three elements of colour in normal vision.

3. That the element of colour, by which the two types of vision differ, is a red, whose relations to vermilion, ultramarine, and emerald-green are expressed by the equation

$$D = 1.198V + 0.078U - 0.276G,$$

where D is the defective element, and V, U and G the three colours named above.

LETTER TO GEORGE GABRIEL STOKES

LATE MAY/EARLY JUNE 1860[1]

From the original in the University Library, Cambridge[2]

Glenlair
Dalbeattie
Dumfries

Dear Stokes

I have just received your letter of the 19th May which went astray owing to its being addressed to Aberdeen. My safest address is always as above but I have written to Aberdeen to put them right.

I wrote an abstract of my postscript[3] & sent it to Dr Sharpey.[4]

In my paper I avoided making any definite statement as to the view taken by Sir D Brewster of the composition of light.[5] I think he regards the three elements of colour as objective realities or actual substances of which the light is compounded.[6] I consider them as properties of homogeneous rays having reference to the human organ of vision whether dichromic or trichromic.

On the one theory it is conceivable that absorption may alter the proportions of the elements in a ray of the pure spectrum. On the other theory it can only have the same effect as weakening the light in any other way.

This is the field of a controversy into which I have not entered as I have not yet sufficient experiments to prove that absorption always acts in the same way as simple weakening of the light.[7]

I also avoid stating Sir D Brewsters method of drawing curves,[8] as he is not very distinct about them and in cases of this sort I think that his methods of observation are sometimes much better than his description seems to indicate so that if you go by the description merely you will find that you are under rating the forces of the enemy and may require to modify a statement.

---

(1) The letter is not dated; but see *infra*.

(2) ULC Add. MSS 7656, M 442. Previously published in Larmor, *Correspondence*, **2**: 21–2.

(3) Number 179.   (4) Number 176 note (3).

(5) David Brewster, 'On a new analysis of solar light, indicating three primary colours, forming coincident spectra of equal length', *Trans. Roy. Soc. Edinb.*, **12** (1834): 123–36. Compare Maxwell, 'On the theory of compound colours', *Phil. Trans.*, **150** (1860): 57–84, esp. 59 (= *Scientific Papers*, **1**: 413).

(6) Supposing the colours of the spectrum to be compounded of mixtures, in various proportions, of red, yellow and blue primary kinds of light; see Number 139 notes (15) and (16).

(7) Compare Maxwell's letter to Forbes of 26 November 1857 (Number 139), and his more circumspect discussion of Brewster in 'On the theory of compound colours': 59 (= *Scientific Papers*, **1**: 413); 'I shall not enter into the very important questions affecting the physical theory of light'. Compare Number 184.   (8) Brewster, 'On a new analysis of solar light': 136.

I shall endeavour to state the difference of the two theories by making one very distinct and stating the other in its own words only.[9]

II J & K are two different observers.[10] I shall try to distinguish them properly when I get proofs.[11]

III Unless the curve of the spectrum has angular points no colour in it represents a primary.[12]

I fear oxen and sheep have no yellow spot at least I never found it. They have plenty of glittering tapetum but I must consult anatomists to see what animals have the yellow spot.[13]

I have not been in Cambridge since you left Pembroke so I am not sure whether your residence derives its name from the lenticular form of its enclosure[14] & if I address right under that impression.

I think it would be worth while to select some portions of the spectrum and determine their heating power also their action on various chemicals and their red green and blue luminous elements and their fluorescent effects on various things so as to prove that if the wavelength and the heating power remain the same everything else will be the same, whatever modifications the light may have undergone.

Yours truly

J. C. MAXWELL

---

(9) See Maxwell, 'On the theory of compound colours': 58–60 (= *Scientific Papers*, **1**: 411–14).

(10) Maxwell himself (J) and his wife Katherine (K); see Maxwell, 'On the theory of compound colours': 69–71, 74–5 (= *Scientific Papers*, **1**: 426–8, 431–2).

(11) Maxwell made only trivial changes to the text of the MS of 'On the theory of compound colours' (Royal Society, PT. 60.4) in seeking to clarify this point.

(12) See Figs. 6 and 7 of 'On the theory of compound colours': opposite 84 (= *Scientific Papers*, **1**: 444).   (13) Compare Number 178.

(14) Stokes' address: Lensfield Cottage, Cambridge.

181

BRITISH ASSOCIATION PAPER ON THE THEORY OF GASES

[JUNE/JULY 1860]

From the *Report of the British Association for 1860*[1]

ON THE RESULTS OF BERNOULLI'S THEORY OF GASES[2]
AS APPLIED TO THEIR INTERNAL FRICTION, THEIR DIFFUSION,
AND THEIR CONDUCTIVITY FOR HEAT

The substance of this paper is to be found in the 'Philosophical Magazine' for January and July 1860.[3] Assuming that the elasticity of gases can be accounted for by the impact of their particles against the sides of the containing vessel, the laws of motion of an immense number of very small elastic particles impinging on each other, are deduced from mathematical principles; and it is shown,[4] – 1st, that the velocities of the particles vary from 0 to ∞, but that the number at any instant having velocities between given limits follows a law similar in its expression to that of the distribution of errors according to the theory of the 'Method of least squares'. 2nd. That the relative velocities of particles of two different systems are distributed according to a similar law, and that the mean relative velocity is the square root of the sum of the squares of the two mean velocities. 3rd. That the pressure is one-third of the density multiplied by the mean square of the velocity. 4th. That the mean *vis viva* of a particle is the same in each of two systems in contact, and that temperature may be represented by the *vis viva* of a particle, so that at equal temperatures and pressures, equal volumes of different gases must contain equal numbers of particles. 5th. That when layers of gas have a motion of sliding over each other, particles will be projected from one layer into another, and thus tend to resist the sliding motion. The amount of this will depend on the average distance described by a particle between successive collisions. From the coefficient of friction in air, as given by Professor Stokes,[5] it would appear that this distance is $\frac{1}{447000}$ of an inch; the mean velocity being 1505 feet per second, so that each

---

(1) *Report of the Thirtieth Meeting of the British Association for the Advancement of Science; held at Oxford in June and July 1860* (London, 1861), part 2: 15. Compare Number 160.

(2) Daniel Bernoulli, *Hydrodynamica, sive de Viribus et Motibus Fluidorum Commentarii* (Strasbourg, 1738): 200–202.

(3) J. C. Maxwell, 'Illustrations of the dynamical theory of gases', *Phil. Mag.*, ser. 4, **19** (1860): 19–32; *ibid.*, **20** (1860): 21–37 (= *Scientific Papers*, **1**: 377–409).

(4) See Number 157.   (5) See Number 152 note (7) and Number 157 note (11).

particle makes 8,077,200,000 collisions per second.[6] 6th. That diffusion of gases is due partly to the agitation of the particles tending to mix them, and partly to the existence of opposing currents of the two gases through each other. From experiments of Graham on the diffusion of olefiant gas[7] into air, the value of the distance described by a particle between successive collisions is found to be $\frac{1}{389000}$ of an inch, agreeing with the value derived from friction as closely as rough experiments of this kind will permit. 7th. That conduction of heat consists in the propagation of the motion of agitation from one part of the system to another, and may be calculated when we know the nature of the motion. Taking $\frac{1}{400000}$ of an inch as a probable value of the distance that a particle moves between successive collisions, it appears that the quantity of heat transmitted through a stratum of air by conduction would be $\frac{1}{10,000,000}$ of that transmitted by a stratum of copper of equal thickness, the difference of the temperatures of the two sides being the same in both cases.[8] This shows that the observed low conductivity of air is no objection to the theory, but a result of it. 8th. That if the collisions produce rotation of the particles at all, the *vis viva* of rotation will be equal to that of translation. This relation would make the ratio of specific heat at constant pressure to that at constant volume to be 1.33, whereas we know that for air it is 1.408. This result of the dynamical theory, being at variance with experiment, overturns the whole hypothesis, however satisfactory the other results may be.[9]

---

(6) Maxwell, 'Illustrations of the dynamical theory of gases. Part I. On the motions and collisions of particles', *Phil. Mag.*, ser. 4, **19** (1860): 19–32 (= *Scientific Papers*, **1**: 377–91).

(7) Ethylene; and see Number 157 note (12).

(8) Maxwell, 'Illustrations of the dynamical theory of gases. Part II. On the process of diffusion of two or more kinds of moving particles among one another', *Phil. Mag.*, ser. 4, **20** (1860): 21–33 (= *Scientific Papers*, **1**: 392–405). Rudolf Clausius criticised Maxwell's calculation of this value in his 'Ueber die Wärmeleitung gasförmiger Körper', *Ann. Phys.*, **115** (1862): 1–56, esp. 54n. Maxwell used an incorrect value for the conductivity of copper given by W. J. M. Rankine in *A Manual of the Steam Engine and other Prime Movers* (London/Glasgow, 1859): 259, which did not correctly reduce the value to English measure. He also used a number which relates to one hour as the unit of time as though it was calculated for one second. He gave a corrected calculation, acknowledging Clausius' result that 'lead should conduct heat 1400 times better than air', in his paper 'On the dynamical theory of gases', *Phil. Trans.*, **157** (1867): 49–88, on 88 (= *Scientific Papers*, **2**: 77). Correcting Maxwell's calculation, following Clausius, copper would conduct heat about 7000 times better than air (see Niven's note in *Scientific Papers*, **1**: 405n).

(9) J. C. Maxwell, 'Illustrations of the dynamical theory of gases. Part III. On the collision of perfectly elastic bodies of any form', *Phil. Mag.*, ser. 4, **20** (1860): 33–7 (= *Scientific Papers*, **1**: 405–9). The paradox of specific heats was a consequence of the equipartition theorem of the distribution of kinetic energy among the rotational and translational motions of the particles: 'The final state...of any number of systems of moving particles of any form is that in which the average *vis viva* of translation along each of the three axes is the same in all the systems, and equal to the average *vis viva* of rotation about each of the three principal axes of each particle.' See also C. Truesdell, 'Early kinetic theories of gases', *Archive for History of Exact Sciences*, **15** (1975): 1–66, esp. 51–2.

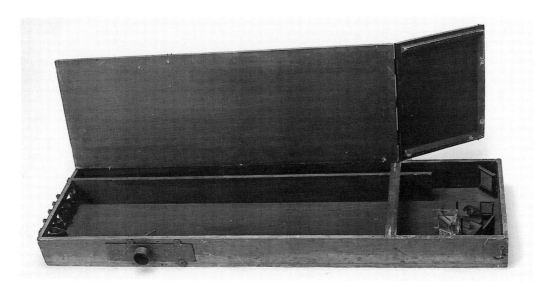

Plate IX. Portable apparatus (1860) for mixing spectral colours, the compound colour being compared with white light (Number 182).

BRITISH ASSOCIATION PAPER ON MIXING THE COLOURS OF THE SPECTRUM

[JUNE/JULY 1860]

From the *Report of the British Association for 1860*[1]

ON AN INSTRUMENT FOR EXHIBITING ANY MIXTURE OF THE COLOURS OF THE SPECTRUM[2]

This instrument consists of a box about 40 inches long by 11 broad and 4 deep. Light is admitted at one end through a system of three slits, of which the position and breadth can be altered and accurately measured. This light, near the other end of the box, falls on two prisms in succession, and then on to a concave mirror, which reflects it back through the prisms, so as to increase the dispersion of colours. The light then falls on a plane mirror inclined 45° to the axis of the instrument, and is reflected on a screen in which is a narrow slit. On this screen are formed three pure spectra, the position and intensity of each depending on the position and breadth of the slit through which the light was admitted. The portions of these spectra which fall on the slit in the screen pass through, and are viewed by the eye placed close behind it. A colour compounded of these three portions of three different spectra is seen illuminating the prisms, and can be compared with white reflected light seen past the edge of the prisms. The advantage of the instrument over that described to the Association in 1859[3] is, that by the principle of reflexion the rays return in the same tube, so as not to require two limbs forming an awkward angle; while at the same time, by doubling the dispersion, the necessary length of the instrument is diminished. By means of this instrument many observations of colours have been taken. Some of these by a colour-blind person are published in the 'Philosophical Transactions' for 1860.[4]

---

(1) *Report of the Thirtieth Meeting of the British Association for the Advancement of Science; held at Oxford in June and July 1860* (London, 1861), part 2: 16.

(2) See Numbers 168 and 169 and plate IX.

(3) Number 161.

(4) The 'Postscript' to Maxwell's paper 'On the theory of compound colours', *Phil. Trans.*, **150** (1860): 78–84 (= *Scientific Papers*, **1**: 436–44). See Number 179.

INAUGURAL LECTURE AT KING'S COLLEGE, LONDON

OCTOBER 1860[1]

From the original in the University Library, Cambridge[2]

Mr Principal and Gentlemen[3]

The study of Natural Philosophy, when once entered on must preclude us from allowing our minds to dwell upon any ideas nobler than those of matter and motion. We must leave on one side all the questions which interest us as social and moral beings, & all the feelings which incline us to take pleasure in what we see, without inquiring into what lies behind; and turn aside into a region where Force reigns supreme, and recognizes matter as its only subject, and where the only theory of action is, that Might makes Right.

When we have commenced our course, it will be too late to speculate on the probable effect of such studies on the ultimate development of our minds, or even on the use of theoretical knowledge, as the prelude to a practical career. We must then give our whole minds to each subject as it comes before us, whether it be the contest of opposing forces, or the paths of moving bodies, and dismiss from our minds everything except what is involved in the problem we have to solve.

---

(1) According to *The Calendar of King's College, London for 1860–61* (London, 1860) the Michaelmas Term began on Tuesday 2 October 1860. Maxwell's lectures were given on Monday, Wednesday and Thursday.

(2) ULC Add. MSS 7655, V, h/3. Previously published in *American Journal of Physics*, **47** (1979): 928–33.

(3) The post of Professor of Natural Philosophy at King's College, together with the professorship of Manufacturing Art and Machinery, both of which had been held by Thomas Minchin Goodeve (see *The Calendar of King's College, London for 1859–60*: 26–7), was advertised on 18 June 1860, the election to be on 13 July 1860. The College Council decided to 'elect to each office separately, [but] they hold themselves at liberty to elect one Gentleman to the two offices combined'. Maxwell applied only for the professorship of Natural Philosophy; seven candidates applied, and five (including G. R. Smalley of King's College School and R. B. Clifton of St John's College, Cambridge) were interviewed. The appointment Committee consisted of Joseph Henry Green (Council member), Thomas Grainger Hall (Professor of Mathematics), and William Allen Miller (Professor of Chemistry), and reported to the College Council on 11 July 1860: 'Mr Green the only member of the Committee who was present saw these Gentlemen, and with the assistance of Professors Hall and Miller, very fully considered their merits…[they] unanimously concur in recommending Mr James Clerk Maxwell Second Wrangler & Second Smith's Prizeman in 1854 late Fellow of Trinity College Cambridge and now Professor of Natural Philosophy in the Marischal College and University of Aberdeen, for the Chair of Natural Philosophy in this College…' (King's College Archives, Special Committees N°.2, ff.232–3 (KA/CS/M2)). Maxwell was placed equal Smith's Prizeman; see Number 45 note (2).

If we require to make up our minds as to the relation of Natural Philosophy to other branches of knowledge, to our own education, or to the progress of mankind, we must do it today; for tomorrow our business will not be *about* Natural Philosophy, but, will be Natural Philosophy itself. I shall now therefore endeavour to point out to you how some of these more general views of the subject appear to me, although I am sure, that after you have yourselves studied the subject, you will confess that what you heard from me at first but faintly indicated what you learnt by experience at last.

Natural Philosophy is the name given in this country to a collection of sciences consisting of two main groups. The first of these consists of Mechanics and includes the general theory of motion and equilibrium, together with the application of mechanical principles to the investigation of the phenomena of nature.[4]

The second group of sciences is commonly called Physics and includes at present the study of Light, Heat, Electricity and Magnetism; and, in general, of those phenomena which we have already referred to more general principles, but which we do not as yet contemplate as the result of known mechanical actions.

Natural Philosophy is bounded on the mechanical side by Mathematics, and on the Physical side by Chemistry.

Mechanics differs from Mathematics only by involving the ideas of matter, time, and force, in addition to those of Quantity and Space. The methods employed are the same as in mathematics, and the axioms, or laws of motion, upon which the science is founded are of the same kind as those of geometry.

Chemistry, the science which bounds us on the opposite or physical side, investigates those properties of matter by which one substance is distinguished from another, and it contemplates these substances as differing not merely in the degree in which they produce similar effects but in kind, in their very essence.[5]

In the physical sciences we investigate general properties of matter, and refer them to causes which we conceive to operate on matter in general.

Thus the formation of water by the combination of oxygen and hydrogen is a chemical phenomenon because it depends on the peculiar nature of those two gases, on the proportion in which they are mixed, and on the facilities for combination. Chemistry professes to describe the properties of substances, to define the proportions in which they combine, and to state the conditions under which the union takes place, but it goes no further, it considers these as ultimate facts relating to each different substance.

---

(4) Compare Number 105.
(5) Compare William Whewell, *Philosophy of the Inductive Sciences*, 2 vols. (Cambridge, ₂1847).

It belongs to Physics to investigate the amount of heat produced by the union of the gases, & the effect of that heat in increasing the pressure, or expanding the combined gases, because the effects of heat, on all substances, though different in degree, are the same in kind, and are subject to general laws which apply to all substances.

The explosion may also produce the mechanical effect of bursting the vessel, blowing off the cover, or setting the air in vibration, and making a noise. These phenomena, considered as motions of various bodies, produced by forces of known amount, are to be investigated by mechanical methods.

I have taken three different classes of phenomena as illustrations of the subject matter of Chemistry Physics and Mechanics. But the difference between these three sciences lies less in the subject studied than in the method of study. In Chemistry we accept a number of elementary substances and study their properties. In Physics we accept certain great natural agencies and study their effects on all kinds of matter. In Mechanics we accept nothing but matter and motion, and recognize no difference in matter except arrangement, and no energy in nature except motion.

If anyone could explain, by means of known actions of heat, electricity, or any other universal agent, the peculiar properties of oxygen and hydrogen, and the results of their combination he would have offered a physical explanation of the nature of these substances, and he would have transferred certain phenomena from Chemistry into Physics.

Those who have attempted to conceive of the chemical elements as different arrangements of particles of one primitive kind of matter, would, if they had been successful, have reduced Chemistry to Mechanics.

No person, however, has hitherto been able to devise arrangements of particles by which chemical phenomena can be explained by the pure science of matter and motion, still less to prove that these arrangements will account for all the known properties of the substances. But though little has been accomplished in our attack on the chemical doctrine of elementary substances, something has been done towards the mechanical explanation of physical phenomena and though we cannot be said as yet to know scientifically the exact kind of motion to which such phenomena as heat and electricity are due, yet we have sufficient evidence to show that any labour we bestow in investigating such subjects by the aid of mechanical ideas will not be in vain. Natural Philosophy, therefore, treats of those properties of matter which do not require us to conceive of different substances as essentially distinct – of the general properties of matter, as distinguished from the special properties of certain substances.

We receive from the Mathematicians the idea of Quantity and all the processes of pure mathematics. We take possession of the field in which the mathematician is first trained in the exercise of his powers, the field of space with all the apparatus of geometry. We then introduce, in addition to the empty forms of geometry, our own idea of matter, and contemplate bodies placed in different parts of space. This arrangement we consider as subject to change, and thus we arrive at the notion of time and motion. In studying the cause of motion, we arrive at the idea of force and its relation to the body moved. We then regard a body not merely as something occupying space and capable of motion, but as something requiring a definite amount of force to produce a definite motion, and capable of producing effects on other bodies depending on the amount of matter in the body moved. Thus we acquire the conception of Mass, or quantity of matter, as a measurable quantity, and also that of Energy, or the amount of work which a body is capable of doing, on account of its motion, or on account of any other state in which it is.

On these conceptions of matter, motion, force, and energy we found the mechanical sciences.

By considering the way in which a body must begin to move, we acquire distinct ideas of the nature of the forces which produce this motion; and we find, that however numerous these forces may be, we can always find a very small number which would produce the same effect. Hence we arrive at the idea of equivalent systems of forces, and the reduction of many forces to a smaller number.

In such cases, the forces which we dismiss from our consideration are such as are so balanced as to produce no effect on the body. These balanced systems of forces are said to be in equilibrium, and the consideration of such systems forms the science of Statics.

In all cases in which the forces are not balanced, the science of Statics gives us the means of reducing them to the equivalent system of the form most convenient for our future operations, so that Statics is a necessary foundation for the more general science of Dynamics.

We may also consider the possible motions of a body apart from the causes of that motion. The science of pure motion is called Cinematics[6] or Kinetics, and forms the other foundation of Dynamics.

Dynamics considers the relation between force and motion. The forces are

---

(6) Compare A. M. Ampère, *Essai sur la Philosophie des Sciences*, 2 vols. (Paris, 1834–43), **1**: 52; '*cinématique, de* χίνημα, *mouvement*'. Ampère's usage is cited by Whewell, *Philosophy of the Inductive Sciences*, **1**: 152, referring to 'kinematics' as the science of motion; and by Robert Willis, *Principles of Mechanism* (Cambridge, 1841): vii, using '*Kinematics* ... to include all that can be said with respect to *motion* in its different kinds, independently of the forces by which it is produced'. See also W. J. M. Rankine, *A Manual of Applied Mechanics* (London/Glasgow, 1858): 15; 'The comparison of motions with each other, without reference to their causes, is the subject of a branch of geometry called "*Cinematics*".'

reduced by statics to their simplest form, and the possible motions are brought into a mathematical form by Kinetics, and then these are brought into relation by Dynamics, which is the science of the motion of matter as produced by known forces.

Force is here considered as the cause of the motion of a body. But Force is always an action between *two* bodies and is the result of some relation between them. The investigation of the particular relations between bodies which give rise to particular manifestations of force in nature forms a large portion of Experimental Physics, but there are certain general laws, regulating the amount of Energy arising from given conditions, and determining the total effect of the forces called into play, which are among the most important conclusions of physical science. The science founded on these laws is called Energetics.[7] The application of these principles to natural phenomena is the special research which the present state of science points out as that from which the greatest results are to be expected in the coming age. The work is only begun, and not till we have measured the energies of all known agents, can we hope to make any progress towards a mechanical explanation of their mode of action. Already in Astronomy & in the theory of Heat and Electricity have the principles of Energetics led to new methods of research, to the discovery of unsuspected relations between known properties of matter, and even to the knowledge of new properties, which subsequent investigation has shown to exist.

The doctrine of the convertibility and equivalence of all forms of Energy[8] may hereafter be made the basis of new inquiries, which, starting with the knowledge of the quantitative relations between mechanical energy and the other forms in which energy exists, such as heat, attraction, &c will proceed with the investigation of these special forms of energy, and discover, not only the *quantity*, but the *quality* of those forms of energy, and to ascertain by what arrangements and motions of matter these different phenomena can be accounted for.

Statics Cinematics Dynamics and Energetics may be regarded as the four great branches of abstract Mechanics. They may be applied to the equilibrium and motion of matter in various states of aggregation. The forms and dimensions of all known bodies are altered by the pressures which may be applied to them, and by the effects of Heat.

The theory of pressures, of changes of form, and of the relation between

---

(7) W. J. M. Rankine, 'Outlines of the science of energetics', *Edinburgh New Philosophical Journal*, **2** (1855): 120–41.

(8) W. J. M. Rankine, 'On the general law of the transformation of energy', *Phil. Mag.*, ser. 4, **5** (1853): 106–17; and Rankine, *Applied Mechanics*: 499–501.

pressure and change of form constitutes an important branch of Mechanics and may be called the general theory of Elasticity. The theory of Pressure and Elasticity as applied to fluids is much more simple than in the case of solids, and forms the sciences of Hydrostatics and Hydrodynamics.

The agency of Heat in producing similar changes is so intimately connected with that of pressure, that we are obliged to treat of the effects of heat and pressure at the same time. We are thus led to those practical applications of the theory of Energetics, which enable us to convert heat into mechanical energy, and form the basis of the theory of the Steam Engine.

The general theory of heat consists of four branches.

1 The laws of the production of heat by mechanical, chemical, or electric action, and the conditions of its transformation into other forms of energy.

2 The theory of the effects it produces on bodies by expanding them and changing their state.

3 The theory of the distribution of heat in bodies by conduction.

4 The theory of Radiant Heat.

The nature of radiant heat appears to differ in nothing from that of light. There is no doubt that the light which we see has all the properties of radiant heat, and that dark heat differs from light only in not being visible to our eyes. Radiant heat, then, being the same thing as light, though perceived by a different sense, we prefer to use that organ which gives us most information, and call the radiation, *light*, and the *science*, Optics.

From a few simple facts about reflexion and refraction, we are able to deduce a systematic science, which rivals the mechanical sciences in precision, and has this great advantage, as an educational science, that the elementary phenomena are easily observed. The science of Light, however, is one in which we have not only explained phenomena by referring them to general laws, but in which those general laws have been explained by a mechanical theory.

To trace the steps by which the nature of the motion which we call Light has been ascertained, is one of the most instructive parts of Physical study, and is most likely to introduce the student into the right path for following out investigations in other parts of science.

Last of all, we have the Electrical and Magnetic sciences, which treat of certain phenomena of attraction, heat, light, and chemical action, depending on conditions of matter of which we have as yet only a partial and provisional knowledge. An immense mass of facts has been collected, and these have been reduced to order and expressed as the results of a number of experimental laws, but the form under which these laws are ultimately to appear as deduced from central principles, is as yet uncertain.

The present generation has no right to complain of the great discoveries already made, as if they left no room for further enterprise. They have only given Science a wider boundary, and we have not only to reduce to order the regions already conquered, but to keep up constant operations on the frontier, on a continually increasing scale.

These are the main divisions of the science of matter and its forces. I must now speak of the method of study. We have seen that Physics differs from pure Mathematics in involving a greater number of ideas, while it agrees with mathematics in using these ideas as the foundation of systematic science.

Our first duty must therefore be to acquire true ideas of the various kinds of quantity with which we have to deal. When we have done this, the application of these ideas to special cases will be comparatively easy.

It is this which gives Physical science its peculiar value as a means of education. In all human knowledge, the acquisition of an idea brings with it, as a logical consequence, a certain system of truths dependent on it, but the mental process by which the idea is acquired is of a different order from that by which deductions are made from it. If a man understands what Force means, I have only to secure his attention, and I can prove to him as many propositions as I please, but if he has not the fundamental idea, no amount of demonstration will give it him. He must think for himself till he gets it.

Now in Natural Philosophy there are a great many different things which must be made our own, before we can have right notions upon what is to follow. And we have this great advantage over the students of many other sciences, that if we once go wrong, errors become manifest as soon as we go a step forward, so that we have no fear of building complacently on a bad foundation, for the whole will go to the ground as soon as we make the first practical application.

We are therefore called upon, during our study of Natural Philosophy, to clear up our ideas of the fundamental truths on which the science is built, and to test the success of this mental process, by comparing the results with facts.

I shall not now enter upon the question whether the fundamental truths of Physics are to be regarded as mere facts discovered by experiment, or as necessary truths, which the mind must acknowledge as true as soon as its attention has been directed to them.[9] Questions of this kind belong to Metaphysics. In this class we do not pretend to study Metaphysics in a formal and direct manner, but if by the careful study of the laws of nature and their dependence on each other we have been trained into watchfulness over the processes of our own minds, and clear habits of thought, we shall come all the better prepared for the study of higher problems, whether they are presented

---

(9) Compare Whewell, *Philosophy of the Inductive Sciences*, **1**: 59; 'necessary truths are those of which we cannot distinctly conceive the contrary'. See P. M. Harman, 'Edinburgh philosophy and Cambridge physics; the natural philosophy of James Clerk Maxwell', in *Wranglers and Physicists*, ed. P. M. Harman (Manchester, 1985): 202–24, esp. 220–3.

to us in a metaphysical shape, or as they occur sooner or later to every thinking man. If we have acquired, not the vague and popular notion that there are laws of Nature, but an acquaintance with some of these laws themselves, in their elementary form, and have been able to form some idea of their complication, when applied even to cases purposely simplified, we shall have learned a lesson of caution, when we examine higher departments of nature, with the expectation of finding there laws of equal simplicity of expression, and equally agreeable to the present state of development of science.

Physical Science affords the exercise which has developed the powers of the greatest and most original thinkers. Bacon, though his supply of physical truth was scanty, had his mind fixed upon the discoveries of the future, and he draws both his wisdom and his eloquence from the contemplation of that new era of which he was himself the prophet.[10] Descartes and Leibnitz need only to be named, to recall systems of metaphysics which were also systems of science. In fact the ideas discussed in metaphysics are so intimately connected with the foundations of Natural Philosophy, that we have only to read a few pages of a metaphysical work, if we wish to ascertain the precise limits of the author's knowledge of physical science.

In the course of your studies here, you will find abundant material for the most abstract speculation, but you must recollect that in physical speculation there must be nothing vague or indistinct. The truths with which we deal are far above the region of mist and storm which conceals them from the undisciplined mind, and yet they are solidly built on the very foundations of the world, and were established of old according to number and measure and weight. Nothing that we can say or think here can escape from the ordeal of the measuring rod and the balance. All quantities must be exact quantities, so that we have a most effectual means of discovering error, and an absolute security against vagueness and ambiguity.

As we proceed in our course we shall see what part has been taken by experiment, demonstration, and hypothesis respectively in the advancement of science; and we shall have to distinguish between demonstrations founded on pure mathematical properties of space or motion, and those which start from a fact determined by experiment or from an hypothetical assumption. We shall also have to distinguish between experiments of illustration, which, like the diagrams of Euclid, serve merely to direct the mind to the contemplation of the desired subject, and experiments of research, in which the thing sought is a quantity, whose value could not be discovered without experiment. We shall also learn the value of hypotheses in the process of discovery, and by

---

(10) Compare Whewell, *Philosophy of the Inductive Sciences*, **2**: 229.

what method hypotheses may be made useful in relation to the present state of science, without forming an obstacle in the way of further discovery.

The student of physical science will find that the method or mode of procedure by which knowledge has been accumulated, and even the process by which he himself masters it from day to day, will furnish him with facts relating to the conditions of human knowledge which he may take with him as guides to the study of other and more complicated subjects.

He will see as he advances that the laws of nature are not mere arbitrary and unconnected decisions of Supreme Power, but that they form essential parts of one universal system, in which infinite Power serves only to reveal unsearchable Wisdom and eternal Truth.

When we examine the truths of science, and find that we can not only say 'This *is* so' but 'This *must be so* for otherwise it would not be consistent with the first principles of truth', or even when we can only say 'This *ought to be* so, according to the analogy of nature', we should think what a great thing we are saying, when we pronounce sentence on the laws of creation, and say they are true, or right when judged by the principles of reason. Man has indeed a very partial knowledge of the simplest real thing, the nature of a drop of water has in it mysteries within mysteries utterly unknown to us at present, but what we do know, we know distinctly and scientifically, and we may expect that more will be understood in due time.

Some facts we know in their first principles, others as experimentally true, we see more before us which we can only guess at as yet, but we are confident that there remains an inexhaustible inheritance of knowledge, not revealed at once, lest we should become proud of our possession, and despise patient enquiry, but so arranged, that as each new truth is unravelled, it becomes a clear, well-established addition to science, quite free from the mystery which involves what lies beyond, and shows that every atom of creation is unfathomable in its perfection.

Objections have sometimes been raised to the study of physical science, on the ground of the supposed effect of exact science in making the mind unfitted to receive truths which it cannot fully comprehend. We shall find that it is the peculiar function of physical science to lead us, by the steps of rigid demonstration, to the confines of the incomprehensible, and to encourage us to apply our minds to that which we do not yet understand, since it is only to those who labour patiently and think steadily, that such mysteries are ever opened. The higher laws of nature are hid from us at present, but we, and those who come after us, will find in the search for them that which will prepare our minds for the next stage of human knowledge; and the discovery of each new law will not only open up new regions of science, but will alter mens expectations of what the laws of nature ought to be. New discoveries must be consistent with the old, for all *facts* are consistent with each other, but there is a vast mass of opinion about scientific facts, which I am certain is susceptible of modification, when the light of new truths is brought to bear upon it. We must beware of 'anticipating nature' and reasoning from the supposed existence in a particular form of laws of which we have no distinct idea, and assuming that the higher laws which we do not yet know are capable of being stated under the same forms as the lower ones which we do know.

When vague ideas are put into the form of physical arguments, we can expect nothing but vague conclusions, and great discredit to the mathematics or physics so desecrated. Vague ideas may possibly give picturesqueness to a declamation, but we must be very careful of them when they are disguised in the forms of exact science.

To avoid this vagueness ourselves, we must eventually make use of that method of expression which, by throwing away every idea but that of quantity, arrives at the utmost limit of distinctness. We cannot express physical facts except in a mathematical form. In this class, as I have said before, we have many *kinds* of quantities to deal with and therefore much of our time must be spent in becoming acquainted with these. This corresponds to that part of mathematics which is put in the form of definitions and axioms. But if geometry, or anything more than pure algebra is mathematics, then Natural Philosophy is, and ought to be, mathematics, that is, the science of quantitative relations.

In this class, I hope you will learn not only the mathematical accuracy of expression of which all physical facts are capable, but the mathematical necessity of their interdependence.

In this way we will carry with us, not merely results, or formulæ applicable to cases that may possibly occur in our practice afterwards, but the principles upon which those formulæ depend, and without which the formulæ are mere mental rubbish.

I know the tendency of the human mind to do anything rather than think. None of us expect to succeed without labour, and we all know that to learn any science requires mental labour, and I am sure we would all give a great deal of mental labour to get up our subjects. But mental labour is not thought, and those who have with great labour acquired the habit of application, often find it much easier to get up a formula than to master a principle. I shall endeavour to show you here, what you will find to be the case afterwards, that principles are fertile in results, but the mere results are barren, and that the man who has got up a formula is at the mercy of his memory, while the man who has thought out a principle may keep his mind clear of formulæ, knowing that he could make any number of them when required.

I need hardly add, that though thought be a process from which the mind naturally recoils, yet, that process once completed, the mind feels a power and an enjoyment which make it think little in future of the pains and throes which accompany the passage of the mind from one stage of development to another.

It is only by that mathematical training which enables us to see the consequences of the introduction of each new principle into science, that we can fully appreciate the value of these principles, and it is only by arithmetical computation that the ultimate results can be compared with facts.

The intermediate portion of mathematical science, which consists of calculation and transformation of symbolical expressions, is most essential to physical science, but it is in reality *pure mathematics*. Everything connected with the original question may be dismissed from the mind during these operations, and the mathematician to whom they are referred may be doubtful whether his results are to be applied to solid geometry, to hydrostatics or to electricity. But as we are engaged in the study of Natural Philosophy we shall endeavour to put our calculations into such a form that every step may be capable of some physical interpretation, and thus we shall exercise powers far more useful than those of mere calculation – the application of principles, and the interpretation of results.

In this class we profess to study Natural Philosophy as an applied science,[11] and we hope in doing so to reap all the benefit to ourselves that can be drawn from the subject. If we studied first principles alone, we should become so familiar with the words and modes of expression of philosophy, that we should think ourselves masters of the subject, because we found no scruples in using language which ought to indicate something more than a talking acquaintance with high subjects.

If, dismissing all concern about principles, as too metaphysical for our taste, we were to accept them as mathematical data, embody them in formulæ, and grind them together, till we got out x as one of the roots of an equation, we should have made ourselves a little more expert in algebra, but not otherwise wiser, especially if it should turn out that we could not determine whether x is a number of square feet or a velocity.

And if we had determined to be practical and instead of working out formulæ, had taken the very case which we wished to solve, and had gone to a standard treatise, and copied out the formula which seemed most suitable, and then got out some result by substituting our numbers, we should *most probably*, by some little oversight get a perfectly useless result, and we should *certainly* leave our minds in a more confused state than when we began.

But if, while here, we can once acquire an intelligent conception of physical principles, and by first investigating the mathematical consequences of these in

---

(11) At King's College the lectures on 'Natural Philosophy and Astronomy' were listed in the 'Engineering Section' of the 'Department of Applied Sciences'; see *Calendar of King's College, London for 1860–61*: 115–17.

a few cases, and then applying the results to actual experiments or observations, convince ourselves that these principles are not mere abstractions made by philosophers, but the *key* by which we ourselves may interpret what we see everyday, and the *charm* by which we may make the forces of nature do our bidding, then our minds will have received an extension and enlargement which will be *permanent*, for we have been forced to pass from philosophy to carpentry, and from the workshop to the *locus principiorum*, till we have learned by experience that the philosophical and the real is the same thing.

If we can repeat the scientific expressions of physical facts in the classroom, we shall have gained little, unless we are able to recognize these facts when we meet them out of doors, not dressed up for the lecture table, but in that *natural retiring* form in which they escaped the notice of so many wise philosophers of old.

But if we can train our minds to see the physical significance of everything that happens, we shall be in the *first place* able to make use of our opportunities in the various professions to which we may be called, *secondly*, we shall never cease to seek, obtain, and enjoy additional knowledge of the world in which we are placed, and *thirdly* all the skill and knowledge we lay up will round itself to a perfect whole of Wisdom when all the elements of Science, from the matter which exhibits its modes of action, to the mind which perceives them, are felt to be mutually related parts of one great whole.

The whole course of history is full of examples showing how the neglect of scientific principles produces, in the first place, the certain failure of every enterprize, secondly, how the unscientific mind has been led from one error to another up to the very pinnacle of absurdity, and thirdly how the want of observation, wrongheadedness, and superstition thus produced have generated systems of philosophy which, by beginning with the contradiction of physical facts, guarantee the thorough unreality of the whole superstructure. We have no time at present to study the history of physical science. We cannot understand the steps by which the human mind has advanced to its present state of knowledge, till we ourselves have some experience of what that state of knowledge is. When we have encountered and overcome the resistance of our minds to the acquisition of new ideas, we shall be the better able to appreciate the labours of those, who for the first time thought out those ideas, and transmitted them to us for a perpetual possession. While we feel the difficulties of acquiring knowledge which has been already discovered, and carefully put into the most convenient form for our acceptance, we must remember that the discoverers of that knowledge had to struggle against established notions, and the force of habits of thought to which they themselves had been trained, and to devote the energy of their lives to labours, the results of which *they* saw but dimly, and *others* not at all.

And now when discoveries have been made and results obtained, when public opinion has changed and everything wearing the garb of science is respected, when, as in this institution, we are directed to study science with a view to success in life – shall we forget the men to whom all this is due, and accept the mere result of their labours, without entering into the spirit of their lives? Shall our descendants be taught to repeat, like the Indian astronomers, statements of facts which had a meaning in former times when men were found who could think? That condition I think need not be ours if we are careful to learn the higher lessons of science, while we study its facts. The conditions of knowledge are always the same, and we, while following out the discoveries of the teachers of science, must experience in some degree the same desire to know and the same joy in arriving at knowledge which encouraged and animated them. Do not check these feelings because you cannot expect mankind to sympathise with your triumph over some elementary proposition. None but yourselves can be partakers of the intellectual fruition which results from the comprehension of a principle, and you should beware how you throw away opportunities of a kind of enjoyment which neither university honours nor worldly reputation can ever afford. Do not be ashamed because the occasion seems small, but cherish any sensation of pleasure which you now feel in the opening up of the mind to the perception of truth. Nothing is easier than for the youngest of us, by repressing such natural feelings, to acquire a permanently contracted mind. The highest intellectual distinction at which man can aim, is to have preserved and nourished to maturity true liberality of thought, in a mind having all its actual knowledge in full and undoubted possession, but always capable of advancing to higher and more comprehensive views of truth.

DRAFT OF THE LECTURE 'ON THE THEORY OF THREE PRIMARY COLOURS'[1]

circa MAY 1861[2]

From the original in the University Library, Cambridge[3]

ON THE THEORY OF THREE PRIMARY COLOURS

During all our waking hours we are deriving knowledge and pleasure from the exercise of sight. Certain colours arranged in a continually varied manner are the indications of outward things from which we learn so much about them in so delightful a way. What do we know about these colours, what relation have they to the object we look at, and the light by which we see it, and by what powers of the eye or of the mind are they perceived, and when we have enjoyed the sight of any object, how may we artificially produce such a resemblance of it as shall enable us to refresh our eyes with it at any time? The Opticians, the Physiologists and the Artists have each taken their own branch of this inquiry and have arrived at results, which though all of them true are independent of each other so that at first sight they seem opposed to each other.

The Opticians have studied Colour as an indication of the nature of Light. Newton, by passing a ray of light through his prism divided white light into its component parts and showed that bodies appear of different colours according to the proportions of these different kinds of light which they send to the eye. The prismatic analysis of light has been improved by Wollaston[4] Fraunhofer[5] and others so that now we can not only distinguish seven shades of colour in the spectrum but in each of these we can distinguish describe and identify many thousand gradations, each of which is as distinct from every other as one chemical element from another. [...] The optician has now done all that is within his power. He has analysed light into its different kinds, he has traced the path of the rays from their source to the sensitive membrane of the eye, on which they fall and are extinguished and he has proved that the

---

(1) J. C. Maxwell, 'On the theory of three primary colours', *Proceedings of the Royal Institution of Great Britain*, **3** (1858–62): 370–4 (= *Scientific Papers*, **1**: 445–50).

(2) The lecture was delivered on 17 May 1861.

(3) ULC Add. MSS 7655, V, b/22ii. Only portions of the MS are extant.

(4) W. H. Wollaston, 'A method of examining refractive and dispersive powers, by prismatic reflexion', *Phil. Trans.*, **92** (1802): 365–80, esp. 378.

(5) Joseph Fraunhofer, 'Bestimmung des Brechungs- und Farben-Zerstreuungs-Vermögens verschiedener Glasarten, in bezug auf die Vervollkommnung achromatischer Fernröhre', *Annalen der Physik*, **56** (1817): 264–313, and Tables III and IV.

different colours which we see are due to the different kinds of light which enter the eye.

The Physiologist has next to reconcile these facts with the theory of Sensation, and to explain how the optic nerve can distinguish not only the *intensity* of the light which falls on every point of the retina, but also its quality.

I believe we have no evidence that any nerve is capable of more than one *kind* of action, so that by exciting a nerve more or less we may increase the intensity of the sensation till it becomes an acute pain or diminish it till it can no longer be felt, but we cannot, by using a different kind of agent to excite the nerve, produce any new sensation. Whatever method be used to excite the nerve, the same kind of sensation will be produced and the character of the sensation will depend only on the function of the nerve so that any excitement of the optic nerve is felt as a colour of the olfactory nerve as a smell, of a nerve of touch as the contact of a body with its extremity &c.

The theory that we have in our optic nerve the power of distinguishing colour by the time of vibration of the ray is almost incredible when we calculate the actual number of vibrations per second for red light which is 477,000000,000000. How are we to distinguish this from green light a vibration of which occupies $\dfrac{1}{577,\text{ billions}}$ of a second insteady of $\dfrac{1}{477\text{ billion}}$ of a second.

The only satisfactory method of explaining our perception of colours is to suppose that we have in our eyes several different sets of nerves one set being most affected by one kind of light and another set by a different kind of light. This theory was first proposed by Dr Thomas Young in his lectures to the Royal Institution.[6] The colours we see, according to this theory depend on the different sets of nerves excited by the light which enters our eyes. If we could excite one set of nerves alone we should see a simple or primary colour. If other sets of nerves were excited we should see a colour compounded of the colours belonging to each set of nerves in proportions depending on their relative amount of excitement. We find that we are unable to recognize, in a compound colour, the elements of which it is compounded. In music we distinguish the notes which are sounded together but in vision, whatever be the colours blended together we have no direct means of perceiving that the resultant tint is not a simple colour.

Anatomists have not hitherto succeeded in demonstrating the existence in the retina of several distinct sets of nerves each distributed over the whole sensitive surface. Even if such a structure were exhibited by dissection, the

(6) Thomas Young, *A Course of Lectures on Natural Philosophy and the Mechanical Arts*, ed. P. Kelland, 2 vols. (London, 1845), **1**: 344–5. See also Young, 'On the theory of light and colours', *Phil. Trans.*, **92** (1802): 12–48.

functions of these nerves in exciting the sensation of colour could only be discovered by one who could feel the sensations themselves. We have therefore to investigate by actually observing various colours, whether all our colour sensations are capable of being produced by the union of a small number of simple colour-sensations, and if this is found to be the case to determine how many simple sensations we have, and what colours they correspond to.

Now the Artists have been pursuing their own investigations into the nature of colour in order to be able to imitate natural colours by means of paint. They have a certain number of pigments at their disposal and they find that by mixing these they can produce any desired tint. Before Newtons discovery of the compound nature of white light white was regarded as a simple colour and other hues as the results of impurity. When the doctrine of Newton became popular those who made experiments with pigments soon adopted the theory that there are three primary colours from which all the rest may be formed by mixture and that these primary colours are Red Yellow and Blue.[7] To fix our ideas let us suppose that Carmine represents Red Chrome yellow Yellow and Ultramarine, Blue.

By mixing Carmine and Chrome Yellow we may obtain all hues of Orange. By mixing Chrome Yellow and Ultramarine we produce various hues of green and Ultramarine & Carmine give all the different purples. By mixing all three colours in proper proportions we obtain a neutral tint having no decided colour, and equivalent to a certain mixture of black and white which we call grey.[8]

By these mixtures we have imitated all the kinds of colour and if any of the tints we have thus formed fall short of any proposed colour in brilliancy it is natural to ascribe the fact to the imperfection of the pigments employed since by getting brighter pigments we might produce a more intense hue by their mixture.

Thus the painters regard all possible colours as composed of three primary colours in various proportions. They have learnt this by their experience in mixing pigments and they have found that red, yellow and blue pigments answer better than any other three colours for making a complete set of tints.

Now let a person with this practical knowledge of painting examine the prismatic spectrum. He will find there Red Yellow and Blue of far more brilliant and intense tone than any pigment in nature. Between the red and yellow he finds orange. Between the yellow and blue he finds green, and beyond the blue he finds violet a colour which he is accustomed to produce by adding a little red to his blue. He is inclined to adopt Sir David Brewsters theory of the

(7) See Number 54 notes (7) and (8).
(8) On pigment mixing see Number 64 note (2).

spectrum, that the red yellow and blue are primary colours while the orange green and violet are formed by their mixture.[9] Nothing is more evident to the eye than that most of the colours of the spectrum are intermediate in quality as well as in position to those on either side of them so that the theory of their being actually compounded of these other colours seems at first sight to be the only natural one.

Now it is very easy to mix two colours of the spectrum and to try whether the mixture is like the intermediate part of the spectrum. We find that the mixture is the same with regard to colour but that if we take a prism and analyze the Mixed light it is divided into its original components, while the part of the original spectrum which has the same colour is undecomposed by the prism. Sir David Brewster has stated that the colours of the spectrum although they cannot be decomposed by the prism, can be analyzed by passing them through certain coloured substances which by stopping one of their component parts allow the colour of the other parts to be distinctly seen.

If the experiments on which the new analysis of light is founded were completely verified we should have to abandon the undulatory theory of light.[10] According to that theory, light of definite refrangibility consists of vibrations of known length, and can therefore vary only in intensity. If the effect of passing a ray of homogeneous light through an absorbing medium is in any respect different from that of diminishing the intensity of that light in any other way we should have a distinct proof that light does not consist of a series of equal undulations. From experiments of Sir John Herschel and others it seems probable that the diminution of intensity of the light in certain parts of the spectrum is sufficient to account for the apparent change of colour.[11]

We must now endeavour to reconcile the independent discoveries of the Opticians, the Physiologists and the Painters.

The Opticians tell us there are innumerable kinds of light all equally simple and equally primary.

---

(9) David Brewster, 'Description of a monochromatic lamp for microscopical purposes, with remarks on the absorption of the prismatic rays by coloured media', *Trans. Roy. Soc. Edinb.*, **9** (1823): 433–44; Brewster, 'On a new analysis of solar light, indicating three primary colours, forming coincident spectra of equal length', *ibid.*, **12** (1834): 123–36. See Numbers 139, esp. note (15) and 180.

(10) See Number 139, esp. note (16).

(11) John Herschel, 'Light', in *Encyclopaedia Metropolitana, Second Division. Mixed Sciences*, Vol. II (London, 1830): 341–586, esp. 432–4; and Herschel, 'On the absorption of light by coloured media, and on the colours of the prismatic spectrum exhibited by certain flames; with an account of a ready mode of determining the absolute dispersive power of any medium, by direct experiment', *Trans. Roy. Soc. Edinb.*, **9** (1823): 445–60. See especially Hermann Helmholtz, 'Ueber Herrn D. Brewster's neue Analyse des Sonnenlichts', *Ann. Phys.*, **86** (1852): 501–23, (trans.) 'On Sir David Brewster's new analysis of solar light', *Phil. Mag.*, ser. 4, **4** (1852): 401–16.

The Physiologists tell us that vision is performed by the optic nerve or nerves and that one nerve can only give us one sensation in greater or less intensity.

The Painters tell us that all colours may be made by properly mixing three primary colours.

Are we to suppose that all the rays of the spectrum except three are compound, and that the Opticians are wrong in saying they are simple? or are we to ascribe to the nerve of vision the power of distinguishing the quality as well as the quantity of the agent which affects it contrary to the opinion of the Physiologists?

By adopting Young's theory, that in the organ of vision there is a three fold sensitive apparatus of nerve fibrils, capable of different kinds of colour-sensation we reconcile the physical discoveries of optics with the practical experience of colourists in a way consistent with the conditions of nervous action.

Each homogeneous ray of the spectrum may act upon one or more of the sensitive systems of nerves, the amount of action on each depending on its wave-length, just as the amount of absorption by different kinds of glass depends on the wave length of the incident ray. If any ray of the spectrum could excite one and one only of the primary colour sensations, that ray of the spectrum would exhibit to us one of the primary colours. We have no direct means of knowing whether the sensation we feel is simple or compound.[12]

---

(12) In the Royal Institution lecture Maxwell performed an experiment to illustrate the principle of colour photography he had outlined in his paper 'Experiments on colour, as perceived by the eye', *Trans. Roy. Soc. Edinb.*, **21** (1855): 275–98, esp. 283–4 (= *Scientific Papers*, **1**: 136–7). 'Three photographs of a coloured ribbon taken through...three coloured solutions respectively, were introduced into the camera, giving images representing the red, the green, and the blue parts separately, as they could be seen by each of Young's three sets of nerves separately. When these were superposed, a coloured image was seen, which, if the red and green images had been as fully photographed as the blue, would have been a truly-coloured image of the ribbon.' (*Scientific Papers*, **1**: 449). The separation plates were prepared by Thomas Sutton; on these plates, and the fortuitous success of Maxwell's exhibition of a colour photograph, see R. M. Evans, 'Maxwell's colour photograph', *Scientific American* (November 1961): 118–227.

185

LETTER TO MICHAEL FARADAY

21 MAY 1861

From the original in the library of the Institution of Electrical Engineers, London[1]

8 Palace Gardens Terrace
W
21 May 1861

Dear Sir

If a sphere were set in rotation about any diameter, it would continue to revolve about that diameter for ever. If the body is of unequal dimensions it will continue to revolve about the same axis provided that axis be a principal axis of the body, of which there are always three.

If the original axis of rotation is not a principal axis then the axis of rotation will change its position both in the body and in space.

In the body, the extremity of the axis of rotation will describe an ellipse (or circle) about the greatest or least axis. In space it describes a complicated curve. The ellipse is described in

$$\frac{a^2}{\sqrt{b^2-a^2}\,\sqrt{c^2-a^2}}$$

revolutions of the body, where $\frac{1}{a^2}$, $\frac{1}{b^2}$, $\frac{1}{c^2}$, are the moments of inertia about the principal axes.

If the body be very nearly spherical like the earth, this motion is very slow. If P be the end of the principal axis of the earth and N that of the actual axis of rotation the rotation being from $W.$ to $E.$ then N will travel round P from $W.$ to $E.$ so as to complete its circuit in about 325.6 solar days.[2]

This phenomenon depends only on the configuration of the earth, and its rotation round an axis which is nearly but not exactly a principal axis, and would be the same if all other bodies were

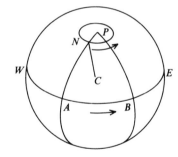

Figure 185,1

absent from the sky. The position of the pole among the stars is not at all affected but the latitude of every place on the earth is alternately increased and diminished, the maximum latitude occurring when N is in the same meridian with the place. Thus our maximum should occur sooner than that of Paris or Poulkova.

---

(1) Institution of Electrical Engineers, London, Special Collection MSS 2. Previously published in *Life of Maxwell* (2nd edn): 241–3.   (2) See Number 116.

The *amount* of the variation can only be determined by observation. I have not the means at present of making reference to the statement that Peters Astronomer at Konigsberg or Berlin has determined the amount at less than $\frac{1}{10}$ of a second, and the period about 312 days. He has also stated the epoch of maximum at his observatory which I do not remember.[3]

An explanation of this motion, illustrated by experiments was published by me in the Edinburgh Transactions Vol XXI pt IV 'On a Dynamical Top'.[4]

The motion without reference to the earth is described in Poinsôts treatise on Rotation.[5]

The Astronomer Royal[6] says he may possibly be able to test it by a long series of observations.

As far as I am concerned you are at liberty to speak to any one on the subject.

I have my dynamical top in London and can show you the motion at any time.[7]

I remain
Yours truly
J. C. MAXWELL

---

(3) C. A. F. Peters, 'Recherches sur la parallax des étoiles fixes', *Recueil de Mémoires présentès à l'Académie des Sciences par les Astronomes de Poulkova*, **1** (1853): 1–180, esp. 146–7. See Number 116 note (7).

(4) J. C. Maxwell, 'On a dynamical top, for exhibiting the phenomena of the motion of a system of invariable form about a fixed point', *Trans. Roy. Soc. Edinb.*, **21** (1857): 559–70, esp. 568–70 (= *Scientific Papers*, **1**: 259–61).

(5) Louis Poinsot, *Théorie Nouvelle de la Rotation des Corps* (Paris, 1852); see Number 116 note (4).

(6) George Biddell Airy; see Numbers 158 and 167.

(7) Faraday acknowledged receipt of this letter in a brief letter to Maxwell of 23 May 1861 (ULC Add. MSS 7655, II/20): 'I am greatly obliged to you for your kindness. It was the amount of the variation & its determination which had caught my mind & which I thought had been recently ascertained but I dare say my imperfect memory has taken up imperfectly what you said.'

LETTER TO ROBERT CAY

11 JUNE 1861

From the original in the library of Peterhouse, Cambridge[1]

8 Palace Gardens Terrace
1861 June 11

Dear Uncle

I enclose the three Receipts.[2] I am glad to hear you are so far better, and I hope the N.E. Wind will abate, for a cessation of that wind often does more good than anything else. Charlie will be all the better for the commencement of the long vacation but three weeks is but a poor substitute for it. I have just lectured for the last time and have only examinations after this. We are going to the coast for a week in the interval before it is my turn to examine.[3]

I hope all your party continue well. Katherine joins me in kind regards to you all.

Your aff^t nephew
J. C. MAXWELL

R. D. Cay Esqre

---

(1) Peterhouse, Maxwell MSS: unnumbered.

(2) Robert Cay, Maxwell's uncle, was responsible for handling some family property in England; there are many receipts signed by Maxwell (Peterhouse, Maxwell MSS). On 19 October 1855 Maxwell wrote separately to Jane and Robert Cay, assuming responsibility on his father's illness, urging that Robert Cay take charge of administering the property (Peterhouse, Maxwell MSS (13), (14)).

(3) According to *The Calendar of King's College, London for 1860–61* (King's College, London, 1860), the Easter term ended on Friday 28 June 1861.

LETTER TO MICHAEL FARADAY

19 OCTOBER 1861

From the original in the library of the Institution of Electrical Engineers, London[1]

8 Palace Gardens Terrace
Kensington W.
19 Oct. 1861

Dear Sir

I have been lately studying the theory of static electric induction, and have endeavoured to form a mechanical conception of the part played by the particles of air, glass or other dielectric in the electric field, the final result of which is the attraction and repulsion of 'charged' bodies.[2]

The conception I have hit on has led, when worked out mathematically to some very interesting results, capable of testing my theory, and exhibiting numerical relations between optical, electric and electromagnetic phenomena, which I hope soon to verify more completely.

What I now wish to ascertain is whether the measures of the capacity for electric induction of dielectric bodies with reference to air have been modified materially since your estimates of them in 'Series XI',[3] either by yourself or others.

[a]I wish to get the numerical value of the 'electric capacity' of various substances especially transparent ones, if formed into a thin sheet of given thickness and coated on both sides with tinfoil.[4] Sir W. Snow Harris has made experiments of this kind[5] but I do not know whether I can interpret them numerically.

Another question I wish to ask is whether any experiments similar to those

---

(a) {Faraday} Magnecrystallic action series.[6]

(1) Institution of Electrical Engineers, London, Special Collection MSS 2. Previously published in *Life of Maxwell* (2nd edn): 243–5.

(2) J. C. Maxwell, 'On physical lines of force. Part III. The theory of molecular vortices applied to statical electricity', *Phil. Mag.*, ser. 4, **23** (1862): 12–24 (= *Scientific Papers*, **1**: 489–502); published in January 1862.

(3) Michael Faraday, 'Experimental researches in electricity. – Eleventh series. On induction', *Phil. Trans.*, **128** (1838): 1–40 (= *Electricity*, **1**: 360–416).

(4) See Maxwell's letter to Thomson of 10 December 1861 (Number 189); and compare *infra*, and his discussion (without numerical values) in 'On physical lines of force. Part III': 22–3 (= *Scientific Papers*, **1**: 500–1) of the relation between the 'inductive power of a dieletric' and the index of refraction. See note (17).

(5) W. Snow Harris, 'On the specific inductive capacities of certain electrical substances', *Phil. Trans.*, **132** (1842): 165–72. On 'specific inductive capacity' see Faraday, 'On induction': 25 (§1252).

in Series XIV on crystalline bodies[7] have yet led to positive results. I expect that a sphere of Iceland spar, suspended between two oppositely electrified surfaces would point with its optic axis transverse to the electric force, and I expect soon to calculate the value of the force with which it should point.[8]

Again, I have not yet found any determination of the rotation of the plane of polarization by magnetism[9] in which the absolute intensity of magnetism at the place of the transparent body was given. I hope to find such a statement by searching in libraries, but perhaps you may be able to put me on the right track.[b]

My theory of electrical forces is that they are called into play in insulating media by *slight* electric displacements, which put certain small portions of the medium into a state of distortion which, being resisted by the *elasticity* of the medium, produces an electromotive force. A spherical cell would, by such a displacement be distorted thus

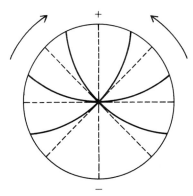

Figure 187,1

where the curved lines represent diameters originally straight, but now curved.

(b) {Faraday} Verdet.[10]

(6) Michael Faraday, 'Experimental researches in electricity. – Twenty-second series. On the crystalline polarity of bismuth (and other bodies), and on its relation to the magnetic form of force', *Phil. Trans.*, **139** (1849): 1–41 (= *Electricity*, **3**: 83–136); and Faraday, 'Thirtieth series. Constancy of differential magnecrystallic force in different media', *Phil. Trans.*, **146** (1856): 159–80.

(7) Michael Faraday, 'Experimental researches in electricity. – Fourteenth series. Nature of the electric force or forces', *Phil. Trans.*, **128** (1838): 265–82 (= *Electricity*, **1**: 533–56).

(8) See Maxwell's discussion, without numerical values, in 'On physical lines of force. Part III': 23–4 (= *Scientific Papers*, **1**: 501–2).

(9) See Michael Faraday, 'Experimental researches in electricity. – Nineteenth series. On the magnetization of light and the illumination of magnetic lines of force', *Phil. Trans.*, **136** (1846): 1–20 (= *Electricity*, **3**: 1–26); the discovery of the rotation of the plane of polarization of linearly polarized light in a magnetic field.

I suppose the elasticity of the sphere to react on the electrical matter surrounding it, and press it downwards.[11]

From the determination by Kohlrausch and Weber of the numerical relation between the statical and magnetic effects of electricity,[12] I have determined the *elasticity* of the medium in air, and assuming that it is the same with the luminiferous ether I have determined the velocity of propagation of transverse vibrations.

The result is

193088 miles per second

(deduced from electrical & magnetic experiments).[13]

Fizeau has determined the velocity of light

= 193118 miles per second

by direct experiment.[14]

This coincidence is not merely numerical.[15] I worked out the formulæ in the

---

(10) See Émile Verdet, 'Recherches sur les propriétés optiques développées dans les corps transparents par l'action du magnétisme', *Ann. Chim. Phys.*, ser. 3, **41** (1854): 370–412; *ibid.*, **43** (1855): 37–44. The problem is considered in detail by Maxwell in 'On physical lines of force. Part IV. The theory of molecular vortices applied to the action of magnetism on polarized light', *Phil. Mag.*, ser. 4, **23** (1862): 85–95 (= *Scientific Papers*, **1**: 502–13). In his letter to Thomson of 10 December 1861 (Number 189) Maxwell gives a more substantive discussion of the issue, by then familiar with Verdet's work. It is possible that Faraday alerted Maxwell to Verdet's papers; see P. M. Heimann, 'Maxwell and the modes of consistent representation', *Archive for History of Exact Sciences*, **6** (1970): 171–213, esp. 193–4.

(11) See Maxwell, 'On physical lines of force. Part III': 14–19 (= *Scientific Papers*, **1**: 491–6). On Maxwell's representation of the dielectric as a structured elastic solid, and his concept of the 'displacement of electricity', see Daniel M. Siegel, 'The origin of the displacement current', *Historical Studies in the Physical and Biological Sciences*, **17** (1986): 99–146; Joan Bromberg, 'Maxwell's displacement current and his theory of light', *Archive for History of Exact Sciences*, **4** (1967): 218-34; Pierre Duhem, *Les Théories Électriques de J. Clerk Maxwell: étude historique et critique* (Paris, 1902): esp. 62; and *Ueber physikalische Kraftlinien. Von James Clerk Maxwell*, ed. L. Boltzmann (Leipzig, 1898): 122–37.

(12) Kohlrausch and Weber determined the number of electrostatic units in one electromagnetic unit of electricity (E), obtaining the value $\frac{1}{2}E = 155{,}370 \times 10^6$ mm/sec; see R. Kohlrausch and W. Weber, 'Elektrodynamische Maassbestimmungen insbesondere Zurückführung der Stromintensitätsmessungen auf mechanisches Maass' *Abhandlungen der Königlichen Sächsischen Gesellschaft der Wissenschaften, math.-phys. Klasse*, **3** (1857): 219–92, esp. 231–2, 260 (= Weber, *Werke*, 6 vols. (Berlin, 1892–4), **3**: 609–76). See Number 189.

(13) J. C. Maxwell, 'On physical lines of force. Part III': 21–2 (= *Scientific Papers*, **1**: 498–9). He demonstrates that the rate of propagation of transverse vibrations through the elastic medium $V = E$, and hence $V = 310{,}740 \times 10$ mm/sec = 193,088 miles per second; see note (12).

(14) See Maxwell, 'On physical lines of force. Part III': 22n (= *Scientific Papers*, **1**: 500n). He refers to the statement of Fizeau's result in J. A. Galbraith and S. Haughton, *Manual of Astronomy* (London, 1855): 39, where 'M. Fizeau's result is stated at 169,944 geographical miles of 1000 fathoms, which gives 193,118 statute miles [per second]'. Compare Number 189, esp. note (16).

country, before seeing Webers number, which is in millimetres, and I think we have now strong reason to believe, whether my theory is a fact or not, that the luminiferous and the electromagnetic medium are one.[16]

Supposing the luminous and the electromagnetic phenomena to be similarly modified by the presence of gross matter, my theory says that the inductive

---

(15) In his letter to Maxwell of 25 March 1857 (ULC Add. MSS 7655, II/6; in *Life of Maxwell*: 519–20; and see Number 133 note (5)) Faraday had stated that: 'I hope this summer to make some experiments on the time of magnetic action or rather on the *time* required for the assumption of the electrotonic state round a wire carrying a current that may help the subject on. The time must probably be short as the time of light, but the greatness of the result if affirmative makes me not despair. Perhaps I had better have said nothing about it for I am often long in realizing my intentions & a failing memory is against me.' In his 'A dynamical theory of the electromagnetic field', *Phil. Trans.*, **155** (1865): 459–512, on 466 (= *Scientific Papers*, **1**: 535–6) Maxwell referred to Faraday's discussion in his paper 'Thoughts on ray-vibrations', *Phil. Mag.*, ser. 3, **28** (1846): 345–50 (= *Electricity*, **3**: 447–52); 'The electromagnetic theory of light, as proposed by him, is the same in substance as that which I have begun to develope in this paper, except that in 1846 there were no data to calculate the velocity of propagation.' Compare Kohlrausch and Weber, 'Elektrodynamische Maassbestimmungen insbesondere Zurückführung der Stromintensitäts-messungen auf mechanisches Maass': 264–5. Stating Weber's electrodynamic force law (see Number 66 notes (3) and (5)) in the form

$$\frac{ee'}{r^2}\left(1 - \frac{1}{c^2}\left(\frac{dr}{dt}\right)^2 + \frac{2r}{c^2}\frac{d^2r}{dt^2}\right)$$

where e and e' are the moving electric charges separated by a distance r, Kohlrausch and Weber state: 'Aus dieser Bestimmung der *Constanten c* ersieht man also, dass zwei elektrische Massen mit sehr grosser Geschwindigkeit gegen einander bewegt werden müssen, weil die *elektrodynamische* Kraft die *elektrostatische* aufheben soll, nämlich mit eine Geschwindigkeit von 439 Millionen Meter oder 59320 Meilen in der Secunde, welche die Geschwindigkeit des Lichts bedeutend übertrifft'. They did not consider this to be physically significant: 'Die Geschwindigkeit des Lichts ist selbst aber nicht die einer Körperbewegung, sondern die einer Wellenbewegung'. Compare also Gustav Kirchhoff, 'Ueber die Bewegung der Elektrizität in Drähten', *Ann. Phys.*, **100** (1857); 193–217 (= *Gesammelte Abhandlungen* (Leipzig, 1882): 131–54), (trans.) 'On the motion of electricity in wires', *Phil. Mag.*, ser. 4, **13** (1857): 393–412; although Kirchhoff found that in the limiting case of vanishing resistance in the wire, electricity is propagated with the velocity of light, he did not develop the implications of this result. Maxwell was apparently unaware of Kirchhoff's paper. The constant c is the ratio of electrostatic and electrodynamic units; conversion from electro-dynamic to electromagnetic units involves a factor of $\sqrt{2}$; see Kohlrausch and Weber, 'Elektro-dynamische Maassbestimmungen': 223, 261–2, 264, where they state the value of c more precisely as $439,450 \times 10^6$ mm/sec. See also Wilhelm Weber, 'Elektrodynamische Maassbe-stimmungen insbesondere über elektrische Schwingungen', *Abhandlungen der Königlichen Sächsischen Gesellschaft der Wissenschaften, math.-phys. Klasse*, **6** (1864): 571–716 (= *Werke*, **4**: 105–241); he deduced that current oscillations in a circuit are propagated with a velocity of $c/\sqrt{2} = 310,740 \times 10^6$ mm/sec, but he again declined to speculate on the physical relationship between optics and electricity. See also Number 189 note (18).

(16) See Maxwell, 'On physical lines of force. Part III': 22 (= *Scientific Papers*, **1**: 500).

capacity (static) is equal to the square of the index of refraction, divided by the coefficient of magnetic induction (air = 1).[17]

I have also examined the theory of the passage of light through a medium filled with magnetic vortices, and find that the rotation of the plane of polarization is in the same direction with that of the vortices, that it varies inversely as the *square* of the wave length (as is shown by experiment)[18] and that its amount is proportional to the *diameter* of the vortices.[19]

The absolute diameter of the magnetic vortices, their velocity and their density, are so involved, that though as yet they are all unknown, the discovery of a new relation among them would determine them all.

---

(17) See Maxwell, 'On physical lines of force. Part III': 22–3, and see note (4). Compare also his subsequent discussion in 'A dynamical theory of the electromagnetic field': 501 (= *Scientific Papers*, **1**: 582–3), where he again concludes that 'the Specific Inductive Capacity is equal to the square of the index of refraction divided by the coefficient of magnetic induction'. In the *Treatise*, **2**: 388–9 (§§ 788–9) he cites values for the dieletric constant and index of refraction of paraffin; arguing that the term for the magnetic capacity could be ignored, he found only approximate experimental support for this result. 'At the same time, I think that the agreement of the numbers is such that if no greater discrepancy were found between the numbers derived from the optical and electrical properties of a considerable number of substances, we should be warranted in concluding that the square root of K [the dielectric constant], though it may not be the complete expression for the index of refraction, is at least the most important term in it.' Ludwig Boltzmann subsequently demonstrated the approximate agreement between the dielectric constant and the square of the index of refraction for several substances, confirming Maxwell's theoretical argument. See especially Boltzmann's 'Über die Verschiedenheit der Dielektricitätsconstante des krystallisirten Schwefels nach Verschiedenen Richtungen', *Sitzungsberichte der Math. Naturwiss. Classe der Kaiserlichen Akademie der Wissenschaften* [Wien], **70**, Abtheilung II (1874): 342–66. In his notes to his translation of *Ueber physikalische Kraftlinien*: 140, he commented that in his work 'wurde die Richtigkeit der *Maxwell*'schen Theorie schon lange vor den klassischen Versuchen *Hertz*' wahrscheinlich gemacht'.

Maxwell attempted to establish the dielectric constant of paraffin in 1870–71 (see his letters to Thomson of 14 April 1870 and 27 March 1871, to be reproduced in Volume II); and he subsequently directed work by J. E. H. Gordon, at the Cavendish Laboratory, on the measurement of the dielectric constants of various substances including glass, paraffin and sulphur. Gordon compared the refractive indices of the transparent dielectrics with the square roots of the dielectric constants, demonstrating only limited agreement. See J. E. H. Gordon, 'Measurements of electrical constants. No. II. On the specific inductive capacities of certain dielectrics. Part I', *Phil. Trans.*, **170** (1879): 417–46. Maxwell's letter to Stokes of 12 August 1878, where reference is made to Gordon's work, will be published in Volume III.

(18) Edmond Becquerel, 'Expériences concernant l'action du magnétisme sur tous les corps', *Ann. Chim. Phys.*, ser. 3, **17** (1846): 437–51; and Gustav Wiedemann, 'Ueber die Drehung der Polarisationsebene des Lichtes durch den galvanischen Strom', *Ann. Phys.*, **82** (1851): 215–32. Compare Number 189 esp. note (29).

(19) See Maxwell, 'On physical lines of force. Part IV', 91–5 (= *Scientific Papers*, **1**: 509–13); and his letters to Thomson of 10 and 17 December 1861 (Numbers 189 and 190).

Such a relation might be obtained by the observation of a revolving electromagnet if our instruments were accurate enough. I have had an instrument made for this purpose, but I have not yet overcome the effects of terrestrial magnetism in masking the phenomena.[20]

When I began to study electricity mathematically, I avoided all the old traditions about forces acting at a distance, and after reading your papers as a first step to right thinking, I read the others, interpreting as I went on, but never allowing myself to explain anything by these forces. It is because I put off reading about electricity till I could do it without prejudice, that I think I have been able to get hold of some of your ideas, such as the electrotonic state,[21] action of contiguous parts &c and my chief object in writing to you is to ascertain if I have got the same ideas which led you to see your way into things, or whether I have no right to call my notions by your names.[22]

I remain
Yours truly
J. C. Maxwell

Professor Faraday

---

(20) See Numbers 143, 189 and 190. Compare Maxwell, 'On physical lines of force. Part II. The theory of molecular vortices applied to electric currents', *Phil. Mag.*, ser. 4, **21** (1861): 281–91, 338–48, esp. 345n (= *Scientific Papers*, **1**: 485–6n); 'The angular momentum of the system of vortices depends on their average diameter; so that if the diameter were sensible, we might expect that a magnet would behave as if it contained a revolving body within it, and that the existence of this rotation might be detected by experiments on the free rotation of a magnet. I have made experiments to investigate this question, but have not yet fully tried the apparatus.' See also his description of apparatus 'which I had constructed in 1861' in the *Treatise on Electricity and Magnetism*, **2**: 202–4 (§575). An electromagnet can rotate about a horizontal axis within an armature which revolves about a vertical axis. If the current in the coil carries momentum, then the coil would precess about the vertical axis. Maxwell found 'no evidence of any change'. He concludes that: 'If therefore, a magnet contains matter in rapid rotation, the angular momentum of this rotation must be very small compared with any quantities which we can measure, and we have as yet no evidence of the existence of the terms...derived from their mechanical action'. See plate X. Maxwell's experiment was subsequently discussed by W. J. de Haas and G. L. de Haas Lorentz, 'Een proef van Maxwell en de moleculaire stroomen van Ampère', *Koninklijke Akademie van Wetenschappen te Amsterdam Verslag*, **24**, 1 (1915): 398–404, who concluded that if electrons were supposed as the cause of the effect, then the change in angle of inclination in Maxwell's apparatus would be unobservably small. See P. Galison, 'Einstein and the gyromagnetic experiments, 1915–1925', *Historical Studies in the Physical Sciences*, **12** (1982): 285–323.

(21) See Faraday, *Electricity*, **1**: 16 (§60), **3**: 420 (§3269); compare Number 87; and see Maxwell, 'On physical lines of force. Part II': 289–91, 338–42 (= *Scientific Papers*, **1**: 475–82).

(22) See Maxwell, 'On physical lines of force. Part III': 14 (= *Scientific Papers*, **1**: 490–1), on dielectric polarization and the concept of electric 'displacement'. Compare Faraday 'On induction': 37–9 (§§1295, 1298, 1304), on the '*polarity*' and 'action of contiguous particles' of the

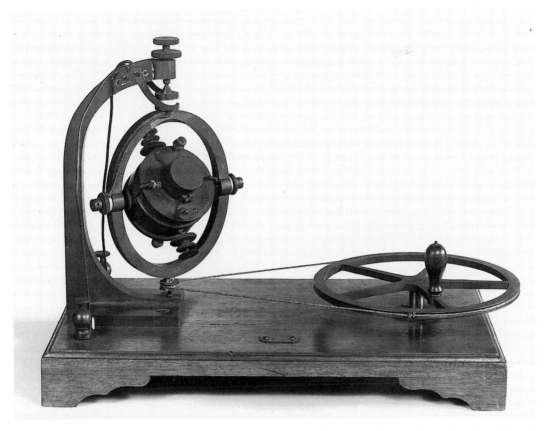

Plate X. Apparatus to detect molecular vortices (1861), by attempting to determine the effect of the angular momentum of revolving vortices on the free rotation of an electromagnet (Number 187).

dielectric medium. Referring to Faraday's 'On induction', Maxwell also makes reference to O. F. Mossotti, 'Discussione analitica sull' influenza che l'azione di un mezzo dielettrico ha sulla distribuzione dell' elettricità alla superfizie di più corpi elettrici disemminati in esso', *Memorie di Matematica e di Fisica della Società Italiana delle Scienze* (Modena), **24** (1850) : 49–74. See also Mossotti, 'Recherches théoriques sur l'induction electrostatique, envisagée d'après les idèes de Faraday', *Supplement à la Bibliothèque Universelle de Genève. Archives des Sciences Physiques et Naturelles*, **6** (1849) : 193–8. On Maxwell's interpretation of Faraday's 'electro-tonic state' see Heimann, 'Maxwell and the modes of consistent representation': 191–3.

188

LETTER TO CECIL JAMES MONRO

circa 20 OCTOBER 1861[1]

From the original in the Greater London Record Office, London[2]

8 Palace Gardens Terrace
Kensington W.

Dear Monro

I was occupied all last week with making preparations for my brother in law who has been very ill and whom we have been urging to come to London for advice,[3] so I have not had much opportunity for other occupations. I send you the papers I promised you about colours[4] and electricity.[5] The velocity of a transverse vibration in a medium having the same elasticity as the electro magnetic medium in my scheme is

193088 miles per second.

The velocity of light in air as found by Fizeau is

193118 miles per second

which is an argument for light being transverse undulations of electro magnetic medium, especially as by hypothesis they are coextensive, that is the luminiferous and electromagnetic media.[6]

I am sorry you have not been able to look me up here but I hope you will be able before you leave. Did you ever see the life of James Halley[7] by Arnott, the man who beat A. C. London[8] at Greek at Glasgow and nearly beat Archibald

---

(1) Monro's reply (addressed from London) is dated 23 October 1861; *Life of Maxwell*: 329.

(2) Greater London Record Office, Acc. 1063/2089.

(3) See Number 194.

(4) Possibly: J. C. Maxwell, 'On the theory of three primary colours', *Proceedings of the Royal Institution*, 3 (1861): 370–4 (= *Scientific Papers*, 1: 445–50).

(5) J. C. Maxwell, 'On physical lines of force' (Parts I and II), *Phil. Mag.*, ser. 4, 21 (1861): 161–75, 281–91, 338–48 (= *Scientific Papers*, 1: 451–88).

(6) See Number 187 esp. notes (13) and (14). In his reply (note (1)) Monro observes: 'The coincidence between the observed velocity of light and your calculated velocity of a transverse vibration in your medium seems a brilliant result. But I must say I think a few such results are wanted before you can get people to think that, every time an electric current is produced, a little file of particles is squeezed along between rows of wheels. But the instances of bodily transfer of matter in the phenomena of galvanism look like it already, and I admit that the possibility of convincing the public is not the question.'

(7) *Memoir of the Late James Halley, A.B., Student of Theology* (Edinburgh, 1842).

(8) Archibald Campbell Tait, Bishop of London 1856–69 (*DNB*); Halley was known as 'the man who beat Tait' (*Memoir*: 5–6n).

Smith[9] at mathematics? He read too hard and had to go to Madeira[10] and his letters show the man very well.

Can you let me hear of you sometime? I expect to be here till June.

Yours truly

J. C. MAXWELL

---

(9) Archibald Smith, Glasgow University and Trinity, Senior Wrangler 1836 (Venn).

(10) Monro took up residence in Madeira, for health reasons; in an undated letter to Maxwell Monro wrote: 'Any 23^{d} of a month that you should feel minded to post a letter addressed "C. J. Monro Esq. Madeira" – that is quite enough, – I should like very much to get it, which I should do about 8 days later' (Greater London Record Office, Acc. 1063/2110).

189

LETTER TO WILLIAM THOMSON

10 DECEMBER 1861

From the original in the University Library, Cambridge[1]

8 Palace Garden Terrace
Kensington W.
1861 Dec 10

Dear Thomson

I have not heard of you for some time except through Balfour Stewart[2] who told me he had seen you lately. I hope you are now well as you are at work.

I was not farther north than Galloway last summer and we spent all our three months vacation there. Since I saw you I have been trying to develope the dynamical theory of magnetism as an affection of the whole magnetic field according to the views stated by you in the Royal Society's proceedings 1856 or Phil Mag 1857 vol 1 p 199 and elsewhere.[3]

I suppose that the 'magnetic medium' is divided into small portions or cells the divisions or cell-walls being composed of a single stratum of spherical particles these particles being 'electricity'.[4] The substance of the cells I suppose to be highly elastic both with respect to compression and distortion and I suppose the connexion between the cells and the particles in the cell walls to be such that there is perfect rolling without slipping between them and that they act on each other tangentially.[5]

---

(1) ULC Add. MSS 7342, M 99. Previously published in Larmor, 'Origins': 728–30.

(2) Balfour Stewart, Edinburgh University 1845, Director of the Kew Observatory 1859–70 (*DNB*).

(3) William Thomson, 'Dynamical illustrations of the magnetic and heliocoidal rotatory effects of transparent bodies on polarized light', *Proc. Roy. Soc.*, **8** (1856): 150–8 (= *Phil. Mag.*, ser. 4, **13** (1857): 198–204). See Maxwell's letter to Faraday of 9 November 1857 (Number 133, esp. note (16)); and his letters to Thomson of 24 November 1857 (Number 137, esp. note (5)) and 30 January 1858 (Number 143). Maxwell developed Thomson's theory of magnetic vortices in his paper 'On physical lines of force. Part I. The theory of molecular vortices applied to magnetic phenomena', *Phil. Mag.*, ser. 4, **21** (1861): 161–75 (= *Scientific Papers*, **1**: 451–66).

(4) J. C. Maxwell, 'On physical lines of force. Part II. The theory of molecular vortices applied to electric currents', *Phil. Mag.*, ser. 4, **21** (1861): 281–91, 338–48, esp. 283 (= *Scientific Papers*, **1**: 468–9). The rotation of contiguous vortices is represented by a mechanism of 'idle wheels': 'in epicyclic trains and other contrivances, as, for instance, in Siemens's governor for steam engines, we find idle wheels whose centres are capable of motion'; referring to Thomas Minchin Goodeve, *The Elements of Mechanism* (London, 1860): 118–20. On the theory of molecular vortices see Daniel M. Siegel, 'Thomson, Maxwell, and the universal ether in Victorian physics', in *Conceptions of Ether. Studies in the History of Ether Theories 1740–1900*, ed. G. N. Cantor and M. J. S. Hodge (Cambridge, 1981): 239–68.

(5) Maxwell, 'On physical lines of force. Part II': 346 (= *Scientific Papers*, **1**: 486). See also Daniel M. Siegel, 'Mechanical image and reality in Maxwell's electromagnetic theory', in

I then find that if the cells are set in rotation the medium exerts a stress equivalent to a hydrostatic pressure combined with a longitudinal tension along the lines of axes of rotation.[6]

Wranglers and Physicists: Studies on Cambridge Physics in the Nineteenth Century, ed. P. M. Harman (Manchester, 1985): 180–202. A draft fragment headed 'Prop XII To find the Angular momentum of a vortex' (ULC Add. MSS 7655, V, c/8), its numbering indicating that it was intended for Part II of the paper, contains a cancelled passage on the mechanical properties of vortices which is omitted from the published Proposition XVIII of 'On physical lines of force. Part IV. The theory of molecular vortices applied to the action of magnetism on polarized light', *Phil. Mag.*, ser. 4, **23** (1862): 85–95, esp. 90–1 (= *Scientific Papers*, **1**: 508–9). 'The Angular Momentum of any system about an axis is the sum of the products of each particle multiplied by the area it describes about that axis in unit of time or if A is the angular momentum about axis of

x $A = \sum dm \left(y\frac{dz}{dt} - z\frac{dy}{dt} \right)$. As we do not know the law of distribution of density ⟨and velocity⟩ in the vortex, we shall determine the relation between the Angular Momentum and the Energy of the vortex ⟨on several different assumptions⟩. ⟨1st⟩ Let the motion of the vortex be such that the angular velocity is the same throughout, and let v be the linear velocity at the circumference, then if A is the angular momentum and E the energy $A = \sum dm\, r^2 \omega$ $E = \frac{1}{2}\sum dm\, r^2 \omega^2$. Making ω constant

we have $A = \dfrac{2E}{\omega} = 2E\dfrac{r}{v}$. ⟨2nd Let the velocity be constant $= v$ then $A = \sum dm\, rv$, $E = \frac{1}{2}\sum dm\, v^2$. If we suppose the vortex spherical and of uniform density then $A = \dfrac{3\pi}{8}\dfrac{r}{v}E = 1.178\dfrac{E}{\omega}$. The more the density increases towards the circumference, the more will the value of A approach to $\dfrac{2E}{\omega}$. As the first supposition is most likely to be true we may suppose $A = 2E\dfrac{r}{v}$ where r is the radius of the vortex and v the velocity at the circumference.⟩ Now we know by Prop VI that $E = \dfrac{1}{4\pi}\mu v^2 V$ where V is the volume of the vortex so that the value of the angular momentum is $A = \dfrac{1}{4\pi}\mu r v V$.'

(6) Maxwell, 'On physical lines of force. Part I': 161–75, esp. 163 (= *Scientific Papers*, **1**: 451–66, esp. 453). In expounding his theory of magnetism as the result of '*stress in the medium*' he refers to W. J. M. Rankine, *A Manual of Applied Mechanics* (London/Glasgow, 1858): 92–3, 113–16, for the mathematical theory of stress. He rejects the supposition, proposed by Challis, that magnetism could be explained by 'undulations issuing from a centre'; see James Challis, 'A theory of galvanic force', *Phil. Mag.*, ser. 4, **20** (1860): 431–41; Challis, 'A theory of magnetic force', *ibid.*, **21** (1861): 65–73, 92–107. Compare Challis' response to Maxwell's comments: Challis, 'On theories of magnetism and other forces, in reply to remarks by Professor Maxwell', *ibid.*: 250–4. Challis wrote to Maxwell on 10 June 1861, enclosing copies of his papers in response to Maxwell's 'desire for a list of my papers relative to the hydrodynamical theorems used in my theories of the dynamic action of undulations and of electricity'; Challis claims that his theories were 'strictly within the rules of the Newtonian principles of Philosophy, which is more than can be said of many theories that have been broached in the present day' (ULC Add. MSS 7655, II/21). On the reference to Newton's third rule of philosophising, the 'analogy of nature', and Maxwell's critique of this principle, see Number 106, esp. note (3).

Maxwell appended a note to the second part of 'On physical lines of force. Part II': 348 (= *Scientific Papers*, **1**: 488), published in the May 1861 number of the *Philosophical Magazine*: 'Since the first part of this paper was written, I have seen in Crelle's *Journal* for 1859, a paper by

If there be two similar systems, the first a system of magnets, electric currents and bodies capable of magnetic induction, and the second composed of cells and cell walls, the density of the cells everywhere proportional to the capacity for magnetic induction at the corresponding point of the other, and the magnitude and direction of rotation of the cells proportional to the magnetic force, then –

1 All the mechanical magnetic forces in the one system will be proportional to forces in the other arising from centrifugal force.[7]

2 All the electric currents in the one system will be proportional to currents of the particles forming the cell walls in the other.[8]

3 All the electromotive forces in the one system, whether arising from changes of position of magnets or currents or from motions of conductors or from changes of intensity of magnets or currents will be proportional to forces urging the particles of the cell walls arising from the tangential action of the rotating cells when their velocity is increasing or diminishing.[9]

4 If in a non conducting body the mutual pressure of the particles of the cell walls (which corresponds to electric tension) diminishes in given direction, the particles will be urged in that direction by their mutual pressure but will be restrained by their connexion with the substance of the cells. They will therefore produce strain in the cells till the elasticity called forth balances the tendency of the particles to move.[10]

Thus there will be a displacement of particles proportional to the electro-

Prof. Helmholtz on Fluid Motion, in which he has pointed out that the lines of fluid motion are arranged according to the same laws as the lines of magnetic force, the path of an electric current corresponding to a line of axes of those particles of the fluid which are in a state of rotation. This is an additional instance of a *physical analogy*, the investigation of which may illustrate both electromagnetism and hydrodynamics.' See H. Helmholtz, 'Über Integrale der hydrodynamischen Gleichungen, welche den Wirbelbewegungen entsprechen', *Journal für die reine und angewandte Mathematik*, **55** (1858): 25–55 (= *Wissenschaftliche Abhandlungen*, 3 vols. (Leipzig, 1882–95), **1**: 101–34). Two folios of notes relating to Helmholtz's paper (ULC Add. MSS 7655, V, c/5,8) may date from 1861, though a later date seems more probable. One of the folios is headed 'On the Condition of the Steady Motion of an Incompressible Fluid' (V, c/5; compare Number 63). Maxwell uses the term 'velocity potential' (Helmholtz's term *Geschwindigkeitspotential*), and refers to the rotation of 'eddy tubes (Helmholtz's Wirbelfäden)' of a fluid; see Helmholtz, 'Über Integrale der hydrodynamischen Gleichungen': 25–6. These notes will be reproduced in Volume II.

(7) Maxwell, 'On physical lines of force. Part I': 168–9 (= *Scientific Papers*, **1**: 458–9).

(8) Maxwell, 'On physical lines of force. Part II': 284–6 (= *Scientific Papers*, **1**: 469–71).

(9) Maxwell, 'On physical lines of force. Part II': 289–91, 338–43 (= *Scientific Papers*, **1**: 475–83).

(10) J. C. Maxwell, 'On physical lines of force. Part III. The theory of molecular vortices applied to statical electricity', *Phil. Mag.*, ser. 4, **23** (1862): 12–24, esp. 13–14 (= *Scientific Papers*, **1**: 490–1).

motive force,[11] and when this force is removed the particles will recover from displacement. I have calculated the relation between the force and the displacement on the supposition that the cells are spherical and that their cubic and linear elasticities are connected as in a 'perfect' solid.[12] I have found from this the attraction between two bodies having given quantities of free electricity on their surfaces and then by comparison with Webers value of the statical measure of a unit of electrical current I have deduced the relation between the elasticity and the density of the cells.[13]

The velocity of transverse undulations follows from this directly and is equal to

193 088 miles per second, very nearly that of light.[14]

Velocity
of light
miles per sec.
$$\begin{cases} 192{,}500 \text{ by aberration}^{[15]} \\ 195{,}777 \text{ by Fizeau}^{[16]} \\ 193\ 118 \text{ Galbraith \& Haughtons statement of Fizeau's} \\ \quad\text{results.}^{[17]} \end{cases}$$

I made out the equations in the country before I had any suspicion of the nearness between the two values of the velocity of propagation of magnetic effects and that of light, so that I think I have reason to believe that the magnetic and luminiferous media are identical and that Weber's number is really, as it appears to be, one half the velocity of light in millimetres per second.[18]

---

(11) Maxwell, 'On physical lines of force. Part III': 14–15 (= *Scientific Papers*, **1**: 491–2); and see Number 187 esp. note (11).

(12) See Maxwell, 'On physical lines of force. Part III': 18–19 (= *Scientific Papers*, **1**: 495–6), where he refers to W. J. M. Rankine, 'Laws of the elasticity of solid bodies', *Camb. & Dubl. Math. J.*, **6** (1851): 47–80, esp. 69, for a definition of a 'perfect solid'; 'the term *perfect solid* is used to denote a body whose elasticity is due entirely to the mutual attractions and repulsions of atomic centres of force'. See Number 34. On the relation between cubic and linear elasticity see Number 26 notes (11) and (44).

(13) See Maxwell, 'On physical lines of force. Part III': 19–22 (= *Scientific Papers*, **1**: 497–9); and Number 187, esp. notes (12) and (13).

(14) Maxwell, 'On physical lines of force. Part III': 22 (= *Scientific Papers*, **1**: 499); see Number 187 esp. note (13).

(15) See Maxwell, 'On physical lines of force. Part III': 22n (= *Scientific Papers*, **1**: 500n), where he states: 'the value deduced from aberration is 192,000 miles [per second]'; see J. A. Galbraith and S. Haughton, *Manual of Optics* (London, ₄1860): 2.

(16) Hippolyte Louis Fizeau, 'Sur une expérience relative à la vitesse de la propagation de la lumière', *Comptes Rendus*, **29** (1849): 90–92. Fizeau gives a value of 70,948 leagues per second (25 leagues to a degree); compare Maxwell, 'On physical lines of force. Part III': 22 (= *Scientific Papers*, **1**: 500), where he calculates Fizeau's value as 195,647 miles per second.

(17) J. A. Galbraith and S. Haughton, *Manual of Astronomy* (London, 1855): 39; see Number 187 note (14).

(18) R. Kohlrausch and W. Weber, 'Elektrodynamische Maassbestimmungen insbesondere Zurückführung der Stromintensitätsmessungen auf mechanisches Maass', *Abhandlungen der König-*

If there is in all media, in spite of the disturbing influence of gross matter, the same relation between the velocity of light and the statical action of electricity, then the 'dielectric capacity', that is, the capacity of a Leyden jar of given thickness formed of it, is proportional to the square of the index of refraction.[19]

Do you know any good measures of dielectric capacity of transparent substances? I have read Faraday & Harris on the subject and I think they are likely to be generally too small.[20] I think Fleeming Jenkin has found that of gutta percha caoutchouc &c.[21] Where can one find his method, and what method do you recommend.

I think I see a way to work with flat plates of different substances along with your divided ring electrometer.[22]

A & E are plates connected with each other and with a source of electricity. C is connected with the ground, B & D are moveable plates connected with the two halves of the ring. F is a dielectric. Then by a proper placing of D things may be arranged so that

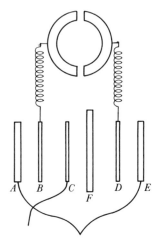

Figure 189,1

lichen Sächsischen Gesellschaft der Wissenschaften, math.-phys. Klasse, **3** (1857): 219–52, esp. 260. See Number 187 notes (12) and (15). Thomson had speculated on the velocity of propagation of electric signals in insulated wires in 1854; see Lord Kelvin, *Baltimore Lectures on Molecular Dynamics and the Wave Theory of Light* (Cambridge, 1904): 45, 688–94); his article on 'Velocity of Electricity' in J. P. Nichol, *A Cyclopaedia of the Physical Sciences* (London, $_2$1860): 274–6 (= *Math. & Phys. Papers*, **2**: 131–7); and his letter to Stokes of 28 October 1854 (ULC Add. MSS 7656, K73), speculating 'that the quantity, σ, by which electrostatical & electrodynamic units are compared, may be determined by finding the velocity of propagation of a regular periodical effect (sin $2nt$) ...'.

(19) Maxwell, 'On physical lines of force. Part III': 22–3 (= *Scientific Papers*, **1**: 500–1). See Number 187 note (17).

(20) Michael Faraday, 'Experimental researches in electricity. – Eleventh series. On induction', *Phil. Trans.*, **128** (1838): 1–40 (= *Electricity*, **1**: 360–416); W. Snow Harris, 'On the specific inductive capacities of certain electrical substances', *Phil. Trans.*, **132** (1842): 165–72; and see Number 187.

(21) Fleeming Jenkin, 'On gutta percha as an insulator at various temperatures', *Report of the Twenty-ninth Meeting of the British Association for the Advancement of Science; held at Aberdeen in September 1859* (London, 1860), part 2: 248–50; Jenkin, 'On the insulating properties of gutta percha', *Proc. Roy. Soc.*, **10** (1860): 409–15.

(22) William Thomson described his divided ring electrometer in his article on 'Atmospheric elasticity', in Nichol, *A Cyclopaedia of the Physical Sciences*: 267–74, esp. 272–3 (= *Electrostatics and Magnetism*: 192–208, esp. 203–5); and in his Royal Institution Lecture of 18 May 1860 'On atmospheric electricity', *Proceedings of the Royal Institution of Great Britain*, **3** (1860): 277–90, esp. 274–6 (= *Electrostatics and Magnetism*: 208–26, esp. 209–12).

the electrification of *A* and *E* produces no difference of potentials in the divided ring, after which the capacity of *F* follows by calculation.[23]

I have also examined the propagation of light through a medium containing vortices[24] and I find that the *only* effect is the rotation of the plane of polarization in the *same* direction as the angular momentum of all the vortices the rotation being proportional to

A   the thickness of the medium
B   the magnetic intensity along the axis[25]
C   the index of refraction in the medium[26]
D   inversely as the square of the wave length in air[27]
F   directly as the radius of the vortices
G   – – – – – the magnetic capacity.[28]

I have been seeking for experiments lately made which I have lost sight of showing that the rotation varies faster than the inverse square of wave length in air,[29] so that C & D give the true law.

(23) See Numbers 190 and 194.

(24) J. C. Maxwell, 'On physical lines of force. Part IV': 85–95, esp. 88 (= *Scientific Papers*, **1**: 506). See Ole Knudsen, 'The Faraday effect and physical theory, 1845–1873', *Archive for History of Exact Sciences*, **15** (1976): 235–81. See Number 187.

(25) Émile Verdet, 'Recherches sur les propriétés optiques développées dans les corps transparents par l'action du magnétisme', *Ann. Chim. Phys.*, ser. 3, **41** (1854): 370–412; *ibid.*, **43** (1855): 37–44. Verdet established that '*Il y a proportionalité entre l'action magnétique et la rotation du plan de polarisation*' produced by a given thickness of a transparent substance (*ibid.*, **41** (1854): 397). See Number 187.

(26) Suggested by A. de la Rive, *Traité d'Électricité Théorique et Appliquée*, 3 vols. (Paris, 1854–8), **1**: 555–7. See note (32).

(27) E. Becquerel, 'Expériences concernant l'action du magnétisme sur tous les corps', *Ann. Chim. Phys.*, ser. 3, **17** (1846): 437–51; Gustav Wiedemann, 'Ueber die Drehung der Polarisationsebene des Lichtes durch den galvanischen Strome', *Ann. Phys.*, **82** (1851): 215–32. See note (29).

(28) Termed by Verdet the '*pouvoir rotatoire magnétique*', a physical constant which depends on the nature of the substance; see Émile Verdet, 'Recherches sur les propriétés optiques développées dans les corps transparentes par l'action du magnétisme', *Ann. Chim. Phys.*, ser. 3, **52** (1858): 129–63, esp. 136.

(29) Émile Verdet, 'Note on the dispersion of the planes of polarization of the coloured rays produced by the action of magnetism', *Report of the Thirtieth Meeting of the British Association for the Advancement of Science; held at Oxford in June and July 1860* (London, 1861), part 2: 54–5. Verdet's full investigation of the variation of the magnetic rotation with the wavelength of the light ray, qualifying the work of Becquerel and Wiedemann, was published subsequently; see Émile Verdet, 'Recherches sur les propriétés optiques développées dans les corps transparentes par l'action du magnétisme. De la dispersion des plans de polarisations des rayons de diverses couleurs', *Ann. Chim. Phys.*, ser. 3, **69** (1863): 415–91.

A & B are proved by Verdet,[30]

F is not yet capable of proof[31]

G is consistent with Verdets result that the rotation in salts of iron is opposite to that in diamagnetic substances.[32] I think that molecules *of iron* are set in motion by the cells and revolve the opposite way so as to produce a very great energy of rotation but an angular momentum in the opposite direction to that of the vortices.[33]

I find that unless the diameter of the vortices is sensible, no result is likely to be obtained by making a magnet revolve freely about an axis perp. to the magnetic axis.[34] I have tried it but have not yet got rid of the effects of terrestrial magnetism which are very strong on a powerful electromagnet which I use.[35] I think it more probable that a coil of conducting wire might be found to have a slight shock tending to turn it about its axis when the electricity is let on or cut off,[36] or that a piece of iron magnetized by a helix might have a similar impulse.

Yours truly

J. C. MAXWELL

---

(30) See note (25).

(31) Compare Maxwell, 'On physical lines of force. Part IV': 89, 95 (= *Scientific Papers*, **1**: 507, 513).

(32) Émile Verdet, 'Recherches sur les propriétés optiques...', *Ann. Phys. Chim.*, ser. 3, **52** (1858): 139–40, 142–63. Verdet rejected de la Rive's suggestion that there was a simple relation between rotatory power and the index of refraction; see also de la Rive, *Traité d'Électricité*, **3**: 715–19.

(33) Compare Maxwell, 'On physical lines of force. Part IV': 89 (= *Scientific Papers*, **1**: 507); and Knudsen, 'The Faraday effect and physical theory': 258. Maxwell concludes that: 'This result agrees so far with that part of the theory of M. Weber which refers to the paramagnetic and diamagnetic conditions'; see Wilhelm Weber, 'Ueber die Erregung und Wirkung des Diamagnetismus nach den Gesetzen inducirter Ströme', *Ann. Phys.*, **73** (1848): 241–56 (= *Werke*, **3**: 255–68); (trans.) 'On the excitation and action of diamagnetism according to the laws of induced currents', *Scientific Memoirs*, ed. R. Taylor, **5** (London, 1852): 477–88. See Number 84 note (40). Maxwell discussed the issue in letters to P. G. Tait (of 23 December 1867) and William Thomson (of 18 July 1868); these letters will be reproduced in Volume II.

(34) On Maxwell's attempt to detect the angular momenta of vortices, by experiments on the free rotation of a magnet, see Numbers 143, 187 and 190, esp. Number 187 note (20) and plate X.

(35) Compare Maxwell's discussion in *Treatise*, **2**: 204 (§575).

(36) Maxwell made reference to such an experiment in a letter to John William Strutt of 18 May 1870 (to be published in Volume II); and compare his discussion in *Treatise*, **2**: 200–1 (§574), concluding that 'No phenomenon of this kind has yet been observed.'

LETTER TO WILLIAM THOMSON

17 DECEMBER 1861

From the original in the University Library, Cambridge[1]

> 8 Palace Gardens Terrace,
> Kensington,
> London, W.
> 1861 Dec 17

Dear Thomson

I have thought of an elementary investigation of the attraction of magnetic poles in a para or dia magnetic substance by means of the theory of electrical images applied to a simple case.[2]

First. Let μ & μ' be the inductive capacities of two media bounded by an infinite plane.

Let A be a magnetic pole, then the force and potential at any point B in the *same* medium are those due to A and its image A' provided

$$A' = A\frac{\mu-\mu'}{\mu+\mu'}.$$

Also at any point B' in the other medium the potential is due to $A\dfrac{2\mu}{\mu+\mu'}$ placed at A.

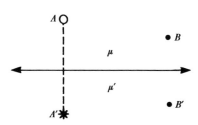

Figure 190,1

For we shall find that these values give the potential the same in both media at the bounding surface, and the variation of potential normal to the surface inversely proportional to μ and μ'.

Second. Let A and B be two magnetic poles distant x from each other and $\frac{1}{2}y$ from the surface and let the law of force be repulsion $= \dfrac{mAB}{x^2}$ where m depends on the medium.

Figure 190,2

---

(1) ULC Add. MSS 7342, M 100. Previously published in Larmor, 'Origins': 731–3.

(2) William Thomson, 'On the mathematical theory of electricity in equilibrium' [Parts III–VI], *Camb. Dubl. & Math. J.*, **3** (1848): 141–8, 266–74; **4** (1849): 276–84; **5** (1850): 1–9; and Thomson, 'Note on induced magnetism in a plate', *ibid.*, **1** (1845): 34–7 (= *Electrostatics and Magnetism*: 52–85, 104–7). See Maxwell, *Treatise*, **1**: 366–9 (§§315–18). On the theory of electrical images see also Number 71 esp. note (18).

The potential of B (with respect to A & the surface) is

$$mB\left\{\frac{A}{x}+\frac{\mu-\mu'}{\mu+\mu'}\left(\frac{B}{y}+\frac{A}{\sqrt{x^2+y^2}}\right)\right\}.$$

Hence if A and B are originally close to the surface at an infinite distance and are then carried by any paths in the first medium so as to be close to the surface and distant a from each other, the work done will be

$$mB\left\{\frac{A}{a}+\frac{\mu-\mu'}{\mu+\mu'}\frac{A}{a}\right\}=\frac{2\mu}{\mu+\mu'}m\frac{AB}{a}.$$

Now if the same thing be done on the other side of the bounding surface we must simply put m' for m μ' for μ & μ for μ' and we find the work

$$=\frac{2\mu'}{\mu'+\mu}m'\frac{AB}{a}.$$

These quantities must be equal, otherwise we might do work by separating two poles putting them into a medium and bringing them near and then taking them out and bringing them to their former position

$$\text{whence } m\mu=m'\mu'$$

or the force is inversely as μ that is greater in dia- than para-magnetics.

This kind of argument does not directly apply to currents because they are doing work even when at rest, but it can be got at by considering

1st The potential of two currents is the one current multiplied by the number of lines of magnetic force *due to the other* which pass through it.

2nd the force on a magnetic pole at given distance from the current depends on the current and is independent of the medium. For the work done on a unit pole by one circuit round the current is $4\pi I$ where I is the current and if it depended on the medium work might be done by a pole going round a closed curve not embracing the current.

3rd The *number* of lines of force in any place depends on the magnetic force multiplied by the magnetic coefficient.

Therefore if the same currents are placed in two different media, the magnetic forces will be the same at corresponding points, but the potential of the currents will be *directly* as μ, and so will the forces and the inductive capacities of the wires if the currents be variable.

It is remarkable that the spontaneous movements of magnets tend to diminish their potentials but the spontaneous movements of currents tend to increase their potentials and the whole energetic value of the field they move in.

Could it be proved that in any piece of machinery containing wheels mounted on moveable axes whose motion alters their velocities: If the machine be set agoing and left the axes will tend to move so as to diminish the actual energy[3] of the machine.

But if the motive power be continually applied to keep up a constant velocity in one wheel, the axles or moveable parts will move so as to increase the actual energy.[4]

I am not sure of the truth of this in general with regard to the last part, but here is an instance.

If a tube be set a spinning about an axis perp. to it and a bullet tied by a string to the axis be allowed to yield to centrifugal force by the elasticity of the string, the tube and ball together will lose vis viva which will be stored in the elastic string.

But if the tube be kept spinning at constant velocity, the yielding of the string to centrifugal force will cause the tube and ball to have greater vis viva.

If this is a general truth it applies to magnets and currents, magnets not having any motive power to back them up like that which currents find in the battery.

I have not been able to ascertain whether any investigations of dielectric capacity have been made besides those of Faraday & Snow Harris[5] and those on different kinds of gutta percha by Jenkin & others.[6]

I think by your electrometer one might eliminate the effects of slow conduction, 'penetration of charge' &c by charging the extreme or middle plates of your instrument instantaneously and then discharging and observing the impulse given to the needle.[7]

Is your instrument such that when the needle is placed at zero the potential of the two halves of the ring may be suddenly changed without disturbing the needle provided the two halves have always the same potential.

I should like to hear from you some time whether you are likely to make any observations on dielectrics, and if you have not the prospect of doing so I should like to have the benefit of the plans you have proposed and to know which of your forms of electrometer you consider most suited for the purpose, for I have more time now to work, and I have several experiments in electricity which I

---

(3) For the term 'actual [kinetic] energy' see W. J. M. Rankine, 'On the general law of the transformation of energy', *Phil. Mag.*, ser. 4, **5** (1853): 106–17, esp. 106; and Number 133 note (11).

(4) Compare Maxwell's discussion of his 'idle wheel' mechanism in 'On physical lines of force. Part II', *Phil. Mag.*, ser. 4, **21** (1861): 281–91, 338–48, esp. 286–91 (= *Scientific Papers*, **1**: 472–6), and see Number 189 note (4).

(5) Number 189 note (20).   (6) Number 189 note (21).

(7) Number 189 note (22).

wish to try, and I hope to get the dielectric coefficients of bodies relative to air as near the truth as I can.[8]

I also, if I can get a good nonconducting liquid, mean to compare the force of two bodies in the liquid and out. If their potentials are given the force ought to vary directly as the dielectric capacity (greater in glass than in air). If their charges are given the force ought to vary inversely as ditto.

It would be difficult to plunge a ball in a liquid without altering its charge but to keep up its potential is easy. I also wish to ascertain whether this property has axes. If it has a sphere of crystal will *set* between two electrified plates.

I shall be glad to know the max^m breadth of atoms.[9] I suppose their length is not so easily limited, but if you could get a minimum breadth, you would go far to establish the existence of the atom.

I know experiments which may determine the maximum breadth of a vortex of magnetism and if we could find the actual breadth we might find the density of ether and the time of revolution of a vortex.[10] This is by a comparison of the rotation of pol. light by a known mag. force with the axis of free rotation of a magnet. I want to put Verdets results[11] into absolute measure, but I need the deflexion produced by a turn of *his bobbin* due to terrestrial mag. at Paris which he has not given so far as I know.[12] I think Arndtsen of Christiania works now at this.[13]

Yours truly

J. C. MAXWELL

(8) Maxwell is concerned to confirm his prediction connecting the dielectric capacity with the square of the index of refraction. See Number 187 notes (4) and (17); and Numbers 189 and 194 for descriptions of the apparatus he planned to use to measure the dielectric constants of substances.

(9) Thomson was interested in the question at the time. See his reference to establishing 'a definite limit for the sizes of atoms' in a letter to James Prescott Joule published in extract in *Proceedings of the Literary and Philosophical Society of Manchester*, **2** (1862): 176–8 (= *Electrostatics and Magnetism*: 317–18).

(10) See Numbers 143, 187 and 189, esp. Number 187 note (20).

(11) See Number 189 notes (25) and (28).

(12) See also Maxwell, 'On physical lines of force. Part IV', *Phil. Mag.*, ser. 4, **23** (1862): 85–95, esp. 94–5 (= *Scientific Papers*, **1**: 512–13).

(13) Adam Arndtsen, 'Magnetische Untersuchungen, angestellt mit dem Diamagnetometer des Hrn. Prof. Weber', *Ann. Phys.*, **104** (1858): 587–611; reported in Arndtsen, 'Expériences magnétiques', *Ann. Chim. Phys.*, ser. 3, **56** (1859): 246–9. In 1875–6, at the Cavendish Laboratory, Maxwell directed research by J. E. H. Gordon on the determination of Verdet's results in absolute (cgs) units; see J. E. H. Gordon, 'Determination of Verdet's constant in absolute units', *Phil. Trans.*, **167** (1877): 1–34. A letter from Maxwell to Gordon on this work will be published in Volume III.

FROM A LETTER TO HENRY RICHMOND DROOP[1]

28 DECEMBER 1861

From Campbell and Garnett, *Life of Maxwell*[2]

Glenlair
Dalbeattie N.B.
28 December 1861

I have nothing to do in King's College till Jany. 20, so we came here to rusticate. We have clear hard frost without snow, and all the people are having curling matches on the ice, so that all day you hear the curling-stones on the lochs in every direction for miles, for the large expanse of ice vibrating in a regular manner makes a noise which, though not particularly loud on the spot, is very little diminished by distance. I am trying to form an exact mathematical expression for all that is known about electro-magnetism without the aid of hypothesis, and also what variations of Ampère's formula are possible, without contradicting his expressions.[3] All that we know is about the action of *closed* currents – that is, currents through closed curves. Now, if you make a hypothesis (1) about the mutual action of the elements of two currents, and find it agree with experiment on closed circuits, it is not proved, for –

If you make another hypothesis (2) which would give *no action* between an element and a *closed* circuit, you may make a combination of (1) and (2) which will give the same result as (1). So I am investigating the most general hypothesis about the mutual action of elements, which fulfils the condition that the action between an element and a closed circuit is null.[4] This is the case if the action between two elements can be reduced to forces between the extremities of those elements depending only on the distance and + or − according as they act between similar or opposite ends of the elements. If the force is an attraction

$$= \phi(r)\, ss'(\cos\omega + 2\cos\theta\cos\theta')$$

where ω is the angle between s and s', r the distance of s and s' and θ and θ' the angles s and s', the elements, make with r, then the condition of no action will be fulfilled.[5]

---

(1) Number 135 note (1).

(2) *Life of Maxwell*: 330; abridged. Maxwell gave an account of the endowment of a church at Corsock, near Glenlair; see Rev. Geo. Sturrock, *Corsock Parish Church: Its Rise and progress &c* (Castle-Douglas, 1899): 10–12.

(3) A. M. Ampère, 'Mémoire sur la théorie mathématique des phénomènes électrodynamiques uniquement déduite de l'expérience', *Mémoires de l'Académie Royale des Sciences*, 6 (1827): 175–388. See Numbers 66 note (3) and 192 note (3).

(4) Compare Maxwell, *Treatise*, 2: 151–61 (§§ 511–27).

(5) Compare Number 192 note (3).

FROM A LETTER TO HENRY RICHMOND DROOP

24 JANUARY 1862

From Campbell and Garnett, *Life of Maxwell*[1]

8 Palace Gardens Terrace
W.
24 January 1862

When I wrote to you about closed currents, it was partly to arrange my own thoughts by imagining myself speaking to you.[2] Ampère's formula containing n and k is the most general expression for an attractive or repulsive force in the line joining the elements;[3] and I now find that if you take the most general expression consistent with symmetry for an action transverse to that line, the resulting expression for the action of a closed current on an element gives a force not perpendicular to that element. Now, experiment 3d (Ampère) shows that the force on a movable element is perp. to the directions of the current, so that I see Ampère is right.[4]

But the best way of stating the effects is with reference to 'lines of magnetic force'. Calculate the magnetic force in any plane, arising from every element of the circuit, and from every other magnetising agent, then the force on an element is in the line perp. to the plane of the element and of the lines of force.[5]

But I shall look up Cellerier[6] and Plana,[7] and the long article in Karsten's

---

(1) *Life of Maxwell*: 331–2.

(2) See Number 191.

(3) A. M. Ampère, 'Mémoire sur la théorie mathématique des phénomènes électrodynamiques', *Mémoires de l'Académie Royale des Sciences*, **6** (1827): 175–388, esp. 204, 252, 323. Ampère's general formula is

$$\frac{ii'\,ds\,ds'}{r^n}(\cos\epsilon + (k-1)\cos\theta\cos\theta')$$

where i and i' are the intensities of the electric currents, ds and ds' the line elements carrying currents, ϵ the angle between the line elements, and θ and θ' the angles between the line elements and the straight line r connecting them. Ampère gives values $n = 2$ and $k = -\frac{1}{2}$.

(4) Compare Maxwell, *Treatise*, **2**: 148–9 (§507), 155–6 (§516); and Ampère, 'Mémoire sur la théorie mathématique des phénomènes électrodynamiques': 194–7.

(5) Compare Maxwell, *Treatise*, **2**: 162–4 (§§528–9).

(6) C. Cellerier, 'De la loi générale des actions électrodynamiques', *Comptes Rendus*, **30** (1858): 693–4.

(7) Jean Plana, 'Mémoire sur la théorie du magnétisme', *Astronomische Nachtrichten*, **39** (1855): col. 225–40, 305–8; *ibid.*, **42** (1856): col. 1–44; and Plana, 'Mémoire sur la distribution de l'électricité à la surface intérieure et sphérique d'une sphère creuse de métal et à la surface d'une autre sphère conductrice électrisée que l'on tient isolée dans sa cavité', *Memorie della Reale Accademia delle Scienze di Torino*, ser. 2, **16** (1857): 57–95.

Cyclopædia.[8] I want to see if there is any evidence from the mathematical expressions as to whether element acts on element, or whether a current first produces a certain effect in the surrounding field, which afterwards acts on any other current.

Perhaps there may be no mathematical reasons in favour of one hypothesis rather than the another.

As a fact, the effect on a current at a given place depends solely on the direction and magnitude of the magnetic force at that point, whether the magnetic force arises from currents or from magnets. So that the theory of the effect taking place through the intervention of a medium is consistent with fact, and (to me) appears the simplest in expression; but I must prove either that the direct action theory is completely identical in its results, or that in some conceivable case they may be different. My theory of the rotation of the plane of polarised light by magnetism is coming out in the *Phil. Mag.*[9] I shall send you a copy.

---

(8) F. C. O. von Feilitzsch, *Die Lehre von den Fernewirkungen des galvanischen Stroms*, in *Allgemeine Encyclopädie der Physik*, Bd.XIX, Hrsg. G. Karsten (Leipzig, 1856).

(9) J. C. Maxwell, 'On physical lines of force. Part IV', *Phil. Mag.*, ser. 4, **23** (1863): 85–95 (= *Scientific Papers*, **1**: 502–13).

FROM A LETTER TO HENRY RICHMOND DROOP

28 JANUARY 1862

From Campbell and Garnett, *Life of Maxwell*[1]

8 Palace Gardens Terrace
Kensington
London W.
28 January 1862

Some time ago, when investigating Bernoulli's theory of gases,[2] I was surprised to find that the internal friction of a gas (if it depends on the collision of particles) should be independent of the density.[3]

Stokes has been examining Graham's experiments on the rate of flow of gases through fine tubes,[4] and he finds that the friction, if independent of density, accounts for Graham's results, but, if taken proportional to density, differs from those results very much. This seems rather a curious result, and an additional phenomenon, explained by the 'collision of particles' theory of gases. Still one phenomenon goes against that theory – the relation between specific heat at constant pressure and at constant volume, which is in air = 1.408, while it ought to be 1.333.[5]

My brother-in-law, who is still with us, is getting better, and had his first walk on crutches to-day across the room.

---

(1) *Life of Maxwell*: 332.
(2) See Number 181 note (2).
(3) See Number 157 esp. note (15).
(4) Thomas Graham, 'On the motion of gases', *Phil. Trans.*, **136** (1846): 573–632; *ibid.*, **139** (1849): 349–401.
(5) See Number 181 esp. note (9).

LETTER TO CECIL JAMES MONRO

18 FEBRUARY 1862

From the original in the Greater London Record Office, London[1]

8 Palace Gardens Terrace
London W.
1862 Feb 18

Dear Monro

I got your letter in Scotland, whither we had gone for the Christmas holidays. I have been brewing Platonic suds but failed, owing I suppose to a too low temperature. I had not read Plateaus recipe then.[2] Some of the bubbles on the surface lasted a fortnight in the air but they were scummy and scaly and inelastic. I shall take more care next time. Elliot of the Strand (30)[3] is going to produce colour-tops with papers from De La Rue[4] and directions for use by me and so I shall be put in competition with the brass Blondin[5] and the Top on the top of the Top. When I wrote you last we were expecting my brother in law whose leg was ailing and we wished him to come here for advice. The advice was as we feared – to have it off – so it was done just 3 months ago by Fergusson[6] and we think the disease is arrested for the recovery has been very steady. He can now go on crutches and is going to be measured for a new leg. He has been a very good patient and we have all been very happy to get rid of the pain and danger. His mother stayed with him during our absence. With regard to Britomarts nurse – I have not Spenser here, but I think Spenser was not a magician himself and got all his black art out of romances and not out of the professional treatises. The notions to be brought out were 1st the unweaving

---

(1) Greater London Record Office, Acc. 1063/2090. Previously published (almost complete) in *Life of Maxwell*: 332–4. The letter is endorsed 'Rec^d March 3' by Monro; see Number 188 note (10).

(2) J. Plateau, 'Recherches expérimentales et théoriques sur les figures d'équilibre d'une masse liquide sans pesanteur (Cinquième [et] Sixième Série)', *Mémoires de l'Académie Royale des Sciences de Bruxelles*, **33** (1861). Compare J. C. Maxwell, 'Plateau on soap-bubbles [review of Joseph Plateau, *Statique Expérimentale et Théorique des Liquides soumis aux seules forces moléculaires*, 2 vols. (Paris/London/Ghent/Leipzig: 1873)]', *Nature*, **10** (1874): 119–21 (= *Scientific Papers*, **2**: 393–9).

(3) Messrs Elliott Bros. subsequently provided apparatus for the Cavendish Laboratory; see ULC Add. MSS 7655, II/80, and ULC, University Archives, Cavendish 1.

(4) Firm of stationers in London; see *Encyclopaedia Britannica*, 11th edn, **7** (Cambridge, 1910): 945.

(5) French tight-rope walker and acrobat; appeared at the Crystal Palace, London in 1861 and 1862; see *Encyclopaedia Britannica*, 11th edn, **4** (Cambridge, 1910): 77.

(6) William Fergusson, Professor of Surgery at King's College, London 1840–70, surgeon extraordinary to the Queen 1855 (*DNB*).

any net in which B. had been caught. 2nd Doing so in witchlike fashion. 3rd Not like a wicked witch but like a well intentioned nurse, unused to the art and therefore blunderlingly.

She believes in the number three and in contrareity and therefore says everything thrice and does everything thrice saying inversions of sentences and doing reversions of her revolutions which are described in similar language. The revolutions begin by $+3\,(2\pi)$ against the visible motion of the sun, then by a revolution -6π she returns all contrary and unweaves the first. Then she goes round $+6\pi$ to make the final result contrary to the natural revolution and to make a complete triad.[7]

Withershins is I believe equivalent to wider die Sonne in High Dutch which I am not aware is a modern or ancient idiom in that language but it may be one in a cognate language.[8] If the 'phamplets' have not turned up in Madeira yet let me know that I may 'replace' them.

I suppose in your equations, when the numbers do not amount to unity, Black has been present.

$$.841 \text{ Brunsw. G.} + .159 \text{ W} = .200 \text{ V} + .423 \text{ U} + .377 \text{ Black.}$$

That is green a little palish and dark mauve, your last eq^n by the young eyes. It is something like a colour-blind eq^n but all those I know say 100 Brunswick G = 100 Vermillion so that this person sees the green darker than the Vermillion or in other words sees much more of the second side of the equation than a colour-blind person would. But in twilight U comes out strong while G does not so that I think the apparent equality arises from suppression of all colours but blue (in U and W) in the twilight so that you may write

$$.841 \text{ Black} + .159 \text{ W} = .577 \text{ Black} + .423 \text{ U.}$$

There is no use going to the 3rd place of decimals unless you spend a good while on each observation and have first rate eyes. But if you can get observations to be consistent to the 3rd place of decimals, glory therein and let me know what the human eye can do.[9]

---

(7) Spenser, *The Faerie Queen*, Book III, Cant. II, 51; 'That sayd, her round about she from her turnd,/She turned her contrarie to the Sunne,/Thrise she turnd contrary, and returnd,/All contrary, for she the right did shunne,/And ever what she did was streight undonne'. See Edmund Spenser, *The Faerie Queen*, 2 vols. (Oxford, 1953), **1**: 368.

(8) Withershins: in a direction contrary to the apparent course of the sun (*OED*); see note (7). The word is used in the novels of Sir Walter Scott; see especially Scott's note (not cited by *OED*) in chap. 24 of *Waverley*: 'To go round a person in the opposite direction [to the sun], or *withershins* (*German wider-shins*), is unlucky, and a sort of incantation'; Sir Walter Scott, *Waverley*, new edn (Edinburgh, 1829), **1**: 255n.

(9) Experiments suggested by Maxwell's work on colour vision: see Number 188.

Donkin[10] gave me tea in Oxford July 1, 1860.

I find that my belief in the reality of state affairs is no greater in London than in Aberdeen, though I can see the clock at Westminster on a clear day. If I went and saw the parks of artillery at Woolwich and the Consols going up and down in the city[11] and the Tuscarora[12] and M[r] Mason[13] I would know what like they were but otherwise a printed statement is more easily appropriated than experience is acquired by being near where things are being transacted.

I am getting a large box made for mixture of colours. A beam of sunlight is to be divided into colours by a prism, certain colours selected by a screen with slits. These gathered by a lens and restored to the form of a beam by another prism and then viewed by the eye directly. I expect great difficulties in getting everything right adjusted but when that is done I shall be able to vary the intensity of the colours to a great extent, and to have them far purer than by any arrangement in which white light is allowed to fall on the final prism.[14]

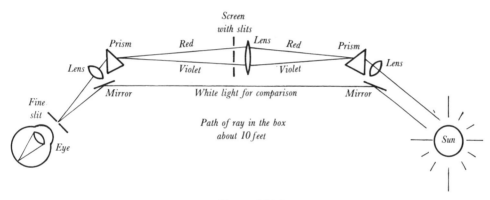

Figure 194,1

---

(10) William Fishburn Donkin, Savilian Professor of Astronomy at Oxford, 1842–70; see Joseph Foster, *Alumni Oxonienses: the Members of the University of Oxford, 1715–1886*, 4 vols. (Oxford, 1888), **1**: 378.

(11) Consolidated Annuities: Government securities of Great Britain.

(12) 'The Tuscarora and the Nashville': Federal and Confederate warships berthed at Southampton; see *The Times*, 9 January to 5 February 1862.

(13) James Murray Mason, a Commissioner from the Confederate States to Great Britain; seized with his colleague John Slidell by the Federal navy from the British mail steamer Trent. The 'Trent affair' led to a diplomatic outcry and threats of war; Mason and Slidell were released, and Mason took up his post in Britain; see *The Times*, 28 November 1861 to 30 January 1862; and *Encyclopaedia Britannica*, 11th edn, **17** (Cambridge, 1911): 839.

(14) Maxwell describes the apparatus in his paper 'On colour-vision at different points of the retina', *Report of the Fourtieth Meeting of the British Association for the Advancement of Science; held at Liverpool in September 1870* (London, 1871), part 2: 40–1 (= *Scientific Papers*, **2**: 230–2). He remarks

I am also planning an instrument for measuring electrical effects through different media, and comparing those media with air.[15] *A* and *B* are two equal metal discs capable of motion towards each other by fine screws. *D* is a metal disc suspended between them by a spring *C*. *E* is a piece of glass, sulphur, vulcanite gutta-percha, &c. *A* and *B* are then connected with a source of +electricity and *D* with −electricity. If everything was symmetrical *D* would be attracted both ways and would be in unstable equilibrium but this is rendered stable by the elasticity of the spring *C*.

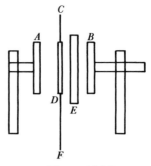

Figure 194,2

To find the effect of the plate *E* you work *A* further or nearer till there is no motion of *D* consequent on electrification. Then the plate of air between *A* and *D* is electrically equivalent to the two plates of air and one of glass (say) between *D* and *B*, whence we deduce the coeff^t for *E*.

I hope to hear from you and that you continue better. We have had no great winter here. One of my cousins is up the Nile but I have not heard how he gets on there. He was very ill in England last winter. Last night we had a tea fight, the first time since we set up hospital and our patient came up stairs very successfully so I think we may repeat the experiment as a man gets nervous who begins with typhus fever and goes on through the summer with an incipient cancer and then has to get the leg off, his attention being necessarily taken from his parish and concentrated on the pain. But now we have great hopes of a complete recovery and of his being an active man with an American leg. We are here till end of June after which probably Glenlair, Dalbeattie NB.

Yours truly

J. C. MAXWELL

that the 'original conception [of the instrument] is due to Sir Isaac Newton, and is described in his *Lectiones Opticae*, though it does not appear to have been actually constructed till the author set it up in 1862, with a solid frame and careful adjustments'. See *Isaaci Newtoni ... Lectiones Opticae, anni MDCLXIX, MDCLXX & MDCLXXI. In scholis publicis habitae; et nunc primum ex MSS. in lucem editae* (London, 1729): 221–6, and Tab. II fig. 28; and *The Optical Papers of Isaac Newton. Volume I. The Optical Lectures 1670–1672*, ed. Alan E. Shapiro (Cambridge, 1984): 516–23. Compare Maxwell's reference to Newton's instrument, as described in Book I, Part II, Prop. XI of the *Opticks*, in his 'Experiments on colour, as perceived by the eye', *Trans. Roy. Soc. Edinb.*, **21** (1855): 290 (= *Scientific Papers*, **1**: 144).   (15) See Numbers 189 and 190.

195

FROM A LETTER TO LEWIS CAMPBELL

21 APRIL 1862

From Campbell and Garnett, *Life of Maxwell*[1]

8 Palace Gardens Terrace
Kensington, W.
21 April 1862

It is now a long time since I wrote half a letter to you, but I have never since had time to write or to find the scrap. I suppose, as it was more than a good intention, but less than a perfect act, it may be regarded as destined to paper purgatory. This is the season of work to you, when folks visit shrines in April and May, but I get holiday this week. I have been putting together a large optical box, 10 feet long, containing two prisms of bisulphuret of carbon, the largest yet made in London, five lenses and two mirrors, and a set of movable slits. Everything requires to be adjusted over and over again if one thing is not quite right placed, so I have plenty of trial work to do before it is perfect, but the colours are most splendid.[2]

I think you asked me once about Helmholtz and his philosophy. He is not a philosopher in the exclusive sense, as Kant, Hegel, Mansel[3] are philosophers, but one who prosecutes physics and physiology, and acquires therein not only skill in discovering any desideratum, but wisdom to know what are the desiderata, e.g., he was one of the first, and is one of the most active, preachers of the doctrine that since all kinds of energy are convertible, the first aim of science at this time should be to ascertain in what way particular forms of energy can be converted into each other, and what are the equivalent quantities of the two forms of energy.[4]

The notion is as old as Descartes (if not Solomon), and one statement of it was familiar to Leibnitz.[5] It was wholly unknown to Comte,[6] but all sorts of people have worked at it of late, – Joule and Thomson for heat and electricals,

---

(1) *Life of Maxwell*: 335–6.

(2) See Number 194 esp. note (14).   (3) See Number 16 note (3).

(4) Hermann Helmholtz, *Über die Erhaltung der Kraft* (Berlin, 1847), and especially his lecture 'On the application of the law of the conservation of force to organic nature', *Proceedings of the Royal Institution*, **3** (1858–62): 347–57.

(5) As the principle of the conservation of *vis viva*. Helmholtz presents his 'principle of the conservation of force' as a generalisation of the 'principle of the conservation of *vis viva*': see also Number 107 note (77).

(6) Compare the first two volumes of Auguste Comte's *Cours de Philosophie Positive*, 6 vols. (Paris, 1830–42).

Andrews for chemical combinations, Dr. E. Smith for human food and labour.[7] We can now assert that the power of our bodies is generated in the muscles, and is not conveyed to them by the nerves, but produced during the transformation of substances in the muscle, which are supplied fresh by the blood.

We can also form a rough estimate of the efficiency of a man as a mere machine, and find that neither a perfect heat engine nor an electric engine could produce so much work and waste so little in heat. We therefore save our pains in investigating any theories of animal power based on heat and electricity. We see also that the soul is not the direct moving force of the body. If it were, it would only last till it had done a certain amount of work, like the spring of a watch, which works till it is run down. The soul is not the mere mover. Food is the mover, and perishes in the using, which the soul does not. There is action and reaction between body and soul, but it is not of a kind in which energy passes from the one to the other, – as when a man pulls a trigger it is the gunpowder that projects the bullet, or when a pointsman shunts a train it is the rails that bear the thrust. But the constitution of our nature is not explained by finding out what it is not. It is well that it will go, and that we remain in possession, though we do not understand it.

Hr. Clausius of Zurich, one of the heat philosophers, has been working at the theory of gases being little bodies flying about, and has found some cases in which he and I don't tally,[8] so I am working it out again.[9] Several experimental results have turned up lately, rather confirmatory than otherwise of that theory.[10]

I hope you enjoy the absence of pupils. I find that the division of them into smaller classes is a great help to me and to them; but the total oblivion of them for definite intervals is a necessary condition of doing them justice at the proper time.

---

(7) J. P. Joule, 'On the mechanical equivalent of heat', *Phil. Trans.*, **140** (1850): 61–82; William Thomson, 'On the dynamical theory of heat', *Trans. Roy. Soc. Edinb.*, **20** (1851): 261–88, 289–98, 475–82; *ibid.*, **21** (1854): 123–71 (= *Math. & Phys. Papers*, **1**: 174–291); Thomas Andrews, 'On the heat disengaged during metallic substitution', *Phil. Trans.*, **138** (1848): 91–104; and Edward Smith, 'Experiments on respiration. On the action of food upon respiration during the primary process of digestion', *Phil. Trans.*, **149** (1859): 715–42.

(8) R. Clausius, 'Ueber die Wärmeleitung gasförmiger Körper', *Ann. Phys.*, **115** (1862): 1–56.

(9) See Number 196.

(10) See O. E. Meyer, 'Ueber die Reibung der Flüssigkeiten', *Ann. Phys.*, **113** (1861): 55–86, 193–228, 383–425, esp. 386, where Meyer remarks that his value for the viscosity of air was compatible with the 'Bernoulli–Clausius'schen Ansicht über die Constitution der Gase'.

196

DRAFTS ON THE MOTIONS AND
COLLISIONS OF PARTICLES

circa 1862[1]

From the originals in the University Library, Cambridge[2]

[1]   [MOLECULAR COLLISIONS][3]

To find the number of collisions in unit of time between n particles having a given velocity v and N particles having velocities distributed so that the number having velocities between u & $u+du$ is $\dfrac{4N}{\alpha^3\sqrt{\pi}}e^{-\frac{u^2}{\alpha^2}}du$ the motion being in one unit of volume.

Let the distance between the centres at collision be s and the relative velocity of a pair of particles r, then the number of collisions in unit of time between that pair of particles will be $\pi s^2 r$.

---

(1) Drafts endorsed 'Collisions' in Maxwell's hand, and 'Cond. Heat in Gases' in pencil. Compare the MS 'On the conduction of Heat in Gases' (ULC Add. MSS 7655, V, f/5; to be published in Volume II), Maxwell's response to Rudolf Clausius' paper 'Ueber die Wärmeleitung gasförmiger Körper', *Ann. Phys.*, **115** (1862): 1–56. These drafts may constitute Maxwell's initial response to Clausius' criticisms of 'Illustrations of the dynamical theory of gases' (1860) to which he alludes in his letter to Campbell of 21 April 1862 (Number 195). Handwriting suggests that these drafts were not written before the early 1860s. Clausius' criticisms focussed on Maxwell's use of an isotropic distribution function in treating the conduction of heat in a gas; see Maxwell, 'Illustrations of the dynamical theory of gases. Part II', *Phil. Mag.*, ser. 4, **20** (1860): 21–33, esp. 31–3 (= *Scientific Papers*, **1**: 403). Compare Clausius, 'Ueber die Wärmeleitung gasförmiger Körper': 13n; 'Maxwell hat in seiner oben erwähnten Abhandlung (*Phil. Mag.* Vol. XX) bei der Bestimmung der Wärmeleitung den Umstand, dass die von einer Schicht ausgesandten Molecüle einer Ueberschuss an positiver Bewegungsgrösse haben, nicht berücksichtigt, sondern hat in seinen Rechnungen stillschweigend vorausgesetzt, dass die Molecüle nach allen Richtungen in gleicher Weise ausgesandt werden.' In his MS 'On the conduction of Heat in Gases' Maxwell identified 'the problem which I neglected to consider in my former paper'. This is: 'When the density and temperature of a gas or the composition of a system of mixed gases vary from one place to another, what is the proportion of particles, which, starting from one given place, arrive at another given place without a collision'.

(2) ULC Add. MSS 7655, V, f/5. Previously published in E. Garber, S. G. Brush, and C. W. F. Everitt, *Maxwell on Molecules and Gases* (Cambridge, Mass./London, 1986): 283–4, 321–32. Garber, Brush and Everitt date these drafts to 1859–61; see notes (1) and (3).

(3) Compare Proposition VIII of 'Illustrations of the dynamical theory of gases. Part I', *Phil. Mag.*, ser. 4, **19** (1860): 26 (= *Scientific Papers*, **1**: 384–5). Garber, Brush and Everitt, *Maxwell on Molecules and Gases*: 283–4, suggest that this draft 'probably formed part of the basis for Prop. VIII of Maxwell's 1860 paper'. Handwriting and similar terminology ('modulus of velocity') suggest that the draft was written with the other drafts in the early 1860s.

Let lines representing the velocities of the N particles be drawn from a point then the number of such lines having their extremities in a space $= 1$ at distance u will be $\dfrac{N}{\alpha^3 \pi^{\frac{3}{2}}} e^{-\frac{u^2}{\alpha^2}}$.

Let us call this the 'density' of the distribution of velocities corresponding to a given direction and magnitude of u. In general it will depend on the direction as well as the magnitude of u.

Now let v be the velocity of the n particles and let r be their velocity relative to a particle of the other system and let θ be the angle between v and r, then

$$u^2 = v^2 + r^2 - 2vr\cos\theta$$

and the number of particles of the second system for which r lies between r & $r+dr$ and θ between θ & $\theta + d\theta$ will be

$$\frac{N}{\alpha^3 \pi^{\frac{3}{2}}} e^{-\frac{u^2}{\alpha^2}} 2\pi r^2 \sin\theta \, dr \, d\theta = \frac{2N}{\alpha^3 \pi^{\frac{1}{2}}} \frac{ru}{v} e^{-\frac{u^2}{\alpha^2}} \, dr \, du$$

if we make r & u the variables instead of r & θ. The number of collisions between these and the n particles whose relative velocity is r, is

$$\frac{2\sqrt{\pi}\, s^2\, Nn}{\alpha^3 v} r^2 u \, e^{-\frac{u^2}{\alpha^2}} \, dr \, du.$$

Integrating with respect to u from $u^2 = (v-r)^2$ to $u^2 = (v+r)^2$

$$\frac{\sqrt{\pi}\, s^2\, Nn}{\alpha v} r^2 \{ e^{-\frac{(r-v)^2}{\alpha^2}} - e^{-\frac{(r+v)^2}{\alpha^2}} \} \, dr.$$

Integrating with respect to r from $r = 0$ to $r = \infty$

$$\sqrt{\pi}\, s^2\, Nn\, \alpha \left\{ e^{-\frac{v^2}{\alpha^2}} + \left(\frac{\alpha}{v} + 2\frac{v}{\alpha} \right) \int_0^v e^{-\frac{v^2}{\alpha^2}} \, dv \right\}.$$

This is the number of collisions in unit of time between n particles moving with velocity v and N particles whose modulus of velocity is α.

[2] [ON THE RELATIVE VELOCITIES OF PARTICLES IN MOTION]

To find the mean relative velocity of particles whose direction makes an angle $\cos^{-1}\lambda$ with the axis with respect to particles the directions of which are distributed according to the law $(1 + F\mu)$ where $\cos^{-1}\mu$ is the angle with the axis.

Let OA be the axis, OB the velocity of the first particle $= v$ $u = OC$ that of any other & BC the relative velocity $= r$.

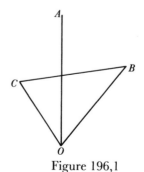

Figure 196,1

Let $\cos AOB = \lambda$ $\cos AOC = \mu$ Let $BOC = \theta$ and angle between planes AOB, $AOC = \phi$.

density of particles at $C = \dfrac{N}{\alpha^3 \pi^{\frac{3}{2}}} e^{-\frac{u^2}{\alpha^2}} (1 + F\mu)$

differential element at $C = u^2 \sin^2 \theta \, du \, d\theta \, d\phi$

relative velocity $r = \sqrt{u^2 + v^2 - 2uv \cos \theta}$ $r \, dr = 2uv \sin \theta \, d\theta$.

Sum of relative velocities $= \displaystyle\iiint \dfrac{N}{\alpha^3 \pi^{\frac{3}{2}}} e^{-\frac{u^2}{\alpha^2}} (1 + F(\lambda \cos \theta +$

$\sqrt{1 - \lambda^2} \sin \theta \cos \phi)) \sqrt{u^2 + v^2 - 2uv \cos \theta} \, u^2 \sin \theta \, du \, d\theta \, d\phi$.

Integrate from $\phi = 0$ to $\phi = 2\pi$

$$\iint \dfrac{2N}{\alpha^3 \pi^{\frac{1}{2}}} e^{-\frac{u^2}{\alpha^2}} (1 + F\lambda \cos \theta) \sqrt{u^2 + v^2 - 2uv \cos \theta} \, u^2 \sin \theta \, du \, d\theta$$

$$\langle \text{from } \theta = 0 \text{ to } \theta = \pi \rangle$$

$$= \iint \dfrac{2N}{\alpha^3 \pi^{\frac{1}{2}}} e^{-\frac{u^2}{\alpha^2}} u^2 \, du \left(1 + F\lambda \dfrac{u^2 + v^2 - r^2}{2uv}\right) \dfrac{r^2}{2uv} \, dr$$

$$= \int \dfrac{2N}{\alpha^3 \pi^{\frac{1}{2}}} e^{-\frac{u^2}{\alpha^2}} u^2 \, du \, \dfrac{1}{4u^2 v^2} \left\{ \dfrac{2uv}{3} r^3 + \tfrac{1}{3} F\lambda (u^2 + v^2) r^3 - \tfrac{1}{5} F\lambda r^5 \right\}.$$

The superior limit of r is $u + v$ and the inferior is $u - v$ or $v - u$ according as u or v is greater. The quantity within brackets becomes

$$\dfrac{4uv^2}{3}(3u^2 + v^2) + \tfrac{4}{15} F\lambda (v^5 - 5u^2 v^3) \text{ when } u \text{ is greater than } v$$

and $\dfrac{4u^2 v}{3}(u^2 + 3v^2) + \tfrac{4}{15} F\lambda (u^5 - 5u^3 v^2)$ when v is greater than u.

If N' is the number of particles projected at an angle λ with the axis & if β is their coeff^t of velocity the number between v & $v + dv$ is $N \dfrac{4}{\beta^3 \sqrt{\pi}} v^2 e^{-\frac{v^2}{\beta^2}} \, dv$. The sum of relative vel^{ies} is $\therefore$.

$$\iint \dfrac{8NN'}{\alpha^3 \beta^3 \pi} e^{-\frac{u^2}{\alpha^2}} e^{-\frac{v^2}{\beta^2}} \, du \, dv \left\{ (u^3 v^2 + \tfrac{1}{3} uv^4) + F\lambda \left(\dfrac{v^5}{15} - \dfrac{u^2 v^3}{3} \right) \right\}$$

when $u > v$.

Integrate the first term with respect to u from $u = v$ to $u = \infty$ and the second with respect to v from $v = 0$ to $v = u$

$$\int \dfrac{8NN'}{\alpha^3 \beta^3 \pi} e^{-\frac{u^2}{\alpha^2}} e^{-\frac{v^2}{\beta^2}} (\tfrac{2}{3} \alpha^2 v^4 + \tfrac{1}{2} \alpha^4 v^2) \, dv +$$

$$F\lambda \int \dfrac{8NN'}{\alpha^3 \beta^3 \pi} e^{-\frac{u^2}{\alpha^2}} e^{-\frac{v^2}{\beta^2}} (\tfrac{2}{15} \beta^2 u^4 + \tfrac{1}{10} \beta^4 u^2 + \tfrac{1}{15} \beta^6) \, dv.$$

Integrating from 0 to ∞

$$\frac{8NN'}{\alpha^3\beta^3}\frac{1}{\sqrt{\pi}}\{(\tfrac{1}{2}\alpha^2\beta^5+\tfrac{1}{4}\alpha^4\beta^3)+F\lambda(\tfrac{1}{10}\beta^2\alpha^5+\tfrac{1}{20}\beta^4\alpha^3+\tfrac{1}{15}\beta^6\alpha)\}$$

for all cases where $u > v$.

Similarly for $v > u$

$$\frac{8NN'}{\alpha^3\beta^3}\frac{1}{\sqrt{\pi}}\{(\tfrac{1}{2}\beta^2\alpha^5+\tfrac{1}{4}\beta^4\alpha^3)+F\lambda(\tfrac{1}{10}\alpha^2\beta^5+\tfrac{1}{20}\alpha^4\beta^3+\tfrac{1}{15}\alpha^6\beta)\}.$$

[3]   [MOLECULAR COLLISIONS]

Let there be n_1 molecules in unit of volume having a relative velocity r to n_2 other molecules in the same unit of volume. Then the number of encounters between molecules of the two kinds in which the alteration of the direction of the relative velocity lies between γ and $\gamma + d\gamma$ will be

$$n_1 n_2 f(r, \gamma) \sin \gamma \, d\gamma$$

where $f(r, \gamma)$ is some function of r and γ depending on the law of action between the molecules.

Now let us assume that there are N_1 molecules of the first kind in unit of volume and that the number of these which have component velocities between ξ and $\xi + d\xi$ η and $\eta + d\eta$ and ζ and $\zeta + d\zeta$ is

$$\frac{N_1}{\alpha^3\pi^{\frac{3}{2}}}e^{-\frac{\xi^2+\eta^2+\zeta^2}{\alpha^2}}\,d\xi\,d\eta\,d\zeta.$$

If we put

$$\frac{N_1}{\alpha^3\pi^{\frac{3}{2}}}e^{-\frac{\xi^2+\eta^2+\zeta^2}{\alpha^2}}=\rho_1$$

we may call ρ_1 the *density* of the velocities which are nearly equal to the given velocity and we may say that $\rho_1 = \dfrac{N_1}{\alpha^3\pi^{\frac{3}{2}}}e^{-\frac{v^2}{\alpha^2}}$ is the density of velocities near to v v being given both in direction and magnitude and if dV be a small element of volume on the diagram of velocities then $\rho\,dV$ will be the number of molecules such that the lines representing their velocities terminate within dV.

Let there be N_2 molecules of another kind in unit of volume and let their modulus of velocity be β. Let us find how many of these have a velocity relative to v lying between r and $r + dr$.

Let θ be the angle between the directions r and v and ϕ the angle which the plane of this angle makes with a fixed plane through v, then we have to integrate $\dfrac{N_2}{\beta^3\pi^{\frac{3}{2}}}e^{-\frac{v^2-2vr\cos\theta+r^2}{\beta^2}}r^2\sin\theta\,dr\,d\theta\,d\phi$ from $\phi = 0$ to $\phi = 2\pi$ and from $\theta = 0$ to $\theta = \pi$. This gives

$$\frac{N_2}{\beta\pi^{\frac{1}{2}}}\frac{r}{v}\{e^{-\frac{(v-r)^2}{\beta^2}}-e^{-\frac{(v+r)^2}{\beta^2}}\}\,dr$$

for the number required.[4] The number of molecules of velocity between v and $v + dv$ which in unit of time have their velocity altered so that γ lies between γ and $\gamma + d\gamma$ is

$$4v\, dV \frac{N_1}{\alpha^3 \pi^{\frac{1}{2}}} e^{-\frac{v^2}{\alpha^2}} \frac{N_2}{\beta \pi^{\frac{1}{2}}} \sin\gamma\, d\gamma \int_0^\infty \{e^{-\frac{(v-r)^2}{\beta^2}} - e^{-\frac{(v+r)^2}{\beta^2}}\} rf(r,\gamma)\, dr.$$

[4]   [ON THE COLLISION OF MOLECULES]

We have next to determine the number of molecules of the first kind which after an encounter with molecules of the second kind in which their relative velocity is altered in direction through an angle between γ and $\gamma + d\gamma$ acquire a velocity between v and $v + dv$.

Let OA, OB be the original velocities of the molecules of the first and second kinds $AB = r$ their relative velocity G their centre of gravity.

$$AG = Ga = r^2 \frac{\alpha^2}{\alpha^2 + \beta^2} \quad Aga = \gamma \quad GaO = \theta$$

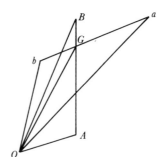

Figure 196,2

Let ϕ be the angle between the plane GaO and a fixed plane through Oa and ψ the angle between the planes aGA and aGO.

Let ρ_1 be the density of distribution of the molecules of the first kind corresponding to the velocity OA and ρ_2 the density of distribution of the second kind corresponding to OB.

To find the number of molecules of the first kind which can acquire a velocity Oa, let AB describe a conical surface about G whose solid angle is

$$\omega = \sin\gamma\, d\gamma \sin\psi.$$

Next let r increase from r to $r + dr$ then the solid element described by A in all its positions will be

$$\omega GA^2(1 - \cos\gamma)\, dGA = \frac{\alpha^6 r^2}{(\alpha^2 + \beta^2)^3}(1 - \cos\gamma)\sin\gamma\, d\gamma\, d\psi\, dr.$$

The number of molecules of the first kind is therefore

$$\frac{N_1}{\pi^{\frac{3}{2}}} \frac{\alpha^3 r^2}{(\alpha^2 + \beta^2)^3} e^{-\frac{OA^2}{\alpha^2}}(1 - \cos\gamma)\sin\gamma\, d\gamma\, d\psi\, dr.$$

Let us next find the number of molecules of the second kind with which each of these molecules may have an encounter so as to acquire such a velocity that a may fall within an element of volume dV described about a. Let A be fixed and

---

(4) Compare Maxwell, 'Illustrations of the dynamical theory of gases. Part I': 26–7 (= *Scientific Papers*, **1**: 385).

let AB describe a solid angle ω' about A Ga continuing parallel to itself then B will describe an area $r^2\omega'$ and a an area $\dfrac{\alpha^2 r^2}{\alpha^2 + \beta^2}\omega'$. Now let r increase from r to $r + dr$ B will advance dr and a will move to a distance $(1 - \cos\gamma)\dfrac{\alpha^2}{\alpha^2 + \beta^2}dr$ from the plane of the small area.

The number of particles which A may meet will therefore be $\rho_2 r^2\omega'\,dr$ and the space dV within which a will be found will be

$$dV = \frac{\alpha^4}{(\alpha^2 + \beta^2)^2}(1 - \cos\gamma)\,r^2\omega'\,dr.$$

[5] [DISTRIBUTION OF VELOCITIES]

Two systems of particles moving in the same space have respectively N_1 & N_2 particles in unit of volume and α_1 α_2 as the modulus of velocity. Required the number of collisions in unit of volume and unit of time after which a particle of the first system has a velocity between v and $v + dv$.

Let OA represent the velocity of the first particle OB that of the second and OG that of the centre of gravity then if we draw $Ga = GA = r_1$ in any direction the probability that Oa will represent the velocity of the first particle after collision will be the same in whatever direction we draw Ga.

We have to determine the number of cases in which the point a will lie within certain limits. If we suppose a volume V described about a then if the number of cases in which a lies within this volume be ρV we may call ρ the 'density' corresponding to the velocity Oa.

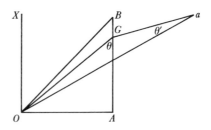

Figure 196,3

Let aG be fixed and let AB describe a conical surface whose solid angle is ω then A will describe an area $= \omega r_1^2$. Now let aG and GA be increased simultaneously by dr_1 the displacement of A in the direction of GA will be $(1 + \cos BGa)\,dr_1$[5] so that the volume at A corresponding to the solid angle ω and the variation dr_1 is $(1 + \cos BGa)\,r_1^2\,\omega\,dr_1$.

If ρ_1 is the 'density' of distribution of particles at A, the number between these limits will be

$$(1 + \cos BGa)\,\rho_1 r_1^2\,\omega\,dr_1. \tag{1}$$

---

(5) Maxwell first wrote: $(1 + \cos BGA)$, correcting this term. Expressions *infra* are modified in conformity with this correction.

Each of these particles, after collision may have a velocity Oa. Let us find the number of particles with which the collision may take place, so that a may lie within the space V. To do this let A be fixed and let AB describe a conical surface whose solid angle is ω and while Ga is always drawn parallel to itself and equal to GA.

Let $GA = r_1$ $GB = r_2$ then B describes an area $= (r_1 + r_2)^2 \omega$ & a describes an area $= r_1^2 \omega'$.

Now let r_1 increase by dr_1 and r_2 by $\dfrac{r_2}{r_1} dr_1'$ then B will advance $\dfrac{r_1 + r_2}{r_1} dr_1'$ and a will advance $(1 + \cos BGa)\, dr_1$ so that if ρ_2 be the 'density' at B the number of particles with which A may meet will be

$$\rho_2 \frac{(r_1 + r_2)^3}{r_1} \omega'\, dr_1' \tag{2}$$

and the volume described about A will be

$$V = (1 + \cos BGa)\, r_1^2 \omega'\, dr_1'. \tag{3}$$

We have hitherto supposed the direction of Ga fixed but if we describe about a a solid angle ω_2 the probability that Ga will lie within this angle is

$$\tfrac{1}{4}\pi\omega_2. \tag{4}$$

By multiplying (1), (2) & (4) we get the number of particles of the first kind which, after collision with particles of the second kind may have velocities comprised within the given limits under the conditions that r_1 is between r_1 & $r_1 + dr_1$, that the direction Ga lies within a solid angle ω_2 and AB within a solid angle ω.

The number of collisions of each of these particles in unit of time is

$$\pi s^2 (r_1 + r_2). \tag{5}$$

We have therefore

$$\delta\rho V = \delta\rho(1 + \cos BGa)\, r_1^2 \omega'\, dr_1'$$

$$= (1 + \cos BGa)\, \rho_1 r_1^2 \omega\, dr_1 \rho_2 \frac{(r_1 + r_2)^3}{r_1} \omega'\, dr_1' \frac{1}{4\pi} \omega_2 s^2 (r_1 + r_2)$$

$$\rho = \sum \tfrac{1}{4} \rho_1 \rho_2 s^2 \frac{(r_1 + r_2)^4}{r_1} \omega\omega_2\, dr_1$$

where ρ_1 is the 'density' of distribution of velocities in the first system corresponding to the velocity OA, ρ_2 that in the second system corresponding to OB, s the distance of centres at collision r_1 the velocity of the 1st particle & r_2 that of the second relative to their centre of gravity ω & ω_2 small solid angles.

Taking the integral of this expression with respect to ω & ω_2 round the whole sphere and with respect to r_1 from 0 to ∞ we obtain ρ the number of particles of the first kind projected in unit of time from unit of volume, after collision with particles of the second kind, with velocities whose extremities lie within unit of volume described about a.

[6] To integrate $\rho = \sum \frac{1}{4}\rho_1\rho_2\, s^2 \dfrac{(r_1+r_2)^4}{r_1}\,\omega\omega_2\, dr_1$

when $\qquad \rho_1 = \dfrac{N_1}{\alpha_1^3\pi^{\frac{3}{2}}}\, e^{-\frac{x^2}{\alpha_1^2}} \quad \rho_2 = \dfrac{N_2}{\alpha_2^3\pi^{\frac{3}{2}}}\, e^{-\frac{y^2}{\alpha_2^2}}$ and $r_1\alpha_2^2 = r_2\alpha_1^2$.

Let $OA = x \quad OB = y \quad OG = z \quad Oa = v \quad OGA = \theta \quad OaG = \theta'$.[6]

Let the angle between the planes XOG and $GOA = \phi$, and that between XOa and $aOG = \phi'$.

then
$$z^2 = v^2 + r_1^2 - 2vr_1\cos\theta'$$
$$x^2 = z^2 + r_1^2 - 2zr_1\cos\theta$$
$$y^2 = z^2 + r_2^2 + 2zr_2\cos\theta$$

$$\frac{x^2}{\alpha_1^2} + \frac{y^2}{\alpha_2^2} = z^2\left(\frac{1}{\alpha_1^2}+\frac{1}{\alpha_2^2}\right) + r_1^2\left(\frac{1}{\alpha_1^2}+\frac{\alpha_2^2}{\alpha_1^4}\right)$$

$$\omega = \sin\theta\, d\theta\, d\phi \qquad \omega_2 = \sin\theta'\, d\theta'\, d\phi'$$

$$\rho = \sum \frac{1}{4}\frac{N_1 N_2 s^2}{\alpha_1^3\alpha_2^3\pi^3}\, e^{-z^2\frac{\alpha_1^2+\alpha_2^2}{\alpha_1^2\alpha_2^2}}\, e^{-r_1^2\frac{\alpha_1^2+\alpha_2^2}{\alpha_1^4}}\left(\frac{\alpha_1^2+\alpha_2^2}{\alpha_1^2}\right)^4 r_1^3\sin\theta\sin\theta'\, d\theta\, d\theta'\, d\phi\, d\phi'\, dr_1.$$

Integrating with respect to ϕ and ϕ' from 0 to 2π we get

$$\rho = \sum \frac{N_1 N_2 s^2}{\alpha_1^3\alpha_2^3\pi}\left(\frac{\alpha_1^2+\alpha_2^2}{\alpha_1^2}\right)^4 e^{-z^2\frac{\alpha_1^2+\alpha_2^2}{\alpha_1^2\alpha_2^2}}\, e^{-r_1^2\frac{\alpha_1^2+\alpha_2^2}{\alpha_1^4}} r_1^3\sin\theta\sin\theta'\, d\theta\, d\theta'\, dr_1.$$

Now let z and x be made independent variables instead of θ & θ'

$$z\, dz = vr_1\sin\theta'\, d\theta \qquad x\, dx = zr_1\sin\theta\, d\theta$$

$$\therefore \rho = \sum \frac{N_1 N_2 s^2}{\alpha_1^{11}\alpha_2^3\pi}(\alpha_1^2+\alpha_2^2)^4\, e^{-z^2\frac{\alpha_1^2+\alpha_2^2}{\alpha_1^2\alpha_2^2}}\, e^{-r_1^2\frac{\alpha_1^2+\alpha_2^2}{\alpha_1^4}}\frac{xr_1}{v}\, dr_1\, dx\, dz.$$

Integrating with respect to x from $x^2 = (z-r_1)^2$ to $x^2 = (z+r_1)^2$

$$\rho = \sum \frac{N_1 N_2 s^2}{\alpha_1^{11}\alpha_2^3\pi}(\alpha_1^2+\alpha_2^2)^4\, e^{-z^2\frac{\alpha_1^2+\alpha_2^2}{\alpha_1^2\alpha_2^2}}\, e^{-r_1^2\frac{\alpha_1^2+\alpha_2^2}{\alpha_1^4}}\frac{2zr_1^2}{v}\, dr_1\, dz.$$

Integrating with respect to z from $z^2 = (v-r_1)^2$ to $z^2 = (v+r_1)^2$

$$\rho = \sum \frac{N_1 N_2 s^2}{\alpha_1^9\alpha_2\pi}(\alpha_1^2+\alpha_2^2)^3\frac{r_1^2}{v}\, e^{-v^2\frac{\alpha_1^2+\alpha_2^2}{\alpha_1^4}}\, e^{-\frac{r_1^2(\alpha_1^2+\alpha_2^2)^2}{\alpha_1^4\alpha_2^2}}\left\{e^{2vr_1\frac{\alpha_1^2+\alpha_2^2}{\alpha_1^2\alpha_2^2}} - e^{-2vr_1\frac{\alpha_1^2+\alpha_2^2}{\alpha_1^2\alpha_2^2}}\right\}.$$

---

(6) See Figure 196,3.

Integrating with respect to r_1 from 0 to ∞

$$\rho = \frac{N_1 N_2}{\alpha_1^3 \pi} \alpha_2 \, e^{-\frac{v^2}{\alpha_1^2}} \left\{ e^{-\frac{v^2}{\alpha_2^2}} + \left(\frac{\alpha_2}{v} + 2\frac{v}{\alpha_2} \right) \int_0^v e^{-\frac{v^2}{\alpha_2^2}} \, dv \right\}.$$

This is the number of particles of the first system projected under the required conditions after striking those of the second system. By putting $\dfrac{N_1}{\alpha_1^3 \pi^{\frac{3}{2}}} e^{-\frac{v^2}{\alpha_1^2}}$ for n in the expression for the number of collisions made by such particles we find the same expression which shows that as many particles of the given velocity are projected as lose their velocity by striking.

[7]   [DISTRIBUTION OF VELOCITIES]

Let a great number of equal particles be in motion in a vessel having two of its sides equal and parallel and let the motion go on till the velocities are distributed according to the law already found. Now let it be supposed that whenever a ball strikes the first of the parallel sides it is turned white and whenever a ball strikes the other side it is turned black and let the motion go on in this way for a sufficient time. The distribution of velocities among the balls will be as before if we pay no regard to colour but if we distinguish the colours we shall find that more white balls are moving in the positive direction than in the negative and that the contrary is the case with the black balls. Let us investigate the number of white balls which in unit of area and at a given part of the vessel have velocities within given limits of direction and magnitude.

Let OX be the positive direction[7] and let us assume that the 'density' of the velocities of the white balls corresponding to OA is $\dfrac{N}{\alpha^3 \pi^{\frac{3}{2}}} e^{-\frac{x^2}{\alpha^2}} \xi \cos AOX = \rho$ where ξ is a function of OA or x and must lie between 0 and 1. Let these white balls strike other balls (either white or black) whose velocity is $OB = y$ and whose 'density' is $\dfrac{N}{\alpha^3 \pi^{\frac{3}{2}}} e^{-\frac{y^2}{\alpha^2}}$ and let the white balls after collision have the velocity $OA = v$. Required the density of the projected balls whose velocity is Oa. We have to sum –

$$\rho = \sum 4 \frac{N^2 s^2}{\alpha^6 \pi^3} e^{-\frac{x^2 + y^2}{\alpha^2}} \xi \cos AOX r_1^3 \, dr_1 \sin\theta \sin\theta' \, d\theta \, d\theta' \, d\phi \, d\phi'.$$

Let $\qquad\qquad XOa = \gamma \quad XOG = \beta$

then $\qquad\qquad \cos AOX = \cos\beta \cos GOA + \sin\beta \sin GOA \cos\phi.$

---

(7) See Figure 196,3.

Integrating with respect to ϕ from 0 to 2π

$$\sum \cos AOX \, d\phi = 2\pi \cos \beta \cos GOA$$
$$= 2\pi \cos \beta \frac{x^2 + z^2 - r_1^2}{2xz}.$$

Again $\cos \beta = \cos \gamma \cos GOa + \sin \gamma \sin GOa \cos \phi'$. Integrating with respect to ϕ' from 0 to 2π

$$\sum \cos \beta \, d\phi' = 2\pi \cos \gamma \frac{v^2 + z^2 - r_1^2}{2vz}.$$

Putting θ & θ' in terms of x & z we find

$$\rho = \sum 4 \frac{N^2 s^2 \cos \gamma}{\alpha^6 \pi} e^{-2\frac{(r_1^2 + z^2)}{\alpha^2}} \xi \frac{r_1}{v^2 z^2} (x^2 + z^2 - r_1^2)(v^2 + z^2 - r_1^2) \, dx \, dz \, dr.$$

In this expression we do not know the form of the function ξ and we therefore cannot integrate with respect to x. We must therefore integrate first with respect to r_1 but we must remember that the limits of r will be different according to the relations of x and z to v.

The conditions are

(1) $r < z + v$ (2) $r > z - v$ (3) $r > v - z$
(4) $r < z + x$ (5) $r > z - x$ (6) $r > x - z$.

By (1) & (6)
$$z > \frac{x - v}{2}.$$

By (4) & (3)
$$z > \frac{v - x}{2}.$$

When $x < v$ the upper limit of r is $z + x$, and the lower limit is

$$r = v - z \text{ when } z \text{ is between } \frac{v - x}{2} \text{ and } \frac{v + x}{2}$$

or $\quad\quad\quad r = z - x \text{ when } z \text{ is between } \frac{v + x}{2} \text{ and } \infty.$

When $x > v$ the upper limit of r is $z + v$ and the lower limit is

$$r = x - z \text{ when } z \text{ is between } \frac{x - v}{2} \text{ and } \frac{x + v}{2}$$

or $\quad\quad\quad r = z - v \text{ when } z \text{ is between } \frac{x + v}{2} \text{ and } \infty.$

The part depending on r is

$$\int e^{-2\frac{r_1^2}{\alpha^2}} \left(r(x^2 + z^2)(v^2 + z^2) - r^3(x^2 + v^2 + 2z^2) + r^5 \right) dr.$$

The indefinite integral is

$$\left\{\frac{\alpha^2}{4}(x^2+z^2)(v^2+z^2)-\left(\frac{\alpha^2}{4}v^2+\frac{\alpha^4}{8}\right)(x^2+v^2+2z^2)+\tfrac14\alpha^2 r^4+\tfrac14\alpha^4 r^2+\tfrac18\alpha^6\right\}e^{-2\frac{r^2}{\alpha^2}}$$
$$=\left\{\tfrac14\alpha^2(x^2+z^2-r^2-\tfrac12\alpha^2)(v^2+z^2-r^2-\tfrac12\alpha^2)+\tfrac14\alpha^6\right\}e^{-2\frac{r^2}{\alpha^2}}.$$

When $r=z\mathrel{\underset{\sim}{+}}v$

$$\frac{N^2 s^2\cos\gamma}{\alpha^4\pi v^2}\xi\left\{4v^2+(\alpha^2+v^2-x^2)\left(\tfrac12\frac{\alpha^2}{z^2}\pm\frac{2v}{z}\right)\right\}e^{-2\frac{v^2}{\alpha^2}}e^{-4\frac{z^2\pm vz}{\alpha^2}}\,dx\,dz. \qquad \begin{matrix}A\\B\end{matrix}$$

When $r=z\mathrel{\underset{\sim}{+}}x$

$$\frac{N^2 s^2\cos\gamma}{\alpha^4\pi v^2}\xi\left\{4x^2+(\alpha^2-v^2+x^2)\left(\tfrac12\frac{\alpha^2}{z^2}\pm\frac{2x}{z}\right)\right\}e^{-2\frac{x^2}{\alpha^2}}e^{-4\frac{z^2\pm xz}{\alpha^2}}\,dx\,dz. \qquad \begin{matrix}C\\D\end{matrix}$$

When $x<v$ the upper limit is C and the lower is B from $z=\dfrac{v-x}{2}$ to $z=\dfrac{v+x}{2}$ and D above this.

When $x>v$ [the upper limit is] A [and the lower is] D [from $z=]\dfrac{x-v}{2}$ $\left[\text{to }z=\dfrac{x+v}{2}\text{ and}\right]$ B [above this].

Integrating these expressions with regard to z and writing

$$\phi\left(\frac{x}{\alpha}\right)\text{ for }\int_0^{\frac{x}{\alpha}}e^{-t^2}\,dt$$

they become

$$-\frac{N^2 s^2\cos\gamma}{\alpha^3\pi v^2}\xi\,e^{-2\frac{v^2}{\alpha^2}}\left\{2(\alpha^2-x^2)\phi\left(\frac{2z\pm v}{\alpha}\right)e^{-\frac{v^2}{\alpha^2}}+\frac12\frac{\alpha}{z}(\alpha^2+v^2-x^2)e^{-4\frac{(z^2\pm vz)}{\alpha^2}}\right\}dx \qquad \begin{matrix}A\\B\end{matrix}$$

and

$$-\frac{N^2 s^2\cos\gamma}{\alpha^3\pi v^2}\xi\,e^{-2\frac{x^2}{\alpha^2}}\left\{2(\alpha^2-v^2)\phi\left(\frac{2z\pm x}{\alpha}\right)e^{\frac{x^2}{\alpha^2}}+\frac12\frac{\alpha}{z}(\alpha^2-v^2+x^2)e^{-4\frac{(z^2\pm xz)}{\alpha^2}}\right\}dx. \qquad \begin{matrix}C\\D\end{matrix}$$

When $x<v$ C is the upper limit from $z=\dfrac{v-x}{2}$ to $z=\infty$. The integral within these limits is

$$\frac{N^2 s^2\cos\gamma}{\alpha^3\pi v^2}\xi\,e^{-\frac{x^2}{\alpha^2}}\left\{2(\alpha^2-v^2)\left(\frac{\sqrt\pi}{2}-\phi\left(\frac{v}{\alpha}\right)\right)-\left(\frac{\alpha^2}{v-x}-v-x\right)\alpha\,e^{-\frac{v^2}{\alpha^2}}\right\}dx.$$

The lower limit is B from $z=\dfrac{v-x}{2}$ to $z=\dfrac{v+x}{2}$.

Taking B within these limits

$$\frac{N^2 s^2\cos\gamma}{\alpha^3\pi v^2}\xi\,e^{-\frac{v^2}{\alpha^2}}\left\{2(\alpha^2-x^2)\left(2\phi\left(\frac{x}{\alpha}\right)\right)-(\alpha^2+v^2-x^2)\frac{2\alpha x}{v^2-x^2}e^{-\frac{x^2}{\alpha^2}}\right\}dx.$$

Between $z = \dfrac{v+x}{2}$ and $z = \infty$ the lower limit is D therefore take D between these limits

$$\frac{N^2s^2\cos\gamma}{\alpha^3\pi v^2}\,\xi\,e^{-\frac{x^2}{\alpha^2}}\left\{2(\alpha^2-v^2)\left(\frac{\sqrt{\pi}}{2}-\phi\left(\frac{v}{\alpha}\right)\right)-\left(\frac{\alpha^2}{v+x}-v+x\right)\alpha\,e^{-\frac{v^2}{\alpha^2}}\right\}dx.$$

Subtracting these integrals at the lower limits from that at the upper one we get

$$\int_0^v \frac{N^2s^2\cos\gamma}{\alpha^3\pi v^2}\,\xi\,e^{-\frac{v^2}{\alpha^2}}\left\{4(\alpha^2-x^2)\,\phi\left(\frac{x}{\alpha}\right)+4\alpha x e^{-\frac{x^2}{\alpha^2}}\right\}dx.$$

for the complete expression to be integrated between $x = 0$ and $x = v$.

Similarly, for the expression between $x = v$ & $x = \infty$ we get

$$\int_0^\infty \frac{N^2s^2\cos\gamma}{\alpha^3\pi v^2}\,\xi\,e^{-\frac{x^2}{\alpha^2}}\left\{4(\alpha^2-v^2)\,\phi\left(\frac{v}{\alpha}\right)+4\alpha v e^{-\frac{v^2}{\alpha^2}}\right\}dx.$$

The sum of these expressions gives the value of ρ, but we cannot integrate them as we do not know the form of the function ξ.[8]

---

(8) There is a preliminary draft of §7, written on the *verso* of the folio here designated §5; this is reproduced in Garber, Brush and Everitt, *Maxwell on Molecules and Gases*: 332–5.

APPENDIX

§1 The following autograph letters written in the period encompassed by this volume, concerned with personal and family matters, are not reproduced.

(a) Letters to Richard Buckley Litchfield

(1) no date [possibly December 1853], Trinity College, Cambridge, Add. MS Letters. c.1[79].

(2) 4 June 1856, Trinity, Add. MS Letters. c.1[82]; reproduced in extract in *Life of Maxwell*: 256–7.

(3) 18 July 1856, Trinity, Add. MS Letters. c.1[84]; reproduced in extract in *Life of Maxwell*: 262.

(4) 23 September 1857, Trinity, Add. MS Letters. c.1[87]; reproduced in extract in *Life of Maxwell*: 280–2.

There is also a typed copy of a letter of 9 September 1856, Trinity, Add. MS Letters. c.1[91]; reproduced in extract in *Life of Maxwell*: 262.

(b) Letter to Elizabeth Cay

27[/28] May 1858, Trinity, Add. MS b.52[1]. To his cousin on the arrangements for his wedding on 2 June 1858; see Number 149.

(c) Letters to Jane Cay

(1) 19 October 1855, Peterhouse, Cambridge, Maxwell MSS (13). To his aunt on the arrangements for family property; see Number 76.

(2) 24 May 1858, Autographen-Sammlung Geigy-Hagenbuch Nr. 2168, Universitätsbibliothek, Basel. On the arrangements for his wedding.

(d) Letters to Robert Cay

(1) 19 October 1855, Peterhouse, Maxwell MSS (14). To his uncle on the arrangements for family property, for which Robert Cay took responsibility.

(2) Notes dated 9 October 1856, 'Saturday', 11 April 1857, 19 May 1857, 21 May 1857, 22 July 1861, and 4 February 1862, Peterhouse, Maxwell MSS (15) to (21); enclosing receipts for cheques for income from property; see Number 186.

(e) Letter to John Maitland

25 February 1862, The Francis A. Countway Library of Medicine, Boston, USA. A note enclosing a cheque.

§2 The following letters, which are not extant as autograph manuscripts, have been reproduced with some cuts from the versions printed in the *Life of Maxwell*.

(a) Letters to Lewis Campbell

[16 November 1847], 26 April 1848, 6 July [1849], October 1849, 26 April 1850, 16 September 1850, 14 July 1853, 22 April 1856, 6 February 1857, 7 August 1857, 4 September 1857, 22 December 1857 (Numbers 7, 9, 19, 20, 30, 33, 41, 99, 112, 127, 130 and 142).

(b) Letters to John Clerk Maxwell

25 November 1855, 3 December 1855 (Numbers 80 and 82).

(c) Letters to Jane Cay

29 May 1853, 27 February 1857 (Numbers 39 and 114).

(d) Letter to Henry Richmond Droop

28 December 1861 (Number 191).

§3 The following documents, printed (generally in partial extract) in the *Life of Maxwell*, have not been reproduced.

(a) Letters to Lewis Campbell

19 October 1849, 18 October 1850, 11 March 1851, 9 June 1851, 10 February 1852, 7 March 1852, 20 February 1853, 15 September 1853, 3 December 1853, 17 October 1855, 20 January 1858, 31 January 1858, 28 April 1858 (*Life of Maxwell*: 125, 148–9, 155–6, 157, 176–7, 178–80, 182–3, 192–3, 194, 218–19, 297–300, 300–1, 308–9).

(b) Letters to John Clerk Maxwell

[April 1842], [1843], [1843], 3 May 1843, 28 March 1844, [19 June 1844], 10 July 1844, 14 October 1844, [1844], 21 April 1855 (*Life of Maxwell*: 57–61, 67, 211).

(c) Letters to Jane Cay

June 1845, July 1845, July 1845, 23 March 1852, 11 March 1853, 7 June 1853, 12 November 1853, 13 January 1854, 3 April 1856, 28 September 1857, 18 February 1858 (*Life of Maxwell*: 70, 71, 180–1, 184, 186–7, 193–4, 195–6, 253, 282, 303–4).

(d) Letters to R. B. Litchfield

4 June 1856, 18 July 1856, 9 September 1856, 23 September 1857, 25 October 1857 (*Life of Maxwell*: 256–7, 262, 280–2, 284–5).

(e) Letters to Miss K. M. Dewar (later his wife)

2 May 1858, 6 May 1858, 9 May 1858, 10 May 1858, 13 May 1858, 16 May 1858, 16 September 1859 (*Life of Maxwell*: 309–13).

(f) Letter to Mrs Blackburn of Killearn

3 April 1856 (*Life of Maxwell*: 253–4).

(g) Letter to J. A. Frere

26 February 1853 (*Life of Maxwell*: 167)

(h) Letter to F. W. Farrar

1854 (*Life of Maxwell*: 200)

(i) Letter to an unidentified friend

1850 (*Life of Maxwell*: 149)

(j) Essays for the 'Apostles' (1853–56)

'Decision', 'Envelopment', 'Language and speculation', 'Autobiography', 'Unnecessary thought' (*Life of Maxwell*: 223–5, 232–4, 235, 244–5, 245–6).

(k) Poems (*Life of Maxwell*: 577–630).

§4 Letters written to Maxwell

Locations of the letters and details of their citation are given.

(a) Letters from James David Forbes

(1)   19 April 1849, ULC Add. MSS 7655, II/1; Number 17 note (5).

(2)   1 May 1849, ULC Add. 7655, II/2; Number 17 note (5).

(3)   4 May 1850, *Life of Maxwell*: 137–8; Number 28 note (2).

(4)   30 January 1851, ULC Add. 7655, II/3.

(5)   4 May 1855, *Life of Maxwell*: 213–14; Number 64 note (2).

(6)   16 May 1855, *Life of Maxwell*: 214–15; Numbers 58 note (6) and 64 note (8).

(7)   13 February 1856, *Life of Maxwell*: 250; Number 92 note (2).

(8)   18 February 1856 (testimonial), Forbes Papers, St Andrews University Library, Letter Book V, 272; Number 92 note (2).

(9)   30 April 1856; *Life of Maxwell*: 256.

(10)   31 March 1857, *Life of Maxwell*: 266–7; Numbers 115 note (8) and 133 note (6).

(11)   13 April 1857, ULC Add. 7655, II/7; Number 116 note (1).

(12)   14 September 1857, ULC Add. 7655, II/10; Number 139 note (19).

(13)   16 March 1858, *Life of Maxwell*: 307–8; Numbers 148 note (2) and 149 note (5).

(14)   16 July 1859, ULC Add. 7655, II/17; Number 159 note (6).

(15)   10 November 1859, Forbes Papers, Letter Book VI, 124; Number 164 note (3).

(16)   6 August 1860, Forbes Papers, Letter Book VI, 292; Number 164 note (3).

(b) Letters from Michael Faraday

(1)   25 March 1857, ULC Add. 7655, II/6; *Life of Maxwell*: 519–20; *The Selected Correspondence of Michael Faraday*, ed. L. P. Williams, 2 vols. (Cambridge, 1971), **2**: 864; Numbers 133 note (5) and 187 note (15).

(2)   7 November 1857, ULC Add. 7655, II/11; *Life of Maxwell*: 288; Number 133 note (2).

(3)   13 November 1857, ULC Add. 7655, II/14; *Life of Maxwell*: 288–9; *Correspondence of Faraday*, **2**: 884–5; Number 133 notes (9), (12) and (13).

(4)   23 May 1861, ULC Add. 7655, II/20; Number 185 note (7).

(c) From George Gabriel Stokes

7 November 1857, ULC Add. 7655, II/12; *Life of Maxwell*: 287–8; Number 139 note (2).

(d) From John Tyndall

7 November 1857, ULC Add. 7655, II/13; *Life of Maxwell*: 288; Number 96 note (12).

(e) From George Biddell Airy

(1)   1 March 1859, Royal Greenwich Observatory Archive, ULC, Airy Papers 6/379, 209; Number 158 note (3).

(2)   30 June 1859, Airy Papers, 6/378, 75 R–V; Number 158 note (3).

(3)   2 December 1859, Airy Papers, 6/378, 78 R–V; Number 167 note (2).

(f) From Robert Baldwin Hayward

(1)   14 November 1857, ULC Add. 7655, II/15; Number 116 note (7).

(2)   26 November 1857, ULC Add. 7655, II/16; Number 116 note (7).

(g) From James Challis

10 June 1861, ULC Add. 7655, II/21; Number 189 note (6).

(h) From Cecil James Monro

(1)   20 January 1855, Greater London Record Office, Acc. 1063/2095; Number 56 note (3).

(2)   8 February 1855, GLRO, Acc. 1063/2096; Number 57 note (2).

(3)   7 March [1855], GLRO, Acc. 1063/2093; Numbers 56 note (7) and 146 note (6).

(4)   15 March 1855, GLRO, Acc. 1063/2094.

(5)   4 November 1856, GLRO, Acc. 1063/2097; Number 104 note (12).

(6)   8 November 1856, GLRO, Acc. 1063/2098; Number 104 note (12).

(7)   11 November 1856, GLRO, Acc. 1063/2099; Number 104 note (13).

(8)   14 November 1856, GLRO, Acc. 1063/2100; Number 118 note (16).

(9)   15 February 1857, GLRO, Acc. 1063/2101; *Life of Maxwell*: 266; Number 118 notes (2) and (3).

(10)   2 June 1857, GLRO, Acc. 1063/2102; Number 124 note (4).

(11)   6 November 1857, GLRO, Acc. 1063/2103.

(12)   18 January 1859, GLRO, Acc. 1063/2104; Number 174 note (4).

(13)   Undated, GLRO, Acc. 1063/2110; Number 188 note (10).

(14)   23 October 1861, *Life of Maxwell*: 329; Number 188 note (6).

(i) From James MacKay Bryson

5 January 1857, ULC Add. 7655, II/4; Number 110 note (19).

(j) From William Pole

4 June 1860, ULC Add. 7655, II/19; Number 169 note (5).

(k) From A. Jardine Lizars

9 February 1857, ULC Add. 7655, II/5; Number 116 note (7).

(l) From A. Milligan

4 July 1857, ULC Add. 7655, II/8; Number 126 note (3).

(m) From J. A. Hunter Paton

16 January 1860, ULC Add. 7655, II/18.

(n) From John Clerk Maxwell

Numerous short extracts are published in the *Life of Maxwell*.

(o) From Vernon Lushington

31 May 1858, *Life of Maxwell*: 312.

INDEX

Bold figures refer to text numbers. Italic figures indicate pages on which biographical details are given.